Astro
67,- €
'18

From Disks to Planets

Michel Blanc • Gregory J. Herczeg •
Veerle Sterken • Helmut Lammer • Willy Benz •
Stéphane Udry • Rafael Rodrigo • Maurizio Falanga
Editors

From Disks to Planets

The Making of Planets and Their Early Atmospheres

Chapters 3–12 and Chapter 2 were previously published in *Space Science Reviews* Volume 205, Issues 1–4, 2016 and Volume 214, Issue 2, 2018, respectively

Springer

Editors
Michel Blanc
International Science Institute in Beijing
Beijing, China

Gregory J. Herczeg
Kavli Institute of Astronomy and Astrophysics
Peking University
Beijing, China

Veerle Sterken
Institute of Astronomy
University of Bern Institute of Astronomy
Bern, Switzerland

Helmut Lammer
Space Research Institute
Austrian Academy of Sciences Space Research Institute
Graz, Austria

Willy Benz
Institute of Physics
University of Bern Institute of Physics
Bern, Switzerland

Stéphane Udry
Observatoire de Genève
Sauverny, Switzerland

Rafael Rodrigo
International Space Science Institute
Bern, Switzerland

Maurizio Falanga
International Space Science Institute
Bern, Switzerland

ISSN 1385-7525 Space Sciences Series of ISSI
ISBN 978-94-024-1517-9

Library of Congress Control Number: 2018942650

Printed on acid-free paper

This Springer imprint is published by the registered company Springer Science+Business Media B.V. part of Springer Nature.
The registered company address is: Van Godewijckstraat 30, 3311 GX Dordrecht, The Netherlands

Contents

Preface
M. Falanga and R. Rodrigo

From Disks to Planets: The Making of Planets and Their Early Atmospheres. An Introduction
H. Lammer and M. Blanc: Space Science Reviews 2018, 2018:60 (14 March 2018) https://doi.org/10.1007/s11214-017-0433-x

Part I From Protoplanetary Disks to Young Planets and Their Proto-atmospheres

The Gas Disk: Evolution and Chemistry
C. Rab, C. Baldovin-Saavedra, O. Dionatos, E. Vorobyov and M. Güdel: Space Science Reviews 2016, 2016:3–40 (7 December 2016) https://doi.org/10.1007/s11214-016-0325-5

Dust Evolution and the Formation of Planetesimals
T. Birnstiel, M. Fang and A. Johansen: Space Science Reviews 2016, 2016:41–75 (17 May 2016) https://doi.org/10.1007/s11214-016-0256-1

Formation, Orbital and Internal Evolutions of Young Planetary Systems
C. Baruteau, X. Bai, C. Mordasini and P. Mollière: Space Science Reviews 2016, 2016:77–124 (12 May 2016) https://doi.org/10.1007/s11214-016-0258-z

Disk Dispersal: Theoretical Understanding and Observational Constraints
U. Gorti, R. Liseau, Z. Sándor and C. Clarke: Space Science Reviews 2016, 2016:125–152 (12 May 2016) https://doi.org/10.1007/s11214-015-0228-x

Formation and Evolution of Protoatmospheres
H. Massol, K. Hamano, F. Tian, M. Ikoma, Y. Abe, E. Chassefière, A. Davaille, H. Genda, M. Güdel, Y. Hori, F. Leblanc, E. Marcq, P. Sarda, V.I. Shematovich, A. Stökl and H. Lammer: Space Science Reviews 2016, 2016:153–211 (27 September 2016) https://doi.org/10.1007/s11214-016-0280-1

Part II The Products of the Coevolution of Disks and Planets

Insights into Planet Formation from Debris Disks: I. The Solar System as an Archetype for Planetesimal Evolution
B.C. Matthews and J. Kavelaars: Space Science Reviews 2016, 2016:213–230 (17 May 2016) https://doi.org/10.1007/s11214-016-0249-0

Insights into Planet Formation from Debris Disks · II. Giant Impacts in Extrasolar Planetary Systems
M.C. Wyatt and A.P. Jackson: Space Science Reviews 2016, 2016:231–265 (22 March 2016) https://doi.org/10.1007/s11214-016-0248-1

The Architecture of Exoplanets
A.P. Hatzes: Space Science Reviews 2016, 2016:267–283 (17 May 2016)
https://doi.org/10.1007/s11214-016-0246-3

Exoplanetary Atmospheres—Chemistry, Formation Conditions, and Habitability
N. Madhusudhan, M. Agúndez, J.I. Moses and Y. Hu: Space Science Reviews 2016, 2016:285–348 (12 May 2016) https://doi.org/10.1007/s11214-016-0254-3

Part III Future Projects and Their Scientific Perspectives

The Way Forward
M. Fridlund, A. Hatzes and R. Liseau: Space Science Reviews 2016, 2016:349–372 (12 May 2016) https://doi.org/10.1007/s11214-016-0247-2

Preface

Maurizio Falanga[1] · Rafael Rodrigo[1]

This book will certainly be a new milestone in the whole series of the ISSI books published since 1996. This is the first joint book between the International Space Science Institute in Beijing (ISSI-BJ) and the International Space Science Institute in Bern (ISSI). In July 2013 a new scientific, and we would say also social, reality appeared in the Chinese capital: The International Space Science Institute in Beijing. ISSI-BJ arose out of an agreement between the International Space Science Institute in Bern, Switzerland, and the National Space Science Center, Chinese Academy of Sciences in Beijing. It was decided, for the first time in the history of ISSI, to share the name, the same scientific committee, the same study tools, and other information of mutual relevance and interest between the two Institutes. At the same time, they remained two fully independent Institutes using independent operational methods and different funding sources. Since its inception, we have seen ISSI-BJ growing well and becoming a wonderful platform for the international space science communities and an important complement to ISSI.

The fast growing space research community in China and Asia is eager to increase its scientific exchanges with their colleagues from the other parts of the world, and at the same time scientists outside of Asia will find new opportunities in Asia/China through ISSI-BJ. The most efficient way to serve these objectives as an open platform was to start with a workshop at ISSI-BJ. Mainly, both institutes agreed in 2013 to have joint activities as milestones for the ISSI-BJ implementation plan, and to have in 2014 one of their workshops organized jointly. It became very clear that the workshop should address timely science topics and should be of broad interest for the different communities, and at the same time we envisaged to address a relatively young but very promising research field. The two Executive directors converged very quickly towards the recommendation made by a previous ISSI Exoplanet Forum on "The Science of Exoplanets and Their Systems", which was held in Bern

✉ M. Falanga
mfalanga@issibern.ch

✉ R. Rodrigo
R.Rodrigo@issibern.ch

1 International Space Science Institute, Bern, Switzerland

Fig. 1 Group picture of the first ISSI-BJ/ISSI joint workshop held in August 2014 in Beijing

on December 5th and 6th, 2012, with the participation of about 25 of the leading experts in the field. The forum discussed in detail what are the avenues for new progress in our understanding of exoplanets, their distribution, their formation and their characteristics. One of the planned outcomes of the forum was to identify key subjects for future ISSI workshops. The outcome of the forum has been the subject of an article published by Lammer et al., Astrobiology, 2013, 793–813. Among the subjects considered for future workshops, the priority was given to the theme of "*The disk in relation to the formation of planets and their protoatmospheres*". Indeed, the interrelationship between circumstellar disks, their evolution and their relation to the formation of planets and their protoatmospheres remain poorly understood, and a subject of intense scientific debate, despite an increasing body of modeling and observational data. The two Executive Directors agreed for the first joint ISSI-BJ/ISSI workshop to address this topic in depth with the participation of the best specialists covering the different aspects of the problem.

The objective of the Workshop was to address one of the greatest challenges of modern astrophysics: understanding the origin and formation of planets around young stars, and the resulting diversity of planetary objects and their protoatmospheres. It is assumed, and for one part observed, that planets form inside the circumstellar disks which are one of the generic characteristics of young stars. But a coherent global picture of how planets and their atmospheres form and evolve inside those disks is yet to emerge. The workshop participants revisited all available scenarios for the joint evolution of disks, planets and atmospheres in the light of the most recent observations and of available models. They tried to establish the most likely scenarios for this co-evolution and, from that, laid the foundations for an improved understanding of the formation and early evolution of the different types of planets.

ISSI-Beijing successfully held its first workshop and brought 45 highly respected Space Science experts from all over the world to Beijing. In addition, ISSI-BJ introduced under its special program for supporting young scientists a young scientists session. Six promising young Chinese scientists were invited to present their work. This opened new perspectives, and was highly appreciated by all participants. Lively debates during the workshop contributed to its very successful outcome: all participants critically yet constructively discussed the status of this very exciting research field and contributed to produce an inspiring outlook into the future, telling us where the synergistic combination of future missions, new instrumentation and observing methodologies, new analysis techniques and the development of theory might lead us.

The first chapter of the book contains a much valuable introductory material about the disk-planets relation, and summarizes some of the main findings of the whole book. The second and third chapters describe gas and dust disks. The fourth chapter describes the formation, orbital and internal evolutions of young planetary systems. The fifth chapter deals with architecture and dynamics of systems. The sixth chapter is about disk dispersal. Chapter seven has been divided in three parts mainly dealing with formation and evolution of protoatmospheres, whereas chapter eight describes exoplanets atmospheres. Chapters nine and ten describe different insights into planet formation from debris disks. The last chapter is devoted to the way forward, with a special emphasis on future space missions devoted to the observation of disks and exoplanets.

We are most pleased and honored to write this foreword for the first ISSI-BJ/ISSI joint book. Overall, we warmly compliment the conveners and organizers of the Workshop who tirelessly conducted the whole workshop and the edition of this book to a successful and fast fruition with enthusiasm and dedication. Let us also thank all those participated in this memorable event and in the writing of this excellent book. Congratulations to all.

Space Sci Rev (2018) 214:5–39
https://doi.org/10.1007/s11214-017-0433-x

From Disks to Planets: The Making of Planets and Their Early Atmospheres. An Introduction

Helmut Lammer[1] · **Michel Blanc**[2,3,4]

Received: 9 November 2015 / Accepted: 16 October 2017 / Published online: 14 March 2018

Abstract This paper is an introduction to volume 56 of the Space Science Series of ISSI, "From disks to planets—the making of planets and their proto-atmospheres", a key subject in our quest for the origins and evolutionary paths of planets, and for the causes of their diversity. Indeed, as exoplanet discoveries progressively accumulated and their characterization made spectacular progress, it became evident that the diversity of observed exoplanets can in no way be reduced to the two classes of planets that we are used to identify in the solar system, namely terrestrial planets and gas or ice giants: the exoplanet reality is just much broader. This fact is no doubt the result of the exceptional diversity of the evolutionary paths linking planetary systems as a whole as well as individual exoplanets and their proto-atmospheres to their parent circumstellar disks: this diversity and its causes are exactly what this paper explores. For each of the main phases of the formation and evolution of planetary systems and of individual planets, we summarize what we believe we understand and what are the important open questions needing further in-depth examination, and offer some suggestions on ways towards solutions.

We start with the formation mechanisms of circumstellar disks, with their gas and disk components in which chemical composition plays a very important role in planet formation. We summarize how dust accretion within the disk generates planet cores, while gas accretion on these cores can lead to the diversity of their fluid envelopes. The temporal evolution of the parent disk itself, and its final dissipation, put strong constraints on how and how far planetary formation can proceed. The radiation output of the central star also plays an important role in this whole story. This early phase of planet evolution, from disk formation to dissipation, is characterized by a co-evolution of the disk and its daughter planets.

From Disks to Planets: The Making of Planets and Their Early Atmospheres
Edited by M. Blanc, G. Herczeg, H. Lammer, V. Sterken, W. Benz, S. Udry, R. Rodrigo and M. Falanga

✉ M. Blanc
Michel.blanc@irap.omp.eu

1 Space Research Institute, Austrian Academy of Sciences, Graz, Austria

2 International Space Science Institute – Beijing, Beijing, China

3 International Space Science Institute, Bern, Switzerland

4 IRAP, Observatoire Midi-Pyrénées, UPS-CNRS, Toulouse, France

During this co-evolution, planets and their protoatmospheres not only grow, but they also migrate radially as a result of their interaction with the disk, thus moving progressively from their distance of formation to their final location. The formation of planetary fluid envelopes (proto-atmospheres and oceans), is an essential product of this planet formation scenario which strongly constrains their possible evolution towards habitability. We discuss the effects of the initial conditions in the disk, of the location, size and mass of the planetary core, of the disk lifetime and of the radiation output and activity of the central star, on the formation of these envelopes and on their relative extensions with respect to the planet core. Overall, a fraction of the planets retain the primary proto-atmosphere they initially accreted from the gas disk. For those which lose it in this early evolution, outgassing of volatiles from the planetary core and mantle, together with some contributions of volatiles from colliding bodies, give them a chance to form a "secondary" atmosphere, like that of our own Earth.

When the disk finally dissipates, usually before 10 Million years of age, it leaves us with the combination of a planetary system and a debris disk, each with a specific radial distribution with respect to their parent star(s). Whereas the dynamics of protoplanetary disks is dominated by gas-solid dynamical coupling, debris disks are dominated by gravitational dynamics acting on diverse families of planetesimals. Solid-body collisions between them and giant impacts on young planetary surfaces generate a new population of gas and dust in those disks. Synergies between solar system and exoplanet studies are particularly fruitful and need to be stimulated even more, because they give access to different and complementary components of debris disks: whereas the different families of planetesimals can be extensively studied in the solar system, they remain unobserved in exoplanet systems. But, in those systems, long-wavelength telescopic observations of dust provide a wealth of indirect information about the unobserved population of planetesimals. Promising progress is being currently made to observe the gas component as well, using millimetre and sub-millimetre giant radio interferometers.

Within planetary systems themselves, individual planets are the assembly of a solid body and a fluid envelope, including their planetary atmosphere when there is one. Their characteristics range from terrestrial planets through sub-Neptunes and Neptunes and to gas giants, each type covering most of the orbital distances probed by present-day techniques. With the continuous progress in detection and characterization techniques and the advent of major providers of new data like the Kepler mission, the architecture of these planetary systems can be studied more and more accurately in a statistically meaningful sense and compared to the one of our own solar system, which does not appear to be an exceptional case. Finally, our understanding of exoplanets atmospheres has made spectacular advances recently using the occultation spectroscopy techniques implemented on the currently operating space and ground-based observing facilities.

The powerful new observing facilities planned for the near and more distant future will make it possible to address many of the most challenging current questions of the science of exoplanets and their systems. There is little doubt that, using this new generation of facilities, we will be able to reconstruct more and more accurately the complex evolutionary paths which link stellar genesis to the possible emergence of habitable worlds.

Keywords Exoplanets · Planetary sciences · Origin and evolution · Planetary systems: protoplanetary disks

Fig. 1 Orbital period in days vs. transit radii for 2842 Kepler planetary candidates. More than 1100 planetary candidates have a transit radius $\leq 1.5\ R_{\mathrm{Earth}}$ (after Rowe et al. 2015)

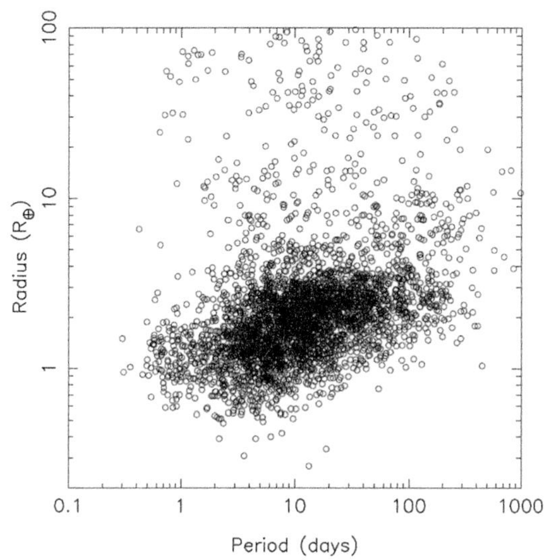

1 Introduction

In our solar system we observe essentially two species of planets: the rocky or terrestrial planets Mercury, Venus, Earth and Mars which have average densities ρ between 3.7–5.5 $\mathrm{g\,cm^{-3}}$ with masses $\leq 1\ M_{\mathrm{Earth}}$, and gas and ice-giants Jupiter, Saturn, Uranus and Neptune which have densities in the range between 0.7–1.6 $\mathrm{g\,cm^{-3}}$, with masses between 14.5–318 M_{Earth}. Prior to the discovery of planets around other stars, it was generally thought that all planets would fall more or less into these two categories.

Since the first discovery of an exoplanet orbiting a main-sequence star via radial velocity measurements in 1995 (Mayor and Queloz 1995), further observations revealed the diversity of exoplanetary characteristics and environments, e.g. the detection of the exotic object Kepler-37b ($R_{\mathrm{pl}} \approx 0.32\ R_{\mathrm{Earth}}$, $M_{\mathrm{pl}} \approx 2.78\ M_{\mathrm{Earth}}$, $\rho \approx 469\ \mathrm{g\,cm^{-3}}$, $d \approx 0.1$ AU), a sub-Mercury sized transiting exoplanet with just unreal planetary parameters located in a multiple planet system close to the host star Kepler-37 (Barclay et al. 2013; Marcy et al. 2014). Further examples representing the wide range of exoplanet discoveries are CoRoT-27b ($R_{\mathrm{pl}} \approx 11.29\ R_{\mathrm{Earth}}$, $M_{\mathrm{pl}} \approx 3302\ M_{\mathrm{Earth}}$, $\rho \approx 12.65\ \mathrm{g\,cm^{-3}}$, $d \approx 0.0476$ AU), a highly massive and dense Jupiter-sized exoplanet in a close orbit to its host star with an orbital period of only ≈ 3.6 days (Parviainen et al. 2014), and GJ 3470 b ($R_{\mathrm{pl}} \approx 4.83\ R_{\mathrm{Earth}}$, $M_{\mathrm{pl}} \approx 13.9\ M_{\mathrm{Earth}}$, $\rho \approx 0.23\ \mathrm{g\,cm^{-3}}$) which has a mass similar to that of Uranus, but is less dense and hotter (Ehrenreich et al. 2014).

The harvest of planet discoveries by Kepler actually unravelled a planetary scene which is very different from the one our Solar System offers to us: the observed distribution of exoplanet radii determined by transit observations is a perfect continuum (Rowe et al. 2015) extending from terrestrial planets with sizes $\leq 1\ R_{\mathrm{Earth}}$ to giants with radii $\geq 4\ R_{\mathrm{Jupiter}}$. These discoveries indicate that exoplanets exist in all sizes and many may have rocky cores that are surrounded by thin H_2/He envelopes. Figure 1 shows radii and periods of 2842 Kepler candidates. Not all of them are confirmed planets, though, and those with very large radii may be eclipsing binaries.

For example, the mean densities of super-Earths or sub-Neptunes such as Kepler-11b and Kepler-11f, of ≈ 3.1 and $\approx 0.7\ \mathrm{g\,cm^{-3}}$ (Lissauer et al. 2011), and the super-Earth GJ 1214b with ≈ 1.87 $g\,cm^{-3}$ (Howe and Burrows 2012) indicate substantial envelopes of light gases such as H_2 and He or H_2O and H. Recent analysis of low-density Neptune-like exoplanets indicate that, if the current estimates of their large radii or small masses are accurate, they should not have maintained their hydrogen envelopes because of extreme high hydrodynamic mass loss rates (Cubillos et al. 2017; Fossati et al. 2017). Because of these recent findings one can expect that in some cases the applied methods may underestimate the planetary masses, or overestimate the planetary radii (due to high-altitude clouds), or infer that hot Neptunes probably have higher albedos than Jupiter-like exoplanets (Cubillos et al. 2017). Misinterpreted parameters can significantly bias the view of populations studies, like the observed radius distribution of exoplanets shown in Fig. 1.

Another example of the diversity discovered in the population of exoplanets can be found in their atmospheric composition. Besides some remaining uncertainties, the discovery of several real low mass H_2 and/or H_2O-rich "rocky" planets indicates that many planets that can be considered as terrestrial may not have lost their initial hydrogen-dominated protoatmospheres (Lammer et al. 2014; Luger et al. 2015; Rogers 2015; Massol et al. 2016 and this volume). In fact there are also arguments that proto-Venus and Earth were surrounded by thin H_2 envelopes after the gas from the disk evaporated, and that significant amounts of noble gases were tapped from the disk and trapped with their initial solar composition in their interiors (e.g., Porcelli and Pepin 2000; Becker et al. 2003). This is plausible if these proto-planets, in agreement with the "Grand Tack" scenario (O'Brien et al. 2014) accreted masses between 0.5–0.75 M_{Earth} (e.g., Walsh et al. 2011; O'Brien et al. 2014) during the life time of the gas disk.

Hence, the recent revelation of this variety of planets that are not present in the Solar system completely changed our current understanding of the origin, formation and evolution of the different classes of solar system planets themselves and showed that observed extrasolar systems can be highly different from our own solar system and display altogether a remarkable diversity. In this chapter, we summarize our understanding of the mechanisms by which the formation of planets and their early atmospheres from their initial protoplanetary disks may generate such diversity.

We start by describing the evolution processes leading from gas and dust circumstellar disks to young planets: dust and gas accretion, disk dispersion, accretion and growth of the different types of planets, and the variety of coupling processes between disk, forming planetesimals and stellar output which largely control the formation and radial migration of the different planetary bodies.

Once the nebula has cleared, we are left with the superposition of a planetary system and a debris disk. We critically examine how well the current evolutionary models of planet formation reproduce the observed diversity of planetary systems architectures, of individual planets and of the volatile contents of each object, drawing some conclusions about the solar system itself. We show how the observations of populations of small bodies and giant impacts in the solar system nicely complement those of dust and gas emissions in protoplanetary disks to produce an integrated understanding of the formation and evolution of debris disks and to make it possible to use them as tracers of the late phases of planetary formation.

We finally focus on the formation of exoplanet atmospheres and show how their observation by the powerful techniques that are progressively becoming available is going to play a key role in understanding their formation processes and in our search for habitable worlds.

2 From Protoplanetary Disks to Young Planets and Their Protoatmospheres

Emanuel Svedenborg (1734) and Immanuel Kant (1755), and later Pierre-Simon Laplace in an independent work (1796) first inferred from the observed geometry of planetary orbits that they must have formed in a "primordial nebula" around the Sun. This visionary "nebular hypothesis" was progressively criticized and abandoned during the XIXth and early XXth centuries, merely because of its inability to explain the observed distribution of angular momentum between the Sun and planets. Several alternative theories competed with one another, until the theory of the primodial nebula was revived by the Soviet astronomer and theoretician Safronov (1972). In the late XXth century, observations of disks around young stars, supported by spectacular progress in protoplanetary disk theory, led astrophysicists to generalize this hypothesis to the formation of all planetary systems.

The accretion of planets from dust to planetesimals, then to planetary embryos and finally to protoplanets requires growth through more than 13 orders of magnitude in size, starting from micron-sized particles of dust and ice up to protoplanetary bodies with radii of thousands or tens of thousands of kilometers. It is convenient to divide up the process into distinct stages in which different physical processes are dominant. In this section we successively describe how the evolution of the two components of circumstellar disks, gas and dust, lead to the formation of early planetary systems and of protoatmospheres.

2.1 Disk Evolution: From Dust and Gas to Planetesimals

Gravitational Collapse and Formation of a Circumstellar Disk Main sequence stars form as the end-products of instabilities within dense molecular clouds composed primarily of molecular hydrogen, leading to a gravitational collapse of cloud cores into protostars within about 10^5 years. Much of the collapsing cloud material initially forms a thick accreting disk around the newly formed star, a T Tauri, or Herbig Ae/Be star, due to excess angular momentum. The rate at which the disk mass is accreted onto the star seems to depend on stellar mass: $M_{\mathrm{acc}} \sim M^2_{\mathrm{star}}$, which demonstrates that the disk lifetime is also stellar-mass dependent (e.g. Alcalá et al. 2000). Conservation of angular momentum results in an increase of the rotation rate with decreasing radius in the disk. Numerical simulations, like the one by Rab et al. (2016 and this volume) illustrated in Fig. 2, reveal the complexity of the joint formation of a young star and a circumstellar cloud/disk: due to fragmentation of the disk, different clumps tend to form in the turbulent cloud within which star and disk are forming simultaneously. They ultimately sometimes accrete onto the central object, or form a separate distant orbiting companion. This process is likely to be at the origin of the formation of stellar binaries and of sub-stellar-mass free floating objects, leading to the observed diversity of the products of stellar genesis.

The disks produced by this process display a broad diversity of sizes and masses. Observations indicate that there are rather strong scaling relations between stellar masses and disk properties. Although variations can be expected, disk masses scale more or less linearly with stellar masses (e.g., Mohanty et al. 2013; Andrews et al. 2013). Their total mass with respect to the star mass is highly variable and weakly constrained by observations. In the Solar system, from estimates of the current mass of planets, the circumstellar disk likely had a minimum mass of a few % of that of the Sun (e.g., Hayashi 1981). However, there was initially most likely more mass than accounted in these estimates.

Once formed, protoplanetary disks are observed to last for only a finite amount of time, far shorter than the typical lifetime of the planetary systems they generate. The lifetimes of

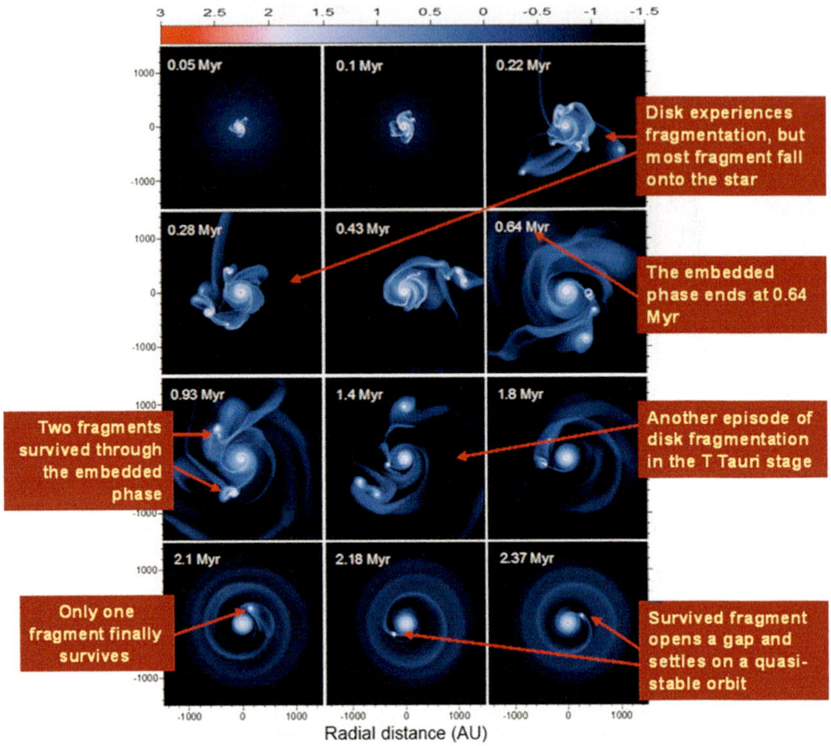

Fig. 2 This numerical simulation of fragmentation inside a forming circumstellar disk shows how a clump can form in the fragmentation and either fall on the central star, generating a luminosity burst, or stabilize as a distant orbiting companion, like the one at 300 AU distance in this example (from Rab et al. 2016 and this volume)

Fig. 3 Inner disk lifetime obtained from IR surveys for different cluster/star-formation regions plotted as a function of their ages (after Haisch et al. 2001; Sicilia-Aguilar et al. 2006; Fedele et al. 2010; Fang et al. 2013). The *red dashed line* represents the best-fit exponential decay for all regions, excluding the sparse associations, and the *red shaded line* represents the best-fit exponential decay for the sparse associations (courtesy of Fang Min)

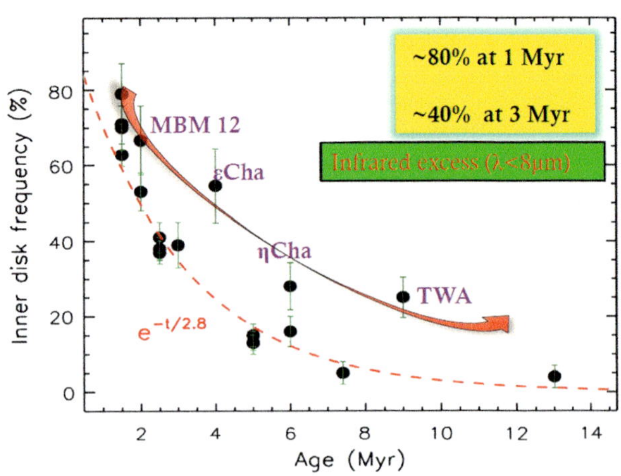

inner disks can be inferred from IR surveys of star-formation regions, as illustrated in Fig. 3, suggesting that disk luminosities decay regularly in a few Myrs, reaching about 40% of their initial values near 3 Myrs of age.

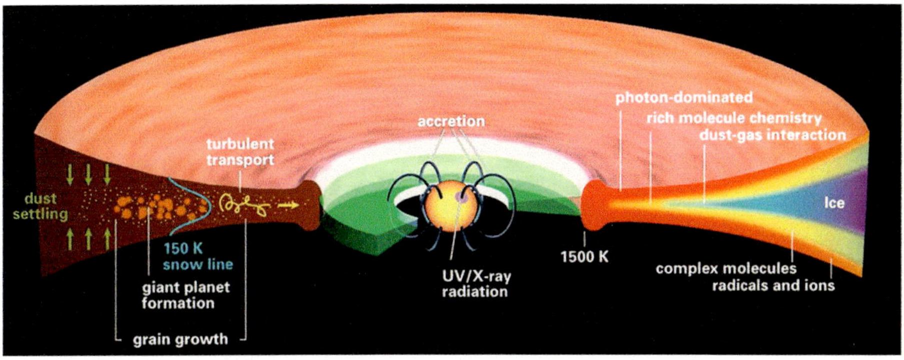

Fig. 4 Sketch of the physical and chemical structure of a protoplanetary disk around a sunlike star (courtesy of Henning and Semenov)

Disk Structure and Composition In the average, disks are composed of about 99% of gas and 1% of dust grains. The distribution between gas and solid particles displays a strong radial variation, however: while light elements like hydrogen and helium are in the gas phase everywhere, heavier molecules associated with the majority heavy elements, water, ammonia and methane meet their condensation temperatures at various radial distances in the disk, forming so-called "snow lines" beyond which they contribute to the solid material of the disk and provide important additional sources for planet formation. The solid component, initially composed of micron-sized dust, is a minor part of the mass budget of the disk, but it is of the utmost importance for its evolution: it is responsible for most of the disk opacity; its thermal radiation spans from IR to sub-mm as grain temperatures vary from about 1000 K to 10s of K between the innermost and outermost regions of the disk; and finally, the solid surfaces of dust grains enable the development of complex heterogeneous chemistry, leading in part to the synthesis of heavy complex molecules. Since the early 90s, ground-based infrared and radio astronomy followed by infrared space observatories such as ISO, Spitzer and Herschel, using dust spectroscopy, have revealed the mineralogical composition of protoplanetary dust particles (e.g., Stahler and Palla 2005; Hartmann 2009), showing that they are in the form of amorphous silicates, crystalline forsterite, water ice and other molecular ices (e.g., Henning and Semenov 2013) and that they can grow in size far beyond the typical submicron sizes of interstellar dust grains (Henning and Meeus 2011).

The overall radial and vertical structure, dynamics and temporal evolution of the disk are a complex result of the interplay of its constituting components: dust, neutral gas, and ionized gas resulting from irradiation by the young stars output. Figure 4 shows a simplified sketch of the disk structure, adapted by Rab et al. (2016 and this issue). A key dynamical characteristics of motions in the disk is that they are near-keplerian, with a tendency to a systematic differential drift between the two main components. The existence of a systematic star-centric pressure gradient in the gas component is at the origin of its sub-keplerian azimuthal motion, and the resulting permanent friction between gas and dust in turn drives dust motions away from keplerian.

The outer layers of the disk are the most subject to stellar irradiation and experience strong erosion and evaporation. In somewhat deeper layers, ionisation by X-ray and UV radiation produces a plasma which, in the presence of the local magnetic field of the disk, enables the development of MHD instabilities, such as the Magneto-Rotational Instability (MRI) which plays an important role in disk dynamics and planet formation. Dust grains shielding from external energetic radiation create a "dead zone" in which the

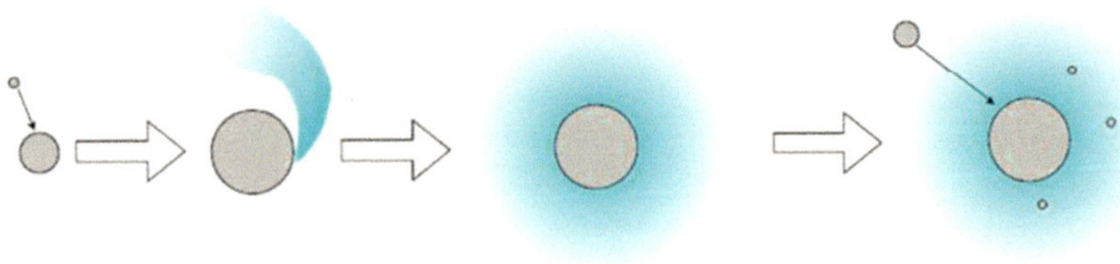

Fig. 5 Illustration of the core accretion scenario (CA) of giant planets formation. The solid components of rocky planets, gas and ice giant cores, first form from dust grains by collisional growth. The primary atmospheres of rocky planets then form by accretion of disk gas. In the case of gas and ice giants, runaway accretion of the gas is triggered as soon as their core masses reach ten Earth masses or more, leading to the formation of a thick mantle of gas. The specific formation processes of each body and its atmosphere are reflected in their composition, which depends on elemental ratios in the disk dust and gas (e.g., Öberg et al. 2011)

magnetorotational instability (MRI) no longer operates (e.g. Balbus and Hawley 1991; Gammie 1996).

Along the radial dimension, one can generally distinguish in a disk an inner/dry and an outer/icy region (e.g., Andrews et al. 2009; Williams and Cieza 2011) separated by a "snowline" whose radial distance scales with stellar intrinsic and accretion luminosity.

Two Main Paths for Planet Formation According to our present understanding, planet formation in circumstellar disks initially follows one of two major formation scenarios: either collisional accretion of solid material, referred to later as "solid accretion" (SA) (from dust to progressively larger particles and bodies), or gravitational instability (GI) (e.g. Boss 2003).

In the solid accretion scenario (SA for short), rocky planets and giant planet cores first form from collisions between dust grains, while gaseous envelopes are accreted separately from the gas disk onto these solid bodies. Rocky proto-planets remain limited in mass inside the snow line due to the limited availability of solid material, while solid cores formed beyond it can reach much larger masses because of the contribution of the different types of ices, mostly water ice. Then, when they reach approximately 10 Earth masses, their gravity field becomes large enough to trigger the so-called run-away accretion of gas from the disk, leading to the formation of gas giants like those of our Solar System (Pollack et al. 1996): this is the description of the so-called Core Accretion (CA) scenario of giant planets formation, illustrated in Fig. 5. Conversely, in the alternative Gravitational Instability (GI) scenario, gas giant planets form via the same mechanism as stars, e.g. the gravitational collapse of a fraction of the disk.

While only SA can account for the formation of rocky planets and their primitive atmospheres, and also of ice giants which are mainly constituted of a thick mantle of ice, the CA and GI mechanisms compete for the generation of gas giants: it is possible that the two mechanisms prevail over two different sub-populations, depending on disk composition, morphology and radial distance, as will be discussed later in Sect. 2.2.

Disk Chemistry Understanding disk chemistry is an important prerequisite to understanding planetogenesis, since it provides the initial conditions determining the chemical composition of planetary bodies, as reviewed in detail by Rab et al. (2016 and this volume).

Protoplanetary disks are characterized by a wide range of temperatures, densities and stellar radiation fields. Thus they display a rich variety of chemical processes that range

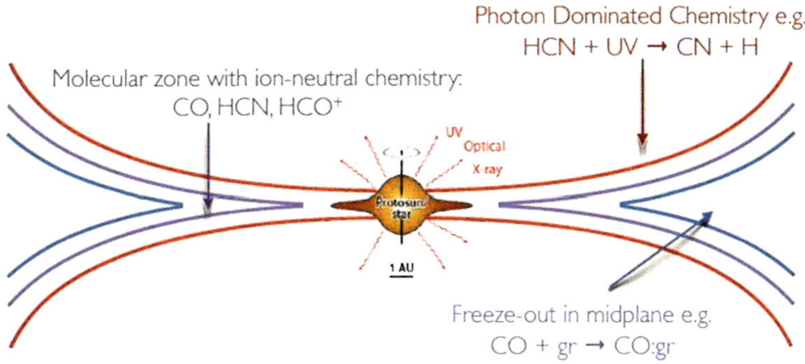

Fig. 6 Illustration showing how disk chemistry is driven by stellar photons and temperature (after Mumma and Charnley 2011)

from high-temperature neutral-neutral reactions in the inner disk to ion-molecular freeze-out up to the outer disk regions (e.g., Henning and Semenov 2013).

X-ray ionization by the young host stars radiation field and the related heating of the disk also play an important role in disk chemistry. Like in a planetary atmosphere, more energetic photons penetrate deeper into the disk. Lower-energy photons have a larger ionization cross section so that they are more easily absorbed either in the wind or in the uppermost disk layers. About 47% of the photoelectron energy is then used for ionization, and a similar amount for excitation and dissociation of H_2 and He. Only $\approx 6\%$ of the energy is used for direct heating of the gas, resulting in inner disk surface heating up to temperatures of ≈ 8000 K (e.g., Aresu et al. 2011).

Chemistry regulates the compositions of nascent planets and the efficiency of planet formation, and provides probes of disk structures and processes. For instance, Spitzer/IRS observations of inner disks around solar mass stars and around brown dwarfs indicate that the HCN/C_2H_2 ratios in their disk atmospheres are different (Pascucci et al. 2009, 2013). Strong vibration-rotation absorption bands of CO_2, C_2H_2 and HCN have also been discovered in disk systems, indicating temperatures between a few 100 K and a few 1000 K and densities $> 10^8$ cm^{-3} necessary for the excitation of rotational-vibrational molecular transitions (Lahuis et al. 2006).

Inner disks have high temperatures and densities. Molecules should be abundant there in the gas phase, until they are destroyed by thermal dissociation ($T \geq 2500$–3500 K). Chemical models of the inner disk chemistry yield high abundances of H_2O and CO vapor at about 1 AU and the lesser presence of nitrogen-bearing molecules (NH_3, HCN, HNC) and a variety of hydrocarbons (e.g. CH_4 and C_2H_2) (e.g., Willacy et al. 1998; Ilgner et al. 2004; Najita et al. 2011).

The sublimation of water ice occurs at about 150 K and defines the so-called snow line for water, which is located at 2–3 AU in the early Solar Nebula (e.g., Morbidelli et al. 2000; Encrenaz 2008). The snow line for CO in the solar system is located at a larger radial distance of about 20 AU where the gas temperature drops below 20 K (e.g., Öberg et al. 2011). Volatile composition and partitioning between ice and gas all depend on disk location and evolutionary stage. Figure 6 shows that disk chemistry depends on disk location, evolutionary time, disk dynamics and the properties of the central star. Compositions of nascent planets will depend on where, when and how planets form (e.g., Mumma and Charnley 2011).

Planetesimals and planetary embryos that originated beyond the water snow line contained ices with trapped NH_3 and CH_4 in the form of clathrates, amino acids and other organic compounds (e.g., Holland 1984; Wordsworth 2016) which can be considered as the main sources of the nitrogen inventories of the inner planets in the solar system. According to Holland (1984) and Kasting et al. (1993), compared to H_2O and CO_2 and/or CO, reduced gases such as NH_3 and CH_4 in the magma ocean-related outgassed early atmospheres of the terrestrial planets were most likely present only in low concentrations of about a few tenths of percent of the total outgassed volatiles. The discovery of highly deuterated amino acids in meteoritic materials and other isotopic anomalies in C, N, and O indicate that their initial synthesis most likely occurred under very cold and dark conditions (e.g., Pizzarello and Huang 2005; Caselli and Ceccarelli 2012; Marty 2012).

Stevenson and Lunine (1988) studied the H_2O-vapor diffusive redistribution and condensation of solid material in the disk near Jupiter's expected formation region. They found that the assumed turbulent disk temperatures decreased inversely with radial distance from the centre. Time scales are related to turbulent viscosities. The length scale for condensation of diffusively-transported water vapour was found to be ≈ 0.4 AU, and the surface density of ice in the Jupiter-formation zone was enhanced by a factor of ≈ 75 (Stevenson and Lunine 1988), which is enough for the rapid accretion of planetesimals into a solid core.

The Solid Accretion Scenario: From Dust to Planetesimals We will now focus on the Solid Accretion scenario, which is likely responsible for the formation of rocky planets and of giant planets cores, when they have one. According to the planetesimal hypothesis, protoplanets form out of dust grains that collide and stick together to form larger bodies. When the growing bodies reach sizes of about one kilometer, they are bound by their own gravity, forming what we call planetesimals; then, in a second step, planetesimals start to attract each other through their mutual gravity, which leads to further growth into planetary embryos of much larger sizes. We focus on the first step here: from dust to planetesimals, the challenge is to understand how solid particles can grow, essentially via mutual sticking, by about 10 orders of magnitude! A general view is that the efficiency of sticking is due first to the sedimentation of solid particles near the mid-plane as the result of friction with the gas. Then, in this medium of increased dust concentration, Brownian or turbulent motions maintain a high enough level of collisions, possibly amplified by a collective gravitational instability in localized regions of increased dust densities. Increased concentrations of this kind can theoretically be generated by several mechanisms: development of pressure bumps by a Magneto-Rotational Instability (MRI), vortex trapping, or concentration and gravitational collapse of swarms of larger particles in the so-called streaming instability.

As discussed in Birnstiel et al. (2016 and this volume) understanding how the population of solid particles accumulated near the disk mid-plane can ultimately build bodies of the size of planetesimal by the mere process of collision and sticking is a difficult scientific challenge: each two-body collision can randomly produce not only sticking, but also mere bouncing, erosion and cratering, mass transfer between the two incident particles, or fragmentation. An accurate calculation of the net statistical cumulative effect of these combined mechanisms on particle growth rests on the use of sophisticated "coagulation algorithms" using appropriate laboratory measurements. Current studies along these lines suggest that the efficiency of coagulation and growth increases with decreasing radial distance in the protoplanetary disk, so that solid particle growth progresses from inside out, fed by the flux of inward drifting particles which is permanently produced by the friction of these particles with the ambient gas. This growth would be efficient up to the size of the so-called "pebble", on the order of a meter or so, producing a "radially expanding pebble propagation front" (Birnstiel et al. 2016 and this issue).

To gain an additional three orders of magnitude, from the size of pebbles to the size of planetesimals for which gravity starts to play an important role, one needs to overcome the "bouncing" and "fragmentation" barriers suggested by simulations. It is unclear at the present time whether these barriers can be overcome simply with the level of collisions induced by turbulence of some kind in the flow, or whether increased local concentrations of particles inducing further gravitational collapse of the solid population are needed. Another suggestion is that a specific additional process, referred to as "pebble accretion", is triggered between particles reaching pebble size. Studies by Ormel and Klahr (2010), Lambrechts and Johansen (2012) and Morbidelli and Nesvorný (2012) indicate that pebble accretion speeds up core formation by a factor ≈ 1000 at 5 AU and up to a factor ≈ 10000 at 50 AU. Pebble accretion would be stopped when the protoplanet grows massive enough ($\approx 20\ M_{\rm Earth}$) to carve a gap in the pebble distribution (Lambrechts et al. 2014). Therefore, protoplanetary cores form well within the lifetime of the protoplanetary gas disk, even at large distances.

Open Questions and Ways Toward Solutions Previous observations and studies of "primitive building blocks" in the solar system have provided mounting evidence for the presence of complex organic matter in the early Solar Nebula. Complex molecules are also most likely present in other protoplanetary disks, but only some species were detected so far using modern (sub-)millimeter interferometers. Present-day submillimeter and millimeter observations provide constraints on the radial and vertical physical and chemical structure, and are delivering information about molecular contents in outer disks. Higher sensitivity and spatial resolution of the Atacama Large Millimeter/submillimeter Array (ALMA) in Chile, and the IR-spectroscopic capabilities of the James Webb Space Telescope will enrich the field of disk chemistry in data in the near future. This data will enhance our understanding related to the global chemical evolution and disk dynamics as well.

While accretion of solid bodies up to the pebble size seems reasonably understood, further growth has to overcome the bouncing and fragmentation barriers identified by detailed collision studies. Determining whether the so-called "pebble accretion" may be the way forward is a subject of current research.

Planet growth and the simultaneous evolution of the disk involve a diversity of phases (stellar radiation and corpuscular output interacting with the dust, gas and plasma components of each circumstellar disk) and of processes (hydrodynamics, gravitational dynamics and MHD, homogeneous and heterogeneous chemistry, collision physics) which are themselves highly dependent on the physical nature and chemical composition of the solid and gas phases of the disk. Our current understanding of the pathways leading to the building of planetesimals out of an initial population of micron-size dust particles is presently hampered by the lack of observational constraints and models capable of representing this diversity. Future progress in our understanding of how planetesimals form will likely come from further sophistication of our modelling tools and from new observational data.

2.2 Disk Dispersal and Planet Formation

IR excess data show that protoplanetary disks disperse in 3 to 5 Myrs. Around 10 Myrs a sharp transition appears in the dust population, while the gas component has essentially vanished. At that time the formation of giant planets stops by lack of a reservoir to feed the growth of their gas envelopes, while terrestrial planet formation may continue for some time. A good understanding of the disk dispersal process and of its timing is therefore essential to better grasp planet formation scenarios and their timelines.

The evolution of protoplanetary disks during the simultaneous phases of protoplanet formation and of disk dispersal is dominated by three competing processes which correspond to the three main sinks of disk material (e.g., Gorti et al. 2016 and this volume). Two of these processes take place essentially near the disk density maximum in the mid-plane: hydrodynamic evolution of the disk, which is dominated by viscous accretion of disk material onto the central star, and growth of protoplanetary bodies, which also contributes to disk dispersal. A third major loss process, photo-evaporation of the external layers of the disk under the effect of stellar X-ray, EUV and FUV illumination, preferentially erodes its outer layers. All three processes contribute to the development of a non-monotonous radial mass distribution in which gaps developing at specific radial distances play an important role, coupling the three disk dispersion processes in a poorly known way.

Internal Disk Evolution Mechanisms Viscous accretion of disk material onto the star appears to be the main loss mechanism for the disk and contributes to the growth of the central star. It is the result of a diversity of hydrodynamic regimes in which the magnetorotational instability (MRI) dominates over a specific radial range (e.g., Baruteau et al. 2016 and this volume). Its inner radial boundary with the quiescent magnetorotational stability zone, or "dead zone", plays an important role in planet formation scenarios and in disk evolution models. Disk accretion combined with the outward radial expansion of disk material required to conserve angular produce a progressive thinning of the disk.

Photo-Evaporation Effects Throughout this accretion process, the disk is illuminated by the photon output of its central star, which is usually dominant, and also in some cases by the radiation of stars belonging to its local cluster. The component of this radiation ranging from X-ray to FUV can efficiently heat and/or ionize the outer layers of the disk, thus contributing to their expansion and evaporation. Some models of disk evolution under photo-evaporation (e.g., Gorti et al. 2015) show that photo-evaporation starts to prevail over accretion in the outer parts of the disk (1 to 10 AU and beyond) after about 1.8 Myrs. The combined effects of inward accretion and outward expansion by photo-evaporation thus generate an expanding gap in disk mass at intermediate distances, which progressively develops into a hole. Once formed, this hole stops further accretion from the outer disk. The inner disk tends to dissipate as it progressively empties onto the star. For solar-type stars, this hole seems to develop in the 1–10 AU radial distance range. One must note, however, that net photo-evaporation rate estimates span from 10^{-11} to 10^{-7} M_{solar}/Yr for stars in the 0.1 to 3 M_{solar} range, implying that, in the present state of our knowledge, we can only say that the role of photo-evaporation in disk evolution lies somewhere between significant and totally negligible!

Planet Formation and Growth, and Associated Coupling Processes The mechanisms governing the formation of protoplanets, first within the gas disk, and later after disk dissipation, are described in detail by Baruteau et al. (2016 and this volume). The growth of a planet within a disk generates density perturbations that create torques and momentum exchanges with the disk and drive changes of its semi-major axis (Kley and Nelson 2012).

The disk also acts as a kind of damping mechanism on eccentricity and inclination. If a planetary core becomes massive enough, it opens up a gap in the disk. If there are several planets forming in the disk, additional mutual interactions develop between them. As a result of these interactions the majority of exoplanets experience significant migrations from their initial formation location to the one where we observe them today: terrestrial planets most likely form in the inner part of the disk; more massive super-Earths could partly have migrated inward from larger distances. Very close-in exoplanets within the Neptune to Jupiter mass domain likely started their formation at larger distances from their host stars and migrated towards closer orbit locations during their formation (Lin and Papaloizou 1986; Lin et al. 1996; Ward 1997a, 1997b; Tanaka et al. 2002; Alibert et al. 2005a, 2005b, 2006;

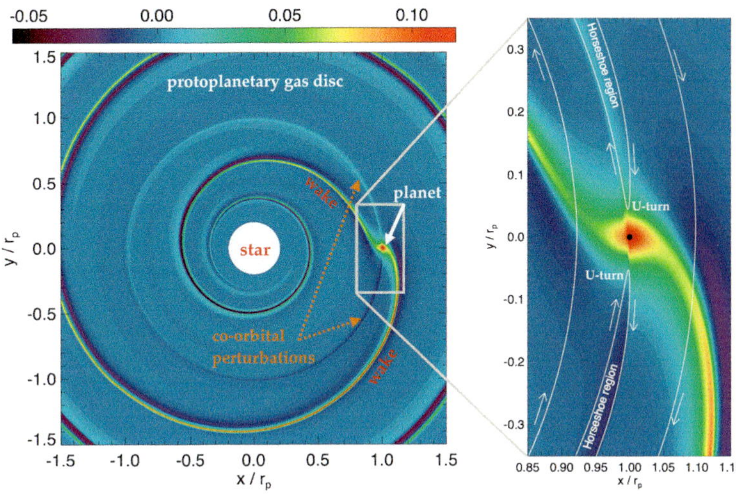

Fig. 7 A numerical simulation by Baruteau et al. (2014) nicely illustrates the different perturbations generated by a forming planet inside a gas disk, and the associated mass and momentum transfer processes

Mordasini et al. 2009a, 2009b). The fact that many of the planets we observe today are found trapped into mean motion resonances is one of the main proofs that they most likely migrated when the gaseous disk was still present. One important result of this large-scale migration is that many planets may also fall onto the star or be ejected from the system. Thus planet migration plays an important role in the shaping of planetary systems architectures, but how well do we understand how migration works?

Planet migration mainly results from the interaction between the growing protoplanet and the gaseous disk. Theoretical models predict the following migration scenarios (e.g. Alibert et al. 2010):

Type I migration: when a protoplanet has a low enough mass compared to the disk mass, it cannot open up a gap along its orbit in the gaseous disk. Due to an imbalance of torques from the inner and outer disks, the protoplanet loses angular momentum and migrates within $\approx 10^5$ years through the disk toward the star. Recent theoretical work by Baruteau et al. (2014) shows that type I migration is very sensitive to disk properties, such as torque-related disk temperature. Both the direction (inward or outward) and speed of migration depend on the disk structure and on its radiative and turbulent properties in the narrow region around the planet (Paardekooper et al. 2011; Dittkrist et al. 2014).

Type II migration: when the protoplanet is heavy enough to disturb significantly the structure of the gaseous disk, it opens up a gap in the disk and drifts with the disk and the gap towards the star. The timescales for type II migration are longer than for type I and lie between 0.1 and 10 Myr (e.g. Alibert et al. 2004; Ida and Lin 2004). Giant planets that are embedded in a protoplanetary disk are usually affected by type II migration. Their final orbit location depends on the disk parameters and particularly on its formation and dispersion time scales (e.g. Alexander and Armitage 2009; Alexander and Pascucci 2012; Rosotti et al. 2013; Baruteau et al. 2016, this volume). For this reason disk dispersal likely has a significant effect on planet formation, as it stops the migration of giant planets and finally shapes the orbital architecture of planetary systems (Bitsch et al. 2015a, 2015b).

Planetary Resonance Processes A direct theoretical consequence of the efficiency of migration mechanisms is that, under their effect, many planets in multiple planets systems migrate until they fall in resonance with another planet and become trapped in the resonance. The orbital configurations of the numerous multi-planet systems observed by Kepler suggest that planet-planet resonances may indeed play a direct role in the shaping of planetary systems, in addition to offering some observational evidence for migration (see Fig. 4, lower left panel, of Baruteau et al. 2016 and this volume). Kepler observations show that planet pairs near first-order mean-motion resonances are preferentially found on orbits just outside the nominal resonance, while avoiding those just inside the resonance. Petrovich et al. (2013) explored this phenomenon with a dynamical planet formation model in which planets grow in mass at a prescribed rate without orbital migration or dissipation. For the 3:2 resonance corresponding to the strongest observed peak-trough structure, their results are consistent with observations for mean planet masses ranging from 20 to 100 M_{Earth}. But this predicted mass range is higher by at least a factor of three than the range obtained from the few Kepler planets with measured masses. Petrovich et al. (2013) thus suggested that part of this discrepancy could be due to oversimplifications in their dynamical model or to uncertainties in the planetary mass-radius relation. The results of their study indicate that the number of planets that are actually in resonance is still uncertain and mitigate the importance of the existence of resonant planetary pairs as a "proof" of migration.

Protoplanet Growth and Initial Formation Our current understanding of the formation of protoplanets and cores in protoplanetary disks can be used to predict the distribution and diversity of these planetary bodies at the end of the disk lifetime (Baruteau et al. 2016 and this volume). In the so-called "evolutionary models" developed to study this phenomenon, the variations in size and temperature of forming planets are calculated by taking into account the three dominant energy sources feeding planet growth according to the standard core accretion scenario: accretion heating deposited into the planet by the two wakes it generates in the disk, internal heating generated by contraction of the gas envelope, and finally radiogenic heating. Overall, evolutionary models are very successful in reproducing many observables of planetary populations, such as the observed mass/radius relationship. They also nicely suggest that a remarkable continuity must exist throughout the diversity of planetary bodies, from "bare rocky" planets like Mercury, through giants, and to planets with masses similar to brown dwarves. The mass distribution of planets probably extends all the way up to the planet/star boundary at 80 R_{J} above which hydrogen fusion takes over.

Another interesting inference of evolutionary model studies concerns the cosmogony of giant planets: because core accretion works very well at distances not too far from the star, but becomes inefficient at farther distances in the disk where solid material and planetesimal densities drop dramatically, it is likely that core accretion is replaced by the classical gravitational instability mechanism at large distances and that giant planets may form there in ways mimicking star formation. Hence planetary formation within disks might well establish a perfect continuity between star and planet populations rather than separating two distinct classes of objects.

Open Questions and Ways Toward Solutions The complex interactions that develop through time between a dynamic accreting disk, growing protoplanetary bodies and the time-variable photon output of the young star ultimately control planet growth and the initial post-disk architecture of planetary systems. These interactions seem to be qualitatively understood, but considerable uncertainties exist on each element of the picture. Progress will come from two directions: improved observations and advanced modelling.

To improve our observational base, we need to know better the temporal evolutions of the stellar photon output, of the radial profile of the dust and gas disks and of the photoevaporation rate. We need to know how and where gaps and holes form in the disk distribution. As one sees, the present lack of a comprehensive description of disk evolution during dispersal calls for broad multi-spectral, high-resolution observations of young-star/disk systems.

Such a comprehensive observational effort must go along with the development of advanced theoretical models, essentially of two types. First, we need models of the coupling between the dust and gas components and the stellar output which will simultaneously take into account hydrodynamics, magnetic fields, chemistry and radiation transfer and will be able to predict the evolution through time of the three sinks of disk material during the dissipation phase: accretion onto the star, photo-evaporation and planet formation. Second, we need evolutionary models of planet growth that can predict the diversity of the population of protoplanets from below the Earth mass to the planet/stars limit of 80 M_J and the respective importance of the core accretion and of the gravitational instability scenarios in giant planets formation.

3 The Products of the Co-evolution of Disks and Planets

When protoplanetary disks finally disperse, they leave behind the superposition of a nascent planetary system and a debris disks. It is not guaranteed that debris survive protoplanetary disk dispersal mechanisms, but it might be expected that, as in the Solar system, the two components are present to varying degrees and in different forms at different distances from their star, since exoplanets are found around 30–50% of Sun-like stars, and debris disks around 20% of them (Wyatt and Jackson 2016 and this volume).

3.1 Exoplanet Systems and Their Architecture

Since the first discovery of an exoplanet via radial velocity measurements (Mayor and Queloz 1995), further observations progressively revealed the diversity of exoplanets and of their planetary systems (e.g. Hatzes 2016 and this volume). With more than 3658 exoplanets, 2742 planetary systems and 614 multiple planet systems discovered (status August 2017: http://exoplanet.eu/catalog/) it is meaningful to compare the architecture of these systems with that of the Solar System, which we describe first (after Winn and Fabrycky 2015):

- The Sun has eight planets, with the four smaller planets ($R_{pl} = 0.4$–$1.0\ R_{Earth}$) interior to the four larger planets (3.9–11.2 R_{Earth}).
- Small low-mass terrestrial planets populate the inner part of the system, while gas and ice giants populate the outer Solar System.
- Planetary orbits are all nearly circular, with a mean eccentricity of 0.06 and individual eccentricities ranging from 0.0068 to 0.21. They are nearly co-planar, with a root-mean-square inclination of 1.9° relative to the plane defined by the total angular momentum of the Solar System (the "invariable plane"), and individual inclinations ranging from 0.33° to about 6.3°.
- The Sun's rotational angular momentum is much smaller than the orbital angular momentum of the planets ($L_{Sun}/L_{orb} \approx 0.5\%$).
- The Sun's equator is tilted by 6.0° relative to the invariable plane.
- The sizes of neighbouring orbits have ratios in the range of 1.4 to 3.4.

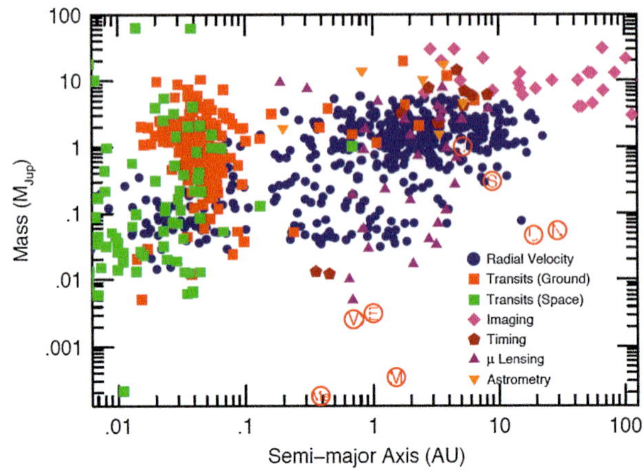

Fig. 8 The mass vs. semi-major axis discovery space for exoplanets with representative discoveries shown. Symbols represent the different detection methods: radial velocity (*dots*), ground-based transit detections (*squares*), space-based transit detections (*green squares*), direct imaging (*diamonds*), timing variations, but excluding transit timing variations (*pentagons*) and micro-lensing (*triangles*). Only transits for which the planet mass has been determined are shown. The planet mass shown is a mixture of true mass (from transits) and minimum mass (from RV surveys). Artie Hatzes, private communication, 2016

Exoplanets Observations Following the 1995 discovery, a first phase of intensive detection of exoplanets was dominated by radial velocity and transit observations by Earth-based telescopes, during which giant, Jupiter-like planets where mostly discovered. Then the advent of satellite missions using the transit technique in space, first CoRoT and then Kepler, extended the sample of discovered objects to a broad continuum of smaller planets which were categorized as super-Earths, sub-Neptunes and Neptune-sized planets. Super-Earths and/or sub-Neptunes are defined to have masses between 1 and 10 M_{Earth}, while exo-Neptunes have masses and radii between $\approx$ 10–30 M_{Earth} and $\approx$ 1.5–5 R_{Earth}.

With the addition of the thousands of planets detected by Kepler, the global mass distribution of exoplanets could be established more accurately. It appeared that planets smaller than gas giants are far more numerous in simulated volume-limited samples around the closest thousand FGK stars (Winn and Fabrycky 2015). More generally, the occurrence probability of exoplanets increases monotonically as mass decreases. Giant planets then account for only 10 to 15% of the sample. The mass vs. semi-major axis distribution of exoplanets as known in 2016 is shown in Fig. 8. Exoplanets populate a broad area in this parameter space, ranging from 0.01 AU (very close-in planets) to about 100 AU in semi-major axis, and from 0.005 to 30 M_{J} in mass. Heaviest among them, the "Super-Jupiters", detected by direct imaging, are located well beyond the outer limits of our Solar System's radial extension. One can also notice that the area of this parameter space which is well covered by detected exoplanets does not really overlap with the locations of our Solar System planets, which are essentially located at lower masses and larger semi-major axes than the bulk of the population of exoplanets: if there was somewhere a planetary system similar to ours, we would not yet have discovered it.

This sample also displays large eccentricities, with an average value of 0.26 far superior to the Solar System value, high-inclination orbits, and even some retrograde ones.

A small fraction of the detected planets belong to binary star systems. They are often found on high-excentricity and sometimes retrograde orbits. While the early detections found them orbiting one of the stars, Kepler managed to detect stars orbiting the binary system at larger distances, suggesting that a broad diversity of planetary orbits must exist in such systems.

The post-Kepler sample also allows a more accurate statistical description of the dependence of planet characteristics on stellar mass: more massive stars display a higher probability of giant planet occurrence, while lower-mass stars tend to have a larger number of planets in their systems. Fressin et al. (2013) analysed the Kepler candidates that were detected during the first six quarters of operation, and found that 16.5% ± 3.6% of main-sequence F, G, and K-type stars have at least one planet with a radius between about 0.8 and 1.25 R_{Earth} and with orbital periods up to 85 days. This result is a significant step toward the determination of η-Earth, the fraction of stars hosting Earth-like planets in their Habitable Zones.

A recent analysis of the planet radius distribution for Kepler stars of different spectral types by Mulders et al. (2015a, 2015b) indicates that low-mass M stars have about 3.5 times more small planets ($\approx$ 1.0–2.8 R_{Earth}) than main-sequence F, G, and K stars, but two times fewer Neptune-sized and larger planets ($>$ 2.8 R_{Earth}). These authors pointed out that the overall higher planet occurrence rates around M stars are not consistent with the redistribution of the same mass into a larger number of smaller planets.

Kepler results also indicate that gas giants are found more frequently around metal-rich stars, while low-mass planets are less frequent there (Buchhave et al. 2014; Wang and Fischer 2015). The trend that metal-poor stars have much less planets compared to metal-rich stars in the Kepler data supports the earlier findings by Santos et al. (2004) and Fischer and Valenti (2005) that stars hosting planets are metal-rich.

A statistical analysis of M-star planetary systems detected by NASA's Kepler mission indicates that, within the sample of M4V to K7V stars observed, 71 host a single planet candidate, 17 host a pair, 12 host three, four host four and two host five candidates (Ballard and Johnson 2016). In the near future M stars are likely to play a very important role in exoplanet research: detection of low-mass (terrestrial) planets close to their stars, for some of them in the Habitable Zone, is relatively easy: their planet-to-star mass ratio is more favourable to detection by the radial velocity technique, and the planet can in principle be observed in the Infrared. PLATO discoveries of small planets in the habitable zones of K and G stars will then widen the comparative study of planets in habitable zones that are further away from their host stars.

Exoplanet Systems Large-Scale Architectures It does not seem possible to define a unique model for planetary systems architectures on the basis of current observations. For instance a model with several planets in nearly coplanar orbits similar to the Solar System does not produce enough single-transit systems, while a model with multiple planets in orbits with large mutual inclinations (i.e. $> 7°$) does not produce enough multiple-transit systems. A similar conclusion drawn from earlier analyses of solar-like G and F-type stars has become known as the "Kepler Dichotomy". It suggests that two different types of architectures for planetary systems may exist (Hansen and Murray 2013): one with a single planet or multiple planets with large mutual inclinations and another one with multiple planets and low mutual inclinations.

Comparison Between Observed Exoplanet Systems and Disk-Planet Interaction Theories The theories of disk-planet interaction and planet formation summarized in Sect. 2.2 mainly predict that:

- Planets should move along circular orbits, because the disk damps eccentricities;
- They should have prograde orbits, because the planets and their star formed from the same rotating cloud;

- Planet orbits should be approximately located in the disk mid-plane;
- Giant planets must have accreted their gas envelope around icy cores that formed beyond the ice line in the disk;
- Rocky planets may form preferentially in the inner disk where there is not enough solid material to form large cores;
- There should be only a few giant planets forming beyond 30 AU, a region depleted in material and where growth timescales are too long compared to planetary formation timescales;
- Inward and outward radial migration of proto-planets and cores may deeply modify the observed planet locations and drive them far away from their formation sites (Bitsch et al. 2015a, 2015b).

Comparison with observations reveals significant differences and strongly suggests that there is more than migration and resonances in the processes controlling the architecture of these systems. Many hot Jupiters are observed on orbits with high eccentricities and/or inclinations, sometimes retrograde, which are difficult to explain by standard migration theories. This suggests that planet-to-planet scattering interactions may play an important additional role. Many planet pairs are observed far from resonance, and when they are closer to it they are usually at some small distance away. This feature, however, may be compatible with the most advanced migration models (Baruteau and Papaloizou 2013) which suggest that planet pairs coupled to the same disk tend to repel each other over time, and consequently to move out of resonance.

Open Questions and Ways Towards Solutions The discrepancies between observations and theoretical predictions of planetary systems architectures should stimulate new observation and modelling studies.

Observations in the coming decades should focus on broadening the coverage of exoplanet populations in parameter space beyond its present limits. As a consequence of important observation biases and limitations associated to each detection technique, the mass/distance parameter space is nearly empty beyond 10 AU, a region where Doppler and transit techniques have very low sensitivities. To fill this large-distance region of parameter space, we will rely on future JWST observations and on specific techniques like astrometry and direct imaging, but Earth-like masses and lower ones will remain poorly accessible to these techniques. A reasonable strategy will be to focus on the characterization of terrestrial planets in the inner systems, a configuration for which radial velocity and transit techniques are sufficiently sensitive.

On the theory side, it is essential to identify, understand and describe the diversity of mechanisms which take part in the build-up of planetary systems architectures in addition to migration and resonance trapping: scattering between planets, interactions with near-by stars and planetary systems, non-standard effects of migration, and some more. The interplay between predictions of models of increasing complexity and a base of observations progressively covering a larger domain of parameter space will in turn provide unique clues to the key processes responsible for the shaping of planetary systems.

3.2 Debris Disks

As discussed in detail by Matthews and Kavelaars (2016 and this volume) the debris disk is made up of these final products of protoplanetary disk evolution, i.e., planetesimals, dust and gas which are not massive enough to be classified as planets. Figure 9 (adapted from

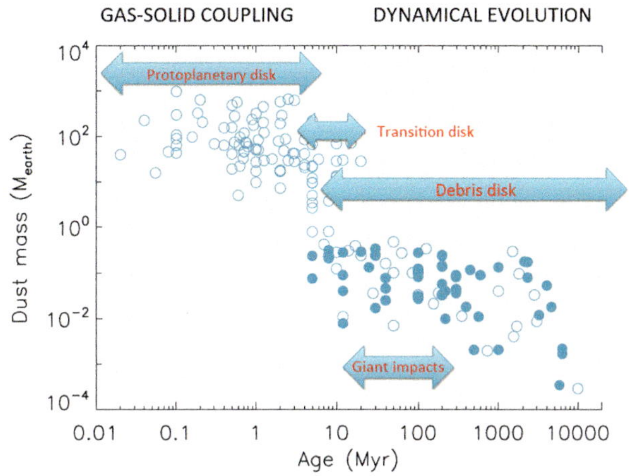

Fig. 9 Periods approximately corresponding to the existence of the successive types of disks along the evolutionary path from initial protoplanetary disks to evolved planetary systems are overlaid to the figure by Paníc et al. (2013) compiling data on the mass of dust disks as a function of time. Adapted from Fig. 1 of Matthews et al. (2014)

Matthews et al., ibid.) places debris disks in the context of the chronology of planet formation: they emerge from protoplanetary disks just after the end of the formation of giant planets, are witnesses and ingredients of the formation of terrestrial planets, and may have lifetimes comparable to the evolution timescales of star-planets systems, so that they may still be there during the late stages of stellar evolution. Unlike protoplanetary disks, whose dynamics is dominated by gas-solid coupling, the evolution of debris disks is governed by gravitational dynamics, collisions and impacts on planets. A secondary, non primordial population of gas and dust forms in debris disks as a result of these collisions and of giant impacts (e.g., Wyatt and Jackson 2016 and this volume). Debris disks are both the last records of the initial conditions of planetary genesis and the remnants of the protoplanetary disk phase: that's what makes them so interesting for our understanding of planetary formation.

The solar system debris disk is mainly composed of the Asteroid belt, Trojan asteroids, the Kuiper belt and the Oort cloud. The zodiacal cloud, which is made of tiny dust particles, is another important component, equivalent to what can be observed as warm/hot debris disk around other stars. These different components are extensively studied by Earth-based telescopes, by in-situ space exploration and via the laboratory analysis of extraterrestrial samples. In contrast, current observational techniques are not sensitive enough to directly detect populations of planetesimals around other planet-forming stars. We see there essentially the emissions of the dust component, e.g. reflected starlight complemented by thermal dust emissions in the near and mid-IR and at longer wavelengths. Recent ALMA observations of the cold molecular gas, which promise to give us access to the gas-dust chemistry of this medium, opened a new window into the study of debris disks. Table 1 compares the different observation techniques of debris disks presently available in the solar system and in exoplanets systems, and shows their unique complementarities.

Two key questions are left open:

- We do not yet have in hand an accurate description of the evolution scenarios from protoplanetary disks to debris disks;
- We do not know the full size distribution of debris disk particles: there is a significant lack of coverage of the size scales intermediate between the small dust particles observed in exoplanets systems and the planetesimals observed in the solar system.

What are the most promising avenues to make progress towards filling these important "knowledge gaps"?

Table 1 A summary of the observational access to the different components of debris disks available in the Solar System and in exoplanet systems, showing how complementary are their observations

Debris disk component	Solar System	Exoplanet Systems
Gas	Lost or undetected	Mostly cleaned or lost
Cold dust	Undetected	Seen in scattered light mm to FIR
Warm dust	Bright in mid-IR	Seen in NIR and mid-IR
Planetesimals	Extensively studied (In situ + remote sensing)	Undetected (may be traced indirectly via gas, dust, dust bursts)

First, observations of the gas phase will play an important role: being strongly coupled to the solid phase before and after the transition from protoplanetary to debris disks, it provides a continuous tracer of the populations of solids throughout that transition. Measurements of the D/H ratio as early as possible after the transition will be particularly important to provide constraints on the formation regions of planetesimals. Detection of the dynamics of the gas and dust phases will be equally important to try and understand the collisional history of this critical age. By studying debris disks in our galactic neighbourhood at a diversity of ages, it is indeed potentially possible to retrieve their temporal evolution during the transition phase. This critical phase can inform us on several important events in planetary formation: giant planets formation by gas accretion; the possible development of large-scale instabilities among the populations of nascent planets, and finally the delivery of volatiles—including water—to terrestrial planets, with its consequences for habitability.

By comparison, the Solar System provides us with only a snapshot of the configuration of a particular debris disk 4.5 Gyrs after its formation. But, in compensation, one can explore in situ the diversity of its populations of objects. The cold population of the Kuiper belt is particularly interesting, as a witness of the primordial population of planetesimals which prevailed before the major instabilities of the outer Solar System developed.

Lessons to be Learnt from Giant Impacts Among all phenomena taking place in debris disks, giant impacts (Wyatt and Jackson 2016 and this volume) should be given a high priority in future observations. They are important both for their role in the late-stage sculpting of planetary bodies and as major sources of the dust and gas populations: Table 1 of Wyatt et al. nicely summarizes the variety of effects of giant impacts on planetary and moon characteristics, how they are observed in the Solar System, and how to observe them indirectly in other planetary systems. Progress in their study will be founded on synergies between observations in the solar system, observations in outer planetary systems and numerical simulations.

Effects of giant impacts can be observed throughout the Solar System, from Mercury to the Kuiper belt and through our own Earth-moon system. By comparing these numerous

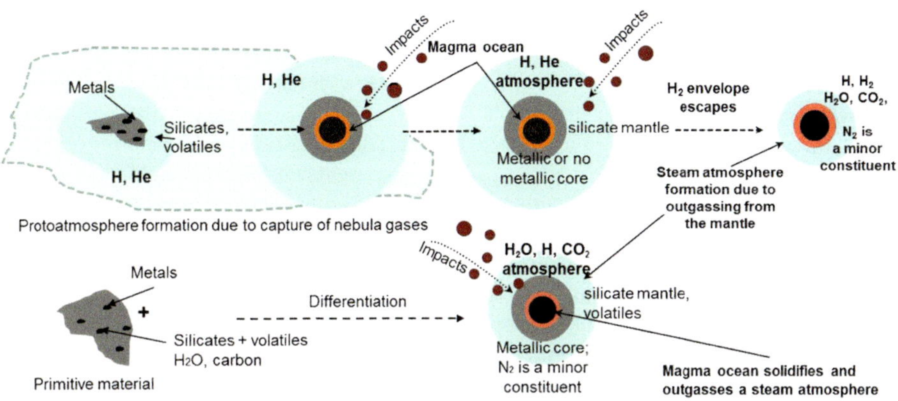

Fig. 10 Illustration of two protoatmosphere formation scenarios for terrestrial planets. The upper scenario illustrates the capture of nebula gases and the accumulation of hydrogen-rich protoatmospheric layers around a protoplanetary core. If the core reaches masses that are $\geq 5\ M_{\mathrm{Earth}}$ a magma ocean forms at the surface. In the case of protoplanet core masses that are $< 1\ M_{\mathrm{Earth}}$, the nebular gas can be lost to space via EUV-driven hydrodynamic escape, the magma ocean solidifies and outgasses a steam atmosphere. This scenario also includes impacts of planetesimals and planetary embryos. The bottom scenario illustrates the "secondary" formation of a protoatmosphere by catastrophic outgassing during accretion and magma ocean solidification if a planet accretes at a slower rate so that no nebular gas is captured. This scenario is also complemented by contributions of volatiles from giant impacts

observations to advanced numerical simulations, one can build a detailed understanding of giant collisions and of their effects on planetesimals, dust and gas. This knowledge can then be used to interpret the growing body of observations of bursts of gas and dust in exoplanet systems. This way, the coupling of modelling and observations of giant impact can potentially give us access to the dynamical history of the populations of extrasolar planetesimals, which would otherwise remain out of reach, and to an improved understanding of their role in the formation of planetary systems.

4 Exoplanet Atmospheres: From Formation to Characterization

4.1 Protoatmospheres: Formation and Evolution

The physical structure and chemical composition of planetary protoatmospheres are the result of a complex interplay between accretion time, nebular gas lifetime and orbit location. If protoplanets grow while the disk gas is still present, they can capture H_2-He envelopes to form a "primordial" atmosphere. Conversely, protoplanets that accrete after dissipation of the disk gas are left with two main sources to form a "secondary" atmosphere: internal (outgassing of the volatile-rich (i.e., H_2O, CO_2) magma ocean and volcanic activity), and external (contribution of volatiles from giant impacts). Following Massol et al. (2016 and this volume), we briefly describe here what we know of these two types of protoatmospheres.

Disk-Captured Protoatmospheres The massive H_2/He atmospheres of large planets such as Jupiter, Saturn, Uranus and Neptune were formed as a result of gas capture onto ice-dominated cores, e.g. in the late stage of the core accretion scenario, as illustrated in Fig. 10 (Mizuno 1980; Stevenson 1982; Wuchterl 1993; Ikoma et al. 2000; Rafikov 2006; Rogers et al. 2011; Ikoma and Hori 2012; Stökl et al. 2015). If a proto-planet accretes to a mass that is $\geq 0.5\ M_{\text{Earth}}$ within the disk lifetime, it captures a H_2 envelope and other disk gases. In this scenario, also illustrated in Fig. 9, a small protoplanet with masses $< 1\ M_{\text{Earth}}$ can lose its H_2 envelope by "boil-off" and stellar XUV powered hydrodynamic escape over tens of Myr (e.g., Lammer et al. 2014; Owen and Wu 2016; Fossati et al. 2017) so that after its escape a magma-ocean-related steam atmosphere can also be outgassed. More massive planets keep their disk-captured gas and may even grow to produce Neptune-type planets or Jupiter-like gas giants.

The type and mass of atmosphere which can be captured around a rocky core depend mainly on core mass, disk conditions, and core formation scenarios.

The capture of disk gas around protoplanetary cores is a complex process which works best for giant planets, being an intrinsic component of the Core Accretion (CA) scenario described previously. Its relevance for low-mass terrestrial planets is most likely different for various systems, and not well understood yet. During recent years the Kepler space observatory has found hundreds of sub-Neptune and Neptune-size ($\approx$ 2–6 R_{Earth}) planet candidates within 0.5 AU of their host stars (e.g., Rogers et al. 2011). The discovery of sub-Neptunes with captured nebula-based atmospheres indicates that there are many planets with H_2/He envelopes around planetary cores with masses between $\approx$ 1–10 M_{Earth}: this was a surprise in the context of our Solar System experience.

The discoveries of sub-Neptunes indicate that many exoplanets within the low mass regime accrete fast within the gas disk to masses large enough for large amounts of H_2 and H_e to be captured, and that only a tiny fraction of these captured primordial hydrogen envelopes has been lost (Lammer et al. 2014; Luger et al. 2015; Rogers 2015; Johnstone et al. 2015; Owen and Mohanty 2016; Cubillos et al. 2017; Rogers et al. 2011; Massol et al. 2016 and this volume). If such a low-mass H_2-dominated exoplanet orbits close to its host star, the observed transit radius can be larger compared to the "true" planetary radius. In such a case the radius-mass relationship and the inferred planetary structure of low mass hot sub-Neptunes can be strongly affected (Lammer et al. 2016; Fossati et al. 2017; Cubillos et al. 2017). For instance, a recent study of the CoRoT-24 planetary system by Lammer et al. (2016) revealed that for the hydrogen-dominated hot sub-Neptune CoRoT-24b with a transit radius R_{T} of $\approx 3.7\ R_{\text{Earth}}$ and a mass of $\leq 5.7\ M_{\text{Earth}}$ the real planetary radius R_{pl} usually considered around the 100 mbar to 1 bar level is located much deeper in the atmosphere at $\approx 2\ R_{\text{Earth}}$. This result was obtained from fitting a set of R_{pl} for CoRoT-24b within $\leq 3.7\ R_{\text{Earth}}$ and comparing the hydrodynamic outflow rate of hydrogen atoms with the maximum possible EUV-driven loss rate. The high thermally-driven hydrodynamic outflow obtained for CoRoT-24b and many other hot sub-Neptunes (Cubillos et al. 2017; Fossati et al. 2017) is caused by the combination of high temperature and low gravity, independent of the XUV flux. Such high escape rates could in principle last for a few Myr after the young planet is disconnected from the gas disk (see also Owen and Wu 2016).

Figure 11 compares the structure of a hot sub-Neptune with a cooler more massive gas giant. The wide separation between the 1 bar level and the observed transit radius R_{T} is most likely due to the presence of Mie scattering caused by high-altitude hazes (i.e. visibility-reducing aerosols) (Bean et al. 2010; Lee et al. 2016; Madhusudhan et al. 2016 and this volume), around the 1 mbar pressure level. For more massive or cooler planets the atmosphere is more compact and the cloud and haze layers that extinct the light during the transit are much closer to the 100 mbar to 1 bar level, so that $R_{\text{T}} \approx R_{\text{pl}}$.

High altitude hazes and clouds (Madhusudhan et al. 2016 and this volume) can therefore introduce a systematic bias in the measured planet radii, especially for the radius distribution of hot sub-Neptunes. A low-mass exoplanetary initial mass function would be extremely valuable to know in order to better constrain planet formation and population models

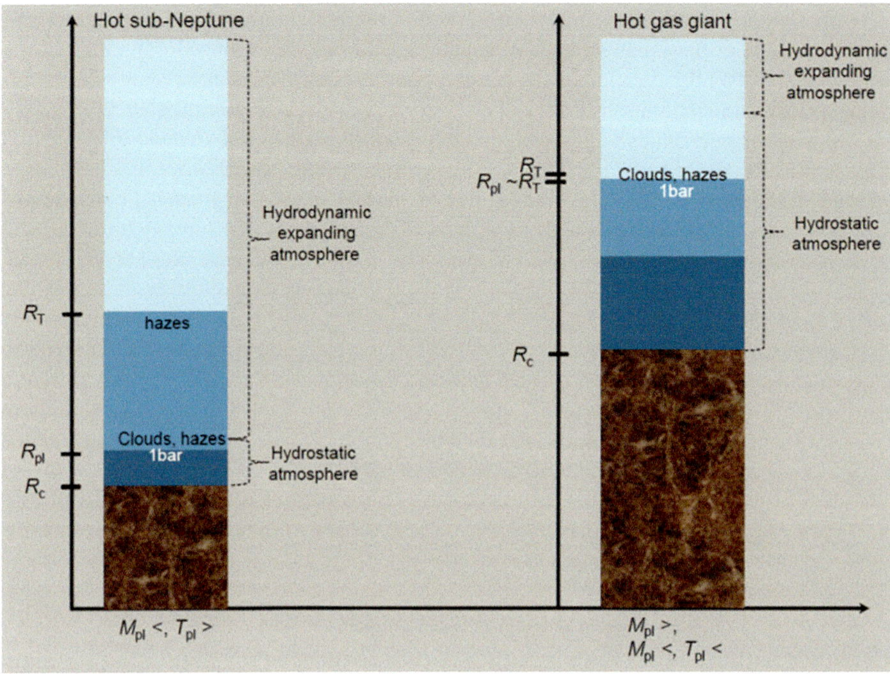

Fig. 11 Illustration of planetary structures of hot gas giants. The *brownish area* at the bottom corresponds to the rocky core, while the *blue areas* above are different layers of the atmosphere with the planet radius $R_{\rm pl}$ (≈ 1 bar level) and transit radius $R_{\rm T}$ indicated. Hot and low-mass hydrogen-dominated exoplanets, such as CoRoT-24b, are less compact compared to cooler and more massive planets. For compact exoplanets, clouds and hazes are located near $R_{\rm pl}$, and $R_{\rm T}$ is therefore close to the core. For hydrogen-dominated close-in hot sub-Neptunes one can expect that high-altitude hazes are more widely distributed and responsible for a $R_{\rm T}$ which can be much larger than $R_{\rm pl}$ (Lammer et al. 2016; Cubillos et al. 2017)

(e.g., Mordasini et al. 2012; Hatzes 2016 and this volume; Matthews and Kavelaars 2016; Baruteau et al. 2016 and this volume).

Studies related to the escape of hydrogen-dominated protoatmospheres indicate that under most protoplanetary core conditions, planets with masses within the sub-Earth to Earth mass domain can lose their captured hydrogen envelopes by thermal escape if the planets orbit a moderately or fast rotating young solar-like host star (Tu et al. 2015) during the first $\leq$ 100 Myr after disk dissipation (Lammer et al. 2014; Johnstone et al. 2015; Luger et al. 2015; Owen and Mohanty 2016; Massol et al. 2016 and this volume). However, depending on the dust depletion factor or low accretion rates that result in low protoplanetary luminosities, it is possible that even protoplanets with Earth-mass cores may keep their hydrogen envelopes during their whole lifetime (Lammer et al. 2014; Stökl et al. 2015). In contrast to lower-mass protoplanetary cores, more massive super-Earth-type cores can accumulate a huge amount of disk gas and lose only tiny fractions of their primordial hydrogen envelopes.

Magma-Ocean-Related Catastrophically Outgassed Steam Atmospheres Smaller Earth-size planets may lose their primordial H_2-H_e envelopes under the effect of irradiation by the high EUV flux of their young host stars, and/or form their atmospheres after dissipation of the gas disk.

As briefly discussed in Rab et al. (2016 and this volume) the initial water inventory and volatile contents of these newly created planets depend mainly on the configuration of the protoplanetary disk, the location of the ice line, orbit location, mass and eccentricity of additional planets, and the impact history of asteroids and comets. Terrestrial planets are likely accreted from planetesimals built from chondrite-dominated material. Wood (2005) reports up to 20 wt% of H_2O in wet chondrites, while achondrites are much drier and have up to ≤ 3 wt% (Jarosewich 1990). The final (oligarchic) accretion of the terrestrial planets most likely proceeded from planetesimals or embryos that were themselves already differentiated and therefore drier (Elkins-Tanton 2008; Odert et al. 2018). A range of initial H_2O inventories of ≈ 0.01–3 wt% in the bulk silicate of terrestrial planets can therefore be expected (Liu 2004; Elkins-Tanton 2008, 2011).

As a result, during the magma ocean solidification process, a fraction of the water and other volatiles which were incorporated into the proto-planet during the main accretion phase and delivered via impacts will degas and produce dense steam atmospheres. Lebrun et al. (2013) and Hamano et al. (2013) studied the thermal evolution of early magma oceans that are connected with hot steam atmospheres at early Venus, Earth and Mars. Hamano et al. (2013) found that Venus- or Earth-mass terrestrial planets that could lose or never captured hydrogen from the protoplanetary disk can be separated into two distinct categories: a type I planet originates beyond a critical distance from its host star, so that the magma ocean solidifies within several million years. Outgassed H_2O then condenses and forms the earliest oceans. In contrast, on a type II planet that originates inside the critical distance, the magma ocean can be sustained during a longer time and its duration can be as long as 100 million years if the planet is formed together with a mass of water comparable to the total inventory of modern Earth. In such a case H_2O remains in vapour form and hydrodynamic escape desiccates type II planets during the slow solidification process (Lammer et al. 2011; Hamano et al. 2013).

According to Lebrun et al. (2013) and Hamano et al. (2013) the planetary environment (i.e., surface and steam atmosphere) cools so that H_2O vapor would begin to condense into liquid water at 1.5 AU on Mars after about 0.1 Myr, at 1 AU on Earth after about 1.5 Myr and at 0.7 AU on Venus after 10.0 Myr or most likely never. In the solar system it is expected that during the late veneer ($\approx$ 80–120 Myr after the beginning of its formation) terrestrial planets were bombarded by large asteroids. A recent analysis of different groups of chondrites, including carbonaceous chondrites and enstatite chondrites of Ruthenium (Ru) isotopes, by Fischer-Gödde and Kleine (2017) points to an inner solar system origin of the late veneer and indicates that it was not composed of carbonaceous chondrites. However, as briefly discussed above, isotope studies show that Earth's volatiles most likely originated from carbonaceous chondrite-like material (Alexander et al. 2012; Marty 2012). One can overcome this contradiction if carbonaceous chondrites were scattered into the inner solar system early during planetary accretion, which is also consistent with the "Grand Tack" scenario (e.g., O'Brien et al. 2014). One should note that the delivery of volatiles by impactors will be system-specific, therefore the processes that lead to the delivery of the initial volatile inventories at terrestrial planets are the results of a complex interplay of protoplanet accretion time, disk lifetime and dynamical effects caused by forming planets.

Open Questions and Ways Towards Solutions While we basically understand there are two main scenarios (disk-captured protoatmospheres and secondary atmospheres) and three sources (disk gas, magma ocean degassing and volatile delivery by giant impacts) for the formation of protoplanets atmospheres, a considerable degree of uncertainty is associated with each of these scenarios.

In the first scenario, the exact mechanisms and quantitative effects of gas accretion from the disk, which is an integral part of planetary accretion, are not fully understood. The stability of the resulting accreted atmosphere depends on the efficiency and time constants of the different atmosphere escape and erosion mechanisms at work, which are poorly known for exoplanets and only partially understood in the case of Solar System planets. In addition, a good understanding of the types of planets for which this scenario may be dominant is hampered by the known biases in planet characterization.

To overcome these difficulties, we need significantly improved observational determinations and models of the structure of exoplanetary atmospheres.

In the second scenario, in which secondary atmospheres are produced by the combined effects of degassing and external bombardment, improved estimates of the magnitudes and chemical compositions of these two sources are urgently needed, but at the present time they are the subject of considerable uncertainties even for Earth. A key parameter controlling the possibility of forming an ocean by condensation of H_2O vapor in the steam atmosphere is the time constant for cooling of the surface and solidification of the primitive magma ocean, which likely displays a strong dependence with respect to distance to the host star.

Given the loose constraints existing presently on our knowledge of how these processes operate at exoplanets, a reasonable way forward is to make the best possible use of the lessons we may learn from Solar System planets, where quantitative studies are easier, and to improve the accuracy of atmosphere models.

4.2 Characterization of Exoplanet Atmospheres

Exoplanet atmospheres in their later stage are a further category of products of planets and disk co-evolution. Observations of these outer shells of exoplanets, which interact strongly with photons over a broad range of wavelengths, made spectacular progress from mere detection to characterization over the last few years, with a strong emphasis on the determination of their chemical composition.

Madhusudhan et al. (2016 and this volume) summarize the expected chemical and physical processes thought to operate and prevail at different altitudes in these atmospheres (Fig. 12), the related recent observations of these properties and their limits, and the major avenues of research which these observations open for the future, driving to a large extent the development of a new generation of observing facilities. Characterizing planetary atmospheres is indeed a challenge, even in the Solar System, because they are complex systems in which a large diversity of processes operate over a broad range of altitudes. To understand exoplanets atmospheres, it is therefore necessary to feed the sparse data we have into more and more sophisticated models, and to make the best possible use of synergies with corresponding Solar System objects.

So far observations of exoplanet atmospheres have been carried out for a variety of exoplanets that were detected via transits, direct imaging and the radial velocity method. Observations with the Spitzer space observatory were carried out for more than 50 exoplanets in 3.6 and 4.5 mm photometric bands and for about 20 exoplanets in the 3.6, 4.5, 5.8 and 8 mm photometric bands (e.g., Madhusudhan et al. 2014, 2016 and this volume). Other photometric observations were carried out in the near-IR by using large ground-based telescopes and also with the Wide Field Camera 3 on the Hubble Space Telescope. These observations led to the discovery of molecular species like CO and H_2O in the vapour phase in the atmospheres of several transiting exoplanets (e.g. de Kok et al. 2013; Brogi et al. 2013; Deming et al. 2013). Temperature inversions were observed in the atmospheres of some exoplanets (e.g., Knutson et al. 2008; Madhusudhan and Seager 2011). They are most likely produced

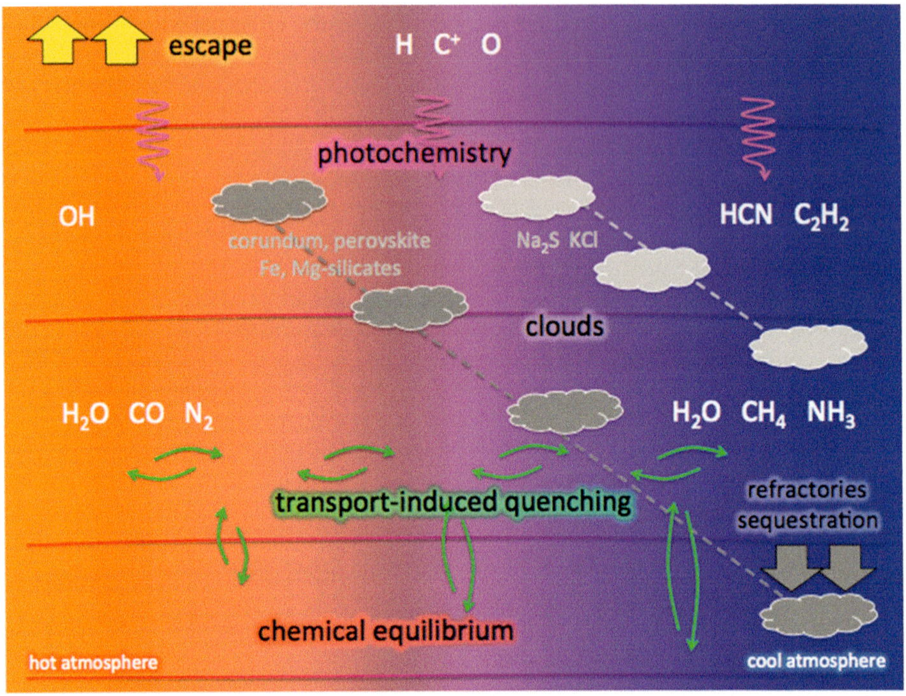

Fig. 12 Illustration of the different physical and chemical mechanisms operating in a H/He dominated atmosphere: from bottom to top, chemical equilibrium, transport-induced quenching, photochemistry and escape (from Madhusudhan et al. 2016 and this volume)

by absorption of stellar radiation in high-altitude absorbing layers (e.g., Hubeny et al. 2003; Knutson et al. 2010). Absorbing molecules may include TiO, VO and sulphur compounds (Hubeny et al. 2003; Fortney et al. 2008; Spiegel et al. 2009).

In the meantime, more than 400 HST orbits were allocated for the acquisition of high signal/noise spectra of more than 12 transiting exoplanets in the visible and IR. Furthermore, FUV HST-STIS transit observations of HD 209458b revealed enhanced transit depths at the positions of Ly-α (H), O I and C II lines, suggesting that this planet has an extended upper atmosphere where oxygen and carbon are dragged by outward-flowing hydrogen atoms (e.g., Vidal-Madjar et al. 2003, 2004; Ehrenreich et al. 2008; Bourrier and Lecavelier des Etangs 2013; Kislyakova et al. 2014). Similar absorption in Ly-α indicating an extended atomic hydrogen atmosphere was recently discovered at the hot Neptune GJ 436b (Kulow et al. 2014; Ehrenreich et al. 2015). Besides detections of atmospheric species in the FUV, observations in NUV have also been carried out (Fossati et al. 2010). These observations indicate significant stellar photospheric absorption related to strong metal lines including Fe II in the spectrum (Haswell et al. 2012).

Information on atmospheric composition of exoplanets can also be obtained from directly imaged exoplanets, which can now be characterized spectroscopically (Barman et al. 2015). Finally, molecules have been detected in the atmospheres of close-in planets using very high-resolution Doppler spectroscopy (Snellen et al. 2010; Madhusudhan et al. 2016 and this volume, and references therein).

Open Questions and Ways Towards Solutions To obtain more data from exoplanet atmospheres, either H_2-dominated atmospheres or outgassed secondary atmospheres, observational efforts are made mainly in two directions: detection of exoplanets around bright nearby stars and high precision observations of exoplanet atmosphere spectra. Bright host stars are important to obtain precise measurements of radii, masses and atmospheric spectra. Their observations also allow to minimize possible uncertainties in input parameters for hydrodynamic models. Because of misinterpreted or uncertain observational exoplanet parameters, the transit-timing variation (TTV) mass determination method could underestimate planetary masses, while optical transit observations could overestimate radii, particularly in the presence of high-altitude aerosols. Lastly, exo-Neptunes could have consistently higher albedos than Jupiter-like exoplanets, leading to an overestimate of their temperature. Many of these uncertainties will be overcome with the next generation of ground-based and space projects.

5 The Way Forward

Success in discovering more atmospheric species in exoplanet atmospheres is strongly related to the technological breakthroughs expected from future ground and space-based observations. Fridlund et al. (2016 and this volume) place these expectations in the broader scientific perspective of the search for alien life, which addresses a multi-millenial question of Humankind: are we alone in the Universe? When we study the birth and evolution of protoplanetary disks, we try to ever better understand how they give birth to a broad variety of planetary objects, how a fraction of them may become habitable, and finally whether a few of them have hosted the emergence of life. In this perspective, the major challenge of the study of disk-planets relationships is to produce a continuous view through time of the evolutionary paths linking disks to habitable planets, thus bridging the most important gaps in our understanding. Two of these gaps must be emphasized.

First, the transition from protoplanetary disks to the (debris disks + planetary systems) couple is still poorly understood. While the Hubble Space Telescope in particular has provided an impressive description of protoplanetary disks, their associated jets and their connection to stellar formation in the visible and UV ranges, space telescopes operating in the infrared can be used to study cold disks and their imbedded planets. Giant millimeter radiotelescopes like ALMA give access to the gas component.

Then, to properly cover the paths leading to habitable planets, among which Earth-like planets in habitable zones are the most promising candidates, we must make major advances in our capacities to first detect these planets and characterize them in terms of mass, size and age, then measure the chemical compositions of their atmospheres, and ultimately search for signatures of life. Most of the plans for advanced exoplanet observations for the coming two decades aim at reaching these objectives step by step.

Building on the major advances in exoplanet transit detection by CoRoT, Kepler and their follow-on, and combining transit and radial velocity detections of the same objects by use of ground-based Radial Velocity experiments of continuously increasing accuracies, the "elementary" characterization of exoplanets is making spectacular progress, while its main limit remains the poor knowledge of stellar masses: hence the importance of deploying in the near future observation platforms that will simultaneously give access to planets and to the characterization of their central star by seismology, resulting in a determination of stellar masses to better than a few percent. Upcoming space observatories like NASA's Transiting

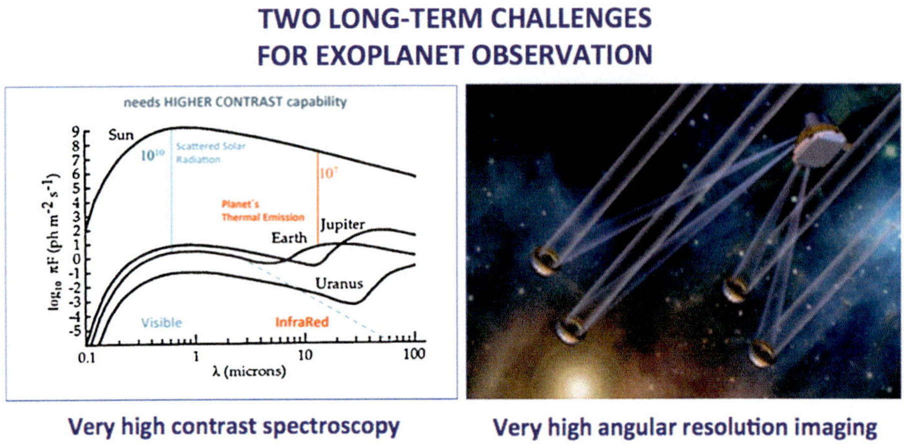

Fig. 13 Illustration of the two types of technical challenges that have to be met in order to observe biosignatures by spectroscopy in an exoplanet's atmosphere (*left panel*) or by interferometric imaging (*right panel*): in the most favourable IR band, the contrast of expected spectral lines to starlight is at best 10^{-6}; imaging the disk of an exoplanet (*right panel*) 10 parsecs away at the modest resolution of 1000 km, using one of the configurations designed for ESA's Darwin project would require 1000 km-long interferometric arms. Adapted from Fridlund et al. (2016 and this volume)

Exoplanet Survey Satellite (TESS) (Ricker et al. 2009), the Swiss-ESA Characterizing ExOPlanets Satellite (CHEOPS) mission (Broeg et al. 2013) and ESA's PLAnetary Transits and Oscillations of stars (PLATO 2.0) (Rauer et al. 2014) satellite will discover and characterize low and high-mass exoplanets by the transit technique. TESS's primary goal will be to identify transit events of terrestrial planets orbiting nearby dwarf stars, while PLATO will mostly focus on the search for transit events of Earth-like planets in the habitable zones of bright solar-like stars.

Major advances are also expected in the spectroscopy of planetary atmospheres. Future space observatories such as the 6.5-m aperture James Webb Space Telescope (JWST) to be launched in 2018, which is optimized for observations at near- and mid-IR, will enable us to derive precise chemical abundances for transiting exoplanets. One major UV telescope beyond HST is the World Space Observatory—UltraViolet (WSO-UV) project (see Shustov et al. 2015), a 2-m class telescope fully devoted to UV photometric and spectral observations, planned to be launched in 2022. Even further, giant ground-based optical telescopes will provide spectacular advances using both spectrometry and imaging, in particular with the first light of the E-ELT around 2024.

In the mid-to-long term future, challenges are well identified, but ways to solutions still have to be found to a large extent: to reach our ultimate goal of detecting biosignatures at Earth-like planets in the habitable zones of M-type or solar-type stars, as illustrated in Fig. 13, we have to solve two major technical challenges: one in high-contrast spectroscopy, and one in very-high-resolution imaging.

In high-contrast spectroscopy first, one needs to subtract the effects of the dominant brightness of stars to look at spectral signatures in close-in planetary atmospheres which are in the best case one million times fainter. Several families of technical solutions, all very challenging, will have to be explored: stellar coronagraphy, interferometry, external occultation techniques... Farther away in time, to reach a 1000-km spatial resolution in imaging in the mid-IR of the disk of an exoplanet 10 parsec away, we will need to reach an angular resolution of 10^{-6} arcsec. This can be done only with an interferometer, using a base 1000-km long at least. There is no doubt that ambitious technology research and development projects are the key to the ultimate discovery of life "elsewhere in the Universe". This objective is well worth the effort!

6 Conclusion

We discussed the role and importance of disk-planet relations throughout the most active period of their host star lifetime, from the initial formation of the disk through its evolution and to the simultaneous generation of planetary systems, planetary atmospheres and debris disks. The initial conditions in the dust and gas disk as well as the temporal evolution of the activity of its host star lead to the emergence of a diversity of protoplanets and planetary orbits which is significantly broader than in our Solar System. The growing protoplanets evolve into gas giants, Neptune-like, sub-Neptune-size or terrestrial planets. The resulting architectures of exoplanetary systems are also very diverse, and none of them up to now were found to resemble the architecture of our own Solar System. Proto-atmospheres, and ultimately atmospheres, are an important final product of disk-planet relations. Much effort will be devoted in the coming decades to characterize them and study their composition, with the ultimate objective of finding signatures of life in an alien world.

Acknowledgements This brief introduction to the role of protoplanetary disks in planetary formation and evolution was conceived during the august 2014 joint ISSI-Beijing/ISSI workshop held in Beijing on "The Disk in Relation to the Formation of Planets and their Protoatmospheres". HL and MB thank ISSI-Beijing for its much-appreciated hospitality during the workshop, and all the participants of the workshop and contributors to this special issue of Space Science Reviews and to the corresponding SSSI book for lively discussions and highly appreciated efforts. HL also thanks the Austrian Science Fund (FWF) NFN project S11601-N16 "Pathways to Habitability: From Disks to Stars, Planets and Life" and the related FWF NFN subprojects S11607-N16 "Particle/Radiative Interactions with Upper Atmospheres of Planetary Bodies Under Extreme Stellar Conditions". We also thank three referees for their detailed suggestions and recommendations which helped us to improve considerably this introductory chapter, and Guest Editor Veerle Jasmin Serken for her patient and efficient handling of the review of this manuscript. Finally, HL thanks P. Cubillos and L. Fossati for discussions related to radius-mass relations of sub-Neptunes and observations.

References

S.M. Andrews, K.A. Rosenfeld, A.L. Kraus, D.J. Wilner, The mass dependence between protoplanetary disks and their stellar hosts. Astrophys. J. **771**, 129–200 (2013)

R. Alexander, P.J. Armitage, Giant planet migration, disk evolution, and the origin of transitional disks. Astrophys. J. **704**, 989–1001 (2009)

R. Alexander, I. Pascucci, Deserts and pile-ups in the distribution of exoplanets due to photoevaporative disc clearing. Mon. Not. R. Astron. Soc. **422**, 82–86 (2012)

C.M.O'D. Alexander, R. Bowden, M.L. Fogel, K.T. Howard, C.D.K. Herd, N.R. Nittler, The provenances of asteroids, and their contributions to the volatile inventories of the terrestrial planets. Science **337**, 721–723 (2012)

J.M. Alcalá, E. Covino, G. Tottres, M.F. Sterzik, M.J. Pfeiffer, R. Neuhäuser, High-resolution spectroscopy of ROSAT low-mass pre-main sequence stars in Orion. Astron. Astrophys. **353**, 186–201 (2000)

Y. Alibert, C. Mordasini, W. Benz, Migration and giant planet formation. Astron. Astrophys. **417**, L25–L28 (2004)

Y. Alibert, C. Mordasini, W. Benz, C. Winisdoerffer, Models of giant planet formation with migration and disc evolution. Astron. Astrophys. **434**, 343–353 (2005a)

Y. Alibert, O. Mousis, C. Mordasini, W. Benz, New Jupiter and Saturn models meet observations. Astrophys. J. **626**, L57–L60 (2005b)

Y. Alibert, I. Baraffe, W. Benz, G. Chabrier, C. Mordasini, C. Lovis, M. Mayor, F. Pepe, F. Bouchy, D. Queloz, S. Udry, Formation and structure of the three Neptune-mass planets system around HD 69830. Astron. Astrophys. **455**, L25–L28 (2006)

Y. Alibert, C. Broeg, W. Benz, G. Wuchterl, O. Grasset, C. Sotin, C. Eiroa, Th. Henning, T. Herbst, L. Kaltenegger, A. Léger, R. Liseau, H. Lammer, C. Beichman, W. Danchi, M. Fridlund, J. Lunine, F. Paresce, A. Penny, A. Quirrenbach, H. Röttgering, F. Selsis, J. Schneider, D. Stam, G. Tinetti, G.J. White, Origin and formation of planetary systems. Astrobiology **10**, 19–32 (2010)

S.M. Andrews, D.J. Wilner, A.M. Hughes, C. Qi, C.P. Dullemond, Protoplanetary disk structures in Ophiuchus. Astrophys. J. **700**, 1502–1523 (2009)

G. Aresu, I. Kamp, R. Meijerink, P. Woitke, W.F. Thi, M. Spaans, X-ray impact on the protoplanetary disks around T Tauri stars. Astron. Astrophys. **526**, A163 (2011). 6 pp.

S.A. Balbus, J.F. Hawley, A powerful local shear instability in weakly magnetized disks. I—Linear analysis. II—Nonlinear evolution. Astrophys. J. **376**, 214–233 (1991)

S. Ballard, J.A. Johnson, The Kepler dichotomy among the M dwarfs: Half of systems contain five or more coplanar planets. Astrophys. J. **816**, 66 (2016)

T.S. Barman, Q.M. Konopacky, B. Macintosh, C. Marois, Simultaneous detection of water, methane, and carbon monoxide in the atmosphere of exoplanet HR8799b. Astrophys. J. **804**, 61 (2015). 10 pp.

T. Barclay, J.F. Rowe, J.J. Lissauer (the Kepler Science Team), A sub-Mercury-sized exoplanet. Nature **496**, 252 (2013)

C. Baruteau, J.C.B. Papaloizou, Disk-planets interactions and the diversity of period ratios in Kepler's multi-planetary systems. Astrophys. J. **778**, A7 (2013). 15 pp.

C. Baruteau, A. Crida, S.-J. Paardekooper, F. Masset, J. Guilet, B. Bitsch, R. Nelson, W. Kley, J. Papaloizou, Planet-disk interactions and early evolution of planetary systems, in *Protostars and Planets VI*, ed. by H. Beuther, R.S. Kessen, C.P. Dullemond, T. Henning (University of Arizona Press, Tucson, 2014), pp. 667–689

C. Baruteau, X. Bai, C. Mordasini, P. Molliere, Formation, orbital and internal evolutions of young planetary systems. Space Sci. Rev. **205**, 77–124 (2016). ibid. (2017, this volume)

J.L. Bean, E. Miller-Ricci Kempton, D. Homeier, A ground-based transmission spectrum of the super-Earth exoplanet GJ 1214b. Nature **468**, 669–672 (2010)

R.H. Becker, R.N. Clayton, E.M. Galimov, H. Lammer, B. Marty, R.O. Pepin, R.O. Wieler, Isotopic signatures of volatiles in terrestrial planets. Space Sci. Rev. **106**, 377–410 (2003)

T. Birnstiel, M. Fang, A. Johansen, Dust evolution and the formation of planetesimals. Space Sci. Rev. **205**, 41–75 (2016). ibid. (2017, this volume)

B. Bitsch, M. Lambrechts, A. Johansen, The growth of planets by pebble accretion in evolving protoplanetary discs. Astron. Astrophys. **582**, A112 (2015a). 24 pp.

B. Bitsch, A. Johansen, M. Lambrechts, A. Morbidelli, The structure of protoplanetary discs around evolving young stars. Astron. Astrophys. **575**, A28 (2015b). 17 pp.

V. Bourrier, A. Lecavelier des Etangs, 3D model of hydrogen atmospheric escape from HD 209458b and HD 189733b: Radiative blow-out and stellar wind interactions. Astron. Astrophys. **557**, A124 (2013). 18 pp.

A.P. Boss, Rapid formation of outer giant planets by disk instability. Astrophys. J. **599**, 577–581 (2003)

C. Broeg, A. Fortier, D. Ehrenreich (CHEOPS the CHEOPS Team), A transit photometry mission for ESA's small mission programme. EPJ Web Conf. **47**, 03005 (2013)

M. Brogi, I.A.G. Snellen, R.J. de Kok, S. Albrecht, J.L. Birkby, E.J.W. de Mooij, Detection of molecular absorption in the dayside of exoplanet 51 Pegasi b? Astrophys. J. **767**, 27 (2013). 10 pp.

L.A. Buchhave, M. Bizzarro, D.W. Latham, D. Sasselov, W.D. Cochran, M. Endl, H. Isaacson, D. Juncher, G.W. Marcy, Three regimes of extrasolar planet radius inferred from host star metallicities. Nature **509**, 593–595 (2014)

P. Caselli, C. Ceccarelli, Our astrochemical heritage. Astron. Astrophys. Rev. **20**, A56 (2012). https://doi.org/10.1007/s00159-012-0056-x

P. Cubillos, E.V. Erkaev, I. Juvan, L. Fossati, C.P. Johnstone, H. Lammer, M. Lendl, P. Odert, K.G. Kislyakova, An overabundance of low-density Neptune-like planets. Mon. Not. R. Astron. Soc. **466**, 1868–1879 (2017)

K.-M. Dittkrist, C. Mordasini, H. Klahr, Y. Alibert, T. Henning, Impacts of planet migration models on planetary populations. Effects of saturation, cooling and stellar irradiation. Astron. Astrophys. **567**, A121 (2014). 18 pp.

D. Deming, A. Wilkins, P.A. McCullough, J. Fortney, E. Agol, I. Dobbs-Dixon, N. Madhusudhan, N. Crouzet, J.-M. Desert, R.L. Gilliland, K. Haynes, H.A. Knutson, M. Line, Z. Magic, A.M. Mandell, S. Ranjan, D. Charbonneau, M. Clampin, S. Seager, A.P. Showman, Infrared transmission spectroscopy of the exoplanets HD 209458b and XO-1b using the wide field camera-3 on the Hubble Space Telescope. Astrophys. J. **774**, 95 (2013). 17 pp.

D. Ehrenreich, A. Lecavelier Des Etangs, G. Hébrard, J.-M. Désert, A. Vidal-Madjar, J.C. McConnell, C.D. Parkinson, G.E. Ballester, R. Ferlet, New observations of the extended hydrogen exosphere of the extrasolar planet HD 209458b. Astron. Astrophys. **483**, 933–937 (2008)

D. Ehrenreich, X. Bonfils, C. Lovis, X. Delfosse, T. Forveille, M. Mayor, V. Neves, N.C. Santos, S. Udry, D. Ségransan, Near-infrared transmission spectrum of the warm-Uranus GJ 3470b with the Wide Field Camera-3 on the Hubble Space Telescope. Astron. Astrophys. **570**, A89 (2014). 24 pp.
D. Ehrenreich, V. Bourrier, P.J. Wheatley, A. Lecavelier Des Etangs, G. Hébrard, S. Udry, X. Bonfils, X. Delfosse, J.-M. Désert, D.K. Sing, A. Vidal-Madjar, A giant comet-like cloud of hydrogen escaping the warm Neptune-mass exoplanet GJ 436b. Nature **522**, 459–461 (2015)
L.T. Elkins-Tanton, Linked magma ocean solidification and atmospheric growth for Earth and Mars. Earth Planet. Sci. Lett. **271**, 181–191 (2008)
L.T. Elkins-Tanton, Formation of water ocean on rocky planets. Astrophys. Space Sci. **332**, 359–364 (2011)
T. Encrenaz, Water in the solar system. Annu. Rev. Astron. Astrophys. **46**, 57–87 (2008)
M. Fang, J.S. Kim, R. van Boekel, A. Sicilia-Aguilar, T. Henning, K. Flaherty, Young stellar objects in Lynds 1641: disks, accretion, and star formation history. Astrophys. J. Suppl. **207**, 5 (2013). 39 pp.
D. Fedele, M.E. van den Ancker, T. Henning, R. Jayawardhana, J.M. Oliveira, Timescale of mass accretion in pre-main-sequence stars. Astron. Astrophys. **510**, A72 (2010). 7 pp.
D.A. Fischer, J. Valenti, The planet-metallicity correlation. Astrophys. J. **622**, 1102–1117 (2005)
M. Fischer-Gödde, T. Kleine, Ruthenium isotopic evidence for an inner solar system origin of the late veneer. Nature **541**, 525–527 (2017)
J.J. Fortney, K. Lodders, M.S. Marley, R.S. Freedman, A unified theory for the atmospheres of the hot and very hot Jupiters: two classes of irradiated atmospheres. Astrophys. J. **678**, 1419–1435 (2008)
L. Fossati, S. Bagnulo, A. Elmasli, C.A. Haswell, S. Holmes, O. Kochukhov, E.L. Shkolnik, D.V. Shulyak, D. Bohlender, B. Albayrak, C. Froning, L. Hebb, A detailed spectropolarimetric analysis of the planet-hosting star WASP-12. Astrophys. J. **720**, 872–886 (2010)
L. Fossati, N.V. Erkaev, H. Lammer, P.E. Cubillos, P. Odert, I. Juvan, K.G. Kislyakova, M. Lendl, D. Kubyshkina, S.J. Bauer, Aeronomical constraints to the minimum mass and maximum radius of hot low-mass planets. Astron. Astrophys. **598**, A90 (2017). 9 pp.
F. Fressin, G. Torres, D. Charbonneau, S.T. Bryson, J. Christiansen, C.D. Dressing, J.M. Jenkins, L.M. Walkowicz, N.M. Batalha, The false positive rate of Kepler and the occurrence of planets. Astrophys. J. **766**, 81 (2013). 20 pp.
M. Fridlund, A. Hatzes, R. Liseau, The way forward. Space Sci. Rev. **205**, 349–372 (2016). ibid. (2017, this volume)
C. Gammie, Layered accretion in T Tauri disks. Astrophys. J. **457**, 355–362 (1996)
U. Gorti, D. Hollenbach, C.P. Dullemond, The impact of dust evolution and photoevaporation on disk dispersal. Astrophys. J. **804**, A29 (2015). 21 pp.
U. Gorti, R. Liseau, Z. Sandor, C. Clarke, Disk dispersal: theoretical understanding and observational constraints. Space Sci. Rev. **205**, 125–152 (2016). ibid. (2017, this volume)
K.E. Haisch Jr., E.A. Lada, C.J. Lada, Disk frequencies and lifetimes in young clusters. Astrophys. J. **553**, L153–L156 (2001)
K. Hamano, Y. Abe, H. Genda, Emergence of two types of terrestrial planet on solidification of magma ocean. Nature **497**, 607–610 (2013)
B.M.S. Hansen, N. Murray, Testing in-situ assembly with the Kepler planet candidate sample. Astrophys. J. **775**, 53 (2013). 17 pp.
L. Hartmann, *Accretion Processes in Star Formation* (Cambridge University Press, Cambridge, 2009)
A.R. Howe, A.S. Burrows, Theoretical transit spectra for GJ 1214b and other "super-Earths". Astrophys. J. **756**, 176 (2012). 14 pp.
C.A. Haswell, L. Fossati, T. Ayres, K. France, C.S. Froning, S. Holmes, U.C. Kolb, R. Busuttil, R.A. Street, L. Hebb, A. Collier Cameron, B. Enoch, V. Burwitz, J. Rodriguez, R.G. West, D. Pollacco, P.J. Wheatley, A. Carter, Near-ultraviolet absorption, chromospheric activity, and star-planet interactions in the WASP-12 system. Astrophys. J. **760**, A9 (2012). 23 pp.
A. Hatzes, The architecture of exoplanet systems. Space Sci. Rev. **205**, 267–283 (2016). ibid. (2017, this volume)
C. Hayashi, Structure of the solar nebula, growth and decay of magnetic fields and effects of magnetic and turbulent viscosities on the nebula. Prog. Theor. Phys. Suppl. **70**, 35–53 (1981)
T. Henning, D. Semenov, Chemistry in protoplanetary disks. Chem. Rev. **113**, 9016–9042 (2013)
T. Henning, G. Meeus, in *Physical Processes in Circumstellar Disks Around Young Stars*, ed. by P.J.V. Grady (Chicago University Press, Chicago, 2011), p. 114
H.D. Holland, *The Chemical Evolution of the Atmosphere and Oceans* (Princeton University Press, Princeton, 1984). 598 pp.
I. Hubeny, A. Burrows, D. Sudarsky, A possible bifurcation in atmospheres of strongly irradiated stars and planets. Astrophys. J. **594**, 1011–1018 (2003)
S. Ida, D.N.C. Lin, Toward a deterministic model of planetary formation. II. The formation and retention of gas giant planets around stars with a range of metallicities. Astrophys. J. **616**, 567–572 (2004)

M. Ikoma, K. Nakazawa, H. Emori, Formation of giant planets: dependences on core accretion rate and grain opacity. Astrophys. J. **537**, 1013–1025 (2000)
M. Ikoma, Y. Hori, In-situ accretion of hydrogen-rich atmospheres on short-period super-Earths: implications for the Kepler-11 planets. Astrophys. J. **753**, A66 (2012). 6 pp.
M. Ilgner, Th. Henning, A.J. Markwick, T.J. Millar, Transport processes and chemical evolution in steady accretion disk flows. Astron. Astrophys. **415**, 643–659 (2004)
E. Jarosewich, Chemical analyses of meteorites: a compilation of stony and iron meteorite analyses. Meteoritics **25**, 323–337 (1990)
C.P. Johnstone, M. Güdel, A. Stökl, H. Lammer, L. Tu, K.G. Kislyakova, T. Lüftinger, P. Odert, N.V. Erkaev, E.A. Dorfi, The evolution of stellar rotation and the hydrogen atmospheres of habitable-zone terrestrial planets. Astrophys. J. Lett. **815**, A12 (2015). 6 pp.
I. Kant, *Universal Natural Theory and Theory of Heaven* (1755)
J.F. Kasting, D.H. Eggler, S.P. Raeburn, Mantle redox evolution and the case for a reduced Archean atmosphere. J. Geol. **101**, 245–257 (1993)
K.G. Kislyakova, M. Holmström, H. Lammer, P. Odert Petra, M.L. Khodachenko, Magnetic moment and plasma environment of HD 209458b as determined from Lyα observations. Science **346**, 981–984 (2014)
W. Kley, R.P. Nelson, Planet-disk interaction and orbital evolution. Annu. Rev. Astron. Astrophys. **50**, 211–249 (2012)
H.A. Knutson, D. Charbonneau, L.E. Allen, A. Burrows, S.T. Megeath, The 3.6–8.0 μm broadband emission spectrum of HD 209458b: evidence for an atmospheric temperature inversion. Astrophys. J. **673**, 526–531 (2008)
H.A. Knutson, A.W. Howard, H. Isaacson, A correlation between stellar activity and hot Jupiter emission spectra. Astrophys. J. **720**, 1569–1576 (2010)
R.J. de Kok, M. Brogi, I.A.G. Snellen, J. Birkby, S. Albrecht, E.J.W. de Mooij, Detection of carbon monoxide in the high-resolution day-side spectrum of HD 189733b. Astron. Astrophys. **554**, A82 (2013). 9 pp.
J.R. Kulow, K. France, J. Linsky, L.R.O. Parker, Lyα transit spectroscopy and the neutral hydrogen tail of the hot Neptune GJ 436b. Astrophys. J. **786**, A132 (2014). 9 pp.
M. Lambrechts, A. Johansen, Rapid growth of gas-giant cores by pebble accretion. Astron. Astrophys. **544**, A32 (2012). 13 pp.
M. Lambrechts, A. Johansen, A. Morbidelli, Separating gas-giant and ice-giant planets by halting pebble accretion. Astron. Astrophys. **572**, A35 (2014). 12 pp.
F. Lahuis, E.F. van Dishoeck, A.C.A. Boogert, K.M. Pontoppidan, G.A. Blake, C.P. Dullemond, N.J.I.I. Evans, M.R. Hogerheijde, J.K. Jørgensen, J.E. Kessler-Silacci, C. Knez, Hot organic molecules toward a young low-mass star: a look at inner disk chemistry. Astrophys. J. **636**, L145–L148 (2006)
H. Lammer, K.G. Kislyakova, P. Odert, M. Leitzinger, R. Schwarz, E. Pilat-Lohinger, Yu.N. Kulikov, M.L. Khodachenko, M. Güdel, A. Hanslmeier, Pathways to Earth-like atmospheres extreme ultraviolet (EUV)-powered escape of hydrogen-rich protoatmospheres. Orig. Life Evol. Biosph. **41**, 503–522 (2011)
H. Lammer, A. Stökl, N.V. Erkaev, E. Dorfi, P. Odert, M. Güdel, Yu.N. Kulikov, K.G. Kislyakova, M. Leitzinger, Origin and loss of nebula-captured hydrogen envelopes from 'sub'- to 'super-Earths' in the habitable zone of Sun-like stars. Mon. Not. R. Astron. Soc. **439**, 3225–3238 (2014)
H. Lammer, N.V. Erkaev, L. Fossati, I. Juvan, P. Odert, P.E. Cubillos, E. Guenther, K.G. Kislyakova, C.P. Johnstone, T. Lüftinger, M. Güdel, Identifying the "true" radius of the sub-Neptunes CoRoT-24b by mass loss modelling. Mon. Not. R. Astron. Soc. **461**, L62–L66 (2016)
T. Lebrun, H. Massol, E. Chassefiére, A. Davaille, E. Marcq, P. Sarda, F. Leblanc, G. Brandeis, Thermal evolution of an early magma ocean in interaction with the atmosphere. J. Geophys. Res., Planets **118**, 1155–1176 (2013)
P.-S. Laplace. *Exposition du Système du Monde* (1796)
G. Lee, I. Doobs-Dixon, Ch. Helling, K. Bognar, P. Woitke, Dynamic mineral clouds on HD 189733b—I. 3D RHD with kinetic, non-equilibrium cloud formation. Astron. Astrophys. **594**, A48 (2016)
D.N.C. Lin, J. Papaloizou, On the tidal interaction between protoplanets and the protoplanetary disk. III—Orbital migration of protoplanets. Astrophys. J. **309**, 846–857 (1986)
D.N.C. Lin, P. Bodenheimer, D.C. Richardson, Orbital migration of the planetary companion of 51 Pegasi to its present location. Nature **380**, 606–607 (1996)
J.J. Lissauer, D. Fabrycky, C. Daniel (the Kepler Science Team), A closely packed system of low-mass, low-density planets transiting Kepler-11. Nature **470**, 53–58 (2011)
L.-G. Liu, The inception of the oceans and CO_2-atmosphere in the early history of the Earth. Earth Planet. Sci. Lett. **227**, 179–184 (2004)
R. Luger, R. Barnes, E. Lopez, J. Fortney, B. Jackson, V. Meadows, Habitable evaporated cores: transforming mini-Neptunes into super-Earths in the habitable zones of M dwarfs. Astrobiology **15**, 57–88 (2015)

N. Madhusudhan, S. Seager, High metallicity and non-equilibrium chemistry in the dayside atmosphere of hot-Neptune GJ 436b. Astrophys. J. **729**, A41 (2011)
N. Madhusudhan, M.A. Amin, G.M. Kennedy, Toward chemical constraints on hot Jupiter migration. Astrophys. J. Lett. **794**, L12 (2014). 5 pp.
N. Madhusudhan, M. Agundez Chico, J.I. Moses, Y. Hu, Exoplanetary atmospheres—chemistry, formation conditions, and habitability. Space Sci. Rev. **205**, 285–348 (2016). ibid. (2017, this volume)
G.W. Marcy, H. Isaacson, A.W. Howard (the Kepler Science Team), Masses, radii, and orbits of small Kepler planets: the transition from gaseous to rocky planets. Astrophys. J. Suppl. **210**, 20 (2014). 70 pp.
B. Marty, The origins and concentrations of water, carbon, nitrogen and noble gases on Earth. Earth Planet. Sci. Lett. **313**, 56–66 (2012)
H. Massol, K. Hamano, F. Tian, M. Ikoma, Y. Abe, E. Chassefière, A. Davaille, H. Genda, M. Güdel, Y. Hori, F. Leblanc, E. Marcq, P. Sarda, V.I. Shematovich, A. Stökl, H. Lammer, Formation and evolution of protoatmospheres. Space Sci. Rev. **205**, 153–211 (2016). ibid. (2017, this volume)
B.C. Matthews, A.V. Krivov, M.C. Wyatt, G. Bryden, C. Eiroa, Observations, modeling, and theory of debris disks, in *Protostars and Planets*, vol. VI (2014), pp. 521–544. https://doi.org/10.2458/azu_uapress_9780816531240-ch023
B.C. Matthews, J.J. Kavelaars, Insights into planet formation from debris disks. I. The solar system as an archetype for planetesimal evolution. Space Sci. Rev. **205**, 213–230 (2016). ibid. (2017, this volume)
M. Mayor, D. Queloz, A Jupiter-mass companion to a solar-type star. Nature **378**, 355–359 (1995)
H. Mizuno, Formation of the giant planets. Prog. Theor. Phys. **64**, 544–557 (1980)
S. Mohanty, J. Greaves, D. Mortlock, I. Pascucci, A. Scholz, M. Thompson, D. Apai, G. Lodato, D. Looper, Protoplanetary disk masses from stars to brown dwarfs. Astrophys. J. **773**, 168 (2013). 33 pp.
A. Morbidelli, J. Chambers, J.I. Lunine, J.M. Petit, F. Robert, G.B. Valsecchi, K.E. Cyr, Source regions and time scales for the delivery of water to Earth. Meteorit. Planet. Sci. **35**, 1309–1320 (2000)
A. Morbidelli, D. Nesvorný, Dynamics of pebbles in the vicinity of a growing planetary embryo: hydrodynamical simulations. Astron. Astrophys. **546**, A18 (2012). 7 pp.
C. Mordasini, Y. Alibert, W. Benz, Extrasolar planet population synthesis I: method, formation tracks and mass-distance distribution. Astron. Astrophys. **501**, 1139–1160 (2009a)
C. Mordasini, Y. Alibert, W. Benz, D. Naef, Extrasolar planet population synthesis II: statistical comparison with observation. Astron. Astrophys. **501**, 1161–1184 (2009b)
C. Mordasini, Y. Alibert, C. Georgy, K.-M. Dittkrist, H. Klahr, T. Henning, Characterization of exoplanets from their formation. II. The planetary mass-radius relationship. Astron. Astrophys. **547**, A112 (2012). 36 pp.
G.D. Mulders, I. Pascucci, A. Dániel, An increase in the mass of planetary systems around lower-mass stars. Astrophys. J. **814**, 130 (2015a). 10 pp.
G.D. Mulders, I. Pascucci, A. Dániel, A stellar-mass-dependent drop in planet occurrence rates. Astrophys. J. **798**, 112 (2015b). 18 pp.
M.J. Mumma, S.B. Charnley, The chemical composition of comets—emerging taxonomies and natal heritage. Annu. Rev. Astron. Astrophys. **49**, 471–524 (2011)
J.R. Najita, M. Ádámkovics, A.E. Glassgold, Formation of organic molecules and water in warm disk atmospheres. Astrophys. J. **743**, A147 (2011). 18 pp.
K.I. Öberg, R. Murray-Clay, E. Bergin, A. Edwin, The effects of snowlines on C/O in planetary atmospheres. Astrophys. J. Lett. **743**, L16 (2011). 5 pp.
D.P. O'Brien, K.J. Walsh, A. Morbidelli, S.N. Raymond, Water delivery and giant impacts in the 'Grand Tack' scenario. Icarus **239**, 74–84 (2014)
P. Odert, H. Lammer, N.V. Erkaev, A. Nikolaou, H.I.M. Lichtenegger, C.P. Johnstone, K.G. Kislyakova, M. Leitzinger, N. Tosi, Escape and fractionation of volatiles and noble gases from Mars-sized planetary embryos and growing protoplanets. Icarus (2018). https://doi.org/10.1016/j.icarus.2017.10.031
C.W. Ormel, H.H. Klahr, The effect of gas drag on the growth of protoplanets. Analytical expressions for the accretion of small bodies in laminar disks. Astron. Astrophys. **530**, A43 (2010). 15 pp.
J.E. Owen, S. Mohanty, Habitability of terrestrial-mass planets in the HZ of M Dwarfs. I. H/He-dominated atmospheres. Mon. Not. R. Astron. Soc. **459**(4), 4088–4108 (2016)
J.E. Owen, Y. Wu, Atmospheres of low-mass planets: the "Boil-off". Astrophys. J. **817**, 107 (2016). 14 pp.
S.-J. Paardekooper, C. Baruteau, F. Meru, Numerical convergence in self-gravitating disc simulations: initial conditions and edge effects. Mon. Not. R. Astron. Soc. **416**, L65–L69 (2011)
O. Paníc, W.S. Holland, M.C. Wyatt, G.M. Kennedy, B.C. Matthews, J.F. Lestrade, B. Sibthorpe, J.S. Greaves, J.P. Marshall, N.M. Phillips, J. Tottle, First results of the SONS survey: submillimetre detections of debris discs. Mon. Not. R. Astron. Soc. **435**, 1037–1046 (2013). https://doi.org/10.1093/mnras/stt1293
H. Parviainen, D. Gandolfi, M. Deleuil (the CoRoT Team), Transiting exoplanets from the CoRoT space mission. XXV. CoRoT-27b: a massive and dense planet on a short-period orbit. Astron. Astrophys. **562**, A140 (2014). 12 pp.

I. Pascucci, D. Apai, K. Luhman, Th. Henning, J. Bouwman, M.R. Meyer, F. Lahuis, A. Natta, The different evolutions of gas and dust in disks around Sun-like and cool stars. Astrophys. J. **696**, 143–159 (2009)
I. Pascucci, G. Herczeg, J.S. Carr, S. Bruderer, The atomic and molecular content of disks around very low-mass stars and brown dwarfs. Astrophys. J. **779**, 178 (2013). 13 pp.
C. Petrovich, R. Malhotra, T. Scott, Planets near mean-motion resonances. Astrophys. J. **770**, 24 (2013). 16 pp.
S. Pizzarello, Y. Huang, The deuterium enrichment of individual amino acids in carbonaceous meteorites: a case for the presolar distribution of biomolecule precursors. Geochim. Cosmochim. Acta **69**, 599–605 (2005)
J.B. Pollack, O. Hubickyj, P. Bodenheimer, J.J. Lissauer, M. Podolak, Y. Greenzweig, Formation of the giant planets by concurrent accretion of solids and gas. Icarus **124**, 62–85 (1996)
D. Porcelli, R.O. Pepin, Rare gas constraints on early Earth history, in *Origin of the Earth and Moon*, ed. by R.M. Canup, K. Richter (University of Arizona Press, Tucson, 2000), pp. 435–458
C. Rab, C. Baldovin-Saavedra, O. Dionatos, E. Vorobyov, M. Güdel, The gas disk: evolution and chemistry. Space Sci. Rev. **205**, 3–40 (2016). ibid. (2017, this volume)
R.R. Rafikov, Atmospheres of protoplanetary cores: critical mass for nucleated instability. Astrophys. J. **648**, 666–682 (2006)
H. Rauer, C. Catala, C. Aerts (the PLATO Team), The PLATO 2.0 mission. Exp. Astron. **38**, 249–330 (2014)
G.R. Ricker, S.W. Latham, R.K. Vanderspek (the TESS Team), The Transiting Exoplanet Survey Satellite (TESS). Bull. Am. Astron. Soc. **41**, 193 (2009)
L.A. Rogers, P. Bodenheimer, J.J. Lissauer, S. Seager, Formation and structure of low-density exo-Neptunes. Astrophys. J. **738**, 59 (2011). 16 pp.
L.A. Rogers, Most 1.6Earth-radius planets are not rocky. Astrophys. J. **801**, 41 (2015). 13 pp.
G.P. Rosotti, B. Ercolano, J.E. Owen, P.J. Armitage, The interplay between X-ray photoevaporation and planet formation. Mon. Not. R. Astron. Soc. **430**, 1392–1401 (2013)
J.F. Rowe, J.L. Coughlin, V. Antoci, T. Barclay, N.M. Batalha, W.J. Borucki, C.J. Burke, S.T. Bryson, D.A. Caldwell, J.R. Campbell, J.H. Catanzarite, J.L. Christiansen, W. Cochran, R.L. Gilliland, F.R. Girouard, M.R. Haas, K.G. Hełminiak, C.E. Henze, K.L. Hoffman, S.B. Howell, D. Huber, R.C. Hunter, H. Jang-Condell, J.M. Jenkins, T.C. Klaus, D.W. Latham, J. Li, J.J. Lissauer, S.D. McCauliff, R.L. Morris, F. Mullally, A. Ofir, B. Quarles, E. Quintana, A. Sabale, S. Seader, A. Shporer, J.C. Smith, J.H. Steffen, M. Still, P. Tenenbaum, S.E. Thompson, J.D. Twicken, C. Van Laerhoven, A. Wolfgang, K.A. Zamudio, Planetary candidates observed by Kepler V: planet sample from Q1–Q12 (36 Months). Astrophys. J. Suppl. **217**, 16 (2015). 22 pp.
V. Safronov, Evolution of the protoplanetary cloud and formation of the Earth and planets (English version 1972)
N.C. Santos, G. Israelian, M. Mayor, New extra-solar planets: the metallicity distribution revised, in *Planetary Systems in the Universe—Observation and Evolution*, ed. by A. Penny, P. Artymovicz, A.-M. Lagrange, S. Russel. Proceedings IAU Symposium, vol. 202 (2004), pp. 118–120
B.M. Shustov, E. Mikhail, D. Bisikalo, V. de Castro, A.-I. Gómez, The World Space Observatory—UV project as a tool for exoplanet science, in *Characterizing Stellar and Exoplanetary Environments*. Astrophysics and Space Science Library, vol. 411 (Springer, Cham, 2015), pp. 275–287
A. Sicilia-Aguilar, L.W. Hartmann, G. Fürész, T. Henning, C. Dullemond, W. Brandner, High-resolution spectroscopy in Tr 37: gas accretion evolution in evolved dusty disks. Astron. J. **132**, 2135–2155 (2006)
I.A.G. Snellen, R.J. de Kok, E.J.W. de Mooij, A. Simon, The orbital motion, absolute mass and high-altitude winds of exoplanet HD209458b. Nature **465**, 1049–1051 (2010)
D.S. Spiegel, K. Silverio, A. Burrows, Can TiO explain thermal inversions in the upper atmospheres of irradiated giant planets? Astrophys. J. **699**, 1487–1500 (2009)
S.W. Stahler, F. Palla, *The Formation of Stars* (Wiley, Weinheim, 2005)
D.J. Stevenson, Formation of the giant planets. Planet. Space Sci. **30**, 755–764 (1982)
D.J. Stevenson, J.I. Lunine, Rapid formation of Jupiter by diffuse redistribution of water vapor in the solar nebula. Icarus **75**, 146–155 (1988)
A. Stökl, E.A. Dorfi, H. Lammer, Hydrodynamic simulations of captured protoatmospheres around Earth-like planets. Astron. Astrophys. **576**, A87 (2015). 11 pp.
E. Svedenborg, Opera philosophica et mineralia (1734)
H. Tanaka, T. Takeuchi, W.R. Ward, Three-dimensional interaction between a planet and an isothermal gaseous disk. I. Corotation and Lindblad torques and planet migration. Astrophys. J. **565**, 1257–1274 (2002)
L. Tu, C.P. Johnstone, M. Güdel, H. Lammer, The extreme ultraviolet and X-ray Sun in time: high-energy evolutionary tracks of a solar-like star. Astron. Astrophys. **577**, L3 (2015). 4 pp.
A. Vidal-Madjar, A. Lecavelier des Etangs, J.-M. Désert, G.E. Ballester, R. Ferlet, G. Hébrard, M. Mayor, An extended upper atmosphere around the extrasolar planet HD209458b. Nature **422**, 143–146 (2003)

A. Vidal-Madjar, J.-M. Désert, A. Lecavelier des Etangs, G. Hébrard, G.E. Ballester, D. Ehrenreich, R. Ferlet, J.C. McConnell, M. Mayor, C.D. Parkinson, Detection of oxygen and carbon in the hydrodynamically escaping atmosphere of the extrasolar planet HD 209458b. Astrophys. J. Lett. **604**, 69–72 (2004)

K.J. Walsh, A. Morbidelli, S.N. Raymond, D.P. O'Brien, A.M. Mandell, A low mass for Mars from Jupiter's early gas-driven migration. Nature **475**, 206–209 (2011)

J. Wang, D.A. Fischer, Revealing a universal planet-metallicity correlation for planets of different sizes around solar-type stars. Astron. J. **149**, 14 (2015). 7 pp.

W.R. Ward, Protoplanet migration by nebula tides. Icarus **126**, 261–281 (1997a)

W.R. Ward, Survival of planetary systems. Astrophys. J. **482**, L211–L214 (1997b)

J.P. Williams, L.A. Cieza, Protoplanetary disks and their evolution. Annu. Rev. Astron. Astrophys. **49**, 67–117 (2011)

K. Willacy, H.H. Klahr, T.J. Millar, Th. Henning, Gas and grain chemistry in a protoplanetary disk. Astron. Astrophys. **338**, 995–1005 (1998)

J.N. Winn, D.C. Fabrycky, The occurrence and architecture of exoplanetary systems. Annu. Rev. Astron. Astrophys. **53**, 409–447 (2015)

J.A. Wood, The chondrite types and their origins, in *Chondrites and the Protoplanetary Disk*, ed. by A.N. Krot, E.R.D. Scott, B. Reipurth. ASP Conference Series, vol. 341 (2005), pp. 953–971

R.D. Wordsworth, Atmospheric nitrogen evolution on Earth and Venus. Earth Planet. Sci. Lett. **447**, 103–111 (2016)

G. Wuchterl, The critical mass for protoplanets revised—massive envelopes through convection. Icarus **106**, 323–334 (1993)

M. Wyatt, A.P. Jackson, Insights into planet formation from debris disks. II. Giant impacts in extrasolar planetary systems. Space Sci. Rev. **205**, 231–265 (2016a). ibid. (2017, this volume)

Part I

From Protoplanetary Disks to Young Planets and Their Proto-atmospheres

Space Sci Rev (2016) 205:43–80
DOI 10.1007/s11214-016-0325-5

The Gas Disk: Evolution and Chemistry

Christian Rab[1] · Carla Baldovin-Saavedra[1] · Odysseas Dionatos[1] · Eduard Vorobyov[1,2] · Manuel Güdel[1]

Received: 12 January 2016 / Accepted: 28 November 2016 / Published online: 7 December 2016

Abstract Protoplanetary disks are the birthplaces of planetary systems. The evolution of the star-disk system and the disk chemical composition determines the initial conditions for planet formation. Therefore a comprehensive understanding of the main physical and chemical processes in disks is crucial for our understanding of planet formation. We give an overview of the early evolution of disks, discuss the importance of the stellar high-energy radiation for disk evolution and describe the general thermal and chemical structure of disks. Finally we provide an overview of observational tracers of the gas component and disk winds.

Keywords Stars: pre-main sequence · Stars: formation · Protoplanetary disks · Accretion, accretion disks · Planet-disk interactions · ISM: jets and outflow · Astrochemistry

1 Introduction

Disks around young stellar objects are the birthplaces of planetary systems. A comprehensive knowledge of their evolution, structure and chemical composition is therefore crucial for our understanding of planet formation.

Low mass stars like our Sun are formed with a disk component. Observations show that the fraction of disk-bearing stars in clusters with an age of $\approx$1 Myr is usually $\gtrsim$80 %. However, the disk fraction drops quite rapidly with cluster age. In clusters with ages of 2–3 Myr still $\approx$50 % of the stars show disks, whereas in older ($\approx$10 Myr) clusters the fraction of disk-bearing stars is lower than 5 % (e.g. Hernández et al. 2008; Fedele et al. 2010). This tells us that disks have a typical lifetime of a few million years. The lifetime of disks defines the timescale on which formation of massive (gaseous) planets can occur.

✉ C. Rab
christian.rab@univie.ac.at

1 Department of Astrophysics, University of Vienna, Türkenschanzstr. 17, 1180 Vienna, Austria

2 Research Institute of Physics, Southern Federal University, Stachki 194, Rostov-on-Don, 344090, Russia

Disks evolve on timescales of a few million years. The evolutionary stages are determined by the balance of accretion onto the disk and different disk dispersal/removal mechanisms (see Williams and Cieza 2011 for a review). There is observational evidence that disks form already within 10^4–10^5 yr after the gravitational collapse of the parental cloud-core (e.g. Murillo et al. 2013) at the phase where the star is still deeply embedded (Class-0 protostellar phase; see review of Li et al. 2014). After the formation disks are growing due to the continuous inflow of mass from their surrounding envelopes during the Class-0 and Class-I (the star becomes visible) protostellar phases. Once this reservoir is drained and the star reaches its Class-II phase (T Tauri star), disk erosion processes due to accretion onto the star, disk photoevaporation, jet acceleration, and planet formation lead to the disappearance of dust and gas disks within typically 2–3 Myr (although in some rare cases disks are still present at ages of $\sim$10 Myr). Observations of disks with large inner holes/gaps, the so called transitional disks, indicate that disk dispersal works from inside out. From the fraction of observed transition disks a timescale of $\gtrsim 10^5$ yr is derived for this period (see Espaillat et al. 2014 for a review). The final result of the gas disk dispersal are debris disks. These disks are mainly made up of asteroid/comet/planet like bodies and dust but contain no or only very little amount of gas. This evolutionary phase can last for Gyr. However, it is still unclear if every gaseous disks ends up as a debris disk (see Matthews et al. 2014 for a review).

During these evolutionary stages also the dust component of the disk evolves. Recent discoveries like the horseshoe-shaped structures of larger dust particles in transition disks (e.g. van der Marel et al. 2013; Pérez et al. 2014a) or the spectacular dark rings in the thermal dust emission of HL Tau (ALMA Partnership et al. 2015; Pinte et al. 2016) and TW Hya (Andrews et al. 2016) clearly show this. The dust component also influences the gas disk. Dust is a strong opacity source in the optical and ultraviolet and therefore absorbs most of the stellar radiation. This significantly affects the disk thermal structure and chemistry (see Sects. 5 and 6). However, our focus here is on the gas disk and we refer the reader to Chap. 3 of this book and to the review by Testi et al. (2014) for more details on dust evolution.

We now briefly summarize the essential steps of disk evolution from formation to dispersal due to various erosion processes. Three major processes: accretion, winds/outflows/jets and planet formation contribute to the slow removal of disk material, determining the typical lifetime of disks.

The first and best studied process is the *continuous accretion* of gas and dust through the disk onto the star. There are several possible mechanisms which can drive matter flows in the disk, of which the most prominent ones are: the magneto-rotational instability (MRI), magnetically launched jet winds, magneto-centrifugally driven winds, gravitational instability and hydrodynamical instabilities (see Turner et al. 2014 for a review). The first mechanism has been the most promising one to explain increased disk viscosity required for the observed accretion rate. However, MRI requires a disk ionization fraction of $\approx 10^{-12}$ which is probably not reached everywhere in the disk (e.g. Cleeves et al. 2015). The second mechanism leads to fast jets and associated molecular outflows, but here the acceleration mechanism of disk material along magnetic fields is poorly understood, and it is also still a matter of debate whether the launching mechanism of molecular outflows is directly associated with jet acceleration (see review of Frank et al. 2014). The other processes (turbulence, winds and instabilities) are sensitive to heating and cooling processes in the disk. Therefore high quality observational data in combination with detailed thermo-chemical models are needed to derive quantities such as the ionization fraction and the gas temperature (see Sects. 4, 5 and 6). In summary, the efficiency of all above processes remains unclear. We also note that disk accretion might affect the chemical composition due to additional heating and transport of material from colder to warmer regions of the disk (see Sect. 6), and jet formation may lead to additional irradiation of disks from above (see Sect. 3.5). Generally, self-consistent thermo-chemical and (magneto-)hydrodynamic simulations including feedback are needed for a full understanding of these processes.

Especially in their earlier phases, most disks probably undergo more *violent accretion phases*. The so-called FU Orionis objects show strong luminosity outbursts (increase in luminosity by factors of 10 to 100s). These outbursts typically last for decades or centuries. During this period the accretion rate rapidly increases from $\approx 10^{-7}$ to $\approx 10^{-4}$ $M_{\odot}\,yr^{-1}$ (see Audard et al. 2014 for a review). The cause of this phenomenon is still unclear. Proposed explanations are disk instabilities (thermal, magneto-rotational and/or gravitational instabilities) and the perturbation of the disk by an external body. Another possibility is disk fragmentation and the infall of the formed fragments onto the star. These outbursts also have an impact on the chemical evolution of the disk (see Sects. 2 and 6). Although rare, repeated FU Ori-type accretion events may be crucial in star and disk evolution as they may bring a significant amount of matter onto the star, and may at the same time dominate disk mass-loss due to accretion on long timescales.

The second major mass-loss processes are via the formation of *disk winds*, either magnetically driven or irradiation driven. We distinguish between fast, probably magnetically driven jets with (at least in the Class-I phase) associated molecular outflows, and radiation-driven slow photoevaporative winds (see Sect. 8, Chap. 6 of this book and the review of Alexander et al. 2014). In the T Tauri phase, mass-loss rates in outflows/jets and photoevaporative winds amount to no more than about 10 % of the stellar accretion rate (Hartigan et al. 1995; Cabrit 2002; Frank et al. 2014). However, in late evolutionary phases the photoevaporation rate may locally (at distances of several au from the star) exceed the disk accretion rate, which leads to the rapid formation of a gap and the subsequent erosion of the entire disk on time scales of only a few hundred kyr or $\approx$10 % of the disk lifetime. This rapid-erosion model is compatible with the relatively rare occurrence of transitional disks with such gaps and eventually inner holes, compared to classical, optically thick disks. An example of gas photoevaporation are winds driven by X-ray or extreme-ultraviolet irradiation of the disk (see Sects. 5 and 8), leading to strong surface heating and ionization and consequently escape of ions. Photoevaporation might have an impact on the final appearance of a planetary systems. In fact the models of Ercolano and Rosotti (2015) suggest that the distribution of the semi-major axes of exoplanet orbits is sensitive to photo-evaporation as it affects planet-migration.

The third process eroding a disk is *planet formation* itself. The formation of gas giants requires an intact gas reservoir still to be present, which suggests that giants form to their full mass during the disk phase. The combined mass of the giant planets in the solar system amounts to about 0.13 % of the solar mass, which provides a strict lower limit to the disk mass loss in the T Tauri phase of our Sun through planetary accretion. Giant planets may open gaps and form spiral-arm patterns, leading to a complex exchange of angular momentum between planet and disk, leading to planet migration. Meanwhile, solids also accumulate and grow up to Mars-sized planetary embryos, adding to the disk mass loss (for more details see the reviews by Baruteau et al. 2014 and Raymond et al. 2014). One spectacular observed example for this process might be the LkCa 15 transitional disks. Recent adaptive optics observations by Sallum et al. (2015) reveal up to three accreting giant proto-planets in the large inner cavity ($r_{cav} \approx 50$ au, Andrews et al. 2011) of this disk. At the same time the outer disk of this system still remains intact and shows similar observational signatures as classical disks (Öberg et al. 2011; Punzi et al. 2015).

As mentioned above, the relative importance of the disk erosion processes are unknown although all of them are observed, and all of them are expected to remove disk material in the range of maybe one tenth and about one percent of the solar mass during a typical disk lifetime (accretion and ejection rates of order 10^{-8} $M_\odot\,yr^{-1}$ for a duration of $\sim 10^6$ yr).

Another important constraint for disk evolution and planet formation is the actual mass of the disk. Disk masses derived from millimeter continuum observations are in the range of $M_{disk} \approx 0.1$ $M_\odot$ to $\lesssim 10^{-3}$ $M_\odot$ (Andrews and Williams 2005). However, disk masses derived from sub-millimeter fluxes can be underestimated by a factor of 2, and in some cases up to a factor of 10 (Dunham et al. 2014).

Williams and Best (2014) proposed a method for a more direct measurement of the disk gas mass using CO isotopologue line ratios. Their results indicate rather low disk gas masses in the range of 10^{-2} to 10^{-3} $M_\odot$ and consequently, lower gas to dust mass ratios of the order of 10, in contrast to the canonical value of 100. However, direct disk gas mass measurements are difficult and are affected by complex chemical processes, which are not fully considered in Williams and Best (2014) (this is also true for CO, see Sect. 6 for more details). In contrast to the results of Williams and Best (2014), Bergin et al. (2013) reported a gas mass of $M_{disk} \gtrsim 5 \times 10^{-2}$ $M_\odot$ for the disk around TW Hya, using the HD molecule as the mass tracer. This result is especially interesting as TW Hya is supposed to be a rather old disk (3–10 Myr, Bergin et al. 2013). The derived disk mass would still allow to form a planetary system similar to our solar system.

To improve our knowledge of the gas disk structure and evolution, further observations are required. However, deriving disk properties directly from observations is not straightforward. Not all regions of the disk (e.g. the midplane) are directly observable and the chemical processing of disk material has to be considered for the interpretation of observations (see Sect. 6). However, the first results of the Atacama Large Millimeter Array (ALMA) have already significantly improved our knowledge of the gaseous disk (see Sect. 7), and most certainly will provide us with more interesting and surprising results in the near future.

From the above introduction of disk evolution and important physical processes it becomes clear that a review like this cannot cover this topic exhaustively. We rather focus on several important aspects and refer the interested reader to dedicated reviews for topics we do not cover in detail.

In Sect. 2 we start with a discussion about the early evolution of gravitationally unstable disks, their potential to form gas giants and their role for the episodic accretion scenario. Section 3 is about the radiative environment of the disk with a focus on stellar high-energy processes as they are in particular interesting for disk-ionization, heating and disk dispersal. In Sect. 4 we discuss in more detail disk ionization and the magneto-rotational instability as a potential driver of accretion. In Sect. 5 we briefly discuss the thermal structure of the disk, heating/cooling processes and disk photo-evaporation as a consequence of gas heating by high-energy radiation. The general chemical structure of classical T Tauri and Herbig Ae/Be disks is described in Sect. 6. In this Section we also discuss in more detail X-ray chemistry, ice-lines and present a simple model for the chemical evolution during episodic accretion events. In Sect. 7 we give an overview of gas disk observations and in Sect. 8 we discuss observational constraints of photoevaporative winds. We conclude this chapter with a summary (Sect. 9).

2 The Early Evolution of Circumstellar Disks

Stars form from the gravitational collapse of dense molecular clouds, a larger fraction of which passes through a circumstellar disk owing to conservation of the net angular momentum of the collapsing cloud. Observational evidence and numerical simulations suggest that circumstellar disks can form as early as a few thousand years after the formation of the protostar (Vorobyov 2011b; Tobin et al. 2013) and sometimes even earlier than the protostar itself (Machida et al. 2010). In this early evolutionary phase, most of the mass reservoir is residing in the collapsing cloud, which lands onto the disk outer regions and drives the disk to the boundary of gravitational instability. Theoretical and numerical studies indicate that circumstellar disks can be prone to gravitational fragmentation in the outer regions if they form from cores of sufficient initial mass and angular momentum (Kratter et al. 2008; Vorobyov 2010; Stamatellos et al. 2011; Zhu et al. 2012). The infall of material from the collapsing core in the early stages of star formation triggers fragmentation in disks that otherwise might have been stable against fragmentation. The location of the fragmentation boundary (usually, at several tens of au from the central star) depends somewhat on the disk physics and is currently a subject of ongoing debate (Rafikov 2005; Boley 2009; Meru and Bate 2012).

In the subsequent protostellar phase of disk evolution, lasting until the parental cloud is accreted onto the star plus disk system or dissipated via star formation feedback, the complex interplay between mass infall onto the disk and disk transport processes, such as gravitational and viscous torques, leads to the formation of gravitationally unstable protostellar disks with radial and azimuthal profiles of density and temperature that are very different from those of the Minimum Mass Solar Nebula (Vorobyov 2011b; Bitsch et al. 2015). Non-axisymmetric structures that emerge in the disk, such as spiral arms, clumps, and vortices, may serve as likely spots for dust accumulation and planetesimal formation, facilitating the formation of solid protoplanetary cores (Rice et al. 2004; Nayakshin 2011; Regály et al. 2012; Gibbons et al. 2015). Gravitational instability and fragmentation in the early evolution of protostellar disks brings about a wealth of phenomena, the most important of which are briefly reviewed below.

2.1 The Burst Mode of Protostellar Accretion

Numerical studies show that fragments forming in the outer regions of protostellar disks can be quickly driven into the inner regions and onto the star thanks to exchange of angular momentum with spiral arms and other fragments. This fast inward migration leads to luminosity outbursts as fragments fall onto the star and release their gravitational energy (Vorobyov and Basu 2006, 2010b; Machida et al. 2011). The magnitude, duration and frequency of the luminosity outbursts is similar to those typically observed in FU-Ori-type eruptions (Vorobyov and Basu 2015). Figure 1 illustrates the burst phenomenon showing the gas surface density distribution in a gravitationally unstable disk.[1] In particular, each row of images presents a time sequence capturing individual luminosity bursts caused by fragments falling onto the star. The left, middle and right panels correspond to the time instances immediately preceding the burst, during the burst, and soon after the burst, respectively. The corresponding times are identified in each panel. The inserts in the left column zoom onto the fragments that are about to fall onto the star and the inserts in the right column present the time evolution of the total stellar luminosity during the shown time sequence. Strong luminosity bursts ranging from several tens to several hundreds of the solar luminosity are evident. Luminosity bursts in the model of disk gravitational instability and fragmentation are a robust phenomenon estimated to take place in at least 40 % of protostellar disks (Vorobyov and Basu 2015).

[1] These data (and those provided in the subsequent sections) were obtained from numerical hydrodynamics simulations of collapsing pre-stellar cores using the method described in detail in Vorobyov and Basu (2015).

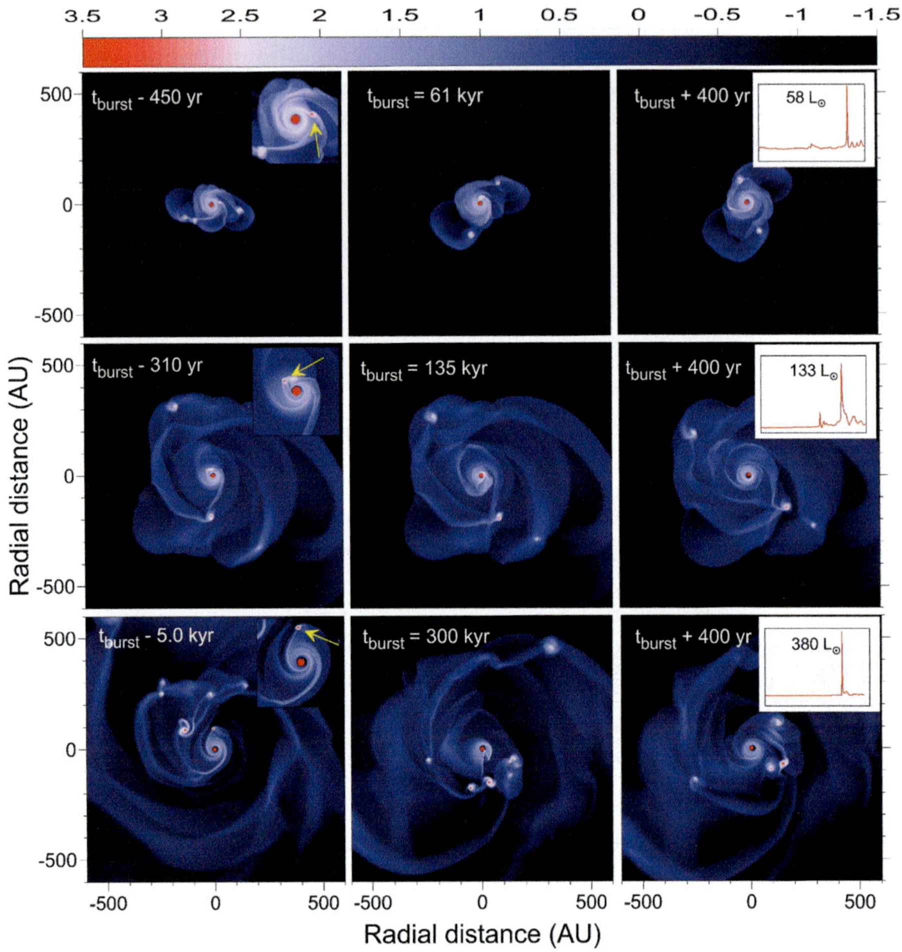

Fig. 1 Sequence of disk images capturing the inward migration of a clump onto the central star. The *scale bar* shows the gas surface density in g cm^{-2}. See the text for more detail

2.2 Formation of Massive Gas Giants and Brown Dwarfs

Another interesting phenomenon associated with disk gravitational fragmentation is the possibility for the formation of massive gas giant planet and brown dwarf companions to solar-type stars (Boss 1997, 2011; Stamatellos and Whitworth 2009; Boley et al. 2010; Nayakshin 2010; Kratter and Murray-Clay 2011; Rogers and Wadsley 2012). Numerical simulations capturing the long-term evolution of circumstellar disks (Vorobyov and Basu 2010a; Vorobyov 2013) show that under favorable conditions, i.e. in sufficiently massive disks that can experience fragmentation not only in the embedded but also T Tauri stages of star formation, some of the fragments may escape fast inward migration and mature into planetary or sub-stellar objects on wide orbits, similar to such systems as Fomalhaut or CT Chameleontis.

Figure 2 illustrates this process showing the evolution of a gravitationally unstable disk leading to fragmentation and ultimate survival of one of the fragments. The time elapsed

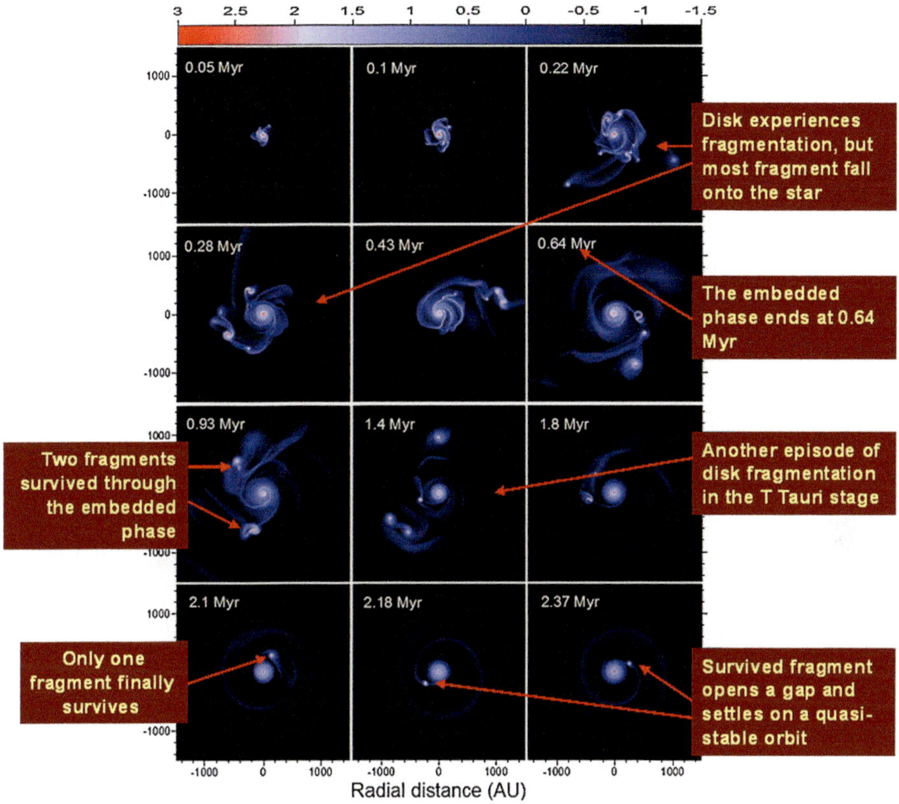

Fig. 2 Formation of a wide-orbit companion to a solar-mass star via disk fragmentation. The *scale bar* shows the gas surface density in $\mathrm{g\,cm^{-2}}$. See text for more detail

since the formation of the central protostar is indicated in each panel. During the early phase $t \leq 0.65$ Myr, when the forming star and disk are both embedded in the parental cloud, the disk experiences vigorous fragmentation, but most fragments are torqued onto the star producing strong luminosity bursts. After the end of the embedded phase, only two fragments are left in the disk. But a new (and last) episode of disk fragmentation takes place around $t = 1.0$ Myr leading to the formation of two more fragments. However, only one fragment survives through the subsequent evolution and finally settles onto a wide-separation ($\sim$300 au) orbit around the host star. The mass of the surviving fragment is 11 M_{Jup}. In most cases, however, companions in the brown dwarf mass regime are formed. The described process operates in massive disks, but has a rather stochastic nature and does not always lead to the formation of a stable companion, even for otherwise similar initial conditions.

2.3 Other Implications of Disk Fragmentation

a) *Ejection of fragments from the disk via many-body interaction.* Close encounters between fragments in a protostellar disk can lead to the ejection of the least massive one into the intracluster medium. The ejected fragments, upon cooling and contraction, can form freely floating brown dwarfs and very-low-mass stars (Basu and Vorobyov 2012; Vorobyov 2016). This phenomenon is distinct from previously suggested scenarios of brown dwarf formation via disk fragmentation, wherein finished point-sized objects are ejected (Bate 2009; Stamatellos and Whitworth 2009), and is more consistent with observations of isolated proto-brown dwarf clumps (André et al. 2012; Palau et al. 2014).

b) *Dust processing in the depths of massive fragments and formation of solid cores.* Theoretical considerations and numerical hydrodynamics simulations of fragmenting disks (Boley 2009; Vorobyov 2011a; Nayakshin et al. 2011) demonstrate that gas temperatures in the depths of massive fragments may exceed 800 K, thus initiating thermal annealing of amorphous silicates. If the tidal destruction timescale of these fragments is shorter than both the migration and dust sedimentation timescales, then the processed dust can be released at distances up to several hundred au, thus explaining the recent detection of crystalline silicates in the solar system comets. The fragments may also form solid cores in their interior before they are destroyed on their approach to the central star, providing a formation mechanism of protoplanetary cores that is alternative to the core accretion mechanism (Nayakshin 2010).

c) *Solution of the luminosity problem.* In the standard model of spherical core collapse (Shu 1977), the mass accretion rate is proportional to the cube of the sound speed. For typical gas temperatures in cloud cores on the order of 10–20 K, the corresponding accretion rate varies in the (1–4) $\times$ 10^{-6} $M_\odot$ range. A significant shortcoming of the standard model is the classic "luminosity problem", whereby accretion at the above rate produces accretion luminosities factors of 10–100 higher than typically observed for embedded protostars (Kenyon et al. 1990; Dunham et al. 2010). Accretion rates in gravitationally unstable disks can provide a solution to this problem. In the recent study, Dunham and Vorobyov (2012) calculated the radiative transfer of the collapsing cores throughout the full duration of the collapse using as inputs the disk, envelope, and stellar properties, as well as stellar mass accretion rates, predicted by the hydrodynamical simulations. They demonstrated that numerical hydrodynamics models that predict gradually declining accretion rates with episodic bursts caused by fragments migrating onto the star can account for the luminosity problem and explain a wide spread in luminosities of embedded sources.

d) *Chemical tracers of recent luminosity bursts.* As simple semi-analytic models show, luminosity outbursts similar in magnitude to those of FU-Orionis-type eruptions can raise the gas and dust temperature and drive significant chemical changes in the surrounding envelope and disk (Lee 2007; Visser and Bergin 2012; Kim et al. 2012; Visser et al. 2015). Numerical hydrodynamics simulations of disk-fragmentation-driven luminosity outbursts (Vorobyov et al. 2013) demonstrate that the CO ice can sublimate into the gas-phase in the inner envelope (a few $\times 10^3$ au) during the luminosity burst, remaining in the gas phase long after the burst has subsided. For the typical conditions in the inner envelope, the e-folding time for CO freeze-out onto dust grains is about 2500 yr. The fact that the CO freeze-out time can be much longer than the burst duration (100–200 yr) opens up a possibility for the observational detection of recent luminosity outbursts (see also Sect. 6).

3 The Radiative Environment of Circumstellar Disks

Radiation from the central star is a crucial driver of disk physics, in particular heating processes, ionization processes, and chemical reactions. The dust component in a disk plays a special role because it adds significant opacity to radiation. The balance between heating mechanisms (e.g., due to chemical reactions, photoionization processes) and cooling processes (line radiation from the gas and continuum radiation from the dust) determines the vertical temperature profile in the upper layers of the disk. Close to the midplane, gas and dust temperatures are equal due to coupling interactions between the two components ("accommodation", Chiang and Goldreich 1997). Both the chemical composition of the gas and the size distribution of dust grains are influencing the heating-cooling balance, so that complex radiation thermo-chemical simulations are required to model temperature profiles. Generally, the disk midplane is relatively cool. However, high-energy radiation leads to excessive ionization and heating of the disk surface layers in which gas and dust thermally decouple due to the low densities. These mechanisms are at the origin of gas-disk erosion and chemical processing of the gas components. We therefore concentrate on high-energy radiation and its effects on the gas disk in the following, while we refer the reader to the chapter on the dust disk for mechanisms involving the dust component (Chap. 3).

Circumstellar disks are immersed in various types of high-energy radiation and particle fluxes. Among them figure most prominently extreme-ultraviolet and X-ray radiation from hot stellar coronae; even some collimated jets have revealed prominent X-ray emission (Güdel et al. 2008); stellar flares are expected to add hard X-ray and gamma radiation, apart from an enhanced flux of energetic particles ("stellar cosmic rays"). High-energy radiation and particles act as ionization agents of disk gas and consequently heat disk gas to several thousand K, thereby driving chemical reactions. Ionization and heating also act toward disk dispersal: the former by possibly inducing the magneto-rotational instability (MRI) through coupling of ionized disk gas to magnetic fields, and the latter by photoevaporating gas from the uppermost layers of the disk. The disk lifetime may thus well be controlled by high-energy processes taking place on or near the central star.

In the subsequent sections, we first briefly review the relevant radiation sources in star-disk systems, and then discuss ionization and heating and their role in gas disk structure and evolution.

3.1 Stellar Coronal X-Rays

X-ray and extreme-ultraviolet radiation from T Tauri stars (TTS) originates predominantly from a magnetized corona, as in the case of the Sun. The X-ray radiation level is much higher than the Sun's, saturating at a maximum level of about 10^{-3} times the stellar bolometric luminosity (Telleschi et al. 2007), i.e., at levels of 10^{29}–10^{31} erg s^{-1}. Similar characteristics apply to Class I protostars. In the Orion region, the X-ray luminosities increase from protostars to TTS by about an order of magnitude, although the situation is unknown below 1–2 keV (Prisinzano et al. 2008).

3.2 Stellar X-Ray Flares

X-ray flares are also common in TTS, producing plasma temperatures of up to $\approx 10^8$ K (Stelzer et al. 2007; Imanishi et al. 2001). Like in main-sequence stars, the flare peak temperature, T_p, correlates with the peak emission measure, EM_p (or, by implication, the peak X-ray luminosity), roughly as (Güdel 2004)

$$EM_\mathrm{p} \propto T_\mathrm{p}^{4.30\pm0.35}. \tag{1}$$

3.3 X-Ray Fluorescence

Photoionization of cool disk gas by X-rays above the Fe K edge at 7.11 keV produces a prominent line feature at 6.4 keV due to fluorescence. This feature has been detected in spectra of several classical TTS, in particular after intense X-ray flaring (Tsujimoto et al. 2005). But in at least one protostellar (Class I) case, strong fluorescence appears to be quasi-steady rather than related to strong X-ray flares (Favata et al. 2005); perhaps, the 6.4 keV line is excited by non-thermal electron impact in relatively dense, accreting magnetic loops.

3.4 Accretion-Induced X-Rays

Accretion streams falling from the circumstellar disk toward the central star reach a maximum velocity corresponding to the free-fall velocity,

$$v_{\mathrm{ff}} = \left(\frac{2GM_*}{R}\right)^{1/2} \approx 620\left(\frac{M}{\mathrm{M}_\odot}\right)^{1/2}\left(\frac{R}{\mathrm{R}_\odot}\right)^{-1/2}\ \mathrm{km\,s^{-1}}. \tag{2}$$

This velocity is an upper limit as the material starts only at the inner border of the disk; a more realistic terminal speed is $v_{\mathrm{m}} \approx 0.8v_{\mathrm{ff}}$ (Calvet and Gullbring 1998). On the stellar surface, a shock develops with a temperature of

$$T_{\mathrm{s}} = \frac{3}{16k}m_{\mathrm{p}}\mu v_{\mathrm{m}}^2 \approx 3.5\times 10^6\frac{M}{\mathrm{M}_\odot}\left(\frac{R}{\mathrm{R}_\odot}\right)^{-1}\ \mathrm{K} \tag{3}$$

(m_{p} is the proton mass and μ is the mean molecular weight, i.e., $\mu \approx 0.62$ for ionized gas). For typical T Tauri stars, $M = (0.1\text{–}1)\ \mathrm{M}_\odot$, $R = (0.5\text{–}2)\ \mathrm{R}_\odot$, and $M/R \approx (0.1\text{–}1)\ \mathrm{M}_\odot/\mathrm{R}_\odot$ and therefore $T_{\mathrm{s}} \approx (0.4\text{–}4)\times 10^6$ K. The ensuing X-rays may be absorbed in the shock gas itself, contributing to its heating, but suggestive evidence for accretion-related X-ray production is available in the very soft X-ray range (e.g., based on flux excess in the O VII and Ne IX line triplets or indications of very high densities; Güdel and Telleschi 2007, Kastner et al. 2002).

3.5 Jets and Herbig–Haro Objects

Terminal shocks between jets and the interstellar medium ("Herbig–Haro (HH) objects"), are obvious candidates for X-ray production. Faint X-ray sources at the shock fronts of HH objects have indeed been found (e.g., Pravdo et al. 2001). However, these sources are very distant and probably of little relevance for the disk. On the other hand, in a few cases X-ray sources have been detected in the launching jet regions no more than a few tens of au from the star; they are likely to form in shocks internal to the jet, and possibly also involve magnetic fields (e.g., Favata et al. 2002, Bally et al. 2003, Güdel et al. 2008). The faint low-resolution spectra are soft and indicate temperatures of a few million K, compatible with shock velocities of a few 100 $\mathrm{km\,s^{-1}}$. These sources may be important as their X-ray luminosities compete with the stellar X-ray luminosity, but they are located vertically above the disk, providing their X-ray radiation relatively unobstructed access to the disk surfaces (in contrast to the grazing-incidence X-rays from the star), where they can drive heating, ionization and therefore photoevaporative flows or chemical reactions.

3.6 Stellar High-Energy Particles

Strong flares on the central star produce high fluxes of accelerated electrons and protons ("stellar cosmic rays") at planetary distances. From solar analogy and flare statistics, Feigelson et al. (2002) estimated a proton flux for the "T Tauri Sun" about 10^5 times higher than at present (i.e., 10^7 protons $\mathrm{cm^{-2}\,s^{-1}}$ at 1 au). This flux is probably present almost continuously given the high flare rate. Protostellar jets may also be excellent accelerators of high-energy particles (Padovani et al. 2015). There is indeed some indirect evidence for high-energy particles ejected by young stars (Ceccarelli et al. 2014).

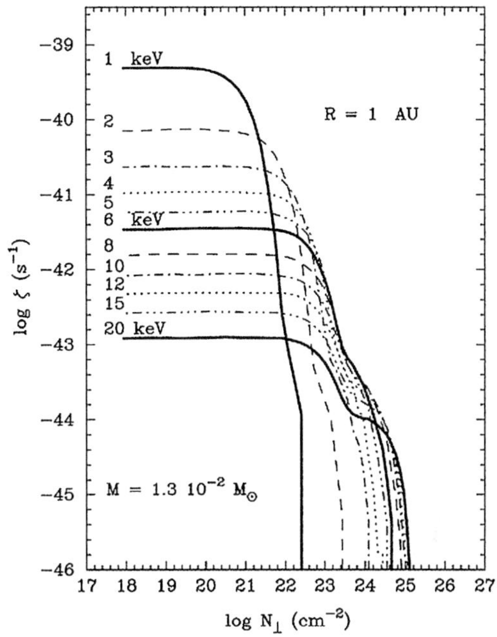

Fig. 3 Ionization rates for monochromatic X-rays as a function of column density vertically into a disk for a gas with standard interstellar composition, at a radius of 1 au. The source X-ray luminosity has been normalized to 1 $\mathrm{erg\,s^{-1}}$. The shoulder at deep levels are due to Compton scattering (from Igea and Glassgold 1999, © AAS. Reproduced with permission)

4 Disk Ionization and the Magneto-Rotational Instability

4.1 The Ionization Structure of the Gas Disk

The stellar X-ray spectrum is continuously being absorbed along the path from the star to the disk surface by the upper, tenuous layers of the disk gas and perhaps a disk wind. The dependence of the photon flux spectrum on photon energy ε and position (r = radius of the disk segment) can be written as, for a geometrically thin, flat disk (Glassgold et al. 1997)

$$F_0(\varepsilon, r) = \underbrace{\frac{R}{r}\frac{1}{2}\frac{1}{4\pi r^2}\frac{L_X}{kT}}_{f(r)}\underbrace{\frac{1}{\varepsilon}e^{-\varepsilon/kT}}_{g(\varepsilon)}, \quad \varepsilon > \varepsilon_0 \tag{4}$$

where the X-ray source is located at the center of the disk but vertically displaced by $R \ll r$; R/r thus is the glancing angle at which the X-rays strike the disk; the $1/\varepsilon$ factor transforms energy flux to photon flux; the $1/(kT)$ factor normalizes the total luminosity to L_X, and ε_0 is introduced to simulate the low-energy cutoff due to absorption along the way from the star to the disk. When arriving at the disk surface, this flux is continuously attenuated according to the photoionization cross section. Essentially all of the initial X-ray energy ε is used up to produce ionization, expending $\Delta\varepsilon \approx 37$ eV per ionization pair. Glassgold et al. (1997) and Igea and Glassgold (1999) performed numerical calculations of disk ionization for realistic disk profiles and shapes, also considering Compton scattering of energetic X-rays inside the disk. They essentially find that (see Fig. 3)

- the ionization rate by the softest X-rays at the disk surface at optical depth $\ll 1$ is approximately constant, owing to the spectral cutoff ε_0;
- the ionization rate rapidly decreases toward larger column depths where harder photons of the spectrum ionize predominantly heavier elements;

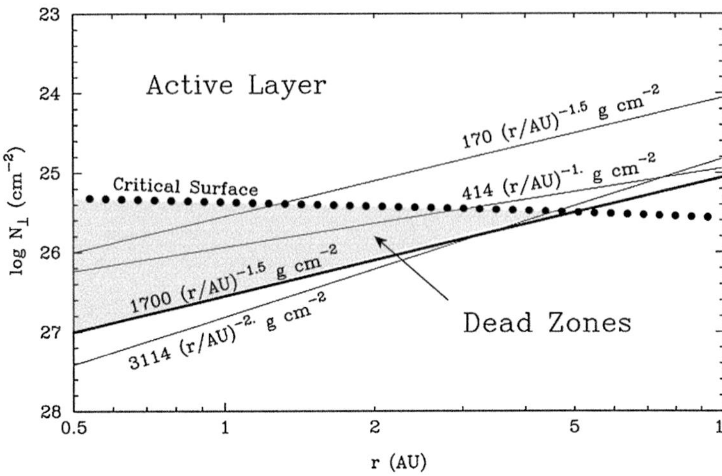

Fig. 4 Electron fraction in a disk due to X-ray ionization, shown in the radius—$N_\perp$ (*vertical column density*) plane. "Critical surface" marks the height below which the ionization fraction is too low to induce the magneto-rotational instability. The *solid lines* give disk mid-planes from different models (from Igea and Glassgold 1999; © AAS. Reproduced with permission)

- harder spectra penetrate deeper into the disk and therefore ionize deeper layers;
- for a typical stellar coronal temperature of $kT = 1$ keV, the ionization degree corresponding to the adopted cosmic-ray ionization degree of 2×10^{-17} s^{-1} is found at depths of $N_\mathrm{H} \approx 10^{23-25}$ cm^{-2} at disk radii of ≈10–1 au.
- At column densities greater than $N_\mathrm{H} \approx 1.5 \times 10^{24}$ cm^{-2}, the ionizing X-rays have energies >10 keV where the Compton cross section becomes larger than the absorption cross section, and the ionization rate begins to be altered by Compton scattering.

Calculations of the resulting ionization fraction $n(\mathrm{e})$ by Glassgold et al. (1997) and Igea and Glassgold (1999) (accounting for sedimentation of condensible heavy elements) show that the electron density $n(\mathrm{e})$ decreases rapidly toward the disk midplane; for a given vertical column density $N_\perp$, $n(\mathrm{e})$ also decreases with distance r; together, the disk ionization structure defines a wedge-shaped region around the inner disk mid-plane with its apex at the mid-plane at some outer radius, within which the ionization degree is very low ("dead zone"; Fig. 4).

4.2 The Magneto-Rotational Instability

Ionization of disk gas couples the disk gas to magnetic fields; this configuration in a differentially rotating disk is subject to the magneto-rotational instability (MRI; Balbus and Hawley 1991). The instability requires a minimum ionization degree; wherever this criterion is fulfilled, the disk is "active", other regions are called "dead zones" (Gammie 1996; see Fig. 4). MRI is held responsible for disk accretion. The criterion is given by the magnetic Reynolds number, Re_M, expressing the relative importance between advection and magnetic diffusion,

$$Re_\mathrm{M} = \frac{Lv}{\eta} \approx 7.4 \times 10^{13} x_\mathrm{e} \alpha^{1/2} \left(\frac{R}{1\ \mathrm{au}}\right)^{3/2} \left(\frac{T}{500\ \mathrm{K}}\right) \left(\frac{M}{\mathrm{M}_\odot}\right)^{-1/2} \quad (5)$$

where L is the characteristic length of variations in B, η is the magnetic diffusivity, v is the plasma flow velocity, and the last numerical expression is due to Gammie (1996), with x_e being the electron fraction and α the Shakura–Sunyaev viscosity parameter ($\alpha^{1/2}$ is of order unity). For MRI to be efficient, the critical Re_M is about 100, and this corresponds to an ionization fraction of $\approx 10^{-12}$ (see discussion in Fromang et al. 2002).

A significant part of the disk surface will couple sufficiently to magnetic fields to drive the instability, while the central layer around the midplane of the inner (few au) disk is insufficiently ionized and is therefore a "dead zone" (Fig. 4; Glassgold et al. 1997 and Igea and Glassgold 1999). However, considering also traces of metallic atoms, Fromang et al. (2002) find for a standard molecular α disk and $Re_{M,\,crit} = 100$ a dead zone extending from a fraction of an au out to 10–100 au. On the other hand, a small admixture of heavy-metal atoms is rapidly ionized by charge exchange with the molecular ions, while atomic ions recombine only slowly by radiation. For standard disks, an atomic abundance of heavy metals of as little as 10^{-6}–10^{-7} of the cosmic value may make the entire disk active, without any dead zone. Newer calculations including complex chemical networks, however, show the presence of a dead zone even if metals are present (Walsh et al. 2012; Cleeves et al. 2013a).

Cosmic rays are also efficient in ionizing disk material. Their importance for driving MRI was recognized in particular by Gammie (1996) who developed a model of layered accretion in T Tauri disks. The crucial question is that of the ionization rate. In the interstellar space, $\zeta_{CR} \approx 10^{-17}$ s^{-1} (Spitzer and Tomasko 1968). The propagation of cosmic rays onto the disk surface is, however, a very significant problem. In the present solar system, low-energy cosmic rays are suppressed because of diffusion, convection and adiabatic cooling in the solar wind. The low-energy cosmic-ray spectrum is therefore highly uncertain for energies $\lesssim 10^9$ eV due to modulation of the interstellar CR spectrum by the magnetized solar wind (filling the "heliosphere"). Various extrapolations down to lower energies provide largely diverging spectra (see summary by Cleeves et al. 2013a). The interstellar ionization rate is therefore unlikely to be appropriate for a protoplanetary disk environment, especially because T Tauri stars are, like the Sun, producing magnetic fields and are very likely to produce ionized winds. Cleeves et al. (2013a) estimate an attenuation of the cosmic ray flux at, say, 1 au from the star by ≈ 3 orders of magnitude compared to the Sun and at least 6 orders of magnitude compared to interstellar conditions, resulting in a CR ionization rate of about $\zeta_{CR} \approx 10^{-22}$–$10^{-20}$ s^{-1}. This may result in much larger "dead zones" in protoplanetary disks, possibly reaching out to distances of 25 au.

5 Heating of Disk Gas and Photoevaporation

5.1 Thermal Structure of a Gas Disk

As laid out in Sect. 3, high-energy radiation strongly modifies the disk surface layers due to their strong heating rates and the decoupling of gas and dust in these low-density regions. In lower regions, dust and chemical processes are pivotal for the temperature profile, but in the upper disk atmospheres most of the heating is due to irradiation by stellar (or jet) ultraviolet, extreme-ultraviolet and X-ray photons.

In Fig. 5 we show the main heating/cooling processes for an exemplary model of a disk around a T Tauri star. The main stellar and disk properties of the model are: stellar mass $M_* = 0.7$ M$_\odot$, photospheric luminosity $L_* = 1$ L$_\odot$, FUV luminosity $L_{FUV} = 2.3 \times 10^{31}$ erg s^{-1}, X-ray luminosity $L_X = 10^{30}$ erg s^{-1} and the disk mass is $M_{disk} = 0.03$ M$_\odot$ (for more details on the model see Sect. 6 and Table 1). In the upper layers of the disk the

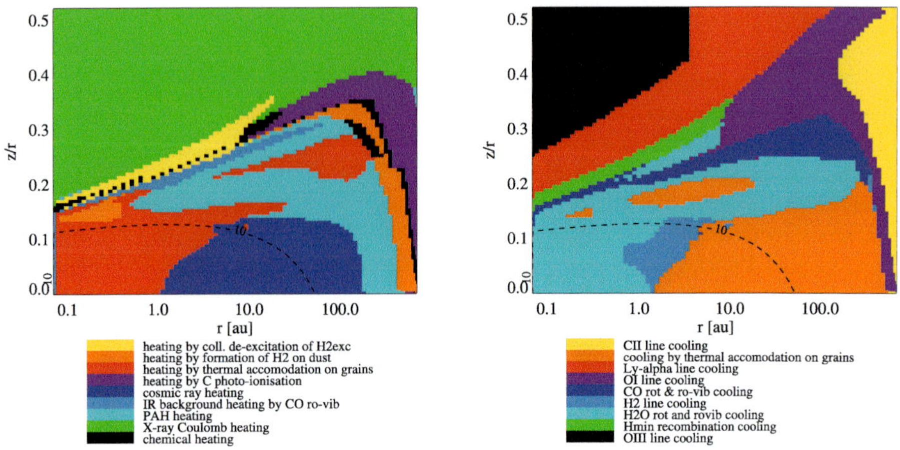

Fig. 5 Dominant heating (*left panel*) and cooling (*right panel*) processes for an exemplary ProDiMo model (Protoplanetary Disk Model, see Sect. 6 for details) of a disk around a T Tauri star. The *plots* show the dominant heating/cooling process at every point in the disk (identified by the different colors and the legend). The height of the disk is scaled by the radius (z/r) to show also the details in the inner region. The *dashed line* indicates a visual extinction of 10

gas heating is dominated by radiation (e.g. X-ray Coulomb heating, C photo-ionisation due to FUV radiation). In the deeper layers the gas temperature is mainly controlled by the dust temperature due to collisions (heating/cooling by thermal accommodation on grains, Burke and Hollenbach 1983). The main cooling process in disks is line cooling. In the upper layers the cooling is dominated by emission of ionized or neutral atomic species like O or C^+. In the transition region where T_{gas} drops rapidly, at $N_\perp \approx (1\text{–}2) \times 10^{21}\ \text{cm}^{-2}$ (see Fig. 6), molecular line cooling is important. Prominent examples are CO and in deeper layers H_2O. CO is formed where H still abounds in its atomic form, and accelerates cooling through CO ro-vibrational and CO rotational emission, generating a strong temperature gradient; corresponding emission lines provide important observational diagnostics of disk heating and cooling.

We want to emphasize that the results shown in Fig. 5 are model dependent. For example the model of Glassgold et al. (2004) includes a viscous heating process which can be dominant in the transition region from the hot upper layers to the cool midplane (see Fig. 6). This process is not included in the model used for Fig. 5. However, all thermo-chemical models show the decoupling of the gas and dust temperature and the resulting high gas temperatures in the upper layers of the disk (e.g. Kamp and van Zadelhoff 2001; Gorti and Hollenbach 2004; Nomura and Millar 2005; Woitke et al. 2009). These results are qualitatively confirmed by Herschel observations (e.g. Bruderer et al. 2012; Tilling et al. 2012).

With the high spatial resolution of ALMA it now becomes possible to measure detailed radial temperature gradients in the disk. Schwarz et al. (2016) uses ALMA observations of several CO transitions (including isotopologues) to derive the radial temperature profile of the TW Hya disk. However, these observations mainly trace the warm molecular layer (≈30–50 K see Sect. 6.1) and not necessarily the midplane of the disk. These kind of observations are required to improve and calibrate thermo-chemical disk models. However, it remains difficult to directly measure disk gas temperatures in particular in the deeper layers of the disk, and derived gas thermal structures of disks remain model dependent.

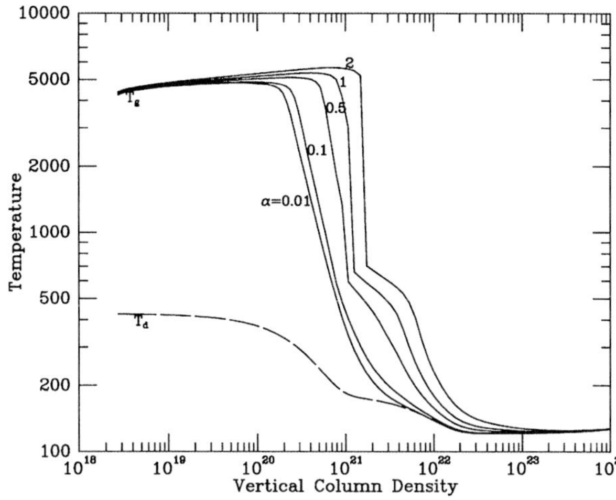

Fig. 6 Gas and dust temperature profiles as a function of vertical column density into a disk, at 1 au from a central T Tauri star; apart from X-ray heating, viscous heating is present, characterized by the various α values ($\alpha = 0$ for no viscous heating; from Glassgold et al. 2004; © AAS. Reproduced with permission)

5.2 X-Ray Heating

X-ray irradiation leads to substantial heating of the upper layers of protoplanetary disks. The heating energy per X-ray ionization event is about $\Delta\epsilon_h \approx 30$ eV (Glassgold et al. 2004). Realistic gas disk calculations irradiated by stellar X-rays show strong heating of the upper, atomic layers of the disk to several thousand K at 1 au (Fig. 6). This surface heating leads to a strong temperature inversion along the vertical direction through the disk, with a steep temperature decline at about $N_\perp \approx 10^{21}$ cm^{-2}. Very close to the star, much hotter, fully ionized "disk coronae" at $T \approx 10^6$ K are formed (Ercolano et al. 2008). X-rays are very important heating agents out to ~35–40 au but disk ionization by stellar X-rays can be seen out to 190 au in such models.

5.3 Disk Photoevaporation

The heating of disk surface layers may produce temperatures at which individual particles escape from the stellar gravitational potential. This occurs once the mean particle energy, $\sim kT$, exceeds the gravitational binding energy of the particles with mass m, GM_*m/R. For a given temperature, this process of *photoevaporation* becomes effective beyond the *gravitational radius*, R_g, where the above criterion is fulfilled, i.e.,

$$R_g = \frac{GM_*}{c_s^2} \tag{6}$$

where $c_s = (kT/m)^{1/2}$ is the sound speed of the ionized gas.

Statistically, protoplanetary dust disks observed in the infrared disperse with a time scale of ≈3 Myr (Mamajek 2009). However, the actual removal of an optically thick disk occurs rapidly at the end of this process, on a time scale of perhaps 0.5 Myr. This "two-time scale problem" was addressed by the "UV switch" model by Clarke et al. (2001), assuming that at some point in time, the disk accretion rate falls below the wind mass loss rate induced by photoevaporation at a certain radius of the disk. Subsequently, the resupply of the inner disk is suppressed and a gap is forming at this distance. The inner disk now disperses on its own, short viscous time scale.

The EUV photoevaporation model of disks was first and extensively developed by Hollenbach et al. (1994) based on the diffuse recombination radiation field that ionizes the disk gas at large radii. The mass-loss rate depends on the flux of EUV photons, Φ, and the mass of the central star, M_*, as

$$\dot{M}_w \approx 4.4 \times 10^{-10} \left(\frac{\Phi}{10^{41}\ \mathrm{s}^{-1}} \right)^{1/2} \left(\frac{M_*}{1\mathrm{M}_\odot} \right)^{1/2} \mathrm{M}_\odot\ \mathrm{yr}^{-1}. \tag{7}$$

If the escaping gas is flowing at sound speed ($c_s \approx 10\ \mathrm{km\,s^{-1}}$), then the total mass loss rate at radius R is

$$\dot{\Sigma}_w(R) = 2\mu m_\mathrm{H} n_0(R) c_s(R) \tag{8}$$

μ being the mean molecular weight of the gas, $\approx$2.3 for a molecular gas, m_H the mass of a hydrogen atom, the factor of 2 accounting for the two sides of the disk, and n_0 being the number density of the disk gas.

Once the inner disk has drained, there is a significant contribution of *direct* radiation supporting the evaporation process in particular at the inner edge of the disk so that the outer disk also rapidly disperses. Alexander et al. (2006a) find a typical escape speed of $0.4c_s$ and a mass-loss formula

$$\dot{M}_w(< R_\mathrm{out}) \approx 2.2 \times 10^{-9} \left(\frac{\Phi}{10^{41}\ \mathrm{s}^{-1}} \right)^{1/2} \left(\frac{R_\mathrm{in}}{3\ \mathrm{au}} \right)^{1/2} \left(\frac{H/R}{0.05} \right)^{-1/2} \times \left[1 - \left(\frac{R_\mathrm{in}}{R_\mathrm{out}} \right)^{0.42} \right] \mathrm{M}_\odot\ \mathrm{yr}^{-1} \tag{9}$$

where R_in and R_out are the inner and outer disk radii, respectively, and H/R is the ratio (assumed to be constant) between disk scale height H and radius R. The effect of direct radiation thus enhances the mass loss rate by an order of magnitude above the estimate in Eq. (7). Numerical simulations confirm this behavior and suggest that at the end of the disk lifetime (several Myr), the disk rapidly disperses within $\gtrsim 10^5$ yr (Alexander et al. 2006b).

EUV (Ly continuum) radiation is problematic because it may not reach the disk at all if disk winds absorb them early on. Also, Pascucci et al. (2014) claim that the EUV contribution is, based on observational evidence, too small to be effective for photoevaporation. Alternatively, soft X-rays may be relevant. Models based on Monte-Carlo methods for radiation show that X-rays may indeed dominate the photoevaporation process, with wind mass-loss rates of order $10^{-8}\ \mathrm{M}_\odot\ \mathrm{yr}^{-1}$ (Ercolano et al. 2008). These studies were qualitatively and quantitatively supported by full hydrodynamic simulations by Owen et al. (2010) and Owen et al. (2011), emphasizing that the soft X-ray part of a stellar spectrum is the dominant driving mechanism for photoevaporation (rather than EUV alone). They find the mass-loss rate to scale linearly with L_X. The important advantage of X-rays here is their much larger penetration depths than for EUV, resulting in a much larger disk area subject to photoevaporation (out to ~40 au) compared to the EUV case (for which the inner cutoff at $R_g \approx 5$ au for 10^4 K gas limits the evaporation efficiency additionally). We will discuss observational evidence for photoevaporation of disks in Sect. 8.

6 Chemical Processing of Disks

Protoplanetary disks are considered to be the birthplaces of planets. It is likely that chemical processes occurring in those disks have an impact on the bodies formed in them. Molecular and atomic species are useful observational tracers of important physical properties of

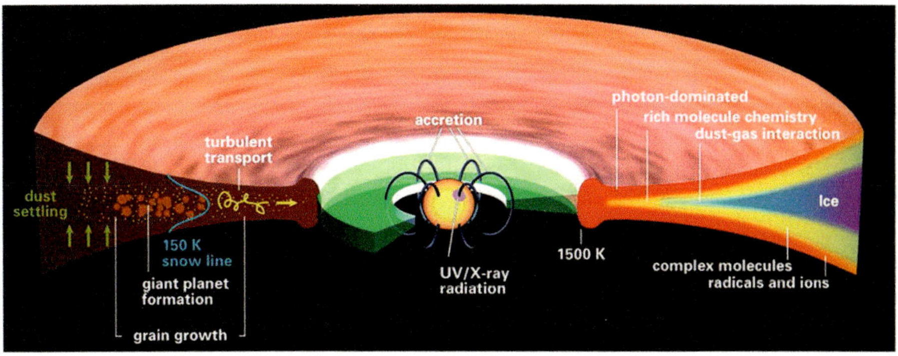

Fig. 7 Sketch of the physical and chemical structure of a protoplanetary disk around a sunlike star (age ≈1–5 Myr). Adapted with permission from Henning and Semenov (2013), ©(2013) American Chemical Society

disks like mass, temperature, disk structure and turbulence. These are important quantities for planet formation theories. For the interpretation of such observations a detailed understanding of chemical processes is required.

This is a challenging task as at least from a chemical point of view, disks are very complex structures. In gas disks temperatures can vary from ≈10 to several 1000 K and densities from $\approx 10^4$–10^{16} cm^{-3}. Also external influences, in particular stellar radiation (e.g. UV, X-rays, see Sect. 3) but also cosmic rays, alter the chemical composition of the disk. Furthermore, the dust component plays an important role (see Chap. 3 on the dust disk). Dust is an important opacity source shielding the deep regions of the disk from stellar radiation. In addition, dust allows for freeze-out of molecules (ice formation) in cold regions and the dust surface can act as a chemical catalyst for the formation of molecules (surface chemistry). Last but not least also dynamical processes like viscous evolution, turbulence, planet-disk interaction (e.g. via spiral arms) and the evolution of the star likely alter the chemical composition of the disk. Figure 7 gives an overview of the physical and chemical structure of a typical protoplanetary disk.

In terms of chemical disk modeling the inclusion of all relevant processes remains challenging. Most chemical disk models commonly assume a fixed (steady state) two dimensional disk structure and fixed stellar properties. So-called *radiation thermo-chemical disk models* also consistently solve for the gas temperature (heating/cooling processes) and often include detailed dust radiative transfer (e.g. Aikawa et al. 2002; Gorti and Hollenbach 2008; Woitke et al. 2009; Walsh et al. 2010; Bruderer et al. 2012; Du and Bergin 2014). Also the impact of turbulent mixing (e.g., Semenov et al. 2006; Furuya and Aikawa 2014) or dust evolution (e.g. Vasyunin et al. 2011; Akimkin et al. 2013) can be studied with such models. An extensive and complete review of this kind of models can be found in Henning and Semenov (2013) and Dutrey et al. (2014). Recently, chemical modeling has also been included in hydrodynamic simulations of disks. Examples are: Ilee et al. (2011) and Evans et al. (2015) who study the chemistry in the context of gravitationally unstable disks (e.g., in spiral arms) or Vorobyov et al. (2013) who study the impact of episodic accretion on the CO gas phase transition (see Sect. 2.3). Such models are computationally expensive and usually make simplifying assumptions about the radiative transfer and/or the chemistry. Nevertheless, such approaches are very powerful and give a first insight on the importance of dynamical processes on the chemical evolution of disks.

Table 1 Main stellar and disk properties for the T Tauri and Herbig Ae/Be (where different) ProDiMo disk models

	T Tauri	Herbig Ae/Be
Stellar mass	0.7 $M_\odot$	2.1 $M_\odot$
Stellar effective temperature	4000 K	6800 K
Stellar luminosity	1.0 $L_\odot$	32.0 $L_\odot$
FUV luminosity	2.3×10^{31} erg s^{-1}	5.2×10^{33} erg s^{-1}
X-ray luminosity	10^{30} erg s^{-1}	10^{29} erg s^{-1}
X-ray emission temperature	2×10^{7} K	
Disk gas mass	0.03 $M_\odot$	
Gas/dust mass ratio	100	
Inner disk radius	0.07 au	0.4 au
Characteristic radius	100 au	
Column density power index	1.0	
Scale height at $r = 100$ au	10 au	
Flaring power index	1.15	

Here we do not discuss chemical models in detail but rather focus on the general chemical structure of the disk. For more complete reviews on disk chemical models and processes see Bergin et al. (2007), Henning and Semenov (2013), Dutrey et al. (2014). To qualitatively illustrate certain aspects discussed in this section, we use the results of two generic disk models, one T Tauri and one Herbig Ae/Be disk (see Table 1 for details). These two models should in particular illustrate the impact of the stellar properties on the chemical structure of the disk. Therefore we use for both models the same parameters for the disk (e.g. disk mass) with the exception of the inner disk radius. The inner disk radius is given by the dust condensation radius, and is therefore larger for the Herbig Ae/Be disk due to the stronger stellar radiation field.

We apply the radiation thermo-chemical disk code ProDiMo (Protoplanetary Disk Model, Woitke et al. 2009, Kamp et al. 2010, Thi et al. 2011, Aresu et al. 2011, Meijerink et al. 2012, Woitke et al. 2016), to calculate the thermal and chemical structure of the disk. We use time-dependent chemistry and evolve the chemistry up to 2 million years (chemical age of the disk). For the chemical reaction network we use the KIDA 2014 (Kinetic Database for Astrochemistry, Wakelam et al. 2015) database supplemented by additional chemical reactions included in ProDiMo. For the adsorption/desorption energies of the chemical species we use the data provided by the UMIST 2012 release (McElroy et al. 2013). More details on this particular model can be found in Woitke et al. (2016). We use the T Tauri and Herbig Ae/Be ProDiMo disk models to discuss the general chemical structure of disks (Sect. 6.1, Figs. 8 and 9) and the importance of ice lines for planet formation (Sect. 6.3, Fig. 10).

6.1 Chemical Structure

The following description of the chemical structure is strictly speaking only valid for T Tauri disks. However, the main chemical processes described are also valid for Herbig Ae/Be disks. We discuss the main differences between T Tauri and Herbig Ae/Be disks at the end of this section.

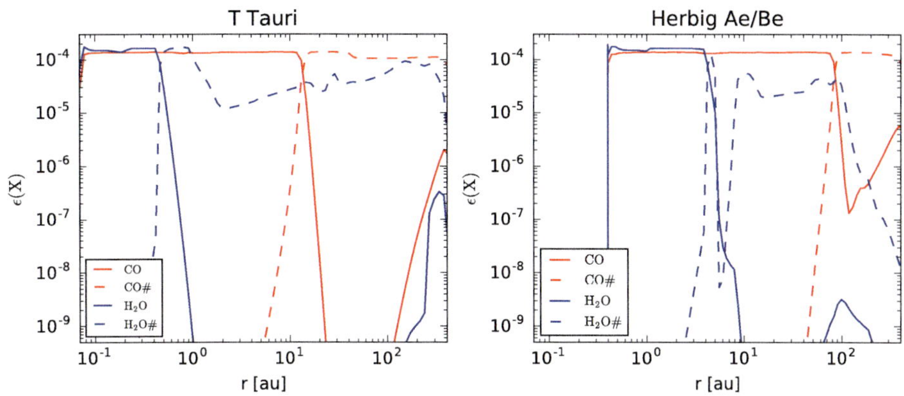

Fig. 8 Abundance of CO and water (relative to the total hydrogen number density) in gas and solid phase (as ice on the dust grains) in the midplane ($z = 0$ au) of the disk for the two generic *ProDiMo* disk models (*left panel*: T Tauri, *right panel*: Herbig Ae/Be, see also Table 1). The *solid lines* show the gas phase, the *dashed lines* the ice-phase abundances. The ice-line for water is at $\approx$0.5(5) au, the one for CO at $\approx$15(80) au for this particular T Tauri (Herbig Ae/Be) disk model. We note that the inner radius of the Herbig Ae/Be disk is at 0.4 au. For easier comparison we use the same scales for both plots

For the discussion of the chemical structure of the disk we divide the disk in different radial zones and vertical layers. From a chemical point of view the midplane of the disk contains three distinct zones:

- *inner zone:* extends up to a distance of a few au from the star; temperatures $T > 100$ K; mostly shielded from radiation due to high optical depths;
- *middle zone:* extends from a few au up to 100 au; $100 \gtrsim T \gtrsim 20$ K; most species are adsorbed onto dust grains (e.g. water ice); shielded from radiation;
- *outer zone:* $T \lesssim 20$ K; most species freeze-out on dust grains; only partly shielded from radiation.

As an example we show in Fig. 8 the abundance of CO and water (for the gas and solid phase) in the midplane as a function of radius. The different locations of the ice-lines due to the different adsorption energies for CO and water are clearly visible. Also the re-increase of the gas phase abundance in the outer disk, in particular for CO, is apparent.

In vertical direction it is common to define three distinct layers:

- *midplane*: see above; ice dominated chemistry (dust–gas interaction, surface chemistry);
- *rich molecular layer*: warm enough ($20 \lesssim T \lesssim 100$ K) so that most molecules are in gas phase; partly shielded from radiation; rich ion-neutral chemistry;
- *photon-dominated layer*: photochemistry (ionization, photo-dissociation); most species are in atomic form and ionized.

These three layers are also indicated in Fig. 7. In Fig. 9 we show, as an example, the two dimensional distribution of the CO molecule. The vertically layered molecular structure is clearly visible and is similar for other molecules.

Due to the strongly varying physical conditions the chemistry can be very different in the zones described above. For a proper treatment of the chemistry in the *inner zone* chemical reactions with high activation barriers can also become important (e.g. Harada et al. 2010), different from the usually colder interstellar conditions. Therefore, it can be expected that

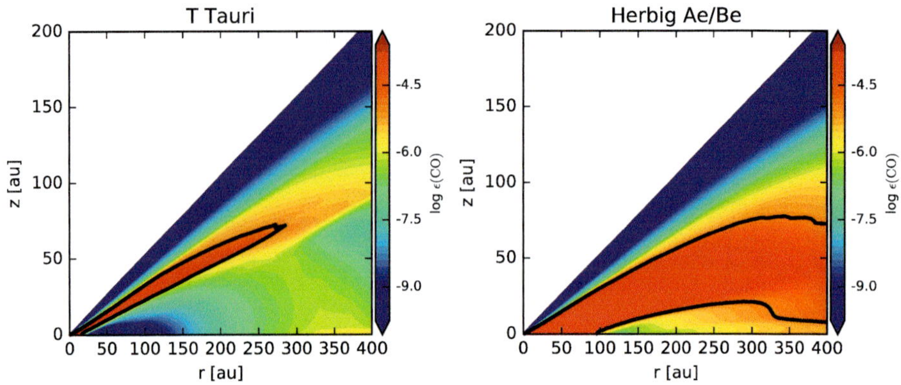

Fig. 9 CO abundance ϵ(CO) (relative to the total hydrogen number density) for the two generic *ProDiMo* disk models (*left panel:* T Tauri, *right panel:* Herbig Ae/Be, see also Table 1). The black contour roughly encircles the warm molecular layer ($\epsilon(\mathrm{CO}) = 10^{-5}$). Above this layer CO is photo-dissociated, below this layer CO freezes-out on dust grains. In the outer region of the midplane ($r \gtrsim 200$ au) CO is also released back into the gas phase due to photo-desorption. The CO layer in the Herbig Ae/Be disk is thicker and more extended compared to the T Tauri disk, as the disk is warmer due to the stronger stellar radiation field

the resulting chemical abundances are also different. Observations presented in Pontoppidan and Blevins (2014) show clear indications for this scenario. This work also shows that thermo-chemical models still have problems to explain the observed abundances. One reason might be that radial migration processes (e.g. large dust particles migrate inwards) are neglected. Migrating dust particles are probably coated in water ice. This ice is sublimates in the warmer conditions of the inner zone, enhancing the oxygen abundance which can subsequently also influence the chemistry. However, the thermo-chemical model of Walsh et al. (2014), for example, takes accretion flows into account, showing that also the abundance of complex molecules (e.g. methanol) can be strongly enhanced owing to the transport of ice-coated dust particles into the inner zone.

In the radiation shielded and much cooler *middle zone* of the midplane the important chemical drivers are cosmic rays and/or radionuclide ionization (and possibly also high-energy X-rays). The ionization rates of these processes are quite low, in the range of $\approx 10^{-20}$–10^{-17} s^{-1} (e.g., Cleeves et al. 2013a,b) compared to the upper layers of the disk. Further photochemistry induced by the secondary UV field from cosmic rays (e.g. Gredel et al. 1989) becomes important in this region. A detailed study by Chaparro Molano and Kamp (2012a,b) has shown that for disks proper dust and gas opacities have to be taken into account for the cosmic-ray induced photo processes. Their derived chemical abundances are compatible with measurements of the chemical composition of solar-system comets. The gas phase abundances of chemical species probably also influences the composition of gas giants that might be formed in this zone. One interesting quantity is the resulting C/O gas phase abundance ratio which can be significantly altered, and therefore directly influences the chemical composition of giant planet atmospheres (e.g Helling et al. 2014).

Different from the *middle zone*, in the cold *outer zone* non-thermal desorption mechanisms (e.g. photo-desorption) may become important. Recent observations of DCO^+ in IM Lup suggest that non-thermal desorption (most likely photo-desorption due to stellar UV radiation) is efficient in the outer disk (Öberg et al. 2015; see also Sect. 7). In addition to the stellar UV radiation also the interstellar background field might play an important role for the DCO^+ abundance (Teague et al. 2015). Surface chemistry processes produce complex molecules on the dust surface (e.g. hydrogenation of CO to form H_2CO, Qi et al. 2013), and these molecules may subsequently be photo-desorbed. One prominent example for this suggested formation process is methanol (Walsh et al. 2010; Dutrey et al. 2014). Gas phase methanol was recently detected in the disk of TW Hya (Walsh et al. 2016). The observationally derived average abundances of methanol are up to two orders or magnitude lower compared to model predictions. This observations probably indicate a surface chemistry dominated formation pathway of methanol, but existing models seem to overestimate the efficiency of methanol formation and/or the efficiency of various desorption processes (Walsh et al. 2016).

Vertical turbulent mixing can transport ice coated dust particles to higher layers where photo-desorption is more efficient, and as a consequence gas phase abundances of complex molecules formed on dust grains are enhanced (Semenov and Wiebe 2011). On the other hand, vertical mixing can also decrease the abundance of complex organic molecules. Enhanced hydrogenation of radicals due to transport of atomic hydrogen to deeper layers can significantly suppress complex molecule formation (Furuya and Aikawa 2014).

In the warm molecular layer above the midplane one finds a very rich chemistry driven by higher temperatures, UV and X-ray radiation (Aikawa et al. 2002; Semenov et al. 2004; Henning and Semenov 2013). In our generic T Tauri disk model the layer is clearly visible in CO (Fig. 9). However, the location and the thickness of this layer probably vary from disk to disk and also depend on the dust properties. Vasyunin et al. (2011) have shown that in chemical models including large dust grains, this layer becomes thicker (e.g. extends deeper into the disk) compared to models considering ISM like dust properties (e.g. only small grains). Similar results are found by Akimkin et al. (2013) who used a self-consistent chemical model including dust evolution. They also find that on average, column densities of neutral molecules (e.g. CO, H_2O) are enhanced in models considering dust evolution.

Another important chemical process in this layer is CO isotopologue chemistry (e.g. Visser et al. 2009), in particular isotopologue selective photo-dissociation. This chemical process may explain the anomalous ^{17}O and ^{18}O ratios measured in meteorites (Visser et al. 2009). CO isotope-selective photo-dissociation must also be considered if CO isotopologue line ratios are used to estimate disk gas masses. Thermo-chemical disk models of Miotello et al. (2014) show that using constant (throughout the disk) CO isotopologue ratios can lead to underestimating disk masses by up to an order of magnitude. Also, the possible depletion of gas phase CO in disks as suggested by Favre et al. (2013) and Bruderer et al. (2012) has to be taken into account. Although this depletion may be explained by isotope-selective photo-dissociation (in particular $C^{18}O$), chemical processes occurring closer to the midplane, like dissociation of CO by He^+ (Aikawa et al. 1998; Bergin et al. 2014; Furuya and Aikawa 2014) and/or surface chemistry (Reboussin et al. 2015) must also be considered if CO and its isotopologues are used as disk mass tracer.

Above the warm molecular layer direct ionization processes of atomic species and dissociation of molecules due to X-ray and UV photons are the main chemistry drivers. This layer shows typical properties of photo-dissociation regions (e.g. Hollenbach and Tielens 1999). A proper treatment of chemical processes in this region is important to determine the gas temperature, which is relevant for e.g. photo-evaporation in disks (see Sect. 5.3 and 8). Also X-rays play a major role in this layer ionizing noble gases like Ne and Ar and heating the gas to temperatures up to 10 000 K (see Sect. 5).

T Tauri Versus Herbig Ae/Be Disk Chemistry A disk around a Herbig Ae/Be disk is on average warmer simply due to its stronger radiation field in the UV and optical wavelength regime. This has an impact on both the radial and vertical chemical structure of the disk.

The locations of midplane ice lines are on average at larger radii in a Herbig Ae/Be disk than in a T Tauri disk (see Fig. 8). But also the vertical structure is affected. Compared to the T Tauri disk the warm molecular layer starts at slightly lower height (due to enhanced photo-dissociation) but extends much deeper towards the midplane of the disk. The warm molecular layer is mainly thicker due to the higher temperatures which prevents freeze-out of molecules (see Fig. 9).

Also molecular observations indicate that T Tauri disks have a larger cold chemistry reservoir compared to Herbig Ae/Be disks. The main indication here is the lower detection rates of cold chemistry tracers like H_2CO (Öberg et al. 2011; Dutrey et al. 2014). Also the measured CO snowline location in HD 163296 of ≈90 au (Qi et al. 2015) is a strong indication for warmer Herbig Ae/Be disks. As ice lines might play an important role for dust evolution (see Sect. 6.3), these chemical differences are most relevant for the study of early planet formation.

6.2 X-Ray Driven Disk Ion Chemistry

Given the importance of X-ray disk ionization, we briefly summarize chemical reactions directly induced by X-rays, in particular those illustrating some principal destruction routes of molecular ions. Among the principal ion destruction paths are the following, leading to "chemical heating" (Glassgold et al. 2012):

$$H_2^+ + e \rightarrow H + H$$
$$H_2^+ + H_2 \rightarrow H_3^+ + H$$
$$H_3^+ + e \rightarrow H_2 + H$$
$$H_3^+ + e \rightarrow H + H + H$$
$$H_3^+ + CO \rightarrow HCO^+ + H_2$$
$$H_3^+ + O \rightarrow OH^+ + H_2$$
$$H_3^+ + H_2O \rightarrow H_3O^+ + H$$
$$HCO^+ + e \rightarrow CO + H$$
$$OH^+ + H_2 \rightarrow H_3O^+$$
$$H_3O^+ + e \rightarrow H_2O + H$$
$$H_3O^+ + e \rightarrow OH + H_2$$
$$H_3O^+ + e \rightarrow OH + 2H$$
$$H^+ + H_2O \rightarrow H_2O^+ + H$$
$$H_2O^+ + H_2 \rightarrow H_3O^+ + H$$
$$He^+ + CO \rightarrow C^+ + O + He$$
$$C^+ + H_2O \rightarrow HCO^+ + H$$

Complex thermo-chemical models of irradiated disks show that the abundances of neutral molecules such as CO, OH, and H_2O are not overly sensitive to X-ray irradiation (a change by less than an order of magnitude over the range of many orders of magnitude in L_X). In contrast, the X-ray sensitive ionization fraction in the upper layers of the disk strongly influences ion chemistry. The following reactions are important (Aresu et al. 2011; M indicates the total mass of the species in the disk):

$H_2 + e \rightarrow H_2^+$ (secondary electrons from direct X-ray ionization)

$H_2^+ + H_2 \rightarrow H_3^+ + H$

$H_3^+ + O \rightarrow OH^+ + H_2$ $M(OH)$ grows approximately linearly with L_X.

$H_3^+ + OH \rightarrow H_2O^+ + H_2$ $M(H_2O^+)$ grows somewhat less than OH^+.

$H_3^+ + H_2O \rightarrow H_3O^+ + H$ $M(H_3O^+)$ grows by 10–100× for 1000× L_X.

Also N^+ reacts very sensitively to the presence of X-rays and its total mass grows about linearly with L_X. These reactions also spread in disk area, being pushed to larger distances as L_X grows. A special case of interest is the fine-structure emission of [O I]. The models clearly predict a linear growth of its line flux with the sum of $L_X + L_{UV}$. Whichever of the two luminosities dominates determines the correlation. Because of the usually appreciable

UV field around T Tauri stars, an X-ray dependence comes into play only above $L_X \approx 10^{30}$ erg s^{-1} (Aresu et al. 2014). Observations support the flux levels approximately although the correlation is not evident, possibly due to some influence of other parameters such as grain size distribution, surface density distribution, etc. in real objects.

6.3 Ice Lines

Ice lines (or condensation fronts) probably play a significant role in planet formation processes (e.g. through dust growth) but are also relevant for the final composition of bodies formed in the disk.

Recently Qi et al. (2013) reported the imaging of the CO snowline in the TW Hya disk via observations of the N_2H^+ ion. N_2H^+ is efficiently destroyed by proton transfer to CO

$$N_2H^+ + CO \rightarrow HCO^+ + N_2. \tag{10}$$

Therefore N_2H^+ can only be abundant in regions with low CO gas phase abundances. The resulting ring like structure visible in images of the N_2H^+ line emissions is also, at least qualitatively, reproduced by thermo-chemical disk models (e.g. Cleeves et al. 2015; Aikawa et al. 2015). The situation is illustrated in Fig. 10 for our generic T Tauri disk model. However, Aikawa et al. (2015) discuss other relevant chemical processes influencing the resulting N_2H^+ distribution. One example is the so called sink effect, where CO is converted to less volatile species in regions with temperatures exceeding the sublimation temperature of CO (e.g. >20 K). Therefore, the peak of the N_2H^+ abundance may be slightly below the CO sublimation radius.

There is a similar mechanism involving water, where HCO^+ is destroyed by water vapor. This mechanism was proposed to explain the ring-like appearance of the $H^{13}CO^+$ emission in young protostars experiencing a recent luminosity outburst (Jørgensen et al. 2013). Such outbursts significantly enhance the gas phase water abundance, leading to subsequent destruction of HCO^+. This chemical process is also relevant for disks; however, more detailed studies are required to investigate its efficiency under disk conditions. Furthermore, direct imaging of the water ice line via HCO^+ would require a much higher spatial resolution, as the water ice line is located much closer to the star (typically around 1 au, see also Fig. 8).

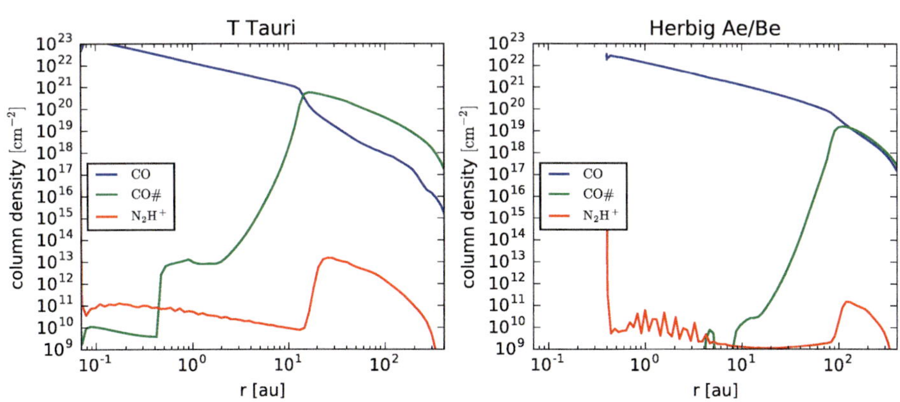

Fig. 10 Vertical column densities for CO, CO ice (CO#) and N_2H^+ as a function of radius for the two generic *ProDiMo* disk models (*left panel:* T Tauri, *right panel:* Herbig Ae/Be, see also Table 1). The jump in the N_2H^+ column density at the CO ice lines (at ≈15 and 80 au) is visible in both models but less pronounced in the Herbig Ae/Be disk

The recent work of Zhang et al. (2015) suggests that the locations of the dark rings visible in ALMA continuum observations of HL Tau (ALMA Partnership et al. 2015) coincide with the expected ice line locations of several molecules. In particular, the locations of the H_2O, NH_3 and CO ice lines agree well with the three most prominent dips in the radial surface brightness distribution. The same authors argue that around the location of the ice lines, enhanced pebble-growth (particles larger than a few cm) quickly reduces the number of mm-sized particles, which would explain the lack of emission inside the rings.

6.4 FU Orions Like Outburst Scenario

As an example for the impact of stellar/disk evolution on the chemical structure we discuss the impact of luminosity outbursts (see Sect. 2.1). During these outbursts the total luminosity of the system can increase by more than a factor of 100 compared to quiescent phases. The most prominent example for such outbursts is probably FU Orionis (e.g., Zhu et al. 2007). As already mentioned, Vorobyov et al. (2013) studied the impact of such outbursts on the gas phase transition of CO in a complex hydrodynamic model with a simplified chemistry (only thermal adsorption/desorption processes for CO and CO_2 were considered).

We present here a simple disk+envelope model using the radiation thermo-chemical disk code ProDiMo. The modeling approach is similar to Visser et al. (2015), who studied the impact of luminosity outbursts on the chemical structure of envelopes around young protostars. Different from the model of Visser et al. (2015), our model considers an outburst scenario at a later evolutionary stage when a prominent Keplerian disk has already formed (e.g. similar to FU Orionis). In addition to the disk component, a remnant envelope structure is included (see Table 2). For the outburst we assume an instantaneous increase of the stellar luminosity by a factor of 100 lasting for 100 years.

Table 2 Main stellar and structure properties for the FU Orions like outburst model

Quantity	Value
Stellar mass	0.5 $M_\odot$
Stellar effective temperature	4000 K
Stellar luminosity	1.0, 100.0 $L_\odot$
FUV luminosity	1.1×10^{31} erg s^{-1}
X-ray luminosity	10^{30} erg s^{-1}
X-ray emission temperature	2×10^7 K
Disk gas mass	0.02 $M_\odot$
Gas/dust mass ratio	100
Inner disk radius	0.6 au
Outer disk radius	200 au
Column density power index	1.0
Scale height at $r = 100$ au	10 au
Flaring power index	1.15
Envelope gas mass	0.02 $M_\odot$
Inner envelope radius	0.6 au
Outer envelope radius	3500 au
Cavity opening angle	30 deg

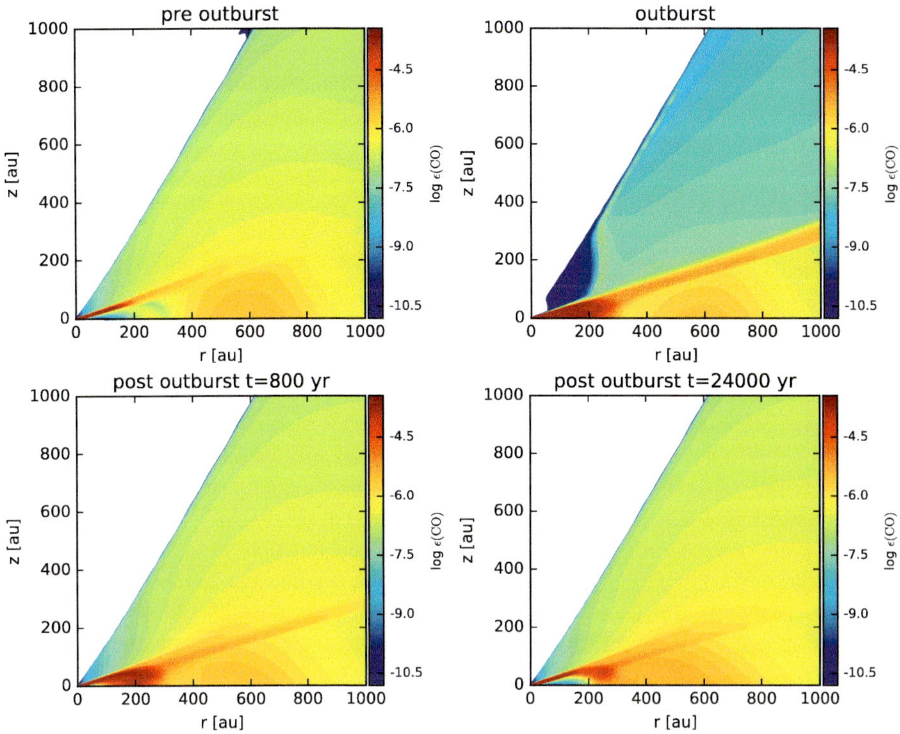

Fig. 11 Time evolution of the CO gas phase abundance $\epsilon(\mathrm{CO})$ for a FU Ori like outburst scenario. The structure includes a disk component and a remnant envelope (see Table 2 for details). The *top left* plot shows $\epsilon(\mathrm{CO})$ for the quiescent phase (stellar luminosity $L_* = 1\ \mathrm{L}_\odot$), the *top right* plot during the outburst phase ($L_* = 100\ \mathrm{L}_\odot$), the *bottom left and right* plots show the abundance 800 yr and 24 000 yr after the outburst, respectively

In Fig. 11 we show the CO gas phase abundance before the outburst, during the outburst, and 800 yr and 24 000 yr after the outburst. Due to the increased luminosity the disk is heated and CO sublimates almost everywhere in the disk. In the upper layers of the disk+envelope structure CO is also dissociated by the enhanced stellar UV field. However, due to the warmer conditions the warm CO layer extends further out into the envelope structure (up to $r \approx 2000$ au). Closer to the midplane the envelope structure ($r > 200$ au) is shielded by the disk and the CO abundance does not change significantly during the outburst (top right plot in Fig. 11).

After the outburst CO freezes out again. In the inner disk this happens quite quickly, whereas in the outer disk ($r \approx 100$ au) the freeze-out timescales are longer ($\approx$1000 yr) due to lower densities. In the envelope structure the gas phase CO abundance is still enhanced after 24 000 yr after the outburst. Approximately 100 000 yr after the outburst even in the outer envelope the gas phase CO abundance reaches again its pre-outburst level.

This is a very simplified approach for modeling an outburst scenario as it ignores the possibly significant impact on the disk structure during these violent phases. However, such models allow us to study the chemical evolution in more detail (e.g. impact on other species than CO). The treatment of detailed dust radiative transfer also allows us to study the impact of different dust properties (size distribution, opacities) that may, for example, change the freeze-out timescales. Although such models are not very realistic in terms of the dynamical evolution, they can identify important chemical processes that can subsequently be incorporated in more complex hydrodynamical models. Furthermore, such models can be used to study the impact on observables in particular molecular line emission.

7 Observations of the Gas Content of Disks

Gas represents the bulk of the mass content in disks. However, from an observational point of view the gas (in particular the gas mass) is difficult to trace. Most molecular gas lines originate within a thin layer, limited by the adsorption of molecules onto dust grains towards the disk midplane, and the influence of high energy radiation in the upper disk layers that dissociates molecules and ionizes atoms (see Sect. 4 and 6).

In the radial direction, the gas content in disks extends closer to the protostar than the dust component and the outer radius is influenced by the disk geometry and the radiation from the central source, although very recent observations suggest that desorption may occur in the outer disk due to non-thermal processes (Öberg et al. 2015). Still, radiation from the protostar is the main source of energy defining the gas excitation, especially in the inner disk, but also as the main catalyst driving diverse chemical pathways. In this respect, higher excitation conditions corresponding to higher kinetic temperatures should occur closer to the protostar, while densities of different species shall probe the vertical structure of disk. In this simplified view the most common molecules detected in disks are represented in Fig. 12, as a function of radius and height in a disk. For more details on gas line diagnostics see also the reviews of Najita et al. (2007) and Carmona (2010) for the inner disk and Dutrey et al. (2014) for the outer disk.

The composition of the gas content in disks is subject to change, and chemical reactions play naturally a pivotal role, along with phase transitions (adsorption/desorption onto dust grains). It is therefore that chemistry in disks can trace different physical conditions but also plays an important role in controlling the disk dynamics. Consequently, understanding the chemical evolution of disks is central to decode the conditions for planet formation and habitability. The study of chemistry in disks in combination with the detailed study of the chemical compositions of solar system bodies is fundamental in understanding the conditions for life on Earth.

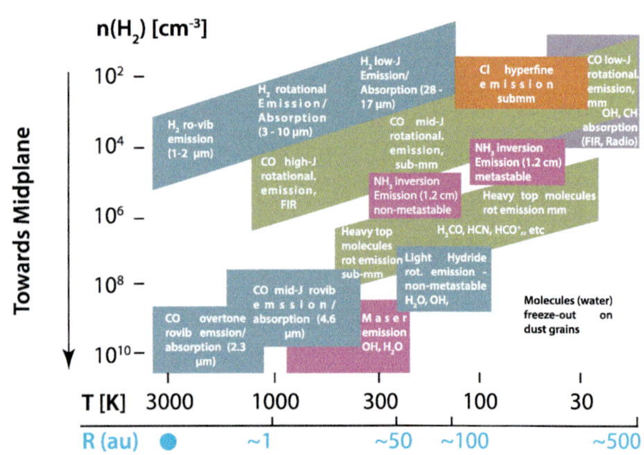

Fig. 12 Commonly observed gas lines, represented as a function of the excitation conditions (temperature, density), corresponding roughly to the radial and vertical structure in a *T Tauri* disk. We note that the temperature scale does not refer to the disk midplane, but is just indicative of the (excitation) temperature as a function of distance from the star (from Dionatos 2015)

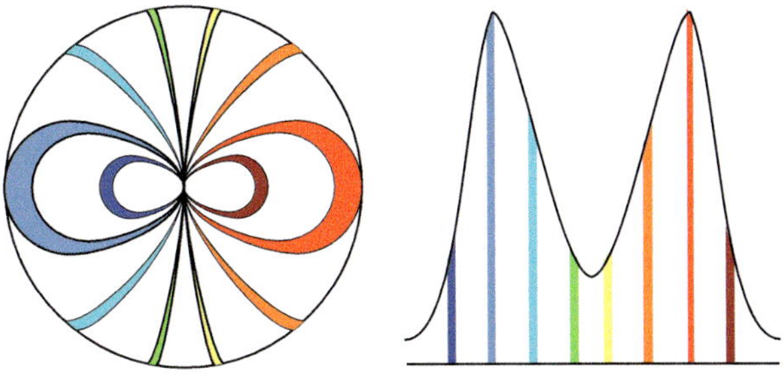

Fig. 13 Contours of constant velocity on a Keplerian disk inclined at an angle θ to the observer and their contributions in forming the characteristic Keplerian line-profile. Line minima come from the regions of the disk that is dissected by the observer's line-of-sight, while maxima are produced for the longest closed iso-velocity contours which occur at R_{out} (from Dionatos 2015)

7.1 Emission-Line Profiles of Disks

Line broadening from gas following the Keplerian rotation of a disk results in a characteristic double-peaked profile (often coined as the Keplerian profile). This is a characteristic signature that can distinguish emission from gaseous disks from other sources of excited gas (envelope cavities for embedded protostars and jets for accreting disk sources).

The line-of-sight loci of gas moving at a constant velocity on a Keplerian rotating disk at an inclination θ to the observer are given by the relation (Beckwith and Sargent 1993):

$$R = R_{out}\left(\frac{u_{out}}{u}\right)^2 \cos^2\theta, \tag{11}$$

where u_{out} is the velocity at the terminating radius R_{out} (Fig. 13). The minimum of the line occurs when the velocity vector of the observer $u \to 0$ and corresponds to the regions in the disk where the gas velocity is directed perpendicularly to the line-of-sight. As long as the velocity remains smaller than u_{out}, the iso-velocity contours on the disk surface are open and truncated at R_{out}. The maximum contour occurs for $u = u_{out}$ that is the first closed (non-truncated) contour which contributes to the line peaks. Further increase in the velocity for $u > u_{out}$ results in shorter, closed contours lying symmetrically on the horizontal disk mid-line that move towards to the center of the disk and contribute to the outer line wings. The contribution of the different curves of constant velocity to the line profile is illustrated in Fig. 13. The limiting u_{out} can be derived from the separation of the line-peaks and Eq. (1) once R_{out} and the inclination θ are known.

7.2 New Input from ALMA

Keplerian profiles are commonly seen in disks of T Tauri and Herbig Ae/Be stars when observed with adequate angular and spectral resolutions, and indeed Keplerian rotation was already observed in such environments even before the advent of ALMA (e.g. Piétu et al. 2005, 2007; Dutrey et al. 2008). Careful analysis and modeling of spectroimaging observations can lead in constraining the physical conditions of specific layers in the disk traced by molecular lines (e.g. review by Dutrey et al. 2007). Most star formation theories predicted that such disks form early on, but it was only very recently that the first disks displaying Keplerian rotation were observed with ALMA around very young, embedded protostars (Tobin et al. 2012; Takakuwa et al. 2012; Murillo et al. 2013; Ohashi et al. 2014).

In disks around T Tauri and Herbig Ae/Be stars, observations of the CO snowline (Qi et al. 2013, see also Sect. 6.3) have for the first time detected the borderlines that define the conditions for the possible formation of rocky and gas-giant planets. High concentrations of dust observed in a number of transitional disks (e.g. van der Marel et al. 2015; Pérez et al. 2014b) may be the result of the interaction between planets or disk fragments (see Sect. 2.2) with the disk. In these cases, the gas appears to fill the dust cavities but at reduced levels (van der Marel et al. 2015), and in some cases show asymmetric distributions that may be related to the dust asymmetries (e.g. HCN and CS, van der Plas et al. 2014).

Large asymmetries and deviations from Keplerian rotation have also been observed in the distribution of the CO emission around the T Tauri star AS 205 (Salyk et al. 2014). However, in that case the influence of a close companion may drive tidal interactions, or a disk wind may be in action (see Sect. 5.3 and 8), or the combination of both.

An unexpected result comes from the observations of IM Lup in DCO^+ (Öberg et al. 2015), showing two co-centric rings at radii of $\sim$90 and 300 au. While the inner ring can be explained in terms of thermal desorption of CO, the outer ring is interpreted in terms of non-thermal desorption of CO which reacts with H_2D^+ to form the observed DCO^+.

8 Observational Constraints of Photoevaporative Disk Winds

8.1 Atomic Tracers

Direct observations of photoevaporative disks winds are needed to test and constrain the different models and ultimately characterize them. In this section we will only cover the observational aspects of disk winds, and refer the reader to the Chap. 6 by Gorti et al. for a complete overview of photoevaporation in the context of disk dispersal and to the review by Alexander et al. (2014). A brief overview of the basic theory of photoevaporation is given in the present chapter (see Sect. 5.3).

Forbidden line emission often observed in T Tauri stars can potentially trace photoevaporative disk winds. Neon has high first and second ionization potentials (21.56 and 41.0 eV, respectively). Ionization can also occur by photoionization of the inner shell, requiring 0.87 and 0.88 keV for ejecting a K-shell electron for neutral and singly ionized neon. Only photons in the EUV range (or beyond) can ionize the outer electrons, while hard X-rays with energies in the range of keV are needed to eject a K-shell electron. Therefore neon can be used to constrain the high-energy stellar radiation reaching the disk. The mid-infrared forbidden transition of [Ne II] (12.81 μm) was first detected with the Spitzer Infrared Spectrograph (IRS, Pascucci et al. 2007, Lahuis et al. 2007). Although the Spitzer spectrograph lacks the spectral resolution needed to confirm or rule out a disk, wind, or jet origin of the line, interesting conclusions were extracted from studies based on statistically significant samples (see for example Lahuis et al. 2007, Flaccomio et al. 2009, Güdel et al. 2010, Baldovin-Saavedra et al. 2011). For example, the line luminosities of sources known to drive outflows and jets are brighter by 1–2 orders of magnitude than those of sources without outflows/jets. Correlations are observed between the line luminosity and different stellar and disk physical parameters, the most prominent ones being with the stellar X-ray luminosity L_X and the disk mass accretion rate.

The [Ne III] line (15.55 μm) on the other hand is detected only in a handful of objects (e.g., Kruger et al. 2013, Espaillat et al. 2013). Espaillat et al. (2013) tested the possibility of the lines being emitted by an X-ray and EUV irradiated disk. The study shows no difference in the X-ray properties between stars with and without [Ne III] emission. In addition the authors used the ratio [Ne III]/[Ne II] to distinguish between X-ray and EUV production of the fine-structure emission. In fact, if the emission is dominated by X-ray production the ratio is expected to be ~0.1; in the case of EUV the line ratio can be much larger. These authors observed one object for which the ratio is ~1 and therefore consistent with EUV production. However, it is not possible to make general conclusions given the small number of detections.

A natural step in the study of [Ne II] emission was to follow-up with ground-based observations those stars that display [Ne II] emission in their Spitzer spectra. Thanks to the high-spectral resolution achievable with ground-based spectrographs in the mid-infrared ($R \sim 30000$, or 10 $\mathrm{km\,s^{-1}}$ for VISIR at VLT), lines could be spectrally resolved, and the nature of the emission (disk, disk wind, or jet) for most of the objects could finally be unveiled. At the time of writing the neon line has been observed with ground-based instruments in ~50 objects (see Herczeg et al. 2007, Najita et al. 2009, Pascucci and Sterzik 2009, van Boekel et al. 2009, Pascucci et al. 2011, Sacco et al. 2012, Baldovin-Saavedra et al. 2012). About 50 % of the [Ne II] lines detected with Spitzer are confirmed with ground based observations. This hints at a high contribution from shocked material in the Spitzer spectra. The lines that are interpreted as emitted in jets present high-velocity blueshifts (30 $\mathrm{km\,s^{-1}}$ or more) and their full width at half maximum (FWHM) can be quite broad (FWHM up to ~140 $\mathrm{km\,s^{-1}}$). Lines consistent with disk wind emission show small velocity blueshifts (~2–20 $\mathrm{km\,s^{-1}}$) and are narrow (FWHM ~10–25 $\mathrm{km\,s^{-1}}$). There is a handful of objects that show no shift in their line peak. However, they cannot be directly associated to disk emission; two of them (AA Tau, CoKu Tau 1) are high inclination objects known to drive bipolar jets. An interesting result from these studies is that the [Ne II] line shows only one velocity component, either a high-velocity jet, a disk wind, or a disk atmosphere. There is no evidence up to now of a neon line with two velocity components as is often observed in other atomic tracers such as [O I] in the optical.

Forbidden lines in the optical range are common in T Tauri stars and have been the subject of studies for quite some time (Jankovics et al. 1983, Edwards et al. 1987, Hamann 1994, Hartigan et al. 1995). Early studies reported the presence of blueshifted line emission and associated it with some sort of disk wind whose redshifted emission is obscured by the protoplanetary disk. Forbidden emission lines often display two-velocity components: a high-velocity component (HVC) blueshifted by hundred kilometers per second with respect to the stellar rest velocity, and a low-velocity component (LVC) blueshifted by a few kilometers per second. Kwan and Tademaru (1988) recognized that these two components are emitted in two different regions and attributed the low-velocity component to a slow-moving wind emitted from the disk, and the high-velocity component to a collimated jet near the star. In addition, Hartigan et al. (1995) observed a trend where the line velocity shifts are higher for lower critical densities, suggesting that the low-velocity component accelerates away from the surface of the disk.

The neutral atomic oxygen lines [O I] at 6300 Å and [O I] at 5577 Å are interesting because they are typically bright, in particular the 6300 Å line. Different authors have attempted to explain the observed properties of the [O I] line with thermal photoevaporative disk wind models considering different types of stellar radiation. The model presented in Font et al. (2004), where photoevaporation is driven by stellar EUV radiation, is difficult to reconcile with the observed [O I] line fluxes because it produces an almost fully ionized layer with a low fraction of neutral gas. Ercolano and Owen (2010) modeled photoevaporation driven by EUV and soft X-ray emission (0.1–10 keV) that can heat the gas to the high temperatures needed to reproduce the observed [O I] luminosity. Gorti et al. (2011) considered FUV and X-rays in their input stellar spectrum and found that the [O I] lines are emitted from dissociation of OH with only a minor contribution from thermal emission.

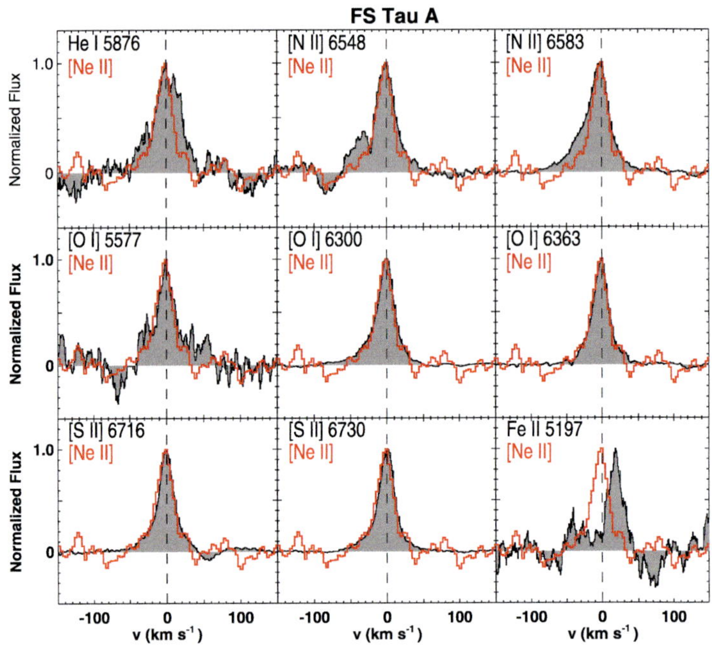

Fig. 14 Comparison between the [Ne II] line (12.81 μm) detected with VISIR (VLT) and the optical lines detected with UVES (VLT) in the young system FS Tau A. In this case the line centroids of the mid-infrared tracer are in good agreement with the optical forbidden lines (Baldovin-Saavedra et al. 2012, ©ESO/A&A. Reproduced with permission.)

Recent observational studies have focused on the study of optical forbidden lines (Baldovin-Saavedra et al. 2012, Rigliaco et al. 2013, Natta et al. 2014, Manara et al. 2014). Baldovin-Saavedra et al. (2012) obtained UVES (VLT) high-resolution spectra for three objects with a [Ne II] line detected from the ground. In two of the objects the optical and the mid-infrared tracers display the same velocity shift and FWHM (see Fig. 14), and in one case the [Ne II] line traces clearly jet emission, while the [O I] lines show two velocity components. Rigliaco et al. (2013) performed a systematic study of the [O I] low-velocity component and compared it to the [Ne II] and CO lines. They find a trend between the line peak velocities, where the [Ne II] line is typically more blueshifted than [O I] which in turn is more blueshifted than CO. The authors obtained a small range in the line ratio of [O I] 6300/5577 Å which they claimed is difficult to reproduce in gas heated by thermal processes. For two objects observed at very high spectral resolution the authors could decompose the line profile in two components: a broad and symmetric component attributed to disk emission and a narrower and blueshifted component attributed to unbound gas. Natta et al. (2014) found a correlation between the [O I] luminosity and both $L_\star$ and $L_{\rm acc}$, the latter being related to the stellar FUV emission. They found that, on average, the low-velocity component peaks at $<20\ \rm km\,s^{-1}$ and determined the physical conditions of the gas, finding high density ($n_{\rm H} > 10^8\ \rm cm^{-3}$) and temperatures ($T \sim 5000$–$1000$ K). The authors warn that the absence of correlation between the luminosity of the [O I] line and $L_{\rm X}$, contradicts an X-ray driven disk wind.

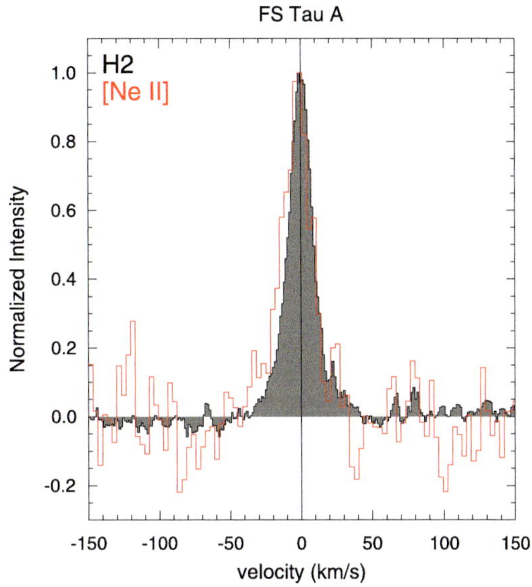

Fig. 15 The *grey area* represents the H_2 1–0 S(1) line at 2.12 μm detected with CRIRES and the *solid line* represents the [Ne II] line (12.81 μm) detected with VISIR in FS Tau A (Baldovin-Saavedra et al. in prep.)

8.2 Molecular Tracers

CO ro-vibrational emission at 4.7 μm is detected in many T Tauri stars. The line probes warm gas in the inner 1 au from the central star. Some of the observed lines show a singly peaked and narrow profile with a broad base that extends to >50 km s^{-1}; they show a small velocity shift of ~ 5 km s^{-1} that cannot be explained by purely Keplerian rotation (e.g., Pontoppidan et al. 2011, Bast et al. 2011, Brown et al. 2013). Pontoppidan et al. (2011) performed spectroastrometry observations of the CO line. This technique allows to obtain spatial and kinematical information at very small scales and very high spectral resolution. In their study, they find that the single-peaked line profiles are consistent with a combination of gas in Keplerian rotation and a non-collimated disk wind. In addition, Brown et al. (2013) detected excess emission on the blue side of the line profile, giving further confirmation for the disk wind hypothesis.

Molecular hydrogen, H_2, is the most abundant component in protoplanetary disks, but this homonuclear molecule is very difficult to observe since its rotational quadrupole transitions from low-energy levels are weak and lie in regions with poor atmospheric transmission. The ro-vibrational transitions observable in the near-infrared trace thermal emission of hot gas at about 1000 K, or gas excited by UV or X-rays (Carmona et al. 2007, 2011). Takami et al. (2004) reported the detection of a slow molecular wind in H_2 for DG Tau. Baldovin-Saavedra et al. (in prep.) performed follow-up observations with CRIRES (VLT) of young systems with detections of [Ne II]. The program aimed at detecting the ro-vibrational transitions of H_2: 1–0 S(1) (2.12 μm), 1–0 S(1) (2.22 μm), and 2–1 S(1) (2.25 μm). Emission from the three transitions is detected in three objects, two of them known to drive high-velocity jets. Two of the sources have line profiles consistent with emission from the disk atmosphere (no shift), and one shows a blueshift of ~ -8 km s^{-1}. The other four objects show detections from the 1–0 S(0) transition only; two of them can be interpreted in terms of gas bound to the disk, and two present lines with small blueshifts (< -10 km s^{-1}). Figure 15 (Baldovin-Saavedra et al. in prep.) shows a comparison between the H_2 1–0 S(1) line and the [Ne II] line detected in the mid-infrared and published in Baldovin-Saavedra et al.

(2012) for one of the stars in the sample. The spectra have been continuum subtracted and the lines normalized to their peak to allow for a better comparison. In this case the [Ne II] line is slightly blueshifted, $v_{peak} = -2.9 \pm 0.7\ \mathrm{km\,s^{-1}}$ with a FWHM of 26.8 $\mathrm{km\,s^{-1}}$ and shows excess emission towards the blue side of the spectrum, while the H_2 line peak is consistent with disk emission, $v_{peak} = -0.1 \pm 0.2\ \mathrm{km\,s^{-1}}$ and FWHM $= 21.2 \pm 0.2\ \mathrm{km\,s^{-1}}$. It is interesting to estimate the distance from the star needed to observe a line in Keplerian rotation that does not show a double peaked profile. Considering the inclination of the system, $i = 35°$, the mass of the star $M_\star = 0.6\ \mathrm{M_\odot}$, and the CRIRES spectral resolution of 3 $\mathrm{km\,s^{-1}}$, the gas should be located at a distance $R > 77$ au.

The trend observed in H_2 is similar to what is observed in CO; the molecular lines tend to be less blueshifted than the [Ne II] line. However, at this stage it is difficult to establish a direct relation between the observed [Ne II] photoevaporative disk winds and their molecular counterpart. Furthermore, the nature of the molecular winds, i.e., magnetically driven or photoevaporation driven, remain uncertain.

9 Summary

We discussed several physical processes which have significant impact on the evolution and the chemical composition of the gas disk. In early stages the disk evolution might be dominated by gravitational instabilities if the initial gas mass is high enough. At this stage most disks probably experience phases of highly increased accretion rates resulting in strong luminosity outbursts. In later phases the disk evolution is dominated by viscous accretion possibly driven by magneto-rotational instabilities. In addition disk winds may play a crucial role in the dispersal of the disk determining the time-scale for planet formation. One example are photo-evaporative winds which are driven by high energy stellar radiation of the star (see also Chap. 6 by Gorti et al.).

Chemical processes in the gas disks determine the chemical composition of planetary atmospheres but also have an impact on the solids due to freeze-out of molecules. Enhanced dust growth at the position of ice lines leads to the formation of larger solid bodies which possibly grow further to planetesimals and planetary cores. The gas mass is another crucial parameter for planet formation. It is still challenging to measure directly the disk gas mass. Using the CO molecular line emission requires detailed chemical models (e.g. CO isotopologue chemistry and surface chemistry) and therefore the derived masses are still model dependent. The HD molecule offers a more direct method but is currently only detected in one disk. Upcoming missions like the James Webb Space Telescope will certainly improve this situation.

The high spatial resolution and sensitivity of ALMA provides detailed observations of the chemical composition of disks (e.g. detection of complex molecules, ice line locations). Future ALMA observations will certainly significantly improve our knowledge of disk chemistry. Interpretations of these observations are a formidable challenge for chemical models. Models consistently treating the dust and gas evolution are necessary. The incorporation of chemical processes in hydrodynamical models is required for a better understanding of the time evolution of the gaseous component of disks. Further detailed knowledge of the evolution of young stars is required as in particular high-energy stellar radiation is one of the main drivers of disk chemistry and evolution.

Acknowledgements We want to thank the anonymous referee for a thorough and constructive report. We are grateful to the organizers of this International Space Science Institute (ISSI) workshop and the editors of this book. C. Rab and M. Güdel acknowledge funding by the Austrian Science Fund (FWF), project number P24790 and project S11601-N16 "Pathways to Habitability: From Disks to Active Stars, Planets and Life". C. Baldovin-Saavedra acknowledges funding by the Austrian Research Promotion Agency (FFG) under grant agreement FA 538022. E. Vorobyov acknowledges support from the RFBR grant 14-02-00719. The research leading to these results has received funding from the European Union Seventh Framework Programme FP7-2011 under grant agreement no 284405. This work was partly supported by the Austrian Science Fund (FWF) under research grant I2549-N27. This publication was partly supported from the FFG ASAP 12 project *JetPro** (FFG-854025). The computational results presented have been achieved in part using the Vienna Scientific Cluster (VSC).

References

Y. Aikawa, T. Umebayashi, T. Nakano, S. Miyama, Faraday Discuss. **109**, 281 (1998)
Y. Aikawa, G.J. van Zadelhoff, E.F. van Dishoeck, E. Herbst, Astron. Astrophys. **386**, 622–632 (2002)
Y. Aikawa, K. Furuya, H. Nomura, C. Qi, Astrophys. J. **807**, 120 (2015)
V. Akimkin, S. Zhukovska, D. Wiebe, D. Semenov, Y. Pavlyuchenkov, A. Vasyunin, T. Birnstiel, T. Henning, Astrophys. J. **766**, 8 (2013)
R.D. Alexander, C.J. Clarke, J.E. Pringle, Mon. Not. R. Astron. Soc. **369**, 216–228 (2006a)
R.D. Alexander, C.J. Clarke, J.E. Pringle, Mon. Not. R. Astron. Soc. **369**, 229–239 (2006b)
R. Alexander, I. Pascucci, S. Andrews, P. Armitage, L. Cieza, in *Protostars and Planets VI* (2014), pp. 475–496
ALMA Partnership, C.L. Brogan, L.M. Pérez, T.R. Hunter, W.R.F. Dent, A.S. Hales, R.E. Hills, S. Corder, E.B. Fomalont, C. Vlahakis, Y. Asaki, D. Barkats, A. Hirota, J.A. Hodge, C.M.V. Impellizzeri, R. Kneissl, E. Liuzzo, R. Lucas, N. Marcelino, S. Matsushita, K. Nakanishi, N. Phillips, A.M.S. Richards, I. Toledo, R. Aladro, D. Broguiere, J.R. Cortes, P.C. Cortes, D. Espada, F. Galarza, D. Garcia-Appadoo, L. Guzman-Ramirez, E.M. Humphreys, T. Jung, S. Kameno, R.A. Laing, S. Leon, G. Marconi, A. Mignano, B. Nikolic, L.-A. Nyman, M. Radiszcz, A. Remijan, J.A. Rodón, T. Sawada, S. Takahashi, R.P.J. Tilanus, B. Vila Vilaro, L.C. Watson, T. Wiklind, E. Akiyama, E. Chapillon, I. de Gregorio-Monsalvo, J. Di Francesco, F. Gueth, A. Kawamura, C.-F. Lee, Q. Nguyen Luong, J. Mangum, V. Pietu, P. Sanhueza, K. Saigo, S. Takakuwa, C. Ubach, T. van Kempen, A. Wootten, A. Castro-Carrizo, H. Francke, J. Gallardo, J. Garcia, S. Gonzalez, T. Hill, T. Kaminski, Y. Kurono, H.-Y. Liu, C. Lopez, F. Morales, K. Plarre, G. Schieven, L. Testi, L. Videla, E. Villard, P. Andreani, J.E. Hibbard, K. Tatematsu, Astrophys. J. Lett. **808**, 3 (2015)
P. André, D. Ward-Thompson, J. Greaves, Science **337**, 69 (2012)
S.M. Andrews, J.P. Williams, Astrophys. J. **631**, 1134–1160 (2005)
S.M. Andrews, D.J. Wilner, C. Espaillat, A.M. Hughes, C.P. Dullemond, M.K. McClure, C. Qi, J.M. Brown, Astrophys. J. **732**, 42 (2011)
S.M. Andrews, D.J. Wilner, Z. Zhu, T. Birnstiel, J.M. Carpenter, L.M. Pérez, X.-N. Bai, K.I. Öberg, A.M. Hughes, A. Isella, L. Ricci, Astrophys. J. Lett. **820**, 40 (2016)
G. Aresu, I. Kamp, R. Meijerink, P. Woitke, W.-F. Thi, M. Spaans, Astron. Astrophys. **526**, 163 (2011)
G. Aresu, I. Kamp, R. Meijerink, M. Spaans, S. Vicente, L. Podio, P. Woitke, F. Menard, W.-F. Thi, M. Güdel, A. Liebhart, Astron. Astrophys. **566**, 14 (2014)
M. Audard, P. Ábrahám, M.M. Dunham, J.D. Green, N. Grosso, K. Hamaguchi, J.H. Kastner, Á. Kóspál, G. Lodato, M.M. Romanova, S.L. Skinner, E.I. Vorobyov, Z. Zhu, in *Protostars and Planets VI* (2014), pp. 387–410
S.A. Balbus, J.F. Hawley, Astrophys. J. **376**, 214–233 (1991)
C. Baldovin-Saavedra, M. Audard, M. Güdel, L.M. Rebull, D.L. Padgett, S.L. Skinner, A. Carmona, A.M. Glauser, S.B. Fajardo-Acosta, Astron. Astrophys. **528**, 22 (2011)
C. Baldovin-Saavedra, M. Audard, A. Carmona, M. Güdel, K. Briggs, L.M. Rebull, S.L. Skinner, B. Ercolano, Astron. Astrophys. **543**, 30 (2012)
J. Bally, E. Feigelson, B. Reipurth, Astrophys. J. **584**, 843–852 (2003)
C. Baruteau, A. Crida, S.-J. Paardekooper, F. Masset, J. Guilet, B. Bitsch, R. Nelson, W. Kley, J. Papaloizou, in *Protostars and Planets VI* (2014), pp. 667–689
J.E. Bast, J.M. Brown, G.J. Herczeg, E.F. van Dishoeck, K.M. Pontoppidan, Astron. Astrophys. **527**, 119 (2011)

S. Basu, E.I. Vorobyov, Astrophys. J. **750**, 30 (2012)
M.R. Bate, Mon. Not. R. Astron. Soc. **392**, 590–616 (2009)
S.V.W. Beckwith, A.I. Sargent, Astrophys. J. **402**, 280–291 (1993)
E.A. Bergin, Y. Aikawa, G.A. Blake, E.F. van Dishoeck, in *Protostars and Planets V* (2007), pp. 751–766. http://arxiv.org/abs/astro-ph/0603358
E.A. Bergin, L.I. Cleeves, U. Gorti, K. Zhang, G.A. Blake, J.D. Green, S.M. Andrews, N.J. Evans II, T. Henning, K. Öberg, K. Pontoppidan, C. Qi, C. Salyk, E.F. van Dishoeck, Nature **493**, 644–646 (2013)
E.A. Bergin, L.I. Cleeves, N. Crockett, G.A. Blake, Faraday Discuss. **168**, 61–79 (2014). http://adsabs.harvard.edu/abs/2014FaDi..168...61B
B. Bitsch, A. Johansen, M. Lambrechts, A. Morbidelli, Astron. Astrophys. **575**, 28 (2015)
A.C. Boley, Astrophys. J. Lett. **695**, 53–57 (2009)
A.C. Boley, T. Hayfield, L. Mayer, R.H. Durisen, Icarus **207**, 509–516 (2010)
A.P. Boss, Science **276**, 1836–1839 (1997)
A.P. Boss, Astrophys. J. **731**, 74 (2011)
J.M. Brown, K.M. Pontoppidan, E.F. van Dishoeck, G.J. Herczeg, G.A. Blake, A. Smette, Astrophys. J. **770**, 94 (2013)
S. Bruderer, E.F. van Dishoeck, S.D. Doty, G.J. Herczeg, Astron. Astrophys. **541**, 91 (2012)
J.R. Burke, D.J. Hollenbach, Astrophys. J. **265**, 223–234 (1983)
S. Cabrit, constraints on accretion-ejection structures in young stars, in *EAS Publications Series*, ed. by J. Bouvier, J.-P. Zahn, EAS Publications Series, vol. 3, 2002, pp. 147–182
N. Calvet, E. Gullbring, Astrophys. J. **509**, 802–818 (1998)
A. Carmona, Earth Moon Planets **106**, 71–95 (2010)
A. Carmona, M.E. van den Ancker, T. Henning, M. Goto, D. Fedele, B. Stecklum, Astron. Astrophys. **476**, 853 (2007). doi:10.1051/0004-6361:20078536. Erratum: **478**, 795 (2008)
A. Carmona, G. van der Plas, M.E. van den Ancker, M. Audard, L.B.F.M. Waters, D. Fedele, B. Acke, E. Pantin, Astron. Astrophys. **533**, 39 (2011)
C. Ceccarelli, C. Dominik, A. López-Sepulcre, M. Kama, M. Padovani, E. Caux, P. Caselli, Astrophys. J. Lett. **790**, 1 (2014)
G. Chaparro Molano, I. Kamp, Astron. Astrophys. **537**, 138 (2012a)
G. Chaparro Molano, I. Kamp, Astron. Astrophys. **547**, 7 (2012b)
E.I. Chiang, P. Goldreich, Astrophys. J. **490**, 368 (1997)
C.J. Clarke, A. Gendrin, M. Sotomayor, Mon. Not. R. Astron. Soc. **328**, 485–491 (2001)
L.I. Cleeves, F.C. Adams, E.A. Bergin, Astrophys. J. **772**, 5 (2013a)
L.I. Cleeves, F.C. Adams, E.A. Bergin, R. Visser, Astrophys. J. **777**, 28 (2013b)
L.I. Cleeves, E.A. Bergin, C. Qi, F.C. Adams, K.I. Öberg, Astrophys. J. **799**, 204 (2015)
O. Dionatos, Gas line observations of disks. Eur. Phys. J. Web Conf. **102**, 00008 (2015)
F. Du, E.A. Bergin, Astrophys. J. **792**, 2 (2014)
M.M. Dunham, E.I. Vorobyov, Astrophys. J. **747**, 52 (2012)
M.M. Dunham, E.I. Vorobyov, H.G. Arce, Mon. Not. R. Astron. Soc. **444**, 887–901 (2014)
M.M. Dunham, N.J. Evans II, S. Terebey, C.P. Dullemond, C.H. Young, Astrophys. J. **710**, 470–502 (2010)
A. Dutrey, S. Guilloteau, P. Ho, in *Protostars and Planets V* (2007), pp. 495–506
A. Dutrey, S. Guilloteau, V. Piétu, E. Chapillon, F. Gueth, T. Henning, R. Launhardt, Y. Pavlyuchenkov, K. Schreyer, D. Semenov, Astron. Astrophys. **490**, 15–18 (2008)
A. Dutrey, D. Semenov, E. Chapillon, U. Gorti, S. Guilloteau, F. Hersant, M. Hogerheijde, M. Hughes, G. Meeus, H. Nomura, V. Piétu, C. Qi, V. Wakelam, in *Protostars and Planets VI* (2014), pp. 317–338
S. Edwards, S. Cabrit, S.E. Strom, I. Heyer, K.M. Strom, E. Anderson, Astrophys. J. **321**, 473–495 (1987)
B. Ercolano, J.E. Owen, Mon. Not. R. Astron. Soc. **406**, 1553–1569 (2010)
B. Ercolano, G. Rosotti, Mon. Not. R. Astron. Soc. **450**, 3008–3014 (2015)
B. Ercolano, J.J. Drake, J.C. Raymond, C.C. Clarke, Astrophys. J. **688**, 398–407 (2008)
C. Espaillat, L. Ingleby, E. Furlan, M. McClure, A. Spatzier, J. Nieusma, N. Calvet, E. Bergin, L. Hartmann, J.M. Miller, J. Muzerolle, Astrophys. J. **762**, 62 (2013)
C. Espaillat, J. Muzerolle, J. Najita, S. Andrews, Z. Zhu, N. Calvet, S. Kraus, J. Hashimoto, A. Kraus, P. D'Alessio, in *Protostars and Planets VI* (2014), pp. 497–520
M.G. Evans, J.D. Ilee, A.C. Boley, P. Caselli, R.H. Durisen, T.W. Hartquist, J.M.C. Rawlings, Mon. Not. R. Astron. Soc. **453**, 1147–1163 (2015)
F. Favata, C.V.M. Fridlund, G. Micela, S. Sciortino, A.A. Kaas, Astron. Astrophys. **386**, 204–210 (2002)
F. Favata, G. Micela, B. Silva, S. Sciortino, M. Tsujimoto, Astron. Astrophys. **433**, 1047–1054 (2005)
C. Favre, L.I. Cleeves, E.A. Bergin, C. Qi, G.A. Blake, Astrophys. J. Lett. **776**, 38 (2013)
D. Fedele, M.E. van den Ancker, T. Henning, R. Jayawardhana, J.M. Oliveira, Astron. Astrophys. **510**, 72 (2010)
E.D. Feigelson, G.P. Garmire, S.H. Pravdo, Astrophys. J. **572**, 335–349 (2002)

E. Flaccomio, B. Stelzer, S. Sciortino, G. Micela, I. Pillitteri, L. Testi, Astron. Astrophys. **505**, 695–706 (2009)
A.S. Font, I.G. McCarthy, D. Johnstone, D.R. Ballantyne, Astrophys. J. **607**, 890–903 (2004)
A. Frank, T.P. Ray, S. Cabrit, P. Hartigan, H.G. Arce, F. Bacciotti, J. Bally, M. Benisty, J. Eislöffel, M. Güdel, S. Lebedev, B. Nisini, A. Raga, in *Protostars and Planets VI* (2014), pp. 451–474
S. Fromang, C. Terquem, S.A. Balbus, Mon. Not. R. Astron. Soc. **329**, 18–28 (2002)
K. Furuya, Y. Aikawa, Astrophys. J. **790**, 97 (2014)
C.F. Gammie, Astrophys. J. **457**, 355 (1996)
P.G. Gibbons, G.R. Mamatsashvili, W.K.M. Rice, Mon. Not. R. Astron. Soc. **453**, 4232–4243 (2015)
A.E. Glassgold, D. Galli, M. Padovani, Astrophys. J. **756**, 157 (2012)
A.E. Glassgold, J. Najita, J. Igea, Astrophys. J. **480**, 344–350 (1997)
A.E. Glassgold, J. Najita, J. Igea, Astrophys. J. **615**, 972–990 (2004)
U. Gorti, D. Hollenbach, Astrophys. J. **613**, 424–447 (2004)
U. Gorti, D. Hollenbach, Astrophys. J. **683**, 287–303 (2008)
U. Gorti, D. Hollenbach, J. Najita, I. Pascucci, Astrophys. J. **735**, 90 (2011)
R. Gredel, S. Lepp, A. Dalgarno, E. Herbst, Astrophys. J. **347**, 289–293 (1989)
M. Güdel, Annu. Rev. Astron. Astrophys. **12**, 71–237 (2004)
M. Güdel, A. Telleschi, Astron. Astrophys. **474**, 25–28 (2007)
M. Güdel, S.L. Skinner, M. Audard, K.R. Briggs, S. Cabrit, Astron. Astrophys. **478**, 797–807 (2008)
M. Güdel, F. Lahuis, K.R. Briggs, J. Carr, A.E. Glassgold, T. Henning, J.R. Najita, R. van Boekel, E.F. van Dishoeck, Astron. Astrophys. **519**, 113 (2010)
F. Hamann, Astrophys. J. Suppl. Ser. **93**, 485–518 (1994)
N. Harada, E. Herbst, V. Wakelam, Astrophys. J. **721**, 1570–1578 (2010)
P. Hartigan, S. Edwards, L. Ghandour, Astrophys. J. **452**, 736 (1995)
C. Helling, P. Woitke, P.B. Rimmer, I. Kamp, W.-F. Thi, R. Meijerink, Life **4**(2), 142–173 (2014)
T. Henning, D. Semenov, Chem. Rev. **113**, 9016–9042 (2013)
G.J. Herczeg, J.R. Najita, L.A. Hillenbrand, I. Pascucci, Astrophys. J. **670**, 509–515 (2007)
J. Hernández, L. Hartmann, N. Calvet, R.D. Jeffries, R. Gutermuth, J. Muzerolle, J. Stauffer, Astrophys. J. **686**, 1195–1208 (2008)
D.J. Hollenbach, A.G.G.M. Tielens, Rev. Mod. Phys. **71**, 173–230 (1999)
D. Hollenbach, D. Johnstone, S. Lizano, F. Shu, Astrophys. J. **428**, 654–669 (1994)
J. Igea, A.E. Glassgold, Astrophys. J. **518**, 848–858 (1999)
J.D. Ilee, A.C. Boley, P. Caselli, R.H. Durisen, T.W. Hartquist, J.M.C. Rawlings, Mon. Not. R. Astron. Soc. **417**, 2950–2961 (2011)
K. Imanishi, K. Koyama, Y. Tsuboi, Astrophys. J. **557**, 747–760 (2001)
I. Jankovics, I. Appenzeller, J. Krautter, Publ. Astron. Soc. Pac. **95**, 883–885 (1983)
J.K. Jørgensen, R. Visser, N. Sakai, E.A. Bergin, C. Brinch, D. Harsono, J.E. Lindberg, E.F. van Dishoeck, S. Yamamoto, S.E. Bisschop, M.V. Persson, Astrophys. J. Lett. **779**, 22 (2013)
I. Kamp, G.-J. van Zadelhoff, Astron. Astrophys. **373**, 641–656 (2001)
I. Kamp, I. Tilling, P. Woitke, W. Thi, M. Hogerheijde, Astron. Astrophys. **510**(26), 260000 (2010)
J.H. Kastner, D.P. Huenemoerder, N.S. Schulz, C.R. Canizares, D.A. Weintraub, Astrophys. J. **567**, 434–440 (2002)
S.J. Kenyon, L.W. Hartmann, K.M. Strom, S.E. Strom, Astron. J. **99**, 869–887 (1990)
H.J. Kim, N.J. Evans II, M.M. Dunham, J.-E. Lee, K.M. Pontoppidan, Astrophys. J. **758**, 38 (2012)
K.M. Kratter, R.A. Murray-Clay, Astrophys. J. **740**, 1 (2011)
K.M. Kratter, C.D. Matzner, M.R. Krumholz, Astrophys. J. **681**, 375–390 (2008)
A.J. Kruger, M.J. Richter, J.S. Carr, J.R. Najita, M.M. Moerchen, G.W. Doppmann, A. Seifahrt, Astrophys. J. **764**, 127 (2013)
J. Kwan, E. Tademaru, Astrophys. J. **332**, L41–L44 (1988). doi:10.1086/185262
F. Lahuis, E.F. van Dishoeck, G.A. Blake, N.J. Evans II, J.E. Kessler-Silacci, K.M. Pontoppidan, Astrophys. J. **665**, 492–511 (2007)
J.-E. Lee, J. Korean Astron. Soc. **40**, 83–89 (2007)
Z.-Y. Li, R. Banerjee, R.E. Pudritz, J.K. Jørgensen, H. Shang, R. Krasnopolsky, A. Maury, in *Protostars and Planets VI* (2014), pp. 173–194
M.N. Machida, S.-i. Inutsuka, T. Matsumoto, Astrophys. J. **724**, 1006–1020 (2010)
M.N. Machida, S.-i. Inutsuka, T. Matsumoto, Astrophys. J. **729**, 42 (2011)
E.E. Mamajek, Initial conditions of planet formation: lifetimes of primordial disks, in *American Institute of Physics Conference Series*, ed. by T. Usuda, M. Tamura, M. Ishii, American Institute of Physics Conference Series, vol. 1158 (2009), pp. 3–10
C.F. Manara, L. Testi, A. Natta, G. Rosotti, M. Benisty, B. Ercolano, L. Ricci, Astron. Astrophys. **568**, 18 (2014)

B.C. Matthews, A.V. Krivov, M.C. Wyatt, G. Bryden, C. Eiroa, in *Protostars and Planets VI* (2014), pp. 521–544
D. McElroy, C. Walsh, A.J. Markwick, M.A. Cordiner, K. Smith, T.J. Millar, Astron. Astrophys. **550**, 36 (2013)
R. Meijerink, G. Aresu, I. Kamp, M. Spaans, W.-F. Thi, P. Woitke, Astron. Astrophys. **547**, 68 (2012)
F. Meru, M.R. Bate, Mon. Not. R. Astron. Soc. **427**, 2022–2046 (2012)
A. Miotello, S. Bruderer, E.F. van Dishoeck, Astron. Astrophys. **572**, 96 (2014)
N.M. Murillo, S.-P. Lai, S. Bruderer, D. Harsono, E.F. van Dishoeck, Astron. Astrophys. **560**, 103 (2013)
J.R. Najita, J.S. Carr, A.E. Glassgold, J.A. Valenti, in *Protostars and Planets V* (2007), pp. 507–522
J.R. Najita, G.W. Doppmann, M.A. Bitner, M.J. Richter, J.H. Lacy, D.T. Jaffe, J.S. Carr, R. Meijerink, G.A. Blake, G.J. Herczeg, A.E. Glassgold, Astrophys. J. **697**, 957–963 (2009)
A. Natta, L. Testi, J.M. Alcalá, E. Rigliaco, E. Covino, B. Stelzer, V. D'Elia, Astron. Astrophys. **569**, 5 (2014)
S. Nayakshin, Mon. Not. R. Astron. Soc. **408**, 36–40 (2010)
S. Nayakshin, Mon. Not. R. Astron. Soc. **413**, 1462–1478 (2011)
S. Nayakshin, S.-H. Cha, J.C. Bridges, Mon. Not. R. Astron. Soc. **416**, 50–54 (2011)
H. Nomura, T.J. Millar, Astron. Astrophys. **438**, 923–938 (2005)
K.I. Öberg, C. Qi, J.K.J. Fogel, E.A. Bergin, S.M. Andrews, C. Espaillat, D.J. Wilner, I. Pascucci, J.H. Kastner, Astrophys. J. **734**, 98 (2011)
K.I. Öberg, K. Furuya, R. Loomis, Y. Aikawa, S.M. Andrews, C. Qi, E.F. van Dishoeck, D.J. Wilner, Astrophys. J. **810**, 112 (2015)
N. Ohashi, K. Saigo, Y. Aso, Y. Aikawa, S. Koyamatsu, M.N. Machida, M. Saito, S.Z. Takahashi, S. Takakuwa, K. Tomida, K. Tomisaka, H.-W. Yen, Astrophys. J. **796**, 131 (2014)
J.E. Owen, B. Ercolano, C.J. Clarke, Mon. Not. R. Astron. Soc. **412**, 13–25 (2011)
J.E. Owen, B. Ercolano, C.J. Clarke, R.D. Alexander, Mon. Not. R. Astron. Soc. **401**, 1415–1428 (2010)
M. Padovani, P. Hennebelle, A. Marcowith, K. Ferrière, Astron. Astrophys. **582**, 13 (2015)
A. Palau, L.A. Zapata, L.F. Rodríguez, H. Bouy, D. Barrado, M. Morales-Calderón, P.C. Myers, N. Chapman, C. Juárez, D. Li, Mon. Not. R. Astron. Soc. **444**, 833–845 (2014)
I. Pascucci, M. Sterzik, Astrophys. J. **702**, 724–732 (2009)
I. Pascucci, D. Hollenbach, J. Najita, J. Muzerolle, U. Gorti, G.J. Herczeg, L.A. Hillenbrand, J.S. Kim, J.M. Carpenter, M.R. Meyer, E.E. Mamajek, J. Bouwman, Astrophys. J. **663**, 383–393 (2007)
I. Pascucci, M. Sterzik, R.D. Alexander, S.H.P. Alencar, U. Gorti, D. Hollenbach, J. Owen, B. Ercolano, S. Edwards, Astrophys. J. **736**, 13 (2011)
I. Pascucci, L. Ricci, U. Gorti, D. Hollenbach, N.P. Hendler, K.J. Brooks, Y. Contreras, Astrophys. J. **795**, 1 (2014)
L.M. Pérez, A. Isella, J.M. Carpenter, C.J. Chandler, Astrophys. J. Lett. **783**, 13 (2014a)
L.M. Pérez, A. Isella, J.M. Carpenter, C.J. Chandler, Astrophys. J. Lett. **783**, 13 (2014b)
V. Piétu, A. Dutrey, S. Guilloteau, Astron. Astrophys. **467**, 163–178 (2007)
V. Piétu, S. Guilloteau, A. Dutrey, Astron. Astrophys. **443**, 945–954 (2005)
C. Pinte, W.R.F. Dent, F. Ménard, A. Hales, T. Hill, P. Cortes, I. de Gregorio-Monsalvo, Astrophys. J. **816**, 25 (2016)
K.M. Pontoppidan, S.M. Blevins, Faraday Discuss. **169**, 49–60 (2014)
K.M. Pontoppidan, G.A. Blake, A. Smette, Astrophys. J. **733**, 84 (2011)
S.H. Pravdo, E.D. Feigelson, G. Garmire, Y. Maeda, Y. Tsuboi, J. Bally, Nature **413**, 708–711 (2001)
L. Prisinzano, G. Micela, E. Flaccomio, J.R. Stauffer, T. Megeath, L. Rebull, M. Robberto, K. Smith, E.D. Feigelson, N. Grosso, S. Wolk, Astrophys. J. **677**, 401–424 (2008)
K.M. Punzi, P. Hily-Blant, J.H. Kastner, G.G. Sacco, T. Forveille, Astrophys. J. **805**, 147 (2015)
C. Qi, K.I. Öberg, D.J. Wilner, P. D'Alessio, E. Bergin, S.M. Andrews, G.A. Blake, M.R. Hogerheijde, E.F. van Dishoeck, Science **341**, 630–632 (2013)
C. Qi, K.I. Öberg, S.M. Andrews, D.J. Wilner, E.A. Bergin, A.M. Hughes, M. Hogherheijde, P. D'Alessio, Astrophys. J. **813**, 128 (2015)
R.R. Rafikov, Astrophys. J. Lett. **621**, 69–72 (2005)
S.N. Raymond, E. Kokubo, A. Morbidelli, R. Morishima, K.J. Walsh, in *Protostars and Planets VI* (2014), pp. 595–618
L. Reboussin, V. Wakelam, S. Guilloteau, F. Hersant, A. Dutrey, Astron. Astrophys. **579**, 82 (2015)
Z. Regály, A. Juhász, Z. Sándor, C.P. Dullemond, Mon. Not. R. Astron. Soc. **419**, 1701–1712 (2012)
W.K.M. Rice, G. Lodato, J.E. Pringle, P.J. Armitage, I.A. Bonnell, Mon. Not. R. Astron. Soc. **355**, 543–552 (2004)
E. Rigliaco, I. Pascucci, U. Gorti, S. Edwards, D. Hollenbach, Astrophys. J. **772**, 60 (2013)
P.D. Rogers, J. Wadsley, Mon. Not. R. Astron. Soc. **423**, 1896–1908 (2012)
G.G. Sacco, E. Flaccomio, I. Pascucci, F. Lahuis, B. Ercolano, J.H. Kastner, G. Micela, B. Stelzer, M. Sterzik, Astrophys. J. **747**, 142 (2012)

S. Sallum, K.B. Follette, J.A. Eisner, L.M. Close, P. Hinz, K. Kratter, J. Males, A. Skemer, B. Macintosh, P. Tuthill, V. Bailey, D. Defrère, K. Morzinski, T. Rodigas, E. Spalding, A. Vaz, A.J. Weinberger, Nature **527**, 342–344 (2015)
C. Salyk, K. Pontoppidan, S. Corder, D. Muñoz, K. Zhang, G.A. Blake, Astrophys. J. **792**, 68 (2014)
K.R. Schwarz, E.A. Bergin, L.I. Cleeves, G.A. Blake, K. Zhang, K.I. Öberg, E.F. van Dishoeck, C. Qi, Astrophys. J. **823**, 91 (2016)
D. Semenov, D. Wiebe, Astrophys. J. Suppl. Ser. **196**, 25 (2011)
D. Semenov, D. Wiebe, T. Henning, Astron. Astrophys. **417**, 93–106 (2004)
D. Semenov, D. Wiebe, T. Henning, Astrophys. J. Lett. **647**, 57–60 (2006)
F.H. Shu, Astrophys. J. **214**, 488–497 (1977)
L. Spitzer Jr., M.G. Tomasko, Astrophys. J. **152**, 971 (1968)
D. Stamatellos, A.P. Whitworth, Mon. Not. R. Astron. Soc. **392**, 413–427 (2009)
D. Stamatellos, A.P. Whitworth, D.A. Hubber, Astrophys. J. **730**, 32 (2011)
B. Stelzer, E. Flaccomio, K. Briggs, G. Micela, L. Scelsi, M. Audard, I. Pillitteri, M. Güdel, Astron. Astrophys. **468**, 463–475 (2007)
S. Takakuwa, M. Saito, J. Lim, K. Saigo, T.K. Sridharan, N.A. Patel, Astrophys. J. **754**, 52 (2012)
M. Takami, A. Chrysostomou, T.P. Ray, C. Davis, W.R.F. Dent, J. Bailey, M. Tamura, H. Terada, Astron. Astrophys. **416**, 213–219 (2004)
R. Teague, D. Semenov, S. Guilloteau, T. Henning, A. Dutrey, V. Wakelam, E. Chapillon, V. Pietu, Astron. Astrophys. **574**, 137 (2015)
A. Telleschi, M. Güdel, K.R. Briggs, M. Audard, F. Palla, Astron. Astrophys. **468**, 425–442 (2007)
L. Testi, T. Birnstiel, L. Ricci, S. Andrews, J. Blum, J. Carpenter, C. Dominik, A. Isella, A. Natta, J.P. Williams, D.J. Wilner, in *Protostars and Planets VI* (2014), pp. 339–361
W.-F. Thi, P. Woitke, I. Kamp, Mon. Not. R. Astron. Soc. **412**, 711–726 (2011)
I. Tilling, P. Woitke, G. Meeus, A. Mora, B. Montesinos, P. Riviere-Marichalar, C. Eiroa, W.-F. Thi, A. Isella, A. Roberge, C. Martin-Zaidi, I. Kamp, C. Pinte, G. Sandell, W.D. Vacca, F. Ménard, I. Mendigutía, G. Duchêne, W.R.F. Dent, G. Aresu, R. Meijerink, M. Spaans, Astron. Astrophys. **538**, 20 (2012)
J.J. Tobin, L. Hartmann, H.-F. Chiang, D.J. Wilner, L.W. Looney, L. Loinard, N. Calvet, P. D'Alessio, Nature **492**, 83–85 (2012)
J.J. Tobin, L. Hartmann, H.-F. Chiang, D.J. Wilner, L.W. Looney, L. Loinard, N. Calvet, P. D'Alessio, Astrophys. J. **771**, 48 (2013)
M. Tsujimoto, E.D. Feigelson, N. Grosso, G. Micela, Y. Tsuboi, F. Favata, H. Shang, J.H. Kastner, Astrophys. J. Suppl. Ser. **160**, 503–510 (2005)
N.J. Turner, S. Fromang, C. Gammie, H. Klahr, G. Lesur, M. Wardle, X.-N. Bai, in *Protostars and Planets VI* (2014), pp. 411–432
R. van Boekel, M. Güdel, T. Henning, F. Lahuis, E. Pantin, Astron. Astrophys. **497**, 137–144 (2009)
N. van der Marel, E.F. van Dishoeck, S. Bruderer, T. Birnstiel, P. Pinilla, C.P. Dullemond, T.A. van Kempen, M. Schmalzl, J.M. Brown, G.J. Herczeg, G.S. Mathews, V. Geers, Science **340**, 1199–1202 (2013)
N. van der Marel, P. Pinilla, J. Tobin, T. van Kempen, S. Andrews, L. Ricci, T. Birnstiel, Astrophys. J. Lett. **810**, 7 (2015)
G. van der Plas, S. Casassus, F. Ménard, S. Perez, W.F. Thi, C. Pinte, V. Christiaens, Astrophys. J. Lett. **792**, 25 (2014)
A.I. Vasyunin, D.S. Wiebe, T. Birnstiel, S. Zhukovska, T. Henning, C.P. Dullemond, Astrophys. J. **727**, 76 (2011)
R. Visser, E.A. Bergin, Astrophys. J. Lett. **754**, 18 (2012)
R. Visser, E.A. Bergin, J.K. Jørgensen, Astron. Astrophys. **577**, 102 (2015)
R. Visser, E.F. van Dishoeck, J.H. Black, Astron. Astrophys. **503**, 323–343 (2009)
E.I. Vorobyov, Astrophys. J. **723**, 1294–1307 (2010)
E.I. Vorobyov, Astrophys. J. Lett. **728**, 45 (2011a)
E.I. Vorobyov, Astrophys. J. **729**, 146 (2011b)
E.I. Vorobyov, Astron. Astrophys. **552**, 129 (2013)
E.I. Vorobyov, Astron. Astrophys. **590**, A115 (2016). doi:10.1051/0004-6361/201628102
E.I. Vorobyov, S. Basu, Astrophys. J. **650**, 956–969 (2006)
E.I. Vorobyov, S. Basu, Astrophys. J. Lett. **714**, 133–137 (2010a)
E.I. Vorobyov, S. Basu, Astrophys. J. **719**, 1896–1911 (2010b)
E.I. Vorobyov, S. Basu, Astrophys. J. **805**, 115 (2015)
E.I. Vorobyov, I. Baraffe, T. Harries, G. Chabrier, Astron. Astrophys. **557**, 35 (2013)
V. Wakelam, J.-C. Loison, E. Herbst, B. Pavone, A. Bergeat, K. Béroff, M. Chabot, A. Faure, D. Galli, W.D. Geppert, D. Gerlich, P. Gratier, N. Harada, K.M. Hickson, P. Honvault, S.J. Klippenstein, S.D. Le Picard, G. Nyman, M. Ruaud, S. Schlemmer, I.R. Sims, D. Talbi, J. Tennyson, R. Wester, Astrophys. J. Suppl. Ser. **217**, 20 (2015)

C. Walsh, T.J. Millar, H. Nomura, Astrophys. J. **722**, 1607–1623 (2010)
C. Walsh, H. Nomura, T.J. Millar, Y. Aikawa, Astrophys. J. **747**, 114 (2012)
C. Walsh, E. Herbst, H. Nomura, T.J. Millar, S.W. Weaver, Faraday Discuss. **168**, 389 (2014)
C. Walsh, R.A. Loomis, K.I. Öberg, M. Kama, M.L.R. van 't Hoff, T.J. Millar, Y. Aikawa, E. Herbst, S.L.W. Weaver, H. Nomura, Astrophys. J. Lett. **823**(1), 10 (2016). http://stacks.iop.org/2041-8205/823/i=1/a=L10
J.P. Williams, W.M.J. Best, Astrophys. J. **788**, 59 (2014)
J.P. Williams, L.A. Cieza, Annu. Rev. Astron. Astrophys. **49**, 67–117 (2011)
P. Woitke, I. Kamp, W.-F. Thi, Astron. Astrophys. **501**, 383–406 (2009)
P. Woitke, M. Min, C. Pinte, W.-F. Thi, I. Kamp, C. Rab, F. Anthonioz, S. Antonellini, C. Baldovin-Saavedra, A. Carmona, C. Dominik, O. Dionatos, J. Greaves, M. Güdel, J.D. Ilee, A. Liebhart, F. Ménard, L. Rigon, L.B.F.M. Waters, G. Aresu, R. Meijerink, M. Spaans, Astron. Astrophys. **586**, 103 (2016)
K. Zhang, G.A. Blake, E.A. Bergin, Astrophys. J. Lett. **806**, 7 (2015)
Z. Zhu, L. Hartmann, N. Calvet, J. Hernandez, J. Muzerolle, A.-K. Tannirkulam, Astrophys. J. **669**, 483–492 (2007)
Z. Zhu, L. Hartmann, R.P. Nelson, C.F. Gammie, Astrophys. J. **746**, 110 (2012)

Space Sci Rev (2016) 205:81–115
DOI 10.1007/s11214-016-0256-1

Dust Evolution and the Formation of Planetesimals

T. Birnstiel[1,2] · **M. Fang**[3] · **A. Johansen**[4]

Received: 6 August 2015 / Accepted: 11 April 2016 / Published online: 17 May 2016

Abstract The solid content of circumstellar disks is inherited from the interstellar medium: dust particles of at most a micrometer in size. Protoplanetary disks are the environment where these dust grains need to grow at least 13 orders of magnitude in size. Our understanding of this growth process is far from complete, with different physics seemingly posing obstacles to this growth at various stages. Yet, the ubiquity of planets in our galaxy suggests that planet formation is a robust mechanism. This chapter focuses on the earliest stages of planet formation, the growth of small dust grains towards the gravitationally bound "planetesimals", the building blocks of planets. We will introduce some of the key physics involved in the growth processes and discuss how they are expected to shape the global behavior of the solid content of disks. We will consider possible pathways towards the formation of larger bodies and conclude by reviewing some of the recent observational advances in the field.

Keywords Accretion disks · Planets and satellites: formation · Protoplanetary disks · Circumstellar matter

✉ T. Birnstiel
tbirnstiel@cfa.harvard.edu

M. Fang
mfang@pmo.ac.cn

A. Johansen
anders@astro.lu.se

1 Harvard-Smithsonian Center for Astrophysics, 60 Garden Street, 02138 Cambridge, MA, USA

2 Present address: Max Planck Institute for Astronomy, Königstuhl 17, 69117 Heidelberg, Germany

3 Purple Mountain Observatory and Key Laboratory for Radio Astronomy, 2 West Beijing Road, 210008, Nanjing, China

4 Lund Observatory, Department of Astronomy and Theoretical Physics, Lund University, Box 43, 22100 Lund, Sweden

1 Introduction

Circumstellar disks consist mainly (99 % by mass) of gas, but the tiny 1 % of condensible material (commonly called solids or dust) is nevertheless a crucial ingredient. After all it is the material out of which planets and minor bodies are formed, but beyond this it is also important for the physical structure and the evolution of the disk: in most parts of the disk, the opacity is dominated by the dust. The dust thus determines the temperature structure of the disk by absorbing the stellar irradiation and reradiating it in the infrared. Consequently, dust determines the observational appearance of the disk: on the one hand through its thermal continuum emission and on the other hand by determining the temperature and density structure and therefore the excitation conditions for gas lines. Furthermore, solids provide the surface area for crucial surface chemical reactions (such as the formation of complex organics) and they influence the ionization levels in the disk by sweeping up free electrons. Finally, dust is also a key observational probe of the gas dynamics as it reacts sensitively to changes in the gas disk (see Sect. 2). Clearly, understanding the evolution of solids is a key element in the puzzle of planet formation.

The evolution of solids in a circumstellar disk is governed by transport processes and by collisional processes. Both categories will be discussed in the following, but it is important to note that both effects strongly depend on each other: transport processes typically depend on the particle size, hence the collisional evolution of the particles, while in turn, the collisions between the particles are driven by the dynamics. Studying transport or particle growth separately may be a useful exercise, but for a self-consistent picture of the global evolution of solids, both processes need to be considered together.

Given the broadness of this review, we can only scratch the surface of these topics. For more in-depth reviews the reader may refer to the relevant reviews in Protostars and Planets VI (Johansen et al. 2014; Testi et al. 2014; Espaillat et al. 2014; Turner et al. 2014, for example), to the reviews by Armitage (2011), Williams and Cieza (2011), and Andrews (2015). In the following we will introduce first some of the basic transport mechanisms (Sect. 2) before discussing the collisional evolution of dust (Sect. 3). We will then put both pieces together to understand the global distribution of dust in the disk (Sect. 4) and how planetesimals, the building blocks of planets can be formed (Sect. 5). Finally, Sect. 6 will review how recent observations can help us constrain the evolution of solids in circumstellar disks and Sect. 7 summarizes this chapter.

2 Dust Dynamics

2.1 Drag Forces

Aerodynamic drag is a phenomenon known from daily life, may it be the head wind experienced on a bicycle or the tail wind that accelerates a sailing boat: whenever there is a difference in velocity between an object and the surrounding gas, the drag force acts towards eliminating the velocity difference. The same principles apply to solids in circumstellar disks that are dynamically coupled to the gas via drag forces. However in most regions of the disk, the drag stems not from the flow around the object as we experience it. Instead, most dust grains are smaller than the mean free path of the gas molecules and they rather feel drag by being bombarded by gas molecules slightly faster from the direction of the flow than behind it, effectively a pressure force. This drag force is called Epstein drag (in contrast to the better known Stokes drag, which can become relevant in the denser inner regions of disks). The Epstein drag force is expressed as (Weidenschilling 1977a)

$$\mathbf{F}_{\mathrm{Ep}} = -\frac{4\pi}{3}\rho_{\mathrm{g}} a^2 \Delta \boldsymbol{w} v_{\mathrm{th}}, \tag{1}$$

where ρ_{g} is the gas density, a the particle radius, $\Delta \boldsymbol{w}$ the particle-gas relative velocity, $v_{\mathrm{th}} = \sqrt{8/\pi}\, c_{\mathrm{s}}$ the mean thermal velocity, and c_{s} the isothermal sound speed. Instead of using this definition of the force, it is much more useful to specify the *stopping time*

$$t_{\mathrm{stop}} = \frac{m \Delta w}{F_{\mathrm{Ep}}} = \frac{\rho_{\mathrm{s}} a}{\rho_{\mathrm{g}} v_{\mathrm{th}}} \tag{2}$$

where we assumed that mass and radius are related as $m = 4\pi\rho_{\mathrm{s}} a^3/3$ with a mean material density of the dust particle ρ_{s}. For fractal particles, the mass is still well defined, but their size or cross section can be defined in various ways (e.g. Ormel et al. 2007; Okuzumi et al. 2009) and the mass-size relation follows a different power-law than 3 (called the fractal dimension). To give an example, for typical disk mid-plane conditions at 1 AU (e.g. $\Sigma_{\mathrm{g}} = 200\ \mathrm{g\,cm^{-2}}$), the stopping time is about a few seconds for a compact micrometer sized particle and about 10 days for a decimeter sized particle. Even more useful than the stopping time is the dimensionless ratio of the stopping to the dynamical time scale, called *Stokes number*,

$$\mathrm{St} = t_{\mathrm{stop}} \Omega, \tag{3}$$

where Ω is the Keplerian angular velocity. The Stokes number is so useful because two particles of different composition, structure, mass, ... behave aerodynamically identical if their Stokes numbers are identical. For a vertically isothermal gas disk with scale height $H_{\mathrm{g}} = c_{\mathrm{s}}/\Omega$ and gas surface density Σ_{g}, the Stokes number at the disk mid-plane becomes

$$\mathrm{St} = \frac{a\rho_{\mathrm{s}}}{\Sigma_{\mathrm{g}}} \frac{\pi}{2}, \tag{4}$$

where we have assumed a gas mid-plane density of $\rho_{\mathrm{g,mid}} = \frac{\Sigma_{\mathrm{g}}}{\sqrt{2\pi} H_{\mathrm{g}}}$. Under these assumptions, the Stokes number is linearly dependent on the particle size, which is true for most but the densest regions of the disk. In the following we will mostly talk about Stokes numbers instead of particle sizes, but the Stokes number can just be viewed as a dimensionless quantity describing the particle size: a small particle with a small Stokes number ($\mathrm{St} \ll 1$) will be adapting to the gas velocity on time scales much shorter than the orbital time scale, while a big particle with very large Stokes number ($\mathrm{St} \gg 1$) will perform several orbits before the drag forces significantly alter its velocity.

From this simple concept of size dependent drag forces, a surprising number of "complications" arise: the velocities of particles in a protoplanetary disk generally become size dependent (unlike the velocity of a dust grain in a free Keplerian orbit). Thus, the trajectory of a particle depends on its size. Different-sized particles having different velocities means that particles collide with each other and those collisions can lead to sticking/growth or to shattering/destruction of the particles. Other effects caused by the drag are: a systematic drift motion of particles (see next section), turbulent mixing of dust particles (Sect. 2.3), and dynamical instabilities caused by the coupling of dust and gas (Sect. 5).

2.2 Dust Drift

It was already found by Whipple (1972) and Weidenschilling (1977a) that dust particles embedded in a gaseous disk should migrate towards the star on short time scales. A derivation of the dust and gas velocities can also be found in Nakagawa et al. (1986). The velocities of dust ($\mathbf{u}_\mathrm{d}$) and of the gas ($\mathbf{u}_\mathrm{g}$) evolve according to two equations

$$\frac{d\mathbf{u}_\mathrm{d}}{dt} = -\frac{1}{t_\mathrm{stop}}(\mathbf{u}_\mathrm{d} - \mathbf{u}_\mathrm{g}) - \frac{GM_\star}{r^3}\mathbf{r}, \tag{5}$$

$$\frac{d\mathbf{u}_\mathrm{g}}{dt} = -\frac{\epsilon}{t_\mathrm{stop}}(\mathbf{u}_\mathrm{g} - \mathbf{u}_\mathrm{d}) - \frac{GM_\star}{r^3}\mathbf{r} - \frac{\nabla P}{\rho_\mathrm{g}}, \tag{6}$$

where ϵ is the dust-to-gas density ratio ($\rho_\mathrm{d}/\rho_\mathrm{g}$), P the gas pressure, G the gravitational constant, and $M_\star$ the stellar mass. Pressure acceleration on the dust particles is negligible (as the dust material density[1] is much smaller than the gas density $\rho_\mathrm{s} \ll \rho_\mathrm{g}$). Rewriting these equations in cylindrical coordinates, assuming a steady state ($d/dt = 0$), a low dust-to-gas ratio, and then solving for the first order deviation from the Keplerian velocity ($\mathbf{v} = \mathbf{u} - \mathbf{V}_\mathrm{k}$) leads to the dust drift speed

$$v_r \simeq -\frac{2}{\mathrm{St} + \mathrm{St}^{-1}}\eta V_\mathrm{k}, \tag{7}$$

$$v_\phi \simeq -\frac{1}{1 + \mathrm{St}^2}\eta V_\mathrm{k}, \tag{8}$$

where

$$\eta = -\frac{1}{2}\left(\frac{H_\mathrm{p}}{r}\right)^2 \frac{\partial \ln P}{\partial \ln r} \tag{9}$$

describes how much slower than Keplerian the gas is orbiting, i.e. $v_{\phi,\mathrm{g}} = \eta V_\mathrm{k}$. Our assumption of low dust-to-gas ratios basically means that the dust feels the drag force by the gas, but the effect on the gas velocity is negligible. The velocities for dust and gas for arbitrary dust-to-gas ratios can be found in Nakagawa et al. (1986). These results have several important consequences:

- Particles with a Stokes number < 1 drift inward with a speed of $v_r \simeq -2\mathrm{St}\eta V_\mathrm{k}$: so small particles move slowly, larger particles move faster.
- For typical disk conditions, η is of the order of a few per mille, which means that the maximum radial drift velocity for $\mathrm{St} = 1$ is a few per mille of the Keplerian speed. In other words, the orbit of a dust particle decays on a time scale of only a few hundred orbits.
- The azimuthal drift velocity for particles with a Stokes number < 1 is ηV_k, so they move along with the gas while the drift speed of larger particles tends towards zero, they move on Keplerian orbits.
- The direction of the radial drift is towards higher pressure, so generally inwards in a disk that is denser and hotter closer to the star.

[1] It should be noted that ρ_d denotes the total mass of dust particles per "volume of space", while ρ_s denotes the specific weight of a dust particle, in other words the mass per "dust volume".

The last point also applies to the vertical dimension: a particle at a height z above the disk mid-plane would orbit on an inclined Keplerian orbit, so effectively oscillate around the mid-plane. The gas disk however will decelerate this motion via gas drag. Let us assume that the gas disk is vertically stable, so $u_{z,\text{gas}} = 0$. A dust particle with a Stokes number St will by definition decelerate on time scales of St $\times$ the orbital time scale. Particles with St > 1 would therefore execute a damped oscillation, while particles with a small Stokes number quickly reach a terminal settling velocity when the vertical acceleration due to the stellar gravity ($\dot{u} = -\Omega^2 z$) and the deceleration from the drag force ($\dot{u} = u/t_\text{stop}$) are in balance. This is the velocity at which particles sediment towards the mid-plane,

$$u_{z,\text{dust}} = -z\Omega\text{St}. \tag{10}$$

Finally, it should be noted, that particles also move azimuthally towards higher pressure. Even slight azimuthal over densities in the gas are thus able to produce very strong dust over-densities (Birnstiel et al. 2013). For example vortex structures are azimuthal over-densities that effectively trap dust particles (e.g., Barge and Sommeria 1995; Klahr and Henning 1997; Lyra and Lin 2013; Raettig et al. 2015). Eccentric gas disks are the exception to the rule: azimuthal over-densities in eccentric disks stem from velocity variations along the eccentric orbit. These velocity (and thus density-) modulations are identical for dust and gas and therefore do not cause strong accumulations of dust particles (Hsieh and Gu 2012; Ataiee et al. 2013).

In addition to this drift motion, dust is also carried along with the radial gas flow. Takeuchi and Lin (2002) derived this component of the dust radial velocity to be

$$u_{r,\text{dust}} = \frac{1}{1+\text{St}^2} u_{r,\text{gas}}. \tag{11}$$

Hence small particles (St < 1) follow the gas flow, while large particles (St $\gg 1$) are unaffected by it.

The ideas presented in this section are based purely on theoretical concepts. Until recently, there has been very little observational evidence in support of radial or azimuthal segregation of dust particles (but some on vertical settling). Recently, this situation has changed dramatically and we will discuss the current observational support of these concepts in Sect. 6.

2.3 Turbulent Mixing

It is widely believed that turbulent effective viscosity is the driver of disk evolution (this picture has recently been put into question by models where the angular momentum is transported by disc winds and the gas motion remains laminar, see Turner et al. 2014, and references therein). If the gas is indeed turbulent, then the dust motion and transport will be affected by it. In the following, we will assume, that the turbulence is described by an effective viscosity (Shakura and Sunyaev 1973)

$$\nu = \alpha \frac{c_\text{s}^2}{\Omega}, \tag{12}$$

where α is the turbulent strength parameter. The random motion induced by this turbulence will act as a diffusivity on the dust, and again, this will depend on the dust particle size. One can imagine that large boulders might be less effectively mixed than micrometer sized dust particles. The ratio of the dust diffusivity to the gas diffusivity is commonly called the Schmidt number[2] Sc which was shown to be (Youdin and Lithwick 2007)

$$\mathrm{Sc} = \frac{D_\mathrm{g}}{D_\mathrm{d}} \simeq 1 + \mathrm{St}^2, \tag{13}$$

and it is commonly assumed that the gas diffusivity D_g equals the gas viscosity ν (but see Johansen and Klahr 2005; Fromang and Papaloizou 2006; Pavlyuchenkov and Dullemond 2007). Turbulent mixing will result in smoothing out concentrations of trace species (dust grains, or molecular species). Such concentrations might be caused by dust drift (in azimuthal, vertical, or radial direction) or by local production. It might also be responsible for mixing thermally processed particles from the hot inner regions throughout the disk (e.g., Bockelée-Morvan et al. 2002; Pavlyuchenkov and Dullemond 2007). We will discuss some of these effects shaping the global appearance of the disk in the following section. Clearly this review can introduce only some of the key concepts. Several other effects, such as radiation pressure (e.g., Dominik and Dullemond 2011) or photophoresis (e.g., Krauss and Wurm 2005) can be relevant in disks as well, but are not discussed here.

3 Dust Growth

The initial stages of planet formation in the core accretion scenario necessarily involve the growth from sub-micrometer sized dust grains to >km sized bodies, called planetesimals, that are gravitationally bound. To understand how particles grow along so many orders of magnitude in mass, we need to understand (1) at which velocities particles collide, (2) how often particles collide, and (3) what the outcome of each collision is. The derivation of the collisional rates is straight forward: a particle i sweeping through a swarm of particles j has a mean free path of

$$l = \frac{1}{n_j \sigma_{ij}}, \tag{14}$$

where n_j is the number density of particles j and σ_{ij} is the cross section of particles i and j. Assuming spherical radii of a_i and a_j respectively, $\sigma_{ij} = \pi(a_i + a_j)^2$. If particles i and j move with a relative velocity of Δv_{ij}, this means, a particle i will on average feel one collision per collisional time scale

$$\tau_\mathrm{col} = \frac{l}{\Delta v_{ij}} = \frac{1}{n_j \sigma_{ij} \Delta v_{ij}}. \tag{15}$$

If there are n_i particles per volume, each feeling one collision with particles j per collisional time scale τ_col, this means

$$R_{ij} = n_i n_j \sigma_{ij} \Delta v_{ij}. \tag{16}$$

This is the rate at which particles i collide with particles j (e.g., forming a new species k). This shows that the particle-particle velocities determine not only the outcome of a collision, but also how often it happens and that the rates depend on density squared and the cross section, both of which are outcomes of the growth process itself. Solving for the evolution

[2] There are several definitions interchanging dust/gas or diffusivity/viscosity. Here we follow the definition of Youdin and Lithwick (2007) who also discusses the different meanings of these notations.

of a particle size distribution means that we need to calculate the rates for all the collisions of all particle sizes and keep book where the results of a collision end up. As an example, let us consider pure sticking: this means that a collision of two particles with masses m_i and m_j will produce one particle with mass $m_k = m_i + m_j$ (a gain term for n_k), while every collision of a particle m_k will lead to a loss term, hence

$$\dot{n}_k = \frac{1}{2} \sum_{i,j} R_{ij} \delta_{i,j-k} - \sum_i R_{i,k}, \tag{17}$$

where the factor of 1/2 comes from double counting collisions. In practice, the analytical or numerical solutions of grain growth are complicated by other collisional outcomes (fragmentation, cratering, ...) and by the fact that the grid spacing of the mass dimension cannot be linear. For example growing a 1 cm sized aggregate out of micrometer sized constituents (monomers) would require a grid with 10^{12} entries, which is computationally unfeasible. Logarithmic grid spacing however comes at the cost of having particles grow to masses where there is no grid point. The mass needs to be distributed over neighboring grid points and the algorithm becomes less accurate. Monte–Carlo methods with discrete particles can overcome these issues, but they suffer from other problems, such as time step constraints (see, Drążkowska et al. 2014, and references within).

3.1 Impact Velocities

If the velocity of a particle depends on its size (as is the case for radial, azimuthal, and vertical drift motion), particles of different sizes will have a relative velocity with respect to each other. For the radial, azimuthal, and vertical velocities above, the relative speed between two particles with Stokes numbers St_i and St_j becomes

$$\Delta v_r(i,j) = \eta V_\mathrm{k} \left| \frac{2}{\mathrm{St}_i + \mathrm{St}_i^{-1}} - \frac{2}{\mathrm{St}_j + \mathrm{St}_j^{-1}} \right| \tag{18}$$

$$\Delta v_\phi(i,j) = \eta V_\mathrm{k} \left| \frac{1}{1 + \mathrm{St}_i^2} - \frac{1}{1 + \mathrm{St}_j^2} \right| \tag{19}$$

$$\Delta v_z(i,j) = z\Omega |\mathrm{St}_i - \mathrm{St}_j|. \tag{20}$$

As we can see, for $\mathrm{St} \ll 1$, the azimuthal velocities vanish, and the radial velocities become proportional to $|\mathrm{St}_i - \mathrm{St}_j|$. We also note that for $\mathrm{St}_i = \mathrm{St}_j$, all velocities vanish, as particles move with the same systematic velocities.

The latter is not the case for random motions. In protoplanetary disks, two effects cause random motion of dust particles: Brownian motion and turbulence. In the case of Brownian motion, the mean thermal kinetic energy $E_\mathrm{kin} = \frac{1}{2} m v_\mathrm{th}^2$ at a temperature T is distributed equally amongst the particles. The average relative velocity of two particles with masses m_i and m_j with a Maxwell–Boltzmann distributed velocity becomes

$$v_\mathrm{BM} = \sqrt{\frac{8 k_\mathrm{B} T (m_i + m_j)}{\pi m_i m_j}}, \tag{21}$$

where k_B is the Boltzmann constant, μ the mean molecular weight and m_p the proton mass. In contrast to the relative velocities discussed above, this one does not vanish for equal mass particles.

The case of turbulent velocities is much more complicated: particle velocities are affected by turbulent eddies through which they move. Turbulent eddies of different sizes cause partial alignment of particle trajectories on short distances or induce random kicks, depending on the eddy turnover time, the particle stopping time, and the time it takes the particle to cross the eddy. The classical picture was introduced by Völk et al. (1980) and allowed Ormel and Cuzzi (2007) to derive closed form expressions that are currently widely used. Pan and Padoan (2010), Pan et al. (2014) and following papers in this series criticized this picture, emphasizing the importance of the particle separation prior to a collision in determining the degree of correlation between particle velocities induced by turbulent eddies of different sizes. Their results deviate somewhat from the results of Ormel and Cuzzi (2007) (predicting a maximum velocity by about a factor of two lower) and they additionally allow the calculation of the distribution of collision velocities. In the following examples, we will use approximations to the velocities derived by Ormel and Cuzzi (2007) for simplicity. In this framework, the collision speed depends on the particle sizes of both particles and on the largest and smallest eddy turn-over times. For equal sized particles with stopping times larger than the smallest eddy turn-over times, the velocities can be approximated by

$$\Delta v_{\mathrm{turb}} \simeq \sqrt{\frac{3\alpha}{\mathrm{St}+\mathrm{St}^{-1}}}\, c_{\mathrm{s}}, \tag{22}$$

where c_{s} is the isothermal sound speed. Similar to relative velocities from dust drift, also turbulent collision velocities increase with Stokes number ($\simeq$ particle size), then reach a maximum at $\mathrm{St}=1$ and decrease for particles with $\mathrm{St}>1$.

Comparing the terms above, we can see that all velocity contributions vanish towards small particle size, apart from Brownian motion (see also Fig. 1). Hence, initial growth is seeded by the Brownian velocities and only once particles have grown larger than a few micrometers, turbulent relative velocities start to become important (Birnstiel et al. 2011). At larger sizes, radial drift velocities are dominating,[3] unless turbulence is quite strong ($\alpha \gtrsim 2(H/r)^2$, see also Testi et al. 2014). For particles with $\mathrm{St} \gtrsim 1$, azimuthal relative velocity contributions are the strongest contribution before gravitational perturbations from the turbulent gas density field come into play (Johansen et al. 2014).

The various contributions to relative particle velocities are plotted in Fig. 1 at 10 AU in a disk with $\alpha = 10^{-3}$, a local temperature of 63 K and a gas surface density of 16 $\mathrm{g\,cm^{-2}}$. The maximum relative velocity from turbulent velocities is given by the gas root mean squared velocity $v_{\mathrm{rms}} = c_{\mathrm{s}}\sqrt{3\alpha/2}$ (see Eq. (22)). These velocities are quite low compared for example to the interstellar medium (for a temperature of 100 K and typical turbulence parameters of $\alpha = 10^{-3} \ldots 10^{-2}$ they range above 20–70 $\mathrm{m\,s^{-1}}$), but they are still high enough to destroy large particles upon collision. Similarly, drift-induced relative velocities reach a few times $c_{\mathrm{s}}H/r$, which also works out to similar numbers ($\lesssim$80 $\mathrm{m\,s^{-1}}$). Collision speeds of micrometer sized bodies, on the other hand, are very small, of the order of millimeters per second. These numbers already indicate that we can expect a wide variety of collisional outcomes and that the outcomes change significantly as particles grow and collide at larger and larger velocities.

[3] As discussed above, for equal sized particles, drift velocities vanish, but typically the particle size dispersion causes particles to collide with particles of similar sizes.

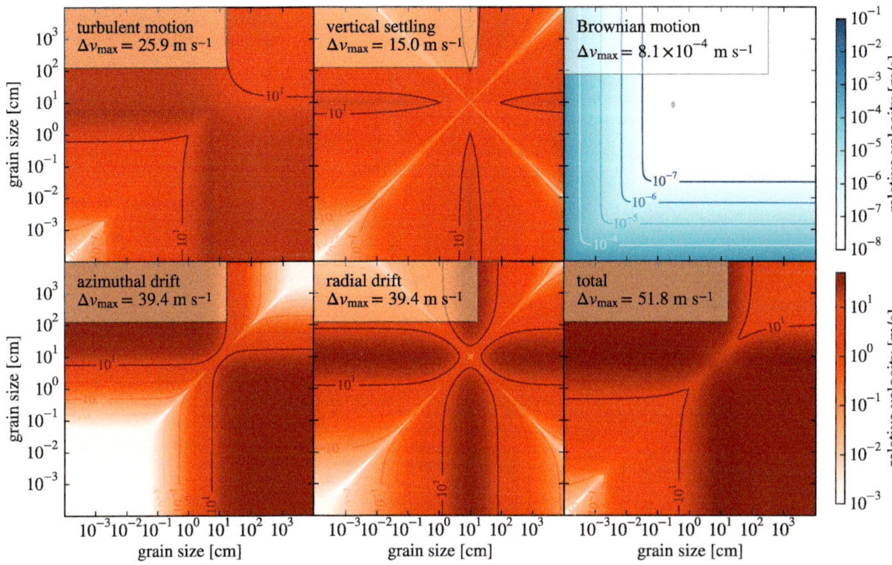

Fig. 1 Various contributions to particle mean relative velocities in protoplanetary disks: turbulence, vertical settling, azimuthal and radial drift yield maximum impact speeds of tens of meters per second (*bottom/red color* scale). Brownian motion is the dominant source of relative velocities for small particles (*top/blue color* scale). The vertical settling velocities plotted here were calculated by using the settling velocity at the respective dust scale height

3.2 Collisional Outcomes

The most important ingredient for models of particle growth are *collisional outcome models*: given two particles and the collision speed, a collisional outcome model predicts the properties of the resulting particle(s). Realistically, this outcome depends not only on the impact velocity, but also on impact parameter, size of the grains, compositional properties of both grains such as porosity, monomer size, fractal dimension, chemical composition (surface ices, hence temperature), and other properties. Obviously this entire parameter space of particle collisions cannot possibly be explored comprehensively with laboratory studies. Still, many studies have shed light on this enormous parameter space and identified possible outcomes and how they depend on or scale with particle properties and collisional parameters (see Blum and Wurm 2008; Testi et al. 2014; Johansen et al. 2014, for recent reviews). Other studies have condensed these results into collision models, using existing laboratory results and experimentally or theoretically constrained scaling relations. Examples include Suyama et al. (2008), Güttler et al. (2010), Zsom et al. (2010), Windmark et al. (2012a), or Krijt et al. (2015). The main collisional outcomes are:

- *sticking:* hit-and-stick collisions
- *bouncing:* particles bouncing off each other without changing their mass, possibly causing compaction
- *erosion/cratering:* smaller projectile removes mass from larger target, possibly shattering itself
- *mass transfer:* smaller projectile fragments upon collision with target, but also deposits some fraction of its mass

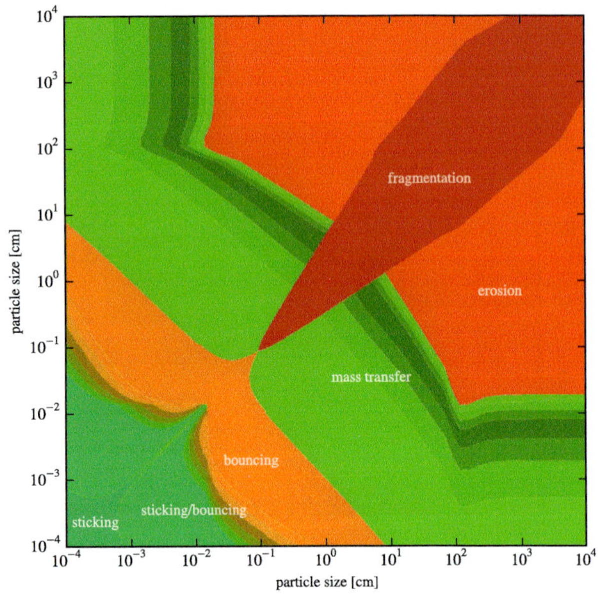

Fig. 2 Mean collisional outcomes for silicate grains from Windmark et al. (2012b) as expected for a minimum-mass solar nebula disk (Weidenschilling 1977b) at 1 AU, including all the contributions of relative velocities discussed in Sect. 3.1. *Green regions* denote net-growth of the larger collision partner, *red* mass loss, and *orange* denotes mass-neutral bouncing collisions

– *fragmentation:* complete destruction of the particle(s), fragments are typically distributed in a power-law fragment size distribution

Figure 2 shows a typical result of the collisional model of Windmark et al. (2012a), where, given two particle sizes, the *mean* impact velocity is calculated (e.g. like in Fig. 1), and a collisional outcome is predicted. A coagulation algorithm can thus evolve a size distribution of particles $n(m)$ by calculating the collision rates according to Eq. (16) and the collisional outcome model then determines how the rates are connected to the gain and loss terms, or in other words: how collisions between two particles i and j affect the number of particles k. Detailed descriptions of different recent astrophysical coagulation codes can be found in Brauer et al. (2008), Laibe et al. (2008), Ormel et al. (2007), Ormel and Spaans (2008), Zsom and Dullemond (2008), Okuzumi et al. (2009), Birnstiel et al. (2010), Okuzumi et al. (2012), and references within.

It should be highlighted, that, as mentioned in the beginning of this section, the collision outcome is not only a function of particle masses and velocities, but also other particle properties, and that the collision velocity of two particles is not a fixed mean velocity, but a distribution of velocities. The effect of the velocity distribution on the outcome can be substantial, as shown in Windmark et al. (2012b) and Garaud et al. (2013). Nevertheless, the mean outcome gives a good estimate for how the bulk of the particle distribution is evolving, as we will see in the following section.

3.3 Simple Estimates

Coagulation/fragmentation processes are not local in mass space. Numerically speaking, particles of one mass bin are not only interacting with neighboring cells (like in hydrodynamics). Instead each particle size can interact with every other particle size and influence the entire distribution, for example by producing a distribution of fragments. This fact makes a full treatment of the growth processes conceptually and numerically difficult. In many cases, however, growth proceeds in a more or less ordered fashion: small particles grow steadily and they grow most efficiently with particles of similar mass. Hence, simple estimates of particle growth are not only very instructive, they can also give reasonable answers, e.g. on how the upper end of the particle size distribution evolves. To derive such estimates, we will consider the case of *monodisperse growth*, i.e. assuming that all particles are of one radius a. If particles only stick to each other, the mass of the newly formed particle is twice the original mass. The rate of collisions of a particle is then one collision per collisional time τ_{col} (Eq. (15)). Upon each collision, the mass of the particle is doubled, so

$$\dot{m} \simeq \frac{m}{\tau_{\mathrm{col}}} = \rho_{\mathrm{d}} \sigma \Delta v. \tag{23}$$

Now the cross section of two spherical particles of size a is given by $\sigma = 4\pi a^2$ and we assume a constant porosity (i.e., $\rho_{\mathrm{s}} = const.$), so $\mathrm{d}m/\mathrm{d}a = 4\pi \rho_{\mathrm{s}} a^2$. This leads to the growth rate

$$\dot{a} = \frac{\rho_{\mathrm{d}}}{\rho_{\mathrm{s}}} \Delta v. \tag{24}$$

Now if the particle collision velocity is set by Brownian motion (Eq. (21)), then

$$\dot{a} = \underbrace{\frac{\rho_{\mathrm{d}}}{\pi} \sqrt{\frac{12 k_{\mathrm{B}} T}{\rho_{\mathrm{s}}^3}}}_{:=A} a^{-3/2}, \tag{25}$$

which can be integrated to (e.g., Dullemond and Dominik 2005)

$$a(t) = \left[a(t_0)^{5/2} + \frac{5}{2} A t \right]^{2/5}. \tag{26}$$

Similar calculations can be done for other relative velocities and they produce trends of numerical simulations well (e.g., Ormel et al. 2009; Birnstiel et al. 2010; Windmark et al. 2012a). These estimates obviously break down at sizes, where other effects come into play, for example when bouncing or fragmentation limits further growth.

4 The Structure of the Dust Disk

In this section, we want to discuss how dust transport and growth shapes the global structure of the disk. For this we have to consider that particle transport depends on the particle size (which is evolving over time) while the particle size evolution in turn depends on local conditions (temperature, gas density, dust density, ...) and the relative velocities, which depend on the dynamical history of the particles. Thus particle growth and transport go hand in hand and to understand global trends, both effects need to be taken into account in a self-consistent way.

4.1 Vertical Structure

As explained in Sects. 2.2 and 2.3, particles sediment towards the mid-plane but turbulence is counteracting this concentration effect. The settling time scale is $t_{\mathrm{sett}} = z/v_{\mathrm{sett}} = (\Omega \mathrm{St})^{-1}$. Even small, micrometer sized particles have a large Stokes number if the density is as low as in the disk atmosphere, thus even the smallest particles sediment quickly. However, if they

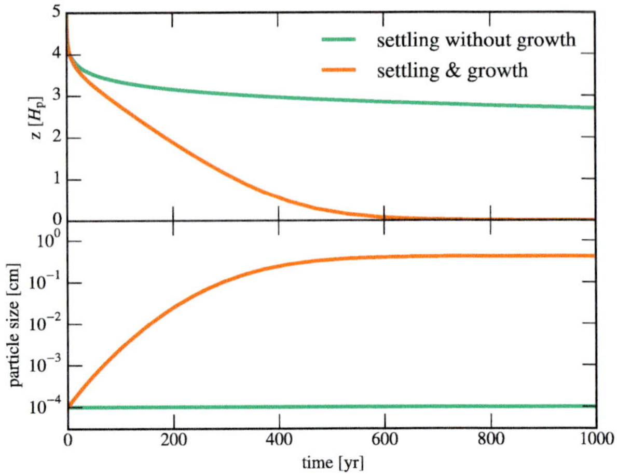

Fig. 3 Trajectory of a settling particle at 1 AU. Figure and parameters after Dullemond and Dominik (2005). The initial particle size is one micrometer and the initial position is 5 H_g above the mid-plane. The *green curve* integrates the trajectory assuming a constant particle size, the *orange curve* assumes sweep-up of other (assumed fixed) particles after Safronov (1969)

do not grow, their Stokes number will decrease as they settle towards higher gas densities and their downward motion will slow down (see green curves in Fig. 3). If they continue to grow while settling down, their increase in radius can partly counteract the increase in gas density and their Stokes number can stay large enough, such that they continue to settle towards the mid-plane, reaching macroscopic sizes along the way (see orange lines in Fig. 3).

Turbulent mixing acts as diffusion on the dust distribution. We can estimate the combined effects of settling and mixing in the following way: the diffusion time scale over a vertical scale z is $t_{\mathrm{diff}} = z^2/D_{\mathrm{d}}$. Setting this in relation to the settling time scale, we find

$$\mathrm{Pe} = \frac{t_{\mathrm{diff}}}{t_{\mathrm{sett}}} = \frac{\mathrm{St}}{\alpha}\left(\frac{z}{H_{\mathrm{g}}}\right)^2, \tag{27}$$

where Pe is called the Péclet number. For large $z > H_{\mathrm{g}}\sqrt{\alpha/\mathrm{St}}$, the Péclet number is > 1, meaning that settling time scales are shorter than diffusion time scales. Close to the mid-plane, $\mathrm{Pe} < 1$ and diffusion is dominating. Thus, we can already expect the vertical dust-to-gas ratio to drop significantly above a dust scale height

$$H_{\mathrm{d}} = H_{\mathrm{g}}\sqrt{\frac{\alpha}{\mathrm{St}+\alpha}}, \tag{28}$$

similar to what was found by Dubrulle et al. (1995). The additional summand α stems from the fact that H_{g} is an upper limit for the dust scale height for the case of perfect mixing, i.e. $\alpha \gg \mathrm{St}$. More generally, the vertical structure and its evolution is described by an advection-diffusion equation (e.g. Dubrulle et al. 1995; Schräpler and Henning 2004; Johansen and Klahr 2005; Fromang and Nelson 2009). After a few settling time scales, an equilibrium will be reached in which the downward flux from sedimentation is balanced by the upward mass flux from diffusion. In this case, the time derivative of the advection-diffusion equation becomes zero and the equation can be integrated analytically. Assuming a vertically isothermal disk and constant diffusivity, Fromang and Nelson (2009) derived the dust density distribution as

$$\rho_{\mathrm{d}}(z) = \rho_{\mathrm{d},0}\exp\left[-\frac{\mathrm{St}_0}{\alpha}\left(\exp\left(\frac{z^2}{2H_{\mathrm{g}}^2}\right)-1\right)-\frac{z^2}{2H_{\mathrm{g}}^2}\right], \tag{29}$$

where a subscript of 0 denotes mid-plane values. Close to the mid-plane, this profile indeed approaches a Gaussian profile with the scale height estimated above, as can be seen by letting $z \ll H$ in Eq. (29).

4.2 Radial Structure

4.2.1 Particle Sizes

In the radial dimension, the same mechanisms are at play, dust drift and turbulent mixing. In addition to that, also the radial flow of the gas transports dust particles. A good intuition of the global transport can again be derived by comparing time scales. The growth time scale is

$$t_{\mathrm{grow}} = \frac{a}{\dot{a}} = \frac{a\rho_{\mathrm{s}}}{\rho_{\mathrm{d}}\Delta v}, \tag{30}$$

where a is the particle radius and we used Eq. (24). For simplicity, we will assume the relative velocities to follow $\Delta v \simeq \sqrt{3\alpha \mathrm{St}}c_{\mathrm{s}}$ (Eq. (22) for $\mathrm{St} \ll 1$) and the mid-plane dust and gas densities to be

$$\rho_{\mathrm{d/g},0} = \frac{\Sigma_{\mathrm{d/g}}}{\sqrt{2\pi}\,H_{\mathrm{d/g}}}. \tag{31}$$

Substituting also the definition of the Stokes number (Eq. (3)) simplifies the growth time scale to

$$t_{\mathrm{grow}} = \frac{1}{\epsilon\Omega}, \tag{32}$$

where ϵ denotes the dust-to-gas mass ratio. Figure 4 depicts this growth time scale and the time scales for radial motion r/u_{r} as arrows, where longer arrows correspond to faster growth/transport. Let us again imagine all particles to be of the same size (monodisperse growth), initially one micrometer. In the beginning the particles will not drift radially, they will only grow to larger sizes. As the particle size increases, the drift motion will start to become significant and the imagined "trajectory" of our dust particle swarm will curve radially inwards. The dashed lines in Fig. 4 show such "streamlines", starting at 1, 10, and 100 AU. Since the growth time scale is proportional to the orbital time scale, we see that growth is slow in the outer regions. Particles in the inner regions will have grown and drifted inwards long before particles in the outer regions have grown significantly. The entire dust disk is therefore sustained by the outermost regions, which act as a reservoir of inflowing mass. The global evolution of the dust disk is thus set by the growth time scales of the outer most disk regions (Garaud 2007; Birnstiel et al. 2012b).

Figure 4 also shows that all the trajectories converge towards the solid black line, which marks the particle sizes at which the local growth time scale equals the local drift time scale (Klahr and Bodenheimer 2006), which is called the drift size limit (Birnstiel et al. 2012b)

$$a_{\mathrm{drift}} \simeq 0.35 \frac{\Sigma_{\mathrm{d}}}{\rho_{\mathrm{s}}\gamma}\left(\frac{H_{\mathrm{g}}}{r}\right)^{-2}, \tag{33}$$

where $\gamma = |d \ln P / d \ln r|$.

Particles below this curve will grow faster than they drift—particles below it will drift faster than they grow. The dashed black line denotes the particle size that corresponds to a Stokes number of unity, i.e. the size where drift rates and turbulent collision velocities

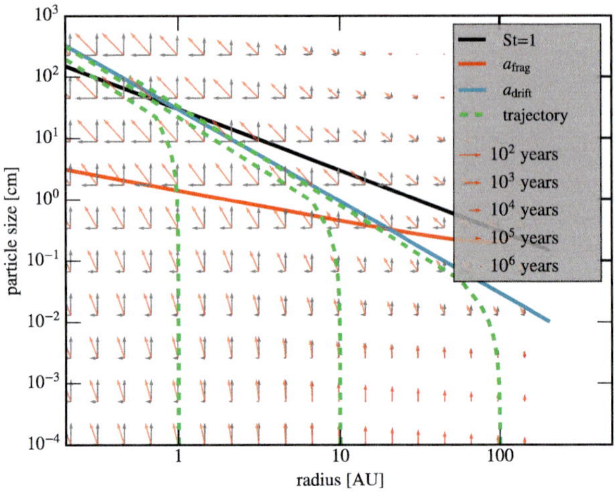

Fig. 4 Comparison of growth and drift time scales. Disk properties follow the model of Birnstiel and Andrews (2014) but using a surface density profile of $\Sigma_g = 0.28\ \mathrm{g\,cm^{-2}} r_d/r$ for $r < r_d$ and $r_d = 200$ AU. The dust-to-gas ratio was chosen to be $\epsilon = 5 \times 10^{-3}$. The arrow length is inversely proportional to the logarithmic time scales, $r/\dot{r}$ in the *horizontal*, and $a/\dot{a}$ in the *vertical direction*. Trajectories were calculated for monodisperse growth and drift, neglecting fragmentation. Fragmentation limits further growth in the inner ~20 AU. The fragmentation barrier (*red line*) was calculated for a fragmentation threshold of $10\ \mathrm{m\,s^{-1}}$ appropriate for icy grains. In the inner disk, it will be at smaller sizes due to the reduced fragmentation threshold of silicate grains

reach a maximum. In this particular plot, a somewhat lower than canonical dust-to-gas ratio of $\epsilon = 5 \times 10^{-3}$ was assumed as evolved disks will have lost dust mass due to drift, or the formation of larger bodies. This also emphasizes the effect that particles not necessarily grow towards St = 1. If they did (as would be the case for a larger dust-to-gas ratios), then radial drift would be very fast, and this would very quickly reduce the dust-to-gas ratio.

So far, we neglected fragmentation and bouncing, but as particles grow, their relative velocities increase. Let us suppose particles fragment at impact velocities above a threshold velocity v_{frag}. Given that turbulent relative velocities increase with size according to Eq. (22), we can derive a maximum size particles can reach before impact velocities become larger than the fragmentation threshold. We find that this size is

$$a_{\mathrm{frag}} \simeq 0.08 \frac{\Sigma_g}{\rho_s \alpha} \left(\frac{v_{\mathrm{frag}}}{c_s} \right)^2. \tag{34}$$

Only if the highest turbulent mean velocity $v_{\mathrm{rms}} = \sqrt{3\alpha/2} c_s$ are smaller than v_f, fragmentation is not happening and Eq. (34) does not apply (but see Windmark et al. 2012b, for the effect of taking the distribution of velocities into account). Birnstiel et al. (2012b) found that fragmentation tends to dominate in the inner regions of the disk, while outer regions, or regions of lowered dust-to-gas ratio are limited by radial drift. Figure 4 shows that inside of about 20 AU, fragmentation (the red line) will limit particle growth, while outside of this radius, the drift limit applies. The trajectories in Fig. 4 ignore the effects of fragmentation—particles are limited to the fragmentation barrier (Eq. (34)) there, and will therefore drift at a lower rate as compared to particles at the drift limit.

This picture of growth, drift, and fragmentation may be quite simplistic, but it was found to match well the behavior seen in numerical simulations (see Birnstiel et al. 2012b), which incorporate many more details. A similar behavior, at least in the outer disk was also found in cases where the porosity of the grains was allowed to change: Okuzumi et al. (2012) considered icy grains and assumed that they do not fragment.[4] In their simulations, particles grow in a fractal way reaching very low filling factors ($\sim 10^{-5}$). Such particles in the inner disk could grow much quicker and thus possibly overcome the drift size limit (Sect. 5.1.2). According to Krijt et al. (2015), however, erosion is expected to limit growth beyond Stokes numbers of unity. But all these simulations find that particle growth beyond about 10–20 AU is limited by the radial drift motion of the grains, irrespective of their porosity.

4.2.2 *Dust Surface Density Profiles*

Once we have derived the maximum particle size as function of distance to the star $a_{\rm max}(r)$ (it could be either due to fragmentation, drift, or other size-limiting effects), we can use this to derive an expected dust surface density profile as follows. Practically all particle size distributions have most of the mass in the largest grains (interstellar medium (ISM): Mathis et al. 1977, circumstellar disks: e.g., Brauer et al. 2008; Birnstiel et al. 2012b; Zsom et al. 2011; Okuzumi et al. 2012, debris disks: e.g., Dohnanyi 1969). In protoplanetary disks, also the transport velocity of grains is largest for the largest grains (as long as $\mathrm{St} \lesssim 1$). This means that the total mass flux (density times velocity), the rate at which the dust is transported is dominated by the largest grains. We can now simply assume that all the mass is near the maximum particle size $a_{\rm max}(r)$, and calculate the dust transport velocity of those particles $v_{\rm r}$. Furthermore, we saw that the mass flux is set by the outer disk regions, which slowly "leak" dust inwards and this dust mass flow is conserved, so the mass accretion rate should approach a constant value. This allows us to write the dust mass accretion rate as

$$\dot{M}_{\rm d} = 2\pi r \Sigma_{\rm d} v(r) = const., \tag{35}$$

and solve for the dust surface density profile

$$\Sigma_{\rm d}(r) = \frac{\dot{M}_{\rm d}}{2\pi r v(r)}. \tag{36}$$

Hence, different prescriptions of $a_{\rm max}(r)$ will yield different dust surface density profiles (Birnstiel et al. 2012b). For typical assumptions ($\Sigma_{\rm g} \propto r^{-1}$, $T \propto r^{-0.5}$), Eq. (36) predicts $\Sigma_{\rm d} \propto r^{-0.75}$ in the drift limit (applicable in the outer disks) and a steeper slope of $\Sigma_{\rm d} \propto r^{-1.5}$ in the inner parts of the disk, where fragmentation dominates (Birnstiel et al. 2012b). This is in agreement with observations of Andrews et al. (2012) and Menu et al. (2014). Observational constraints on $a_{\rm max}(r)$ and $\Sigma_{\rm d}(r)$ thus enable us to probe the collisional processes in protoplanetary disks (see Sect. 6.2).

The fact that the growth time scales become longer in the outer regions means that the dust surface density is draining from the inside-out, where the mass flux is supplied by regions lying further and further outside. In this framework (see, Garaud 2007; Birnstiel et al. 2012b), particle growth can be interpreted as a radially expanding "pebble formation" front, which allows the dust mass accretion rate to be estimated analytically (Lambrechts and Johansen 2014). As Ormel and Klahr (2010) have demonstrated, pebbles can efficiently be accreted by growing proto-planets as the drag force around them effectively increases

[4]Water-ice particles fragment only at velocities beyond $v_{\rm f} \gtrsim 10\ \mathrm{m\,s^{-1}}$, see Gundlach and Blum (2015); silicate or carbonaceous grains fragment at velocities $\lesssim 1\ \mathrm{m\,s^{-1}}$, see Blum and Wurm (2008).

their cross section. With growth and drift providing a constant supply of inward drifting grains of the right sizes to be accreted, the growth time scale of cores, particularly in the outer regions of disks can be reduced significantly Lambrechts and Johansen (2012, 2014).

At this point, it should be mentioned that the overall grain sizes and surface density profiles seem to be in reasonable agreement with observations (see Sects. 6.2.2 and 6.2.3). However the radial drift mechanism seems too efficient—for typical disk parameters, and particles of millimeter sizes, the expected disk life times are at least one order of magnitude too short (Brauer et al. 2007), as compared to observational constraints (see Sect. 6.2.1).

5 Planetesimal Formation

The previous sections have already achieved the first step towards planet formation: the initial growth from sub-micrometer sized dust grains towards macroscopic solid particles. However this is only the first step of many. The next step is the formation of the building blocks of planets, so called planetesimals. They are defined as being bound by their own gravitational attraction (as opposed to their surface and material binding forces), which typically happens at sizes above several kilometers (e.g., Benz and Asphaug 1999). How planetesimal are assembled into planets is the subject of the next chapter in this book. In the following sections, we will discuss the step from dust to planetesimals, i.e., from particles that are strongly affected by gas drag to large bodies that are basically unaffected by gas drag. We will start by reviewing collisional planetesimal formation, i.e. particles growing successively larger via collisions in Sect. 5.1. Section 5.2 will then discuss various ideas how planetesimals form via gravitational instabilities and Sect. 5.3 will review recent synthesis models.

5.1 Collisional Growth

The basic physics of collisional growth was already briefly discussed in Sect. 3: particle collisions are driven by relative velocities between grains (Sect. 3.1). The collision rates depend on the relative velocity, but also the number of colliding particles as well as their cross sections. Assuming perfect sticking upon impact and compact particles, we defined a growth time scale (Eq. (24)) and already saw that the fast inward drift tends to be more effective in removing particles than particle growth can grow them. At the same time, collision velocities of larger grains can become too large and fragmenting collisions potentially prevent further growth. In the following, we will discuss how these obstacles could be overcome.

5.1.1 Compact Growth

Figure 2 shows the collisional outcome for conditions at 1 AU (where water is not frozen out on the grain surfaces) according to the model of Windmark et al. (2012a). For collisions of equal-sized particles (the diagonal of Fig. 2), we see that as particles grow, their sticking probability decreases and bouncing becomes the dominant collisional outcome. The situation where particles stop growing and become stuck at the sticking/bouncing transition was termed the bouncing barrier (Zsom et al. 2010). In Fig. 2, this is approximately below 10^{-2} cm.

Figure 2 also shows that even if the bouncing barrier could be overcome, collisions of equal-sized particles will become more violent: fragmentation and erosion come into play (e.g. Weidenschilling 1997; Dullemond and Dominik 2005; Brauer et al. 2008), limiting further growth at the fragmentation barrier (e.g. Birnstiel et al. 2009). However, there exists a pathway towards growing larger bodies: suppose even only a single particle exists that has a size of about a centimeter, then its collisions with the small particles at the bouncing barrier lie in the green region of Fig. 2, where "mass transfer" leads to a net growth of the larger particle. The effect is similar to throwing a snow ball against a wall: the impactor is fragmented but still able to deposit some of its mass onto the target. This type of collisional outcome was first seen in laboratory experiments by Wurm et al. (2005) and subsequently in Paraskov et al. (2007), Teiser and Wurm (2009) as well as in Kothe et al. (2010). Windmark et al. (2012a) showed that growth of larger bodies is possible in such a way, but it is slow: 10 m sized boulders could grow within about 2×10^5 years, 100 m sized boulders in about 10^6 years.

The question where these "seed particles" beyond the bouncing barrier could come from was answered in Windmark et al. (2012b) (see also Garaud et al. 2013): the results so far assumed that two particles of a given size always collide with their *mean* relative velocity. However, particle velocities and thus also their relative velocities will have a non-zero dispersion around the mean. Thus, two particles whose mean collision outcome might be bouncing or fragmentation have in fact a non-zero probability of colliding either at high velocity (causing fragmentation) or a low velocity (causing sticking). This way, some particles might be "lucky" enough to avoid disruptive collisions.

But this is both a cure and a curse. Let us consider the positive effect first: "lucky" particles can continue to grow into regions where the mean collision outcome would not allow growth. The bouncing barrier thus becomes permeable: particles either bounce (= no growth, but no destruction) or they grow.[5] Thus, the bouncing barrier will only slow down growth, but it will not stop it. Particles will continue to grow beyond the bouncing barrier, as found by Windmark et al. (2012b).

The negative effect is that also fragmentation becomes relevant already at smaller sizes than before. This also slows down growth and unlike bouncing, fragmentation leads to a reduction of the particle size. For a particle to become twice as massive as the fragmentation limit $a_{\rm frag}$, one "lucky" collision with an equal sized particle is enough. To double its size, it already needs 8 lucky collisions, and so on. The chances of the particle to grow become smaller and smaller. Fortunately, there are many particles: for the minimum mass solar nebula (Weidenschilling 1977b) at 1 AU, a region of width $H_{\rm g}$ would contain a mass corresponding to 10^{29} particles of millimeter sizes. We could therefore expect some particles to be "lucky enough" to continue to grow to larger sizes. But still: only a very small fraction of the entire mass would be in those large bodies—if there were more of them, collisions among them would be frequent and destroy them. And finally, as discussed above the growth time scale for sweep-up growth are long, longer than the radial drift time scale, so particles would still drift inwards before they could grow to larger sizes.

5.1.2 Porosity Evolution

A solution to this might lie in the internal structure of the particles: so far, we considered particle growth inside the snow line, where the constituents of dust aggregates, the so called *monomers* consist of silicates or carbonaceous particles. These particles are sticking to each other due to the weak van der Waals force. Further outside, where temperatures in the disk are lower, water will freeze out on the grain surfaces. Water ice develops hydrogen bounds

[5] The onset of fragmentation is still so far away in velocity space that the probability of a fragmenting impact velocity is negligibly low.

leading to surface energies that are about 10 times higher (Heim et al. 1999; Gundlach et al. 2011). These stiffer bounds between monomers might help overcome the radial drift barrier or destructive collisions: Wada et al. (2008) simulated collisions of icy aggregates of sizes of a few micrometers[6] and derived a fragmentation threshold velocity of around 50 $\mathrm{m\,s^{-1}}$. This is indeed higher than the $\sim\mathrm{m\,s^{-1}}$ measured for silicate dust (e.g., Wurm and Blum 1998), as was expected, but recent laboratory studies found threshold velocities that were not quite as high, with about 10 $\mathrm{m\,s^{-1}}$ (Gundlach and Blum 2015). Still, higher threshold velocities of icy aggregates allow growth to proceed to larger sizes or possibly avoid destructive fragmentation completely.

As another consequence of strong surface energy, compression of aggregates becomes more difficult. This prevents bouncing and the bouncing barrier (Wada et al. 2011; Seizinger and Kley 2013) and means that aggregates become more porous. For particles in the Epstein drag regime, the growth time scale does not depend on the surface-to-mass ratio (see Eq. (32)). But in the inner regions of disks, where gas densities are high, particles may be in the Stokes drag regime. In this case, the growth time scale at a given Stokes number becomes inversely proportional to the particle size (Okuzumi et al. 2012). At the crucial regime around Stokes numbers of unity, extremely fluffy particles (internal densities of $\ll 10^{-3}\ \mathrm{g\,cm^{-3}}$) have a very large size resulting in very short growth time scales, short enough to overcome the radial drift barrier (Okuzumi et al. 2012; Kataoka et al. 2013). This mechanism could thus form relatively compact planetesimals by avoiding the growth barriers via extremely low internal densities, followed by compression via gas drag and self-gravity to reach cometary densities (Kataoka et al. 2013).

There are however some caveats: first, the gas densities have to be high enough for particles to be in the Stokes drag regime, which is only expected to be the case inward of $\sim$10 AU in the model of Okuzumi et al. (2012). Second, sintering can cause the break-up of the aggregates. The sintering time scales are shorter than the growth time scales inside of a few AU (Sirono 2011b,a), preventing the formation of planetesimals in such a model. Finally, Krijt et al. (2015) have considered the additional effect of erosion, which was not included in the study by Okuzumi et al. (2012) and Kataoka et al. (2013). The relative velocity between unequal sized particles can be much larger than between equal sized particles (see Fig. 1). Even if equal size particles do not destroy each other, small and large bodies can collide at velocities above the erosion threshold velocity. If this is the case, erosion stops further growth at Stokes numbers around unity also resulting in somewhat more compact aggregates (Krijt et al. 2015). Recently measured erosion threshold velocities of around 15 $\mathrm{m\,s^{-1}}$ Gundlach and Blum (2015) seem to support this result.

Even if erosion prevents the collisional growth of planetesimals, not all is lost: particles are still expected to grow to Stokes numbers large enough to participate in gas induced particle concentrations which can lead to gravitationally bound bodies, as we will discuss in the following.

5.2 Formation of Planetesimals by Gravitational Collapse

As discussed in Sect. 2, growth of dust aggregates also changes the aerodynamic properties of the aggregates. The increased stopping time leads to a partial decoupling of the motion of the dust aggregates from the motion of the gas and this in turn allows particles to sediment to the mid-plane of the protoplanetary disc, drift radially towards the central star and to become concentrated in the turbulent gas flow. In the following, we will discuss some of

[6] These results strongly depend on the monomer size. These works assumed a monomer size of 0.1 μm.

the dynamical effects that are able to cause strong local concentrations of dust particles. We will not discuss here the turbulent concentration mechanism proposed by Cuzzi et al. (2008), where chondrule-sized particles are concentrated on the smallest scales of the turbulent flow, as the relevance of this mechanism for protoplanetary discs is still under exploration. We refer the reader to the recent review paper by Johansen et al. (2015a), and references therein, for a detailed description of the latest development in the study of small-scale turbulent concentration.

5.2.1 Pressure Bumps

One of the most generic flow structures that can concentrate particles is the so-called pressure bump. Pressure bumps are axisymmetric overpressure regions that prevail in perfect balance between the outwards-directed pressure gradient force and the inwards-directed Coriolis force. Particles do not react to the pressure gradient directly, but the surrounding zonal flow envelope is super-Keplerian on the inside of the pressure bump and sub-Keplerian on the outside. Dust particles sense a tailwind on the inside of the bump and a headwind on the outside of the bump, resulting in their migration to the centre of the bump (cf. Eq. (7)). The convergence region will move slightly inwards towards the star in the presence of a global, radial pressure gradient.

Pressure bumps can arise spontaneously in protoplanetary discs by differential transport of angular momentum. While simplified models of protoplanetary accretion discs with a constant value of the turbulent viscosity coefficient α always display outwards transport of angular momentum, any variation in the strength of the turbulence can lead to a local pile-up of angular momentum—this is the zone that in turn forms the pressure bump through the convergence of the radial flow induced by the Coriolis force.

Simulations of turbulence driven by the magnetorotational instability (Balbus and Hawley 1991) show that large-scale variations in the strength of the turbulence lead to the spontaneous formation of pressure bumps at the largest scales of the turbulent flow (Johansen et al. 2007; Simon et al. 2012; Dittrich et al. 2013; Simon and Armitage 2014). These pressure bumps are quasi-stable and live for hundreds of orbits before they disassemble and reform. Particle concentrations inside pressure bumps reach values above the Roche density, and very large planetesimals can form by the subsequent gravitational contraction and collapse phases (Johansen et al. 2011).

Turbulence is nevertheless a two-edged sword. Collision speeds between m-sized particles moving in turbulence caused by the magnetorotational instability reach tens to hundreds of meters per second. Hence it is not clear that fully developed turbulence is a good environment for forming planetesimals—despite the emergence of pressure bumps that concentrate particles.

The differential angular momentum transport needed to form pressure bumps can also arise from radial variations in the very nature of the turbulence. The transition from regions of fully developed turbulence caused by the magnetorotational instability and the "dead zone" whose ionization fraction is too low to sustain the magnetorotational instability has been shown to lead to the formation of pressure bumps at the inside and outside of the dead zone (Lyra et al. 2008, 2009; Kretke et al. 2009). Such sweet spots could experience planetesimal formation ahead of the bulk parts of the protoplanetary disc and hence have important implications for the formation sequence of planets in planetary systems.

5.2.2 Vortex Trapping

Azimuthally elongated vortices may form directly from a flow instability, such as the baroclinic instability, or result from Rossby wave instabilities that are triggered by high-amplitude pressure bumps.

The baroclinic instability is a process analogous to radial convection (Klahr and Bodenheimer 2003; Lesur and Papaloizou 2010; Raettig et al. 2013). The non-linear state of the instability is characterized by the emergence of large-scale, slowly overturning vortices. Such vortices can trap dust particles (e.g., Barge and Sommeria 1995; Meheut et al. 2012; Lyra and Lin 2013; Raettig et al. 2015) in a similar way to pressure bumps (Paardekooper and Mellema 2004; Pinilla et al. 2012; Zhu et al. 2012; Birnstiel et al. 2013; Lyra and Lin 2013). Recent observations show a remarkable resemblance with expected signatures of these models (see Sect. 6). Dust feedback (the impact of the dust dynamics on the gas dynamics) nevertheless limits the particle concentration obtainable inside such vortices, as the vortex flow is destroyed by particle feedback for a dust-to-gas ratio above unity (Johansen et al. 2004; Fu et al. 2014). Still, the achieved over densities in vortices might trigger planetesimal formation via the streaming instability (Raettig et al. 2015), which will be discussed in the following section.

5.2.3 Streaming Instability

The sedimentation of dust particles paves the way for a concentration mechanism in which the particles play an active role through their friction on the gas. As the dust-to-gas ratio in the mid-plane approaches unity, the back-reaction friction force from the particles onto the gas becomes strong enough to accelerate the gas to orbit at closer to the Keplerian speed together with the particles. Thus the headwind on the particles is reduced and the particles fall more slowly towards the star. An overdense structure in the mid-plane (such as an axisymmetric particle filament) would fall even more slowly than the surrounding particles. This is now a run-away situation because these faster-drifting particles continue to drift at their full rate and thus pile up in the filament—this is the essence of the so-called streaming instability (Youdin and Goodman 2005).

The characteristic length scale of the streaming instability is approximately 5 % of a gas scale-height H, although the separation between filaments in the non-linear state is higher, close to 0.2 gas scale-heights (Yang and Johansen 2014).

Particle concentration by the streaming instability can reach at least several thousand times the local gas density (Bai and Stone 2010; Johansen et al. 2012). The particle filaments appears to obtain a fractal structure where smaller-scale filaments embedded within the main filament allow very high particle densities at the smallest, resolvable scales of the simulations (Johansen et al. 2012). The high mass loading in the gas could be an additional factor needed to reach very high densities, as the turbulent diffusion of the gas is drastically reduced in such particle dominated flows (Johansen et al. 2009).

The characteristic planetesimal size forming by the gravitational collapse of the overdense regions formed by the streaming instability depends sensitively on the column density of the particles. Models with a few times the minimum mass solar nebula worth of particles in the asteroid belt give rise to Ceres-mass planetesimals (Johansen et al. 2012), while the characteristic planetesimal size falls with decreasing particle column density, down to 100 km for the more realistic case where the column density is slightly lower than in the minimum mass solar nebula (see Fig. 5).

An important prerequisite for the formation of dense particle filaments by the streaming instability appears to be a sufficiently high ratio of particle column density to gas column

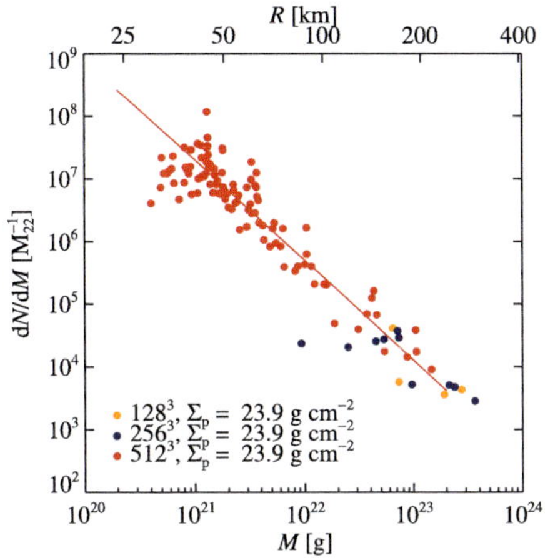

Fig. 5 Differential mass distribution of planetesimals formed by the streaming instability at three different numerical resolutions, from Johansen et al. (2015b), as a function of planetesimal mass (*lower axis*) and size (*upper axis*). The mass distribution is given per 10^{22} g in an annulus of 1 AU width situated at 2.5 AU from the central star. A fit of $\mathrm{d}N/\mathrm{d}M \propto M^{-1.6}$ is overplotted (*full line*). Such a differential mass distribution is dominated in mass by the few largest bodies, but the number is dominated by the smallest bodies. The results here are for a column density of approximately six times the value in the minimum mass solar nebula; reducing the column density to the nominal value leads to a characteristic planetesimal size closer to 100 km in radius (Johansen et al. 2015b)

density, $Z = \Sigma_\mathrm{p}/\Sigma_\mathrm{g}$. Strong particle concentration is triggered above a threshold value of $Z \approx 0.015$ (Bai and Stone 2010). This threshold is approximately constant for particle sizes from $\mathrm{St} = 0.03$ to $\mathrm{St} = 0.3$ (corresponding to approximately cm-dm sized particles in the asteroid belt and to mm-cm sized particles in the outer Solar System). Smaller and larger particles require a higher mass-loading to trigger particle concentration (Youdin and Goodman 2005; Carrera et al. 2015).

5.3 Synthesis: Combined Models

Clearly dust coagulation and particle concentration are in no way mutually exclusive processes. In fact, the two are intimately coupled. Dust growth to mm and cm sizes is key for allowing particles to decouple from the gas and be concentrated, while the dense environment caused by particle concentration incites high collision rates, where the outcomes of the collisions in turn determine the continued contraction of the cloud.

The growth of dust aggregates to sizes that can undergo concentration by the streaming instability is a subject of intense scrutiny at the moment. Drążkowska and Dullemond (2014) showed that the appropriate cm-sized particles can form (at least outside of the ice line where sticking is easier), but that production of sufficient amounts of large dust aggregates to trigger the streaming instability requires a super-solar metallicity (or photoevaporation of the gas). The challenge is that only a fraction of the total dust mass reaches the particle sizes necessary to trigger the streaming instability and that the streaming instability becomes self-limiting by removing the dust mass and lowering the metallicity below the necessary values. Similar conclusions were reached by Krijt et al. (2016) for the growth of very fluffy ice particles. It thus appears that there remains a gap in our understanding between the amount of large particles that can be produced by direct growth and the amount needed to trigger strong particle concentration.

Pebbles in the asteroid belt appear to be represented by the mm-sized chondrules found in primitive meteorites. Typical chondrules have Stokes numbers around 0.001 when assuming a background gas density like the minimum mass solar nebula. Sedimentation of such small particles to the mid-plane is not very efficient, as even a moderate degree of turbulent stirring is enough to lift them up to the disk atmosphere. Particle concentration by the streaming instability also appears problematic for mm-sized particles in the asteroid belt (Carrera et al. 2015). The situation would be better if the gas density was lower than in the minimum mass solar nebula, as motivated by observations (e.g. Andrews et al. 2012). A reduction in gas density by a factor of 10 would increase the friction time correspondingly and put chondrules within the range of particles that can be concentrated by the streaming instability. It is also possible that chondrules formed chondrule aggregates, the sticking facilitated by acquiring porous dust rims (Ormel et al. 2008). Alternatively, other hydrodynamical concentration mechanisms may be are able to bridge the gap, e.g. the baroclinic instability/convective overstability (Klahr and Bodenheimer 2003) or the vertical shear instability (Nelson et al. 2013). Such instabilities do not necessarily rely on an initially high dust-to-gas ratio to further concentrate the dust. The concentrations formed by these mechanisms could then trigger the streaming instability even in cases where the overall metallicity would be too low.

Hydrodynamical simulations of planetesimal formation by gravitational collapse typically employ a fixed grid for the dynamics and hence are not able to follow the collapse of the particle clouds down to planetesimal sizes. This collapse phase is nevertheless so decoupled from the remaining disc that it can be modelled in a separate N-body simulation. Nesvorný et al. (2010) showed that the collapse typically leads to the formation of binary planetesimals, in good agreement with the prevalence of binaries in the classical cold component of the Kuiper belt (e.g., Noll et al. 2008). Wahlberg Jansson and Johansen (2014) included a detailed particle collision model in the collapse and found that a large fraction of the pebbles survive the collapse without undergoing fragmentation, leading to the formation of pebble-pile planetesimals consisting of primordial pebbles from the protoplanetary disc.

This "pebble paradigm" for planetesimal formation will be put further to the test in the future with the observations of minor bodies in the Solar System. The presence of "goosebumps" on the comet 67P has been interpreted as the primordial pebbles of the solar protoplanetary disc (Sierks et al. 2015)—although if this interpretation is true, then that poses a challenge for coagulation theory and experiments to explain how the dominant pebble sizes can be more than one meter, in contrast to drift dominated models that predict sizes closer to 1 cm in the outer disc (Birnstiel et al. 2011).

6 Observations

6.1 Methods

The dust particles in disks span a large range of temperatures from over one thousand to only tens of K, decreasing with the distance to the central star. The grains emit thermal emission from the near-infrared band where the hot inner disk dominates, through the mid-infrared where the warm disk atmosphere contributes most, to far-infrared and (sub-)millimeter wavelengths where the optical depth is lower and we can see deeper into the disk. Apart from the inner $\sim$10 AU, disks are normally optically thin in the millimeter continuum and the emission thus probes the cool disk interior where most of the mass resides.

Various techniques have been developed to study disks based on dust emission. The infrared imaging of star-forming regions of young clusters is one of the most efficient ways to characterize disks properties of a large sample of young stars with disks. Our understanding of disk structure and evolution has been advanced progressively with the advent of various space infrared telescopes, e.g. IRAS, ISO, Spitzer, and Herschel, and complemented with a lot of ground-based telescopes with infrared instruments. From these observations, in combination with (sub-)millimeter imaging, the entire spectral energy distributions (SED) of disks can be constructed. Modeling the SEDs of disks has become a classical way to investigate general disk properties. However, because of the lack of spatial resolution, the results from SED modeling are strongly model dependent. The resolved images from (sub-)millimeter interferometric observations in dust continuum emission and molecular line emission can constrain the surface density distribution of dust and gas, as well as dust grain sizes, disk inclinations, et cetera. A combination of modeling both SEDs and resolved images from (sub-)millimeter interferometric observations has proven to be the best way to study disks (Andrews et al. 2009).

6.1.1 Infrared Observations

Mendoza V (1966, 1968) presented the first discovery of near- and mid-infrared excess emission of T Tauri stars. At that moment it was unclear where the excess emission came from. The launching of IRAS in 1983 extended infrared observations to the far-infrared bands. Observations with IRAS revealed that many T Tauri stars, showing near- and mid-infrared excess, also show significant far-infrared excesses, indicating the existence of cold circumstellar dust (Rucinski 1985). Various models have been proposed to explain the infrared excess emission of T Tauri stars. An optically thick but physically thin circumstellar disk was believed to be the best model for explaining these observations (Myers et al. 1987; Adams et al. 1987). ISO, launched in 1995, opened a new infrared window for spectroscopy and discovered the presence of crystalline silicates in disks (Waelkens et al. 1996). Thanks to the incredible sensitivity of the Spitzer Space Telescope, our knowledge of disks has been significantly improved in the last ten years. Due to the high efficiency in observations, many star-forming regions have been surveyed with Spitzer in both imaging and spectroscopic modes, which advanced our understanding of disk lifetimes, disk evolution with local environments, dust mineralogy in disks around different types of young stars from intermediate mass stars to brown dwarfs (Apai et al. 2005; Furlan et al. 2006; Sicilia-Aguilar et al. 2006; Balog et al. 2007; Hernández et al. 2007; Guarcello et al. 2009; Sicilia-Aguilar et al. 2011; Kraus et al. 2012; Fang et al. 2012, 2013b). The Spitzer imaging survey also largely extended the sample of "transition disks". This class of disks (first seen by Strom et al. 1989) have optically thin, dust depleted (or even dust-free) inner regions. They were thought to be in the act of clearing the disk from the inside out, therefore called transition disks (Marsh and Mahoney 1993; Fang et al. 2009; Cieza et al. 2010; Muzerolle et al. 2010; Cieza et al. 2012; Fang et al. 2013a; Espaillat et al. 2014). Recently, the Herschel space telescope, launched in 2009, has presented incredible sensitivity at far-infrared wavelengths, which makes it possible to study dust contents of large samples of disks and to investigate the disks around brown dwarfs at these wavelengths (Meeus et al. 2012; Dent et al. 2013; Harvey et al. 2012, and also see sensitivities of various instruments in Fig. 6).

6.1.2 SED Modeling

Before circumstellar disks could be spatially resolved, the observed infrared excess of young stars was explained using simple disk models. Modeling the observed SEDs became a classical way to characterize global properties of disks. Initially simple geometrically thin and

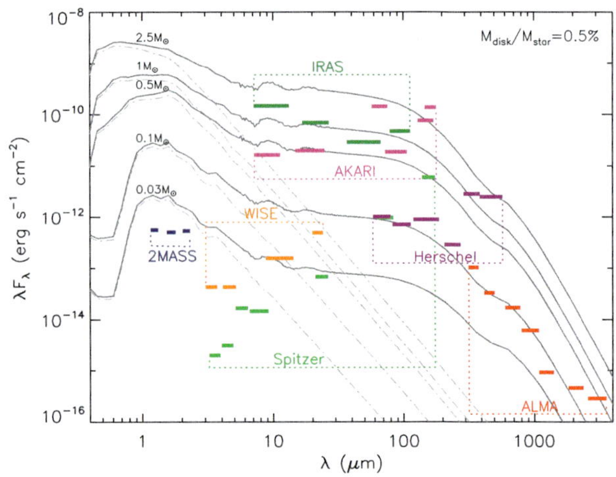

Fig. 6 The 3σ detection limits of various instruments overlaid on the model SEDs of young disk systems, located at a distance of 300 pc, with stellar masses of 2.5, 1.0, 0.5, 0.1, and 0.03 $M_{\odot}$, and a disk-to-star mass ratio of 0.5 %. The SEDs are calculated with a Monte Carlo radiative transfer code RADMC-3D (Dullemond 2012). The detection limits of IRAS, 2MASS, AKARI, and WISE are typical values for these all-sky surveys. The ones of Spitzer, Herschel, and ALMA are for observations with on source integration time of 1 minute

optically thick disks were used in SED modeling (Adams et al. 1987; Kenyon and Hartmann 1987), which was refined by including an optically thin upper layer above the simple disk models (Chiang and Goldreich 1997). The SED modeling was further improved by 2D/3D radiative transfer disk models in which the temperature and vertical structure were calculated self-consistently (Whitney et al. 2003, 2013; Dullemond and Dominik 2004; Dullemond 2012). Even web-based SED fitting tools, built upon those codes are available (Robitaille et al. 2006, 2007). In the typical SED modeling approach, the dust properties and disk geometry are parameterized. The SEDs computed from these models are then compared against the observed SEDs to find the best-fitting parameters. The advantage of SED modeling is that one can investigate a large sample of disks since the SEDs can be easily constructed from various imaging surveys. The physical properties of disks can be inferred using this technique and studied statistically, rather than simply quantifying the SED shapes. However, the disadvantage of this technique is that the results are strongly model dependent and highly degenerate because of the spatially unresolved and often sparse observations.

6.1.3 (Sub-)Millimeter Interferometry

The resolved images in dust continuum emission and molecular line emission do not only directly indicate the disk sizes, but also provide strong constrains on the distribution of dust and gas in circumstellar disks, which cannot be achieved by SED modeling alone. While the infrared excess gave indirect evidence of disks, the gas kinematics inferred from early interferometric observations provided strong evidence that the circumstellar material around T Tauri stars was in disks with a Keplerian rotation profile (Weintraub et al. 1987; Sargent and Beckwith 1987). The first large interferometric survey was carried out by Dutrey et al. (1996) with the IRAM Plateau de Bure Interferometer, in which many disks in Taurus were resolved with typical angular sizes of 1–2″, or 140–280 AU at a distance of 140 pc. A substantial improvement was provided by the Sub-millimeter Array (SMA) with high sensitivities and angular resolutions (Andrews and Williams 2007a,b; Andrews et al. 2010). A combination of modeling SEDs and the high spatial resolution images with SMA significantly improved our understanding of the surface density distribution in disks (Andrews et al. 2009, 2010). These SMA data also directly resolved gaps/holes in transition disks which were already inferred from SED modeling (e.g., Brown et al. 2009; Andrews et al. 2011). The typical hole size from these SMA observations are around several tens AUs, which may be due to the limited angular resolutions. From these high angular resolution observations, the typical disk size is around 100 AU, and the typical disk-to-host mass ratio is around 0.2–0.6 % (Andrews et al. 2009, 2010, 2013).

Recently (sub-)millimeter interferometry revealed radial variations in the dust-to-gas ratio in several disks (e.g., Panić et al. 2009; Andrews et al. 2012; Isella et al. 2012; de Gregorio-Monsalvo et al. 2013; Rosenfeld et al. 2013) and resolved multi-wavelength observations indicate also radial variations in the dust properties, see Sect. 6.2.3.

There are several challenges associated with interpreting the (sub-)millimeter interferometric data. For the dust continuum emission, the primary issue are the unknown optical properties of the dust, which mainly depend on the unknown size distribution and internal structure (i.e. porosity, composition, etc.) of the grains. The observations at several (sub-)millimeter wavelengths may give some constraints on the size distribution of dust grains (Ricci et al. 2010b). However, a considerable amount of dust mass could be locked up in larger particles or planetesimals, in which case the measurements of dust mass from (sub-)millimeter observations would only be lower limits. Moreover, for a conversion from the (sub-)millimeter fluxes into disk masses, one needs to know the gas-to-dust ratio, which is expected to change with disk radii and with disk evolution (Takeuchi and Lin 2002; Birnstiel et al. 2010). In circumstellar disks, the bulk of dominant mass constituent, H_2, is unobservable due to the lack of a dipole moment. Other molecular lines, e.g. CO and its isotopes, are used to probe disks. The line emissions are then converted to disk masses assuming abundances for the molecules which depend on the complex chemistry, freeze-out, and optical depth and can vary from region to region in disks.

6.2 Results

6.2.1 Disk Lifetimes

The disk lifetimes are observationally constrained by counting in star-forming regions or clusters the numbers of young stars that show disk infrared excess. The resulting drop of disk fraction with the age of the cluster/star-forming region then indicates the average disk lifetime. The early study of the nearby Taurus star-forming region at near-infrared (2.2 μm) wavelength suggest disk lifetimes of $\lesssim$3–10 Myr (Strom et al. 1989), which was confirmed later by the study at 10 μm (Skrutskie et al. 1990). The first systematic and extensive study of disk lifetimes comes from Haisch et al. (2001). They studied a sample of 8 star-forming regions/clusters with ages ranging from $\lesssim$1–30 Myr at L band (3.4 μm), and derive an overall disk lifetime of ~6 Myr. With the advent of the Spitzer Space Telescope, our ability to image star-forming regions/clusters at mid-infrared bands (3.5–24 μm) has been dramatically increased. A large sample of star-forming regions/clusters with ages from 1 Myr to tens of Myrs have been surveyed by various projects with Spitzer (e.g. Allen et al. 2004; Megeath et al. 2004; Luhman 2007; Evans et al. 2009; Gutermuth et al. 2009; Fang et al. 2009; Luhman et al. 2010; Megeath et al. 2012; Fang et al. 2013a; Sicilia-Aguilar et al. 2015). The Spitzer data generally confirm the previous results on disk lifetimes (Sicilia-Aguilar et al. 2006; Hernández et al. 2007). Furthermore, due to the excellent sensitivity and high observing efficiency, the Spitzer data can be used to characterize the timescale of "transition disks", which is around 0.1–1 Myr (Muzerolle et al. 2010; Currie and Sicilia-Aguilar 2011).

The Spitzer observations are not able to resolve the binary systems with separations $\lesssim 2''$ even in the nearest star-forming regions like Taurus. Thus the disk lifetimes concluded from these data are actually for both single stars and stellar systems. When correcting for stellar systems, Kraus et al. (2012) find that the fraction of disks surrounding the single stars can be around 80 % in Taurus, while less than 40 % of binary systems with separations $\lesssim$40 AU have disks. A combination of investigations in other nearby star-forming regions indicates that ~2/3 of close (<40 AU) binary systems disperse their disks within $\lesssim$1 Myr (Kraus et al. 2012). Thus, the lifetimes of disks surrounding single stars should be longer than those of close binary systems, which could potentially explain why exoplanet host stars perfer single stars (Roell et al. 2012). Beside in close binary systems, disk evolution can be also affected by other factors. One example are nearby massive stars as found in some massive clusters, e.g. Orion and Pismis 24, where a type of objects, named "proplyds", show tails pointing away from the massive stars. This is interpreted as the outer disks of young stars that are being photoevaporated by ultraviolet radiation from the massive stars (O'dell et al. 1993; Störzer and Hollenbach 1999; Fang et al. 2012). Spitzer observations have revealed a clear anti-correlation between the frequencies of circumstellar disks and the presence of massive stars in several massive clusters harboring extremely massive stars (typically earlier than O5) (e.g., NGC2244, NGC6611, Pimis24, and etc. Balog et al. 2007; Guarcello et al. 2009; Fang et al. 2012). The disks in these massive clusters can be dissipated roughly twice as quickly as in clusters/star-forming regions without extremely massive stars (Fang et al. 2012). The disk lifetimes also show dependence on the density of the cluster environment. In sparse stellar associations disks are dissipated more slowly than those in denser (cluster) environments (Fang et al. 2013b). Metallicity could also affect disk evolution in a way which is still controversial (Yasui et al. 2010; Spezzi et al. 2012).

There are several issues on disk lifetimes that need to be addressed. The infrared observations can only be useful when the disks still have hot/warm inner regions. Thus, the disk fractions counted from these observations may miss some sources with only cold outer disks, i.e. transition disks. The second is that the disk fractions in star-forming regions/clusters could be also biased to higher values since it is hard to identify the diskless young stars. The age estimates of star-forming regions/clusters are also problematic since there are substantial systematic differences on ages between the different sets of tracks (Hillenbrand et al. 2008). An improved age estimates of young clusters indicate the disk lifetimes could be much longer (10–12 Myr) than we thought before (Bell et al. 2013, see also Pfalzner et al. 2014).

6.2.2 Density Profiles, Disk and Hole Sizes

Our knowledge of surface density profile of a circumstellar disk was initiated by the minimum-mass solar nebula (MMSN) model, which was constructed by adding enough light elements to the present planetary bodies to reach solar composition, and then spreading this material out over the regions half-way to their neighboring planets. The resulting total surface density (dust and gas) can be fitted by a power-law function, $\Sigma \propto r^{-3/2}$ (Weidenschilling 1977b; Hayashi 1981). A similar model was constructed for extrasolar planets, resulting in a similar slope of around 1.4 (Chiang and Laughlin 2013). The observed features of disks, e.g. the SEDs and the distribution of dust emission, can be generally reproduced assuming the power-law form ($\Sigma_g \propto r^{-p}$) for the surface density profile (Beckwith et al. 1990; Andrews and Williams 2007a). The power-law index (p) can be only constrained by fitting the resolved images of disks from (sub-)millimeter interferometric observations, which suggest that the p values are typically less than 1.5 (Andrews and Williams 2007a). The typical disk size from these high spatial resolution observations is around several hundred AU (Andrews and Williams 2007a). Though succeeding in explaining many observed properties of disks, the power-law form for surface density profiles in disks hardly reconcile the observational fact that disk sizes from CO rotational line emission are much larger than those from dust emission (Piétu et al. 2005; Panić et al. 2009). A similarity solution of a simple accretion disk with time-independent viscosity (ν) and $\nu \propto r^{\gamma}$ solve the issue by providing a surface density profile (Lüst 1952; Lynden-Bell and Pringle 1974; Hartmann et al. 1998; Andrews et al. 2009),

$$\Sigma_{\mathrm{g}} = \Sigma_{\mathrm{g,c}} \left(\frac{r}{r_{\mathrm{c}}}\right)^{-\gamma} \exp\left[-\left(\frac{r}{r_{\mathrm{c}}}\right)^{2-\gamma}\right], \tag{37}$$

where $\Sigma_{\mathrm{g,c}}$ is the normalization parameter, r_{c} the characteristic scaling radius, and γ is the gradient parameter. When $r \ll r_{\mathrm{c}}$, Σ_{g} is approximately a power-law, and when $r \gtrsim r_{\mathrm{c}}$, Σ_{g} starts to be dominated by the exponential factor. The profile has been widely used to reproduce the resolved images of the dust emission or the CO rotational line emission of disks (Andrews et al. 2009; Isella et al. 2009; Andrews et al. 2011, 2012). When the above exponentially tapered surface density profiles are used in the disk models, physical disk sizes will be less direct than using the power law form, and must be defined explicitly by an intensity threshold. Since r_{c} is a characteristic scaling radius beyond which the surface density profiles decrease significantly in an exponential form, its values can probably used to compare the sizes of different disks. From the investigation of a sample of resolved images of disks, r_{c} shows a wide range from tens to several hundred AU (Hughes et al. 2008; Andrews et al. 2009, 2010). The power-law parameter γ was found mostly between 0.5 and 1, much less steep than in the MMSN model. It should be noted that those observations were most sensitive to the disk regions beyond the size of the solar system (on which the MMSN model and the model of Chiang and Laughlin 2013 is based). It is very well possible, that the inner regions of disks show changes in the dust surface density profile. This is expected theoretically when the particle size distribution transitions from being drift-dominated to being fragmentation dominated, see Sect. 4.2.2 and Birnstiel et al. (2012b). Such a behavior was recently observed for the disk around TW Hya disk by Menu et al. (2014).

Advances in sensitivity, however showed that the discrepancy of dust and gas disk sizes remained even if the surface density profile of Eq. (37) was used (Andrews et al. 2012; Isella et al. 2012; Rosenfeld et al. 2013; de Gregorio-Monsalvo et al. 2013). At the low gas densities in the outer disk, even micron-sized particles can decouple from the gas and start drifting inwards. This drift of small particles can thus leave the outermost regions of the disk devoid of dust, causing steep decreases in the dust-to-gas ratio (Birnstiel and Andrews 2014). The observed discrepancy between dust and gas disk sizes might thus be a finger-print of the radial drift process.

Another advantage of high spatial resolution images from (sub-)millimeter interferometric observations is to directly resolve gaps or holes in some transition disks. The typical gap/hole sizes from the high angular resolution SMA observations are around ten to tens of AU (Brown et al. 2009; Andrews et al. 2011). However, there are still many transition disks with gap/hole sizes less than 10 AU, inferred from the SED fitting (Espaillat et al. 2012; Kim et al. 2013), which could be resolved with ALMA. The causes for the gaps and cavities are still unknown. Proposed mechanisms include clearing by a unseen companion, photoevaporation, opacity effects, or instabilities. In the recent years, it has become clear, that opacity effects from grain growth can be excluded (Birnstiel et al. 2012a), and also photoevaporation models alone can currently not account for most of the gap sizes and accretion rates seen in those objects (Owen et al. 2011), but possibly for a subset of them (Owen and Clarke 2012). The currently favored mechanism is "dust trapping": a bump in the radial gas pressure profile is able to stop radial drift, thus trapping large particles (e.g., Rice et al. 2006; Pinilla et al. 2012; Zhu et al. 2012), while only a small fraction of small dust particles are following the gas flow into the inner disk. The cause of such a pressure bump could be either a planet embedded in the disk, or variations in the viscosity profile (e.g., Kretke and Lin 2007; Regály et al. 2012).

6.2.3 Spectral Indices, Grain Sizes

The thermally emitted intensity of the vertically isothermal, face-on dust disk at a distance r from the star is approximately

$$I_\nu(r) = B_\nu\big(T(r)\big)\cdot\big(1-\exp\big(-\tau_\nu(r)\big)\big), \tag{38}$$

where $B_\nu(T(r))$ is the Planck spectrum at a temperature T and $\tau_\nu(r)\simeq\Sigma_\mathrm{d}\kappa_\nu$ is the optical depth (ν now denotes the frequency). If the optical depth is high, the spectral index is entirely set by the Planck spectrum. If the optical depth is low, we can expand the term in brackets to first order and write

$$I_\nu(r) = B_\nu\big(T(r)\big)\cdot\kappa_\nu(r)\cdot\Sigma_\mathrm{d}(r). \tag{39}$$

In this optical thin case, the spectral index of the flux at (sub-)millimeter wavelength, α_mm, defined as $F_\nu \propto \nu^{\alpha_\mathrm{mm}}$ depends on the dust opacity and can be used to characterize the dust grain sizes (Beckwith and Sargent 1991; Andrews and Williams 2005; Ricci et al. 2010b). The disk-integrated flux is often roughly approximated as

$$F_\nu \approx \kappa_\nu B_\nu(T_\mathrm{d}) M_\mathrm{d} D^{-2}, \tag{40}$$

where T_d is the temperature of the region dominating the dust emission, M_d is the dust mass, and D is the distance from the source. Under the assumption that dust emission at (sub-)millimeter wavelengths is in the Rayleigh-Jeans limit,

$$F_\nu \approx \kappa_\nu 2k_\mathrm{b}\nu^2 T_\mathrm{d} M_\mathrm{d} c^{-2} D^{-2}, \tag{41}$$

where k_b is the Boltzmann constant and c the speed of light. At (sub-)millimeter wavelengths, the dust opacity (κ_ν) can be approximated as $\kappa_\nu \propto \nu^\beta$, where β is sensitive to dust properties, in particular to the grain size (Draine 2006; Ricci et al. 2010b; Testi et al. 2014) since large dust particles have smaller β. The observed flux density can be written as $F_\nu \propto \nu^{2+\beta}$. The dust properties in disks can be characterized using the observed spectral indices (α_mm) since $\beta = \alpha_\mathrm{mm} - 2$. Regions outside of a few AU are expected to be optically thin, apart from local dust concentrations, such as seen in transition disks. The derived values of β from the spectral indices of disks are usually much smaller than for ISM-like dust grains, indicating that dust grains in disk have grown to considerably larger sizes (Beckwith and Sargent 1991; Testi et al. 2003; Natta et al. 2004; Rodmann et al. 2006; Andrews and Williams 2007a; Ricci et al. 2010b,a). While the (sub-)millimeter wavelengths reveal dust growth in the disk interior, dust growth in the disk atmosphere is also evidenced by the broad and weak silicate 10 μm emission features which are expected from silicates with larger sizes than those in ISM (van Boekel et al. 2005).

In recent years, the increasing resolution of (sub-)millimeter interferometers enabled radially resolved measurements of α_mm, which can put constrains on the radial distribution of grain sizes (Isella et al. 2010; Guilloteau et al. 2011; Pérez et al. 2012; Trotta et al. 2013). It was generally found that the spectral index is radially increasing, which indicates larger grains in the inner disk, in agreement with theoretical expectations.

6.2.4 Recent Imaging Results

The field of protoplanetary disks is currently being revolutionized by observational advances in near-infrared imaging (Subaru, NACO, and the upcoming Sphere and GPI instruments) and in (sub-)millimeter interferometry with ALMA. These observations of disks can provide images with comparable angular resolution (of the order of $0.03''$), but are sensitive to dust particles of very different sizes (μm vs. mm) and temperatures (hot vs. cold). In near-infrared bands, imaging can resolve the inner regions of circumstellar disks with inner working angles of $< 0.1''$, sensitive to the μm-sized dust grains. Observations with Subaru confirm the gaps/holes in some transition disks, which are resolved with (sub-)millimeter interferometric observations (Hashimoto et al. 2012; Mayama et al. 2012; Tsukagoshi et al. 2014). It is even more interesting that some transition disks that show gaps/holes in the (sub-)millimeter interferometric data lack evidence of evolved inner disks in H-band polarized intensity images (Muto et al. 2012; Follette et al. 2013), which may be due to the decoupling between the spatial distributions of the small (μm) and big (mm) dust grains inside the holes/gaps caused by dust filtration (Rice et al. 2006; Dong et al. 2012; Pinilla et al. 2012; Zhu et al. 2012). Toward the transition disk, HD 135344B, the Subaru polarized H-band intensity image reveals the spiral structures in the disk, which may be due to the gravitational perturbation from an embedded planet (Muto et al. 2012). Imaging with VLT/NACO revealed gaps, brightness asymmetries, or spiral patterns that might be indicative of planets (e.g., Quanz et al. 2013a,b; Avenhaus et al. 2014). Similar features have recently been imaged in the first results from GPI (Rapson et al. 2015), and SPHERE (Benisty et al. 2015).

With its extremely high angular resolution and sensitivity, ALMA has provided many ground breaking findings already. van der Marel et al. (2013) discovered that the entire emission from large (mm-sized) dust particles, trapped in the outer regions of a transition disk Oph IRS 48 comes from one side of the disk, a region spanning less than one third of the orbit. A possible explanation for this is the combination of a radial dust trap and an azimuthal asymmetry such as a vortex. This would cause dust to be radially drifting towards the pressure bump where it further concentrates in the azimuthal asymmetry. Such a dust trap in Oph IRS 48 seems consistent with theoretical predictions (Pinilla et al. 2012; Birnstiel et al. 2013), and may support the formation of analogues to Kuiper Belt objects in this object. Similar asymmetries in the ring-like dust emission from transition disks have been detected in Casassus (2013) and Pérez et al. (2014). Casassus (2013) also detected faint stream-like features in HCO^+ and dust continuum at 345 GHz in the transition disk, HD 142527, which could be explained as funnel flows into the inner disk through planets (Zhu et al. 2011).

Recently ALMA observations revealed an intriguing image of a circumstellar disk in HL Tau (ALMA Partnership et al. 2015) which is a Class I young stellar object, earlier than the typical Class II T Tauri stars. In the disk of HL Tau, several concentric dust gaps and rings can be clearly identified, possibly sculpted by young planets (Tamayo et al. 2015), which hint that planets could already be formed at very young disk ages. Other proposed explanations include pressure bumps formed due to magnetised turbulence (Johansen et al. 2009; Dittrich et al. 2013; Flock et al. 2015) or condensation fronts (Cuzzi and Zahnle 2004; Ros and Johansen 2013; Zhang et al. 2015; Okuzumi et al. 2015). Whether such sub-structure is a common, yet currently unresolved feature of disks remains to be seen.

7 Summary

Small dust particles, starting out with sizes of a micrometer or less, are not only the material out of which planets form, they are also a key ingredient to protoplanetary disks: they are the main source of opacity, thus determining the temperature structure, the evolution of the disk, and ultimately the observational appearance. They also provide the surface area for chemical reactions that produce complex organic molecules, the pre-biotic building blocks of life. Recent years have seen a true revolution in the observational capabilities and with facilities such as ALMA, Sphere, or GPI just getting started, the pace is not going to slow down. Our ability to interpret these observations of disks rests on a good understanding of the interaction of solids and gas and therefore also on the evolution of particle sizes in disks. In this chapter, we reviewed some of the current ideas about how particles grow, how they are transported, and how they eventually make up the building blocks of planets. The key findings are summarized as follows.

Growth and transport processes are driven by interaction between the gas and the dust through drag forces. The drag forces depend on the particle size, hence, they induce collisions between the particles. The collisions can result in a wide variety of outcomes that depend on the collision speed and particle properties. The coarsest classification of collision outcomes would be growth (mass gain of the larger particle), fragmentation/erosion (mass loss of at least one particle, resulting in small fragments), or bouncing collisions (growth neutral, but possibly compaction). The change in size or mass of the particle in turn alters its aerodynamic behavior, hence the size evolution and the global transport of particles cannot be treated separately.

Small grains are well coupled to the gas motion, so they mostly follow the motion of the gas along (e.g., accretion, viscous spreading, turbulent mixing). But as their surface to mass ratio decreases with particle growth, particles start to decouple from the gas flow resulting in substantial velocity differences between dust and gas. This drift motion of the dust relative to the gas tends to move dust towards higher pressure regions, where the dust-gas relative velocity is reduced or vanishes. In a typical model of a protoplanetary disk, this means that particles sediment towards the mid-plane and spiral inwards. The time scale for this orbital decay is quite short, of the order of a few hundred orbits. This motion tends to deplete the surface layers and the outer regions of disks, but regions of high pressure, particularly pressure maximums are able to collect or "trap" the dust particles and cause accumulations of dust.

But as particles grow not only do their drift speeds increase, but also the collision they experience happen at higher velocities. This tends to lead to bouncing, fragmentation, or erosion instead of continued growth. Even if particles continued to grow, the increasing drift motion removes them faster than growth is able to resupply them. One way or another, particles are expected to reach only a finite maximum size, typically less than a meter, unless some mechanism is able to circumvent or avoid these issues. Possible pathways to continued growth include the following:

- Stochastic effects ("lucky particles") together with the effect that large particles gain mass by being bombarded by many small grains allow growth of a small number of large bodies. But the time scales for this growth are still longer than the drift motion.
- Icy particles are expected to be much harder to compress. Their increased surface to mass ratio combined with aerodynamic effects in the inner regions of disks can lead to very short growth time scales. However erosion or sintering might limit this mechanism.
- Other solutions to this problem may be the aforementioned particle traps, for example vortices that can efficiently accumulate large amounts of dust. Alternatively local two-fluid instabilities, as discussed in Sect. 5.2.3 may lead to gravitationally bound clumps of particles.

The latter accumulation mechanisms, however, rely on particle sizes being much larger than the micrometer sized dust inherited from the interstellar medium. Hence, the most likely pathway towards the building blocks of planets seems to be a hybrid approach where particle growth to macroscopic sizes is able to provide particle sizes that can efficiently be accumulated by aerodynamic effects to build up gravitationally bound clumps.

Observations over the last few years have started to provide the crucial input for theoretical models of disk evolution and planetesimal formation. Until recently, the radial drift motion of dust grains was merely a theoretical expectation. But now, observed structures in transition disks, extreme dust over-densities, and different sizes of dust and gas disks seem impossible to explain without it. Radially resolved multi-wavelength observations indicate larger grains being more centrally confined than smaller grains. The seemingly upper limit of the grain size distribution and its radial variations are in reasonable agreement with theoretical models.

Still crucial questions remain unanswered. There is no consistent picture of disk evolution or how the observed morphology of disks can be categorized into one or several evolutionary sequences. The role of various disk evolution mechanisms such as photoevaporation, disk winds, and even the driver of disk accretion remains unknown, let alone their impact on the solids in the disk. Several recent high-resolution observations in both near-infrared and in (sub-)millimeter interferometry have revealed striking morphologies in disks that seem to pose more questions than they answer. Our future understanding of planet formation and disk evolution will crucially depend on larger samples at high resolution and multiple wavelengths that are able to reveal correlations, frequencies, and the full diversity in disk morphology. At the same time, theoretical models will need to evolve to include all relevant physical effects, not just a subset. Furthermore, they need to provide clear predictions that can be tested observationally. With a number of facilities providing a wealth of observations, at resolutions and sensitivities orders of magnitude better than before, the time seems ripe to answer some of the most fundamental questions of disk evolution and the early stages of planet formation.

Acknowledgements We like to thank Liubin Pan, Christian Lenz, and the referees, Willy Benz and Alessandro Morbidelli for providing helpful comments that have significantly improved the manuscript. T.B. acknowledges support from the NASA Origins of Solar Systems grant NNX12AJ04G and from the DFG through grant KL 1469/13-1. M.F. acknowledges support by the NSFC through grant 11203081. A.J. is grateful for support from a Starting Grant from the European Research Council (278675-PEBBLE2PLANET) and the Knut and Alice Wallenberg Foundation and the Swedish Research Council (2014-5775).

References

F.C. Adams, C.J. Lada, F.H. Shu, Astrophys. J. **312**, 788 (1987)
L.E. Allen, N. Calvet, P. D'Alessio et al., Astrophys. J. Suppl. Ser. **154**, 363 (2004)
S.M. Andrews, Publ. Astron. Soc. Pac. **127**, 961 (2015)
S.M. Andrews, J.P. Williams, Astrophys. J. **631**, 1134 (2005)
S.M. Andrews, J.P. Williams, Astrophys. J. **659**, 705 (2007a)
S.M. Andrews, J.P. Williams, Astrophys. J. **671**, 1800 (2007b)
S.M. Andrews, D.J. Wilner, A.M. Hughes, C. Qi, C.P. Dullemond, Astrophys. J. **700**, 1502 (2009)
S.M. Andrews, D.J. Wilner, A.M. Hughes, C. Qi, C.P. Dullemond, Astrophys. J. **723**, 1241 (2010)
S.M. Andrews, D.J. Wilner, C. Espaillat et al., Astrophys. J. **732**, 42 (2011)
S.M. Andrews, D.J. Wilner, A.M. Hughes et al., Astrophys. J. **744**, 162 (2012)
S.M. Andrews, K.A. Rosenfeld, A.L. Kraus, D.J. Wilner, Astrophys. J. **771**, 129 (2013)

D. Apai, I. Pascucci, J. Bouwman et al., Science **310**, 834 (2005)
P.J. Armitage, Annu. Rev. Astron. Astrophys. **49**, 195 (2011)
S. Ataiee, P. Pinilla, A. Zsom et al., Astron. Astrophys. **553**, L3 (2013)
H. Avenhaus, S.P. Quanz, H.M. Schmid et al., Astrophys. J. **781**, 87 (2014)
X.-N. Bai, J.M. Stone, Astrophys. J. **722**, 1437 (2010)
S.A. Balbus, J.F. Hawley, Astrophys. J. **376**, 214 (1991)
Z. Balog, J. Muzerolle, G.H. Rieke et al., Astrophys. J. **660**, 1532 (2007)
P. Barge, J. Sommeria, Astron. Astrophys. **295**, L1 (1995)
S.V.W. Beckwith, A.I. Sargent, Astrophys. J. **381**, 250 (1991)
S.V.W. Beckwith, A.I. Sargent, R.S. Chini, R. Guesten, Astron. J. **99**, 924 (1990)
C.P.M. Bell, T. Naylor, N.J. Mayne, R.D. Jeffries, S.P. Littlefair, Mon. Not. R. Astron. Soc. **434**, 806 (2013)
M. Benisty, A. Juhász, A. Boccaletti et al., Astron. Astrophys. **578**, L6 (2015)
W. Benz, E. Asphaug, Icarus **142**, 5 (1999)
T. Birnstiel, S.M. Andrews, Astrophys. J. **780**, 153 (2014)
T. Birnstiel, C.P. Dullemond, F. Brauer, Astron. Astrophys. **503**, L5 (2009)
T. Birnstiel, C.P. Dullemond, F. Brauer, Astron. Astrophys. **513**, A79 (2010)
T. Birnstiel, C.W. Ormel, C.P. Dullemond, Astron. Astrophys. **525**, A11 (2011)
T. Birnstiel, S.M. Andrews, B. Ercolano, Astron. Astrophys. **544**, A79 (2012a)
T. Birnstiel, H. Klahr, B. Ercolano, Astron. Astrophys. **539**, A148 (2012b)
T. Birnstiel, C.P. Dullemond, P. Pinilla, Astron. Astrophys. **550**, L8 (2013)
J. Blum, G. Wurm, Annu. Rev. Astron. Astrophys. **46**, 21 (2008)
D. Bockelée-Morvan, D. Gautier, F. Hersant, J.-M. Huré, F. Robert, Astron. Astrophys. **384**, 1107 (2002)
F. Brauer, C.P. Dullemond, A. Johansen et al., Astron. Astrophys. **469**, 1169 (2007)
F. Brauer, C.P. Dullemond, T. Henning, Astron. Astrophys. **480**, 859 (2008)
ALMA Partnership, C.L. Brogan, L.M. Pérez et al., Astrophys. J. Lett. **808**, L3 (2015)
J.M. Brown, G.A. Blake, C. Qi et al., Astrophys. J. **704**, 496 (2009)
D. Carrera, A. Johansen, M.B. Davies, Astron. Astrophys. **579**, A43 (2015)
S. Casassus, Nature **493**, 191 (2013)
E. Chiang, P. Goldreich, Astrophys. J. **490**, 368 (1997)
E. Chiang, G. Laughlin, Mon. Not. R. Astron. Soc. **431**, 3444 (2013)
L.A. Cieza, M.R. Schreiber, G.A. Romero et al., Astrophys. J. **712**, 925 (2010)
L.A. Cieza, M.R. Schreiber, G.A. Romero et al., Astrophys. J. **750**, 157 (2012)
T. Currie, A. Sicilia-Aguilar, Astrophys. J. **732**, 24 (2011)
J.N. Cuzzi, K.J. Zahnle, Astrophys. J. **614**, 490 (2004)
J.N. Cuzzi, R.C. Hogan, K. Shariff, Astrophys. J. **687**, 1432 (2008)
I. de Gregorio-Monsalvo, F. Ménard, W. Dent et al., Astron. Astrophys. **557**, 133 (2013)
W.R.F. Dent, W.F. Thi, I. Kamp et al., Publ. Astron. Soc. Pac. **125**, 477 (2013)
K. Dittrich, H. Klahr, A. Johansen, Astrophys. J. **763**, 117 (2013)
J.S. Dohnanyi, J. Geophys. Res. **74**, 2531 (1969)
C. Dominik, C.P. Dullemond, Astron. Astrophys. **531**, A101 (2011)
R. Dong, R. Rafikov, Z. Zhu et al., Astrophys. J. **750**, 161 (2012)
B.T. Draine, Astrophys. J. **636**, 1114 (2006)
J. Drążkowska, C.P. Dullemond, Astron. Astrophys. **572**, A78 (2014)
J. Drążkowska, F. Windmark, C.P. Dullemond, Astron. Astrophys. **567**, 38 (2014)
B. Dubrulle, G.E. Morfill, M.F. Sterzik, Icarus **114**, 237 (1995)
C. Dullemond, ASCL, -1, 02015 (2012)
C.P. Dullemond, C. Dominik, Astron. Astrophys. **417**, 159 (2004)
C.P. Dullemond, C. Dominik, Astron. Astrophys. **434**, 971 (2005)
A. Dutrey, S. Guilloteau, G. Duvert et al., Astron. Astrophys. **309**, 493 (1996)
C. Espaillat, L. Ingleby, J. Hernández et al., Astrophys. J. **747**, 103 (2012)
C. Espaillat, J. Muzerolle, J. Najita et al., *Protostars & Planets VI* (2014), p. 497
N.J. Evans, M.M. Dunham, J.K. Jørgensen et al., Astrophys. J. Suppl. Ser. **181**, 321 (2009)
M. Fang, R. van Boekel, W. Wang et al., Astron. Astrophys. **504**, 461 (2009)
M. Fang, R. van Boekel, R.R. King et al., Astron. Astrophys. **539**, A119 (2012)
M. Fang, J.S. Kim, R. van Boekel et al., Astrophys. J. Suppl. Ser. **207**, 5 (2013a)
M. Fang, R. van Boekel, J. Bouwman et al., Astron. Astrophys. **549**, A15 (2013b)
M. Flock, J.P. Ruge, N. Dzyurkevich et al., Astron. Astrophys. **574**, A68 (2015)
K.B. Follette, M. Tamura, J. Hashimoto et al., Astrophys. J. **767**, 10 (2013)
S. Fromang, R.P. Nelson, Astron. Astrophys. **496**, 597 (2009)
S. Fromang, J. Papaloizou, Astron. Astrophys. **452**, 751 (2006)
W. Fu, H. Li, S. Lubow, S. Li, E. Liang, Astrophys. J. Lett. **795**, L39 (2014)

E. Furlan, L. Hartmann, N. Calvet et al., Astrophys. J. Suppl. Ser. **165**, 568 (2006)
P. Garaud, Astrophys. J. **671**, 2091 (2007)
P. Garaud, F. Meru, M. Galvagni, C. Olczak, Astrophys. J. **764**, 146 (2013)
M.G. Guarcello, G. Micela, F. Damiani et al., Astron. Astrophys. **496**, 453 (2009)
S. Guilloteau, A. Dutrey, V. Piétu, Y. Boehler, Astron. Astrophys. **529**, 105 (2011)
B. Gundlach, J. Blum, Astrophys. J. **798**, 34 (2015)
B. Gundlach, S. Kilias, E. Beitz, J. Blum, Icarus **214**, 717 (2011)
R.A. Gutermuth, S.T. Megeath, P.C. Myers et al., Astrophys. J. Suppl. Ser. **184**, 18 (2009)
C. Güttler, J. Blum, A. Zsom, C.W. Ormel, C.P. Dullemond, Astron. Astrophys. **513**, 56 (2010)
K.E. Haisch, E.A. Lada, C.J. Lada, Astrophys. J. **553**, L153 (2001)
L. Hartmann, N. Calvet, E. Gullbring, P. D'Alessio, Astrophys. J. **495**, 385 (1998)
P.M. Harvey, T. Henning, Y. Liu et al., Astrophys. J. **755**, 67 (2012)
J. Hashimoto, R. Dong, T. Kudo et al., Astrophys. J. Lett. **758**, L19 (2012)
C. Hayashi, Prog. Theor. Phys. Suppl. **70**, 35 (1981)
L.-O. Heim, J. Blum, M. Preuss, H.-J. Butt, Phys. Rev. Lett. **83**, 3328 (1999)
J. Hernández, L. Hartmann, T. Megeath et al., Astrophys. J. **662**, 1067 (2007)
L.A. Hillenbrand, A. Bauermeister, R.J. White, in *14th Cambridge Workshop on Cool Stars*, vol. 384 (2008), p. 200
H.-F. Hsieh, P.-G. Gu, Astrophys. J. **760**, 119 (2012)
A.M. Hughes, D.J. Wilner, C. Qi, M.R. Hogerheijde, Astrophys. J. **678**, 1119 (2008)
A. Isella, J.M. Carpenter, A.I. Sargent, Astrophys. J. **701**, 260 (2009)
A. Isella, J.M. Carpenter, A.I. Sargent, Astrophys. J. **714**, 1746 (2010)
A. Isella, L.M. Pérez, J.M. Carpenter, Astrophys. J. **747**, 136 (2012)
A. Johansen, H. Klahr, Astrophys. J. **634**, 1353 (2005)
A. Johansen, A.C. Andersen, A. Brandenburg, Astron. Astrophys. **417**, 361 (2004)
A. Johansen, J.S. Oishi, M.-M. Mac Low et al., Nature **448**, 1022 (2007)
A. Johansen, A. Youdin, H. Klahr, Astrophys. J. **697**, 1269 (2009)
A. Johansen, H. Klahr, T. Henning, Astron. Astrophys. **529**, 62 (2011)
A. Johansen, A.N. Youdin, Y. Lithwick, Astron. Astrophys. **537**, A125 (2012)
A. Johansen, J. Blum, H. Tanaka et al., *Protostars & Planets VI* (2014), p. 547
A. Johansen, E. Jacquet, J.N. Cuzzi, A. Morbidelli, M. Gounelle (2015a). arXiv:1505.2941
A. Johansen, M.-M. Mac Low, P. Lacerda, M. Bizzarro, Sci. Adv. **1**, 1500109 (2015b)
A. Kataoka, H. Tanaka, S. Okuzumi, K. Wada, Astron. Astrophys. **557**, L4 (2013)
S.J. Kenyon, L. Hartmann, Astrophys. J. **323**, 714 (1987)
K.H. Kim, D.M. Watson, P. Manoj et al., Astrophys. J. **769**, 149 (2013)
H.H. Klahr, P. Bodenheimer, Astrophys. J. **582**, 869 (2003)
H. Klahr, P. Bodenheimer, Astrophys. J. **639**, 432 (2006)
H.H. Klahr, T. Henning, Icarus **128**, 213 (1997)
S. Kothe, C. Güttler, J. Blum, Astrophys. J. **725**, 1242 (2010)
A.L. Kraus, M.J. Ireland, L.A. Hillenbrand, F. Martinache, Astrophys. J. **745**, 19 (2012)
O. Krauss, G. Wurm, Astrophys. J. **630**, 1088 (2005)
K.A. Kretke, D.N.C. Lin, Astrophys. J. **664**, L55 (2007)
K.A. Kretke, D.N.C. Lin, P. Garaud, N.J. Turner, Astrophys. J. **690**, 407 (2009)
S. Krijt, C.W. Ormel, C. Dominik, A.G.G.M. Tielens, Astron. Astrophys. **574**, A83 (2015)
S. Krijt, C.W. Ormel, C. Dominik, A.G.G.M. Tielens, Astron. Astrophys. **586**, A20 (2016)
G. Laibe, J.-F. Gonzalez, L. Fouchet, S.T. Maddison, Astron. Astrophys. **487**, 265 (2008)
M. Lambrechts, A. Johansen, Astron. Astrophys. **544**, A32 (2012)
M. Lambrechts, A. Johansen, Astron. Astrophys. **572**, A107 (2014)
G. Lesur, J.C.B. Papaloizou, Astron. Astrophys. **513**, 60 (2010)
K.L. Luhman, Astrophys. J. Suppl. Ser. **173**, 104 (2007)
K.L. Luhman, P.R. Allen, C. Espaillat, L. Hartmann, N. Calvet, Astrophys. J. Suppl. Ser. **186**, 111 (2010)
R. Lüst, Z. Naturforsch. Teil A, Phys. Phys. Chem. Kosmophys. **7**, 87 (1952)
D. Lynden-Bell, J.E. Pringle, Mon. Not. R. Astron. Soc. **168**, 603 (1974)
W. Lyra, M.-K. Lin, Astrophys. J. **775**, 17 (2013)
W. Lyra, A. Johansen, H. Klahr, N. Piskunov, Astron. Astrophys. **479**, 883 (2008)
W. Lyra, A. Johansen, H. Klahr, N. Piskunov, Astron. Astrophys. **493**, 1125 (2009)
K.A. Marsh, M.J. Mahoney, Astrophys. J. **405**, L71 (1993)
J.S. Mathis, W. Rumpl, K.H. Nordsieck, Astrophys. J. **217**, 425 (1977)
S. Mayama, J. Hashimoto, T. Muto et al., Astrophys. J. Lett. **760**, L26 (2012)
G. Meeus, B. Montesinos, I. Mendigutía et al., Astron. Astrophys. **544**, A78 (2012)
S.T. Megeath, L.E. Allen, R.A. Gutermuth et al., Astrophys. J. Suppl. Ser. **154**, 367 (2004)

S.T. Megeath, R. Gutermuth, J. Muzerolle et al., Astron. J. **144**, 192 (2012)
H. Meheut, Z. Meliani, P. Varniere, W. Benz, Astron. Astrophys. **545**, 134 (2012)
E.E. Mendoza V, Astrophys. J. **143**, 1010 (1966)
E.E. Mendoza V, Astrophys. J. **151**, 977 (1968)
J. Menu, R. van Boekel, T. Henning et al., Astron. Astrophys. **564**, 93 (2014)
T. Muto, C.A. Grady, J. Hashimoto et al., Astrophys. J. Lett. **748**, L22 (2012)
J. Muzerolle, L.E. Allen, S.T. Megeath, J. Hernández, R.A. Gutermuth, Astrophys. J. **708**, 1107 (2010)
P.C. Myers, G.A. Fuller, R.D. Mathieu et al., Astrophys. J. **319**, 340 (1987)
Y. Nakagawa, M. Sekiya, C. Hayashi, Icarus **67**, 375 (1986)
A. Natta, L. Testi, R. Neri, D.S. Shepherd, D.J. Wilner, Astron. Astrophys. **416**, 179 (2004)
R.P. Nelson, O. Gressel, O.M. Umurhan, Mon. Not. R. Astron. Soc. **435**, 2610 (2013)
D. Nesvorný, A.N. Youdin, D.C. Richardson, Astron. J. **140**, 785 (2010)
K.S. Noll, W.M. Grundy, D.C. Stephens, H.F. Levison, S.D. Kern, Icarus **194**, 758 (2008)
C.R. O'dell, Z. Wen, X. Hu, Astrophys. J. **410**, 696 (1993)
S. Okuzumi, H. Tanaka, M. Sakagami, Astrophys. J. **707**, 1247 (2009)
S. Okuzumi, H. Tanaka, H. Kobayashi, K. Wada, Astrophys. J. **752**, 106 (2012)
S. Okuzumi, M. Momose, S.-I. Sirono, H. Kobayashi, H. Tanaka (2015). arXiv:1510.03556
C.W. Ormel, J.N. Cuzzi, Astron. Astrophys. **466**, 413 (2007)
C.W. Ormel, H.H. Klahr, Astron. Astrophys. **520**, A43 (2010)
C.W. Ormel, M. Spaans, Astrophys. J. **684**, 1291 (2008)
C.W. Ormel, M. Spaans, A.G.G.M. Tielens, Astron. Astrophys. **461**, 215 (2007)
C.W. Ormel, J.N. Cuzzi, A.G.G.M. Tielens, Astrophys. J. **679**, 1588 (2008)
C.W. Ormel, D. Paszun, C. Dominik, A.G.G.M. Tielens, Astron. Astrophys. **502**, 845 (2009)
J.E. Owen, C.J. Clarke, Mon. Not. R. Astron. Soc. **426**, L96 (2012)
J.E. Owen, B. Ercolano, C.J. Clarke, Mon. Not. R. Astron. Soc. **412**, 13 (2011)
S.-J. Paardekooper, G. Mellema, Astron. Astrophys. **425**, L9 (2004)
L. Pan, P. Padoan, J. Fluid Mech. **661**, 73 (2010)
L. Pan, P. Padoan, J. Scalo, Astrophys. J. **791**, 48 (2014)
O. Panić, M.R. Hogerheijde, D. Wilner, C. Qi, Astron. Astrophys. **501**, 269 (2009)
G.B. Paraskov, G. Wurm, O. Krauss, Icarus **191**, 779 (2007)
Y. Pavlyuchenkov, C.P. Dullemond, Astron. Astrophys. **471**, 833 (2007)
L.M. Pérez, J.M. Carpenter, C.J. Chandler et al., Astrophys. J. Lett. **760**, L17 (2012)
L.M. Pérez, A. Isella, J.M. Carpenter, C.J. Chandler, Astrophys. J. Lett. **783**, L13 (2014)
S. Pfalzner, M. Steinhausen, K. Menten, Astrophys. J. **793**, L34 (2014)
V. Piétu, S. Guilloteau, A. Dutrey, Astron. Astrophys. **443**, 945 (2005)
P. Pinilla, M. Benisty, T. Birnstiel, Astron. Astrophys. **545**, A81 (2012)
S.P. Quanz, A. Amara, M.R. Meyer et al., Astrophys. J. Lett. **766**, L1 (2013a)
S.P. Quanz, H. Avenhaus, E. Buenzli et al., Astrophys. J. Lett. **766**, L2 (2013b)
N. Raettig, W. Lyra, H. Klahr, Astrophys. J. **765**, 115 (2013)
N. Raettig, H. Klahr, W. Lyra (2015, in press). Astrophys. J. arXiv:1501.05364v1
V.A. Rapson, J.H. Kastner, S.M. Andrews et al., Astrophys. J. Lett. **803**, L10 (2015)
Z. Regály, A. Juhász, Z. Sándor, C.P. Dullemond, Mon. Not. R. Astron. Soc. **419**, 1701 (2012)
L. Ricci, L. Testi, A. Natta et al., Astron. Astrophys. **512**, 15 (2010b)
L. Ricci, L. Testi, A. Natta, K.J. Brooks, Astron. Astrophys. **521**, 66 (2010a)
W.K.M. Rice, P.J. Armitage, K. Wood, G. Lodato, Mon. Not. R. Astron. Soc. **373**, 1619 (2006)
T.P. Robitaille, B.A. Whitney, R. Indebetouw, K. Wood, P. Denzmore, Astrophys. J. Suppl. Ser. **167**, 256 (2006)
T.P. Robitaille, B.A. Whitney, R. Indebetouw, K. Wood, Astrophys. J. Suppl. Ser. **169**, 328 (2007)
J. Rodmann, T. Henning, C.J. Chandler, L.G. Mundy, D.J. Wilner, Astron. Astrophys. **446**, 211 (2006)
T. Roell, R. Neuhäuser, A. Seifahrt, M. Mugrauer, Astron. Astrophys. **542**, A92 (2012)
K. Ros, A. Johansen, Astron. Astrophys. **552**, 137 (2013)
K.A. Rosenfeld, S.M. Andrews, D.J. Wilner, J.H. Kastner, M.K. McClure, Astrophys. J. **775**, 136 (2013)
S.M. Rucinski, Astron. J. **90**, 2321 (1985)
V.S. Safronov, Evolution of the protoplanetary cloud and formation of the earth and planets (1969). English translation (1972)
A.I. Sargent, S. Beckwith, Astrophys. J. **323**, 294 (1987)
R. Schräpler, T. Henning, Astrophys. J. **614**, 960 (2004)
A. Seizinger, W. Kley, Astron. Astrophys. **551**, A65 (2013)
N.I. Shakura, R.A. Sunyaev, Astron. Astrophys. **24**, 337 (1973)
A. Sicilia-Aguilar, L. Hartmann, N. Calvet et al., Astrophys. J. **638**, 897 (2006)
A. Sicilia-Aguilar, T. Henning, C.P. Dullemond et al., Astrophys. J. **742**, 39 (2011)

A. Sicilia-Aguilar, V. Roccatagliata, K. Getman et al., Astron. Astrophys. **573**, A19 (2015)
H. Sierks, C. Barbieri, P.L. Lamy et al., Science **347**, 440 (2015)
J.B. Simon, P.J. Armitage, Astrophys. J. **784**, 15 (2014)
J.B. Simon, K. Beckwith, P.J. Armitage, Mon. Not. R. Astron. Soc. **422**, 2685 (2012)
S.-I. Sirono, Astrophys. J. Lett. **733**, L41 (2011a)
S.-I. Sirono, Astrophys. J. **735**, 131 (2011b)
M.F. Skrutskie, D. Dutkevitch, S.E. Strom et al., Astron. J. **99**, 1187 (1990)
L. Spezzi, G. de Marchi, N. Panagia, A. Sicilia-Aguilar, Mon. Not. R. Astron. Soc. **421**, 78 (2012)
H. Störzer, D. Hollenbach, Astrophys. J. **515**, 669 (1999)
K.M. Strom, S.E. Strom, S. Edwards, S. Cabrit, M.F. Skrutskie, Astron. J. **97**, 1451 (1989)
T. Suyama, K. Wada, H. Tanaka, Astrophys. J. **684**, 1310 (2008)
T. Takeuchi, D.N.C. Lin, Astrophys. J. **581**, 1344 (2002)
D. Tamayo, A.H.M.J. Triaud, K. Menou, H. Rein, Astrophys. J. **805**, 100 (2015)
J. Teiser, G. Wurm, Mon. Not. R. Astron. Soc. **393**, 1584 (2009)
L. Testi, A. Natta, D.S. Shepherd, D.J. Wilner, Astron. Astrophys. **403**, 323 (2003)
L. Testi, T. Birnstiel, L. Ricci et al., *Protostars & Planets VI* (2014), p. 339
F. Trotta, L. Testi, A. Natta, A. Isella, L. Ricci, Astron. Astrophys. **558**, A64 (2013)
T. Tsukagoshi, M. Momose, J. Hashimoto et al., Astrophys. J. **783**, 90 (2014)
N.J. Turner, S. Fromang, C. Gammie et al., *Protostars & Planets VI* (2014), p. 411
R. van Boekel, M. Min, L.B.F.M. Waters et al., Astron. Astrophys. **437**, 189 (2005)
N. van der Marel, E.F. van Dishoeck, S. Bruderer et al., Science **340**, 1199 (2013)
H. Völk, F.C. Jones, G.E. Morfill, S. Röser, Astron. Astrophys. **85**, 316 (1980)
K. Wada, H. Tanaka, T. Suyama, H. Kimura, T. Yamamoto, Astrophys. J. **677**, 1296 (2008)
K. Wada, H. Tanaka, T. Suyama, H. Kimura, T. Yamamoto, Astrophys. J. **737**, 36 (2011)
C. Waelkens, L.B.F.M. Waters, M.S. de Graauw et al., Astron. Astrophys. **315**, L245 (1996)
K. Wahlberg Jansson, A. Johansen, Astron. Astrophys. **570**, A47 (2014)
S.J. Weidenschilling, Mon. Not. R. Astron. Soc. **180**, 57 (1977a)
S.J. Weidenschilling, Astrophys. Space Sci. **51**, 153 (1977b)
S.J. Weidenschilling, Icarus **127**, 290 (1997)
D.A. Weintraub, B. Zuckerman, C.R. Masson, Astrophys. J. **320**, 336 (1987)
F.L. Whipple, *From Plasma to Planet* (1972), p. 211
B.A. Whitney, K. Wood, J.E. Bjorkman, M. Cohen, Astrophys. J. **598**, 1079 (2003)
B.A. Whitney, T.P. Robitaille, J.E. Bjorkman et al., Astrophys. J. Suppl. Ser. **207**, 30 (2013)
J.P. Williams, L.A. Cieza, Annu. Rev. Astron. Astrophys. **49**, 67 (2011)
F. Windmark, T. Birnstiel, C. Güttler et al., Astron. Astrophys. **540**, A73 (2012a)
F. Windmark, T. Birnstiel, C.W. Ormel, C.P. Dullemond, Astron. Astrophys. **544**, L16 (2012b)
G. Wurm, J. Blum, Icarus **132**, 125 (1998)
G. Wurm, G. Paraskov, O. Krauss, Icarus **178**, 253 (2005)
C.-C. Yang, A. Johansen, Astrophys. J. **792**, 86 (2014)
C. Yasui, N. Kobayashi, A.T. Tokunaga, M. Saito, C. Tokoku, Astrophys. J. **723**, L113 (2010)
A.N. Youdin, J. Goodman, Astrophys. J. **620**, 459 (2005)
A.N. Youdin, Y. Lithwick, Icarus **192**, 588 (2007)
K. Zhang, G.A. Blake, E.A. Bergin, Astrophys. J. Lett. **806**, L7 (2015)
Z. Zhu, R.P. Nelson, L. Hartmann, C. Espaillat, N. Calvet, Astrophys. J. **729**, 47 (2011)
Z. Zhu, R.P. Nelson, R. Dong, C. Espaillat, L. Hartmann, Astrophys. J. **755**, 6 (2012)
A. Zsom, C.P. Dullemond, Astron. Astrophys. **489**, 931 (2008)
A. Zsom, C.W. Ormel, C. Güttler, J. Blum, C.P. Dullemond, Astron. Astrophys. **513**, 57 (2010)
A. Zsom, C.W. Ormel, C.P. Dullemond, T. Henning, Astron. Astrophys. **534**, 73 (2011)

Space Sci Rev (2016) 205:117–164
DOI 10.1007/s11214-016-0258-z

Formation, Orbital and Internal Evolutions of Young Planetary Systems

Clément Baruteau[1] · Xuening Bai[2] · Christoph Mordasini[3] · Paul Mollière[4]

Received: 22 September 2015 / Accepted: 27 April 2016 / Published online: 12 May 2016

Abstract The growing body of observational data on extrasolar planets and protoplanetary disks has stimulated intense research on planet formation and evolution in the past few years. The extremely diverse, sometimes unexpected physical and orbital characteristics of exoplanets lead to frequent updates on the mainstream scenarios for planet formation and evolution, but also to the exploration of alternative avenues. The aim of this review is to bring together classical pictures and new ideas on the formation, orbital and internal evolutions of planets, highlighting the key role of the protoplanetary disk in the various parts of the theory. We begin by briefly reviewing the conventional mechanism of core accretion by the growth of planetesimals, and discuss a relatively recent model of core growth through the accretion of pebbles. We review the basic physics of planet-disk interactions, recent progress in this area, and discuss their role in observed planetary systems. We address the most important effects of planets internal evolution, like cooling and contraction, the mass-luminosity relation, and the bulk composition expressed in the mass-radius and mass-mean density relations.

Keywords Planets and satellites: formation · Planets and satellites: interiors · Protoplanetary disks · Planet-disk interactions

1 Introduction

Planet formation and evolution is a fast-moving field, stimulated by the rapid increase in the number of exoplanets and their great diversity. Despite the wealth of observational

✉ C. Baruteau
clement.baruteau@irap.omp.eu

1 Institut de Recherche en Astrophysique et Planétologie, CNRS/Université de Toulouse, Toulouse, France

2 Harvard-Smithsonian Center for Astrophysics, Cambridge, USA

3 Physikalisches Institut, Universität Bern, Bern, Switzerland

4 Max-Planck-Institut für Astronomie, Heidelberg, Germany

data on planetary systems, including our own, it is difficult to have a general theory for planet formation and evolution as it involves a broad range of physical processes that happen at widely different length and time scales. The protoplanetary disk where planets form embodies many of the difficulties and uncertainties related to planet formation and evolution. Protoplanetary disks are made of dust and poorly-ionized gas governed by non-ideal magneto-hydrodynamics. The disk's structural and turbulent properties are thus influenced by complex radiative and chemical processes, which set the thermodynamical behavior and ionization state of the gas. They have critical implications in every part of planet formation and evolution: (i) they condition the location in protoplanetary disks where dust grains grow to planetesimals and to planets, (ii) they play a prominent role in the early orbital evolution of low-mass planetary systems by changing the direction and speed of planetary migration, (iii) they impact the early internal evolution of planets by determining the amount of heat retained from the formation process. The formation and evolution of planetary systems is obviously affected by gravity and by the local stellar environment. This broad range of physical processes echoes the diversity of extrasolar planetary systems, and suggests that there is likely no unique theory for planet formation and evolution, but several.

Dramatic progress in understanding planet formation and evolution in the past few years has motivated a number of recent reviews (e.g., Raymond et al. 2014; Helled et al. 2014; Baruteau et al. 2014; Chabrier et al. 2014; Baraffe et al. 2014). The aim of this chapter is to provide an update. It brings together the formation, the orbital and internal evolutions of young planetary systems as all three are intimately linked. For instance, the radius and luminosity of a young planet depends on its formation scenario, and so does its orbital migration. The energy released by a forming planet as it cools and contracts can change the properties of the protoplanetary disk in the planet's vicinity and thus affect its orbital migration as much as its growth. This chapter thus puts an accent on, but is not restricted to, the early stages of planetary formation and evolution, before the dispersal of the protoplanetary disk. It begins with a review on the growth of planetary cores by planetesimal and pebble accretions in Sect. 2. It continues with the orbital evolution of planets driven by planet-disk interactions, which is presented in Sect. 3. The internal evolution of planets follows in Sect. 4. Summary points are provided in Sect. 5.

2 Growth of Planetesimals to Protoplanets

The journey of planet formation begins with the growth of dust grains in protoplanetary disks and the formation of planetesimals, as reviewed in the previous chapter by Birnstiel et al. Due to major uncertainties in our understanding of planetesimal formation, most studies of the later phase of planet formation assume a population of planetesimals are readily in place to start with, and follow the growth of planetesimals into planetary mass objects, or protoplanets. If the protoplanets reach sufficiently high mass within the lifetime of their parent protoplanetary disk, they can further accrete gas and become the cores of giant planets (i.e., the "core accretion" paradigm for giant planet formation). For this reason, the terms "protoplanet" and "core" are sometimes used interchangeably in the literature.

2.1 Planetesimal Accretion

We begin by briefly reviewing the conventional mechanism of planetesimal growth, where planetesimals grow by accreting other planetesimals. The theory has been developed and refined over decades. Given the limited space of this chapter and the maturity of this subject, we simply introduce the fundamental concepts of planetesimal accretion. A pedagogical introduction can be found in the book by Armitage (2010), and a more in-depth review was given by Goldreich et al. (2004). Applications to terrestrial and extrasolar planetary systems are summarized in the recent PPVI chapters by Raymond et al. (2014) and Helled et al. (2014).

2.1.1 Gravitational Focusing and Runaway Growth

Consider a swarm of planetesimals, with surface density $\Sigma_{\rm p}$, characteristic radial size R (corresponding mass M), and velocity dispersion $v_{\rm p}$. They are located around disk radius a, where the disk's angular frequency is $\Omega_{\rm K}$ (equal to the Keplerian angular frequency). These planetesimals are thus distributed over a vertical thickness of $H \approx v_{\rm p}/\Omega_{\rm K}$, with corresponding midplane number density $n_{\rm p} \approx \Sigma_{\rm p}/MH \approx \Sigma_{\rm p}\Omega_{\rm K}/Mv_{\rm p}$.

Assuming every collision among the planetesimals leads to coagulation, the rate of planetesimal growth is given by, within order unity,

$$\frac{dM}{dt} \approx n_{\rm p} M \sigma v_{\rm rel} \approx \Sigma_{\rm p}\Omega_{\rm K}\sigma, \tag{1}$$

where $v_{\rm rel} \approx \sqrt{2} v_{\rm p}$ is the typical relative velocity of planetesimal encounters, and the corresponding collisional cross section σ is

$$\sigma = \pi R^2 \left(1 + \frac{v_{\rm esc}^2}{v_{\rm rel}^2}\right), \tag{2}$$

where $v_{\rm esc}^2 = 2GM/R$ is the escape velocity. The second term is called *gravitational focusing*.

When the planetesimals' mass is small, or when the population of planetesimals are "hot" (with large velocity dispersions $v_{\rm p}$), we have $v_{\rm esc} \ll v_{\rm rel}$, and hence the collisional cross section is simply geometric. Correspondingly, $dM/dt \propto R^2 \propto M^{2/3}$, or $d(\ln M)/dt \propto M^{-1/3} \propto R^{-1}$. This is called *ordered growth*. If we rewrite the mass growth rate to growth rate in planetesimal radius, it becomes $dR/dt \approx$ const: planetesimals of all sizes grow at the same rate.

As planetesimals become more massive, or when their relative velocity is sufficiently low so that $v_{\rm esc} \gtrsim v_{\rm rel}$, gravitational focusing significantly enhances the cross section. At fixed $v_{\rm rel}$, gravitational focusing is more effective for more massive planetesimals. This leads to the so-called *runaway growth*, where we have $dM/dt \propto R^2 \times (M/R) \propto M^{4/3}$, or $d\ln M/dt \propto M^{1/3} \propto R$. In this regime, the growth of larger bodies dramatically runaways over smaller bodies, and as a result, few massive bodies stand out over the rest of the planetesimals population, as firstly found from numerical studies (Greenberg et al. 1978; Wetherill and Stewart 1989; Kokubo and Ida 1996).

A key requirement for runaway growth is that the velocity dispersion $v_{\rm p}$ for the bulk of the (small) planetesimal population is kept small. This is a result of the balance among various dynamical heating and cooling processes. In general, larger bodies resulting from runaway growth are cooled by dynamical friction with small bodies, which contain most of the mass. The small body populations are heated during this process, but they cool by mutual collisions and gas drag.

2.1.2 Oligarchic Growth, Isolation Mass and Timescales

At late stages of the runaway growth, planetary embryos become sufficiently large and start to interact with each other, and the overall dynamics is dominated by these few bodies, called the *oligarchs*. They tend to heat up the *neighboring* planetesimals, increasing their velocity dispersion approximately as $v_p \propto M^{1/3}$. More precisely, N-body simulations suggest that the root mean square planetesimal eccentricity e and inclination i scale with embryo mass as (e.g., Ida and Makino 1993; Kokubo and Ida 2000)

$$\langle e^2 \rangle^{1/2} \sim 6R_H/a, \qquad \langle i^2 \rangle^{1/2} \sim 3.5R_H/a, \tag{3}$$

where

$$R_H = \left(\frac{GM}{3\Omega_K^2}\right)^{1/3} = \left(\frac{M}{3M_*}\right)^{1/3} a \tag{4}$$

is the embryo's Hill radius, and M_* is the stellar mass. This enhancement of velocity dispersion reduces the efficiency of gravitational focusing, and hence slows down the growth of planetary embryos, a regime called *oligarchic growth* (Kokubo and Ida 1998, 2000), with growth rate scaled as $dM/dt \propto R^2 \times (M/R)/M^{2/3} \propto M^{2/3}$, or $d\ln M/dt \propto M^{-1/3} \propto R^{-1}$. Therefore, the growth mode of the large embryos reduces to orderly. In other words, neighboring oligarchs growth at similar rates and maintain similar masses. On the other hand, the collisional cross section continues to be dominated by gravitational focusing, thus large embryos runaway over small planetesimals. Therefore, the overall outcome of runaway growth followed by oligarchic growth is a bi-modal distribution of an embryo-planetesimal system.

During oligarchic growth, each planetary embryo establishes its own domain of dominance, and maintains certain separation with neighboring embryos. Empirically, it was found that the separation Δa is about $10R_H$ (Kokubo and Ida 1995, 1998). If it were closer, planet embryos would experience close encounters, and such strong scattering would increase their eccentricity and expand their separation. Circularization is then achieved due to dynamical friction from the planetesimals. Therefore, the maximum mass a planet embryo can achieve by planetesimal accretion is limited by the amount of planetesimals available in its *feeding zone*. This mass is called the *isolation mass*, given by $M_{iso} \approx 2\pi a \Delta a \Sigma_p$. Assuming $\Delta a = 10R_H$ and $M_* = M_\odot$, this mass is found to be

$$M_{iso} \approx 50\left(\frac{\pi a^2 \Sigma_p}{M_*}\right)^{3/2} M_* \approx 0.16\left(\frac{\Sigma_p}{10\,\mathrm{g\,cm^{-2}}}\right)^{3/2}\left(\frac{a}{1\,\mathrm{AU}}\right)^3 M_\oplus. \tag{5}$$

The isolation mass mainly depends on radius and surface density of planetesimals as $(\Sigma_p a^2)^{3/2}$. Although the scaling with a is not well known, we generally expect Σ_p to decrease with radius slower than a^{-2} (e.g., $\Sigma_p \propto a^{-1.5}$ in the Minimum-Mass Solar Nebula model – hereafter MMSN), and hence M_{iso} increases with a.

We can roughly estimate the growth timescale during the oligarchic growth phase. Using Eqs. (3), we get the typical velocity dispersion $v_p \sim 5R_H\Omega_K$, and we substitute it into Eqs. (1) and (2) to obtain

$$\begin{aligned}\tau_{grow} &\approx \frac{M}{dM/dt} \approx 8\Omega_K^{-1}\left(\frac{M_*}{\Sigma_p a^2}\right)\left(\frac{R_H^2}{aR}\right) \\ &\approx 1.2\left(\frac{M}{0.1M_\oplus}\right)^{1/3}\left(\frac{\Sigma_p}{10\,\mathrm{g\,cm^{-2}}}\right)^{-1}\left(\frac{a}{1\,\mathrm{AU}}\right)^{1/2}\ \mathrm{Myrs}.\end{aligned} \tag{6}$$

The results above suggest that Mercury and Mars may be the leftover embryos from oligarchic growth, and they form within a few million years, most likely within the lifetime of the Solar nebular. Faster growth is possible when considering the accretion of small planetesimals where gas drag efficiently damps their eccentricities (Rafikov 2004; Kenyon and Bromley 2009), the overall result is a reduction of the pre-factor in Eq. (6) but not the scaling.

Upon accreting most planetesimals from their feeding zones to achieve the isolation mass, there is insufficient damping of random velocities of the planetary embryos from dynamical friction (Kenyon and Bromley 2006). This marks the end of oligarchic growth, and the system becomes chaotic, with eccentricity growth, orbits crossing, etc. Over secular timescales, some embryos may collide, while some may be ejected from the system. This *chaotic* growth characterizes the final stage of planet formation, and sets the mass, composition, and orbital architecture of the planetary system.

The planetesimal accretion scenario suffers from a major difficulty when applied to the formation of giant planet cores towards larger separation. In the case of Jupiter and Saturn at $a \sim 5$–10 AU, reduction of Σ_p and increase of the dynamical time at larger radii make the growth timescale substantially longer than in the inner disk, as can be seen from Eq. (6). Correspondingly, the timescale to build up a sufficiently massive core (~ 10–$15M_\oplus$, Pollack et al. 1996) to trigger runaway accretion of a gaseous envelope would well exceed the typical lifetime of a protoplanetary disk (~ 3 Myrs, Haisch et al. 2001). This issue can be alleviated when considering accretion of smaller planetesimals (Rafikov 2004), enhancement of Σ_p by a factor of a few beyond the snow line (e.g., Kennedy and Kenyon 2008), and probably resolved when type-I migration of the embryos is considered (Alibert et al. 2005). In the latter scenario, orbital migration via planet-disk interaction allows the embryos to sweep up planetesimals over the course of their migration, and avoids severe depletion of the feeding zone. Incorporating inward planet migration into planet population synthesis models (e.g., Ida and Lin 2008; Mordasini et al. 2009a), it is found that, while forming giant planets at the location of Jupiter and Saturn is plausible, the overall outcome depends sensitively on the strength of planetary migration, which remains to be better understood (see Sect. 3).

Recently, a number of giant planets have been discovered via direct imaging surveys to reside at large orbital separations. Notable examples include β-Pic *b* (Lagrange et al. 2010) and the HR 8799 system (Marois et al. 2010), where orbital separation reaches up to ~ 70 AU. The conventional core accretion theory via planetesimal accretion simply fails in such cases due to the rapid decline of planetesimal density and rapid increase of dynamical timescale towards large orbital distances. One possible alternative to form giant planets at large orbital separation is by the gravitational instability (GI, Boss 1997), which likely operates in the outer disk during the early stages of disk evolution. We will discuss GI further in Sect. 4.2.

2.2 Pebble Accretion

Recently, a new model of planetesimal/core growth has been proposed, where it was found that accretion of mm- to cm-sized grains, or pebbles, can be a much more efficient mode of core growth (Lambrechts and Johansen 2012; Ormel and Klahr 2010; Johansen and Lacerda 2010; Perets and Murray-Clay 2011; Morbidelli and Nesvorny 2012). A large fraction of protoplanetary disks have been observed to contain a significant fraction of dust mass in mm-sized grains (Natta et al. 2007; Ricci et al. 2010), some also reveal the presence of cm sized pebbles (Testi et al. 2003; Wilner et al. 2005; Rodmann et al. 2006). These solids couple strongly with disk gas via aerodynamic drag, the relative velocity can be strongly damped during gravitational encounters, leading to substantially enhanced cross section. Because of its relative novelty, we devote more effort discussing this new picture of pebble accretion.

2.2.1 Pebble Aerodynamics

Due to partial pressure support, the gas in protoplanetary disks rotates at a velocity slightly slower than the Keplerian velocity. The velocity difference, denoted by Δv_K, is typically a very small fraction of the Keplerian velocity ($\sim 10^{-3}$–10^{-2} depending on location), and is commonly normalized by the local sound speed c_s. In the standard MMSN disk, it is given by $\Delta v_K \approx 0.1(R/10\ \mathrm{AU})^{1/4} c_s$ at disk midplane.

The pressureless dust grains, on the other hand, tend to follow Keplerian orbits. They thus experience a headwind from the gas and drift radially inward. The aerodynamic friction between gas and pebbles is characterized by the stopping time t_s, which is commonly normalized to the orbital time to give $\tau_s \equiv \Omega_K t_s$. This dimensionless stopping time is also called the Stokes number. Strong and weak aerodynamic coupling correspond to $\tau_s \ll 1$ and $\tau_s \gg 1$, respectively. For an MMSN disk, a spherical grain with size s has a stopping time in the disk midplane (in the Epstein regime) equal to

$$\tau_s = 0.14 \left(\frac{s}{1\ \mathrm{cm}} \right) \left(\frac{R}{10\ \mathrm{AU}} \right)^{3/2}. \tag{7}$$

Thus, pebble-sized grains in the outer disk are close to be marginally coupled with the gas. Assuming pebbles are passive in a laminar disk, particles experience a radial drift, with sub-Keplerian rotation, whose velocity components are given by

$$v_r = -\frac{2\tau_s}{1+\tau_s^2} \Delta v_K, \qquad v_\phi - v_K = -\frac{1}{1+\tau_s^2} \Delta v_K, \tag{8}$$

where v_K is the Keplerian velocity. Radial drift is most efficient for marginally coupled particles with $\tau_s \sim 1$. In any case, the difference between the particles velocity and the local Keplerian velocity is still on the order of Δv_K.

In the vertical direction, particles experience vertical gravity $g_z = -\Omega_K^2 z$, leading to settling towards the disk midplane. This is balanced by turbulent diffusion. Denoting by $D_{p,z}$ the particles diffusion coefficient in the vertical direction, particles tend to adopt a Gaussian density profile of the form $\rho_p = \rho_{p0} e^{-z^2/2H_p^2}$ that is characterized by a particle scale height

$$H_p \approx \sqrt{\frac{D_{p,z}}{\Omega_K \tau_s}}. \tag{9}$$

The turbulent diffusion coefficient of the gas, D_g, can be generally written as $D_g \sim \delta v_g^2 \times \tau_{corr}$, where δv_g is the turbulent velocity of the gas, and τ_{corr} is the correlation time of the gas turbulence, which is expected to be on the order of Ω_K^{-1}. The particles' diffusion coefficient D_p is generally comparable to D_g for marginally to strongly coupled particles $\tau_s \lesssim 1$ (Youdin and Lithwick 2007). The level of turbulence in protoplanetary disks is uncertain, but is expected to be weak (e.g., Bai 2015). For $\delta v_g \sim 10^{-2} c_s$, we have $D_g \approx D_p \sim 10^{-4} c_s H_g$, with H_g the gas pressure scale height ($H_g = c_s/\Omega_K$). The particles scale height is therefore $H_p \sim 0.01 H_g$ for $\tau_s = 1$, and $H_p \sim 0.1 H_g$ for $\tau_s = 0.01$.

2.2.2 Two Regimes of Pebble Accretion

Accurate determination of pebble accretion rates requires detailed analysis of particle orbits as they approach the planetesimals/embryos (e.g., Ormel and Klahr 2010), and the orbits can become fairly complex especially for loosely coupled large solids (τ_s well above 1). For more tightly coupled particles ($\tau_s \lesssim 1$), which are relevant in the pebble accretion scenario, simple order-of-magnitude analysis proves to be sufficient, as we describe here.

We consider particle trajectories as they approach the planetesimal/core in the background gas. The gas is assumed to be unperturbed by the core.[1] Therefore, if particles are very strongly coupled to the gas, they are largely entrained by the gas without being accreted. More generally, particles can decouple from the gas on timescales of their stopping time $\sim t_s$. They can be accreted if strong gravitational deflection can be achieved within t_s, which depends on the impact parameter r. This is the basic physics of pebble accretion. Key in calculating the pebble accretion rate is to estimate the maximum impact parameter (r_a) or the accretion radius, within which pebbles can be accreted.

Before calculating the rate of pebble accretion, it is useful to distinguish two regimes of pebble accretion, determined by the mass M_c of the planetesimal/core. To begin with, we can conveniently define a Bondi radius as[2]

$$R_{\rm B} = \frac{GM_c}{\Delta v_{\rm K}^2}. \tag{10}$$

This definition of $R_{\rm B}$ marks the length scale at which pebbles can be significantly deflected by the planetesimal/core by two-body interactions. Additionally, this deflection can not be effective beyond the Hill radius $R_{\rm H}$ (Eq. (4)), where three-body effects become important and the core loses gravitational dominance.

By equating $R_{\rm B}$ and $R_{\rm H}$, the *transition mass* can be defined as

$$M_t = \frac{\Delta v_{\rm K}^3}{G\Omega_{\rm K}} \approx 0.16 M_\oplus \left(\frac{R}{30\ \rm AU}\right)^{3/4} \left(\frac{\Delta v_{\rm K}}{0.1 c_s}\right)^3, \tag{11}$$

where in the second equality we use the MMSN scaling for the sound speed. Note that $R_{\rm B} \propto M_c$ while $R_{\rm H} \propto M_c^{1/3}$, thus $(R_{\rm B}/R_{\rm H}) \approx (M_c/M_t)^{2/3}$. Therefore, the distance from the planetesimal/core within which its gravitational influence is effective is given by $R_{\rm B}$ for low-mass cores, and by $R_{\rm H}$ for high-mass cores. Correspondingly, the transition mass defined above separates the two regimes of pebble accretion, which are termed *drift regime* for $M_c < M_t$, and *Hill regime* for $M_c > M_t$. Estimates of the accretion radius r_a differ between both regimes, as we address below.

2.2.3 Drift Regime

In the drift regime where $M_c < M_t$ (or, equivalently, the Bondi radius $R_{\rm B} < R_{\rm H}$), the accretion radius $r_a \le R_{\rm B}$. Within this radius, particles approach the core with relative velocity $\Delta v \sim \Delta v_{\rm K}$. The timescale for a particle to pass by the core is

$$t_B \approx \frac{R_{\rm B}}{\Delta v_{\rm K}} = \frac{1}{\Omega_{\rm K}} \frac{M_c}{M_t}. \tag{12}$$

[1] As long as M_c is much smaller than the "thermal mass" $M_{\rm th} = c_s^3/G\Omega_{\rm K}$, which is about the mass scale of Jupiter and is well beyond the mass scale for core growth.

[2] Note this definition given in Lambrechts and Johansen (2012) is different from the conventional definition of the Bondi radius, where $\Delta v_{\rm K}$ is replaced by the sound speed.

Particles with stopping time $t_s \approx t_B$ decouple from the gas during the encounter, and are expected to spiral into the core once their impact parameter is within R_B. Therefore, the accretion radius $r_a \approx R_B$ for $t_s \approx t_B$.

For more strongly coupled particles with $t_s < t_B$, they can only be captured when their impact parameter is smaller than R_B, so that they experience stronger gravity with shorter gravitational deflection time (and hence they can decouple from the gas and spiral in). Otherwise, they will be entrained by the gas flow without being accreted. The gravitational deflection time for given impact parameter r is

$$t_g(r) \approx \frac{\Delta v_K}{GM_c/r^2} = \left(\frac{r}{R_B}\right)^2 t_B. \tag{13}$$

By equating $t_g(r)$ with t_s, the accretion radius r_a can be obtained as

$$r_a \approx \left(\frac{t_s}{t_B}\right)^{1/2} R_B \approx \tau_s^{1/2} \left(\frac{M_c}{M_t}\right)^{1/6} R_H. \tag{14}$$

This radius can equivalently be understood as the radius at which the gravitational acceleration is equal to the acceleration due to gas drag (also termed as the wind-shearing radius, Perets and Murray-Clay 2011).

For more weakly coupled particles with $t_s > t_B$, they behave more similarly to the gas-free scenario, where particles follow hyperbolic orbits as they encounter the core. Instead of the Bondi radius, it is the physical size of the core (much smaller) that determines whether accretion takes place. This results in a rapid decline of r_a as t_s increases beyond t_B, and it becomes negligible compared with the $t_s \sim t_B$ case.

We note that in the drift regime, $\Omega_K t_B \approx M_c/M_t$. Therefore, efficient accretion of dust grains in the drift regime always takes place for particles with $\tau_s < 1$. The smaller the core mass, the smaller the optimal particle size.

2.2.4 Hill Regime

In the Hill regime where $M_c > M_t$ (or, equivalently, the Bondi radius $R_B > R_H$), we expect the accretion radius $r_a \leq R_H$. Within this radius, the relative velocity with which particles approach the core can be dominated by the Keplerian shear, $\Delta v_{sh}(r) \equiv (3/2)\Omega_K r$. To see this, note that

$$\frac{\Delta v_{sh}(r)}{\Delta v_K} = \frac{3}{2}\left(\frac{M_c}{M_t}\right)\left(\frac{r}{R_B}\right). \tag{15}$$

Therefore, the shear velocity dominates for impact parameters $r \gtrsim R_B$. For a given impact parameter r, the relative velocity is $\Delta v(r) \approx \max[\Delta v_K, \Delta v_{sh}(r)]$.

Following the same spirit as before, we write the gravitational deflection time for particles as

$$t_g(r) = \frac{\Delta v}{GM_c/r^2} \approx \frac{\Omega_K r}{GM_c/r^2} \approx \frac{1}{\Omega_K}\left(\frac{r}{R_H}\right)^3, \tag{16}$$

where we have taken $\Delta v \approx \Omega_K r$. Efficient pebble accretion requires $t_g \lesssim t_s$, which leads to an accretion radius of

$$r_a \approx (\Omega_K t_s)^{1/3} R_H = \tau_s^{1/3} R_H. \tag{17}$$

This result again holds[3] for particles with $\tau_s \lesssim 1$. Particles with lager stopping times would undergo 3-body scattering whose orbits can become fairly chaotic. While gas drag still enhances particle accretion, accretion is less efficient with increasing τ_s, and more over, we do not expect a large population of particles with stopping times far exceeding 1 to be present in protoplanetary disks.

The above scaling needs to be revised when $r_a < R_{\rm B}$, at which point we should have taken $\Delta v = \Delta v_{\rm K}$. The revised r_a then becomes the expression in Eq. (14). Therefore, the overall result is

$$r_a \approx \min\left[\tau_s^{1/2}\left(\frac{M_c}{M_t}\right)^{1/6}, \tau_s^{1/3}\right] R_{\rm H}. \tag{18}$$

The fact that the accretion radius approaches the Hill radius in the Hill regime makes pebble accretion an extremely efficient mechanism for planetesimal/core growth. Additionally, the weak dependence of the accretion radius on τ_s in Eq. (18) indicates efficient accretion takes place for a wide range of particle sizes (say $\tau_s = 0.01\text{–}1$).

2.2.5 Timescale for Core Growth

The rate of pebble accretion is determined by the accretion radius r_a, and the radial flux of pebbles approaching the core. To calculate it, we need to distinguish the 2D and 3D cases, depending on whether the particles' scale height $H_{\rm p}$ is larger or smaller than r_a. In the following, we denote by $\Sigma_{\rm p}$ the surface density of the pebbles.

When $H_{\rm p} < r_a$, we are in the 2D regime where particles settle strongly to the midplane so that the entire particle column approaching the core can be accreted, with accretion rate

$$\dot{M}_{2D} \approx 2\Sigma_{\rm p} r_a \Delta v. \tag{19}$$

When $H_{\rm p} \gtrsim r_a$, we are in the 3D regime where particles are suspended, and only those close to the midplane with $|z| \lesssim r_a$ can be accreted. The accretion rate is given by

$$\dot{M}_{3D} \approx \frac{\Sigma_{\rm p}}{H_{\rm p}} \pi r_a^2 \Delta v. \tag{20}$$

In both cases, we have $\Delta v \approx \max[\Delta v_{\rm K}, \Delta v_{\rm sh}(r_a)]$.

Now we can discuss the planetesimal/core growth rate due to pebble accretion. While multiple combinations of 2D/3D and drift/Hill regimes are possible, here we simply discuss two most representative cases.

In the early stage of planet formation when the planetesimal/core is small, pebble accretion is in the drift regime with $R_{\rm B} < R_{\rm H}$, and likely proceeds in 3D. In that case, we get

$$\dot{M}_c \approx \pi \frac{\Sigma_{\rm p}}{H_{\rm p}} \left(\frac{t_s}{t_B} R_{\rm B}^2\right) \Delta v_{\rm K}. \tag{21}$$

In the optimistic case where the contribution from $t_s/t_B \sim 1$ particles is considered (t_s increases with M_c), growth can achieve super-runaway with $\dot{M}_c \sim M_c^2$, although the base rate (i.e., proportional coefficient) can be very low.

[3] By integrating individual particle trajectories, r_a is found to be about $R_{\rm H}$ for particles with $\tau_s \sim 0.1$ (Lambrechts and Johansen 2012), thus $r_a \approx (\tau_s/0.1)^{1/3} R_{\rm H}$ is probably more accurate for $\tau_s \lesssim 0.1$, which still agrees with the order-of-magnitude derivation within a factor of order unity.

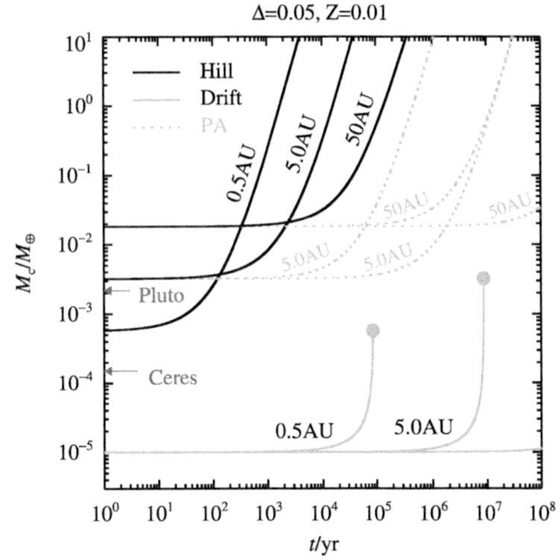

Fig. 1 Core growth as a function of time for various orbital distances for the pebble accretion scenario in the Hill and drift regimes, as well as the conventional planetesimal accretion (denoted by "PA"), taken from Lambrechts and Johansen (2012). This calculation assumes parameter $\Delta \equiv \Delta v_K/c_s = 0.05$, and dust-to-gas mass ratio $Z = 0.01$ in a minimum-mass solar nebular disk

Towards later stages of planet formation, when the core has grown substantially across the transition mass, accretion proceeds in the Hill regime and can operate in 2D if turbulence is not too strong. The corresponding accretion rate is

$$\dot{M}_c \approx 2\Sigma_p \tau_s^{2/3} R_H^2 \Omega_K. \tag{22}$$

In terms of scaling, we obtain $\dot{M}_c \propto M_c^{2/3}$, which is the same rate as with oligarchic growth. However, the proportional coefficient is much larger than in the planetesimal accretion scenario, because of the greatly enhanced cross section of the order R_H^2.

To evaluate the core growth timescale via pebble accretion (in the Hill regime), we first rewrite Eq. (22) into the more intuitive form

$$\frac{1}{M_c}\frac{dM_c}{dt} \approx \frac{\Sigma_p r^2}{M_*}\left(\frac{M_*}{M_c}\right)^{1/3} \Omega_K, \tag{23}$$

where we consider the most optimistic case that all particles entering the Hill sphere get accreted. This roughly corresponds to the case where most dust particles have stopping time $\tau_s \sim 0.1$–1 (and 2D accretion), which is probably roughly satisfied in the outer regions of protoplanetary disks. We note that the growth rate is independent of Δv_K (but the transition mass does). Assuming a Solar mass star, the growth time is

$$t_{\text{grow}} \approx 2.2 \times 10^3 \left(\frac{2\pi}{\Omega_K}\right)\left(\frac{0.005 M_\odot}{\Sigma_p r^2}\right)\left(\frac{M_c}{M_\oplus}\right)^{1/3}. \tag{24}$$

Assuming a typical disk mass of $\sim 0.01 M_\odot$, and that the planet is formed half-way, then for an earth-mass core the growth time is about 2000 orbits, which is only $\sim 3.6 \times 10^5$ years at 30 AU.

Figure 1 shows a simple model for the time evolution of the core's mass resulting from the conventional planetesimal accretion as well as from pebble accretion, taken from Lambrechts and Johansen (2012). The calculations correspond to the optimistic 2D accretion case with maximum accretion rate. The general conclusions are that pebble accretion in the Hill regime is extremely efficient, especially at large orbital distances. On the other hand, in the drift regime, pebble accretion is slow and core growth generally proceeds more efficiently via planetesimal accretion.

2.2.6 Global Models and Uncertainties

The general formulation of pebble accretion has recently been incorporated in global models of planet formation (Chambers 2014; Kretke and Levison 2014; Lambrechts and Johansen 2014; Bitsch et al. 2015a). In general, these models confirm that pebble accretion is indeed very efficient and can well account for the formation of the cores of not only the gas and ice giant planets in the Solar system, but also the extrasolar giant planets at large separations. The growth might be fast enough to avoid large-scale type I planetary migration (see Sect. 3.1.3). Additionally, a large fraction ($\sim 50\ \%$) of pebbles undergoing radial drift can be intercepted by the embryos/planetesimals, leading to very efficient conversion of solids into planets (Lambrechts and Johansen 2014). On the other hand, formation of gas giants is less favored in systems with relatively small disk masses and/or metallicities, consistent with the metallicity trend in the occurrence rate of giant exoplanets (Fischer and Valenti 2005; Bitsch et al. 2015a; Johnson et al. 2010).

Applying to the Solar system, criticisms arise concerning the fact that it may be too efficient: the system tends to form a large number of planetary embryos inside-out, which would lead to violent dynamical interactions and can hardly be made consistent with the planetary architecture of the inner Solar system (Kretke and Levison 2014). On the other hand, Morbidelli et al. (2015) pointed out that pebble accretion is likely very inefficient within the snow line due to reduced grain size, which readily explains the dichotomy of low-mass terrestrial planets in the inner Solar system and gas giants in the outer parts.

At the present stage, studies from different groups have adopted different methodologies in treating various physical processes during planetesimal/core growth, and inevitably involve simplified assumptions in one way or another. As also noticed in these works, major uncertainties exist in these modeling efforts, which include: (1) The size distribution and evolution of pebbles, which were either treated as the end product of grain growth process, or as fragments from planetesimal collision. (2) The initial mass function of planetesimals, the starting point of planetesimal growth. (3) Global disk structure and evolution, and level of turbulence, particularly relevant for the dynamics and transport of pebbles. None of these factors are well understood, and they are interrelated with each other.

The key to unravel these uncertainties probably lies in better understanding the internal structure and evolution of protoplanetary disks from first principle. Recent local 3D MHD simulations including realistic non-ideal MHD physics have suggested a paradigm shift from the conventional viscous evolution picture towards a magnetized disk wind driven evolution of the inner disk (Bai and Stone 2013; Gressel et al. 2015), and a weakly turbulent outer disk with substructures due to magnetic flux concentration (Bai and Stone 2014; Bai 2015). Extension of these local studies to global simulations is essential and will likely help establish a most realistic picture of protoplanetary disks. Such a picture will provide essential input for both grain growth calculations and necessary conditions for planetesimal formation (see chapter by Birnstiel et al.), and altogether, for modeling the entire processes of planet formation.

3 Orbital Evolution of Planets in Their Protoplanetary Disk

As planets form in their protoplanetary disk, the gravitational interaction between the disk and the planets change the semi-major axis, eccentricity and inclination of the planets. Planet-disk interactions usually damp eccentricities and inclinations efficiently, but as we will see throughout this section, they can either decrease or increase the planets semi-major axes, an effect usually termed planetary migration or disk migration. Planet-planet interactions tend to pump eccentricities and inclinations, and can significantly change semi-major axes during the disk's lifetime, for example through scattering events due to the disk migration of two or more planets. After the protoplanetary disk is cleared by photo-evaporation, by typically 1 to 10 Myrs, the orbital evolution of planets can continue through interaction with the central star (via tides or stellar evolution for close-in planets), with distant stars (e.g., via Kozai cycles), with a disk of remnant planetesimals (debris disk), or via planet-planet interactions.

As this chapter is about the early evolution of planetary systems, we will focus on the orbital evolution of planets driven by planet-disk interactions. The long-term evolution of planetary systems, after dispersal of the protoplanetary disk, will be presented in the chapter by Zhou et al. The aim of this section is to give a simple, practical overview of the physical processes that drive planet-disk interactions. The case of low-mass planets (typically Earth-mass planets) is described in Sect. 3.1, that of massive planets (typically Jovian planets) follows in Sect. 3.2. We then discuss in Sect. 3.3 what models of planet-disk interactions can and cannot explain about hot Jupiters (Sect. 3.3.1), and about the many super-Earths in multi-planet systems discovered by *Kepler* (Sect. 3.3.2). More in-depth reviews of planet-disk interactions can be found in Kley and Nelson (2012), Baruteau and Masset (2013) and Baruteau et al. (2014).

3.1 Disk Migration of Low-Mass Planets

Since planetary migration is driven by the gravitational interaction between a planet and its parent protoplanetary disk, the basics of planetary migration can be drawn through inspection of the gas density perturbation caused by a protoplanet. Figure 2 shows the relative perturbation of the gas surface density of a protoplanetary disk where a 5 Earth-mass planet forms. The figure is taken from a 2D hydrodynamical simulation of planet-disk interactions for a non-magnetized radiative disk model. As can be seen in the figure, the planet induces two kinds of density perturbations in the disk: (i) spiral density waves, called the planet's wakes, which propagate throughout the disk, and (ii) density perturbations very close to the planet's orbital radius, called co-orbital density perturbations, which are confined in the planet's horseshoe region (see right panel), where gas describes horseshoe trajectories relative to the planet.

The basic idea underlying planet-disk interactions is the law of action-reaction: the disk reacts to the planet's gravity by exerting a gravitational force on the planet which changes the planet's semi-major axis, eccentricity and inclination. Considering the simplest case of a planet on a circular and coplanar orbit (with zero eccentricity and inclination), the planet's angular momentum is $J_\mathrm{p} = M_\mathrm{p}\sqrt{GM_\star r_\mathrm{p}}$, where M_p is the mass of the planet, $M_\star$ that of the star, G is the gravitational constant, and r_p is the distance between the star and the planet. Denoting by Γ the torque exerted by the disk on the planet, and further assuming the planet mass is stationary, $\Gamma = dJ_\mathrm{p}/dt$ implies $dr_\mathrm{p}/dt = \Gamma \times 2r_\mathrm{p}/J_\mathrm{p}$. This relation simply means that, in order to know the direction and speed of planetary migration, the sign and magnitude of the disk torque need to be determined. This is why most numerical studies of planet-disk interactions take planets on fixed orbits and calculate the disk torque on the planet. However, as we will see later, the disk torque can actually be a function of the migration rate and migration history, with the consequence that, under some circumstances, the migration can runaway. We briefly outline below the different components of the disk torque.

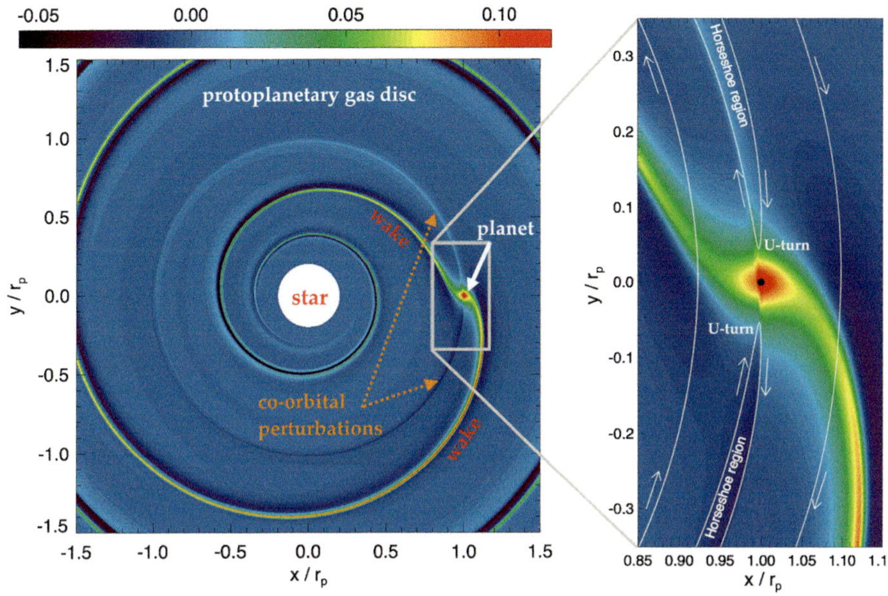

Fig. 2 Relative perturbation of the gas surface density of a protoplanetary disk perturbed by a 5 Earth-mass planet. The planet's location is spotted by the *white arrow* in the *left panel*. The planet generates a one-armed spiral density wave – the wake – that propagates throughout the disk, as well as co-orbital density perturbations within the planet's horseshoe region. Gas trajectories with respect to the planet are depicted by *white curves* and *arrows* in the *right panel*. Figure adapted from Baruteau et al. (2014)

3.1.1 Wake Torque

First, there is the torque due to the inner wake, which is the wake that propagates between the planet and the star. Most of the inner wake's torque comes from the excess of gas just in front of the planet in the azimuthal direction, and which is distant from the planet by a few pressure scale heights at most. This gas excess exerts a gravitational force on the planet whose azimuthal component F_φ is positive, which corresponds to a torque $\sim r_\mathrm{p} F_\varphi$ on the planet that is therefore positive. Similarly, the outer wake's torque is mostly due to the excess of gas immediately behind the planet in the azimuthal direction. That local gas excess now has $F_\varphi < 0$ and thus exerts a negative torque on the planet. The total wake torque then has two opposite contributions: the (positive) torque due to the inner wake that tends to move the planet outwards, and the (negative) torque due to the outer wake that tends to move the planet inwards. Inspection of the inset panel in the right-hand side of Fig. 2 shows that the gas density perturbation in the outer wake is slightly larger and closer to the planet, which indicates that the outer wake's torque is stronger. The total wake torque is therefore negative in that case and favors inward migration. While this is indeed the general expectation, the reader should be aware that this result depends on the radial gradients of the gas temperature and density near the planet's location, as they affect the location of the wakes relative to the planet as well as their density enhancement. These dependencies can be found by solving the linear perturbation equations numerically, either in 2D (Korycansky and Pollack 1993; Ward 1997; Paardekooper et al. 2010; Masset 2011) or in 3D (Tanaka et al. 2002). The reader is referred to Sect. 2.1.1 of Baruteau et al. (2014) for more details on the wake torque and its expression as a function of disk gradients.

3.1.2 Corotation Torque

The disk gas in the planet's horseshoe region follows horseshoe trajectories as seen from the planet. At its closest approach to the planet, the gas undergoes a gravitational kick from the planet. The gas just behind the planet in the azimuthal direction and inside the planet's orbital radius moves radially outward by taking angular momentum from the planet. This gas takes outward U-turns relative to the planet and exerts a negative torque on the planet. Meanwhile, gas just in front on the planet in the azimuthal direction and outside the planet's orbital radius moves radially inward by giving angular momentum to the planet. That gas embarks on inward U-turns and exerts a positive torque on the planet. Much like the wake torque, the torque exerted by the horseshoe region, which is called the corotation torque or horseshoe drag, has two opposite contributions. The sign and magnitude of the corotation torque then depend on the angular momentum difference between gas doing inward and outward U-turns.

Complexity arises as the angular momentum of the gas in the planet's horseshoe region evolves in time due to the advection-diffusion of two hydrodynamical quantities: (i) the gas specific vorticity, sometimes called vortensity, which in 2D is the vertical component of the velocity curl (vorticity) divided by the surface density, and (ii) the gas specific entropy, which in 2D is basically the quantity $P\Sigma^{-\gamma}$ with P the thermal pressure, Σ the surface density and γ the adiabatic index. Advection of vortensity and specific entropy along horseshoe streamlines implies that the sign and magnitude of the corotation torque depend on the density and temperature gradients across the horseshoe region. Diffusion of both quantities implies that the corotation torque also depends on the nature and efficiency of turbulent diffusion mechanisms taking place near or inside the horseshoe region, which for low-mass planets is a small fraction of the disk's pressure scale height. Modeling of turbulence by a viscosity and a thermal diffusivity shows that the corotation torque is very sensitive to both diffusion parameters. Hence the importance of better understanding how turbulent transport of angular momentum operates in regions of planet formation (e.g., Lesur et al. 2014; Bai 2015; Gressel et al. 2015). 2D and 3D models of viscous disks show that the corotation torque is usually positive and therefore favors outward migration (e.g., Paardekooper and Mellema 2006; Baruteau and Masset 2008; Paardekooper and Papaloizou 2008; Kley and Crida 2008; Kley et al. 2009; Masset and Casoli 2010; Paardekooper et al. 2010, 2011; Lega et al. 2014). The reader is referred to Sect. 2.1.2 of (Baruteau et al. 2014) for more details on the corotation torque. Its expression in 2D as a function of disk gradients, viscosity and thermal diffusivity can be found in Masset and Casoli (2010) or Paardekooper et al. (2011). A derivation of the corotation torque in 3D globally isothermal disk models has been recently presented by Masset and Benítez-Llambay (2016). Dynamical corotation torques, which arise as the planet drifts relative to the disk, will be described in Sect. 3.1.4.

3.1.3 Type I Migration

The migration of low-mass planets (typically up to Neptune's mass), known as type I migration, is driven by the sum of the wake torque (Sect. 3.1.1) and the corotation torque (Sect. 3.1.2). The former is usually negative, the latter usually positive. Once more, we have two competing effects, and the balance between them is very sensitive to the disk model. This is illustrated in Fig. 3, which displays the type I migration torque with varying planet mass and planet's orbital radius. In this disk model taken from Bitsch et al. (2013b), opacity transitions near the evaporation line of silicates (at $r \approx 0.7$ AU) and the water ice line (at $r \approx 5$ AU) boost the corotation torque beyond these transitions, where a broad range

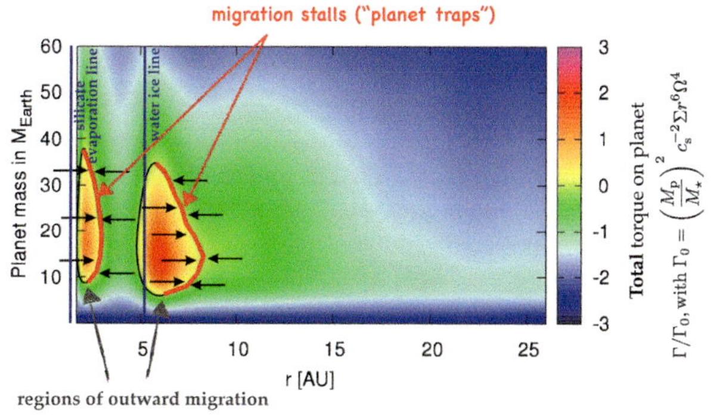

Fig. 3 Type I migration torque on planets of different masses (*y-axis*) and orbital radii (*x-axis*). *Horizontal arrows* show the direction of migration. Figure adapted from Bitsch et al. (2013b), where the disk model sets the water ice line at $r \approx 5$ AU, and the evaporation line of silicates at $r \approx 0.7$ AU

of planet masses undergoes outward migration. Note that type I migrating cores converge towards the outer edge of these regions of outward migration; these locations are known as "planet traps".

The torque value in Fig. 3 can be used to estimate the type I migration timescale. The factor Γ_0 by which the torque is normalized in the color bar sets indeed a typical timescale for type I migration, τ_0, given by $\tau_0 = \Omega_K M_p r_p^2 / 2\Gamma_0$, and which can be recast as

$$\tau_0 \approx 1.3\ \mathrm{Myr} \times \left(\frac{h}{0.05}\right)^2 \left(\frac{\Sigma}{200\ \mathrm{g\,cm^{-2}}}\right)^{-1} \left(\frac{M_\star}{M_\odot}\right)^{3/2} \left(\frac{r_p}{5\ \mathrm{AU}}\right)^{-1/2} \left(\frac{M_p}{M_\oplus}\right)^{-1}, \qquad (25)$$

with $h = c_s/v_K$ the disk's aspect ratio, and where all gas quantities are to be calculated at the planet's orbital radius. Above estimate shows that an Earth-mass planet located at 5 AU from a Solar-mass star, with local gas surface density of 200 g cm^{-2} and disk's aspect ratio of 5 %, and which experiences a total disk torque $\Gamma \approx -2\Gamma_0$ (see Fig. 3), would migrate inwards on a characteristic timescale of ~ 0.7 Myr. This should be compared with the typical growth timescale of the planet. In the Hill regime of pebble accretion, the growth timescale of such a planet could be as short as $\sim 2 \times 10^4$ yr (see the black bold curves in Fig. 1), which suggests that Earth-mass planets could grow fast enough to avoid significant inward migration and reach planetary traps (the red bold curves in Fig. 3). Planetesimal accretion, however, would result in a growth timescale at least comparable to the aforementioned migration timescale (see the dashed curves in Fig. 1), in which case significant inward migration would occur.

There are several points to bear in mind about Fig. 3. First is that this disk model assumes a turbulent disk viscosity associated with a constant alpha viscous parameter $\alpha \sim$ a few $\times\ 10^{-3}$. Different values of α, or perhaps more realistically a radial profile of α inspired from simulations of MHD turbulent disks, would result in different locations for regions of outward migration (Bitsch et al. 2014). Second is that this picture is time-dependent: as the disk evolves, it loses mass and cools with the consequence that regions of outward migration progressively shift towards smaller orbital radii and smaller planet masses (Bitsch et al. 2015b).

3.1.4 Some Recent Progress on Type I Migration

The predicted rapid inward migration of planets of few Earth masses has stimulated much work on type I migration. A selection of recent works on type I migration is detailed below.

(i) Disk's magnetic field. The recent work by Guilet et al. (2013) has shown that there is an additional corotation torque on low-mass planets when the disk has a weak toroidal magnetic field. This new torque comes about because of advection-diffusion of the magnetic flux inside the planet's horseshoe region. The horseshoe trajectories of the gas cause an accumulation of the magnetic flux contained in the horseshoe region downstream of the U-turns. The magnetic pressure increases locally, and the thermal pressure decreases to maintain a mechanical equilibrium with the surrounding gas. This, in turn, decreases the local gas surface density if the disk's temperature is stationary. An azimuthal asymmetry of the horseshoe U-turns relative to the planet, which is related to the radial pressure gradient, makes this new corotation torque generally positive and favors outward migration. Its magnitude depends sensitively on the strength of the magnetic field and the magnetic resistivity (Guilet et al. 2013). We point out that the surface density perturbation obtained in the 2D laminar disk models of Guilet et al. (2013), where the effects of turbulence are modeled by a viscosity and a magnetic resistivity, is in excellent agreement with the time-averaged density perturbation found in the 3D MHD turbulent simulations of Baruteau et al. (2011a), which were for Saturn-mass planets in high aspect ratio disks (see the comparison in Fig. 3 of Baruteau et al. 2014). More work is needed in this area, in particular for more deeply embedded planets. When the disk has a strong toroidal magnetic field, horseshoe streamlines disappear and the corotation torque is replaced by a so-called MHD torque that is driven by angular momentum carried away by slow MHD waves excited in the planet's vicinity. The recent work by Uribe et al. (2015) shows that this MHD torque can be positive and also reverse type I migration, in agreement with earlier analytical and numerical works (Terquem 2003; Fromang et al. 2005).

(ii) Heating torque. Material accreted by a planet releases energy that heats up the disk gas in the planet's vicinity (located typically by a few Hill radii from the planet; the Hill radius is defined at Eq. 4). The gas flow near the planet implies that this released energy forms two hot lobes: (i) one in front of the planet in the azimuthal direction and inside the planet's orbit, and (ii) a second one behind the planet in the azimuthal direction and outside the planet's orbit (Benítez-Llambay et al. 2015). The gas density decreases at the location of the lobes to maintain a mechanical equilibrium with the surrounding gas. The under-dense lobe in front of the planet favors inward migration (its gravitational force on the planet has a negative azimuthal component), whereas the lobe behind the planet favors outward migration. The flow asymmetry near the planet, which is related again to the radial pressure gradient, makes the lobe behind the planet hotter and more under-dense. The resulting torque on the planet, termed "heating torque" by Benítez-Llambay et al. (2015), therefore favors outward migration. It is not a corotation torque since it is not associated with horseshoe U-turns. The magnitude of the heating torque depends sensitively on the planet's growth timescale and the disk opacity. The heating torque increases with the accretion rate and with increasing disk opacity (since higher opacities inhibit cooling). The results of Benítez-Llambay et al. (2015) show that the heating torque can reverse type I migration of planets of 1 to 3 Earth masses that are located near 5 AU if their mass doubling time is shorter than a few $\times 10^4$ years. Such short growth timescales could be obtained through pebble accretion.

(iii) Eccentricity. Despite the relative efficiency of the disk gas at damping eccentricities, a planet can maintain some level of eccentricity if it is forced by another planet, in particular if planets are in or near mean-motion resonances. Fendyke and Nelson (2014) have shown that the radial width of the planet's horseshoe region decreases with increasing planet eccentricity, and therefore so does the magnitude of the corotation torque. This clearly has important implications for the strength of type I migration in multi-planetary systems that contain Earth- to Neptune-mass planets.

(iv) Dynamical corotation torque. We have examined so far the various contributions to the type I migration torque for planets on fixed orbits, which assumes that the migration rate is directly given by the disk torque, and that the latter is independent of the migration rate. However, as a planet drifts in the disk, gas outside the horseshoe region eventually executes a unique horseshoe U-turn relative to the planet and thus contributes to the corotation torque. The corotation torque imparted by orbit-crossing gas scales with the mass flow rate across the orbit and therefore depends on the migration rate. This torque is negative if the planet moves inward, positive if it moves outward. It therefore causes a positive feedback on migration. It adds to the corotation torque due to the gas trapped inside the planet's horseshoe region, which has two contributions. First is the horseshoe drag caused by the gas in the horseshoe region executing U-turns relative to the planet, as in the case when there is no radial drift between the disk and the planet. Second is the torque the horseshoe region has to apply on the planet so that both migrate at the same rate. This torque, which we call the trapping torque, is perhaps more easily understood as the opposite of the torque the planet exerts on its horseshoe region to have it migrate at its own drift rate. The trapping torque is positive (negative) if the planet moves inward (outward); it thus yields a negative feedback on migration.

Adding all contributions together, we see that the corotation torque has in general two components: (i) the *static corotation torque*, which is the horseshoe drag, and (ii) the *dynamical corotation torque*, which is the sum of the trapping torque and the orbit-crossing torque (Masset and Papaloizou 2003; Paardekooper 2014). As this terminology implies, only the dynamical corotation torque depends on the migration rate and causes a feedback on migration.

The key quantity to assess whether feedback is negative or positive is a quantity called "vorticity-weighted coorbital mass deficit" (Masset and Papaloizou 2003). This quantity is well known for massive planets that open a gap around their orbit (see Sect. 3.2). Since low-mass planets do not or hardly deplete their horseshoe region, above quantity should be thought of as a vorticity or vortensity deficit, which comes about because orbit-crossing gas has a vortensity different from the averaged vortensity of the horseshoe region (Paardekooper 2014). For type I migrating planets, the vortensity deficit can be either negative or positive depending on the radial gradient of disk vortensity in isothermal disk models (where the gas temperature is kept stationary). Extension to non-isothermal, radiative disk models (including viscous heating, stellar irradiation and radiative cooling) has been carried out by Pierens (2015), where the vortensity deficit is primarily a surface density deficit brought about by radial variation of the gas entropy in the disk; the vortensity deficit can be positive or negative depending mostly on the radial gradient of disk entropy. Conditions for type I migration to runaway are described in Paardekooper (2014) for isothermal disks, and in (Pierens 2015) for non-isothermal disks. For this review, it is sufficient to take away that, in massive low-viscosity disks, the dynamical corotation torque can dramatically slow down inward type I migration (negative feedback), and possibly lead to runaway outward type I migration (case where positive feedback causes a runaway). In any case, the dynamical corotation torque can largely widen the range of orbital radii where outward migration can occur (Paardekooper 2014; Pierens 2015).

3.2 Disk Migration of Massive Planets

We provide in this section a short overview of planet-disk interactions for massive planets that open a gap around their orbit, their migration properties, and touch upon some aspects of their observable signatures.

3.2.1 Gap Opening

The more massive a planet, the stronger its wakes, and the closer to the planet will wakes turn into shocks and deposit their energy and angular momentum. As the inner wake carries negative fluxes of energy and angular momentum, it will push gas inward, away from the planet's orbit. Similarly, the outer wake carries positive fluxes of energy and angular momentum and will push gas outward, away from the orbit. Massive planets thus progressively deplete their co-orbital horseshoe region, which becomes an annular gap. The final gap width is determined by the balance between the gap-opening gravity torque of the planet and the gap-closing viscous and pressure torques (Crida et al. 2006). In other words, the ability of a planet to carve a gap does not only depend on the planet's mass, or, actually, the planet-to-star mass ratio ($q = M_{\rm p}/M_{\star}$). It also depends on the disk's aspect ratio, h, as q/h^3, and on the disk's turbulent alpha viscosity, α, as $\alpha h^2/q$ (Lin and Papaloizou 1986; Crida et al. 2006).

We stress that there is no accurate definition or criterion as to when a planet opens a gap. The widely used gap-opening criterion formulated by (Crida et al. 2006) is for when the planet's co-orbital surface density drops to 10 % its initial value. Of course this threshold value is arbitrary, and planets less massive than predicted by Crida et al's criterion can still open a partial gap or dip. This can be the case for planets down to few Earth masses in weakly turbulent disks (e.g., Rafikov 2002; Muto et al. 2010; Dong et al. 2011; Duffell and MacFadyen 2013; Duffell 2015). Also, this criterion is for planets on fixed orbits. The gap-opening criterion is modified by migration, particularly in massive disks, as recently highlighted by Malik et al. (2015). An illustration of this is the rapid inward migration expected for planets formed by disk fragmentation (e.g., Baruteau et al. 2011b), which is initially so rapid that planets have no time to carve a gap, unless or until significant gas accretion occurs (Zhu et al. 2012; Nayakshin and Cha 2013; Stamatellos 2015). We also point out the recent work by (Fung et al. 2014) who estimated via hydrodynamical simulations the gas surface density contrast inside and outside a planet gap. Note that simulations of MHD turbulent disks show that the width and depth of planet gaps can be somewhat different from viscous disk models (Papaloizou et al. 2004; Zhu et al. 2013).

Gap-opening planets are often invoked to explain the large cavities observed in transition disks (see, e.g., the review by Espaillat et al. 2014). It should be borne in mind that the width of a planetary gap varies with dust size. Dust grains up to few tens of microns are tied to the gas, so the width of their gap is just as narrow as in the gas. Dust grains in the mm/cm range (depending on the local gas density) are generally not tied to the gas: they undergo significant radial drift and tend to concentrate in regions of pressure maxima (see Sect. 2.2.1). Simulations of planet-disk interactions show that the outer edge of a planet gap is generally a robust pressure maximum but the inner edge is not. This implies that mm/cm-grains initially beyond the planet's orbit can be trapped efficiently at the gap's outer edge, and those initially inside the planet's orbit will migrate inward (e.g., Zhu et al. 2012). A single massive planet could therefore simultaneously open a narrow gap in μm-grains and a large cavity in mm-grains. We point out that large cavities in μm-grains are not a natural expectation of planet-disk interactions, unless one invokes the presence of several massive planets in the disk (Zhu et al. 2011). Finally, we stress how uncertain it would be to apply predictions of gap widths and depths for the gas to observed gap structures in protoplanetary disks imaged in the mm/submm, like for example the (suggestive) gap structures seen in the HL Tau disk imaged by ALMA (ALMA Partnership et al. 2015).

Quite similarly, massive, yet unseen planets have been widely invoked to explain the spiral density waves observed in near-infrared images of protoplanetary disks. While spiral waves could be generated by other means, for example by gravitational instability, studies on the observability of the spiral wakes induced by a massive planet, coupling hydrodynamical simulations and radiative transfer calculations, are emerging (Juhász et al. 2015; Zhu et al. 2015). The work by Dong et al. (2015) provides convincing evidence that the two spirals seen in MWC 758 in polarized scattered light (Benisty et al. 2015) can be due indeed to an unseen outer planet companion of a few Jupiter masses.

3.2.2 *Formation of a Circumplanetary Disk*

Part of the gas that is progressively expelled from the planet's co-orbital region during the opening of a gap forms a circumplanetary disk. The recent discoveries of young, massive planets on wide orbits by direct imaging, like HD 100546 b (recently confirmed by Currie et al. 2014 and Quanz et al. 2015), has stimulated a number of recent works that have examined the structure, accretion rate and observability of circumplanetary disks. Accretion onto the protoplanetary disk is found to be inherently 3D and to proceed from high latitudes (e.g., Tanigawa et al. 2012; Szulágyi et al. 2014), can be stochastic and even launch outflows (Gressel et al. 2013). The observability of circumplanetary disks has been assessed by the prediction of distinct features in the gas kinematics (Perez et al. 2015) and spectral energy distribution (Zhu 2015; Eisner 2015; Montesinos et al. 2015).

3.2.3 *Type II Migration*

The migration of massive, gap-opening planets is driven by the wake torque and the corotation torque (both its static and dynamical components). The balance between both torques depends on how deep and wide the gap is, which, as we have seen above, depends on the planet's mass, the disk's aspect ratio and viscosity. For planets that open a deep gap around their orbit, which, roughly speaking, correspond to planets more massive than Jupiter, the corotation torque is largely suppressed and the wake torque is weakened. Still, the wake torque remains the main driver of migration. This corresponds to the so-called type II migration regime. It is directed inwards and runs on timescales longer than 10^{4-5} yrs. Recently, Dürmann and Kley (2015) have shown through numerical simulations that type II migration does not depend on the viscous inflow speed of the disk gas. This implies that type II migration is not tied to the disk's viscous evolution, as often thought, and that gas can cross the gap during the planet's migration. Type II migration is considerably slowed down when the planet mass becomes larger than the mass of the gas outside the planet gap.

3.2.4 *Type III Migration*

For planets that open a partial gap around their orbit, which is typically the case of Saturn-mass planets, the corotation torque still plays a major role in the migration. At low to moderate gas surface densities at the planet's location, that is when the local Toomre parameter $Q \equiv c_s \Omega_K / \pi G \Sigma$ typically exceeds 10, the migration regime is intermediate between the type I and type II regimes (e.g., Masset and Papaloizou 2003; Crida and Morbidelli 2007; D'Angelo and Lubow 2008). At larger gas surface densities at the planet's location ($Q \lesssim 10$), the dynamical corotation torque becomes important and leads to a different migration regime termed type III migration (Masset and Papaloizou 2003). The co-orbital vortensity deficit featured by the dynamical corotation torque (see Sect. 3.1.4) becomes a surface density deficit or mass deficit (it is the mass that should be added to the planet's horseshoe region so that it gets the averaged surface density of the orbit-crossing gas). When the mass deficit becomes larger than the mass of the planet plus its circumplanetary disk, migration runs away, otherwise the dynamical corotation torque accelerates migration but does not cause a runaway (Masset and Papaloizou 2003). The planet's circumplanetary disk can have a significant impact on migration (D'Angelo et al. 2005; Pepliński et al. 2008; Crida et al. 2009).

3.3 Disk Migration and the Observed Exoplanets

This section discusses the role of planet-disk interactions in some of the observed properties of planetary systems. Sect. 3.3.1 deals with spin-orbit (mis-)alignments among hot Jupiters and their constraints on planet migration scenarios. Sect. 3.3.2 discusses the relevance of planet-disk interactions for the near- and non-resonant architecture of many of the multi-planet systems discovered by *Kepler*. For a detailed statistical comparison between the orbital properties of exoplanets and those predicted by global models of planetary formation and evolution, the reader is referred to, e.g., Benz et al. (2014) and Mordasini et al. (2015).

3.3.1 Aligned and Misaligned Hot Jupiters: Constraints on Planet Migration Scenarios

Observations of sky-projected planet obliquities have gained considerable importance in the past few years (the obliquity of a planet is the angle between the star's spin axis and the planet's orbital axis). Projected obliquities have been primarily obtained via the Rossiter MacLaughlin effect (e.g., Winn et al. 2005), which is a time-varying distortion in the spectral lines of the star caused by a transiting planet. Other techniques, like the analysis of starspots occultations by a planet, have also been used (e.g., Sanchis-Ojeda et al. 2011). Projected obliquities have been determined for a little more than 60 planets at the time of writing, the vast of majority of them are hot Jupiters with orbital periods below 10 days. Surprisingly, a fair number of hot Jupiters have a large projected obliquity (see Fig. 4). When their projected obliquity exceeds 90°, planets are said to be retrograde. It is seen that the projected obliquities seem to depend on the star's effective temperature, with hot Jupiters around hot stars being more likely misaligned (see, e.g., the review by Winn and Fabrycky 2015). These observations have considerable impact on evolutionary models of planetary systems, since they indicate that dynamical processes may commonly lead planets to acquiring large obliquities, which is often thought of as planets getting large inclinations at some point of their evolution. We mention that, under some circumstances, measuring the stellar inclination relative to the line of sight, $i_\star$, for example through asteroseismology, can constrain the true obliquity of a planet (a small $i_\star$ implying a large misalignment; Huber et al. 2013; Chaplin et al. 2013) and even determine it (Benomar et al. 2014). At the time of writing, true obliquities have been estimated for 11 planets,[4] they exceed 30° for 5 of them.

How to explain these observations? Hot Jupiters are thought to form at typically several AU from their central star, reaching orbital periods of a few days either by disk migration or by high-eccentricity migration. Disk migration is expected to preserve zero obliquity (e.g., Bitsch et al. 2013a) unless (i) the disk becomes misaligned with the stellar equator due to gravitational interactions with nearby stars (Bate et al. 2010; Batygin 2012; Picogna and Marzari 2015), or (ii) the star itself becomes misaligned, which might occur

[4]See, e.g., http://www.astro.keele.ac.uk/jkt/tepcat/rossiter.html.

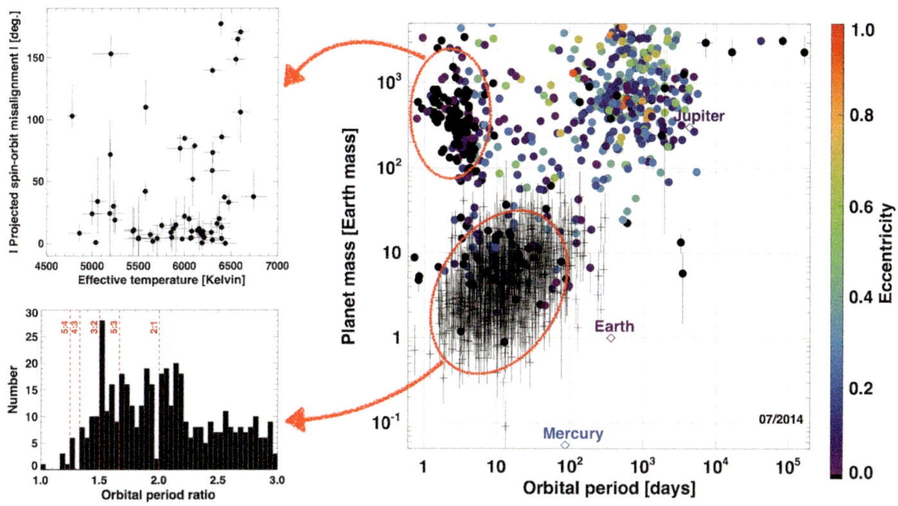

Fig. 4 *Right*: Mass, orbital period and eccentricity of observed exoplanets (*filled circles*), and for Mercury, Earth and Jupiter. Planets labeled by *plus signs* are planets detected by transit only, for which the mass is inferred from the mass-radius relationship $M \propto R^{2.06}$ for low-mass planets with both mass and radius determined. A 50 % error bar on the mass has been added to these planets detected by transit only. *Upper-left*: projected obliquity versus stellar effective temperature. *Lower-left*: period ratio histogram for confirmed multi-planet systems, the period ratio of a few mean-motion resonances is shown by vertical dashed lines. All data were extracted from http://exoplanets.org

through star-planet tidal interactions (tidal flip of stellar axis, Cébron et al. 2013; Barker and Lithwick 2014). Note that the delivery of hot Jupiters by disk migration does not have to proceed solely by smooth type II migration. It may be assisted by the inward scattering of a giant planet resulting from interactions with one or more giant planet companions in the protoplanetary disk (e.g., Marzari et al. 2010). In the high-eccentricity migration scenario, a Jupiter-mass planet begins on an inclined, highly eccentric orbit that shrinks and circularizes due to tidal interactions with the central star. Mechanisms that may pump planet eccentricities and/or inclinations in the first place include (i) planet-planet scattering between two or more massive planets, a scenario that has been proposed to explain the broad distribution of exoplanet eccentricities (e.g., Rasio and Ford 1996; Chatterjee et al. 2008; Jurić and Tremaine 2008), and (ii) Kozai cycles with an inclined, probably stellar-mass companion (e.g., Wu and Murray 2003; Fabrycky and Tremaine 2007; Naoz et al. 2012). High-eccentricity migration can produce a broad range of planet obliquities depending on the excitation mechanism and the effectiveness of tidal interactions at damping obliquities (which depends on the star's mass, rotation period, the planet's mass and orbital separation etc.).

Interestingly, it has been suggested that perhaps *all* hot Jupiters formed by high eccentricity migration with a broad range of initial obliquities, and that tidal interactions with the central stars were more or less efficient at reducing obliquities over the age of the systems (that is, at aligning the stellar spin axis with the planet's orbital axis, Triaud et al. 2010; Albrecht et al. 2012). (Albrecht et al. 2012) argued that aligned systems would be those where the timescale for spin-orbit alignment is shorter than the age of the system. They used the estimates of (Zahn 1977) for the tidal synchronization timescales in close binary stars, showing that tidal alignment should be more efficient for cooler stars that harbor a convective envelope. While Zahn's theory does predict an increase in the alignment timescale with the star's effective temperature, in practice it gives alignment timescales that are at least one or two orders of magnitude longer than the age of the systems where projected obliquities have been measured (Albrecht et al. 2012; Ogilvie 2014). This shows the limitations in applying Zahn's theory of tidal evolution of binary stars to hot Jupiters. This result has nevertheless stimulated much work on star-planet tidal interactions. In particular, the conditions for tides to produce alignment faster than orbital decay lead to the prediction that there should be as many retrograde aligned planets as prograde aligned ones, which is not what is observed (Lai 2012; Rogers and Lin 2013).

Could measurements of projected obliquities help distinguish and/or constrain the mechanisms of hot Jupiter migration? Answer to this important question probably requires a statistical comparison between model predictions and observations. We mention the recent work by Crida and Batygin (2014), who compiled the predictions of high-eccentricity migration models in terms of projected obliquities, and compared them with observations. They find that the models successfully reproduce the distribution of projected obliquities beyond 40°, but they all underestimate the proportion of systems with low obliquities. An alternative scenario mentioned above is the possibility that the disk can be misaligned relative to the star's equator due to the interaction with a binary companion star. Crida and Batygin (2014) use Monte-Carlo simulations to predict the distribution of projected obliquities according to this scenario. They find that, (i) just like for the combined high-eccentricity and tidal friction scenario, the misaligned disk migration scenario can robustly reproduce the distribution of projected obliquities beyond 40°, but (ii) only the aligned disk migration scenario can account for the number of aligned and nearly aligned hot Jupiters, depending on the semi-major axis distribution of the binary companions. Inclusion of magnetic interactions between the disk and the central star allows the misaligned disk migration scenario to reproduce the observed trend between misalignment and stellar mass (Spalding and Batygin 2015). These comparisons highlight that (i) current observations cannot disentangle between the high-eccentricity migration scenario and the disk migration scenario in a misaligned disk to explain misaligned hot Jupiters, and that (ii) delivery of aligned hot Jupiters via disk migration is necessary to account for the number of aligned hot Jupiters. Chemical depletions (e.g., O and C abundances) in the atmosphere of hot Jupiters could be a way forward in constraining their migration scenario (Madhusudhan et al. 2014). We also point out that the fraction of observed aligned hot Jupiters does not seem to vary significantly with orbital period[5] which, in our opinion, is a challenge to the above tidal alignment theory. We finally mention the recent work by Mazeh et al. (2015), who show that the trend of lower obliquity around cooler stars extends to *Kepler* objects of Interest with orbital periods up to at least 50 days, for which star-planet tidal re-alignment should be largely negligible.

3.3.2 Architecture of Kepler's Multi-Planetary Systems

Kepler has uncovered a large population of Earth- to Neptune-mass planets in compact multi-planetary systems with orbital periods between 1 and 100 days. The distribution of period ratios between all pairs of planets of a given system is displayed in the lower-left corner of Fig. 4 for *Kepler*'s confirmed multi-planet systems. There are two salient points.

[5]This is inferred from the sample of planets listed in http://exoplanets.org that have a minimum mass greater than $0.5M_J$. Defining as aligned a planet with projected obliquity less than about 40°, we find that, in the range of orbital periods between 1 and 3 days, there are 18 aligned and 7 misaligned planets, while for orbital periods between 3 and 5 days, there are 16 aligned and 8 misaligned planets.

First, there are many planet pairs far from resonances, while the usual expectation from disk migration models is that *Kepler*-mass planets should form resonant planet pairs, that is, one would a priori expect only spikes in this distribution at the location of the resonant period ratios, like 2 : 1, 3 : 2 etc. Second, planet pairs near resonances tend to have period ratios somewhat greater than resonant. It is worth noting that, despite a much smaller number statistics, a similar trend can be seen for planetary systems detected by RV techniques only. We point out that some features in the period ratio distribution may be artifacts of unseen additional companions (Steffen 2013). For example, a two-planet system with a period ratio ~ 1.9 could actually be a three-planet resonant system where the middle planet, missed by the transit observation, is in 4 : 5 mean-motion resonance with the innermost planet, and in 3 : 2 mean-motion resonance with the outermost planet.

How to explain these observed features? One model that has been recently put forward is in-situ growth. Here, it is important to stress that this is not in-situ growth all the way from micrometer-sized grains to planetary-sized objects, but in-situ growth of pre-formed and pre-evolved Mars- to Earth-sized embryos in a gas-free disk, and that these embryos may well have been delivered previously by disk migration, before the protoplanetary disk gets cleared out. The period ratio distribution predicted by the in-situ growth model of Hansen and Murray (2013) tends to produce too few close and resonant planet pairs compared to observations. The recent work by Ogihara et al. (2015) shows that planet embryos actually form very rapidly in the inner parts of protoplanetary disks, and that therefore disk migration cannot be neglected. Inclusion of disk migration in the "in-situ" growth calculations of Ogihara et al. (2015) shows that disk migration leads this time to an excess of close pairs of planets compared to observations.

Another model that has been highlighted is that a resonant planet pair formed by disk migration may evolve away from resonance due to star-planet tidal interactions, which tend to slowly increase the planets' period ratio over time, at a rate that depends on the star-planet separation (Papaloizou 2011; Lithwick and Wu 2012). While tides may indeed account for some planet pairs being just wide of resonance, we note that a similar trend is observed for planet pairs above 10 days, for which tides should be inefficient (Baruteau and Papaloizou 2013).

More recently, disk migration of planet pairs has been revisited. Baruteau and Papaloizou (2013) have shown that, at the typical location of Kepler planets, the wakes of the planets should be shock waves, which implies that: (i) the planets should open partial gaps around their orbits, and (ii) the wakes deposit energy and angular momentum in the co-orbital region of each planet, with the consequence that the planets tend to repel each other over time. This stresses that disk migration models do not necessarily predict *Kepler*-mass planets to form resonant pairs. Turbulence in the disk is another way to explain why many low-mass multiple planet systems are near- or non-resonant (Pierens et al. 2011; Rein 2012), and may play a prominent role in the formation of closely packed systems (Paardekooper et al. 2013).

4 Internal Evolution of Planets

We now turn our attention to individual planets, treating them no more just as point-like masses, but as geophysical objects that are characterized by a radius, a luminosity, an atmospheric and a bulk compositions, which can all evolve in time.

A small fraction of self-luminous objects like β Pic b (Lagrange et al. 2009) among the known extrasolar planets has been directly detected at young ages, using coronagraphs and dedicated data reduction procedures, and a handful of objects has even been observed during their formation, which occurs on a timescale of $\sim 10^7$ years (e.g., Quanz et al. 2015). However, the large majority of known exoplanets orbits stars that are at least $\sim$100 Myrs old with a typical age of several Gyrs, meaning that we usually observe (exo)planets at an epoch much after their formation. Many of these objects are then too faint or too close to their star to be detected directly. More often their mass can, however, be inferred from radial velocity measurements. Further, if the planets are transiting their star, it becomes possible to measure their radius if the radius of the host star is known. Transits also allow to probe the planetary atmospheres and to get hints at the planet's abundances and day-side emission. The combination of radial velocity and transit measurements finally yields the mean density of a planet, which is its first rough but important geophysical characterization (e.g., Dressing et al. 2015; Santos et al. 2015).

The number of planets for which various observational data is available has increased enormously in the past two decades (e.g., Mayor et al. 2011; Borucki et al. 2011b; Marcy et al. 2014). One of the major goals of the research on (exo)planets is to better understand the physics of planet formation thanks to this data, including in particular the Solar system. A typical approach is to compare the results from theoretical planet formation models and the observational constraints from the aforementioned techniques. However, in order to draw conclusions about the formation of planets from properties that we observe today like orbital parameters, mass, radius, luminosity and composition, one needs to understand not only how planets form, but also how they evolve from formation until the epoch of observation. The reason for this is that the evolution spans Gyr-long timescales, and the link between observation and formation can be non-trivial and potentially mask the original imprints of formation.

Close-in low-mass planets are an illustrative example. Their formation mechanism is currently debated (see, e.g., Sect. 3.3.2). Additional constraints that can be derived from their geophysical characteristics like for instance the H/He envelope mass are therefore very valuable to distinguish different formation scenarios (Sect. 4.5). However, because of envelope evaporation (atmospheric escape), some planets might have started with a gaseous envelope after formation but then lost it during the evolutionary phase.

Directly imaged planets are another example. Even for these young planets, which are typically only a few Myrs old, the question of understanding their temporal evolution since formation is of central importance when it comes to infer their mass based on their luminosity. This mass-luminosity relationship is in turn of high interest for formation theories as will be discussed in Sect. 4.2: it constrains the physics of gas accretion and potentially even the mode of formation, namely core accretion (Perri and Cameron 1974; Mizuno 1980; Bodenheimer and Pollack 1986) or gravitational instability (Toomre 1981; Boss 1997; Durisen et al. 2007). Not only must evolutionary models be well-suited to describe the governing physics, but also the correct initial conditions for the planet's evolution must be known, meaning that formation and evolution must be linked self-consistently.

In this section, we mostly concentrate on how to derive constraints on formation models from studying the long-term internal (mainly thermodynamical) evolution of planets after the dissipation of the protoplanetary disk, which is the link between formation and observation. Concentrating mainly on planets with primordial H/He, we address some of the most important aspects of planetary evolution like cooling and contraction, the mass-luminosity relation, and the bulk composition as expressed in the mass-radius and mass-mean density relations. We also quickly touch upon other topics that can play important roles like atmospheric escape, radius inflation and deuterium burning. For further aspects of the theory of planetary evolution, including crucial elements like high-pressure physics and equations of state, the internal rotation of planets and their shape, the specific properties and cooling histories of the Solar system planets, or potential phase separations of elements in planetary interiors (to name just a few), the reader is referred to the reviews of Stevenson (1982); Guillot (2005); Fortney and Nettelmann (2010); Chabrier et al. (2014); Guillot and Gautier (2014) or Baraffe et al. (2014).

4.1 Thermodynamical Evolution: Cooling and Contraction

The thermodynamical evolution of giant planets after formation at constant mass is controlled by cooling and contraction. They evolve from a comparatively hot post-formation state characterized by high entropy, high luminosity, low degeneracy, and large radius to a colder state with low entropy, low luminosity, higher degeneracy, and smaller radius (e.g., Guillot 2005). In contrast to stars, due to their comparatively low internal temperature, giant planets do not radiate away energy released by thermonuclear fusion as they are not massive enough to allow for fusion in their interior. The only exception may be a short stage of deuterium burning in planets with masses in excess of 12–13 M_J, as we will see in Sect. 4.3. But, for the bulk of giant planets, cooling and contraction are the main source of internal luminosity (Hubbard 1980), which can be seen from the fact that Jupiter, Saturn and Neptune emit more energy than they receive from the Sun.

The total amount of energy that can be radiated away by a giant planet during its lifetime is primarily determined by its post-formation entropy (Sect. 4.2). For terrestrial planets, such as Earth, another important contributor to the internal luminosity, besides delayed secular cooling, is the radioactive decay of unstable heavy element isotopes (e.g., Urey 1955; Hofmeister and Criss 2005). For the Earth, the total intrinsic heat flux is around 47 TW (which is only about 0.03 % the solar irradiation flux), of which about half is estimated to arise from radiogenic heating (Kamland Collaboration et al. 2011). For Jupiter, the intrinsic luminosity (which is of the same order as the solar irradiation flux, Guillot and Gautier 2014) is largely dominated by the cooling and contraction of the fluid H/He envelope; a possible solid core could contribute a few percent, while the radiogenic contribution is tiny ($\sim 10^{-4}$).

The cooling of gas giant planets is caused by the radiation escaping through their atmosphere, which decreases the total energy of the planet. Because of virial equilibrium, for self-gravitating objects composed mainly of ideal gas, radiative escape causes somewhat counter-intuitively the interior to heat up (e.g., Guillot 2005). Such objects are therefore said to have a negative heat capacity: the loss of energy at their surface leads to an increase in their internal energy and central temperature, which is compensated for by the object's contraction. This regime applies to low-mass pre-main-sequence stars, as well as extremely young hot giant planets at ages less than 10^{5-6} years they are not yet significantly degenerate (Graboske et al. 1975).

The situation is quite different for older giant planets however, for which cooling and contraction cause the interior temperature to decrease. This is because the thermal pressure needed to balance the planet's gravity is established by degenerate electrons, which increase their mean kinetic energy, in contrast to ions, which decrease their kinetic energy and cool, yielding most of the intrinsic luminosity (Guillot 2005). In this phase, giant planets behave qualitatively like degenerate brown dwarfs (Graboske et al. 1975).

The time evolution of the cooling of a planet, and thus its internal evolution, depend on how energy is transported from the interior to the top and then through the atmosphere of the planet. The atmosphere can be thought of as a blanket which, with increasing thickness (i.e., optical depth), diminishes the planet's ability to cool, making it stay hot and have a larger physical radius for a longer period of time at high opacity (Burrows et al. 2007). For strongly irradiated planets the cooling timescale also increases: the incoming radiation makes the atmospheres more isothermal, reducing the effectiveness of radiative energy transport, as

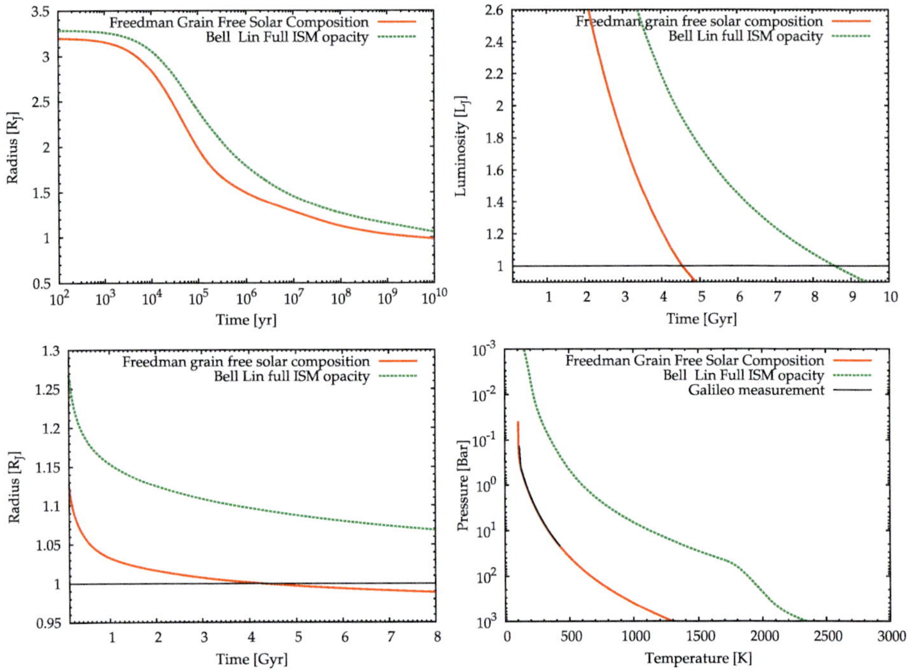

Fig. 5 Calculation of the internal evolution of a Jupiter-mass planet for two different input opacities: full ISM grain opacities (Bell and Lin 1994, *green dotted curves*), and Solar abundance molecular opacities (Freedman et al. 2008, *red solid curves*). The time evolution of the planet's $\tau = 2/3$ radius is shown in the *upper-* and *lower-left panels* with two different time windows. The time evolution of the planet's luminosity is in the *upper-right panel*, and the pressure-temperature profiles of the planet's atmosphere at Jupiter's current age is displayed in the *lower-right panel*. The *black solid curve* in the *lower-right panel* shows the pressure-temperature profile measured by the Galileo probe (Seiff et al. 1996)

the atmospheric temperature gradient decreases (Burrows et al. 2000; Guillot and Showman 2002).

An example of planetary evolution calculation illustrating the impact of opacities is shown in Fig. 5 (Mordasini et al. 2012a). For such calculations, the fundamental equations for a planet's internal structure (equations of mass conservation, hydrostatic equilibrium, energy conservation, and energy transport by radiation or convection) are solved (e.g., Guillot and Gautier 2014) which allows the construction of evolutionary sequences based on the principle that the luminosity is equal to minus the temporal change of the planet's total energy. The post-formation evolution of a Jupiter-like planet is simulated assuming (ad hoc) either very high opacities caused by ISM grains, or more realistic grain-free molecular opacities for a solar composition gas which are about three orders of magnitude lower (see Bell and Lin 1994 and Freedman et al. 2008 for the grain and molecular opacities, respectively). For both sets of opacities, the planet has the same initial entropy. Note that planets with the same set of opacities, but different initial entropies, should still get the same structure in the end because the initial evolutionary timescales are short, at least if the post-formation entropies are not very low (Sect. 4.2). Figure 5 displays the time evolution of the planet's radius and luminosity, as well as the pressure-temperature profiles of the planet's atmosphere at Jupiter's current age. One clearly sees that for ISM grain opacities (in reality the grains would fall out of the atmosphere on a very short timescale, Mordasini 2014), the planet's radius remains larger than for molecular opacities. Also, after a few Gyrs, the planet luminosity obtained with ISM grain opacities is about a factor 2 larger than with molecular opacities. This is expected, as with ISM opacities the planet cools much more slowly, so that at late evolutionary times the planet has retained more heat than if it had cooled down with low molecular opacities. Only molecular opacities can reproduce Jupiter's present day radius, luminosity, and the pressure-temperature profiles measured by the Galileo probe (Seiff et al. 1996).

Energy transport in the interior of gaseous planets is commonly assumed to be due to convection. This is because the interior is usually characterized by high opacities, such that Schwarzschild's criterion for convection is satisfied (but see also Guillot et al. 1995). However, different means of energy transport might make planetary interiors considerably deviate from the picture of efficient convective energy transport. If planetary envelopes have a compositional gradient with mean molecular weight increasing towards the interior, the large-scale energy transport by convection could be shut down. The reason for this is that blobs of gas perturbed to upper layers of the atmosphere must sink back downward because the surrounding material has a lower density due to a smaller mean molecular weight. The possible implications for planetary structures were discussed by Stevenson (1985) and recently revisited by Leconte and Chabrier (2012). Following these studies, layers which are unstable according to Schwarzschild's criterion, but stable when including the compositional gradient, are thought to become semi-convective, exhibiting many convective layers separated by thin conductive layers with steep compositional gradients. The conductive energy transport is established by the thermal motion of electrons and ions. Planets with such semi-convective envelopes are thought to cool much slower, as the conductive boundary layers hamper efficient energy transport. It has been suggested that this phenomenon could be responsible for Saturn's high luminosity (Leconte and Chabrier 2013), which is too high when compared to classical, fully convective models (Pollack et al. 1977). Semi-convection is thus an alternative to the classical explanation that Saturn's excess luminosity is caused by the phase separation of hydrogen and helium (He insolubility) and a subsequent helium "rain" into the deeper layers, which releases gravitational potential energy (e.g., Stevenson and Salpeter 1977).

There are two important processes to assess the extent to which strong compositional gradients, and therefore semi-convection, are present in planetary envelopes. First, the solid core present in planets thought to have formed by core accretion might dissolve, leading to an increasing mean molecular weight in the layers close to the core-envelope boundary. This has been studied by Guillot et al. (2004) and Wilson and Militzer (2012), who found that a dissolution and subsequent mixing of the core might be possible. Second, during the planet formation phase, the destruction and heavy element deposition by planetesimals breaking up in the envelope similar to comet Shoemaker-Levy 9 could also lead to compositional variations. The ablated material could either stay suspended near the location where it has been deposited, settle towards deeper regions, or get mixed through the envelope by convection (Mordasini et al. 2006; Iaroslavitz and Podolak 2007). Clearly, this can lead to compositional gradients in the planet's envelope that are determined by the formation process. In this way the planetesimal accretion history might influence the planet's evolution even at late times.

4.2 Initial Conditions for Planetary Evolution

The early evolution of gaseous planets depends crucially on the amount of heat retained from the formation process. Depending on whether they form with a low, average or high initial

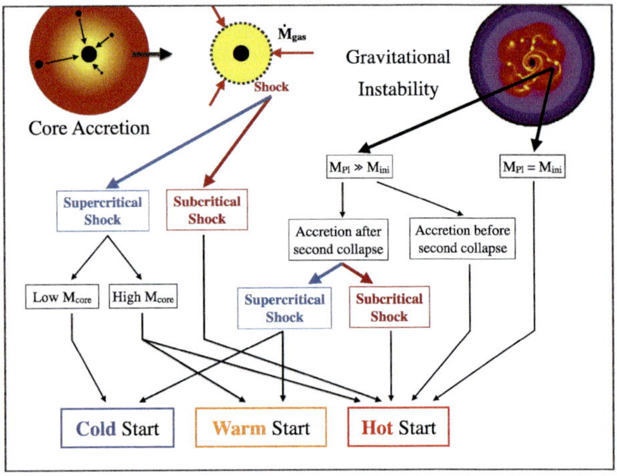

Fig. 6 Summary of the various pathways to giant planet formation leading to different post-formation entropies (Mordasini et al. 2012a). The gas surface density plot in the *upper-right corner* illustrating gravitational instability is taken from Boley (2009). M_{ini} denotes the initial mass of the clump formed by gravitational instability. Note that several pathways have to date not been simulated quantitatively

entropy (with a correspondingly low, average and high luminosity, radius and temperature), planets are said to have "cold", "warm" or "hot" start initial conditions. A typical cold start post-formation entropy is ~ 8–$9k_{\mathrm{B}}/\mathrm{baryon}$ (k_{B} denotes the Boltzmann constant) whereas a typical hot start post-formation entropy is ~ 11–$15k_{\mathrm{B}}/\mathrm{baryon}$ (see, e.g., Marley et al. 2007; Spiegel et al. 2011; Mordasini 2013; Marleau and Cumming 2014). Given the significant impact that such initial conditions can have on the early internal evolution of giant planets, we first review the expected range of initial entropies for the two mainstream scenarios of giant planet formation: core accretion and disk gravitational instability. This is also summarized in Fig. 6, but it should be noted that several assumptions have not been validated yet with quantitative simulations, especially in the gravitational instability scenario.

(i) Core accretion. Traditionally, planet formation via core accretion is thought to produce gaseous planets with a low initial entropy (Fortney et al. 2005; Marley et al. 2007). The reason for this is the assumption that the accretion shock produced on the planet's surface by the infalling gas is radiatively efficient (or "super-critical"). For such a shock structure (found, e.g., for gas accretion on the first stellar core, Commerçon et al. 2011), the gravitational potential energy released at the shock is fully radiated away and none of it is incorporated in the convective interior of the planet. By analogy with the stellar case (e.g., Hartmann et al. 1997), planets undergoing such "cold" accretion will start their evolution with a low initial entropy and low internal temperature, dubbed "cold start".

However, the assumption that the shock is radiatively efficient is quite uncertain. If, on the contrary, the accretion shock is radiatively inefficient (or "sub-critical"), that is if the energy released at the shock is not radiated away (as found for gas accretion on the second stellar core, see Vaytet et al. 2013), planets will have high post-formation entropies, actually similar to those assumed in classical purely evolutionary simulations with a hot start (Mordasini et al. 2012a). The intermediate case where the accretion shock is neither radiatively efficient nor inefficient will lead to warm starts (Spiegel and Burrows 2012). Furthermore, and somewhat surprisingly, the formation of a massive solid core increases the post-formation entropy (Mordasini 2013; Bodenheimer et al. 2013). The requirement for this is an increased solid surface density (Σ_{s}) in the protoplanetary disk, which results in a higher solid accretion luminosity within the planet and a higher final core mass. Planets forming in disks with a high Σ_{s} will go into runaway gas accretion earlier, as the core also forms more rapidly. Therefore, gas is accreted onto a protoplanet with a larger radius, as the planet has had less time to cool and is more strongly stabilized against contraction owing to the increased planetesimal accretion luminosity. A larger radius means a weaker shock, therefore less gravitational potential energy released by the gas before it is accreted onto the planet. The gas is then incorporated into the planet at a larger entropy, also leading to warm starts (Mordasini 2013; Bodenheimer et al. 2013).

(ii) Disk gravitational instability. Another pathway to the formation of gaseous planets is gravitational instability (GI) in the outer parts of protoplanetary disks (typically beyond 50 to 100 AU; see, e.g., Boley et al. 2010). The possibility that GI occurs, the location where fragmentation sets in, and the subsequent internal and orbital evolutions of the clumps depend on the disk's mass, its structure and, critically, its cooling time scale (see, e.g., the reviews by Helled et al. 2014 and Kratter and Lodato 2016). Analytical arguments based on the Toomre (Toomre 1981) and cooling criteria (Gammie 2001) show that at smaller distances from the star, in a disk that is dense enough to be gravitationally unstable, it is not possible to cool rapidly enough to fragment, so that the two criteria cannot be simultaneously fulfilled (Rafikov 2005). In the outer parts of the disk, gravitational instability could instead occur early in the disk's evolution when it is still massive and accreting rapidly from the protostellar cloud (D'Angelo et al. 2011). But it is then unclear if bound clumps forming at this stage survive migration, evolve into planets, massive brown dwarfs or stellar companions (e.g., Stamatellos and Whitworth 2009), or if they get tidally disrupted (e.g., Nayakshin 2010; Forgan and Rice 2013).

Because in the case of a gravitational collapse the material is thought to not process through a shock, GI should yield hot starts. This has been found indeed in the numerical simulations of Galvagni et al. (2012), who report an entropy of $\sim 15k_{\mathrm{B}}/\mathrm{baryon}$. However, if the collapsed protoplanet continues to accrete gas after it has fully collapsed (called "second collapse", analogous to stellar formation, a second collapse denotes the collapse after the molecular hydrogen has been dissociated) an accretion shock can potentially develop. The more mass is added through a super-critical shock onto the initial GI unstable object, the lower the final post-formation entropy of the planet (Mordasini et al. 2012a). If, however, the initially unstable object accretes its mass before the second collapse, it should rather have a hot start as the large radius of the clump should prevent the formation of strong shocks.

We point out that in "classical" planetary evolution calculations, which do not treat the formation phase, it is often simply assumed that planets start with an arbitrarily high initial entropy (that is, a hot start; see, e.g., Burrows et al. 1997; Baraffe et al. 2003). Because the Kelvin-Helmholtz (KH) timescale is short for large entropies, the initial condition is quickly erased such that after $\sim 10^7$–10^9 years for 1 and $10M_{\mathrm{J}}$ planets, respectively, the luminosities in the cold and hot start cases converge to the same evolutionary tracks (see, e.g., Marley et al. 2007; Spiegel et al. 2011; Marleau and Cumming 2014). Therefore, the choice of an arbitrarily high initial entropy does not affect predictions on the planet appearance at ages $> 10^7$–10^9 years. But it will do so for younger planets. This is particularly important for directly imaged planets in young stellar systems, for which the deduction of the planet's mass based on the planet's luminosity depends on the underlying assumptions of the utilized planetary evolutionary tracks.

A comparison between hot and cold start calculations is illustrated in Fig. 7 (updated from Marleau and Cumming (2014)). It shows the time evolution of the luminosity of planets with a mass of 1, 2, 3, 5, 10, 15, 20 and $25M_{\mathrm{J}}$ for cold (blue curves) and hot initial conditions (red curves). Observational data of directly imaged planets are overplotted (updated from Neuhäuser and Schmidt 2012). One can see that for the specific evolutionary tracks shown in the figure, cold and hot start tracks converge after $\sim 10^8$ years. The mass degeneracy for a given age and luminosity is obvious when looking at the figure: at an age of ~ 4 Myrs, a

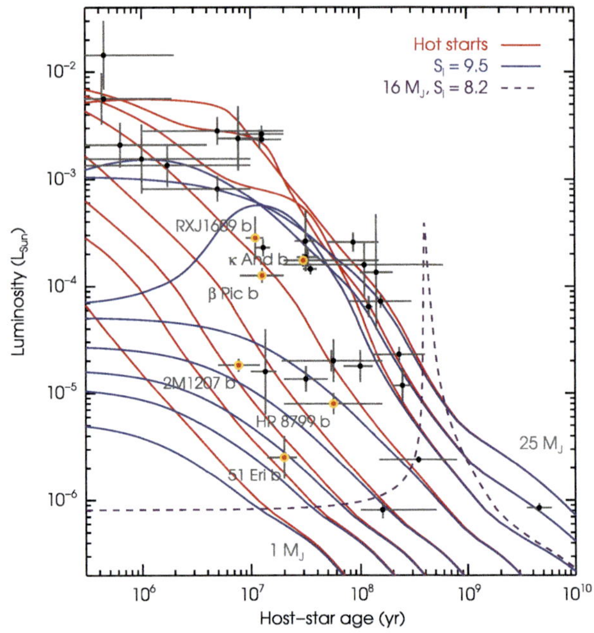

Fig. 7 Luminosity as a function of time from theoretical models (Marleau and Cumming 2014) compared with directly imaged objects with a hot-start mass below 25 Jovian masses (data compiled by Neuhäuser and Schmidt 2012). The *lines* show cooling curves for $M = 1, 2, 3, 5, 10, 15, 20$ and $25 M_J$ (bottom to top) assuming a cold start with a post-formation entropy of $9.5 k_B$/baryon (*blue lines*) or a hot start (*red lines*). Figure updated from Marleau and Cumming (2014)

cold-start planet of $10 M_J$ and a hot-start planet of $3 M_J$ have the same luminosity. For cooler starts than assumed in the figure, the time after which the evolutionary tracks of cold and hot starts converge becomes longer and the difference in masses for a given age and luminosity increases.

4.3 Deuterium Burning in Massive Planets

Traditionally, gaseous objects which are able to fuse deuterium, but which are not massive enough to burn hydrogen on a stable main sequence, are called "brown dwarfs" (see, e.g., Boss 2007). They were first theorized to exist in the 1960s (Kumar 1963; Hayashi and Nakano 1963). The distinction between massive gaseous planets and brown dwarfs has been based on the hypothesized formation mode, with formation in a protoplanetary disk for giant planets, and gravoturbulent fragmentation of a molecular clump (i.e., a star-like formation mode) for brown dwarfs (Chabrier et al. 2014). In principle, it is well possible that, in massive protoplanetary disks, planets form with masses in excess of 13 Jupiter masses via core accretion (Mordasini et al. 2009b), and it has been shown that such planets are able to burn deuterium in the layers on top of their solid core (Baraffe et al. 2008).

Classical studies in the brown dwarf regime include Burrows and Liebert (1993); Saumon et al. (1996); Chabrier and Baraffe (2000); Burrows et al. (2001); Baraffe et al. (2003). They put a mass limit for the onset of deuterium burning at roughly $13 M_J$. Deuterium burning in forming planets that harbor a solid core has been studied systematically in Mollière and Mordasini (2012) and Bodenheimer et al. (2013), and the mass threshold for deuterium burning is found to be $\sim 13 M_J$ as well. For planets, the exact threshold value can vary with the helium content, the metallicity and deuterium fraction (Spiegel et al. 2011; Mollière and Mordasini 2012). It also varies with the core mass (Mollière and Mordasini 2012;

Bodenheimer et al. 2013), since a more massive core leads to a higher post-formation entropy (see Sect. 4.2) and therefore a higher internal temperature, which increases the deuterium burning rate (Mordasini 2013; Bodenheimer et al. 2013).

In general, deuterium burning proceeds via the reaction

$$\mathrm{H} + \mathrm{D} \rightarrow {}^{3}\mathrm{He} + \gamma, \tag{26}$$

with an energy release of 5.494 MeV (Fowler et al. 1967). Since the hydrogen and deuterium ions need to be brought close enough to overcome the Coulomb repulsion and fuse, the reaction rate depends very strongly on the temperature. It is limited by the fact that, at the central temperatures in the objects of interest, only a small fraction of ions have kinetic energies large enough to tunnel through the ions' repulsive Coulomb barrier. Screening plays an important role too, as the electrons that are present in the fully ionized interiors of the planet effectively shield the ions from each other during their approach, decreasing the kinetic energy needed for the tunneling probability. As the objects are already considerably degenerate in their interior, classical Debye-Hückel shielding is not sufficient for these objects and screening theories for degenerate material need to be invoked (see, e.g., Dewitt et al. 1973; Graboske et al. 1973).

Deuterium burning can have a very strong dependence on temperature. The deuterium burning rate (ϵ_{D}, energy per unit mass and time) can be expressed as a power-law function of temperature T (see, e.g., Kippenhahn and Weigert 1990):

$$\epsilon_{\mathrm{D}} \propto \rho X_{\mathrm{D}} X_{\mathrm{H}} T^{\nu} \tag{27}$$

with ρ the gas density, X_{H} and X_{D} the mass fractions of hydrogen and deuterium, respectively, and

$$\nu = \frac{80}{3}\left(\frac{10^5\ \mathrm{K}}{T}\right)^{1/3} - \frac{2}{3}. \tag{28}$$

The power-law exponent of the temperature thus depends itself on the temperature. For deuterium burning in low-mass protostars, it is often assumed that $\nu = 11.8$ (Stahler and Palla 2005), which corresponds to temperatures of $\sim 10^6$ K. The coolest deuterium burning objects (with masses of $\sim 13 M_{\mathrm{J}}$) burn deuterium at central temperatures of roughly 4×10^5 K (see, e.g., Mollière and Mordasini 2012), corresponding to $\nu \sim 16$. Note, however, that electron screening in these objects may somewhat decrease ν (Kippenhahn and Weigert 1990).

The outcome of deuterium burning is quite different whether a hot start or a cold start is assumed. Hot start objects, which might be rather related to "classical" brown dwarfs, start their evolution with a large radius and a high temperature (see Sect. 4.2), and then contract slowly with time, starting the deuterium burning process once their central density is high enough. The main effect of deuterium burning is to slow down the contraction. Cold start objects, however, have quite low post-formation entropies, resulting in small radii and small internal temperatures (see Sect. 4.2). The fact that their internal temperature is smaller than in hot start objects, while having smaller radii, is due to the partial degeneracy of their interior. When cold start planets reach internal temperatures that are high enough for the onset of deuterium burning, they have much higher densities than hot start objects. The effect of deuterium burning for cold start planets is not a decreased contraction speed, but a re-inflation to larger radii. The increase in planetary luminosity for these objects therefore does not exclusively stem from the deuterium burning itself, but also from the later re-contraction of the planets inflated by deuterium burning.

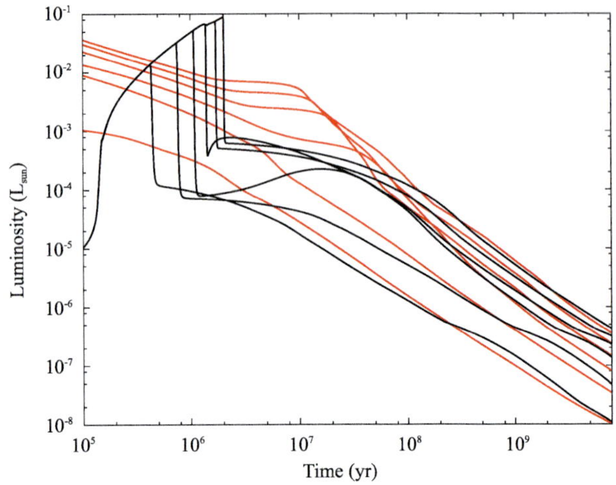

Fig. 8 Luminosities of cold and hot start objects with a mass of 5, 10, 15, 20, 25 and $30M_J$ (from bottom to top). The *red solid curves* show evolutionary calculations with a high initial entropy (hot start), taken from Burrows et al. (1997). The *black solid curves* show combined formation and evolutionary calculations of objects formed by core accretion with a low post-formation entropy (cold accretion). The high luminosities at around 1 Myr are due to the gas accretion shock, they rapidly fall as gas accretion is (artificially) terminated (Mollière and Mordasini 2012)

In general, the deuterium number fraction in the hydrogen and helium dominated atmospheres is quite low, $N_D/N_H \sim 2 \times 10^{-5}$ (Prodanović et al. 2010). Therefore, the phase of deuterium burning is rather short, between 1 and 100 Myrs, depending on the mass and formation mode (cold/hot).

Illustrative luminosity tracks for hot start and cold accretion objects are shown in Fig. 8. The hot start tracks are taken from Burrows et al. (1997), while the cold start formation and evolution calculations for core accretion planets use the model of Mollière and Mordasini (2012). The objects shown in the figure have a mass ranging from 5 to $30M_J$. Note that the low-entropy tracks in Fig. 7 do not agree with those in Fig. 8: Fig. 7 assumes arbitrarily low post-formation entropies for cold start objects, whereas in Fig. 8 the low initial post formation entropies are determined self-consistently from the formation process. The point of Fig. 8 is that even objects formed by core accretion produce deuterium burning rates similar (but not identical) to those obtained by classical calculations which normally neglect the formation process, while the point of Fig. 7 is to show the differences in evolution between high and low entropy objects and how long it takes for both branches to converge. The formation is neither considered nor shown to focus on the effects of evolution.

The effect of the deuterium burning delaying the hot start's luminosity decrease for masses $> 13M_J$ can clearly be seen in Fig. 8, as well as the increase in luminosity for cold start objects (in order to compare the hot and cold start calculations more easily, the cold start calculations were shifted in time by roughly 2 Myrs). For the cold start objects with masses $> 20M_J$, the inflation of the planetary radius happens already during the formation phase of the planet, such that the increase in luminosity cannot be seen in the figure as it lies within the period where the powerful accretion shock luminosity dominates ($<$ 3 Myrs in the figure). The accretion shock luminosity decreases a bit because of deuterium burning as the infalling gas is added at larger radii.

4.4 Mass-Radius Relation

A diagram that is of particular interest for planet formation and evolution theories is the planetary mass-radius diagram, as it contains information about the internal structure of (extrasolar) planets (e.g., Guillot 1999; Chabrier et al. 2009). It constrains the processes

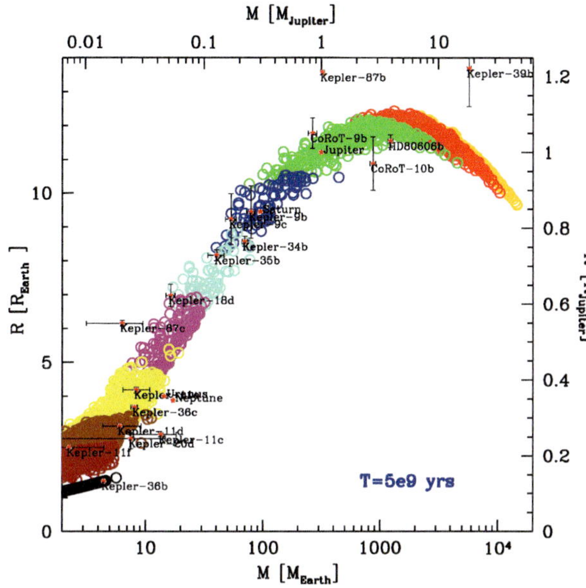

Fig. 9 Mass-radius diagram of observed and synthetic planets. The synthetic planets have an age of 5 Gyr and a semi-major axis a between 0.1 and 1 AU. Observed planets with $a > 0.1$ AU are included. The colors give the mass fraction of H/He ranging from solid planets without H/He (*black*) over *ocher* (< 1 %) to *dark orange* (> 99 %)

that determine the structure during the formation period (solid/gas accretion, orbital migration) as well as during the evolutionary phase (like atmospheric opacities, energy transport mechanisms, atmospheric escape, outgassing, or radius inflation).

Figure 9 compares the mass-radius relation obtained with a planet population synthesis calculation at an age of 5 Gyr with observational data (updated from Mordasini et al. 2012b). The metallicity (core mass fraction) of the synthetic planets is color-coded in the figure. The chemical composition of the heavy elements depends on the planet's formation track. Heavy elements accreted beyond the iceline in the parent protoplanetary disk are assumed to consist of 50 % water ice and 50 % "rocky" material, while material accreted inside the iceline is fully "rocky", which is itself assumed to be made of 2/3 silicates and 1/3 iron. Such mass fractions are approximately expected for the condensation of solar composition gas (Lodders 2003; Santos et al. 2015). Although the observed exoplanets shown in Fig. 9 are not at an exact age of 5 Gyr the observational errors on the host star's age often allow for ages of about 5 Gyr. Furthermore, the planetary radius evolution in the Gyr time range is slow, such that the 5 Gyr snapshot should represent planets of various ages within the Gyr regime reasonably well.

The population synthesis model used in Fig. 9 self-consistently combines a planet formation model based on the core accretion paradigm (Alibert et al. 2005) with the long-term thermodynamical evolution of the planets after the dissipation of the protoplanetary disk (Mordasini et al. 2012a). The simulation includes the effect of atmospheric escape (see Sect. 4.5) but does not model radius inflation (see Sect. 4.6). To avoid planets which could be significantly inflated (Demory and Seager 2011), the minimum semi-major axis of the synthetic and observed planets included in the figure is 0.1 AU. mechanisms that could lead to inflated close-in exoplanets will be discussed in Sect. 4.6.

In general the synthetic mass-radius diagram has a specific shape that is characterized by an increase in radius with mass, which levels off and turns over at roughly $4 M_J$. This maximum is due to an increase in the compressibility of the matter as the electron degeneracy increases. Furthermore, one sees that neither the upper-left corner in Fig. 9 (i.e., large low-mass planets) nor its lower-right corner (i.e., small massive planets) seem to be populated by

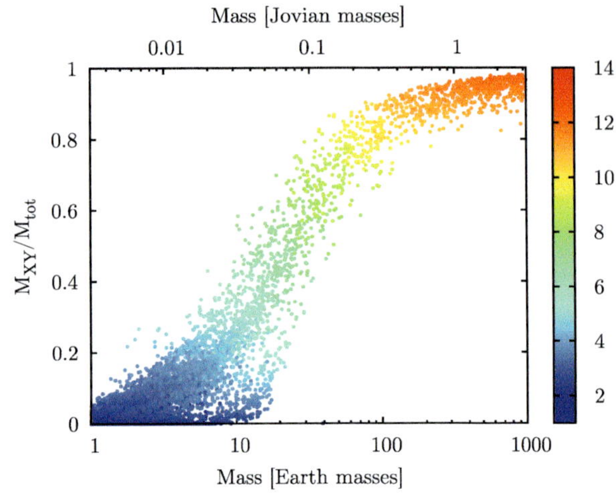

Fig. 10 Mass fraction of H/He relative to the total mass of a planet as a function of its total mass. The color-code gives the planetary radius in Earth radii. The larger the planet mass, the higher the mass fraction of the gas envelope. The change in slope at about 100 $M_\oplus$ results from the transition between the Kelvin-Helmholtz contraction limited regime and the disk-limited gas accretion regime. This figure can be compared with observational constraints (see, e.g., Fig. 9 in Lopez and Fortney 2014)

synthetic or observed planets. Synthetic planets with a low metal fraction have larger radii and masses than high-metallicity synthetic planets. At a given mass, the lower the radius, the more enriched is a synthetic planet, while planets of fixed enrichment have larger radii when their mass is higher, leading to diagonally colored bands in the synthetic mass-radius diagram.

The shape of the mass-radius diagram is a consequence of the formation process, in particular the resulting core and H/He envelope masses, as well as material properties. The fraction of H/He envelope mass as a function of the planet's total mass is displayed in Fig. 10. As the population synthesis is calculated under the core accretion paradigm, the planets start as low-mass solid cores, which can bind only tenuous gaseous envelopes due to long KH cooling timescales of their protoplanetary envelope during the nebular phase (Ikoma et al. 2000; Lee and Chiang 2015). Additionally, if the protoplanets are close to their star during formation, they will not accrete large amounts of gas due to the high local disk temperature (Ikoma and Hori 2012). Whatever gaseous envelope they accrete during formation can then be reduced again by envelope evaporation due to the stellar XUV insolation during evolution (e.g., Lammer et al. 2003; Owen and Jackson 2012; Lopez and Fortney 2013; Jin et al. 2014). Therefore, the lowest mass objects will have the highest heavy element content (essentially they are solid rocky or icy planets). As rocky or icy planets are only weakly compressible, their radius increases as their mass increases.

More massive planets can bind a larger gaseous envelope on top of their solid core. As they grow in mass, the amount of planetesimals which can be accreted increases as well due to the expansion of the feeding zone, but the net effect is still a decrease in the planetary metallicity (heavy element mass fraction) with mass M roughly as $M^{-0.7}$ in the giant planet regime (Mordasini et al. 2014), in agreement with observational data (Miller and Fortney 2011).

Although the compressibility of gas is much higher than of solids, the planets still grow in radius for a given metallicity if the mass is increased. For ever higher solid core masses, the growing planets will eventually try to accrete and bind more gas from the protoplanetary disk than is locally available due to their very short KH timescales. This leads to a quasi-static collapse of their gaseous envelope and the so-called stage of runaway gas accretion (Bodenheimer et al. 2000). In this phase, the growing planets will accrete a large fraction of the gas delivered to their location by the viscous evolution of the protoplanetary disk (Lubow et al. 1999). The disk's lifetime and gas supply at the location of the planet then become the sole determining factors of the planets' final mass. The transition from the regime where the planet's KH-contraction controls gas accretion, to the one where the disk limits it, is visible in Fig. 10 by a change in slope at about 100 $M_\oplus$. This figure can be compared to observational data where the H/He mass fraction of actual exoplanets is derived from internal structure modeling (Fig. 9 in Lopez and Fortney 2014).

As it is inherently impossible to form low-mass gaseous planets without a core in the core accretion paradigm, no very large low-mass planets exist in the synthetic mass-radius diagram (nor actually in the observed diagram). Such planets would populate the upper-left corner in Fig. 9 (and Fig. 10). The presence of a massive core always leads to comparatively smaller radii for low-mass planets, even if there is a gaseous envelope on top of the core. On the other hand, massive solid planets will always undergo runaway gas accretion in the disk, such that it is impossible to form small (i.e., bare rocky or icy) but massive planets in the core accretion picture. This explains why the lower-right corner of Fig. 9 (and Fig. 10) also remains empty.

We see that the majority of the observed exoplanets fall within the mass-radius diagram obtained with the population synthesis. Kepler-87b and Kepler-39b are notable exceptions, being larger than the predicted synthetic radii of any given mass. The structure of these planets remains to be explained (Bouchy et al. 2011). The large radius of Kepler-39b could be caused by rapid rotation, its rotation period being estimated at only 1.6 ± 0.4 hours (Zhu et al. 2014). It is interesting to note that some objects similar to Kepler-39b in terms of mass, but at even higher insolation, follow the prediction of population syntheses much better (like CoRoT 3b, which has a mass of $21.7 \pm 1.0 M_\mathrm{J}$ and a radius of $1.01 \pm 0.07 R_\mathrm{J}$, see Moutou et al. 2013; Deleuil et al. 2008).

The fact that some of the observed exoplanets with a mass comparable to that of Neptune seem to be smaller than predicted by the model needs to be explained as well. There are two potential explanations. A first possibility is a higher opacity in the protoplanetary envelope during the formation phase. The accretion of H/He during formation is regulated by the envelope's ability to cool, which is in turn mostly controlled by the opacity caused by small grains suspended in the planet's upper radiative zone. If the opacity is higher than assumed in this population synthesis (it is taken here equal to the ISM opacity multiplied by a factor of 0.003, Mordasini et al. 2014), the planets would have smaller radii at a given mass due to a lower H/He fraction. This shows how the planetary mass-radius relation can be used to constrain the microphysical models of the grain dynamics during formation like coagulation, settling and evaporation (Podolak 2003; Ormel 2014; Mordasini 2014). A second possibility is a more efficient envelope evaporation during evolution. If the loss of the primordial H/He envelopes is actually more efficient than in the simple evaporation model taken here (e.g., because of different evaporation regimes applying to planets of different types, Owen and Alvarez 2016), this would also lead to smaller radius at a given mass. These two scenarios could potentially be disentangled by their different dependence on the planet's orbital distance, the planet's atmospheric composition, or on time. But in order to see such features, a high number of well-characterized planets is likely necessary, highlighting the importance of future missions like CHEOPS (Broeg et al. 2013) or PLATO (Rauer et al. 2014).

4.5 The Formation of Close-in Low-Mass Planets, Envelope Evaporation, and the Planetary Mass-Density Diagram

Close-in low-mass planets are extremely common around Sun-like stars with about 45–55 % of FGK stars having a planet with $M \lesssim 30 M_\oplus$ and a period $P < 100$ days, a result from

Table 1 Predictions of three scenarios for the formation of close-in low-mass planets in terms of the resulting core composition of the planets, whether planets attain a primordial H/He envelope, are expected to be in mean motion resonances, and the disk mass necessary to form the planets. The term isolation refers to the situation where the planetesimals' orbits are sufficiently circularized by the protoplanetary gas disk so that they can never grow enough in mass by collisions and cannot attract a large envelope of gas during their formation

Quantity	Scenario		
	Strictly in situ/ no migration	Weak migration, then final assembly	Assembly at $a \gtrsim a_{\rm ice}$, large scale migration
Core composition	Rocky	Rocky to icy with potential gradient	Icy
Primordial H/He	Yes (w/o isolation) No (w isolation)	Yes (w/o isolation) No (w isolation)	Yes
Mean-Motion Resonance	No	No/Yes	Yes/No (see Sect. 3.3.2)
Necessary disk mass	High	Intermediate	Low

high-precision RV surveys (Udry and Mayor 2008; Mayor et al. 2011) that was subsequently confirmed and extended by the *Kepler* survey (Borucki et al. 2011a).

Despite their high frequency, the formation mechanism of this class of planets (without analogue in the Solar system) is currently debated in the literature (e.g., Hansen and Murray 2012; Chiang and Laughlin 2013; Ida et al. 2013; Raymond and Cossou 2014; Schlichting 2014; Hands et al. 2014; Chatterjee and Tan 2014; Ogihara et al. 2015; Inamdar and Schlichting 2015 and Sect. 3.3.2). Some models propose an in situ formation at the current location of the planets close to the star, other models propose that the planets have originally formed further away, potentially in the colder regions of the protoplanetary disk beyond the water ice line, while some other models finally are of a hybrid type where at least some migration of the planets themselves or their building blocks has occurred. As summarized in Table 1, these models differ in their predictions not only regarding the architecture of the planetary systems (e.g., mean motion resonances, concentration towards the inner edge, relative size or mass of the planets), but also in their predicted bulk composition (ice mass fraction in the planetary core, H/He mass fraction).

The table makes clear that knowing the bulk composition (H/He content; rocky, icy, mixed core composition) and also the atmospheric composition would be of very high interest to disentangle the different formation mechanisms. The different bulk compositions express themselves in different mean planetary densities, even if degeneracies unfortunately prevent a unique conversion of the observed mean density into an underlying bulk composition, at least for some parameters (e.g., Adams et al. 2008; Dorn et al. 2015).

Figure 11 shows the planetary mass-mean density diagram obtained in a population synthesis calculation based on the core accretion paradigm and assuming Sun-like stars (Swoboda et al., in prep.). During the formation phase, the concurrent formation of 10 embryos was calculated using the model of Alibert et al. (2013), while during the subsequent long-term evolution, the effects of cooling and contraction as well as atmospheric escape were included (Mordasini et al. 2012a; Jin et al. 2014). Atmospheric escape is an important process for low-mass close-in planets because (i) they are more susceptible to it than giant planets due to their weaker gravitational potential, and (ii) the loss of even a tenuous H/He envelope strongly decreases their radius (e.g., Lopez and Fortney 2013). Envelope evaporation has been observed both for giant and low-mass planets (Vidal-Madjar et al. 2008; Ehrenreich et al. 2015). The mass-density diagram is interesting because it reveals more clearly than the mass-radius diagram the imprint of several prominent physical processes

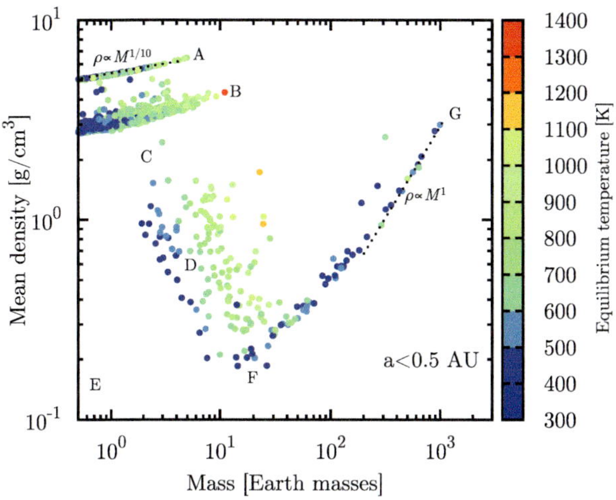

Fig. 11 Mean planetary density $\bar{\rho}$ as a function of mass M in a synthetic population around Sun-like stars at an age of 5 Gyr. Planets have a semi-major axis $a < 0.5$ AU. The *color bar* gives the equilibrium temperature of the planets. Important features are: (A) bare rocky planets; (B) bare icy planets; (C) the evaporation valley; (D) low-mass core-dominated planets with H/He; (E) a forbidden zone because of evaporation ($a < 0.5$ AU); (F) the transition point to giant planets; (G) the giant planet branch. The analytic scaling $\bar{\rho} \propto M^q$ for a polytropic index of $n = 1/7$ ($q = 1/10$) applying to terrestrial planets and $n = 1$ ($q = 1$) for partially degenerate giant planets are overplotted by *dotted lines*

during planet formation and evolution (for a comparison with an observational M–$\bar{\rho}$ diagram, see Hatzes and Rauer 2015). They can be identified with seven features in Fig. 11:

A: bare rocky planets. These are low-mass rocky planets which have either started without an H/He envelope (very low-mass planets) or which have lost it completely due to atmospheric escape. These planets have formed inside the iceline but have migrated to their current orbital distance. In the simulation, they fall on one single mass-density relation as a universal mass fraction of iron (1/3) was assumed in the synthesis, whereas in reality a certain spread is expected due to varying stellar compositions (Dorn et al. 2015) and to giant impacts which can partially remove the mantle (Benz et al. 2007). The colors show that at smaller orbital distances (corresponding to higher equilibrium temperatures T_{eq}), more massive planets have lost all their primordial H/He. This illustrates that the dividing line between solid and gaseous planets depends on orbital distance if it is only due to escape, which is however unlikely (Schlichting et al. 2015; Rogers 2015).

B: bare icy planets. These are low-mass planets formed outside the iceline which have migrated close to their host star and then have lost their H/He due to evaporation. In the syntheses of Alibert et al. (2013), the planetary ice-mass fraction increases with increasing orbital distance and planetary mass, as more massive planets migrate faster in type I migration, and get more easily lost out of type I migration convergence zones ("traps") due to saturation of the corotation torque (e.g., Baruteau et al. 2014; Dittkrist et al. 2014 and Sect. 3.1.2). The mass-density relation is less narrow than for the bare rocky planets indicating a spread in the ice mass fraction, despite the fact that the increase in the planetesimals' surface density at the iceline (a factor of 2) was the same in all synthetic disks. This means that all planets accreting outside of the iceline start with an identical ice mass fraction in the core. The final spread is then a consequence of the accretion of rocky planetesimals during the migration through the inner disk (see Figueira et al. 2009 for the case of GJ436b).

C: evaporation valley. The evaporation valley separates planets without H/He from low-mass planets that still have retained some H/He (Owen and Wu 2013; Lopez and Fortney 2013; Jin et al. 2014). It comes about as a small amount of H/He (just 0.1–1 % in mass) is sufficient to decrease $\bar{\rho}$ by a factor 2 or so, and because the timescale over which this last H/He is lost is short ($\sim 10^5$ years) compared to the planet's lifetime. This means that it is not likely that at a given moment in time, like here 5 Gyrs, many planets are observed during the process of losing the last H/He, leading to the depletion of points in the valley.

D: low-mass core-dominated planets with H/He. These are planets in the super-Earth mass range. They can have both icy or rocky cores depending on their formation history. A clear imprint of evaporation is seen: the hotter a planet, the larger its density at a given total mass as more primordial H/He is lost for closer-in planets.

E: forbidden zone because of evaporation. This lower left triangle remains empty as only planets with $a < 0.5$ AU are included, and at such distances, low-mass, very low-density planets would undergo intense evaporation. Therefore, these planets have already moved to higher mean densities by 5 Gyr. This means that the parameter space is in principle not only three-dimensional (M, $\bar{\rho}$, $T_{\rm eq}$), but that time should be an extra dimension.

F: transition point to giant planets. The point of lowest mean density in the diagram can be defined as the transition point from solid-dominated to gas-dominated planets. It occurs at a mass of 10–30$M_\oplus$, which corresponds to the mass where rapid gas accretion starts in the core-accretion model (Pollack et al. 1996). In the planetary mass function, a break in the slope can be seen at such a mass in both the theoretical and observational data (Mordasini et al. 2009b; Mayor et al. 2011).

G: giant planet branch. With further increasing mass, planets become denser again as the self-gravitational compaction becomes stronger for these gas-dominated, more compressible planets. In this regime, closer-in planets have a lower density (even if this is not clearly visible in this specific synthesis due to the low number of giant planets inside of 0.5 AU), which is the opposite to the core-dominated planets with H/He. For giant plants, the change of the atmospheric boundary conditions for more strongly irradiated planets delays somewhat the contraction (Guillot and Showman 2002), resulting in lower densities at higher $T_{\rm eq}$. Additional radius inflation mechanisms (not included in this synthesis) tend to render this correlation even clearer (e.g., Laughlin et al. 2011).

Other features in the mass-mean density diagram, in particular the mass scalings, can be understood by considering polytropic models for planetary interiors for which the pressure P depends on the density ρ as $P \propto \rho^{1+1/n}$, with n the polytropic index (Chandrasekhar 1939; Chabrier et al. 2009). For such models, the mean density scales with mass as $\bar{\rho} \propto M^{2n/(3-n)}$.

In the giant planet regime the electrons are partially degenerate due to the relatively low temperatures and high densities. But the degeneracy is not complete so that both the quantum electron-degeneracy pressure and the classical electrostatic contribution from the non-degenerate ions matter. The polytropic index is therefore $n = 1$ which corresponds to a (mean) density that increases linearly with mass ($\bar{\rho} \propto M$) or, equivalently, to a radius that is independent of mass. Indeed, such a behavior is seen in the numerical simulations, as shown by the right dotted line close to Feature G (the giant planet branch). At lower masses, the degeneracy decreases and the composition changes from gas-dominated to solid-dominated planets. These planets are described by equations of state which are much less compressible.

For completely incompressible matter ($n = 0$), the mean density would be independent of mass, corresponding to a horizontal line in Fig. 11. Such a regime is approached in the case

Table 2 Power-law indices governing the structure of various types of planets described by polytropic models (see text): $n = 1/(\gamma - 1)$ is the polytropic index, γ the adiabatic index, $p = (1-n)/(3-n)$ describes the power-law dependence between radius and mass, $q = 2n/(3-n)$ that between mean density and mass, and $\chi_{\rm norm} = 1/\gamma$ that between mean density and pressure

Planet type	Quantity			
	Polytropic index n	$p = \partial \log R/\partial \log M$	$q = \partial \log \bar{\rho}/\partial \log M$	$\chi_{\rm norm} = \partial \log\bar{\rho}/\partial \log P$
Incompressible	0	1/3	0	0
Terrestrial planets	1/7	3/10	1/10	1/8
Giant planets	1	0	1	1/2
Low mass BDs	6/5	−1/9	4/3	6/11
High mass BDs	3/2	−1/3	2	3/5

of low-mass solid planets. The small, but still non-zero compressibility of rocks, ices, and iron however causes a weak increase of the density with mass also in this regime (see Features A and B). For low-mass solid planets, the mass-radius relation scales approximately as $R \propto M^{3/10}$ (Grasset et al. 2009), corresponding to $n = 1/7$, and therefore $\bar{\rho} \propto M^{1/10}$ which is a weak increase. Such scaling is shown by the left dotted line in Fig. 11. It approximately reproduces the results found by solving directly the interior structure equations of bare rocky or icy planets (Mordasini et al. 2012b).

The scaling relations for polytropic models are summarized in Table 2. The results are shown for planets that are approximately incompressible (very low-mass planets), terrestrial planets, partially degenerate giant planets with masses roughly between 1 and $10M_{\rm J}$, more strongly degenerate high-mass giant planets and low-mass brown dwarfs with masses from about 10 to $30M_{\rm J}$, and finally high-mass brown dwarfs which can be approximated with a fully degenerate non-relativistic equation of state ($n = 3/2$). The table gives the polytropic index n, the power-law exponent p for the radius R as function of mass M ($R \propto M^p$), which, from the polytropic Lane-Emden equation, is linked to n via $p = (1-n)/(3-n)$. The table also gives the power-law exponent q for the mean density as function of mass ($\bar{\rho} \propto M^q$), given by $q = 2n/(3-n)$, and finally the dimensionless power-law compressibility $\chi_{\rm norm} = \partial \log\bar{\rho}/\partial \log P$, i.e. $\chi_{\rm norm} = n/(n+1) = 1/\gamma$ (γ denotes the adiabatic index). This is a good measure of the compressibility because it shows that for small values of $\chi_{\rm norm}$ the density will only increase weakly even if the pressure increases strongly. For large values of $\chi_{\rm norm}$, small changes in the pressure can change the density substantially. The quantity $\chi_{\rm norm}$ increases with increasing mass. In reality, the transition from one regime to the other is smooth, as can be seen in Fig. 9. In this figure we see for example that, for low-mass giant planets, the radius increases with mass indicating an effective $n < 1$, followed by the maximum at about $4M_{\rm J}$ where $n = 1$, and finally a smooth increase to even higher n for more massive objects that only ends when hydrogen burning sets in at about 80 Jovian masses.

4.6 Radius inflation of Hot Jupiters

Radius inflation is a distinctive feature of close-in giant planets which was not expected theoretically and which is still not fully understood. With a radius of about $1.38R_{\rm J}$, HD 209458 b, the first detected transiting extrasolar planet (Henry et al. 2000; Charbonneau et al. 2000), is significantly larger than expected from theoretical evolution models, even when the effects of stellar irradiation are taken into account (Guillot and Showman 2002). This can

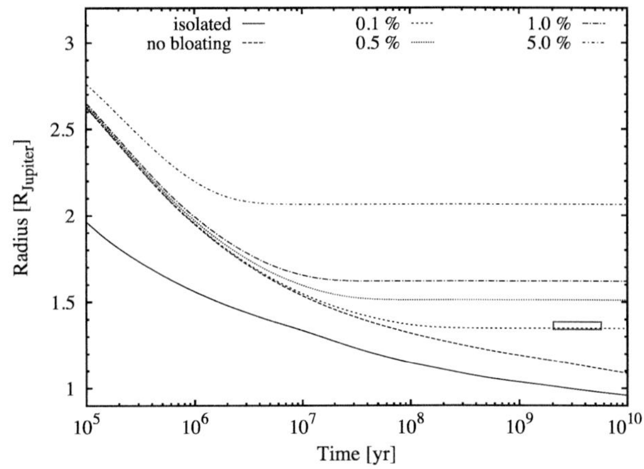

Fig. 12 Time evolution of the radius of a HD 209458 b-like planet. Results are shown for an isolated planet without stellar irradiation, an irradiated planet without inflation, and for four irradiated planet models where a fraction between 0.1 and 5 % of the incoming stellar flux is added to the planet's luminosity in the deep convective zone. The *red box* indicates the observed radius and age of HD 209458 b

be seen in Fig. 9 which shows that standard cooling models do not predict the existence of planets with radii larger than about $1.1R_J$ (or up to $1.3R_J$ if smaller orbital distances are included) at ages of a few Gyrs. Yet, a significant fraction of hot Jupiters are larger than above threshold values, with some extreme cases such as HAT-P-32 b ($R = 1.79 \pm 0.03R_J$, Hartman et al. 2011) and WASP 17 b ($R = 1.93 \pm 0.08R_J$, Anderson et al. 2010). The latter has a mean density about 10 times smaller than Jupiter's.

At the time of writing, thanks to a fairly large sample of transiting giant planets with a wide range of orbital distances, it is fairly well established that there exists a correlation between inflation and the intensity of stellar irradiation (Laughlin et al. 2011). It is found that modestly irradiated planets do not have inflated radii and that below a threshold value for the incident flux of about 2×10^8 erg s^{-1} cm^{-2} (corresponding to a solar-like star at an orbital distance of about 0.08 AU) radius inflation ceases (Demory and Seager 2011).

The observation that radius inflation, also called radius *anomaly*, is correlated with the incident stellar flux may therefore indicate that some of the incident stellar energy is deposited deep inside the planets' interior, thereby delaying or even halting their contraction. For this to work, part of the stellar irradiation flux must be converted into some other form of energy that is then deposited in the deeper layers, as pure irradiation of the planet's atmosphere by the star does not penetrate deep enough into the planet to cause a large enough inflation (Burrows et al. 2000) – it is instead radiated back to space. From evolutionary models, one finds that 0.1–1 % of the stellar flux impinging on the planet's atmosphere is sufficient to explain the radius anomaly at the current age of the planets if it can penetrate to the deep convective zone of the planets (Guillot and Showman 2002).

This is illustrated in Fig. 12, which displays the time evolution of the radius of a planet with parameters similar to HD 209458 b ($M = 0.69 \pm 0.02M_J$, $a = 0.0475 \pm 0.0006$ AU, age $= 4.0 \pm 2.0$ Gyr). The results of six models are illustrated in the figure: an isolated planet (without stellar irradiation), an irradiated planet for which stellar irradiation does not come as an additional energy source for the planet (no bloating, that is no radius inflation), and for four irradiated planets for which a fraction of the incoming stellar flux is added to the planet's intrinsic luminosity at the center (thereby inflating the planet). According to these calculations, which use the planet evolution model of Mordasini et al. (2015), stellar irradiation increases the radius relative to the isolated case at late times by about $0.13R_J$, bringing it to $1.13R_J$ at 4 Gyr, which is clearly less than the observed radius (marked by the red box in Fig. 12). These calculations indicate that about 0.1 % of the incoming flux must be added to the planet luminosity in the deep convective zone in order to reproduce approximately the observed radius (for comparison, Guillot and Showman 2002 found about 0.08 %).

We note that a difference in radius of about $0.25 R_J$ between the standard irradiated model and the aforementioned inflated model corresponds to a much hotter thermodynamics state of the planet as the partially degenerate equation of state governing the planet's interior means that the radius is only weakly dependent on temperature. This is visible from the result that the irradiated planet without additional energy sources contracted to the observed radius when the planet was almost 100 times younger than the actual age of the system. The 0.1 % fraction of the incoming stellar irradiation necessary to bring the planet to its observed radius corresponds to a luminosity that is about 90 times bigger than the present-day luminosity of a planet undergoing standard cooling without inflation.

There are two main mechanisms that can turn stellar irradiation into an energy flux reaching the deep interior of planets, but note that none of them can so far explain the radius anomaly of all extrasolar planets *quantitatively* (Baraffe et al. 2014):

- The first one is the so-called ohmic heating mechanism, for which ionized alkali atoms in the winds caused by stellar irradiation of the planet produce an electric current that is dissipated in the deep layers of the planet (Batygin and Stevenson 2010; Perna et al. 2010). However, the study by Huang and Cumming (2012), who calculated the heating rate more self-consistently than earlier works, indicates that ohmic dissipation might not be strong enough to explain the radius anomaly of the very massive and very hot Jupiters.
- The second mechanism is the so-called *weather noise* mechanism (Showman and Guillot 2002). The strong winds resulting from the large temperature contrasts between the day- and night-sides of tidally locked planets are found in the numerical simulations of atmospheric circulation of Showman and Guillot (2002) to lead to a significant downward transport of kinetic energy. This energy is then dissipated in the deep layers, the details of which are currently not well understood. Also, the feasibility of a sufficiently large downward transport of kinetic energy depends on the details of the employed atmospheric circulation model and is subject to significant uncertainties (Baraffe et al. 2014). Lastly, the combination of tides together with the temperature contrast between the planetary day- and night-sides may produce work in the planet which could again result in an energy flux to the planet's deep layers (thermal tides, see Arras and Socrates 2010).

Other mechanisms have been proposed to explain the observed radius anomaly of hot Jupiters, which we briefly outline below:

- An increased metallicity in the planet's atmosphere which could lead to a slower contraction, because the opacity is expected to increase along with the metallicity of the planet and therefore to slow down its cooling (Burrows et al. 2007),
- A decrease in the efficiency of convective energy transport in the deep interior of the planet due to compositional gradients and the associated onset of semi-convection, which would also slow down the planets' cooling (Chabrier and Baraffe 2007),
- Tidal dissipation in the planets' interior resulting from star-planet tidal interactions (e.g., Bodenheimer et al. 2001). While tidal circularization of a giant planet (whose eccentricity could arise from interactions with unseen planetary companions) could provide significant heating in the early evolution of the planet, it is unlikely to remain a significant source of radius inflation at the typical ages where inflated hot Jupiters are observed (Leconte et al. 2010). Also, tidal dissipation does not account for the observed dependency between radius anomaly and stellar irradiation flux.

In conclusion of this paragraph, the correlation between the occurrence of inflated hot Jupiters and stellar irradiation is well established observationally, but the mechanisms behind it remain poorly understood. Theorists have yet to come up with a convincing theory which can explain the observations and quantitatively work when coupled to planetary structure models.

5 Summary

The main points to take away from this chapter are the following:

- The planetesimal accretion scenario of planetary growth is dictated by self-gravity, and proceeds through runaway growth and oligarchic growth. It can account for terrestrial planet formation in the inner Solar system. Growth becomes inefficient in the oligarchic phase, and growth timescales become exceedingly long toward large orbital separations.
- Accretion of small pebbles is assisted by gas drag, and is most efficient for marginally coupled particles with stopping time to orbital period ratios between roughly 0.1 and 1. Once a substantial fraction of solid mass resides in such pebbles, embryos that have grown across the transition mass accrete most pebbles entering their Hill sphere, and growth timescales are well within the lifetime of protoplanetary disks even at large orbital separations.
- Both planetesimal and pebble accretions likely play important roles in the formation of the Solar system as well as for exoplanets, and they may dominate at different phases and in different regions of protoplanetary disks. Global models of planet formation have begun to take both accretion scenarios into account. Pinning down the uncertainties requires better understandings of disk physics.
- The gravitational interaction between planets and their protoplanetary disk plays a prominent role in the early orbital evolution of planetary systems. Planet-disk interactions usually damp the eccentricity and inclination, but can lead to inward or outward migration depending on the planet's mass and, crucially, the disk's physical properties near the planet (radial profiles of the disk's density and temperature, magnetic field, turbulence). While tremendous progress has been made in the past few years in understanding the physics of planetary migration, especially that of low-mass planets, predictive scenarios of planet migration require more knowledge of protoplanetary disks in regions of planet formation. Pursuing detailed observations of protoplanetary disks as well as comprehensive models of the dynamics of such disks will help theories of disk migration become more predictive.
- While disk migration is certainly inevitable, it is one among many mechanisms that shape the architecture of planetary systems. The great diversity of extrasolar planetary systems shows that disk and high-eccentricity migrations, interactions with host and nearby stars all play some role. (Mis-)aligned hot Jupiters and the many multiple super-Earths systems illustrate the plurality of mechanisms shaping planetary systems.
- Evolutionary models establish the link between formation and observations. They are keys to estimate the mass of young directly imaged planets for which only the age and luminosity are known, and to learn about the interior composition and structure of planets when only their mass and radius are known like for close-in low-mass planets. For this class of planets, the bulk composition like the H/He mass fraction and the ice mass fraction in the core represent important constraints for their formation mode(s). Atmospheric compositional measurements (using transit spectra) could help break degeneracies between composition and structure in the future, and help constrain the planet's formation mechanism and migration history.

- The transition between giant planets and brown dwarfs is smooth. Planets with a mass of $\sim 13M_J$ can burn deuterium in their interior, just like brown dwarfs. After a rapid phase of deuterium burning, a non deuterium-burning planet of $12M_J$ and a planet that has depleted its deuterium (say, with a mass of $14M_J$) will have very similar structures, comparable to that of low-mass brown dwarfs after their deuterium is depleted (except, for a solid core which can, however, dissolve into the H/He in time).
- Like for models of planet migration, predictions of evolutionary models are tied to the formation process (e.g., hot or cold start models), especially for young planets. While many exoplanets are well described by the mass-radius or the mass-mean density distributions predicted by core accretion linked to evolutionary models, there are several ways such models can be improved, in particular regarding the planet's internal structure (e.g., the occurrence of semi-convection), its material properties (via updated equations of states at high pressures and temperatures), and models for special evolutionary effects like atmospheric escape or radius inflation.

Acknowledgements We thank the organizers of the ISSI workshop "The Disk in Relation to the Formation of Planets and their Protoatmospheres", which was held in Beijing in August 2014. We thank D. Lin, N. Madhusudhan, Z. Sandor and S. Udry for stimulating discussions at the workshop. We thank Gabriel Marleau for interesting discussions and for providing Figure 7, and David Swoboda and Yann Alibert for providing the starting data for Figure 11. We also thank Aurélien Crida and Bertram Bitsch for detailed comments on an earlier draft of this paper, and the referee for a detailed and constructive report. XNB acknowledges support from Hubble and ITC Fellowships. CM acknowledges the support from the Swiss National Science Foundation under grant BSSGI0_155816 "PlanetsInTime".

References

E.R. Adams et al., Astrophys. J. **673**, 1160 (2008)
S. Albrecht et al., Astrophys. J. **757**, 18 (2012)
Y. Alibert et al., Astron. Astrophys. **434**, 343 (2005)
Y. Alibert et al., Astron. Astrophys. **558**, A109 (2013)
ALMA Partnership et al., Astrophys. J. Lett. **808**, L3 (2015)
D.R. Anderson et al., Astrophys. J. **709**, 159 (2010)
P.J. Armitage, *Astrophysics of Planet Formation* (2010)
P. Arras, A. Socrates, Astrophys. J. **714**, 1 (2010)
X.-N. Bai, Astrophys. J. **798**, 84 (2015)
X.-N. Bai, J.M. Stone, Astrophys. J. **769**, 76 (2013)
X.-N. Bai, J.M. Stone, Astrophys. J. **796**, 31 (2014)
I. Baraffe et al., Astron. Astrophys. **402**, 701 (2003)
I. Baraffe et al., Astron. Astrophys. **482**, 315 (2008)
I. Baraffe et al., in *Protostars and Planets VI* (2014), pp. 763–786
A.J. Barker, Y. Lithwick, Mon. Not. R. Astron. Soc. **437**, 305 (2014)
C. Baruteau, F. Masset, Astrophys. J. **672**, 1054 (2008)
C. Baruteau, F. Masset, in *Lecture Notes in Physics*, vol. 861, ed. by J. Souchay, S. Mathis, T. Tokieda (Springer, Berlin, 2013), p. 201
C. Baruteau, J.C.B. Papaloizou, Astrophys. J. **778**, 7 (2013)
C. Baruteau et al., Astron. Astrophys. **533**, A84 (2011a)
C. Baruteau et al., Mon. Not. R. Astron. Soc. **416**, 1971 (2011b)
C. Baruteau et al., in *Protostars and Planets VI* (2014), pp. 667–689
M.R. Bate et al., Mon. Not. R. Astron. Soc. **401**, 1505 (2010)
K. Batygin, Nature **491**, 418 (2012)
K. Batygin, D.J. Stevenson, Astrophys. J. Lett. **714**, L238 (2010)
K.R. Bell, D.N.C. Lin, Astrophys. J. **427**, 987 (1994)
M. Benisty et al., Astron. Astrophys. **578**, L6 (2015)
P. Benítez-Llambay et al., Nature **520**, 63 (2015)
O. Benomar et al., Publ. Astron. Soc. Jpn. **66**, 94 (2014)

W. Benz et al., Space Sci. Rev. **132**, 189 (2007)
W. Benz et al., in *Protostars and Planets VI* (2014), pp. 691–713
B. Bitsch et al., Astron. Astrophys. **555**, A124 (2013a)
B. Bitsch et al., Astron. Astrophys. **549**, A124 (2013b)
B. Bitsch et al., Astron. Astrophys. **570**, A75 (2014)
B. Bitsch et al., Astron. Astrophys. **582**, A112 (2015a)
B. Bitsch et al., Astron. Astrophys. **575**, A28 (2015b)
P.H. Bodenheimer, J.B. Pollack, Icarus **67**, 391 (1986)
P.H. Bodenheimer et al., Icarus **143**, 2 (2000)
P. Bodenheimer et al., Astrophys. J. Lett. **548**, 466 (2001)
P. Bodenheimer et al., Astrophys. J. **770**, 120 (2013)
A.C. Boley, Astrophys. J. Lett. **695**, L53 (2009)
A.C. Boley et al., Icarus **207**, 509 (2010)
W.J. Borucki et al., Astrophys. J. **728**, 117 (2011a)
W.J. Borucki et al., Astrophys. J. **736**, 19 (2011b)
A.P. Boss, Science **276**, 1836 (1997)
A.P. Boss, Astrophys. J. **661**, L73 (2007)
F. Bouchy et al., Astron. Astrophys. **533**, A83 (2011)
C. Broeg et al., in *European Physical Journal Web of Conferences*, vol. 47 (2013), p. 3005
A. Burrows, J. Liebert, Rev. Mod. Phys. **65**, 301 (1993)
A. Burrows et al., Astrophys. J. **491**, 856 (1997)
A. Burrows et al., Astrophys. J. Lett. **534**, L97 (2000)
A. Burrows et al., Rev. Mod. Phys. **73**, 719 (2001)
A. Burrows et al., Astrophys. J. **661**, 502 (2007)
D. Cébron et al., Icarus **226**, 1642 (2013)
G. Chabrier, I. Baraffe, Annu. Rev. Astron. Astrophys. **38**, 337 (2000)
G. Chabrier, I. Baraffe, Astrophys. J. Lett. **661**, L81 (2007)
G. Chabrier et al., Am. Inst. Phys. Conf. Ser. **1094**, 102 (2009)
G. Chabrier et al., in *Protostars and Planets VI* (2014), pp. 619–642
J.E. Chambers, Icarus **233**, 83 (2014)
S. Chandrasekhar, *An Introduction to the Study of Stellar Structure* (The University of Chicago Press, Chicago, 1939)
W.J. Chaplin et al., Astrophys. J. **766**, 101 (2013)
D. Charbonneau et al., Astrophys. J. Lett. **529**, L45 (2000)
S. Chatterjee, J.C. Tan, Astrophys. J. **780**, 53 (2014)
S. Chatterjee et al., Astrophys. J. **686**, 580 (2008)
E. Chiang, G. Laughlin, Mon. Not. R. Astron. Soc. **431**, 3444 (2013)
B. Commerçon et al., Astron. Astrophys. **530**, 13 (2011)
A. Crida, K. Batygin, Astron. Astrophys. **567**, A42 (2014)
A. Crida, A. Morbidelli, Mon. Not. R. Astron. Soc. **377**, 1324 (2007)
A. Crida et al., Icarus **181**, 587 (2006)
A. Crida et al., Astron. Astrophys. **502**, 679 (2009)
T. Currie et al., Astrophys. J. Lett. **796**, L30 (2014)
G. D'Angelo, S.H. Lubow, Astrophys. J. **685**, 560 (2008)
G. D'Angelo et al., Mon. Not. R. Astron. Soc. **358**, 316 (2005)
G. D'Angelo et al., Giant planet formation, in *Exoplanets*, ed. by S. Seager (University of Arizona Press, Tucson, 2011), pp. 319–346
M. Deleuil et al., Astron. Astrophys. **491**, 889 (2008)
B.-O. Demory, S. Seager, Astrophys. J. Suppl. Ser. **197**, 12 (2011)
H.E. Dewitt et al., Astrophys. J. **181**, 439 (1973)
K.-M. Dittkrist et al., Astron. Astrophys. **567**, A121 (2014)
R. Dong et al., Astrophys. J. **741**, 57 (2011)
R. Dong et al., Astrophys. J. Lett. **809**, L5 (2015)
C. Dorn et al., Astron. Astrophys. **577**, A83 (2015)
C.D. Dressing et al., Astrophys. J. **800**, 135 (2015)
P.C. Duffell, Astrophys. J. Lett. **807**, L11 (2015)
P.C. Duffell, A.I. MacFadyen, Astrophys. J. **769**, 41 (2013)
R.H. Durisen et al., in *Protostars and Planets V* (2007), pp. 607–622
C. Dürmann, W. Kley, Astron. Astrophys. **574**, A52 (2015)
D. Ehrenreich et al., Nature **522**, 459 (2015)
J.A. Eisner, Astrophys. J. Lett. **803**, L4 (2015)

C. Espaillat et al., in *Protostars and Planets VI* (2014), pp. 497–520
D. Fabrycky, S. Tremaine, Astrophys. J. **669**, 1298 (2007)
S.M. Fendyke, R.P. Nelson, Mon. Not. R. Astron. Soc. **437**, 96 (2014)
P. Figueira et al., Astron. Astrophys. **493**, 671 (2009)
D.A. Fischer, J. Valenti, Astrophys. J. **622**, 1102 (2005)
D. Forgan, K. Rice, Mon. Not. R. Astron. Soc. **432**, 3168 (2013)
J.J. Fortney, N. Nettelmann, Space Sci. Rev. **152**, 423 (2010)
J.J. Fortney et al., Astron. Nachr. **326**, 925 (2005)
W.A. Fowler et al., Annu. Rev. Astron. Astrophys. **5**, 525 (1967)
R.S. Freedman et al., Astrophys. J. Suppl. Ser. **174**, 504 (2008)
S. Fromang et al., Mon. Not. R. Astron. Soc. **363**, 943 (2005)
J. Fung et al., Astrophys. J. **782**, 88 (2014)
M. Galvagni et al., Mon. Not. R. Astron. Soc. **427**, 1725 (2012)
C.F. Gammie, Astrophys. J. **553**, 174 (2001)
P. Goldreich et al., Annu. Rev. Astron. Astrophys. **42**, 549 (2004)
H.C. Graboske et al., Astrophys. J. **181**, 457 (1973)
H.C. Graboske et al., Astrophys. J. **199**, 265 (1975)
O. Grasset et al., Astrophys. J. **693**, 722 (2009)
R. Greenberg et al., Icarus **35**, 1 (1978)
O. Gressel et al., Astrophys. J. **779**, 59 (2013)
O. Gressel et al., Astrophys. J. **801**, 84 (2015)
J. Guilet et al., Mon. Not. R. Astron. Soc. **430**, 1764 (2013)
T. Guillot, Science **286**, 72 (1999)
T. Guillot, Annu. Rev. Astron. Astrophys. **33**(1), 493 (2005)
T. Guillot, D. Gautier, ArXiv e-prints (2014)
T. Guillot, A.P. Showman, Astron. Astrophys. **385**, 156 (2002)
T. Guillot et al., Astrophys. J. **450**, 463 (1995)
T. Guillot et al., *The Interior of Jupiter* (Cambridge University Press, Cambridge, 2004), pp. 35–57
K.E. Haisch Jr. et al., Astrophys. J. Lett. **553**, L153 (2001)
T.O. Hands et al., Mon. Not. R. Astron. Soc. **445**, 749 (2014)
B.M.S. Hansen, N. Murray, Astrophys. J. **751**, 158 (2012)
B.M.S. Hansen, N. Murray, Astrophys. J. **775**, 53 (2013)
L. Hartmann et al., Astrophys. J. **475**, 770 (1997)
J.D. Hartman et al., Astrophys. J. **742**, 59 (2011)
A.P. Hatzes, H. Rauer, Astrophys. J. Lett. **810**, L25 (2015)
C. Hayashi, T. Nakano, Prog. Theor. Phys. **30**, 460 (1963)
R. Helled et al., in *Protostars and Planets VI* (2014), pp. 643–665
G.W. Henry et al., Astrophys. J. Lett. **529**, L41 (2000)
A.M. Hofmeister, R.E. Criss, Tectonophysics **395**, 159 (2005)
X. Huang, A. Cumming, Astrophys. J. **757**, 47 (2012)
W.B. Hubbard, Rev. Geophys. **18**(1), 1 (1980)
D. Huber et al., Science **342**, 331 (2013)
E. Iaroslavitz, M. Podolak, Icarus **187**, 600 (2007)
S. Ida, D.N.C. Lin, Astrophys. J. **673**, 487 (2008)
S. Ida, J. Makino, Icarus **106**, 210 (1993)
S. Ida et al., Astrophys. J. **775**, 42 (2013)
M. Ikoma, Y. Hori, Astrophys. J. **753**, 66 (2012)
M. Ikoma et al., Astrophys. J. **537**, 1013 (2000)
N.K. Inamdar, H.E. Schlichting, Mon. Not. R. Astron. Soc. **448**, 1751 (2015)
S. Jin et al., Astrophys. J. **795**, 65 (2014)
A. Johansen, P. Lacerda, Mon. Not. R. Astron. Soc. **404**, 475 (2010)
J.A. Johnson et al., Publ. Astron. Soc. Pac. **122**, 905 (2010)
A. Juhász et al., Mon. Not. R. Astron. Soc. **451**, 1147 (2015)
M. Jurić, S. Tremaine, Astrophys. J. **686**, 603 (2008)
Kamland Collaboration et al., Nat. Geosci. **4**, 647 (2011)
G.M. Kennedy, S.J. Kenyon, Astrophys. J. **673**, 502 (2008)
S.J. Kenyon, B.C. Bromley, Astron. J. **131**, 1837 (2006)
S.J. Kenyon, B.C. Bromley, Astrophys. J. Lett. **690**, L140 (2009)
R. Kippenhahn, A. Weigert, *Stellar Structure and Evolution* (1990)
W. Kley, A. Crida, Astron. Astrophys. **487**, L9 (2008)
W. Kley, R.P. Nelson, Annu. Rev. Astron. Astrophys. **50**, 211 (2012)

W. Kley et al., Astron. Astrophys. **506**, 971 (2009)
E. Kokubo, S. Ida, Icarus **114**, 247 (1995)
E. Kokubo, S. Ida, Icarus **123**, 180 (1996)
E. Kokubo, S. Ida, Icarus **131**, 171 (1998)
E. Kokubo, S. Ida, Icarus **143**, 15 (2000)
D.G. Korycansky, J.B. Pollack, Icarus **102**, 150 (1993)
K.M. Kratter, G. Lodato, *ArXiv e-prints* (2016)
K.A. Kretke, H.F. Levison, Astron. J. **148**, 109 (2014)
S.S. Kumar, Astrophys. J. **137**, 1121 (1963)
A.-M. Lagrange et al., Astron. Astrophys. **493**, L21 (2009)
A.-M. Lagrange et al., Science **329**, 57 (2010)
D. Lai, Mon. Not. R. Astron. Soc. **423**, 486 (2012)
M. Lambrechts, A. Johansen, Astron. Astrophys. **544**, A32 (2012)
M. Lambrechts, A. Johansen, Astron. Astrophys. **572**, A107 (2014)
H. Lammer et al., Astrophys. J. Lett. **598**, L121 (2003)
G. Laughlin et al., Astrophys. J. Lett. **729**, L7 (2011)
J. Leconte, G. Chabrier, Astron. Astrophys. **540**, A20 (2012)
J. Leconte, G. Chabrier, Nat. Geosci. **6**, 347 (2013)
J. Leconte et al., Astron. Astrophys. **516**, A64 (2010)
E.J. Lee, E. Chiang, Astrophys. J. **811**, 41 (2015)
E. Lega et al., Mon. Not. R. Astron. Soc. **440**, 683 (2014)
G. Lesur et al., Astron. Astrophys. **566**, A56 (2014)
D.N.C. Lin, J. Papaloizou, Astrophys. J. **307**, 395 (1986)
Y. Lithwick, Y. Wu, Astrophys. J. Lett. **756**, L11 (2012)
K. Lodders, Astrophys. J. **591**, 1220 (2003)
E.D. Lopez, J.J. Fortney, Astrophys. J. **776**, 2 (2013)
E.D. Lopez, J.J. Fortney, Astrophys. J. **792**, 1 (2014)
S.H. Lubow et al., Astrophys. J. **526**, 1001 (1999)
N. Madhusudhan et al., Astrophys. J. Lett. **794**, L12 (2014)
M. Malik et al., Astrophys. J. **802**, 56 (2015)
G.W. Marcy et al., Astrophys. J. Suppl. Ser. **210**, 20 (2014)
G.-D. Marleau, A. Cumming, Mon. Not. R. Astron. Soc. **437**, 1378 (2014)
M.S. Marley et al., Astrophys. J. **655**, 541 (2007)
C. Marois et al., Nature **468**, 1080 (2010)
F. Marzari et al., Astron. Astrophys. **514**, L4 (2010)
F.S. Masset, Celest. Mech. Dyn. Astron. **111**, 131 (2011)
F.S. Masset, P. Benítez-Llambay, Astrophys. J. **817**, 19 (2016)
F.S. Masset, J. Casoli, Astrophys. J. **723**, 1393 (2010)
F.S. Masset, J.C.B. Papaloizou, Astrophys. J. **588**, 494 (2003)
M. Mayor et al., arXiv, astro-ph.EP (2011)
T. Mazeh et al., Astrophys. J. **801**, 3 (2015)
N.K. Miller, J.J. Fortney, Astrophys. J. Lett. **736**, L29 (2011)
H. Mizuno, Prog. Theor. Phys. **64**, 544 (1980)
P. Mollière, C. Mordasini, Astron. Astrophys. **547**, A105 (2012)
M. Montesinos et al., Astrophys. J. **806**, 253 (2015)
A. Morbidelli, D. Nesvorny, Astron. Astrophys. **546**, A18 (2012)
A. Morbidelli et al., Icarus **258**, 418 (2015)
C. Mordasini, Astron. Astrophys. **558**, A113 (2013)
C. Mordasini, Astron. Astrophys. **572**, A118 (2014)
C. Mordasini et al., in *Tenth Anniversary of 51 Peg-b: Status of Prospects for Hot Jupiter Studies*, ed. by L. Arnold, F. Bouchy, C. Moutou (2006) pp. 84–86
C. Mordasini et al., Astron. Astrophys. **501**, 1139 (2009a)
C. Mordasini et al., Astron. Astrophys. **501**, 1161 (2009b)
C. Mordasini et al., Astron. Astrophys. **547**, 111 (2012a)
C. Mordasini et al., Astron. Astrophys. **547**, 112 (2012b)
C. Mordasini et al., Astron. Astrophys. **566**, A141 (2014)
C. Mordasini et al., Int. J. Astrobiol. **14**, 201 (2015)
C. Moutou et al., Icarus **226**, 1625 (2013)
T. Muto et al., Astrophys. J. **724**, 448 (2010)
S. Naoz et al., Astrophys. J. Lett. **754**, L36 (2012)
A. Natta et al., in *Protostars and Planets V*, ed. by B. Reipurth, D. Jewitt, K. Keil (2007), pp. 767–781

S. Nayakshin, Mon. Not. R. Astron. Soc. **408**, L36 (2010)
S. Nayakshin, S.-H. Cha, Mon. Not. R. Astron. Soc. **435**, 2099 (2013)
R. Neuhäuser, T.O.B. Schmidt, *ArXiv e-prints* (2012)
M. Ogihara et al., Astron. Astrophys. **578**, A36 (2015)
G.I. Ogilvie, Annu. Rev. Astron. Astrophys. **52**, 171 (2014)
C.W. Ormel, Astrophys. J. Lett. **789**, L18 (2014)
C.W. Ormel, H.H. Klahr, Astron. Astrophys. **520**, A43 (2010)
J.E. Owen, M.A. Alvarez, Astrophys. J. **816**, 34 (2016)
J.E. Owen, A.P. Jackson, Mon. Not. R. Astron. Soc. **425**, 2931 (2012)
J.E. Owen, Y. Wu, Astrophys. J. **775**, 105 (2013)
S.-J. Paardekooper, G. Mellema, Astron. Astrophys. **459**, L17 (2006)
S.-J. Paardekooper, J.C.B. Papaloizou, Astron. Astrophys. **485**, 877 (2008)
S. Paardekooper et al., Mon. Not. R. Astron. Soc. **401**, 1950 (2010)
S. Paardekooper et al., Mon. Not. R. Astron. Soc. **410**, 293 (2011)
S.-J. Paardekooper et al., Mon. Not. R. Astron. Soc. **434**, 3018 (2013)
S. Paardekooper, Mon. Not. R. Astron. Soc. **444**, 2031 (2014)
J.C.B. Papaloizou, Celest. Mech. Dyn. Astron. **111**, 83 (2011)
J.C.B. Papaloizou et al., Mon. Not. R. Astron. Soc. **350**, 829 (2004)
A. Pepliński et al., Mon. Not. R. Astron. Soc. **386**, 164 (2008)
H.B. Perets, R.A. Murray-Clay, Astrophys. J. **733**, 56 (2011)
S. Perez et al., Astrophys. J. Lett. **811**, L5 (2015)
R. Perna et al., Astrophys. J. **724**, 313 (2010)
F. Perri, A.G.W. Cameron, Icarus **22**, 416 (1974)
G. Picogna, F. Marzari, Astron. Astrophys. **583**, A133 (2015)
A. Pierens, Mon. Not. R. Astron. Soc. **454**, 2003 (2015)
A. Pierens et al., Astron. Astrophys. **531**, A5 (2011)
M. Podolak, Icarus **165**, 428 (2003)
J.B. Pollack et al., Icarus **30**, 111 (1977)
J.B. Pollack et al., Icarus **124**, 62 (1996)
T. Prodanović et al., Mon. Not. R. Astron. Soc. **406**, 1108 (2010)
S.P. Quanz et al., Astrophys. J. **807**, 64 (2015)
R.R. Rafikov, Astrophys. J. **572**, 566 (2002)
R.R. Rafikov, Astron. J. **128**, 1348 (2004)
R.R. Rafikov, Astrophys. J. Lett. **621**, L69 (2005)
F.A. Rasio, E.B. Ford, Science **274**, 954 (1996)
H. Rauer et al., Exp. Astron. **38**, 249 (2014)
S.N. Raymond, C. Cossou, Mon. Not. R. Astron. Soc. **440**, L11 (2014)
S.N. Raymond et al., in *Protostars and Planets VI* (2014), pp. 595–618
H. Rein, Mon. Not. R. Astron. Soc. **427**, L21 (2012)
L. Ricci et al., Astron. Astrophys. **512**, A15 (2010)
J. Rodmann et al., Astron. Astrophys. **446**, 211 (2006)
L.A. Rogers, Astrophys. J. **801**, 41 (2015)
T.M. Rogers, D.N.C. Lin, Astrophys. J. Lett. **769**, L10 (2013)
R. Sanchis-Ojeda et al., Astrophys. J. **733**, 127 (2011)
N.C. Santos et al., Astron. Astrophys. **580**, L13 (2015)
D. Saumon et al., Astrophys. J. **460**, 993 (1996)
H.E. Schlichting, Astrophys. J. Lett. **795**, L15 (2014)
H.E. Schlichting et al., Icarus **247**, 81 (2015)
A. Seiff et al., Science **272**, 844 (1996)
A.P. Showman, T. Guillot, Astron. Astrophys. **385**, 166 (2002)
C. Spalding, K. Batygin, Astrophys. J. **811**, 82 (2015)
D.S. Spiegel, A.S. Burrows, Astrophys. J. **745**, 174 (2012)
D.S. Spiegel et al., Astrophys. J. **727**, 57 (2011)
S.W. Stahler, F. Palla, *The Formation of Stars* (2005)
D. Stamatellos, Astrophys. J. Lett. **810**, L11 (2015)
D. Stamatellos, A.P. Whitworth, Mon. Not. R. Astron. Soc. **392**(1), 413 (2009)
J.H. Steffen, Mon. Not. R. Astron. Soc. **433**, 3246 (2013)
D.J. Stevenson, Annu. Rev. Earth Planet. Sci. **10**, 257 (1982)
D.J. Stevenson, Icarus **62**, 4 (1985)
D.J. Stevenson, E.E. Salpeter, Astrophys. J. Suppl. Ser. **35**, 221 (1977)
J. Szulágyi et al., Astrophys. J. **782**, 65 (2014)

H. Tanaka et al., Astrophys. J. **565**, 1257 (2002)
T. Tanigawa et al., Astrophys. J. **747**, 47 (2012)
C.E.J.M.L.J. Terquem, Mon. Not. R. Astron. Soc. **341**, 1157 (2003)
L. Testi et al., Astron. Astrophys. **403**, 323 (2003)
A. Toomre, The structure and evolution of normal galaxies, in *Proceedings of the Advanced Study Institute* (1981), p. 111
A.H.M.J. Triaud et al., Astron. Astrophys. **524**, A25 (2010)
S. Udry, M. Mayor, in *Extreme Solar Systems*, ed. by D. Fischer, F.A. Rasio, S.E. Thorsett, A. Wolszczan. Astronomical Society of the Pacific Conference Series, vol. 398 (2008), p. 13
H.C. Urey, Proc. Natl. Acad. Sci. **41**, 127 (1955)
A. Uribe et al., Astrophys. J. **802**, 54 (2015)
N. Vaytet et al., Astron. Astrophys. **557**, A90 (2013)
A. Vidal-Madjar et al., Astrophys. J. Lett. **676**, L57 (2008)
W.R. Ward, Icarus **126**, 261 (1997)
G.W. Wetherill, G.R. Stewart, Icarus **77**, 330 (1989)
D.J. Wilner et al., Astrophys. J. Lett. **626**, L109 (2005)
H.F. Wilson, B. Militzer, Astrophys. J. **745**, 54 (2012)
J.N. Winn, D.C. Fabrycky, Annu. Rev. Astron. Astrophys. **53**, 409 (2015)
J.N. Winn et al., Astrophys. J. **631**, 1215 (2005)
Y. Wu, N. Murray, Astrophys. J. **589**, 605 (2003)
A.N. Youdin, Y. Lithwick, Icarus **192**, 588 (2007)
J.-P. Zahn, Astron. Astrophys. **57**, 383 (1977)
Z. Zhu, Astrophys. J. **799**, 16 (2015)
Z. Zhu et al., Astrophys. J. **729**, 47 (2011)
Z. Zhu et al., Astrophys. J. **755**, 6 (2012)
Z. Zhu et al., Astrophys. J. **768**, 143 (2013)
W. Zhu et al., Astrophys. J. **796**(1), 67 (2014)
Z. Zhu et al., Astrophys. J. **813**, 88 (2015)

Space Sci Rev (2016) 205:165–192
DOI 10.1007/s11214-015-0228-x

Disk Dispersal: Theoretical Understanding and Observational Constraints

U. Gorti[1] · R. Liseau[2] · Z. Sándor[3] · C. Clarke[4]

Received: 18 May 2015 / Accepted: 8 December 2015 / Published online: 12 May 2016

Abstract Protoplanetary disks dissipate rapidly after the central star forms, on time-scales comparable to those inferred for planet formation. In order to allow the formation of planets, disks must survive the dispersive effects of UV and X-ray photoevaporation for at least a few Myr. Viscous accretion depletes significant amounts of the mass in gas and solids, while photoevaporative flows driven by internal and external irradiation remove most of the gas. A reasonably large fraction of the mass in solids and some gas get incorporated into planets. Here, we review our current understanding of disk evolution and dispersal, and discuss how these might affect planet formation. We also discuss existing observational constraints on dispersal mechanisms and future directions.

Keywords Protoplanetary disks · Planet formation · Accretion · Winds

1 Introduction

Disks provide the raw material for planet formation and the timescales on which they are dispersed therefore greatly affect their potential to form planets. Observations of disks, viz., the cluster age-disk frequency plot, suggest that dust disk lifetimes are $\lesssim 3$–5 Myr (e.g., Haisch et al. 2001; Hernández et al. 2007; Ribas et al. 2015), and are only a fraction of the typical stellar lifetime. In such studies, which were initially conducted in the near-infrared (NIR), the fraction of young star cluster members with emission indicative of the presence

✉ U. Gorti
uma.gorti-1@nasa.gov

1 NASA Ames Research Center/SETI Institute, Mountain View, CA, USA

2 Department of Earth and Space Sciences, Onsala Space Observatory, Chalmers University of Technology, 439 92 Onsala, Sweden

3 Konkoly Observatory of the Hungarian Academy of Sciences, P.O. Box 67, 1525 Budapest, Hungary

4 Institute of Astronomy, University of Cambridge, Madingley Road, Cambridge, CB3 0HA, United Kingdom

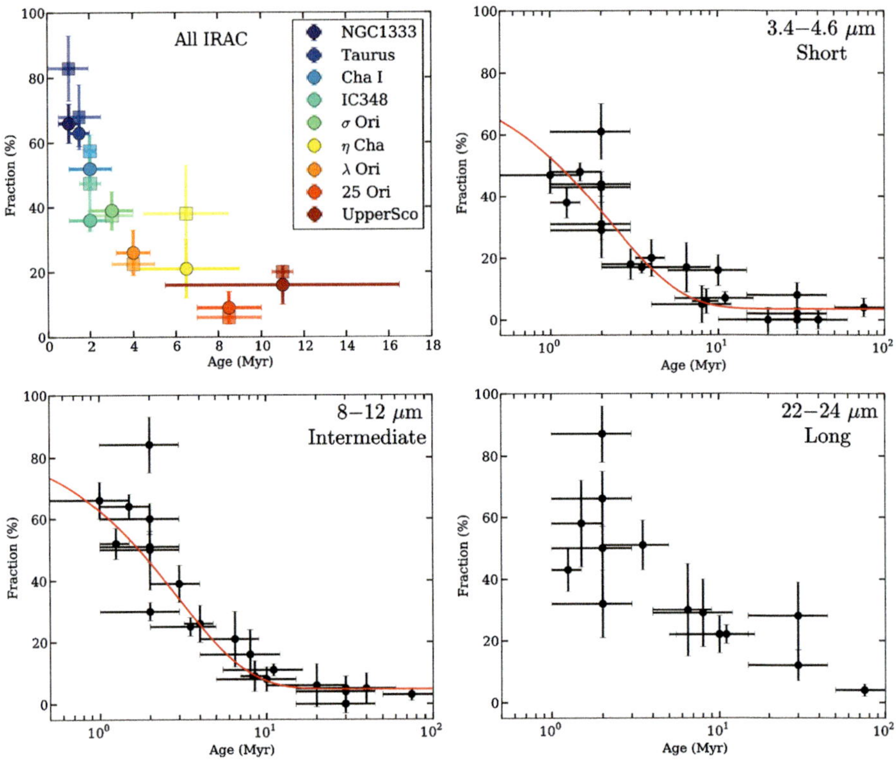

Fig. 1 The disk frequency as a function of cluster age at different wavelengths as probed by *Spitzer* photometry. The inferred disk lifetime varies from ~ 2–3 Myr at short wavelengths to ~ 4–6 Myr at longer wavelengths (figure reproduced with permission from Ribas et al. 2015 © ESO)

of disk material is seen to dramatically decline with cluster age. Although the reliability of the derived lifetimes is limited by the uncertainty in determining ages (e.g., Soderblom et al. 2014; Bell et al. 2013), there is clear evidence that NIR excess fractions decline with age from the classical T Tauri stage to the non-accreting, weak line T Tauri stage—a transition that spans a few Myr at the most. Most disks seem to survive just long enough to allow planet formation (Lissauer et al. 2009), and only in a small fraction (~ 10 %, Winn and Fabrycky 2015) may the formation of giant planets be possible. If gas does manage to persist late into planet formation epochs, it can further affect planetary dynamics. Even small amounts of gas can influence the dynamics of young planetary systems: causing migration, damping eccentricities and mitigating the effects of planetesimal collisions.

Disk depletion lifetimes at longer wavelengths, tracing dust in regions farther from the star, are similar (e.g., see review by Williams and Cieza 2011). Sub-millimetre emission is rarely seen from disks without NIR excesses (Andrews and Williams 2005), indicating either that the entire disk is depleted simultaneously or that the larger grains are lost earlier due to some combination of drift, fragmentation and/or planetesimal formation. Dust at mid-infrared wavelengths appears to last slightly longer (e.g., Wahhaj et al. 2010; Hardy et al. 2015; see Fig. 1), however, debris disks may contaminate emission statistics at these wavelengths. Nevertheless, the transition from optically thick to optically thin disks—believed to be represented by a class of objects called transition disks—appears to proceed from inside-out, i.e., the inner dust is depleted first. Transition disks have lower accretion rates and dust holes in their inner radii and constitute about $\sim 10\ \%$ of the disk population (Williams and Cieza 2011). While they are often believed to be a result of planet formation (e.g., Najita et al. 2015), these disks may represent a mixed class of objects (see recent review by Espaillat et al. 2014) with only the lower mass disks on the verge of dispersal (Ercolano et al. 2011; Owen and Clarke 2012; Koepferl et al. 2013).

Despite the fact that dust disk depletion timescales are fairly well known, disk dispersal mechanisms are not yet well constrained. In fact, the cluster age vs. disk frequency plots are also consistent with the interpretation that the $\sim$ 3–5 Myr dust disk lifetime tracks the process of dust agglomeration into planetesimals. Planet formation from planetesimals, and even giant planet formation if the gas reservoir is still present, may then proceed over substantially longer periods. In this interpretation, which holds if transition disks were mainly caused by planet formation, dust disk depletion does not necessarily imply dispersal of the disk material. Gas, which dominates the disk mass through most of its evolution, could in this case be removed on different timescales.

However, observationally inferred timescales for the dispersal of gas in disks are less certain, mainly due to the fact that gas emission is intrinsically very faint. The earliest study was the seminal CO survey of $\sim$ 10 disks by Zuckerman et al. (1995) which set a loose constraint on the gas disk dispersal time (at large $\sim$ 100 AU radii) of $\sim$ 10 Myr. The FEPS legacy survey on the Spitzer Science Telescope, based on non-detections of H_2, set a dispersal timescale of the order of about 5–30 Myr at radii 1–40 AU (Pascucci et al. 2006). A more recent [OI] 63 μm survey using the Herschel Space Observatory (GASPS program) derived a similar timescale for the dispersal of gas ($\sim$ 5–200 AU). Dent et al. (2013) quote that $\sim 18\ \%$ of stars retain more than $1 M_J$ worth of gas at ages of 4 Myr, and that all disks are dispersed by $\sim$ 10–20 Myr. Thus, gas dispersal times could, in principle, be longer than the $\sim$ 3–5 Myr dust disk depletion time.

Different gas tracers have different sensitivity thresholds, making it difficult to compare gas and dust disk lifetimes. In the inner disk ($\sim$ 1 AU), however, there appears to be clearer evidence of simultaneous dispersal. The fraction of accreting disks (with $\dot{M}_{acc} > 10^{-11}\ M_\odot\,yr^{-1}$) in stellar groups declines on timescales similar to those of NIR excesses (Fedele et al. 2010), with some non-accreting sources (with $\dot{M}_{acc} < 10^{-11}\ M_\odot\,yr^{-1}$) still showing IR excesses (also Ingleby et al. 2013; Hardy et al. 2015). This may indicate that gas in the inner disk is removed first, consistent with dispersal scenarios (see Alexander et al. 2014).

The main processes believed to disperse disks are a combination of viscous accretion and photoevaporation (see reviews by Hollenbach et al. 2000; Armitage 2011; Clarke 2011; Alexander et al. 2014) and to some extent, planet formation.

Protostellar disks build the central star, hence it is to be expected that much of the disk gas is channeled into the central object. Evidence that disk evolution is largely driven by accretion during the bulk of the disk lifetime is provided by the fact that disk accretion rates approximately decline with age as expected from viscous accretion theory (e.g., Hartmann et al. 1998). To begin with, disks form due to the rotation of a gravitationally collapsing cloud core. Gravitational instabilities in the initially massive disk lead to strong accretion onto the central star (e.g., Laughlin and Bodenheimer 1994). At early stages, magnetic fields drive powerful jets and winds that carry away angular momentum from close to the star (e.g., Koenigl 1991; Shu et al. 1994, see reviews by Li et al. 2014; Turner et al. 2014). As star formation proceeds, infall decreases and the disk becomes gravitationally stable. Viscous accretion through the disk continues to transport mass inwards to the central star and angular momentum outwards.

As the star reaches close to its eventual final mass, the rate of accretion declines (Hartmann et al. 1998; Mendigutía et al. 2012). However, accretion does not proceed indefinitely and appears to abruptly halt with the disappearance of the inner gas and dust (e.g., Haisch et al. 2001; Fedele et al. 2010). Dispersal by pure accretion alone implies an indefinite expansion of the outer (gas) disk as angular momentum is re-distributed; there is no observational evidence to support such expansion. Further, relatively short (~ 10 Myr) gas disk lifetimes call for an additional dispersal mechanism that removes gas from the system.

Photoevaporation, whereby gas is heated to escape temperatures by the central star, is thought to disperse gas at later stages of evolution. Although observational evidence only requires the operation of some such mechanism at late stages, irradiation due to accretion-generated high energy photons should be even stronger and proceed to remove disk material even at earlier stages of disk evolution. When the magnetically-driven outflow/jet weakens along with the accretion rate (e.g., Hartmann et al. 1998; White and Hillenbrand 2004) and becomes more transparent, stellar high energy photons begin to penetrate the outflow column and irradiate the disk to heat its surface. Penetration is first by hard X-ray[1] and FUV photons when accretion rates fall to $\dot{M}_{acc} < 10^{-6}\ \mathrm{M}_\odot\ \mathrm{yr}^{-1}$, and later by EUV and soft X-ray photons ($\dot{M}_{acc} < 10^{-8}\ \mathrm{M}_\odot\ \mathrm{yr}^{-1}$) (Gorti et al. 2009). This irradiation marks the beginning of disk photoevaporation, where surface gas is heated to temperatures sufficient to overcome gravity and mass is lost in a slow, thermal wind. Viscous accretion and photoevaporation subsequently work together to disperse the disk with time (Clarke et al. 2001). Theoretical estimates for a disk evolving under the influence of viscous accretion and photoevaporation, including parametrised population synthesis models, agree with observationally inferred values ($\sim$ 1–10 Myr) and can reproduce a range of observational diagnostics of disk dispersal (e.g., Alexander et al. 2006; Gorti et al. 2009, 2015; Ercolano et al. 2009; Owen et al. 2010, 2012).

The continuous removal of gas by photoevaporation may hold consequences for the evolution of the disk as it forms planets. While viscosity depletes gas mass uniformly at all radii and spreads the outer disk as it transports angular momentum, photevaporation preferentially removes gas at specific radii (inner 1–10 AU and outer $\gtrsim 50$ AU, depending on which of EUV, FUV and X-rays dominate). If the mass loss rate due to photoevaporation ($\dot{M}_{pe}$) is low, then planet formation proceeds unaffected by photoevaporation, except at late stages where the presence or absence of gas influences planet dynamics (e.g., Baruteau et al. 2014; Coleman and Nelson 2014). If $\dot{M}_{pe}$ is high, then photoevaporation can influence early stages of planet formation by altering the gas/solids ratio (e.g., Gorti et al. 2015) and the type of planet formed (rocky vs. gaseous).

Exoplanet statistics to date appear to indicate a relative paucity of gas giants (estimated frequency of $\sim 10\ \%$ in solar-mass stars, Winn and Fabrycky 2015), but an abundance of Super-Earths ($M_\mathrm{p} \sim 3$–$10\ M_\oplus$) with some gas in an envelope. There must be insufficient gas present after the formation of planetesimals and rocky cores (a process thought to last $\sim$ 1 Myr, Connelly et al. 2012) or gas giants would be more common. On the other hand, gas must necessarily be present at planetary core formation epochs to explain the frequency of Super-Earths. The rate at which gas is dispersed is thus closely aligned to planet formation timescales.

Photoevaporation may also influence the final architectures of planetary systems due to the migration of planets in a disk with gaps cleared during dispersal, leading to preferred semi-major axis distributions for exoplanets (Alexander and Pascucci 2012; Ercolano and Rosotti 2015).

[1] Hard X-rays: $E \gtrsim 1$ keV; Soft X-rays: ~ 0.1–0.3 keV; EUV: 13.6 eV–100 eV; FUV: 6–13.6 eV.

This chapter, in keeping with the theme of this book, mainly deals with these above connections or links between disk dispersal and planet formation. We are interested not only in the lifetime of gas disks but also in the radial distribution of material at different stages of disk evolution. Dust evolution is discussed only as it pertains to gas dispersal mechanisms. (We refer the reader to chapters by Birnstiel et al. and Wyatt et al. for more in-depth reviews of dust.) We also do not discuss MHD winds which might deplete some disk mass especially at early stages. The structure of the chapter is as follows: we first describe early stages of disk evolution and accretion (Sect. 2), then photoevaporation due to the central star (Sect. 3) and in a cluster environment (Sect. 4), planet formation and dispersal (Sect. 5), and in Sect. 6 describe observational constraints on disk dispersal. We end with a discussion on future directions (Sect. 7).

2 Accretion

It is widely accepted that the formation of stars and planetary systems is fundamentally governed by the action of gravity and angular momentum. Whereas the overall picture is quite clear, the details are still far from fully understood. This concerns both theory and observation as both suffer from resolution problems. In addition, the highly complex interplay between physics and chemistry in the dusty plasmas of interstellar clouds leaves a complete description essentially intractable, at least for quite some time to come.

We focus here on the formation of low and intermediate-mass stars with masses $\lesssim 8\ M_\odot$, for which Kelvin-Helmholtz time scales are longer than other time scales of relevance and consequently, for which we can follow the time evolution of the formation process. These less massive stars form in density enhancements of rotating molecular clouds. At supercritical density, the dense cores collapse, conserving angular momentum and forming flattened structures (Terebey et al. 1984; Walch et al. 2009, 2010) that eventually develop into disks and rings.

The hallmark of the dynamical phase of star formation, i.e. the infall phase, would be spectrally resolved molecular transitions with very high optical depths, as illustrated in Fig. 2. There, the spectral infall signature is qualitatively shown by a radiative transfer model, depicted as a smooth red line. Observed excess emission in the blue and red wings is attributed to outflowing gas. It is now firmly established that both early gravitational infall and later accretion are accompanied by mass loss phenomena, and these most often exhibit a bipolar geometry.

2.1 Accretion Processes in Disks

It is believed that the infall occurs onto the disk, and that the matter is accreted onto the central object through the disk. This, however, needs removal of a significant fraction of the angular momentum that is carried by the disk to prevent the break-up of the central object. Magnetic fields that thread through the core and the disk are invoked to act as a lever arm to brake the rotation. Outside magnetic dead zones, the fields are capable of providing the necessary viscosity due to the combination of magnetic and Reynolds stresses in a turbulent shearing flow (e.g., Balbus 2003; Cao and Spruit 2013 and references therein).

The theoretical foundation of accretion disks was laid by the works of Shakura and Sunyaev (1973) and Lynden-Bell and Pringle (1974). The viscosity is described parametrically by the product of a turbulent eddy size H (of the order the pressure scale height) and its

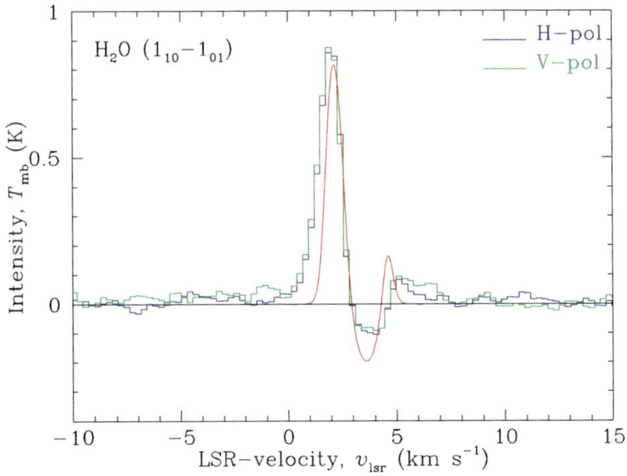

Fig. 2 Observed line profiles with *Herschel*-HIFI of the optically thick ground-state line of ortho-H_2O of a dense protostellar core, displaying, within the 38″-beam (0.02 pc), signs of both infall and outflow at the same time. All data are continuum subtracted around the zero-baseline, and the two polarisations are individually shown (*blue* and *green*) to demonstrate the high quality of the HIFI data. The *red curve* is a qualitative example from a radiative transfer model with centre optical depth over a hundred, generating the central absorption. The *blue-red asymmetry of the line core* is due to the infalling gas in the unstable Bonnor-Ebert sphere, and the excess emission in the line wings is due to the outflow

sound speed c_s, i.e. $\nu = \alpha c_s H$, where α is typically 10^{-4} to 10^{-2}. The time evolution of the surface density $(\Sigma(r))$ of the disk is given by

$$\frac{\partial \Sigma}{\partial t} = \frac{3}{r}\frac{\partial}{\partial r}\left(\sqrt{r}\frac{\partial}{\partial r}(\nu \Sigma \sqrt{r})\right) \tag{1}$$

the solution of which describes the basic characteristics of an accretion disk, viz. that angular momentum is transported outward through the disk as matter is accreted into the inner regions. Disk observations have revealed rotational Keplerian signatures (Sargent and Beckwith 1987; Dutrey et al. 1994; Olofsson et al. 2001; Guilloteau et al. 2014) but predicted radial accretion drift velocities are too small to be measurable (on the order of some cm s^{-1}).

Another observable would be the Spectral Energy Distribution (SED). The SED of a classical accretion disk is essentially that of a multi-temperature broadened blackbody. Recent models still exhibit these basic features (e.g., MCFOST, Pinte et al. 2006). A useful quantity is the integrated SED, i.e., the accretion luminosity $L_{acc} = \eta G \dot{M}_{acc}(M/R)_{star}$, where $0.5 \leq \eta < 1$ is an energy conversion efficiency. For typical parameters one finds $L_{acc} = 10\ L_\odot$, which is far above what had been determined from observations (Hartmann et al. 1997). It was concluded that the accretion luminosity most likely is not steady in time, i.e. $dL_{acc}/dt \propto d\dot{M}_{acc}/dt \neq 0$, but variable within $\dot{M}_{acc} \approx 10^{-8}$–$10^{-4}\ M_\odot\,yr^{-1}$. The intermittent high states would be reached during FU Orionis type outbursts (Hartmann and Kenyon 1996), whereas the low states would correspond to the typical T Tauri phase. Rise and decay times are of the order of 1 yr and 100 yr, respectively. Shorter time scales for $d\dot{M}_{acc}/dt$ have been examined by Costigan et al. (2014).

There are observational signatures of accretion. Optical emission lines from T Tauri stars, e.g. Hα, have been modeled as excited by shocks at the foot-prints of magnetised funnel flows (Muzerolle et al. 2001). However, it appears that the geometry and magnetic field

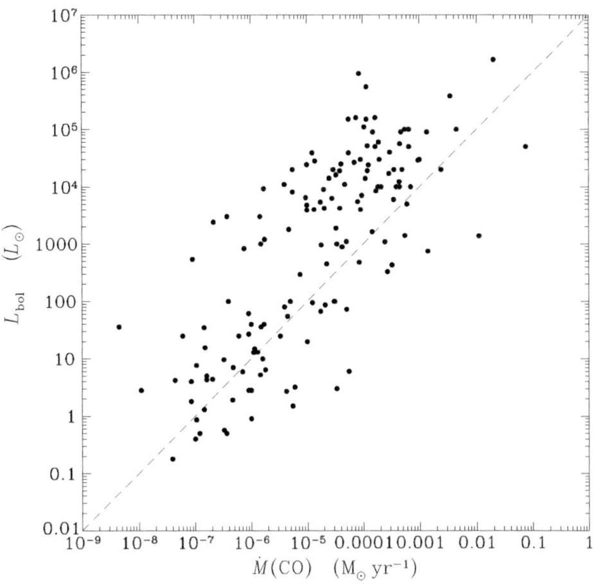

Fig. 3 Mass loss rate, as determined from CO rotational line observations, viz. $\dot{M}$(CO), versus the bolometric luminosity L_{bol}, of the outflow driving sources (data from Wu et al. 2004). The *dashed line* has unit slope and is for reference only

topology are much more complex than envisaged in these one-dimensional models (e.g., see Fig. 1 of Günther et al. 2013).

2.2 Mass Loss Accompanying Accretion

Jets Optically visible HH-objects (Haro 1950; Herbig 1951; Reipurth et al. 2000) and jets (e.g., Mundt 1985; Shang et al. 2007) are the cooling radiation from fast interstellar shock waves in star forming regions. Observations reveal jets on many length scales, viz. micro-jets (sub-arcsec) to pc-scales. As the name indicates, the collimation of jet flows is very high. The absence of detectable [OIII] but prominent [SII] emission often indicates that the excitation (or density) is not very high, consistent with jet velocities not exceeding about 80 km s^{-1}. However, a number of jets are now known to emit in X-rays, implying jet velocities of the order of 500 km s^{-1} or higher (Liseau 2006 and references therein). In many cases, but not all, these jets are seen together with generally much less collimated molecular outflows (Bally et al. 1983).

Molecular Outflows Figure 3 is based on the compilation of literature data of CO-outflows by Wu et al. (2004), and shows the relation of the mass loss rate, as determined from CO-line mapping, and the bolometric luminosity of the driving sources, over seven orders of magnitude, and obtained from their infrared SEDs. The plot exhibits a large scatter that is due to the heterogeneity of the sample. However, in spite of this, it seems pretty clear that there is a dichotomy between low-luminosity (< 100 L$_\odot$) and high-luminosity (≥ 100 L$_\odot$) stellar sources. However, in both cases, the data can be fit by power laws, viz. $L_{bol} \propto \dot{M}^{a}_{loss}$. While the low-luminosity distribution (where the luminosity is dominated by accretion) is consistent with $a = 1$, the distribution steepens at the higher end, with $a = 2.5$ (see also Beuther et al. 2002 and references therein). In the latter case, the luminosity most certainly is due to nuclear burning (objects already on the main-sequence). This power-law behaviour strongly suggests that the underlying physics have common grounds and that the same physical laws govern these processes.

Theories of jet acceleration all invoke the presence of relatively strong magnetic fields, whether for protostellar X-winds (Shu et al. 1994) or for disk-anchored disk winds (Pudritz et al. 2007; Li et al. 2014). The nomenclature "wind" describes the idea that the flows are initially poorly collimated. The precise nature of the interplay between disk-jet-molecular flow is difficult to determine observationally, primarily due to insufficient spatial resolution capability.

However, there are a few clues: for instance, Hartigan et al. (1995) derived an outflow mass loss-to-accretion rate $\dot{M}_{loss}/\dot{M}_{acc} \lesssim 0.01$ from optical observations, while White and Hillenbrand (2004) derive a value ~ 0.05–0.1. Intriguingly, theoretical estimates of this parameter are $0.1 \lesssim \dot{M}_{loss}/\dot{M}_{acc} < 1$. Hartigan et al. (1995) concluded that these flows traced by [OI] forbidden lines may not carry enough momentum to drive the heavy CO-outflow. This was also the conclusion arrived at by Liseau et al. (2005) in their detailed study of the protostellar object L1551-IRS, its jets and its CO flow. For this young binary, the dynamical mass is known (Rodríguez et al. 2003). For the ratio of the rates, Liseau et al. (2005) found $\dot{M}_{loss}/\dot{M}_{acc} = 0.23 \pm 0.10$ for the primary, and 0.7 ± 0.3 for the secondary, which were based on the large-scale CO outflow. These values are more in agreement with the theoretical prediction.

Since both the observed and theoretical values of the outflow mass loss rate are lower than the accretion rate, outflows cannot overwhelm accretion and hence do not play a major role in dispersing the disk at later stages.

3 Photoevaporation: Central Star

Brief History Photoevaporation was first studied in the context of massive stars by Hollenbach et al. (1994), who examined the effects of the rather strong radiation fields of massive stars on their disks (as suggested by Bally and Scoville 1982). The basic premise is that the heating of the surface gas drives thermal winds from the disk ($c_s^2 > GM_*/r$) which then results in mass loss and a steady depletion of the disk material. Clarke et al. (2001) combined viscous evolution with photoevaporation to find that gaps open in disks at a preferred inner location (the gravitational radius, $r_g = GM_*/c_s^2$). Viscosity depletes matter interior to the gap, leading to inner holes. Adams et al. (2004) concluded that angular momentum support against gravity leads to the launching of flows at smaller radii (~ 0.1–$0.2 r_g$, also Begelman et al. 1983; Liffman 2003; Font et al. 2004). Alexander et al. (2006) recognised that the creation of a hole leads to the direct irradiation of the inner rim and results in a rapid dispersal of the outer disk. These theories were more recently extended to include the heating effects of FUV and X-ray irradiation (Ercolano et al. 2009; Gorti and Hollenbach 2009; Owen et al. 2012). Overall, photoevaporation and viscous evolution together lead to the dispersal of gas on observed timescales ($\sim$ 1–10 Myrs). For a more complete account of earlier work, we refer the reader to existing reviews of this topic (Hollenbach et al. 2000; Dullemond et al. 2007; Clarke 2011; Alexander et al. 2014).

3.1 Overview of Photoevaporation

The gravitational pull exerted by the central star decreases with distance, and so does the gas temperature; hence the ease with which flows can be launched depends on disk radius. The rate of change of surface density $\dot{\Sigma}_{pe}(r,t) = \rho_b v_{flow} \propto \rho_b c_s$, where ρ_b is the density of gas at the base of the flow and the flow velocity v_{flow} is proportional to the sound speed c_s at this location. $\dot{\Sigma}_{pe}$ is therefore sensitive to the density and temperature of the heated disk

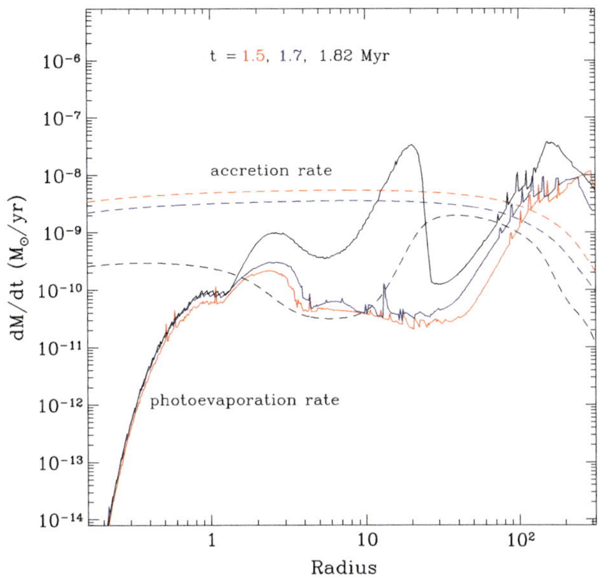

Fig. 4 The increase in photoevaporation rate (*solid lines*) and the decrease in accretion rate (*dashed lines*) near the gap opening epoch is shown for a EUV+FUV+X-ray photoevaporation model. Here, the gap opens at $r \sim 10$ AU, at about 1.8 Myr. After gap opening, rim irradiation increases the photoevaporation rate, lowering Σ and lowering the accretion rate further. Also note that photoevaporation rates are higher than the accretion rate in the outer disk ($\gtrsim 100$ AU) where significant mass loss occurs

surface. In order to escape the system, the critical temperature needed at a given radius is $\sim 18000/r_{AU}$ K for a neutral atomic flow and $\sim 9500/r_{AU}$ K for fully ionised gas to escape with a flow velocity equal to the sound speed. Typical launch speeds are slightly subsonic, ~ 0.5–$1c_s$ (Owen et al. 2012; Gorti et al. 2015).

At early stages of evolution, accretion rates ($\dot{M}_{acc} \sim 3\pi\nu\Sigma$) are high compared to photoevaporative mass loss ($\dot{M}_{pe}$) and although photoevaporation may remove mass, its effects on the radial surface density distribution in the disk are minimal. As the surface density Σ decreases with time due to viscous accretion, $\dot{M}_{acc}$ declines with it. $\dot{M}_{pe}$, on the other hand, stays fairly constant with time; disk mass and its depletion is concentrated at the midplane whereas the density and temperature at the surface (and hence $\dot{\Sigma}_{pe}$) stay relatively unaffected. Eventually $\dot{M}_{acc}$ drops below $\dot{M}_{pe}$ and this is when photoevaporation begins to play a dominant role in determining the evolution of the disk surface density with radius and time (Fig. 4).

Gaps open in the inner disk when accretion can no longer replenish photoevaporative mass loss (e.g., Clarke et al. 2001). This happens at inner radii of ~ 1–10 AU for solar-mass stars due to the strong heating by the high energy photon flux. Gap opening halts the advection of mass from the outer disk, the inner disk drains rapidly and a hole is created. The disk continues to photoevaporate from the irradiated inner rim outward (e.g., Alexander et al. 2006). We note that holes can be created only if the high energy radiation field has a significant non-accretion generated component, i.e., it is mainly chromospheric/coronal in origin. If not, then the cessation of accretion chokes photoevaporation and gaps and holes cannot be sustained (see Gorti et al. 2015). In the case of FUV photoevaporation, flows are also launched from the outer disk where the surface temperature is still high but escape speeds are relatively low (e.g., Gorti and Hollenbach 2009; Owen et al. 2012). Under some conditions, gaps may also open in the outer disk. Since disks have shallow surface density distributions (e.g., Andrews et al. 2011), most of their mass is at large radii and photoevaporation here can affect the evolution of the entire disk. Viscous expansion in the outer disk is curtailed and the disk evolves into a shrinking torus of material (Gorti et al. 2009, 2015).

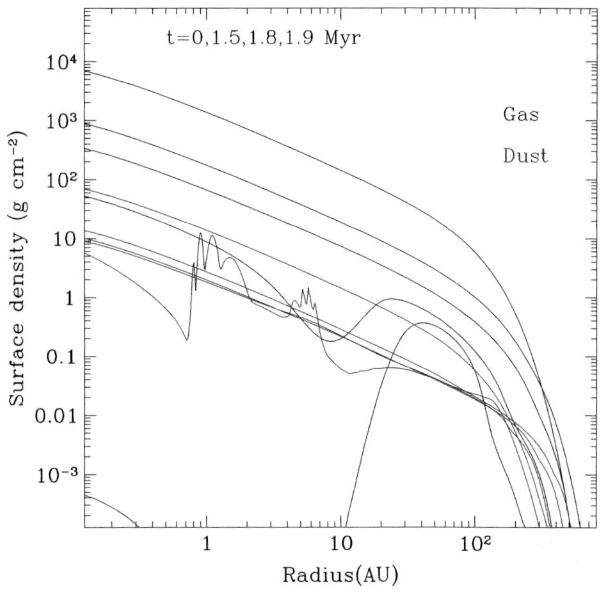

Fig. 5 The gas and dust surface density evolution of a viscously evolving disk, with FUV, EUV and X-ray photoevaporation, $M_d(0) = 0.1\ \mathrm{M}_\odot$, and $\alpha = 0.01$. FUV photoevaporation leads to the creation of a gap at ~ 3–10 AU and the gas disk disperses in ~ 2 Myr. After the dispersal of the gas disk a substantial amount of dust is retained in the disk ($\sim 3 \times 10^{-4}\ \mathrm{M}_\odot$). In these models, the largest solids are 1 cm in size, and no planetesimal formation is taken into account (see Gorti et al. 2015)

Concurrent dust evolution plays an additional role for FUV photoevaporation. FUV heating of gas is due to collisions with energetic electrons ejected by small dust grains (small grains here include polycyclic aromatic hydrocarbons or PAHs) that absorb FUV photons (e.g., Bakes and Tielens 1994). The evolving abundance of small grains in the disk therefore affects the heating. However, small grains also attenuate FUV photons and their depletion increases penetration and shifts the base of the flow to higher densities. Since as noted earlier, $\dot{\Sigma}_{pe}(r,t) \propto \rho_b \sqrt{T_{gas}}$, the depletion of small grains simultaneously decreases T_{gas} and increases ρ_b resulting in a smaller net effect on $\dot{M}_{pe}$. Overall, 2-fluid models of the evolution of the gas and dust (with a range of sizes) show that FUV mass loss rates are not significantly affected by dust evolution. See Fig. 5 for the evolution of one such photoevaporating, viscous disk model. Interestingly, the gas/solids ratio in the disk is reduced by photoevaporation (first noted by Throop and Bally 2005) because dust grains are not coupled to the low-density gas in the wind and preferentially leave dust particles behind (Gorti et al. 2015). Owen et al. (2011) describe how wind entrainment of dust could be observationally detected via their emission in edge-on disks.

3.2 Photoevaporative Mass Loss Rates

As discussed in Sect. 1, the principal determinant of the relevance of photoevaporation for disk evolution and planet formation is the rate at which the disk loses mass. The mass loss rate also dictates how early on during evolution photoevaporation becomes important. As long as accretion dominates (i.e., $\dot{M}_{acc} \gtrsim \dot{M}_{pe}$), viscous diffusion and advection will smear out any radial effects and replenish regions where photoevaporative mass loss has occurred.

While there is reasonable agreement on the qualitative behaviour of disk photoevaporation between different models, the calculated mass loss rates vary by over two orders of magnitude, from $10^{-8} - 10^{-10}\ \mathrm{M}_\odot\,\mathrm{yr}^{-1}$. At the high end, $\dot{M}_{pe} > \dot{M}_{acc}$ during the Class II stage and photoevaporation determines the radial distribution of disk material and can significantly affect planet formation. Rapid dispersal may even preclude the formation of planets.

For low $\dot{M}_{pe}$, the role of photoevaporation may be limited to clearing the disk of small amounts of remnant gas and facilitating the circularization of planetary orbits (Kominami and Ida 2002).

Some of the differences in estimated $\dot{M}_{pe}$ can be attributed to the high energy photons under consideration. For pure EUV models, $\dot{M}_{pe}$ can be low ($< 10^{-10}$ $M_\odot\,yr^{-1}$). Although ionised gas is heated to $\sim 10^4$ K, EUV is absorbed at very small column densities and the low ρ_b results in low mass loss rates. High EUV luminosities can yield higher mass loss rates, but recent studies suggest photon luminosities $\lesssim 10^{40}$–10^{41} s^{-1} ($L_{EUV} \sim 10^{30}$ erg s^{-1}) and hence that the associated $\dot{M}_{pe}$ is low (Pascucci et al. 2014).

X-ray and FUV models result in higher mass loss rates, $\sim 10^{-8}$–10^{-9} $M_\odot\,yr^{-1}$ for typical stellar radiation fields. The calculated $\dot{M}_{pe}$ is in general sensitive to the density and temperature structure of the disk which now has to be determined unlike in the EUV case. The disk structure is in turn based on disk chemistry and calculated cooling rates that are all highly model-dependent (e.g., see Röllig et al. 2007). For X-ray photoevaporation models, Owen et al. (2012) however state that the resulting flow properties are insensitive to the detailed thermal and density structure of the upper disk layers but are instead set by a criticality condition at height $\sim R$ above the disk plane where the flow makes a subsonic to supersonic transition. They further argue that provided the flow structure is optically thin to the X-rays dominating the heating at this surface, the mass loss rate is independent of any complex thermochemical effects at greater depth in the flow. If the optically thin condition is not met however, then the flow instead makes a sonic transition in regions where heating is dominated by FUV and hard X-rays and then it is essential to calculate the disk vertical structure.

The X-ray spectrum assumed also impacts disk temperatures and hence $\dot{M}_{pe}$ (Gorti et al. 2009); Gorti, Hollenbach et al. assume that soft X-rays (0.1–0.3 keV) are mostly absorbed in accretion and outflow columns before they reach the disk surface, while Ercolano, Owen et al. assume a small covering factor of the accretion columns and no absorption in the column, to attain much higher temperatures in their disk models. The latter do not consider molecular cooling, but Gorti et al. find that molecular cooling can be important for regions penetrated by hard X-rays ($\gtrsim 1$ keV) and their model disks have cooler temperatures.

Gorti et al. further treat the flow dynamics using simple analytical estimates drawn from previous work on thermal winds (Begelman et al. 1983; Liffman 2003; Adams et al. 2004; Waters and Proga 2012), but conduct detailed thermo-chemical modeling. Owen et al. claim that the flow structure is unimportant for soft X-rays, and adopt the opposite approach to solve for $\dot{M}_{pe}$ using full radiation hydrodynamics models with simpler thermal physics. However, Gorti et al. include FUV photoevaporation along with X-rays and EUV, and in spite of a smaller role for XEUV photons, get comparable mass loss rates. The high mass loss is partly due to the time-dependent accretion FUV luminosity, which can be substantial in disks (e.g., Gullbring et al. 1998). More recent models by Owen et al. (2012) also find that FUV photoevaporation can dominate if the FUV luminosities are high and better reconcile the differences between the two groups.

Disk mass loss rates can vary depending on a number of parameters, e.g. stellar mass, initial disk mass and radius, viscosity in the disk, EUV, FUV and X-ray luminosities, and the time-dependent XEFUV spectrum (e.g., Ercolano et al. 2009; Gorti et al. 2009, 2015; Owen et al. 2012), many of which are known to vary widely—often by an order of magnitude or more—in young stars. This diversity results in photoevaporation rates that can vary widely depending on the system, and $\dot{M}_{pe}$ can generally range from 10^{-11} to 10^{-7} $M_\odot\,yr^{-1}$ for $M_* \sim 0.1$–3 $M_\odot$.

The disk lifetime for a disk of initial mass $M_d(0)$ and a time-averaged photoevaporation rate $\langle \dot{M}_{pe} \rangle$ can be approximately written as

$$\tau_{disk} \approx 10^7 \left(\frac{M_d(0)}{0.1\ \mathrm{M}_\odot} \right)^{2/3} \left(\frac{\langle \dot{M}_{pe} \rangle}{10^{-9}\ \mathrm{M}_\odot\ \mathrm{yr}^{-1}} \right)^{-2/3} \mathrm{yr}, \tag{2}$$

for a linear viscosity profile and assuming $\alpha = 0.01$ (e.g. Clarke et al. 2001; Gorti et al. 2015). For the fiducial photoevaporation rates discussed above of 10^{-8} to 10^{-9} $\mathrm{M}_\odot\ \mathrm{yr}^{-1}$ and an initial disk mass of 0.1 $\mathrm{M}_\odot$, the corresponding disk lifetimes are thus $\sim$ 2–10 Myr. We note that, in principle, $\dot{M}_{pe}$ can change with time as the disk evolves, and the equation above represents an average rate over the disk lifetime (see Gorti et al. 2015).

4 Photoevaporation in the Cluster Environment

So far we have considered three flavours of disk photoevaporation driven by the EUV, FUV or X-ray radiation from the disk's central star. In the crowded environments of young clusters, however, there is also the possibility of 'external' photoevaporation by the radiation field produced by (more massive) neighbouring stars. This is particularly to be expected in the case of the EUV and FUV where stars' photospheric outputs are a strong function of stellar mass (Diaz-Miller et al. 1998) and where, even taking into account the relative rarity of higher mass stars, the integrated contribution to the EUV and FUV backgrounds peaks at masses in the range 10–55 $\mathrm{M}_\odot$ (Fatuzzo and Adams 2008). This last point is important in assessing the types of cluster environments in which one expects external photoevaporation to be important. At high cluster membership number (N), even the top end of the IMF is statistically well populated and thus the distribution of total UV luminosity at a given N is sharply peaked at a value that simply scales with N. In the case of low N clusters, by contrast, there are large stochastic variations in the population of the upper IMF and the distribution of UV luminosities at given N is broad with a median that is well below the mean. This just means that external photoevaporation is unimportant in low N clusters, partly because the over-all number of stars is lower and partly because, in consequence, the IMF often ends up not containing the most massive stars that dominate the UV budget: see Fatuzzo and Adams (2008) for a detailed analysis of this issue. The behaviour of the X-ray background is more complex because X-ray luminosity does not increase monotonically with stellar mass, attaining a minimum in the case of fully radiative stars in the A star range. Measurements of diffuse X-ray emission in rich clusters such as M17 or the Rosette Nebula (e.g. Townsley et al. 2003) suggest that this emission is largely dominated by early type O stars (Feigelson et al. 2003) since the ONC (for which the most massive star is of spectral type O5) lacks a comparable diffuse field.

4.1 External X-Ray Photoevaporation

It is easy to demonstrate that external X-ray photo-evaporation is negligible compared with internal X-ray photo-evaporation. The X-ray driven mass loss rate at each radius scales linearly with the X-ray flux (since, in the ionisation parameter formulation, the density corresponding to the local escape temperature is proportional to F_X). Even the X-ray flux reported in the ONC near θ_1C Ori is orders of magnitude less than the X-ray flux of an average T Tauri star at a radius of 100 AU. We therefore do not consider this possibility further.

4.2 EUV+FUV Photoevaporation from a Single Star

We now consider the simplest case where the external UV field in a cluster is dominated by a single star. This is approximately the case in the ONC, where θ_1C Ori has an ionising output that substantially exceeds that of the other O stars in the cluster core. We will not for now concern ourselves with the fact that there are other O stars such as θ_1A or θ_1B that are the major contributors of FUV flux to their nearest neighbours. If we consider the luminosity from a single source, rather than an isotropic background, then we do not expect the resulting wind structures to be spherically symmetric and indeed the photoevaporating discs in the ONC show generally cometary morphologies, often being brighter on the side facing θ_1C and with a tail pointing away from this star (Tsamis et al. 2013). Detailed modeling of proplyd morphology assumes that the hemisphere of the ionised flow is directly illuminated whereas the far-side receives a diffuse EUV field derived from recombinations in the nebula (see Henney and Arthur 1998). We will however follow Johnstone et al. (1998) in setting out a simplified spherically symmetric photoevaporation model, an approach that yields mass loss rates that agree to order unity with more complex modeling.

EUV radiation impinges on low density gas above the disk and sets up an ionisation front in which the integrated number of recombinations per unit area matches the star's ionising flux (one can readily justify a posteriori the neglect of additional consumption of EUV photons in ionising neutral material flowing through the ionisation front since this turns out to be a small addition for typical flow parameters; Churchwell et al. 1987). The ionisation front represents a contact discontinuity of the flow for which imposition of mass and momentum conservation on either side results in a flow that is transonic in the ionised region (i.e. the ionised gas flows out at around 10 km s^{-1}) but enters the ionisation front sub-sonically. Naturally there must be heating processes (*not* involving ionising photons) that produce the pressure gradients throughout the neutral region, but provided this flow is subsonic throughout the neutral region it is in causal contact with the ionisation front and it is thus conditions at the ionisation front that set the mass flow rate. This situation is known as EUV photoevaporation, even though FUV heating must contribute also in setting up the neutral flow that feeds the ionisation front. There is however a qualitatively different situation if the solution contains a sonic point in the neutral flow. In this case the flow has to undergo shock deceleration before it can enter the ionisation front and thus neutral flow below the shock is causally decoupled from the ionisation front. This means that the agent heating the neutral flow (the FUV) now *sets* the wind mass loss rate.

For EUV driven flows, ionisation equilibrium at the ionisation front implies that $\int n^2 dr$ scales with the ionising flux so that, for a spherical transonic flow one obtains:

$$\dot{M}_{EUV} = 10^{-8} \Phi_{49}^{1/2} d_{17}^{-1} r_{d14}^{3/2} \, \mathrm{M}_\odot \, \mathrm{yr}^{-1} \tag{3}$$

(Johnstone et al. 1998) where Φ_{49} is the ionising luminosity of θ_1C in units of 10^{49} s^{-1}, d_{17} is the distance of the disk from θ_1C in units of 10^{17} cm and r_{d14} is the radius of the ionisation front in units of 10^{14} cm. Note that in the case of EUV driven winds (where by definition the neutral flow is thin and sub-sonic), r_{d14} roughly equates with the disk radius.

In the case of FUV flows, by contrast, the wind mass flux is set by the maximum density of the flow for which FUV heating is effective. For the strong FUV fields in the centre of Orion, this condition is set by dust absorption (rather than molecular self-shielding) and therefore imposes a maximum column in the FUV heated neutral flow. Since the scale height of FUV heated gas in the outer disk is of order r_d (i.e. the FUV heats to around the escape temperature of the outer disk), the number density in the flow is set by $n r_d \sim 10^{21}$ cm^{-2}.

Putting this together (again assuming spherical symmetry and that the flow velocity is transonic in the neutral gas, i.e. with velocity $\sim 3\ \mathrm{km\,s^{-1}}$) we obtain the relation:

$$\dot{M}_{FUV} = 10^{-8} r_{d14}\ \mathrm{M_\odot\,yr^{-1}} \tag{4}$$

(Johnstone et al. 1998). Note that, unlike, the expression for $\dot{M}_{EUV}$, this rate does not depend explicitly on the strength of the FUV field. However, this expression does not apply at arbitrarily low FUV levels because at some point self-shielding by molecular hydrogen becomes more important than dust absorption in setting the column of neutral gas (also see discussion in Sect. 4.5). Störzer and Hollenbach (1999) argued that the critical flux level was $G_0 \sim 5 \times 10^4$ which corresponds to distances from θ_1C of around 0.3 pc.[2]

We thus have a schematic picture which would imply three radial zones with distinct disk photoevaporation properties. Noting that $\dot{M}_{EUV}$ scales inversely with distance from θ_1C while $\dot{M}_{FUV}$ is constant over a range extending out to a limiting G_0, we then have (i) a small inner region with very fierce EUV photoevaporation, (ii) an intermediate FUV zone with spatially constant mass loss rate and (iii) an outer EUV zone which resumes at the point that FUV photoevaporation is no longer effective. The latter is at ~ 0.3 pc while the interface between (i) and (ii) is set by equality of the EUV and FUV mass loss rates at that point, implying a radius of $\sim 0.15 r_{d14}^{1/2}$.

4.3 The Observational Evidence for External Photoevaporation in Orion

An immediate implication of the FUV model described above is that the ionisation front is spatially offset from the disk because it is separated from the disk by a spatially thick supersonic neutral flow. Such a structure corresponds well to the 'proplyds' first imaged in Orion by O'dell et al. (1993). (Note that whereas O'Dell applied the term 'proplyd' to all the circumstellar structures imaged in Orion, including the pure silhouette discs, we will here restrict the definition to those showing offset ionisation fronts, regardless of whether a central silhouette disk is also detected.) The spatial distribution of the bulk of the proplyds accords well with the 'FUV zone' described above. (Note that the innermost EUV zone is plausibly filled in by projection effects.)

Although the bulk of the proplyds indeed lie within the 0.3 pc radius predicted above, there are several tens of proplyds at larger radius, only some of which are explicable as being instead powered by the more modest FUV heating provided by θ_1A and B (Vicente and Alves 2005). It is possible that FUV winds may be driven at lower values of G_0 than argued above but in this case it is unclear why the number of proplyds at large radii is small, though non-zero. Alternatively, Clarke and Owen (2015) suggested that these far flung proplyds result from external EUV ionisation of a neutral flow driven by internal X-ray photoevaporation. They showed that the numbers and sizes of far flung proplyds are consistent with statistical expectations based on the X-ray luminosity function (though the correlation between instantaneous X-ray luminosity and resolvable proplyd structure on a source by source basis is weak, an effect that might be attributed to variability). Similarly there are far flung proplyds detected in Carina (Smith et al. 2003), NGC 3603 (Brandner et al. 2000) and Cyg OB2 (Wright et al. 2012) although some of these are likely to instead be ionised clumps of molecular gas (Sahai et al. 2012a, 2012b).

Having established that theory more or less correctly predicts the spatial distribution of proplyds, we now turn to the mass loss rates in these objects. These were first deduced from

[2] G_0 is the ratio of the FUV flux to its ambient value in the solar neighbourhood which is $1.6 \times 10^{-3}\ \mathrm{erg\,cm^{-2}\,s^{-1}}$ (Habing 1968).

resolved radio free-free observations (Churchwell et al. 1987) combined with the assumption of transonic expansion in the ionised flow. As the resulting flow rates (comparable with $\dot{M}_{FUV}$ as given above) imply problematically short disk lifetimes (see below) O'dell (1998) instead suggested that the structures might be pressure confined. These high mass loss rates were however subsequently confirmed through emission line modeling (Henney and O'Dell 1999) which showed that the kinematics were clearly incompatible with a static, pressure confined structure but involved free expansion.

These high mass loss rates, both predicted and observed, imply that the total mass photo-evaporated over the cluster lifetime is of order a solar mass. This is far more than the initial gas reservoir in discs and implies that we would expect to see a distinct deficit of discs at the present epoch in the centre of Orion. The most troubling aspect of the 'proplyd lifetime problem' is that the disk fraction in the core of Orion is actually high; in fact, when projection effects are taken into account, the disk fraction is ~ 100 % within the central FUV region. Störzer and Hollenbach (1999) suggested that the unexpected survival of discs at small radii could be explained if stars were on very radial orbits and just 'lit up' as proplyds during their brief sorties through the core region. However, such an orbital configuration was found to be unsustainable in N-body simulations (Scally and Clarke 2001). Another effect that may mitigate the very short predicted survival times is that FUV photoevaporation is considerably less efficient in the case of "sub-critical" discs (i.e., those in which the escape velocity at the disk outer edge exceeds the sound speed in the heated gas). Adams et al. (2004) have shown that the mass loss rate in this case declines with decreasing radius much more steeply than implied by Eq. (4); although Clarke (2007) suggested that this might enhance the survivability of small disks in the ONC, the important of this effect is limited by the fact that even in the case of negligible FUV mass loss, the rate reverts to the EUV rate (Eq. (3)).

This then only leaves two possibilities (or a combination of the two): unexpectedly large disk masses and/or a recent 'switch-on' time for θ_1 1C.

Large disk masses are however not indicated by any of the sub-mm studies in Orion (see Mundy et al. 1987; Bally et al. 1998; Eisner and Carpenter 2006; Mann et al. 2014) and although estimation of gas disk masses is traditionally beset by systematic uncertainties regarding the gas to dust ratio and grain opacity, it is clear that in a comparative sense the discs in Orion are not unusually massive. This instead encourages the 'last gasp' interpretation, in which θ_1C has only been 'switched on' (or perhaps, more correctly, been optically revealed) over a timescale that is a small fraction of the cluster lifetime. In this scenario, the Orion proplyds (and indeed all disk emission in the inner parts of Orion) will be gone within a timescale of $> 10^5$ years (also see Clarke 2007).

Although it is slightly uncomfortable to argue that we are witnessing our nearest massive star formation region at a 'special epoch' there is some circumstantial evidence that this is indeed the case. The disk mass distribution derived by Mann et al. (2014) does show some hints that preferential depletion of disk mass is indeed starting to occur at the very small radii characterising the inner EUV zone (where the mass loss rates are even higher than in the FUV zone: Eq. (3)). Secondly, proplyds indeed seem to be rather rare in other star forming regions (Yusef-Zadeh et al. 2005; Balog et al. 2006; Koenig et al. 2008), although this result has to be interpreted with caution given that resolving prolyds is more challenging in more distant environments than in Orion. Perhaps the most conclusive argument indicating that Orion is being observed at a special epoch is that it is unusual among massive clusters in not showing a deficit of disk bearing stars in the proximity of massive stars. It is to the disk demographics in other clusters that we now turn.

4.4 The Distribution of Disc Bearing Stars in Massive Star Clusters

A number of studies have now conducted disk censuses in star forming regions where one can reasonably expect that external photoevaporation will be important: for example Guarcello et al. (2007, 2009, 2010) in NGC 6611, Fang et al. (2012) in Pismis 24 and Guarcello et al. (2015) in Cygnus OB2. In all cases it is seen that the disk fraction is lower in the proximity of massive stars. Guarcello et al. (2015) have quantified this effect by generating an estimated map of the ambient FUV field across Cyg OB2 and have demonstrated that the disk fraction declines monotonically with increasing FUV flux, being around a factor two lower in the highest FUV versus lowest FUV category. There is also some evidence from this analysis that the (anti-)correlation between disk frequency and FUV flux is more convincing than that between disk frequency and the expected frequency of star-disk collisions. This is to be expected given that in any environment where star-disk collisions are at all significant, the effect of external photoevaporation is likely to be much more important (Scally and Clarke 2001).

4.5 The Effect of Mild FUV Fields in Sparse Star Forming Environments

Environments like the ONC are highly atypical in terms of the strength of the ambient FUV field. The population synthesis exercise of Fatuzzo and Adams (2008), which assembled clusters according to the observationally inferred spectrum of cluster richness, demonstrated that regions which—like the core of Orion—have G_0 values in excess of 10^4 are environments in which at most a few percent of star formation occurs. On the other hand, other star forming environments span an enormous dynamic range in G_0 values (4–6 orders of magnitude). Adams et al. (2004) studied flow solutions in a range of low G_0 environments, arguing that in this case photoevaporation is predominantly in the (cylindrical) radial direction from the outer edge of the disk. Recently Facchini et al. (2015) have revisited the problem and obtained solutions over a wider range of parameter space, additionally iterating on the thermal solution to take account of the fact that only small grains are entrained in the flow. Preliminary solutions indicate rather significant mass fluxes for discs larger than 100 AU even at low G_0. A generic feature of these solutions is a cliff in the gas surface density at the disk outer edge and then a low density plateau of (nearly dust free) gas at larger radii. Such solutions offer the prospect of possibly constraining the properties of such flows through deep molecular line imaging.

Regardless of the numbers that emerge from this exercise, it is also worth noticing that the interplay between viscosity and outer edge can have some unexpected outcomes. A viscous disk without a photoevaporative flow evolves towards a viscous similarity solution in which the disk outer edge grows in such a way that the viscous time is always of order the disk age. Consequently the surface density in such a disk declines as a power law in time and—as the viscous time gets longer and longer as the disk expands—the time required for the disk to clear (e.g. become optically thin in the near infrared) is extremely long.

It might be thought that the addition of a disk flow from the outer edge might at best reduce the disk lifetime to a value equal to t_{evap}, the ratio of the present disk mass to the photoevaporation rate. In fact, however, the effect is much more dramatic: when external photoevaporation is coupled to the viscous flow then the disk at some stage stops growing and then shrinks from the outside in (an effect also seen in the internal FUV photoevaporation models of Gorti et al. 2009, 2015). In this case, the viscous timescale decreases with time and hence the clearing is accelerated. This effect results in the disk clearing an order of magnitude faster than t_w (Clarke 2007; Anderson et al. 2013) and it is obvious that in this case most of the disk mass is cleared by accretion rather than photoevaporation. Nevertheless photoevaporation is playing a decisive role in preventing disk spreading and thus keeping the viscous timescale short.

5 Planet Formation and Photoevaporation

5.1 Inventory of Disk's Solid Material and Its Depletion

For disks that form planetary systems, the planets themselves form an important sink for the initial mass in solids. If giant planets are present, then a fraction of the disk gas is also consumed. In our solar system, there was a minimum of about $\sim 1\text{–}3 \times 10^{-4}$ $M_{\odot}$ of solids and a little more than a Jupiter mass or $\sim 1\text{–}3 \times 10^{-3}$ $M_{\odot}$ of gas that was "depleted" into planet formation. We use the term "depletion" because this disk component is dark for most external systems. A disk-less object may very well have a nascent planetary system that is undetected. The protosolar disk therefore lost 10 times as much gas as the dust, and this is a fact that dispersal theories must necessarily explain.

Exoplanetary disks, especially the birth-sites of the compact Kepler systems, may efficiently convert even more mass into rocky planets (e.g., the Minimum Mass Extrasolar Nebula models of Hansen and Murray 2012; Chiang and Laughlin 2013). The low frequency of gas giants ($\lesssim 5\text{–}10$ %, Winn and Fabrycky 2015) and the abundance of Super-Earths (~ 50 %) indicates that in most disks gas is depleted relative to dust even more than in the solar nebula.

Unless migration is not as efficient as theories predict, accretion cannot preferentially remove gas. Small dust is well coupled and must be carried along with the gas. Radial drift, in fact, may result in the rapid loss of mm-size grains where most of the solids mass is initially contained (e.g., Testi et al. 2014). If larger planetesimals form prior to dispersal, it is not clear that they survive migration. Hasegawa and Ida (2013) conclude that even gas giants may not survive migration in a massive disk. In a recent study, Coleman and Nelson (2014) modeled migration and planet formation via oligarchic growth (high relative planetesimal velocities) and dynamical evolution. They find that as long as the gas disk is present, the formation and retention of a giant planet ($\gtrsim 10\ M_{\oplus}$) is difficult because it would rapidly migrate. In their scenario, low mass disks form close-packed systems, and in high mass disks planets continuously form and migrate into the star and the last generation survives. All of these processes deplete solids relative to gas, and hence do not provide an explanation for the preferential removal of gas, at least in our solar system. On the other hand, gas in photoevaporative flows is not dense enough to lift any dust but the smallest particles; since the mass is typically concentrated in larger sized particles this mechanism leaves most of the dust behind.

5.2 Disk Dispersal in the Classic Core Accretion Scenario

Dispersal of protoplanetary disks by photo-evaporation has a strong effect on planet formation since it limits the time needed to build planetary systems. We recall that planet formation in the core accretion scenario (Lissauer and Stevenson 2007) should explain a growth process through 12 orders of magnitude from the micron sized dust to Jupiter-like giant planets within the very limited lifetime of a protoplanetary disk. Planet formation in the core accretion scenario has different, well separated phases: (i) dust coagulation and formation of planetesimals, (ii) formation of planetary embryos and growth of the solid cores of giant planets, (iii) runaway gas accretion by the solid cores to form giant planets, and finally, (iv) the assembly of the planetary embryos to terrestrial planets. In the following we overview the timescales of the above phases of planet formation, and investigate how they are influenced by photoevaporation of the disk.

Formation of planetesimals due to dust coagulation is presently very uncertain due to several barriers to planetesimal growth: the bouncing barrier (Zsom et al. 2010), the charge barrier (Okuzumi et al. 2009), and the meter-size barrier (due to drift and fragmentation) (Blum and Wurm 2000). New approaches, such as formation in pressure maxima (Lyra et al. 2009), particle concentration due to streaming instability and gravoturbulent planetesimal formation (Johansen et al. 2014), have attempted to clarify these issues. In spite of many uncertainties, the timescale of particle growth to km-sized planetesimals is believed to be quite short; thus planetesimal formation takes place in the gas-rich protoplanetary disk and is unaffected by the disk's photoevaporation.

The next step is the further growth of planetesimals toward terrestrial planets leading to runaway (Wetherill and Stewart 1989) and oligarchic growth (Kokubo and Ida 1998). These initial stages of the early phase of terrestrial planet formation in the classic core-accretion paradigm are rapid, lasting between 0.01 and 1 Myr. As the result of these processes, $\sim$ 100–1000 bodies between Moon and Mars size are formed in the relevant region of terrestrial planet formation still in a gaseous environment. Planetesimals are affected by aerodynamic drag (due to the slightly sub-Keplerian motion of gas), while the formed oligarch feel torques leading to *type I migration*. Drag and torques both result in angular momentum loss for the formed bodies, which spiral in. Large planetesimals ($\gtrsim$ 100 km in size) are less sensitive to drag force, but gravitationally perturb the ambient disk creating dense spiral wakes. Gravity due to these over-dense regions results in a net torque which modifies the planetesimal orbit. In earlier studies assuming isothermal disk models the net torque calculated was negative (Ward 1997) causing a significant loss of angular momentum and therefore inward migration of the planetesimal. More recent studies (Paardekooper et al. 2010) suggest the possibility of outward migration as well. We note that the speed of type I migration is linearly proportional to the mass of the migrating body, therefore the rapid inward or outward migration of relative massive objects will result in their loss from the region of terrestrial planet formation.

Formation of giant planets clearly occurs in the presence of gas. Therefore, one of the most pressing issues in planet formation is building a giant planet within the few Myr disk lifetime. According to core accretion theory, first a solid core forms beyond the water snowline, where the abundance of solids is increased due to ice condensation on dust grains. The higher surface density of solids leads to efficient oligarchic growth allowing fast formation of a planetary core. The characteristic mass of a solid core able to capture a gaseous atmosphere at Jupiter's orbit is roughly $M_{crit} = 1\ M_\oplus$. Both the core and envelope slowly grow, until the core reaches a critical mass of $\sim 10\ M_\oplus$ and runaway gas accretion ensues leading to the rapid formation of a giant planet (on a 10^5 year timescale). This rapid gas accretion slows down when the giant planet opens a gap in the gas disk. According to hydrodynamic simulations (Kley 1999) the planet is not entirely isolated from the ambient gas, and accretion continues along tidally generated spiral arms (Lubow and D'Angelo 2006). Therefore, gas accretion onto the planet's surface stops only when the surrounding gas is dispersed and formation of a giant planet ends with the photoevaporation of the gas disk.

In summary, the presence of gas significantly influences the early phase of terrestrial planet formation; the effect of migration diminishes if much of the disk gas is dispersed before the runaway and oligarchic growth phases. Giant planets can form only if gas is present. However, the final assembly of terrestrial planets takes place in a few tens of million

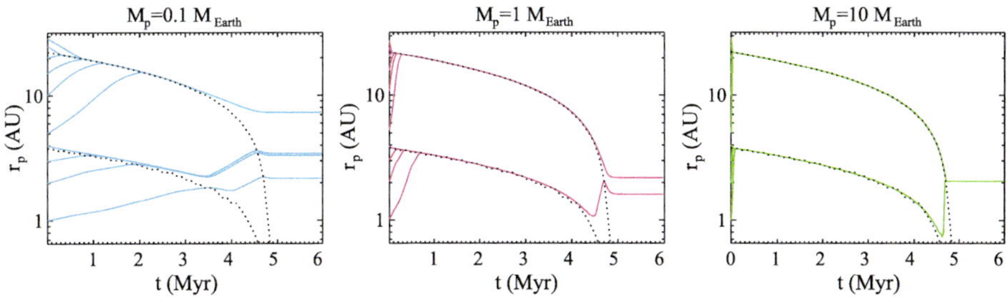

Fig. 6 Migration of different mass planets captured in the two planet traps in the model of Lyra et al. (2010). The orbital distances of planets are displayed with *solid lines*, while the *dotted lines* show the locations of the planet traps as functions of time (figure reproduced with permission from Lyra et al. 2010, The Astrophysical Journal Letters)

years, thus certainly in a gas free environment with no migration. We note that in more recent studies terrestrial planet formation may happen at *planet traps* (places where the torque felt by a planet becomes zero) on much shorter timescales, well before the disk's final dispersal. In such models the effect of gas disk dispersal should be taken into account. These models will be briefly described in the next subsection.

5.3 Disk Dispersal and Type I Migration of Terrestrial Planets and Planetary Cores

Planets and massive cores, after they form, can migrate, and type I migration can be very fast resulting in a rapid loss of the formed cores or planets. Planet traps can halt migration: traps can be caused by a change in disk thermal properties due to sudden changes in dust opacity (Lyra et al. 2010), or by the formation of a large scale vortex at the outer edge of the dead zone, which stops or reduces the migration speed of massive planetary cores (Regály et al. 2013).

Lyra et al. (2010) showed that opacity and temperature jumps in the disk can help prevent migration of planets (with $M_\mathrm{p} \sim 0.1\text{–}10\ M_\oplus$) using an evolving radiative 1D disk model with photoevaporation. In this model, the external photoevaporative disk wind prescription of Veras and Armitage (2004) is used, the wind is effective only outside a critical radius of 5 AU. The locations of the equilibrium radii with zero torque migrate due to evolution of the disk and the planet migrates toward these zero torque regions. As Σ decreases, the radii of the zero torque locations (the traps) move faster toward the star than the migrating planets. As a consequence the planets are decoupled from equilibrium radii. Σ at these times is too low to cause further migration. This effect is shown in Fig. 6: in these simulations the evolution of the disk due to the combined effects of gas accretion and photoevaporation keep terrestrial planets from migrating into the star.

5.4 Effect of the Disk Dispersal on the Type II Migration of Giant Planets

A massive Jupiter-like giant planet interacts with the disk to open an annular gap depleted of material, and then migrates on nearly viscous timescales (*type II migration*, see chapter by Lin et al. 2015, this issue). This migration timescale can be written as (see e.g. in Baruteau and Masset 2013):

$$\tau_\mathrm{II} = \frac{2r_0^2}{3\nu(r_0)}\left(1 + \frac{M_\mathrm{p}}{4\pi\Sigma(r_0)r^2}\right), \tag{5}$$

where $r_0 = r_p + 2.5r_{Hill}$, and r_{Hill} is the Hill radius. The first term in the above formula is the effect of gas accretion and the second term is the ratio between the planet's mass and the local disk mass and can be interpreted as the "resistance" of the inner disk to the inward migrating giant planet. In equilibrium, the gap moves at the accretion velocity during type II migration.

Recent hydrodynamical simulations by Dürmann and Kley (2015) show that the migration of the giant planet is determined by the torques exerted by the disk. In general, they find that the migration of the giant planet does not follow the disk's viscous evolution and gas can flow through the gap. An important result of this research that if $M_D/M_P < 0.2$ (where $M_D = \Sigma(r_p)r_p^2$ is interpreted as the local disk mass) the migration speed becomes significantly lower than the viscous speed (the type II migration rate). (We note that this behaviour is also reflected in Eq. (5) which accounts for the disk mass.) While there is no simple analytical formulation for torque-based migration, these slower rates are more consistent with planet synthesis models that can reproduce observations. Since a low disk mass is necessary to slow migration, the disk mass must be reduced at this stage by the combined effect of gas accretion and photoevaporation.

The final location of a giant planet is determined by the migration and disk dispersal times. Ignoring the effects of gravitational scattering in multi-planet systems, one can estimate the distribution of the semi-major axes of the randomly formed giant planets in a given disk model. Alexander and Armitage (2009) investigated the effects of gas accretion, EUV photo-evaporation, and the migration of the giant planet with a gap opening criterion (as in Lin and Papaloizou 1986). With plausible disk conditions and a range of planet masses ($0.5M_{Jup} < M_p < 5M_{Jup}$), the orbital distribution of planets were found to be comparable to data from the Lick radial velocity survey (Fischer and Valenti 2005); the two distributions are qualitatively similar (see Alexander and Armitage 2009). Considering the effects of gap opening due to EUV photoevaporation, Alexander and Pascucci (2012) found more recently that there would be deficit of planets at semi-major axis values close to the gap radius at $r_p = 1$–3 AU can be seen. This deficit is accompanied by a corresponding increase in the number of planets just outside these radii (see Fig. 2 of Alexander and Pascucci 2012). We note, however, that model results could be affected by many uncertainties due to migration rates which are linked to unknown aspects of planet formation (see discussion in Ercolano and Rosotti 2015).

5.5 Rapid Photoevaporation of a Disk with a Giant Planet

The formation of planets may in turn influence or accelerate disk dispersal. Conventional photoevaporation theory relies on the formation of a gap and then a hole, whose rim is irradiated directly to enhance mass loss subsequently. If a planet forms a gap and creates a similar rim, photoevaporation may be accelerated and trigger disk clearing (Alexander and Armitage 2009). In a recent investigation (Rosotti et al. 2013) used the hydrodynamic code FARGO (Masset 2000) coupled with a 1D code used for the initial ~ 2 Myr evolution of the disk, including X-ray photoevaporation: these models yield a rapid dispersal of the inner disk (interior to the planet) compared to the case without X-ray photoevaporation where the inner disk persisted for the duration of the simulation.

6 Observational Constraints

We next examine observational insights into the process of disk dissipation. Ideally, we would like to be able measure gas and dust masses, understand how their accretion rates decline, measure photoevaporative mass loss rates, detect disks on the verge of dispersal, and assess their planet formation activity. Rapid strides are being made in this area especially with newer, sensitive facilities capable of high spatial and spectral resolution. The broad scenario drawn from observations is consistent with the dispersal theory outlined so far, although details are not well understood. We first summarise what we currently know from observations and then discuss each of these:

- Dust and gas disk lifetimes are $\lesssim$ 10 Myr, very few disks survive beyond this timescale.
- Disks evolve through a transition phase where the infrared excesses decrease; the inner disk probably clears first in most cases to form dust cavities.
- Photoevaporative winds have been detected in [NeII] 12.8 μm emission, and possibly also in [OI] forbidden line and CO emission; mass loss rates are yet to be determined.
- Exoplanet studies indicate that gas is present at late epochs of planet formation in most disks.

6.1 Disk Lifetimes

As discussed in Sect. 1, infrared excesses in disks appear to decline with an e-folding time of $\sim$ 3 Myr (e.g., Mamajek 2009). Inner gas also disappears on similar timescales (Fedele et al. 2010), as is to be expected—gas if present would drag along small dust which causes the NIR excess. Moreover, Ribas et al. (2014) find that the fraction of sources with excesses increases at longer wavelengths, suggesting that the disk evolves inside-out. Inferred lifetimes are $\sim$ 4–6 Myr at 24 μm compared to $\sim$ 2–3 Myr at 3–12 μm. The disk/star flux ratio at 24 μm shows a sharp change at $\sim$ 10 Myr, suggesting that the nature of dust perhaps changes at these ages (Ribas et al. 2014). There is a peak in the fraction of evolved disks (lower disk/star 24 um flux) at $\sim$ 10 Myr, marking the transition from primordial to debris disk stage and the dispersal of gas (cf. Ercolano et al. 2011; Koepferl et al. 2013 for a theoretical perspective). Similarly, Wyatt (2008) argues that the disk mass derived from the sub-mm remains more or less constant (albeit with a wide dispersion, see Carpenter et al. 2014) and shows a sharp decline at 10 Myr, perhaps again indicating this transition to the debris disk phase (see also Wyatt et al. 2014). Since debris disks are almost always gas-free, the 10 Myr time also serves as an upper limit on the gas disk dispersal time, which is also consistent with gas observations (Zuckerman et al. 1995; Dent et al. 2013).

Photoevaporation can explain the above general behaviour of dissipating disks quite well, and timescales are roughly in accordance with observations. The inside-out dispersal of disks observed is, in fact, a main predicted characteristic of XEUV photoevaporation theory. Although FUV photoevaporation causes mass loss in the outer disk, gap opening in the inner disk (at larger radii, $\sim$ 10 AU) typically precedes eventual disk dispersal here as well (Gorti et al. 2015). The 2–10 Myr timescales inferred above indicate photoevaporative mass loss rates that are at least of the order $\sim 10^{-9}$ to 10^{-8} $M_\odot\,yr^{-1}$, in reasonable agreement with theoretical rates as discussed in Sect. 3 (Eq. (2)).

The sudden and simultaneous removal of dust along with gas is harder to explain. Alexander and Armitage (2007) propose that as photoevaporation sweeps across the disk to remove gas, it may cause planetesimal formation at the expanding inner rim and deplete dust. This conversion needs to be rapid and highly efficient, if not, substantial amounts of dust may remain after gas disk removal. Debris disk processes such as PR drag could remove the remnant primordial dust (see chapter by Wyatt et al.), but these mechanisms act on Myr timescales and are not consistent with the rapid transition to the debris disk stage (e.g., Luhman et al. 2010; Wyatt et al. 2014). The most likely scenario is that planet formation has already removed most of the dust before gas disk dispersal, although this suggests a causal link between these two processes (see discussion in Gorti et al. 2015).

6.2 Transition Disks

Transition disks are believed to represent one of possibly multiple pathways from primordial to debris disks (e.g. Williams and Cieza 2011). Hence, this special class of objects are particularly relevant for disk dispersal theories. While we conceptually understand how disks evolve and disperse, the origin and nature of transition disks is still under considerable debate. (There are also several definitions of what constitutes a transition disk, we adopt the definition of Espaillat et al. 2014, i.e., disks with a clear deficit in short wavelength emission.) Transition disks have larger dust grains (Pinilla et al. 2014), are typically millimetre-bright (suggesting high disk mass) and accrete at rates a factor of $\sim$ 3–10 lower than full primordial disks (e.g. Espaillat et al. 2014; Najita et al. 2015). They also show differences in their gas emission, with higher line ratios of HCN/H_2O in the infrared (Najita et al. 2013) and lower [OI] 63 μm line luminosities (Howard et al. 2012), trends that have both been explained as a result of their evolved dust content.

The two main mechanisms proposed to explain transition disks are photoevaporation and planet-disk interactions. XEUV photoevaporation predicts that about $\sim$ 10 % of disks should be caught in the act of viscously draining their inner gas after gap formation (e.g., Owen et al. 2010), in agreement with the observed fraction of transition disks. However, photo evaporating disks are also predicted to be low in mass (note that this is the gas mass, however that it is the dust mass that is measured). FUV-dominated photoevaporation can open gaps at higher disk masses, but does not predict a large fraction of observable disks in the viscous draining phase ($\sim 10^4$–10^5 yr) because of longer dispersal times ($\sim$ 3–5 Myr). Gaps opened by planets in the cavities of transition disks is a more popular explanation, but not without difficulties. On the one hand, the planet needs to be massive to open a gap, $\gtrsim 1 M_J$, and massive disks may be required to form giant planets, naturally explaining the higher mass of transition disks (Najita et al. 2015). On the other hand, Jupiters are believed to be rare (Winn and Fabrycky 2015), and moreover, the formation time of giant planets is on the order of the disk dispersal time (e.g., D'Angelo et al. 2010), making the likelihood of observing an embedded Jovian mass planet too low to explain the $\sim$ 10 % frequency of transition disks. Multiple planetary systems are more common, but loss to migration and maintaining the observed stellar accretion rate past several planets pose problems (Zhu et al. 2011; Coleman and Nelson 2014). Owen and Clarke (2012) propose that there are two classes of transition disks, low mass disks compatible with photoevaporation theory and higher mass ones perhaps better explained by planet-disk interactions. Rosotti et al. (2013) further consider the interaction between planet formation and photoevaporation and suggest that planets when they form could accelerate disk dispersal.

While transition disks are believed to be an evolutionary stage that most disks go through, we note that a significant number of objects have, in fact, turned out to be unresolved binaries (e.g., CoKu Tau/4, CS Cha, HD142527 among others). Future, high resolution observations using facilities such as ALMA may shed light on the true nature of some of these disks (e.g., van der Marel et al. 2015).

6.3 Gas Diagnostics of Photoevaporation

Emission from winds is the best method to directly measure the mass loss rates and assess the efficiency with which photoevaporation can deplete gas. The most promising such detection is the blue-shifted emission in [NeII] 12.8 μm line from slow, thermal winds seen from a few objects (Herczeg et al. 2007; Pascucci and Sterzik 2009). Neon can be ionised by EUV (~ 21 eV photons) and X-rays (~ 1 keV), and gas temperatures need to be $\gtrsim 500$ K to excite the observed line. Line luminosities and profiles are well reproduced by photoevaporation models (Alexander 2008; Ercolano and Owen 2010), but mass loss rates using [NeII] are hard to determine without knowledge of the ionisation level of the gas. For EUV-heated, fully ionised winds, data is consistent with mass loss rates of $\sim 10^{-10}$ $M_\odot$ yr^{-1}, while partly neutral X-ray heated gas implies higher rates $\sim 10^{-9}$ to 10^{-8} $M_\odot$ yr^{-1} (Hollenbach and Gorti 2009; Pascucci et al. 2012). Other low velocity wind tracers such as [OI] forbidden line emission and CO rovibrational emission are more difficult to interpret with contributions from multiple components (Rigliaco et al. 2013). For a more in-depth discussion, see Alexander et al. (2014).

Additional constraints on the ionisation in the disk come from free-free emission, indicated by cm-excesses in a few disks (e.g., Pascucci et al. 2012; Owen et al. 2013). Recent studies by Pascucci et al. (2014) and Galván-Madrid et al. (2014) indicate that the EUV luminosities in disks are low based on observed free-free emission fluxes. These studies also find that the observed [NeII] emission is too high to arise from EUV-ionised gas at the inferred EUV luminosities. Therefore they conclude that the [NeII] emission must trace a wind ionised by hard ~ 1 keV X-rays, indirectly implying higher mass loss rates—due to either a X-ray driven photoevaporative wind, or a FUV-driven wind that is partly ionised by X-rays.

Tracers of FUV photoevaporation are more difficult. Since the flows are launched subsonically and are considerably cooler, emission (for e.g., CO rotational lines) is dominated by the base of the flow which is at higher densities. The molecules further get dissociated higher up in the wind. The blue-shifts and asymmetries in the line profiles expected here are small for detection with current facilities and hard to disentangle from other non-Keplerian sources like turbulence (Gorti et al. in preparation). With the higher resolution and sensitivity of full ALMA and probes of the higher surface layers such as the weak [CI] 609 μm line becoming accessible (Tsukagoshi et al. 2015), these flows may be detected in the near future.

6.4 Gas at Late Stages and Planet Formation

Exoplanet properties indicate that gas is present at late stages of planet formation in most disks. The most direct evidence stems from the detection of Super-Earths or mini-Neptunes—planets with masses $\gtrsim 2$–$3M_E$ and gaseous envelopes (Winn and Fabrycky 2015). Close-in systems detected by Kepler transit surveys require gas both for in-situ and migration theories of their formation (e.g., Laughlin and Lissauer 2015). The eccentricities of these systems are low, indicating that there were at least small quantities of gas present after the giant impact stage of forming terrestrial planets. On the other hand, the paucity of gas giants and the low gas masses of the Super-Earths indicate that there could not have been too much gas present. In that case, the planetary cores would have accreted more of the disk gas to form gas giants.

Exoplanet masses and compositions therefore indicate that while gas was present at the epochs of planet formation, it dispersed shortly thereafter. This is particularly true for gas giant planet formation, with the final mass of the giant planet closely linked to the gas dispersal time (e.g., Lissauer et al. 2009; Movshovitz et al. 2010; Rogers et al. 2011).

Contrary to the all the observational evidence presented so far, gas appears to persist in at least some debris disks (e.g., HD 21997, Kóspál et al. 2013), well past planet formation epochs. Although it is still unknown if the gas is primordial in origin, it is worth noting that all of the debris disks with gas detected are A stars, pointing to either longer disk dispersal times for intermediate-mass stars (however, see Ribas et al. 2014), or a detection bias.

7 Future Directions

To summarise, the evolution of protoplanetary disks is initially dominated by viscous accretion, but at later critical planet-forming epochs internal and external photoevaporation by high energy photons (UV and X-rays) dictate the radial distribution of disk gas with time. Photoevaporation sets gas disk lifetimes and through its influence on gas disk evolution can impact all stages of planet formation, from planetesimal growth to the formation of giant planets. Although we qualitatively understand how disks evolve and disperse, photoevaporation rates are still not measured. Determining the decrease in gas mass with time and quantifying disk mass loss rates are essential toward developing a comprehensive theory of disk evolution that includes accretion, planet formation and disk dispersal. We end with a list of possible future directions that may help resolve many outstanding issues:

- Some of the biggest uncertainties pertain to the stellar high energy spectrum. The flux of the accretion-generated X-ray and UV components, the relative strengths of the soft and hard X-ray fluxes and the strength of the Lyman α contribution to the FUV flux, and the evolution of all of these with time (along with relation to other variables such as accretion rates, disk and stellar masses) are some of the less well characterised inputs that need further investigation.
- Tracers of subsonic flow are almost non-existent, they may be needed to actually determine mass loss rates. Ideally, a measure of the gas mass for disks of different ages is desirable to quantify the rate at which disks dissipate. Gas emission line observations probe the density and temperature structure which are important for setting flow conditions; emission line modeling further indirectly measures disk irradiation. Future observations from high sensitivity facilities like ALMA will inform disk heating and cooling physics and help calibrate disk models.
- More sophisticated disk models, which treat gas and dust separately and include hydrodynamics, radiative transfer and chemistry are needed. With rapid advances in supercomputing facilities and techniques, such models may soon become possible. Disk evolution models need to self-consistently account for planet formation and dynamical evolution along with disk dispersal. As comprehensive studies become more common (e.g., Coleman and Nelson 2014), future work may allow for advanced population synthesis models that can simultaneously explain the diversity in disk and exoplanet properties.

Ultimately, we would like to be able to connect disk evolution to planet formation and understand the close, and perhaps causal, correspondence between timescales for planet formation and disk dispersal.

Acknowledgements We thank ISSI for their kind invitation and our Chinese hosts for their hospitality during the workshop. U. Gorti acknowledges support from the National Science Foundation (NSF AST-1313003). Zs. Sándor thanks the support of the János Bolyai Research Scholarship of the Hungarian Academy of Sciences.

References

F.C. Adams, D. Hollenbach, G. Laughlin, U. Gorti, Astrophys. J. **611**, 360 (2004)
R.D. Alexander, Mon. Not. R. Astron. Soc. **391**, L64 (2008)
R.D. Alexander, P.J. Armitage, Mon. Not. R. Astron. Soc. **375**, 500 (2007)
R.D. Alexander, P.J. Armitage, Astrophys. J. **704**, 989 (2009)
R.D. Alexander, I. Pascucci, Mon. Not. R. Astron. Soc. **422**, L82 (2012)
R.D. Alexander, C.J. Clarke, J.E. Pringle, Mon. Not. R. Astron. Soc. **369**, 216 (2006)

R. Alexander, I. Pascucci, S. Andrews, P. Armitage, L. Cieza, *Protostars and Planets VI* (University of Arizona Press, Tucson, 2014), p. 475
K.R. Anderson, F.C. Adams, N. Calvet, Astrophys. J. **774**, 9 (2013)
S.M. Andrews, J.P. Williams, Astrophys. J. **631**, 1134 (2005)
S.M. Andrews, D.J. Wilner, C. Espaillat et al., Astrophys. J. **732**, 42 (2011)
P.J. Armitage, Annu. Rev. Astron. Astrophys. **49**, 195 (2011)
E.L.O. Bakes, A.G.G.M. Tielens, Astrophys. J. **427**, 822 (1994)
S.A. Balbus, Annu. Rev. Astron. Astrophys. **41**, 555 (2003)
J. Bally, N.Z. Scoville, Astrophys. J. **255**, 497 (1982)
J. Bally, R.L. Snell, R. Predmore, Astrophys. J. **272**, 154 (1983)
J. Bally, L. Testi, A. Sargent, J. Carlstrom, Astron. J. **116**, 854 (1998)
Z. Balog, G.H. Rieke, K.Y.L. Su, J. Muzerolle, E.T. Young, Astrophys. J. Lett. **650**, L83 (2006)
C. Baruteau, F. Masset, in *Lecture Notes in Physics, vol. 861* (Springer, Berlin, 2013), p. 201
C. Baruteau, A. Crida, S.-J. Paardekooper et al., *Protostars and Planets VI* (University of Arizona Press, Tucson, 2014), p. 667
M.C. Begelman, C.F. McKee, G.A. Shields, Astrophys. J. **271**, 70 (1983)
C.P.M. Bell, T. Naylor, N.J. Mayne, R.D. Jeffries, S.P. Littlefair, Mon. Not. R. Astron. Soc. **434**, 806 (2013)
H. Beuther, P. Schilke, T.K. Sridharan et al., Astron. Astrophys. **383**, 892 (2002)
J. Blum, G. Wurm, Icarus **143**, 138 (2000)
W. Brandner, H. Zinnecker, J.M. Alcalá et al., Astron. J. **120**, 950 (2000)
X. Cao, H.C. Spruit, Astrophys. J. **765**, 149 (2013)
J.M. Carpenter, L. Ricci, A. Isella, Astrophys. J. **787**, 42 (2014)
E. Chiang, G. Laughlin, Mon. Not. R. Astron. Soc. **431**, 3444 (2013)
E. Churchwell, M. Felli, D.O.S. Wood, M. Massi, Astrophys. J. **321**, 516 (1987)
C.J. Clarke, Mon. Not. R. Astron. Soc. **376**, 1350 (2007)
C. Clarke, *Physical Processes in Circumstellar Disks Around Young Stars* (2011), p. 355
C.J. Clarke, J.E. Owen, Mon. Not. R. Astron. Soc. **446**, 2944 (2015)
C.J. Clarke, A. Gendrin, M. Sotomayor, Mon. Not. R. Astron. Soc. **328**, 485 (2001)
G.A.L. Coleman, R.P. Nelson, Mon. Not. R. Astron. Soc. **445**, 479 (2014)
J.N. Connelly, M. Bizzarro, A.N. Krot et al., Science **338**, 651 (2012)
G. Costigan, J.S. Vink, A. Scholz, T. Ray, L. Testi, Mon. Not. R. Astron. Soc. **440**, 3444 (2014)
G. D'Angelo, R.H. Durisen, J.J. Lissauer, *Exoplanets* (University of Arizona Press, Tucson, 2010), p. 319
W.R.F. Dent, W.F. Thi, I. Kamp et al., Publ. Astron. Soc. Pac. **125**, 477 (2013)
R.I. Diaz-Miller, J. Franco, S.N. Shore, Astrophys. J. **501**, 192 (1998)
C.P. Dullemond, D. Hollenbach, I. Kamp, P. D'Alessio, *Protostars and Planets V* (University of Arizona Press, Tucson, 2007), p. 555
C. Dürmann, W. Kley, Astron. Astrophys. **574**, A52 (2015)
A. Dutrey, S. Guilloteau, M. Simon, Astron. Astrophys. **286**, 149 (1994)
J.A. Eisner, J.M. Carpenter, Astrophys. J. **641**, 1162 (2006)
B. Ercolano, J.E. Owen, Mon. Not. R. Astron. Soc. **406**, 1553 (2010)
B. Ercolano, G. Rosotti, Mon. Not. R. Astron. Soc. **450**, 3008 (2015)
B. Ercolano, C.J. Clarke, J.J. Drake, Astrophys. J. **699**, 1639 (2009)
B. Ercolano, N. Bastian, L. Spezzi, J. Owen, Mon. Not. R. Astron. Soc. **416**, 439 (2011)
C. Espaillat, J. Muzerolle, J. Najita et al., *Protostars and Planets VI* (University of Arizona Press, Tucson, 2014), p. 497
S. Facchini, C. Clarke, T. Bisbas, Mon. Not. R. Astron. Soc. (2015, submitted)
M. Fang, R. van Boekel, R.R. King et al., Astron. Astrophys. **539**, A119 (2012)
M. Fatuzzo, F.C. Adams, Astrophys. J. **675**, 1361 (2008)
D. Fedele, M.E. van den Ancker, T. Henning, R. Jayawardhana, J.M. Oliveira, Astron. Astrophys. **510**, A72 (2010)
E.D. Feigelson, J.A. Gaffney III, G. Garmire, L.A. Hillenbrand, L. Townsley, Astrophys. J. **584**, 911 (2003)
D.A. Fischer, J. Valenti, Astrophys. J. **622**, 1102 (2005)
A.S. Font, I.G. McCarthy, D. Johnstone, D.R. Ballantyne, Astrophys. J. **607**, 890 (2004)
R. Galván-Madrid, H.B. Liu, C.F. Manara et al., Astron. Astrophys. **570**, L9 (2014)
U. Gorti, D. Hollenbach, Astrophys. J. **690**, 1539 (2009)
U. Gorti, C.P. Dullemond, D. Hollenbach, Astrophys. J. **705**, 1237 (2009)
U. Gorti, D. Hollenbach, C.P. Dullemond, Astrophys. J. **804**, 29 (2015)
U. Gorti, D. Hollenbach, in preparation
M.G. Guarcello, L. Prisinzano, G. Micela et al., Astron. Astrophys. **462**, 245 (2007)
M.G. Guarcello, G. Micela, F. Damiani et al., Astron. Astrophys. **496**, 453 (2009)
M.G. Guarcello, G. Micela, G. Peres, L. Prisinzano, S. Sciortino, Astron. Astrophys. **521**, A61 (2010)

M.G. Guarcello, J.J. Drake, N.J. Wright et al. (2015). arXiv:1501.03761
S. Guilloteau, M. Simon, V. Piétu et al., Astron. Astrophys. **567**, A117 (2014)
E. Gullbring, L. Hartmann, C. Briceño, N. Calvet, Astrophys. J. **492**, 323 (1998)
H.M. Günther, P.C. Schneider, Z.-Y. Li, Astron. Astrophys. **552**, A142 (2013)
H.J. Habing, Bull. Astron. Inst. Neth. **19**, 421 (1968)
K.E. Haisch Jr., E.A. Lada, C.J. Lada, Astrophys. J. Lett. **553**, L153 (2001)
B.M.S. Hansen, N. Murray, Astrophys. J. **751**, 158 (2012)
A. Hardy, C. Caceres, M.R. Schreiber et al. (2015). arXiv:1504.05562
G. Haro, Astron. J. **55**, 72 (1950)
P. Hartigan, S. Edwards, L. Ghandour, Astrophys. J. **452**, 736 (1995)
L. Hartmann, S.J. Kenyon, Annu. Rev. Astron. Astrophys. **34**, 207 (1996)
L. Hartmann, P. Cassen, S.J. Kenyon, Astrophys. J. **475**, 770 (1997)
L. Hartmann, N. Calvet, E. Gullbring, P. D'Alessio, Astrophys. J. **495**, 385 (1998)
Y. Hasegawa, S. Ida, Astrophys. J. **774**, 146 (2013)
W.J. Henney, S.J. Arthur, Astron. J. **116**, 322 (1998)
W.J. Henney, C.R. O'Dell, Astron. J. **118**, 2350 (1999)
G.H. Herbig, Astrophys. J. **113**, 697 (1951)
G.J. Herczeg, J.R. Najita, L.A. Hillenbrand, I. Pascucci, Astrophys. J. **670**, 509 (2007)
J. Hernández, N. Calvet, C. Briceño et al., Astrophys. J. **671**, 1784 (2007)
D. Hollenbach, U. Gorti, Astrophys. J. **703**, 1203 (2009)
D. Hollenbach, D. Johnstone, S. Lizano, F. Shu, Astrophys. J. **428**, 654 (1994)
D.J. Hollenbach, H.W. Yorke, D. Johnstone, *Protostars and Planets IV* (University of Arizona Press, Tucson, 2000), p. 401
A.W. Howard, G.W. Marcy, S.T. Bryson et al., Astrophys. J. Suppl. Ser. **201**, 15 (2012)
L. Ingleby, N. Calvet, G. Herczeg et al., Astrophys. J. **767**, 112 (2013)
A. Johansen, J. Blum, H. Tanaka et al., *Protostars and Planets VI* (University of Arizona Press, Tucson, 2014), p. 547
D. Johnstone, D. Hollenbach, J. Bally, Astrophys. J. **499**, 758 (1998)
W. Kley, Mon. Not. R. Astron. Soc. **303**, 696 (1999)
X.P. Koenig, L.E. Allen, R.A. Gutermuth et al., Astrophys. J. **688**, 1142 (2008)
A. Koenigl, Astrophys. J. Lett. **370**, L39 (1991)
C.M. Koepferl, B. Ercolano, J. Dale et al., Mon. Not. R. Astron. Soc. **428**, 3327 (2013)
E. Kokubo, S. Ida, Icarus **131**, 171 (1998)
J. Kominami, S. Ida, Icarus **157**, 43 (2002)
Á. Kóspál, A. Moór, A. Juhász et al., Astrophys. J. **776**, 77 (2013)
G. Laughlin, P. Bodenheimer, Astrophys. J. **436**, 335 (1994)
G. Laughlin, J.J. Lissauer, (2015). arXiv:1501.05685
Z.-Y. Li, R. Banerjee, R.E. Pudritz et al., *Protostars and Planets VI* (University of Arizona Press, Tucson, 2014), p. 173
K. Liffman, Publ. Astron. Soc. Aust. **20**, 337 (2003)
D.N.C. Lin, J. Papaloizou, Astrophys. J. **309**, 846 (1986)
Lin, et al. Space Sci. Rev. (2015, this issue)
R. Liseau, Astron. Astrophys. **459**, 843 (2006)
R. Liseau, C.V.M. Fridlund, B. Larsson, Astrophys. J. **619**, 959 (2005)
J.J. Lissauer, D.J. Stevenson, *Protostars and Planets V* (2007), p. 591
J.J. Lissauer, O. Hubickyj, G. D'Angelo, P. Bodenheimer, Icarus **199**, 338 (2009)
S.H. Lubow, G. D'Angelo, Astrophys. J. **641**, 526 (2006)
K.L. Luhman, P.R. Allen, C. Espaillat, L. Hartmann, N. Calvet, Astrophys. J. Suppl. Ser. **189**, 353 (2010)
D. Lynden-Bell, J.E. Pringle, Mon. Not. R. Astron. Soc. **168**, 603 (1974)
W. Lyra, A. Johansen, A. Zsom, H. Klahr, N. Piskunov, Astron. Astrophys. **497**, 869 (2009)
W. Lyra, S.-J. Paardekooper, M.-M. Mac Low, Astrophys. J. Lett. **715**, L68 (2010)
E.E. Mamajek, in *American Institute of Physics Conference Series, vol. 1158* (2009), p. 3
R.K. Mann, J. Di Francesco, D. Johnstone et al., Astrophys. J. **784**, 82 (2014)
F. Masset, Astron. Astrophys. Suppl. Ser. **141**, 165 (2000)
I. Mendigutía, A. Mora, B. Montesinos et al., Astron. Astrophys. **543**, A59 (2012)
N. Movshovitz, P. Bodenheimer, M. Podolak, J.J. Lissauer, Icarus **209**, 616 (2010)
R. Mundt, *Protostars and Planets II* (1985), p. 414
L.G. Mundy, N.J. Evans II, R.L. Snell, P.F. Goldsmith, Astrophys. J. **318**, 392 (1987)
J. Muzerolle, N. Calvet, L. Hartmann, Astrophys. J. **550**, 944 (2001)
J.R. Najita, J.S. Carr, K.M. Pontoppidan et al., Astrophys. J. **766**, 134 (2013)
J.R. Najita, S.M. Andrews, J. Muzerolle (2015). arXiv:1504.05198

C.R. O'dell, Astron. J. **115**, 263 (1998)
C.R. O'dell, Z. Wen, X. Hu, Astrophys. J. **410**, 696 (1993)
S. Okuzumi, H. Tanaka, M.-a. Sakagami, Astrophys. J. **707**, 1247 (2009)
G. Olofsson, R. Liseau, A. Brandeker, Astrophys. J. Lett. **563**, L77 (2001)
J.E. Owen, C.J. Clarke, Mon. Not. R. Astron. Soc. **426**, L96 (2012)
J.E. Owen, B. Ercolano, C.J. Clarke, R.D. Alexander, Mon. Not. R. Astron. Soc. **401**, 1415 (2010)
J.E. Owen, B. Ercolano, C.J. Clarke, Mon. Not. R. Astron. Soc. **411**, 1104 (2011)
J.E. Owen, C.J. Clarke, B. Ercolano, Mon. Not. R. Astron. Soc. **422**, 1880 (2012)
J.E. Owen, A.M.M. Scaife, B. Ercolano, Mon. Not. R. Astron. Soc. **434**, 3378 (2013)
S.-J. Paardekooper, C. Baruteau, A. Crida, W. Kley, Mon. Not. R. Astron. Soc. **401**, 1950 (2010)
I. Pascucci, M. Sterzik, Astrophys. J. **702**, 724 (2009)
I. Pascucci, U. Gorti, D. Hollenbach et al., Astrophys. J. **651**, 1177 (2006)
I. Pascucci, U. Gorti, D. Hollenbach, Astrophys. J. Lett. **751**, L42 (2012)
I. Pascucci, L. Ricci, U. Gorti et al., Astrophys. J. **795**, 1 (2014)
P. Pinilla, M. Benisty, T. Birnstiel et al., Astron. Astrophys. **564**, A51 (2014)
C. Pinte, F. Ménard, G. Duchêne, P. Bastien, Astron. Astrophys. **459**, 797 (2006)
R.E. Pudritz, R. Ouyed, C. Fendt, A. Brandenburg, *Protostars and Planets V* (University of Arizona Press, Tucson, 2007), p. 277
Z. Regály, Z. Sándor, P. Csomós, S. Ataiee, Mon. Not. R. Astron. Soc. **433**, 2626 (2013)
B. Reipurth, S. Heathcote, K.C. Yu, J. Bally, L.F. Rodríguez, Astrophys. J. **534**, 317 (2000)
Á. Ribas, B. Merín, H. Bouy, L.T. Maud, Astron. Astrophys. **561**, A54 (2014)
Á. Ribas, H. Bouy, B. Merín, Astron. Astrophys. **576**, A52 (2015)
E. Rigliaco, I. Pascucci, U. Gorti, S. Edwards, D. Hollenbach, Astrophys. J. **772**, 60 (2013)
L.F. Rodríguez, S. Curiel, J. Cantó et al., Astrophys. J. **583**, 330 (2003)
L.A. Rogers, P. Bodenheimer, J.J. Lissauer, S. Seager, Astrophys. J. **738**, 59 (2011)
M. Röllig, N.P. Abel, T. Bell et al., Astron. Astrophys. **467**, 187 (2007)
G.P. Rosotti, B. Ercolano, J.E. Owen, P.J. Armitage, Mon. Not. R. Astron. Soc. **430**, 1392 (2013)
R. Sahai, R. Güsten, M.R. Morris, Astrophys. J. Lett. **761**, L21 (2012a)
R. Sahai, M.R. Morris, M.J. Claussen, Astrophys. J. **751**, 69 (2012b)
A.I. Sargent, S. Beckwith, Astrophys. J. **323**, 294 (1987)
A. Scally, C. Clarke, Mon. Not. R. Astron. Soc. **325**, 449 (2001)
N.I. Shakura, R.A. Sunyaev, Astron. Astrophys. **24**, 337 (1973)
H. Shang, Z.-Y. Li, N. Hirano, *Protostars and Planets V* (University of Arizona Press, Tucson, 2007), p. 261
F. Shu, J. Najita, E. Ostriker et al., Astrophys. J. **429**, 781 (1994)
N. Smith, J. Bally, J.A. Morse, Astrophys. J. Lett. **587**, L105 (2003)
D.R. Soderblom, L.A. Hillenbrand, R.D. Jeffries, E.E. Mamajek, T. Naylor, *Protostars and Planets VI* (University of Arizona Press, Tucson, 2014), p. 219
H. Störzer, D. Hollenbach, Astrophys. J. **515**, 669 (1999)
S. Terebey, F.H. Shu, P. Cassen, Astrophys. J. **286**, 529 (1984)
L. Testi, T. Birnstiel, L. Ricci et al., *Protostars and Planets VI* (University of Arizona Press, Tucson, 2014), p. 339
H.B. Throop, J. Bally, Astrophys. J. Lett. **623**, L149 (2005)
L.K. Townsley, E.D. Feigelson, T. Montmerle et al., Astrophys. J. **593**, 874 (2003)
Y.G. Tsamis, N. Flores-Fajardo, W.J. Henney, J.R. Walsh, A. Mesa-Delgado, Mon. Not. R. Astron. Soc. **430**, 3406 (2013)
T. Tsukagoshi, M. Momose, M. Saito et al., Astrophys. J. Lett. **802**, L7 (2015)
N.J. Turner, S. Fromang, C. Gammie et al., *Protostars and Planets VI* (University of Arizona Press, Tucson, 2014), p. 411
N. van der Marel, E.F. van Dishoeck, S. Bruderer, L. Pérez, A. Isella, Astron. Astrophys. **579**, A106 (2015)
D. Veras, P.J. Armitage, Mon. Not. R. Astron. Soc. **347**, 613 (2004)
S.M. Vicente, J. Alves, Astron. Astrophys. **441**, 195 (2005)
Z. Wahhaj, L. Cieza, D.W. Koerner et al., Astrophys. J. **724**, 835 (2010)
S. Walch, A. Burkert, A. Whitworth, T. Naab, M. Gritschneder, Mon. Not. R. Astron. Soc. **400**, 13 (2009)
S. Walch, T. Naab, A. Whitworth, A. Burkert, M. Gritschneder, Mon. Not. R. Astron. Soc. **402**, 2253 (2010)
W.R. Ward, Astrophys. J. Lett. **482**, L211 (1997)
T.R. Waters, D. Proga, Mon. Not. R. Astron. Soc. **426**, 2239 (2012)
G.W. Wetherill, G.R. Stewart, Icarus **77**, 330 (1989)
R.J. White, L.A. Hillenbrand, Astrophys. J. **616**, 998 (2004)
J.P. Williams, L.A. Cieza, Annu. Rev. Astron. Astrophys. **49**, 67 (2011)
J.N. Winn, D.C. Fabrycky, Annu. Rev. Astron. Astrophys. **53**, 409 (2015)
N.J. Wright, J.J. Drake, J.E. Drew et al., Astrophys. J. Lett. **746**, L21 (2012)

Y. Wu, Y. Wei, M. Zhao et al., Astron. Astrophys. **426**, 503 (2004)
M.C. Wyatt, Annu. Rev. Astron. Astrophys. **46**, 339 (2008)
M.C. Wyatt, O. Panic, G.M. Kennedy, L. Matra (2014). arXiv:1412.5598
F. Yusef-Zadeh, J. Biretta, T.R. Geballe, Astron. J. **130**, 1171 (2005)
Z. Zhu, R.P. Nelson, L. Hartmann, C. Espaillat, N. Calvet, Astrophys. J. **729**, 47 (2011)
A. Zsom, C.W. Ormel, C. Güttler, J. Blum, C.P. Dullemond, Astron. Astrophys. **513**, A57 (2010)
B. Zuckerman, T. Forveille, J.H. Kastner, Nature **373**, 494 (1995)

Space Sci Rev (2016) 205:193–251
DOI 10.1007/s11214-016-0280-1

Formation and Evolution of Protoatmospheres

H. Massol[1] · K. Hamano[2,3] · F. Tian[4] · M. Ikoma[2] · Y. Abe[2] · E. Chassefière[1] · A. Davaille[5] · H. Genda[6] · M. Güdel[7] · Y. Hori[8] · F. Leblanc[9] · E. Marcq[9] · P. Sarda[1] · V.I. Shematovich[10] · A. Stökl[7] · H. Lammer[11]

Received: 10 November 2015 / Accepted: 16 August 2016 / Published online: 27 September 2016

Abstract The origin and evolution of planetary protoatmospheres in relation to the protoplanetary disk is discussed. The initial atmospheres of planets can mainly be related via two formation scenarios. If a protoplanetary core accretes mass and grows inside the gas disk, it can capture H_2, He and other gases from the disk. When the gas of the disk evaporates, the core that is surrounded by the H_2/He gas envelope is exposed to the high X-ray and extreme ultraviolet flux and stellar wind of the young host star. This period can be considered as the onset of atmospheric escape. It is shown that lower mass bodies accrete less gas and

✉ H. Massol
helene.massol@u-psud.fr

K. Hamano
keiko@elsi.jp

F. Tian
tianfengco@mail.tsinghua.edu.cn

M. Ikoma
ikoma@eps.s.u-tokyo.ac.jp

Y. Abe
ayutaka@eps.s.u-tokyo.ac.jp

E. Chassefière
eric.chassefiere@u-psud.fr

A. Davaille
anne.davaille@u-psud.fr

H. Genda
genda@elsi.jp

M. Güdel
manuel.guedel@univie.ac.at

Y. Hori
yasunori.hori@nao.ac.jp

F. Leblanc
francois.leblanc@latmos.ipsl.fr

E. Marcq
emmanuel.marcq@latmos.ipsl.fr

depending on the host stars radiation environment can therefore lose the gaseous envelope after tens or hundreds of million years. Massive cores may never get rid of their captured hydrogen envelopes and remain as sub-Neptunes, Neptunes or gas giants for their whole life time. Terrestrial planets which may have lost the captured gas envelope by thermal atmospheric escape, or which accreted after the protoplanetary nebula vanished will produce catastrophically outgassed steam atmospheres during the magma ocean solidification process. These steam atmospheres consist mainly of water and CO_2 that was incorporated into the protoplanet during its accretion. Planets, which are formed in the habitable zone, solidify within several million years. In such cases the outgassed steam atmospheres cool fast, which leads to the condensation of water and the formation of liquid oceans. On the other hand, magma oceans are sustained for longer if planets form inside a critical distance, even if they outgassed a larger initial amount of water. In such cases the steam atmosphere could remain 100 million years or for even longer. Hydrodynamic atmospheric escape will then desiccate these planets during the slow solidification process.

1 Introduction

Stellar systems originate from the accumulations of gas and dust scattered in space. Such aggregations of matter form huge disks with the active young star in their center. The protoplanets grow from the dust to planetesimals and planetary embryos over timescales of several million to tens of million years. During the growth of the protoplanets in the primordial nebula, massive hydrogen/helium envelopes can be captured (e.g. Mizuno et al. 1978;

P. Sarda
philippe.sarda@u-psud.fr

V.I. Shematovich
shematov@inasan.ru

A. Stökl
alexander.stoekl@univie.ac.at

H. Lammer
helmut.lammer@oeaw.ac.at

[1] GEOPS, Univ. Paris-Sud, CNRS, Université Paris-Saclay, Orsay, France

[2] Department of Earth and Planetary Science, Graduate School of Science, The University of Tokyo, Tokyo, Japan

[3] Present address: Earth-Life Science Institute, Tokyo Institute of Technology, Tokyo, Japan

[4] National Astronomical Observatories, Chinese Academy of Sciences, Beijing, China

[5] Lab. FAST, UMR 7608 CNRS-Université de Paris-Sud, Orsay, France

[6] Earth-Life Science Institute, Tokyo Institute of Technology, Tokyo, Japan

[7] Institute of Astronomy, University of Vienna, Vienna, Austria

[8] National Astronomical Observatory of Japan/Astrobiology Center, National Institutes of Natural Sciences, Tokyo, Japan

[9] Lab. LATMOS, UMR 9190 CNRS, Guyancourt, France

[10] Department of Solar System Research, Institute of Astronomy of the Russian Academy of Sciences, Moscow, Russia

[11] Austrian Academy of Sciences Space Research Institute, Graz, Austria

Wuchterl 1993; Ikoma et al. 2000; Rafikov 2006). The capturing and retaining portions of the local nebular gas depend on the conditions of the nebula and the evolution of the host star's stellar radiation and plasma environment. This process builds up the protoplanet's primordial atmosphere as long as solar wind and solar radiation have not depleted the disk's local matter resources. In the inner zones of an early stellar system, the removal of disk gas and matter may happen within three million years, while farther out, where dispersal most likely takes more time, planetary embryos have more time to collect matter and hence may become more massive.

In the above mentioned pioneering studies, it was shown that the capture of nebula gas can reach a stage where gas accretion exceeds the accretion of solid bodies (here called cores), so that the protoplanet evolves very quickly to a Jovian-like gas giant (e.g. Pollack 1985; Lissauer 1987; Wuchterl 1993, 2010). The results of these studies indicate that the onset of core instability occurs when the mass of the captured hydrogen/helium atmosphere around the core becomes comparable to the core mass (e.g. Mizuno et al. 1978, 1980; Hayashi et al. 1979; Stevenson 1982; Wuchterl 1993, 1994, 2010; Ikoma et al. 2000; Rafikov 2006).

However, the results and conclusions of the majority of these pioneering studies, that have been carried out decades ago, are based on very rough assumptions, because, no reliable observational data from discs and related nebula lifetimes, as well as no astrophysical observation of the radiation environment of very young solar-like stars have been available.

The recent discoveries of low mass exoplanets with known radii and masses by the Kepler and CoRoT satellites, combined with ground-based radial velocity follow-up and dynamical analyses of transit timing variations, indicate that there is a transition between small Earth-like rocky planets and Neptune-like planets, where many rocky planets may have kept their captured hydrogen envelopes during their lifetime (Marcy et al. 2014; Rogers 2015). Rogers (2015) found that the majority of so-called 'super-Earths' with radii of about 1.6 Earth-radius are too low in density to be composed of Fe and silicates alone. For discovered planets with larger radii, the constraints on their rocky fraction are even more strict. These findings agree with Lammer et al. (2014) who modeled the capture of nebula-based hydrogen envelopes and their escape from rocky protoplanetary cores of 'sub-' to 'super-Earths' within the mass range from Mars—like bodies up to cores with 5 Earth-masses. These authors found that protoplanets inside the habitable zone with core masses less than 1 Earth-mass, that are exposed to stellar soft X-ray and EUV fluxes (XUV) of an average young solar-like host star, can lose their captured hydrogen envelopes during the active stellar XUV saturation phase, but rocky cores within the so-called super-Earth domain most likely cannot get rid of their nebula captured hydrogen envelopes during their whole lifetime.

Planets that have too low masses during growth, when the nebula gas is not evaporated, but finish their accretion a few tens of million years later will produce catastrophically outgassed steam atmospheres during the solidification of their magma oceans (e.g. Elkins-Tanton 2008; Lebrun et al. 2013; Hamano et al. 2013). The same scenario can be expected for rocky planets that could lose their nebular captured hydrogen envelopes. The energy sufficient for melting the planet's mantle, thereby creating an ocean of liquid magma, originates from collisions with large planetesimals or planetary embryos or the decay of short lived radioactive isotopes. Over time, the planet's heat is continuously lost to space allowing the magma to cool down and eventually to solidify. The liquid magma contains large amounts of volatiles that were acquired during the accretion process. Since H_2O and CO_2 cannot be trapped in large amounts in the solid phase, these volatiles are outgassed during solidification together with minor amounts of other volatiles such as CH_4, N_2 and NH_3. All these volatiles enter and thereby modify the pre-existing atmospheres (e.g., Lebrun et al. 2013; Hamano et al. 2013).

The major motivation of this work is to review the processes that affect and control timescales of atmospheric formation. Since atmospheres are an essential criterion for habitability and determine planetary evolution (e.g. towards super Earth or mini Neptune), it is important to investigate the capture and loss of nebula-based hydrogen envelopes, and magma ocean related outgassed steam atmospheres from protoplanets. In the first section, the origin of the captured hydrogen-dominated envelopes of nebular gas around accreting protoplanets is discussed. The second section presents the state of the art of thermal evolution processes taking place within magma oceans and surrounding degassed steam atmospheres at the end of accretion. These degassed atmospheres can be considered as the initial atmospheres of Earth-like terrestrial planets. In the third section, we discuss the atmospheric escape processes, including relevant processes such as the heating efficiency in upper atmospheres that are exposed to high XUV fluxes related to the young host stars. Finally, we conclude and address open questions that need some further investigations within the context of future missions and habitability on exo-planets.

2 Nebular Origin of Protoatmospheres

2.1 Introduction

Planets are born in protoplanetary disks composed predominantly of hydrogen and helium. The mass of Mars is large enough for a solid planet to attract hydrogen gas gravitationally, provided that the gas temperature is lower than 2000 K, since the gravitational-potential energy then dominates the thermal-motion energy of the gas. Thus, planetary atmospheres of disk origin should be an inevitable by-product of planet formation.

In our Solar System, the four giant planets have massive H + He envelopes, which must have come from the disk that surrounded the proto-sun (i.e., the solar nebula). The fractions of H + He differ from planet to planet: those of the two gas giants, Jupiter and Saturn, are more than about 85 % and 60 %, respectively, whereas those of the two ice giants, Uranus and Neptune, are both about 10 % (e.g., Fortney and Nettelmann 2010). In contrast, the four terrestrial planets, the Earth, Venus, Mars, and Mercury, retain no such atmospheres; instead they have thin hydrogen-poor atmospheres.

Beyond our Solar System, there are plenty of objects intermediate in size between the Earth and Uranus/Neptune, as revealed by observations with the Kepler space telescope (see Batalha 2014, for a recent review). Those objects are often called super-Earths or mini-Neptunes. Figure 1 shows the observed relationships between masses and radii of exoplanets of less than 20 $M_\oplus$, together with the theoretical relationships for rocky bodies with and without H/He atmospheres (see the caption for details). As seen from Fig. 1, many of the super-Earths are significantly larger in radius than they would be if they had an Earth-like core-mantle ratio (dashed line), indicating the possibility that they retain H/He atmospheres (solid lines) or have a smaller core-mantle ratio than the Earth. Furthermore, the radius differs largely among the planets of similar mass, which may mean that super-Earths are diverse in hydrogen content. What causes the diversity in hydrogen content of planets?

The gravitational capture of disk gas by protoplanets was first explored in the context of the formation of the solar-system giant planets. Pioneering studies were done by Perri and Cameron (1974) and Mizuno et al. (1978, 1980). They assumed that a growing solid

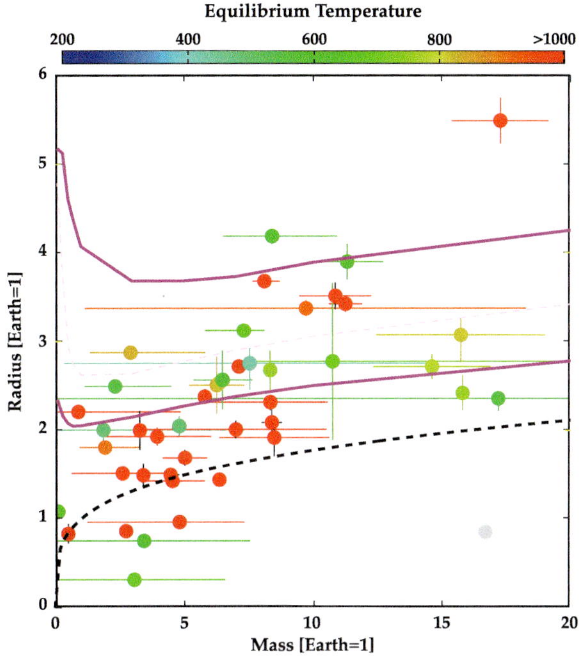

Fig. 1 The mass-radius relationships for exoplanets with mass less than 20 $M_\oplus$. The observational data are represented by *symbols*, which are *colored* according to the estimated zero-albedo equilibrium temperature. The *two solid lines* represent the theoretical relationships for rocky planets with 10 % (upper) and 1 % (lower) hydrogen contents (Lopez and Fortney 2014), while the *dashed line* is for rocky bodies with the terrestrial bulk composition. *Symbols with error bars* are from www.exoplanets.org. The *solid lines* were drawn based on Lopez and Fortney (2014)

protoplanet (or proto-core) was embedded in a protoplanetary disk and had a hydrostatic-equilibrium atmosphere (or proto-envelope) that connected smoothly to the ambient disk at the protoplanet's Hill radius (see below). They demonstrated that the proto-envelope mass increased with the proto-core mass and found a critical value of the proto-core mass beyond which no purely-hydrostatic solution was found. By performing not purely-hydrostatic but instead quasi-static simulations, Bodenheimer and Pollack (1986) demonstrated that run-away accretion of disk gas occurred after the core mass became comparable to the envelope mass (i.e., the crossover mass). As shown analytically (Stevenson 1982; Wuchterl 1993) and numerically (Ikoma et al. 2000), these two characteristic masses always have similar values. While the basic behavior of the critical core mass is well-studied, several issues remain unresolved such as the timescale and location of the formation of Jupiter and Saturn, the consistency of the inferred small core mass of Jupiter with formation theories, the difference between the solar-system giants and extrasolar gas giants with short orbital periods (i.e., hot Jupiters), and so on.

The hypothesis about the Earth's primordial atmosphere of nebular origin was developed by Hayashi, Nakazawa, and their colleagues. The primary aim was to explain the differentiated structure of the terrestrial planets, namely the presence of metallic cores surrounded by rocky mantles. As demonstrated by Hayashi et al. (1979), the blanketing effect of the primordial atmosphere is so strong that a large part of the interior of the proto-Earth is molten, making a global magma ocean, where molten iron can separate from silicate and settle down toward the planetary center. This hypothesis, however, yields several outcomes that seem to be inconsistent with geochemical facts such as the paucity of noble gases in the atmosphere and mantle. While the presence of a primordial atmosphere on the proto-Earth is controversial, this possibility should be revisited in the context of the diversity of exo-Earths.

In this section, we review our current understanding of the disk (nebular) origin of atmospheres of solar-system and extrasolar planets. First, we summarize the basic properties of the atmospheres of protoplanets embedded in protoplanetary disks in Sect. 2.2. In Sect. 2.3, we discuss the relation between physics and observational constraints. Then, in Sect. 2.4, we review several important issues about low-density super-Earths (Sect. 2.4.1), exo-Earths (Sect. 2.4.2), and the primordial atmospheres of the Earth and Mars (Sect. 2.4.3).

2.2 Properties of the Protoatmosphere

Concerning the atmosphere of growing protoplanets embedded in protoplanetary disks, there are two characteristic radii, namely the Hill and Bondi radii, which are defined respectively as

$$R_{\mathrm{H}} = \left(\frac{M_p}{3M_*}\right)^{1/3} a \tag{1}$$

and

$$R_{\mathrm{B}} = \frac{\gamma - 1}{\gamma} \frac{G M_p \mu m_u}{kT}, \tag{2}$$

where M_p and M_* are the masses of the planet and its host star, respectively, a is the planet's orbital distance, μ, γ and T are the mean molecular weight, specific heat ratio and temperature of the disk gas, respectively, and G, m_u, and k are the gravitational constant, the atomic mass unit, and the Boltzmann constant, respectively. Outside the Hill radius, the stellar tidal force is greater than the protoplanet's gravity; outside the Bondi radius, the thermal motion of gas molecules dominates the protoplanet's gravity. Thus, only inside of both the Hill and Bondi radii can the atmosphere be in a gravitationally bound, hydrostatic equilibrium state. For the ranges of the protoplanet's bulk density and orbital distance of interest (i.e., $\bar{\rho} \gtrsim 0.1\ \mathrm{g/cm^3}$ and $a \gtrsim 0.01$ AU), the Hill radius is always larger than the protoplanet's radius; in contrast, the Bondi radius is smaller than the protoplanet's radius when the protoplanet's mass is smaller than a certain value. Thus, it sets a condition on the protoplanetary mass for the existence of a hydrogen atmosphere, which is:

$$M_p \gtrsim 1.1 \times 10^{23} \left(\frac{T}{300\ \mathrm{K}}\right)^{3/2} \left(\frac{\mu}{2}\right)^{-3/2} \left(\frac{\bar{\rho}}{5.5\ \mathrm{g/cm^3}}\right)^{-1/2} \mathrm{kg}, \tag{3}$$

where we have assumed $\gamma = 7/5$.

The embedded atmosphere is thus defined as the gas contained inside the smaller of the Bondi and Hill radii. The atmosphere is usually assumed to connect smoothly to the ambient disk, which means that the outer boundary conditions for the atmospheric structure are determined from the disk properties, namely,

$$P = P_d \quad \text{and} \quad T = T_d \quad \text{at } r = R_p, \tag{4}$$

where P is pressure, T is temperature and P_d and T_d are the pressure and temperature of the ambient disk gas, respectively, r is the radial distance from the protoplanet's center, R_p is the protoplanet's outer radius, which is defined by $R_p = \min(R_{\mathrm{H}}, R_{\mathrm{B}})$. This boundary condition (4) assumes that the planetary mass is sufficiently low for the radius of the embedded atmosphere to be smaller than the vertical thickness of the protoplanetary disk. Using $0.01a$ as a rough estimate for the latter, comparison with the Hill radius yields a mass limit of $M_p \leq 3 \times 10^{-6} M_*$ (i.e. $\lesssim 1\ M_\oplus$) for a solar mass star. This corresponds to the assumption that the gravitational perturbation to the disk structure by the planet's mass remains negligible and that the influx of disk gas into the planetary atmosphere does not exhaust the disk

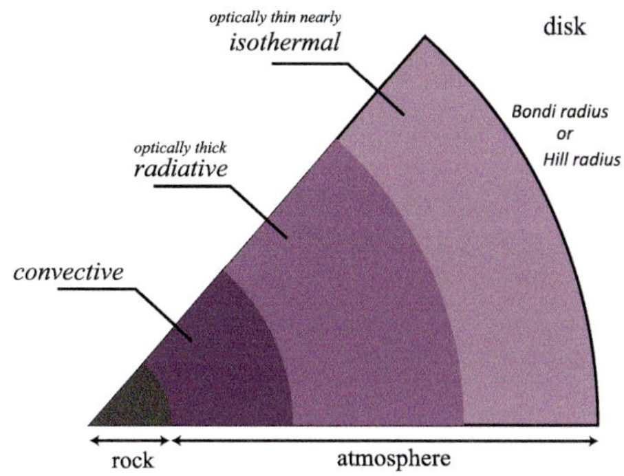

Fig. 2 Schematic illustration of the thermal structure of the protoatmosphere that connects to the ambient disk

gas reservoir or gas transport processes (viscous accretion) within the disk. For more massive planets, these assumptions eventually break down, at the latest when the atmospheric mass exceeds the critical crossover mass and the atmosphere undergoes runaway accretion that will quickly outstrip the gas reservoir of the disk leading to the opening of an annular gap in the disk.

The volume of the atmosphere is given by $4\pi(R_p^3 - R_s^3)/3$, R_s being the outer radius of the solid part of the protoplanet. It is a function of the mass and radius of the solid protoplanet and the properties of the ambient disk, provided the mass of the atmosphere is quite small relative to that of the solid protoplanet. The mass of the atmosphere depends on the structure of the atmosphere. In particular, the thermal structure is important, because it determines the atmospheric scale height. Due to the low mean molecular weight and high temperature, the pressure scale-height ($\propto T/\mu$) is large and usually comparable to the radius of the solid protoplanet, which is in clear contrast to the current atmospheres of the Earth and other terrestrial planets. Thus, as the lower atmosphere becomes hotter, the atmosphere becomes less massive due to enhanced escape.

When modeling embedded planetary atmospheres, it is usually assumed that the atmosphere is in dynamical and thermal equilibrium. Except for violent events such as the impact of a large planetesimal, the former is well justified as the typical sound crossing time, even for expanded atmospheres, is on the order of days at most. The assumption of thermal equilibrium is less well justified (Stökl et al. 2015), it is, however, probably still a good approximation during most phases of planetary evolution and therefore a frequently adopted simplification. Thermal equilibrium corresponds to the assumption of a constant luminosity throughout the atmosphere, namely at different atmospheric depths, which is, in the strict sense, incompatible with some of the atmospheric energy sources discussed in Sect. 2.3.1.

The atmosphere consists of three characteristic regions (see Fig. 2). The outer atmosphere is optically thin and nearly isothermal. Below the optically thin region, the thermal structure is governed by radiative transport and, in the stationary limit, can be described by (Kippenhahn and Weigert 1994):

$$\frac{dT}{dr} = -\frac{3\kappa\rho L}{64\pi\sigma r^2 T^3}, \tag{5}$$

where κ is the opacity, L is the energy flux passing through the sphere of radius r, ρ is the gas density, and σ is the Stefan-Boltzmann constant. In the deep atmosphere, the gas density is so high that convective transport dominates. The structure in that region is well approximated to be iso-entropic and thus described by

$$\frac{dT}{dr} = \left(\frac{\partial T}{\partial P}\right)_{\rm ad} \frac{dP}{dr}, \tag{6}$$

where $(\partial T/\partial P)_{\rm ad}$ is the adiabatic temperature gradient. Note that the structure of the convective part of the atmosphere is independent of the planetary luminosity (as long as it is high enough for convective instability). This is one reason why thermal equilibrium is still a useful approximation, even if the luminosity is not constant in the deep atmosphere. Usually most of the atmospheric mass is contained in the convective region and of special importance is the entropy of the region, which is determined by the structure of the intermediate radiative region.

As seen from Eq. (5), the structure of the inner radiative region depends on the energy flux L and the opacity κ. The sources of the energy flux include the accretion energy of the solid protoplanet (i.e., the kinetic energy of incoming planetesimals), the accretion energy of the atmosphere, the radiogenic energy, and the stellar insolation. Below the critical core mass (see Introduction), the accretion energy of the solid protoplanet is much greater than the other three energy sources and is given approximately by

$$L = \frac{GM_s}{R_s} \dot{M}_s, \tag{7}$$

where M_s is the mass of the solid part of the protoplanet, $\dot{M}_s$ is the total mass of incoming planetesimals per unit time and G the gravitational constant. The sources of the opacity are absorption by gas molecules such as H_2O and scattering and absorption by dust grains floating in the atmosphere. For a given composition of gas, the gas opacity does not include any large uncertainty, whereas the grain opacity is quite uncertain, because we do not have enough knowledge of the amount and size distribution of the dust grains. The details of the sources of the energy flux and the grain opacity are discussed in Sect. 2.3.

Simple analytical argument helps us to know the basic properties of such an atmosphere. If the atmosphere is entirely radiative and $\kappa L/M_r$ and μ are radially constant (M_r being the mass inside a sphere of radius r), Eq. (5) combined with the equation of hydrostatic equilibrium (Eq. (1) from Ikoma and Genda 2006), can be integrated analytically with the ideal-gas equation of state (see Ikoma and Genda 2006). The pressure P_s at the bottom of the atmosphere is calculated to be

$$P_s \approx \frac{\pi \sigma G^5 m_u^4}{48k^4} \frac{M_s^5 \mu^4}{R_s^4 \kappa L} = 750\ \text{MPa} \left(\frac{M_s}{M_\oplus}\right)^{11/3} \left(\frac{\kappa}{10^{-3}\ \text{m}^2/\text{kg}}\right)^{-1} \left(\frac{L}{10^{17}\ \text{W}}\right)^{-1}, \tag{8}$$

where we have assumed the solid protoplanet's bulk density $\bar{\rho}_s = 5500$ kg/m^3 and $\mu = 2$. Under the same assumptions, one can obtain the density as a function of radial distance from the protoplanet center. Then, integrating the density profile, the mass of the radiative atmosphere M_a, is given by:

$$M_a \approx \frac{\pi^2 \sigma}{3\kappa L} \left(\frac{GM_s \mu m_u}{k}\right)^4 \ln\left(\frac{R_{\rm B}}{R_s}\right). \tag{9}$$

The atmospheric mass depends strongly on the mass of the solid protoplanet (i.e., $M_a \propto M_s^4$). Also, the atmospheric mass increases linearly with decreasing κ and L.

The temperature at the bottom of the radiative atmosphere (or the surface temperature) is given by

$$T_s \approx \frac{GM_s \mu m_u}{4kR_s} = 3700\ \mathrm{K}\left(\frac{M_s}{M_\oplus}\right)^{2/3}, \tag{10}$$

where we have assumed $\bar{\rho}_s = 5500\ \mathrm{kg/m^3}$ and $\mu = 2$ again. In the case of an Earth-mass protoplanet, the surface temperature is found to be higher than 3000 K. Note that Eq. (10) does not include radiative properties of the atmosphere such as κ and L. In reality, the opacity varies with radial distance, and the deep atmosphere is not radiative, but convective in most cases. Detailed numerical simulations (Ikoma and Genda 2006) show that T_s weakly depends on κ and L; nevertheless, the surface temperature is always above the liquidus temperature of basalt ($\sim$ 1500 K).

In the above discussion, we have implicitly assumed that the atmosphere is sufficiently bound; in other words, the surface pressure P_s and temperature T_s are sufficiently high relatively to the local pressure P_d and temperature T_d of the disk gas. The degree of gravitational binding can be measured by the escape parameter defined as

$$\lambda \equiv \frac{GM_s \mu m_u}{R_s k T_d} = 54\left(\frac{M_s}{M_\oplus}\right)^{2/3}\left(\frac{T_d}{280\ \mathrm{K}}\right)^{-1} \tag{11}$$

with $\mu = 2$ and $\bar{\rho}_s = 5500\ \mathrm{kg/m^3}$. Ikoma and Genda (2006) showed that the atmospheric structure and mass of the Earth-mass protoplanet ($\lambda = 56$) is insensitive to the outer boundary conditions, whereas those of the Mars-mass protoplanet ($\lambda = 12$) are quite sensitive to changes in the latter. Thus, the protoatmosphere of a Mars-mass protoplanet, if any, never survives the dispersal of the protoplanetary disk. The disk temperature may be estimated to be $T_d = 280(a/\mathrm{AU})^{-1/2}$ K around solar-type stars (Hayashi 1981), with which $\lambda = 17(M_s/M_\oplus)^{2/3}(a/0.1\ \mathrm{AU})^{1/2}$. This indicates that Earth-mass planets orbiting at 0.1 AU from solar-type stars are not massive enough to retain the protoatmospheres. The λ-values for super-Earths with short-orbital periods ($\sim$ 0.1 AU) are intermediate between those of Mars and Earth. This may cause the diversity in density of currently detected low-density super-Earths.

2.3 Constraints on Involved Processes

As described above, the structure and mass of the protoatmosphere depends on the mass of the solid protoplanet, the energy supply rate and opacity, the temperature and density of the ambient disk gas, and the composition (e.g., mean molecular weight) of the atmosphere. Of these, the most important factors are likely to be the energy supply rate and opacity. Below we describe our current understanding of these factors.

2.3.1 Energy Sources

Energy sources for maintaining the atmospheric pressure include accretion energy of solid, that of gas, radiogenic energy, and stellar insolation. The third and fourth sources are always negligible in the accretion era. For example, the radiogenic luminosity is estimated to be $\sim 10^{21}$ erg/s for the early Earth (O'Neill and Palme 1988), which corresponds to the solid accretion rate of $\sim 10^{-11}\ M_\oplus$/yr. Also, stellar light is unavailable for heating the atmosphere, since it is blocked by dust in the protoplanetary disk. Thus, the solid accretion energy is the major source in the early stages of planet formation, namely, the runaway-growth (Wetherill and Stewart 1989; Kokubo and Ida 1996) and oligarchic-growth stages (Lissauer 1987; Kokubo and Ida 1998). The processes of solid accretion and its impact on the atmospheric accumulation are outlined briefly below.

The early runaway growth slows down relatively early, because of the enhanced velocity dispersion of planetesimals (Ida and Makino 1993). In the stage characterized by such a relatively slow accretion, the accretion rate is estimated to be approximately $2 \times 10^{-6}\ M_\oplus$/yr (or 2×10^{26} erg/s) at 1 AU and $1 \times 10^{-7}\ M_\oplus$/yr at 5 AU (or 1×10^{25} erg/s) for a 1 $M_\oplus$ protoplanet in the minimum-mass solar nebula (Tanaka and Ida 1999). Because of energy supply via solid accretion, the atmosphere is relatively less massive (see Eqs. (7) and (9)). For example, for the luminosity of 1×10^{26} erg/s, the atmospheric mass is calculated to be up to 3×10^{24} g (see Fig. 3a of Ikoma and Genda 2006), which corresponds to 0.05 % of the Earth mass.

N-body simulations, however, show that the runaway/oligarchic growth ends before protoplanets become massive enough, unless their feeding zones are replenished with planetesimals via gravitational perturbation by other protoplanets or sweeping during orbital migration. The protoplanets then become isolated. According to Kokubo and Ida (1998), in the minimum-mass solar nebula, the isolation mass $M_{\rm iso}$ is approximately as small as Mars' mass at 1 AU and $\sim 2\ M_\oplus$ at 5 AU, which is well below the conventional critical core mass (e.g., 10 $M_\oplus$) (e.g. Mizuno et al. 1980 or Ikoma et al. 2000). This conventional critical core mass might be affected by the thermal state in the envelope of a planet. For instance, lower opacities in the envelope (e.g. Mizuno et al. 1980) and a lower rate of planetesimal accretion (e.g. Ikoma et al. 2000) result in a smaller critical core mass. Also, the density and temperature of an ambient disk gas (i.e., a distance from a central star) are thought to be key factors for determining a critical core mass especially in the case of fully-convective envelopes (Wuchterl 1993). Specifically, the higher density of a disk gas is and the lower temperature of a disk gas is, the smaller a critical core mass becomes (e.g. Ikoma et al. 2001).

Once a protoplanet becomes isolated, the major source is the energy supplied by the gravitational contraction of the atmosphere itself. In this stage, the atmospheric accumulation is dictated by cooling of the atmosphere; Namely, it proceeds on the Kelvin-Helmholtz timescale (Ikoma et al. 2000). The accumulation timescale $\tau_{\rm KH}$ is strongly dependent on the mass of the solid planet M_s and moderately dependent on the opacity $\bar{\kappa}$; $\tau_{\rm KH}$ decreases almost with the cube of M_s (the exact power being dependent on the opacity) and linearly with $\bar{\kappa}$. For 1 $M_\oplus$ planets, for example, $\tau_{\rm KH}$ is as long as more than 10 Myr except for the metal-free atmosphere (Hori and Ikoma 2010; Ikoma and Genda 2006).

Actual accretion processes including orbital migration are much more complicated. The details are delegated to Sect. 4.

2.3.2 Opacity

A planet radiates heat into space and continues to cool down until it reaches thermodynamic equilibrium. The heat transfer in a planetary atmosphere is controlled by conduction, radiation or convection. In the upper atmosphere, radiation carries energy away from it. The efficiency of energy transfer via radiation is regulated by the opacity of an ambient medium, as shown in Eq. (5). Dust grains are dominant absorbers for temperatures below the sublimation temperature (~ 1500 K), and for higher temperatures, gaseous species are the main sources of the opacity (Semenov et al. 2003). Optical properties of dust grains strongly depend on their particle sizes and structures, e.g., the shape of monomers and porosity of dust aggregates. Thus, a detailed prescription of the microphysical properties of dust grains in the atmosphere is essential to understanding the thermal evolution of a planet.

The growth of dust grains in a planetary atmosphere is mainly governed by six microscopic processes: the vertical transport via the eddy diffusion, collisions via the Brownian motion and sedimentation, electrostatic effects, condensation, and vaporization. The evolution of the size distribution of dust grains is described by a coagulation equation, the so-called Smoluchowski's equation (e.g. Wetherill 1990). Solving the coagulation equation that includes the microphysical effects mentioned above, Podolak (2003) computed the grain opacity in the atmospheres of proto-gas giants with cores of 11.5 $M_\oplus$ and 21.09 $M_\oplus$. He found that the size distribution of dust grains is significantly different from that of the interstellar medium (ISM), and concluded that the grain opacity in the planetary atmosphere is less than 0.1 $\mathrm{cm^2\,g^{-1}}$ because of the grain settling associated with the grain growth.

In the context of Jupiter formation, the grain opacity in the atmosphere is also found to be of the order of 0.01 $\mathrm{g\,cm^{-2}}$, which is much lower than the ISM value (Movshovitz and Podolak 2008; Movshovitz et al. 2010). Such a low grain opacity accelerates gas accretion onto protoplanets and reduces the formation timescale of giant planets (e.g. Ikoma et al. 2000; Hubickyj et al. 2005) because the gravitational contraction of the entire planet, i.e., the Kelvin-Helmholtz contraction, compensates for the energy loss via radiation. The conclusion of Movshovitz and Podolak (2008) was confirmed by two ensuing studies on the grain growth: analytical models of grain opacities Mordasini (2014) and grain growth models including effects of the mass deposition by incoming planetesimals and the porosity of dust grains (Ormel 2014). Although Podolak (2003) and Movshovitz and Podolak (2008) considered porous dust aggregates with a fractal dimension of 2.11, Ormel (2014)'s work adopted a detailed recipe for the porosity evolution of dust aggregates based on direct N-body collision simulations of collisions (see Sect. 2 in Okuzumi et al. 2012).

Recently, Mordasini (2014) introduced a constant reduction factor of ISM grain opacities (f), into their population synthesis model of planet formation; $\kappa = f\kappa_{\mathrm{grain}} + \kappa_{\mathrm{gas}}$, where κ is the opacity and the subscripts of "grain" and "gas" mean the ISM grain opacity and gas opacity, respectively. They concluded that an extremely-low grain opacity model with $f = 0.003$ is favorable in order to explain the observed population of exoplanets. This fact suggests that the atmosphere of a planet is close to being a grain-free environment, as discussed in Hori and Ikoma (2010).

2.4 Relevant Issues

2.4.1 Low-Density Super-Earths

As described in the introduction, many super-Earths larger in radius than their "naked" rocky counterparts have been found beyond the Solar System (see Fig. 1). Those large radii may be consistent with the presence of H/He atmospheres that constitute up to 10 wt% of the planetary total masses (see Lopez and Fortney 2014).

The in situ accretion of disk gas onto rocky super-Earths was examined by Ikoma and Hori (2012) in order to explain the origin of low-density, multiple super-Earths orbiting the star Kepler-11. They assume that disk gas dissipation triggers giant collisions of multiple protoplanets, followed by the cooling and accumulation of H/He atmospheres. They demonstrated that the accretion of disk gas occurring concurrently with disk dissipation results in an intermediate-mass atmospheres. One of their results is that atmospheric erosion occurs because of depressurization and cooling due to disk dissipation. Also, the thermal contribution of the rocky body to the atmospheric evolution was found to have a significant impact on the atmospheric erosion. As discussed in Sect. 2.2, the sensitivity to a change in disk gas density is significant in relatively hot environments ($\gtrsim 500$ K). They concluded that the in situ accumulation of super-Earths with intermediate-mass H/He atmospheres is in principle possible; however, the radius of Kepler-11e is too large to conclude that its H/He atmosphere was accreted in a hot environment, implying that the planet migrated from a cooler region. Recently, Lee et al. (2014) made a more extensive study on the in situ accretion of H/He atmospheres for different stellar metallicities, and confirmed that the suppression of gas accretion onto rocky cores in metal-rich environment leads to super-Earths with tiny H/He atmospheres (less than a few wt%), as well as grain-rich cases as suggested in Ikoma and Hori (2012).

Bodenheimer and Lissauer (2014) simulated the concurrent accretion of solid and gas that results in the formation of super-Earths of 2.2 to 2.5 $M_\oplus$ at 0.5, 1, 2, and 4 AU, adopting the numerical model developed by Pollack et al. (1996). They concluded that low-temperature environments are needed for such low-mass objects to have atmospheres with 2–7 wt% of planet total masses, and concluded that the in-situ accretion of Kepler-11f in a warm environment (i.e., 0.1 AU) cannot account for its large radius, which is also consistent with the conclusion of Ikoma and Hori (2012). This suggests that those super-Earths likely migrated from beyond the snowline.

The migration scenario has been examined via theoretical population synthesis. Mordasini et al. (2012a,b) presented their population synthesis models of planetary accretion, and compared the resultant relationships between the masses and radii of short-period super-Earths ($a \leq 0.1$ AU and $1R_\oplus \leq R_p \leq 2R_\oplus$). They found that super-Earths have proto-atmospheres of 0 to 20 wt% of their total masses at the end of migration. The theoretical radii are consistent with (but somewhat larger than) the observed radii of Kepler planets orbiting FGK host stars. Mordasini et al. (2012a,b) predict that the occurrence rate of planets decreases with planetary radius for $R_p < 2R_\oplus$, which, however, seems to be inconsistent with the Kepler results.

Formation of low-density super-Earths orbiting M stars (of 0.1 to 0.7 $M_\oplus$) was investigated recently via population synthesis modeling by Hori et al. (2016). They demonstrated that super-Earths of 1–10 $M_\oplus$ gain relatively massive atmospheres of 0.1–20 wt%, that more massive planets of more than 10 $M_\oplus$ obtain more massive atmospheres, and that sub-Earths (i.e., less than 1 $M_\oplus$) gain less massive atmospheres of less than 1 wt% of the planetary total masses.

Several effects that can modify the atmospheric mass and composition have been investigated, including heavy element enrichment due to planetesimal bombardment during migration (Fortney et al. 2013), which was followed by Dittkrist et al. (2014) who improved the migration model, the hydrodynamic escape of the atmosphere (Jin et al. 2014), reduction in opacity due to grain settling in the atmosphere (Mordasini 2014), and gravitational interaction among multiple planets (Alibert et al. 2013; Pfyffer et al. 2015). However, no detailed comparison with observation was provided.

While the above studies consider the 1D accretion of the atmosphere, Ormel et al. (2015a,b) examined the 2D and 3D effects of the accumulation of H/He atmospheres onto low-mass planets (i.e., the ratio of the Bondi radius to the disk scale height being 0.01), simulating the flow of the surrounding disk gas. They found the timescale for the disk gas to enter the Bondi radius is shorter than the Kelvin-Helmholtz timescale for the atmospheric contraction. Thus, super-Earths are unable to gain massive atmospheres from protoplanetary disks. This may be consistent with the fact that there are many super-Earths with moderate amounts of H/He gases.

2.4.2 Exo-Earths in Habitable Zone

The accumulation and evolution of a primordial, nebula-accreted planetary atmosphere has consequences for the habitability of a planet. Atmospheres are an essential criterion for habitability by moderating and steadying the surface conditions on a planet. A very massive and optically thick atmosphere, on the other hand, can also produce hot and inhospitable surface conditions as on present day Venus. Clearly, planetary habitability depends on various influences (see, e.g., Güdel et al. 2014).

There are good arguments to expect life on other planets also being based on a carbon based chemistry in a watery environment (Council 2007). The conventional limits of the habitable zone (Kasting et al. 1993) are therefore connected to the existence of liquid water on the planetary surface. This is certainly not a sufficient condition, but it allows an unambiguous definition of a circumstellar habitable zone, the radii of which depend on the type of host star and will change due to stellar evolution. The definition of the habitable zone fixes the position of a potentially habitable, embedded planet in the structure of the protoplanetary disk and thus allows a deduction of a range for the temperature (and to a lesser degree, also for the density) of the disk gas around the planet. Therefore, if a planetary core forms early enough to be embedded in the protoplanetary disk for a significant time, one can estimate, based on typical disk temperatures and densities for the habitable zone, that an Earth mass planetary core can accumulate $\sim 5 \times 10^{22}$–1.5×10^{26} g of steam atmosphere (Ikoma and Genda 2006; Lammer et al. 2014), corresponding to 0.3 to 1000 Earth oceans of water. As outlined above, the accumulation of a primordial atmosphere strongly depends on core mass; for planetary cores with Mars-mass, atmospheric masses between $\sim 3 \times 10^{19}$–6×10^{22} g can be captured from the nebula. For planetary cores more massive than 1 $M_{\oplus}$, correspondingly more gas is accreted from the disk, quickly reaching amounts of atmosphere that are unlikely to be lost in the later evolution. According to Lammer et al. (2014), planets with masses exceeding about 2 $M_{\oplus}$ will not lose their disk-accreted primordial atmospheres during the about 100 Myr phase of intensive extreme UV and soft X-ray irradiation in the early evolution of a sun-like star and retain a thick hydrogen envelope, eventually more resembling a "mini-Neptune" than an Earth-like planet.

2.4.3 Primordial Earth and Mars

Obviously, none of the terrestrial planets currently possess a massive hydrogen-dominated atmosphere that would be expected (see Sect. 2.4.2) around a planetary core of Mars to Earth mass embedded in the likely environment of the protoplanetary disk around the early sun (e.g. Hayashi 1981). Some insight into the formation and early history of the terrestrial planets can be gained from the abundance of the tungsten isotope ^{182}W. ^{182}W is produced from the decay of hafnium ^{182}Hf with a half-life of 8.9 Myr. During differentiation, the siderophile W sinks to the planetary core whereas the lithophile Hf remains in the mantle. The ratio of the abundance of ^{182}W in the mantle with respect to other W isotopes therefore allows some conclusions about the timescale of core-mantle differentiation. This and the similar, yet somewhat more complex, U-Pb systematics indicate that Earth's formation took about 30 Myr to 100 Myr to complete (Halliday 2004; Allègre et al. 2008; Kleine et al. 2009). This is consistent with numerical simulations of Earth's formation through the collision of planetesimals also predicting a typical formation time scale of up to 100 Myr (O'Brien et al. 2006; Raymond et al. 2006, 2014; Thommes et al. 2008; Morishima et al. 2010), though the formation time scale could have been shorter in the presence of a planetary nebula (Hayashi et al. 1985).

Confronted with the observed typical disk life time of a few Myr (Haisch et al. 2001; Hillenbrand 2005), these results indicate that Earth most likely formed only after the evaporation of the gas disk and therefore was unable to accumulate a primordial atmosphere from the disk.

For Mars, ^{182}W measurements from Martian meteorites indicate a faster accretion and core formation process lasting between 1 Myr and 10 Myr (Nimmo and Kleine 2007; Dauphas and Pourmand 2011; Brasser 2013). This suggests that Mars could be a leftover planetary embryo that escaped later collisions in the planetary growth process. Accordingly, Mars is a much more likely candidate for having had a primordial disk-accreted atmosphere. However, due to Mars' low gravity atmospheric escape driven by the intensive extreme UV and soft X-ray radiation of the young sun after the evaporation of the gas disk would have been very efficient, causing a complete loss of the primordial atmosphere after a few Myr, at the latest (Erkaev et al. 2014).

3 Magma Ocean Outgassed Atmospheres

3.1 Observational Constraints

It has been known for decades that the chemical compositions of the external gas layers surrounding the inner planets and the outer planets are completely different. Jupiter and Saturn's gas is essentially hydrogen with 10 % helium, and is therefore a sample of the protosolar nebula (e.g. Lewis and Prinn 1983). Envelopes of such gas around giant planets are thus called primary atmospheres. In contrast, the volatile envelopes around Venus and Mars are made of $\sim 95\ \%$ CO_2 and 5 % N_2, excluding water, and this would be true for the Earth as well, if one considers that O_2 was produced by photosynthetic vegetal life along geologic times and that CO_2 is now stored in the carbonated rocks that we observe in sedimentary basins (Turekian and Clark 1975). Water is very variable, because it appears under various phases, and certainly had a more complex history. The atmospheres of the inner planets are thus called secondary atmospheres. That terrestrial air is very different from solar gas is recorded in a geochemical way by the drastically different abundances of the noble gases in both reservoirs, as observed by Brown (Brown et al. 1949) and Suess (Suess 1949). Such big differences again imply that the atmospheres of telluric planets have a different origin compared to primary atmospheres. Two main end-member possibilities then arise: either they come from the interior and this is planetary degassing, or they were brought in by a late bombardment with comets or meteorites. A combination of both is probable.

The idea that atmospheric volatiles (CO_2, N_2, H_2O) could originate from volcanic degassing was first conjectured in the XIXth century (e.g. Crawford 1997). W. Rubey (Rubey 1951) showed that alteration and erosion of surface rocks cannot account for the present atmospheric inventory of the various volatiles that make up the atmosphere. Degassing occurs at present mainly at mid-ocean ridges, as indicated by measuring the helium isotopic ratio, $^3He/^4He$ in sea water (e.g. Clarke et al. 1969; Jenkins et al. 1978; Craig and Lupton 1981) and in rocks collected on the seafloor (e.g. Kurz et al. 1982): the isotope ^3He indeed reveals gas that was never at the surface of the Earth. Degassing is recorded in the isotopic compositions of argon and xenon found in gas released by heating or crushing of mid-ocean ridge basalt glasses. Such volcanic rocks come directly from the convective upper mantle and enable the determination of the argon and xenon isotopic compositions of this reservoir because such heavy isotopes cannot easily fractionate. The data show enormous isotopic anomalies compared to atmospheric values (e.g. Ozima and Kudo 1972; Fisher 1975; Staudacher and Allègre 1982; Sarda et al. 1985). These isotopic anomalies are a geochemical trace of degassing. Comparison of these isotopic compositions with those in the atmosphere indicate that degassing occurred extremely early in terrestrial history and that degassing was extremely strong, i.e. a volume comparable to the Earth's upper mantle, at least, was almost completely degassed (e.g. Hamano and Ozima 1978; Ozima 1975; Staudacher and Allègre 1982; Sarda et al. 1985; Allègre et al. 1987). Indeed, 90 % of primitive Argon was degassed within the first 100 Myr (Allègre et al. 1987). The measurement of noble gas isotopes in meteorites gives a consistent picture of atmosphere degassing on the Earth (see review by Moreira 2013). Even if neon possibly tells a more complex story, with hydrodynamic escape of part of the degassed atmosphere (Sarda et al. 1988), the geochemical picture given by neon, argon and xenon is robust and points to degassing to have been both very extensive and very early, sometimes called "early catastrophic degassing" (e.g. Fanale 1971). The exact way such early degassing occured was not clear. It was once envisioned that pressure increase upon impacts would devolatilize hydrous minerals, hence releasing water (Lange and Ahrens 1982; Abe and Matsui 1985). It was also realized that the energy of impacts is such that it should have led to melting of the silicate part of planetary embryos (e.g. Safronov 1978; Tonks and Melosh 1993). Consideration of noble gas diffusion also led to the conclusion that degassing should have demanded molten silicate, either through volcanism or during a magma ocean stage (Allègre et al. 1987).

Today, degassing at mid-ocean ridges occurs via melting of mantle peridotite, magma ascent, vesiculation during ascent, and vesicle bursting at eruption. But diffusion of volatiles through crystals and convective upwelling velocities in a solid-state mantle are far too slow to match the measured early degassing. The latter requires upwellings in a molten body. And so does iron cores segregation in planetesimals and planets (e.g. Fujii and Uyeda 1966; Yoshino et al. 2003; Wood et al. 2006; Li and Agee 1996; Righter and Drake 2000).

This is consistent with the early melting of planetesimals owing to radiogenic heating from short-lived radioisotopes such as ^{26}Al and ^{60}Fe (e.g. Urey 1955; Reynolds et al. 1966; MacDonald 1959; La Tourette and Wasserburg 1998; Monteux et al. 2007), and with the energetics of accretionary impacts, where conversion of kinetic energy to heat could induce the repeated formation of surface magma oceans on planets (e.g. Safronov 1978; Canup 2004b; Tonks and Melosh 1993; O'Brien et al. 2014; Monteux et al. 2007; Golabek et al. 2011). Moreover, a magma ocean would also be produced by gravitational energy released deep in the planet by core formation (Sasaki and Nakazawa 1986).

We define here "magma ocean" (hereafter "MO") to occur when at least 10 % of the silicate mantle is molten and when the magma behaves rheologically like a liquid, i.e. for a volume fraction of liquid above 40 % (e.g. Taylor and Norman 1992; Solomatov 2000). The evolution of primordial MOs, which involves cooling, solidification and degassing, is therefore of fundamental importance because it sets initial conditions for (1) the chemical stratification of a planet, (2) the composition of its degassed atmosphere, as well as (3) its subsequent tectonic regime. As such, it has been studied since the mid 60's, first to produce accurate thermal evolution models of the Earth capable of generating present-day heat flow measurements (Reynolds et al. 1966; MacDonald 1959), and then to determine the conditions for habitability—namely the existence of liquid water on a planet surface—(Abe 1997; Solomatov 2000; Elkins-Tanton 2008; Lebrun et al. 2013; Hamano et al. 2013). As will be shown below, all these studies concur that MO and atmosphere evolutions cannot be considered separately (Fig. 3).

Indeed, once an MO is formed, the atmospheric degassing rate depends on the degree of enrichment of volatile species in the magma. It is likely that gaseous species would be

Fig. 3 Relationships between the magma ocean, the atmosphere and space

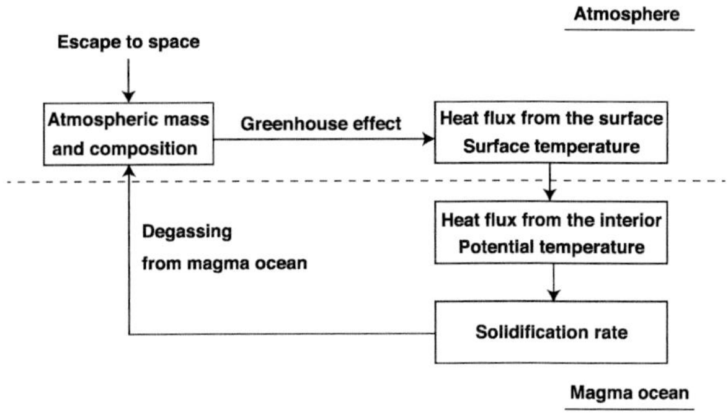

partitioned into a magma ocean according to their solubility into silicate melt at an early stage. As the solidification of the magma ocean proceeds, the residual magma is increasingly enriched in volatiles, which are usually incompatible with silicate cumulates. Volatiles in excess of the saturation limit are therefore exsolved and expelled out of the MO to form an outgassed atmosphere. In turn the rapid growth of the outgassed atmosphere slows down the MO cooling and solidification, due to enhancement of the thermal blanketing and greenhouse effects which prevent heat escape from the MOs. The final state of the planet will then be modulated by the atmospheric escape and by the sequence of accretion.

3.2 Heat and Mass Transfers in a Magma Ocean

Earlier and more complete reviews can be found in Solomatov (2007) and Elkins-Tanton (2012). Figure 4a presents the typical structure of an MO. Giant impacts should produce very deep and hot MO with an initial temperature distribution which is adiabatic and mostly above the liquidus (e.g. Lejeune and Richet 1995). Crystals grow as soon as the magma temperature is lower than the liquidus T_{liq}, and the liquid fraction at temperature T is usually taken as:

$$\phi = \frac{T - T_{sol}}{T_{liq} - T_{sol}}, \tag{12}$$

where T_{sol} is the solidus temperature. However, for liquid fractions lower than $\phi_c \sim 40$ %, the crystals are too closely-packed to move freely anymore and a solid skeleton starts to develop. A corresponding crystal volume fraction of 60 % therefore corresponds to a "rheological" front (e.g. Solomatov 2000). The end of the MO stage is usually defined as the time when the initially molten mantle has everywhere reached a liquid fraction lower than ϕ_c (e.g. Solomatov 2009, "soft magma ocean" of Abe in Abe 1993).

Given the pressure/depth-dependence of the adiabats, T_{sol} and T_{liq} (Fig. 4b), a solidification front (corresponding to temperature lower than T_{sol}) is expected to start from the MO bottom and to move up if the crystals density is greater than the residual liquid density. This is the most common case (e.g. Elkins-Tanton et al. 2003). However, as different mineral phases crystallize depending on temperature, pressure and magma composition, this situation depends on planetary radius and composition (see Elkins-Tanton 2012 for a review). For example, recent measurements on iron-partioning at pressures of the Earth's deep mantle seem to confirm that a liquid becomes denser than the coexisting solid at $\sim$ 1800 km-depth in the lower mantle (Nomura et al. 2011). If the Earth's mantle had initially been completely molten, that would imply that solidification started at a septum around 1800 km depth, soon separating an upper MO solidifying upwards from a bottom MO solidifying downwards

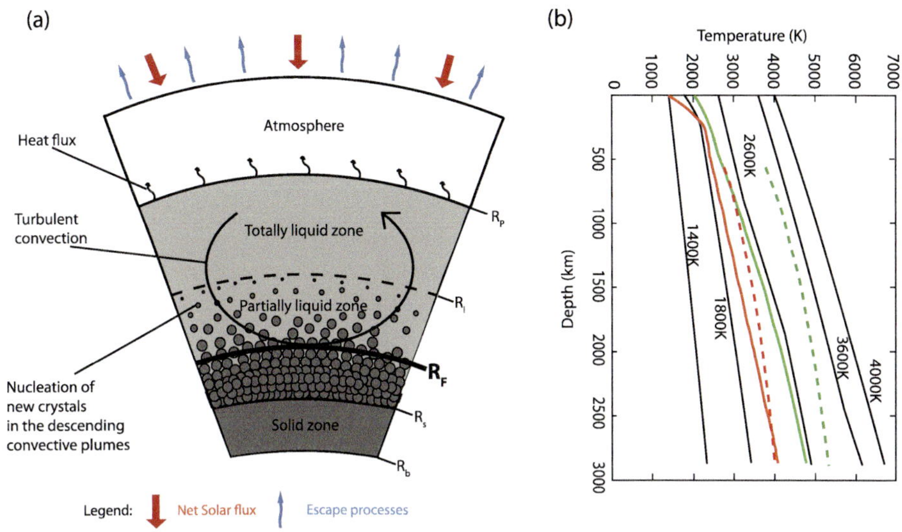

Fig. 4 (**a**) Schematics of magma ocean solidification adapted from Solomatov (2000). Radius R_b designates the base of the upper MO (see text), and R_F the position of the rheological front. (**b**) Selected adiabats (*solid black lines*), and the solidus (*red line*) and liquidus (*green line*) measured by Andrault et al. (2011) for a fertile peridotite and by Fiquet for a synthetic chondritic composition, Fiquet et al. (2010) (*dashed lines*) as a function of depth

(Labrosse et al. 2007). Early MO degassing therefore would only involve the upper MO. On the other hand, the anorthosites discovered on the lunar surface suggest that they formed by flotation of plagioclase crystals on top of the primordial lunar MO (Smith et al. 1970; Wood et al. 1970; Elkins-Tanton 2012). In that case, the late stages of the lunar MO cooling and degassing occurred in the presence of a solid lid at the surface.

3.2.1 MO Cooling

MO cooling is determined by the conservation of energy which can be written as (e.g. Abe 1997; Solomatov 2000; Lebrun et al. 2013):

$$\mathrm{SC} = \int_{R_b}^{R_p} \rho C_p \frac{dT_p}{dt} r^2 dr = R_P^2 Q_S + R_B^2 Q_B + H_{int} + H_{sol}. \tag{13}$$

SC represents the MO secular cooling, ρ is the melt + solid averaged density, C_p the specific heat, T_p designates the potential temperature, t is time and r radius. Q_S and Q_B are the heat flows getting out of the top and bottom surfaces, respectively. H_{int} is the internal heating due to radioactive elements, and H_{sol} is due to the latent heat release of solidification. H_{int} is usually negligible compared to Q_S, except for very small bodies (first planetesimals) cooling within ~ 3 Myr, i.e. when ^{26}Al and ^{60}Fe are still strongly active (Elkins-Tanton 2012, and references therein). Indeed the contribution of U, K and Th amounts at most to 140 mW/m^2 (e.g. Labrosse and Jaupart 2007) compared to the typical 10^2–10^6 W/m^2 going out of the MO (Fig. 5d; Abe and Matsui 1986; Elkins-Tanton 2008; Lebrun et al. 2013; Hamano et al. 2013; Zahnle et al. 1988). Similarly, the approximation $Q_B = 0$ is usually taken since heat transport in the solid bottom of the MO is much slower than convective transfer in the magma. The MO cools from above by losing its heat to outer space, either

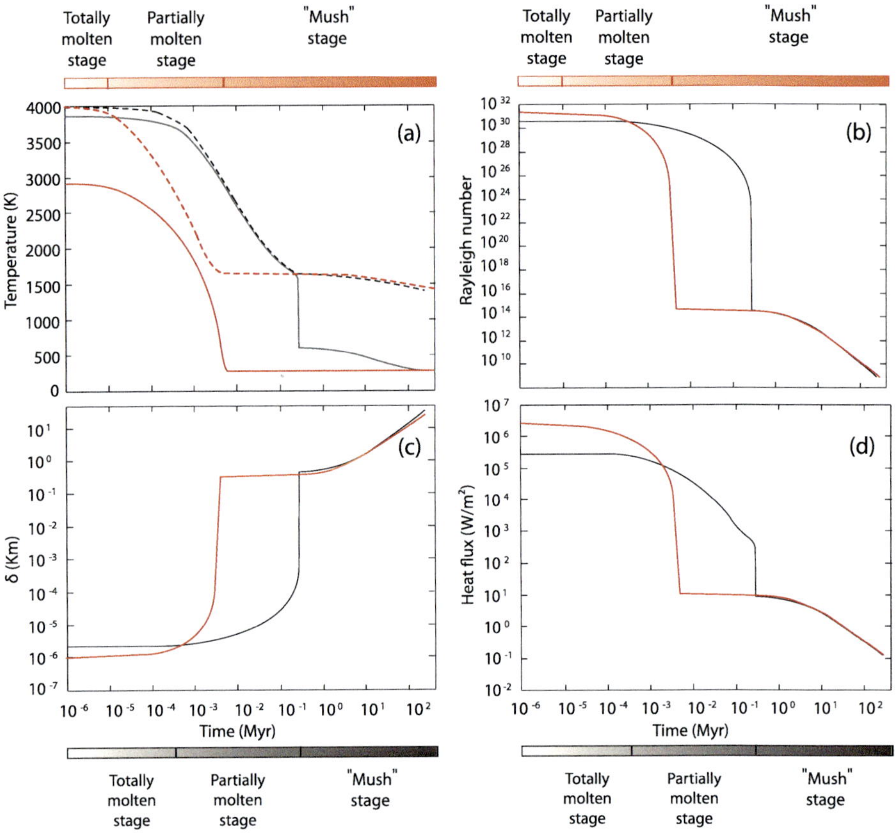

Fig. 5 Typical time evolution of (**a**) temperature (surface temperatures T_S are represented by the *solid lines*, and the potential temperatures T_p by the *dashed lines*), (**b**) Rayleigh number, (**c**) upper cold thermal boundary layer thickness given by $\delta = k(T_p - T_S)/Q_S$ and (**d**) surface heat flux, when the magma ocean is cooling directly through space (*red lines*) or through a grey atmosphere (*black lines*) (adapted from Lebrun et al. 2013)

directly or through an atmosphere. In the MO, heat transfer proceeds through thermal convection, with cold plumes sinking from the planet surface and a diffuse hot return flow. The convection intensity is characterized by the Rayleigh number:

$$Ra = \frac{\alpha g (T_p - T_S) D^3}{\kappa \nu} \tag{14}$$

where T_S is the surface temperature, α thermal expansion coefficient, g gravity acceleration, D the MO thickness (i.e. $R_p - R_F$), κ the heat conductivity, and ν the magma kinematic viscosity. Typical values of MO Ra range between 10^{20} and 10^{30} (Fig. 5b), similar to the values encountered in our present-day atmosphere (e.g. Priestley 1959). The surface heat flux out of the convective system then scales as:

$$Q_S = C_0 \frac{k(T_p - T_S)}{D} Ra^n \tag{15}$$

where k is the heat conductivity. Coefficients C_0 and n are constants which depend on the regime and pattern of convection (Malkus 1954; Siggia 1994; Grossmann and Lohse 2000; Solomatov 2007). These are still debated at very high Ra. An exponent $n = 1/3$ is usually taken which suggests that heat losses are governed solely by local instabilities of the upper thermal boundary layer and does not depend explicitly on the MO thickness (for a more complete discussion, see Solomatov 2007; Lebrun et al. 2013; Davaille and Limare 2015). Equations (14) and (15) show that the convective heat flux in the MO decreases when viscosity increases. Viscosity itself strongly increases when temperature or liquid fraction decrease. The low viscosity of basaltic magmas (as low as 0.01 Pa s, Shaw 1972) induces Q_S values as high as 10^6 W/m^2 (Fig. 5d) in the liquid MO stage. The value of Q_S however strongly decreases through time as crystallization proceeds (Fig. 5d). Besides, it could be strongly limited by the development of a high viscosity region ("lid") at the surface of the MO (like in the lunar case).

On the other hand, conservation of energy of the planet also requires that the MO convective heat flux Q_S be equal to the heat flux directly radiated through space (and given by a black body law) in absence of atmosphere, or to the heat flux transported through the atmosphere to space. The second limitation of the convective heat flux out of the MO is therefore the existence and dynamics of the atmosphere. Figure 5 shows that MO cooling is much faster in the first case than in the second case. The latter will be discussed in more details in Sect. 3.4.

3.2.2 Crystallization and MO Segregation

Crystallization influences MO cooling in two ways: through viscosity and therefore Q_S, and through the MO structure, and therefore SC, H_{sol}, and Q_S in Eq. (13). Of critical importance will especially be the development of a buoyant solid lid on top of the MO, which would drastically slow down MO cooling. There are two end-members: in batch solidification, liquid and solid remain in contact and in equilibrium throughout the whole MO, so that the final solid-state mantle composition will be homogeneous. In fractional solidification, segregation between solid and liquid occurs as the growing mineral grains settle; and therefore the residual liquid composition evolves with the removal of each package of solidified material. This leads to compositional stratification of the resulting solid, as is encountered in magma chambers. Hence, whether the MO can segregate, or on the contrary retain a well-mixed state while cooling, depends on the competition between crystal settling (due to gravity and the density difference $\Delta\rho$ between melt and crystals under mantle P and T conditions) and crystal entrainment due to the viscous stresses generated by the turbulent convective flow (Martin and Nokes 1988; Tonks and Melosh 1990; Solomatov and Stevenson 1993a,b). This simple balance gives an estimate of the critical crystal size above which settling occurs (Solomatov and Stevenson 1993a):

$$a_c = \frac{1}{\Delta\varrho g}\left(\frac{0.1\eta\alpha g Q_S}{C_p}\right)^{0.5}. \tag{16}$$

Typical density differences $\Delta\rho \sim$ 100–300 kg/m^3 and MO surface heat flux $Q_S \sim 10^2$–10^6 W/m^2 give critical sizes between a few μm and 1 mm. According to the 10^{-3}–1 cm-crystal size estimates (e.g. Solomatov and Stevenson 1993b), both settling and re-entrainment are therefore expected in an MO. This question of crystal settling in a turbulent magma flow is still not fully understood and is the subject of active research (e.g. Höink et al. 2005; Suckale et al. 2012).

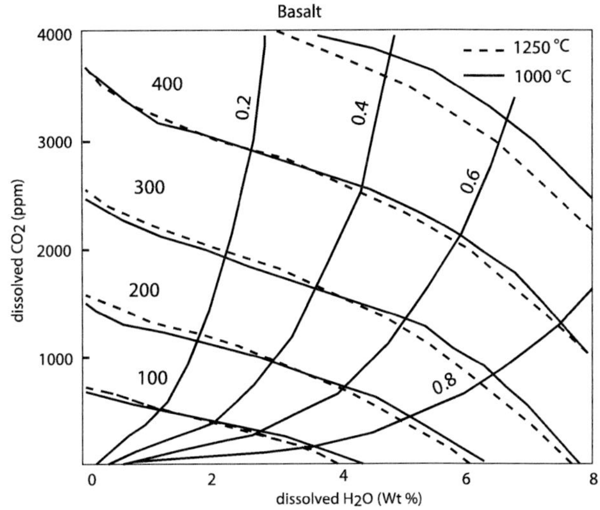

Fig. 6 Solubility curves redrawn from Berlo et al. (2011) and Papale (1999)

Meanwhile, assuming fractional crystallization leads to some interesting predictions for planetary mantle structures (for a review, see Elkins-Tanton 2012). In order to float, solid phases such as plagioclase on the lunar MO require the development of sufficiently iron-rich and dry, denser liquids, which can only form at the late stages of MO cooling, and for small planets (Elkins-Tanton 2012). On the other hand, larger and wetter planets like Mars or Earth are not expected to develop buoyant solid lids in the MO stage. Moreover, as fractional crystallization would render the solidifying liquids progressively denser, the inherited solid-state mantle stratification would be gravitationally unstable (with the top denser than the bottom). This would result in large-scale mantle overturn in the first Myr after the MO (Hess and Parmentier 1995; Elkins-Tanton et al. 2003, 2005; Elkins-Tanton 2008), before thermal mantle convection started. This scenario would be an efficient way to generate a denser reservoir at the bottom of the mantle, which could remain partially "hidden" from the surface for a long time, in agreement with geochemical mass balances and the presence of large seismic velocity heterogeneities at the base of the mantle (e.g. Brown and Lesher 2014). However, such an overturn could also establish too stable a layering to be overcome by thermal convection (e.g. Le Bars and Davaille 2004), as recent numerical simulations seem to suggest (Plesa et al. 2014; Tosi et al. 2013). Since plate tectonics on Earth, and volcanism on Mars both require solid-state mantle convection, this is an indication that fractional crystallization was not the only mode of MO solidification, but that batch solidification played also a significant role and that the onset of solid state convection, well before complete solidification, may have reduced strong mantle layering (Maurice et al. 2015; Solomatov 2015).

3.2.3 Volatiles and Degassing

A magma can contain dissolved volatiles only up to its saturation limit. The latter depends on the species (H_2O, CO_2, ...) considered, on temperature and pressure (Fig. 6), on the composition of the magma, and on the phase (liquid or solid). For example, in an Earth-scale MO with an initial whole-mantle bulk composition, saturation will be reached after 80 % and 40 % by volume of pure fractionation for an initial 500 ppm concentration in H_2O and CO_2, respectively (e.g. Elkins-Tanton 2012).

As the magma crystallizes, the volatiles go preferentially into the residual liquid, which becomes enriched in volatiles. Degassing can then occur by two processes: (i) exsolution of all volatiles in excess of the saturation limit in bubbles which burst at the MO surface, and (ii) diffusion of the volatiles through the liquid magma thermal boundary layer. This second process will dominate if the bubbles are small enough to be re-entrained in the MO by the vigorous convective motions, and the resulting diffusion-limited volatile flux constitutes a lower bound for degassing. Hamano et al. (2013) showed that even this lower bound is sufficient to generate a water atmosphere out of a solidifying MO.

More realistically, a number of authors (e.g. Zahnle et al. 1988; Elkins-Tanton 2008; Lebrun et al. 2013) consider that bubbles are indeed growing fast and will be able to escape from the MO to burst at the surface, so that the first mechanism is dominant. Indeed, by analogy to the fate of bubbles in terrestrial volcanism, small bubbles are likely to coalesce into larger gas pockets when upward melt flow rates decrease on approaching the surface. In low viscosity magmas (10 Pa s), ascent velocities of gas pockets given by the Stokes velocity range between 1 and 100 ms^{-1} for bubble diameters between 1 cm and 1 m. Compared to the typical MO convection velocity of 0.5 ms^{-1}, it is clear that such gas pockets will easily separate from the melt. Besides, with such convective velocities, the whole MO can circulate near to the surface in a few weeks: when a batch of melt travels upward, pressure decreases as well as volatile solubility, and a gas phase appears. All volatiles in excess of the saturation limit will be exsolved as bubbles, which, upon reaching the MO surface, will burst to the growing atmosphere (Zahnle et al. 1988; Elkins-Tanton 2008; Lebrun et al. 2013; Hamano et al. 2013). It is thus often assumed that the MO and the atmosphere are in equilibrium, and their volatiles partial pressures are given by (e.g. Carroll and Holloway 1994; Pan et al. 1991; Jendrzejewski et al. 1997):

$$pH_2O = \left(\frac{x_{H_2O}}{6.810^{-8}}\right)^{1/0.7}, \tag{17}$$

$$pCO_2 = \frac{x_{CO_2}}{4.410^{-12}}, \tag{18}$$

with x_{H_2O} and x_{CO_2} being respectively the mass fraction of water and CO_2 dissolved in the melt respectively expressed in wt% and ppm.

More recent solubility laws might also be used, such as in Liu et al. (2005), Witham et al. (2012), Burgisser et al. (2015) or Papale (1999) where a mixed CO_2-H_2O volatile phase is considered (see Fig. 6 in Berlo et al. 2011). Further, recent studies show that total pressure may be very important in changing completely the solubility law in the case of CO_2 at pressure higher than 1 GPa (e.g. Sarda and Guillot 2005; Guillot and Sarda 2006).

This avenue should be further pursued in the future generation of models.

3.3 Structure of Outgassed Atmospheres

3.3.1 History and Heritage of MO Atmospheric Models

The outgassed volatiles are able to form a secondary atmosphere, which will in turn impact the radiative cooling rate of the whole planet, and will be subject to the escape processes. All coupled magma-atmosphere models, even the most recent ones (Lebrun et al. 2013; Hamano et al. 2013) have used a global mean column (1D) atmospheres since the surface conditions for magma ocean planets are not thought to depend strongly on the location on the planet (e.g. in latitude or local solar time). Moreover, the numerical solution of 1D models is much faster and therefore more convenient for further coupling.

Due to their common typical composition (usually including at least H_2O and most often CO_2 and/or N_2) and total inventory (from several tens to several hundred bars), these atmospheric models are inspired by similar models used to study the runaway greenhouse phenomenon (Ingersoll 1969) that some telluric planets experience at a later stage of their evolution. Thus, the atmospheric module of Lebrun et al. (2013), described in more detail by Marcq (2012), heavily draws on the runaway greenhouse model from Kasting (1988). Similarly, Hamano et al. (2013) have used a steam atmosphere model based on the runaway greenhouse studies from Nakajima et al. (1992).

There is nevertheless a major difference between these two kinds of models. Runaway atmospheres are in global radiative balance and are chiefly interested in the *greenhouse* effect: the absorbed stellar flux integrated over the whole planetary surface must be equal to the integrated outgoing radiative flux radiated by the planet back to space. This absorbed stellar flux can be described by the *equilibrium temperature* T_{eq} defined such as $\sigma T_{\mathrm{eq}} = \bar{F}_{\mathrm{abs}}$ where $\bar{F}_{\mathrm{abs}}$ is the spatially averaged absorbed stellar flux at shorter wavelengths. In a similar manner, the outgoing longwave radiation (OLR) can be described by the *effective temperature* T_{eff} defined such as $\sigma T_{\mathrm{eff}}^4 = \overline{OLR}$. The global radiative balance condition is then equivalent to $T_{\mathrm{eff}} = T_{\mathrm{eq}}$, and the temporal evolution of such runaway atmospheres is driven by the steady increase of T_{eq} caused by the aging of the main sequence star.

In contrast, atmospheres surrounding MO planets are out of global radiative balance, since they have to radiate not only the absorbed stellar flux, but also the internal heat flux originating from the magma ocean. Their main focus is therefore on the *blanketing effect* (Matsui and Abe 1986; Abe and Matsui 1988) of such atmospheres with respect to the internal heat loss.

The blanketing effect can be defined as the screening effect of a radiatively thick atmosphere on the magma ocean heat loss, and quantified using the difference between the magma ocean surface temperature T_{surf} and the effective temperature T_{eff}: an hypothetical atmosphereless magma ocean planet would have $T_{\mathrm{surf}} = T_{\mathrm{eff}}$, whereas the blanketing effect implies that $T_{\mathrm{surf}} > T_{\mathrm{eff}}$. Nevertheless, the internal energy source ensures that $T_{\mathrm{eff}} > T_{\mathrm{eq}}$ in such atmospheres, since we have $F = \sigma(T_{\mathrm{eff}}^4 - T_{\mathrm{eq}}^4)$ where F stands for the internal heat flux (e.g. from the magma ocean). The secular cooling of the MO and the fact that T_{eq} remains quite constant during the early history of these telluric planets implies a secular decrease of T_{eff} towards a limit equal to T_{eq} when global radiative balance is finally reached. The temporal evolution of MO atmospheres therefore appears "reversed" compared to a runaway greenhouse. Depending on various parameters (chiefly the distance to parent star), the final stage can result in a permanent runaway greenhouse state (type II planets from Hamano et al. 2013) whose further evolution is ruled by atmospheric escape, or in a geologically fast exit of the runaway greenhouse state through an atmospheric water collapse. This results in the formation of a global water ocean (type I planets from Hamano et al. 2013), which mostly shelters H_2O from the atmospheric escape encountered by type II planets.

3.3.2 Vertical Structure

Due to the aforementioned heritage, the atmospheric models from Lebrun et al. (2013) and Hamano et al. (2013) both adopt a hydrostatic, three vertical regions, namely from bottom to top: (1) Dry (unsaturated) troposphere; (2) Moist (saturated) troposphere and (3) mesosphere (purely radiative). The first two layers may or may not exist, depending on criteria detailed below. Thus, the following vertical structures are possible: $(1+2+3)$, $(1+3)$ and $(2+3)$.

The dry troposphere is sometimes absent, provided the partial pressure of H_2O exceeds the saturation pressure of water (properly defined only below the critical point, e.g. for $P_S < P_c \approx 221$ bar and $T_S < T_c \approx 648$ K). This never occurs in the model of Hamano et al. (2013) since their simulations stop at a temperature of about 1370 K, but it sometimes does in the model of Lebrun et al. (2013). In such a case, $P_{\mathrm{surf}}(H_2O)$ is set to the saturation pressure, and the extra water condenses into a water ocean that may coexist with an underlying magma ocean. Both are separated only by a thin, transient crust, that might break and recycles back in the MO at the beginning of its formation (Elkins-Tanton et al. 2005). This crust is supposed not to hamper heat transfers. Water clouds are included by Lebrun et al. (2013) but not by Hamano et al. (2013). They are supposed to extend throughout the moist tropospheric layer (provided it exists), with optical and microphysical properties similar to present day Earth's clouds. The radiative layer is isothermal in Lebrun et al. (2013) with a local radiative equilibrium condition in the longwave spectrum at the top of the atmosphere (null vertical divergence of spectrally integrated OLR at vanishing thermal opacities). On the other hand, Hamano et al. (2013), following Nakajima et al. (1992), adopt an analytic two-stream radiative temperature profile in their mesosphere $T(\tau)^4 = \frac{T_{\mathrm{eff}}^4}{2}(1 + \frac{3}{2}\tau)$. This model is at radiative equilibrium in the mesosphere, but not spectrally resolved in the long wave range (grey model).

If the pressure and temperature profiles reach the top of atmosphere temperature without crossing the water saturation, then the moist tropospheric layer is absent, and clouds cannot exist. The atmospheric model from Marcq (2012) used by Lebrun et al. (2013) treated CO_2 as an ideal gas, but not H_2O whose thermodynamical properties were instead derived from thermodynamical steam tables like in Kasting (1988)'s model. The humidity profile is also constant in the dry troposphere as well as in the mesosphere, but decreases with increasing height in the moist troposphere. Atmospheres without water saturation therefore exhibit a vertically uniform humidity profile, without any vertical sequestration of water.

3.3.3 Radiative Transfer

The computation of OLR is performed only for the thermal radiation in the studies of Hamano et al. (2013) and Lebrun et al. (2013). The separation between stellar and planetary thermal radiation is justified in most situations, since thick outgassed atmospheres are optically thick enough so that radiating layers are usually below 700 K in most situations provided enough volatiles are available in the interior (Marcq 2012). Lebrun et al. (2013) use primarily a non-grey model, using a k-correlated code in 36 spectral bands in the long-wave from 0 to 10^4 cm^{-1}, although they could revert to a grey model for easier comparison with previous works and accelerated computations. Conversely, Hamano et al. (2013) used primarily a grey radiative transfer model as Nakajima et al. (1992) did, although they discussed the effects of including water lines opacity in the visible and near IR based on the k-correlated model from Kopparapu et al. (2013). They could also use a non-grey, older, random band model from Abe and Matsui (1988) for further investigations; these did not undermine the main conclusions about the critical stellar distance separating type I from type II planets.

The non grey calculations from Kopparapu et al. (2013) also show that weak H_2O lines in the visible and near IR included in the HITEMP spectral database could significantly lower the planetary albedo below the usual value of 0.4–0.5 to about 0.3. Lebrun et al. (2013) prescribed different values for planetary albedos depending on the existence of a cloud cover (0.3 for a clear atmosphere and up to 0.7 for a cloudy atmosphere). In any case, precise determinations of globally averaged planetary albedos are out of reach of 1D models,

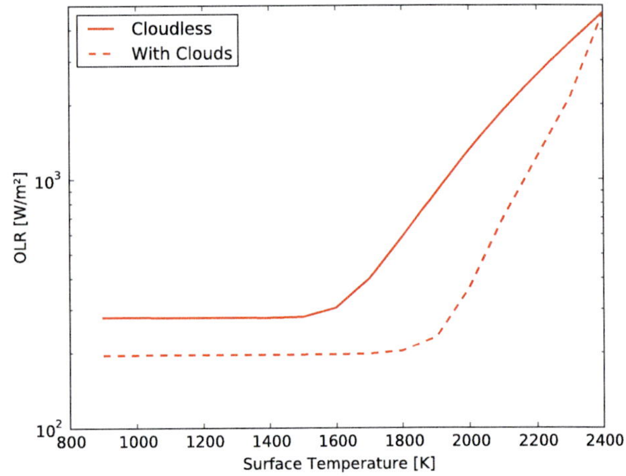

Fig. 7 OLR vs. T_{surf} for an Earth-like planet with an atmospheric inventory of 300 bar of H_2O and 100 bar of CO_2. Here, NL is reached for surface temperatures lower than 1500 to 1800 K (updated from Marcq 2012). Clouds are supposed to be present throughout the moist troposphere, and their optical properties follow Kasting (1988)

and would require the use of 3D General Circulation Model for further progress (see also Sect. 3.3.5): the cloud cover could not encompass the whole planet, and the planetary mean cannot simply be obtained with a mean 1D cloud opacity. In other words, the effect of cloud cover is highly non linear, and exact determinations of planetary albedos would require the horizontally-resolved determination of the stellar flux backscattered to space.

3.3.4 Atmospheric Regimes

The qualitative behavior of OLR vs. T_{surf} in both studies is very similar: for relatively high surface temperatures, the whole atmosphere is hot enough so that it radiates very effectively. In this high temperature regime, OLR increases rapidly with increasing surface temperature. Conversely, at relatively low surface temperatures, OLR is nearly independent of the surface temperature until values low enough for the steam to condense into a water ocean (see Fig. 7). This asymptotic OLR limit is already known from runaway greenhouse studies and is called *Nakajima's limit* (NL) following its discussion in Nakajima et al. (1992). NL for thermal opacities dominated by H_2O vapor (and neglecting clouds) has been estimated between 280 and 300 W/m^2 (equivalent to 265 K $< T_{eff} <$ 270 K) in a number of studies (Nakajima et al. 1992; Goldblatt et al. 2013; Leconte et al. 2013; Kopparapu et al. 2013). The atmospheric model from Marcq (2012) wrongly estimated a much lower NL value near 220 W/m^2, but has been corrected since. Depending on the cloud thickness and location within the moist troposphere, its radiative effect can be negligible, or could lower the NL value by about 30 %. Here also, 3D GCMs could lead to much better estimations of this Nakajima's limit, which is of uttermost importance for runaway greenhouse studies (Leconte et al. 2013). The threshold surface temperature value between these two regimes is primarily related to the atmospheric composition and inventory; it typically lies between 1500 and 2500 K for inventories comparable of H_2O, $\sim$ 100 bar of CO_2 and $\sim$ 1 bar of N_2.

High Temperature Regime This high-temperature regime can be labelled as *true runaway atmospheres*. It is defined by a high OLR, and characterized as follows (see Fig. 8): no water condensation within the atmosphere nor at the surface. The atmospheric humidity is vertically uniform, there is no moist troposphere, and temperatures are relatively warm

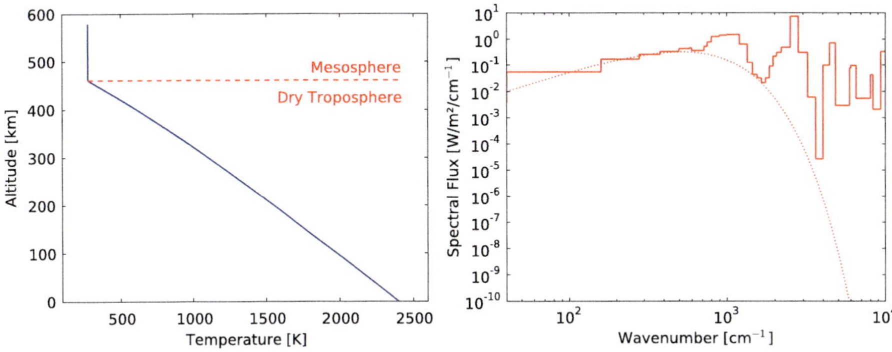

Fig. 8 *Left*: typical vertical $T(z)$ profile in a true runaway atmosphere (Planck curves corresponding to $T_{surf} = 2400$ K) from updated Marcq (2012) model. *Right*: typical spectrum of the thermal emission in a runaway atmosphere according to updated Marcq (2012) model, for $T_{surf} = 2400$ K

even in the radiative layers. These atmospheres are very thick due to their large hydrostatic scale heights and usually high surface pressures. This makes them particularly vulnerable to escape processes (see Sect. 4).

The spectrum of their thermal radiation is characterized by a several peaks in the near IR between 1 and 10 μm, where brightness temperatures can sometimes exceed 600 K. This would yield a good contrast with their host star in this spectral range, leading to good detectability prospects by future dedicated exoplanet observatories such as the ARIEL (Atmospheric Remote-Sensing Infrared Exoplanet Large-survey) ESA M4 mission candidate, based on the former EChO ESA M3 candidate (Tinetti et al. 2012).

The duration of this stage is highly variable depending on the planetary classification of Hamano et al. (2013). Type-II planets are characterized by a $T_{eq} > 270$ K, so that they can reach global radiative balance during this stage. Cooling of the magma ocean becomes then extremely slow, and further atmospheric evolution is driven not by radiative cooling, but by escape processes that will slowly deplete the atmospheric inventory. This will progressively result in a cooler T_{surf} at a given OLR (now fixed by global radiative balance with the parent star) until the magma ocean can solidify. No water ocean is ever formed on this kind of planet, whose typical evolution is described in Hamano et al. (2013, their Fig. 2). For type-I planets, this stage is only transitory since they reach global radiative balance only for much lower OLR ($T_{eq} < 270$ K), which lasts less than 1 Myr (Hamano et al. 2013, their Fig. 1).

Low Temperature Regime This (relatively) low-temperature regime can be labelled as *marginally runaway atmospheres*. They are defined by an OLR approximately equal to the NL value (±5 W/m^2 at most), and can be characterized as follows (see Fig. 9). Water condensation occurs in a moist troposphere located above a dry (unsaturated) troposhere. This would result in massive water clouds, whose detailed study would require a microphysical modeling effort beyond the scope of both Lebrun et al. (2013) and Hamano et al. (2013). The uppermost layers (top of the clouds, mesosphere) are relatively dry and cool, and comparable to the mesospheres of planets like Venus. The moist layer therefore acts as an efficient cold trap, making the water inventory of these planets much more stable with respect to escape processes. The fact that the OLR remains constant with surface temperature can be qualitatively explained following the discussion in Goldblatt et al. (2013): a hotter surface results in a thicker unsaturated troposphere, so that the moist layers lie further up in the atmosphere.

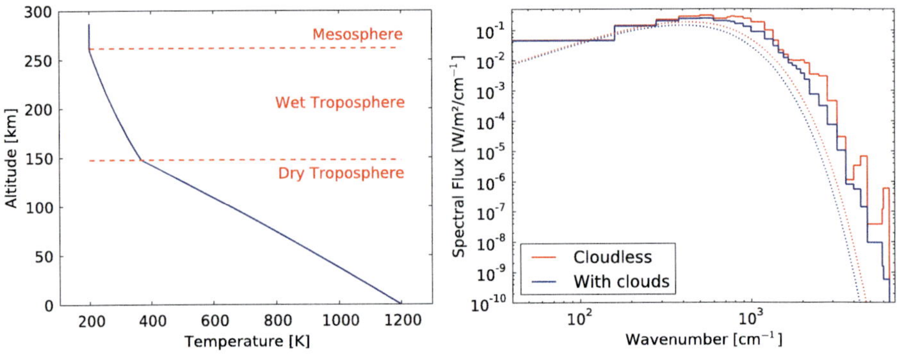

Fig. 9 *Left*: typical vertical $T(z)$ profile in a marginally runaway atmosphere ($T_{surf} = 1200$ K) from updated Marcq (2012) model. *Right*: typical spectrum of the thermal emission in a marginally runaway atmosphere according to updated Marcq (2012) model. Clouds are supposed to be present throughout the moist troposphere, and their optical properties follow Kasting (1988)

But the temperature of these moist layers is unchanged, and since outgoing thermal radiation originated mostly from these layers, the spectrum and total radiated power exhibits very little dependence with respect to surface temperature in this regime. The lower end of possible surface temperatures in this regime results in thermal opacity windows reaching the ground level. This usually leads to the formation of a water ocean between the MO and the moist layer, provided enough water is present.

The spectrum of their thermal radiation is characterized by globally low brightness temperatures (around 200 K) except in a few narrow near-IR windows (provided the cloud layer is not too optically thick). This situation is very reminiscent of present-day near IR windows in Venus' thick CO_2 atmosphere (Kamp et al. 1988), which may have experienced a runaway greenhouse state, and is now in a dessicated state following the escape of H_2O evidenced by the very high D/H fractionation of remaining water vapor (Grinspoon 1993). The detectability prospect of exoplanets in this atmospheric regime is therefore much lower than for true runaway atmospheres, and virtually undistinguishable from more evolved planets like Venus.

As mentioned above, this stage (formation of a water ocean) is never reached by type-II planets. Type-I planets enter this stage after a short full runaway stage, and reach almost complete solidification after about 4 Myr according to Hamano et al. (2013): when most of the water has condensed into the oceans, the atmosphere becomes more and more transparent to the infrared radiation originating from the lowermost atmospheric layers or even the surface itself; this results again in a decreasing OLR with decreasing surface temperature. The atmosphere then leaves the marginally runaway state and enters a final stage similar to present day Earth provided enough water is present. Whether this stage is then reached depends on the incoming energy from the parent star ($T_{eq} = T_{eff}$). As a final note, we emphasize the fact that due to the independence of OLR versus surface temperature, a marginally runaway atmosphere is highly unstable: small increases in internal heat flux, impact energy or stellar radiation can force the atmosphere to reach a true runaway state with surface temperatures higher by several hundred Kelvins. Conversely, very small decreases in heat flux or stellar radiation can force the atmosphere to a partial collapse and an Earth-like state. This instability of surface temperature with respect to small variations in the internal or external heat fluxes may lead to chaotic alternations of condensed/vaporized water before stable oceans can form.

3.3.5 Limits and Desirable Improvements

Although the above atmospheric models include most of the relevant physical phenomena occurring during the evolution of MO planets, their limits of validity are often reached, and one must be careful when extrapolating their results out of their validity range. We already discussed one of these limits previously, namely the impossibility to predict with a good accuracy the planetary albedo with 1D models. Furthermore, a more sophisticated inclusion of stellar radiation would be desirable for other reasons: the presence of stellar light absorbers (whether gaseous or particulate) could significantly alter the temperature profile, including the possible formation of true stratospheres (defined as atmospheric layers where $dT/dz > 0$) that could act as effective seals with respect to the escape of condensible volatiles such as water vapor. Such modifications of the thermal profile of radiative layers could also in turn alter the thermal radiation of the upper layers, thus refining the determination of the various critical thresholds discussed by e.g. Hamano et al. (2013) and Lebrun et al. (2013).

There are other model limitations related to the earliest (and therefore hottest) stages of atmospheres surrounding MO's, especially for smaller, Mars-sized planets. Preliminary modeling using the Marcq (2012) model suggest that the 0.1 Pa mesopause level could be reached at altitudes reaching several thousands of kilometers in such cases. This evidently jeopardizes the plane-parallel assumption common to all 1D models discussed here, and even most 3D GCMs. The required adaptations would range from trivial (atmospheric mass calculations by integration on spherical shells, spherical instead of Cartesian divergence of the heat flux vector in the atmospheric layers) to very difficult problems (radiative transfer including multiple scattering in a curved atmosphere). Fortunately, these departures from plane-parallel geometry do not last long during the evolution of MO atmospheres, since the rapid initial cooling leaves much thinner atmospheres (mesopause below 1000 km) after about ten thousands years.

3.4 Coupled Evolution

The thermal evolution of an MO is closely linked to the formation of an outgassed atmosphere (Fig. 3). In the early stage of solidification, the cooling rate of an MO is limited by thermal blanketing and the greenhouse effect by the planetary atmosphere. It is because the heat transport in the vigorously convective MO is extremely efficient as long as the surface temperature is high enough for the liquid fraction in the magma to exceed the critical value for rheological transition.

Along with crystallization, an outgassed atmosphere would form by degassing from the interior. As water is highly soluble in silicate melt, a large fraction of water would be dissolved in the MO at an early stage. CO_2 has a moderate solubility, and would degas at an earlier stage of solidification, compared to H_2O, which is exsolved at a relatively later stage, when the magma becomes highly enriched in dissolved water (Elkins-Tanton 2008; Lebrun et al. 2013). Other gaseous species such as CO and N_2 are thought to be less soluble in silicate melt. The atmospheric amount of these gaseous components remains less affected by degassing from the interior, and therefore the coupling to the MO solidification is relatively weak.

Once the degassed atmosphere forms, it would slow the solidification process of the MO by preventing heat escaping from the planet. The gaseous species less soluble in silicate melt such as CO_2 and CO would be responsible for the thermal blanketing at an early stage when the largest fraction of water is dissolved in the MO.

As the magma volume decreases, the water vapor would degas and take CO_2 over as the main blanketing species. A rapidly solidifying MO becomes highly enriched in volatile components, leading to a large degassing rate. In turn, the rapid growth of the outgassed atmosphere makes solidification slower due to the enhancement of the thermal blanketing and greenhouse effects.

In parallel with the formation of an outgassed atmosphere, atmospheric escape would occur on early terrestrial planets. It is well known that young stars emit strong XUV radiation due to their high activities. After the nebula gas dissipates out, terrestrial proto-planets and planets would be exposed to extremely strong XUV radiation from the young host star. The stellar XUV radiation is strongly absorbed in the upper atmosphere, thus heating it up and dissociating atmospheric molecules and driving hydrodynamic escape (see Sect. 4 for more details). The outgassed atmosphere would then be partially lost from the planet. Hydrogen-bearing gaseous species such as H_2O are more susceptible to the hydrodynamic escape, because hydrogen is the lightest element. As long as the surface temperature is high enough, no condensation of water vapor occurs and atomic hydrogen is abundant in the thermosphere due to the high mixing ratio of water. The atmosphere can thus escape efficiently at a rate limited by the energy available to heat up the thermosphere. Based on the classification by Tian et al. (2008), the thermosphere is in a hydrodynamic regime in this case. As the surface temperature decreases, the temperature at the tropopause decreases as well so that water vapor starts to condense at the tropopause. Then, the tropopause starts to act as a cold trap, which limits the mixing ratio of water vapor entering the mesosphere and dehumidifies it. The escape rate of hydrogen then becomes diffusion-limited and starts to decrease. As the atmospheric escape rate decreases, the planetary thermosphere shifts from the hydrodynamic regime to a hydrostatic regime. The loss of water associated with the hydrodynamic escape of hydrogen plays a role in reducing the greenhouse effect by water vapor and accelerate the solidification of the magma ocean.

Recently, the MO thermal evolution of an Earth-sized planet along with the formation of a degassed atmosphere has been examined with simple coupled models for a H_2O-CO_2 atmosphere (Elkins-Tanton 2008; Lebrun et al. 2013) and a H_2O atmosphere (Hamano et al. 2013). Results with radiative-convective equilibrium calculations for a steam atmosphere indicate that two disparate types of planets can develop during solidification, depending on their orbital distance from the host star (Fig. 10): wet and potentially habitable planets like Earth (type I), and dried-out planets like Venus (type II) (Hamano et al. 2013). The orbital regions of the two distinct types of planets are limited by a critical distance from the host star (see below).

Type I planets are planets which formed further than a certain critical distance from the star. This type of planet monotonically cools and solidifies on a timescale of less than several million years. Condensation of water vapor starts to occur at the tropopause when the surface temperature becomes low and the amount of water vapor becomes large. Thereafter, the thermal radiation outgoing to space is fixed to 280–300 $\mathrm{W m^{-2}}$, which is called the radiation limit or runaway greenhouse limit of steam atmospheres (Kasting 1988; Abe and Matsui 1988; Kopparapu et al. 2013; Goldblatt et al. 2013). The fact that a radiation limit is present places an upper limit on the MO crystallization timescale.

In contrast, a type II planet, which formed closer to the star, reaches a radiative energy balance during the course of solidification and remains in a runaway greenhouse state by receiving a net stellar flux larger than the radiation limit. Indeed, the planet emits a slightly larger radiative flux than the net incoming stellar insolation by losing its water via hydrodynamic escape. In other words, the planet is desiccated by the hydrodynamic escape during solidification. The initial total water inventory determines the overall solidification time,

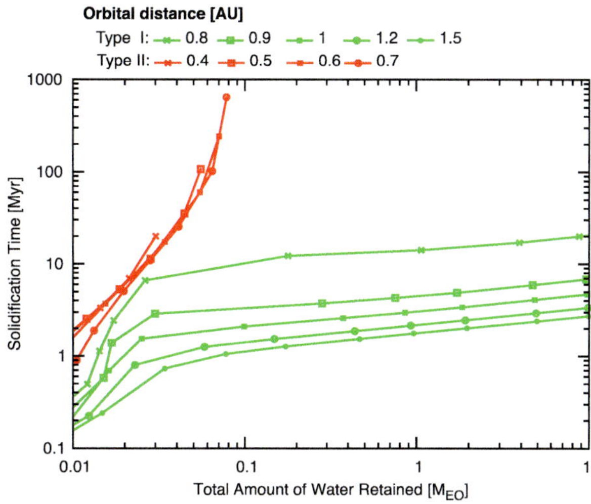

Fig. 10 Relation between final amount of water retained on the planet and its solidification time. The calculations are done with the evolutionary model by Hamano et al. (2013) in a range of initial water inventory from 0.01 to 10 times the Earth's ocean mass (M_{EO}). A critical distance is ~ 0.76 AU in their model

which could exceed 100 Myr for water endowment comparable to the total inventory of the current Earth. When a large amount of other gaseous species is added in the atmosphere, the diffusion-limited escape rate of hydrogen would become lower, due to the decrease of the mixing ratio of water vapor in the upper atmosphere. As a result, the loss rate of water would become lower so that the MO period would be prolonged.

A critical distance which separates the orbital region of the two distinct types of planets is defined as the distance at which the net stellar flux equals the radiation limit. The net stellar flux that a planet receives depends on the planetary albedo and the luminosity of the host star, as well as the planetary orbital distance. For a thick steam atmosphere, the clear-sky planetary albedo becomes small at lower effective stellar temperatures due to the shift of the stellar energy spectrum toward the longer wavelengths (Kopparapu et al. 2013). The stellar luminosity becomes small with decreasing stellar mass as well. For both reasons, smaller stars have a smaller critical distance. Although the stellar luminosity varies with time, the stellar age at which its influence is highest (time of giant impact) can be used to obtain a critical distance of low—and intermediate—mass stars such as the Sun. It is because the solidification timescale of type I planets is typically shorter than the timescale of the evolution of their central star. The critical distance can range from approximately 0.6 to 0.8 AU under solar system condition. This critical distance provides the innermost edge of the habitable zone in the sense that it determines whether water could remain and first form earliest oceans. On the other hand, the threshold orbital distance for ocean re-evaporation due to luminosity enhancement of the host star is larger than the critical distance and depends on various factors such as the distribution of surface water (Abe et al. 2011), the distribution of relative humidity in the atmosphere (Leconte et al. 2013), and the planetary rotation rate (Yang et al. 2014).

Earth is representative of type I planets. The time required for cooling to the sufficiently low surface temperature at which the liquid fraction at the surface reaches the critical value for the rheological transition would have been typically 1 Myr (Lebrun et al. 2013). The subsequent cooling history depends on the development time of a thermal boundary layer at the surface and solid-state convection of a 'hard magma ocean'. Even with a steam atmosphere equivalent to more than several times the Earth's ocean, the overall crystallization time of Earth would not have exceeded 5 Myr due to the presence of the radiation limit (Hamano et al. 2013). The ocean formation after solidification can be estimated to occur roughly in the order of 1000 years for a 270-bar steam atmosphere. This short cooling time would have allowed Earth to maintain a certain amount of water until the ocean formation, whereas a modest loss could have partly oxidized the abundant surface magma.

Venus is on the borderline between the orbital regions of these two types of planets. If Venus is a type I planet, it would have once possessed water oceans, probably allowing formation of carbonate rocks. In accordance with the luminosity enhancement of the Sun, it would have lost its water in a moist greenhouse state, because of photodissociation of water vapor followed by escape of hydrogen (Kasting 1988). If the surface temperature becomes sufficiently high for carbonate decomposition, the carbonate rocks would have been decomposed and released CO_2 into the atmosphere. Although this type I scenario likely explains the dry and CO_2-rich atmosphere of the current Venus, an issue of leftover oxygen still remains. Selective escape of hydrogen could have left Venus with oxygen, which is a relatively heavy element compared to hydrogen. If Venus started with one Earth ocean of water, which is equivalent to the surface atmospheric pressure of ~ 270 bar, approximately 240-bar of oxygen would have been left on Venus. Although an energetic hydrogen wind would drag some of the oxygen to space, this could account for 60 bar of the oxygen at most (Chassefière 1996b), still leaving an oxygen-rich atmosphere on Venus. On the other hand, if Venus is a type II planet, this issue of leftover oxygen can be avoided, because abundant surface magma can serve as a massive oxygen sink. Venus would have lost most of the water during a long magma ocean period by escaping hydrogen to space and taking up oxygen in the interior. In this type II scenario, the final amount of water vapor which could have remained in the atmosphere would depend on the amount of the background gas. Under a 90-bar CO_2 background atmosphere, the hydrogen escape rate would have become extremely low, when the partial pressure of water vapor became lower than about 10 bar (Kasting et al. 1984). The evolution of the Venusian atmosphere based on both the type I and type II scenarios certainly deserves full investigation with detailed calculations of multi-component thermal and non-thermal escape processes.

The vast difference in solidification timescales between type I and type II planets suggests that occurrence rates of hot molten planets could show an orbital dependence reflecting the critical distance from the host star. According to spectral calculations on steam-bearing atmospheres, intense thermal emission from the hot surface could leak through near-infrared atmospheric windows and provide a diagnostic sign of its molten surface, depending on the atmospheric mass (Miller-Ricci et al. 2009; Lupu et al. 2014). Recent calculations predict that terrestrial planets remain molten over the main sequence of the G-type host star in the type-II orbital region if the initial bulk content of water exceeds ~ 1 wt% (Hamano et al. 2015). On the other hand, in the type-I orbital region, which is located further from the star than the critical distance, the lifetime of the magma ocean is typically less than several million years, irrespective of the initial amount of water. Detecting the critical distance could provide observational evidence that water is one of major volatile species on terrestrial planets in extrasolar planetary systems.

3.5 Effects of Multiple Impacts on Protoatmospheric Loss

When a giant impact occurs, there is a possibility that proto-atmospheres on protoplanets are lost through two processes: mechanical (or shock-induced) loss and thermal loss. In the mechanical loss process, a giant impact causes a strong shock wave traveling in the planetary interior, and then the globally expanding motion of the ground pushes out the proto-atmosphere. The thermal loss of the proto-atmosphere occurs after the mechanical loss event. Protoplanets and protoatmospheres can achieve extremely high temperatures due to the energy release of a giant impact. Such a high-temperature proto-atmosphere would expand into space and would be lost through hydrodynamic escape. The fraction of the atmosphere lost by the mechanical process is sensitive to the protoplanet surface conditions. In particular, the existence of liquid water on the surface enhances the fraction lost. Genda and Abe (2003) considered giant impacts between solid protoplanets with various types of atmospheres but without an ocean, and calculated the relationship between atmospheric loss and global motion of the ground induced by a giant impact. They found that a significant loss of the atmosphere occurs only when the ground motion is close to the escape velocity. A typical impact velocity between protoplanets during terrestrial planet formation is approximately that of the two-body escape velocity of colliding protoplanets (Agnor et al. 1999), which is about 5 km/s and 10 km/s for Mars-sized and Earth-sized protoplanets, respectively. Therefore, the global motion of the ground should not exceed roughly half that of the escape velocity. Genda and Abe (2003) showed that such a global motion of the ground can blow off a maximum of about 30 % of the atmosphere by a single giant impact. Subsequently, Genda and Abe (2005) investigated the effect of a water ocean. They found that the presence of an ocean on the surface significantly enhances the atmospheric loss due to the following two effects. One is the vaporization of the ocean, which effectively pushes out the atmosphere. The other is lower shock impedance of the ocean compared to the ground, which causes an increase in the shock velocity at the interface between ocean and ground. Both effects accelerate the velocity at the ocean atmosphere interface close to the escape velocity. However, it should be noted that an entire ocean can survive a giant impact of a Mars-sized impactor.

The proto-atmosphere (including a steam atmosphere) that survives the mechanical loss would undergo thermal loss. The temperature of proto-planets becomes extremely high (5000–10000 K) just after a giant impact, according to the numerical simulations of giant impacts (e.g. Canup 2004a). A high-temperature proto-atmosphere would expand adiabatically, and would be lost through hydrodynamic escape.

The criterion for the onset of the hydrodynamic escape of the atmosphere is given by (e.g. Parker 1963)

$$\lambda_0 < \frac{\gamma}{\gamma - 1}, \tag{19}$$

where γ is the polytropic exponent of the atmosphere and λ_0 is the escape parameter defined by:

$$\lambda_0 = \frac{GMm}{kT_0 r_0}, \tag{20}$$

where G, M, m, k, T_0, and r_0 are the gravitational constant, the mass of the planet, the mass of the molecule of the atmosphere, the Boltzmann constant, the temperature at the planetary surface, and the planetary radius, respectively. In the case of adiabatic flow, the polytropic exponent is equal to the specific heat ratio. If we consider a pure hydrogen atmosphere, which is the lightest gas, the hydrodynamic escape occurs when $T_0 > 4200$ K. Here we assume hydrogen molecules with $\gamma = 1.4$ and an Earth-sized planet. Therefore, if the proto-atmosphere is composed of pure hydrogen just after a giant impact, the adiabatic expansion of the atmosphere results in rapid escape from the planetary surface. However, rock vapor should inevitably exist in very hot planetary atmosphere. Since rock vapor has typically a large molecular weight ($m = 30$ g/mol), the escape requires extremely high temperature ($T_0 > 21000$ K in the case of $\gamma = 1.1$). Moreover, the proto-atmosphere would be composed of various species with large molecular weight other than hydrogen, such as H_2O,

CO_2 and N_2. For example, H_2O with $m = 18$ g/mol and $\gamma = 1.33$ requires extremely high temperature ($T_0 > 30000$ K) for hydrodynamic escape. Therefore, it is impossible for the bulk of the proto-atmosphere to escape adiabatically.

The loss fraction of the atmosphere via mechanical loss is quite sensitive to the surface conditions of the protoplanet as discussed before. We consider four cases i.e. case 1: where the surface is covered by an MO, case 2: dry rock, case 3: an ocean and case 4: water ice.

Cases 2 and 3 were discussed by Genda and Abe (2003, 2005) as summarized above. In case 2, the lost fraction is insensitive to the atmospheric properties, and is approximately proportional to the ground motion velocity. A significant loss occurs only when the ground motion is close to the escape velocity. One third of original atmospheres are expected to survive the sequence of typical impacts among protoplanets (Genda and Abe 2005). In case 3, the atmosphere is efficiently blown away by the impact, while the water ocean survives (Genda and Abe 2005). Hence, giant impacts on such planets result in relative enrichment of water compared to other gases.

Though cases 1 and 4 have not been discussed yet, we can roughly estimate them based on cases 2 and 3. The efficiency of the atmosphere blow off in case 1 is expected to be similar to that of case 2, because the material behavior upon shock wave propagation is not much different between molten and solid silicate. However, the effect on the composition of volatile materials is likely different, because the atmospheric composition is affected by the solubility of gases in the MO. In particular, since the solubility of water vapor is large compared to other gases, such as CO_2, it is likely retained in the MO. Hence, the atmospheric blow off in case 1 likely causes relative enrichment of water vapor over other gases. The enrichment factor depends on the depth of the MO and the amount of atmosphere itself, because the solubility coefficient of water vapor in silicate melt is not constant but depends on the partial pressure of water vapor.

The efficiency of the atmosphere blow off in case 4 is expected to be intermediate between cases 2 and 3, because the mechanical material behavior upon shock wave propagation is not much different between ice and rock, although ice clearly vaporizes much more easily than rock. Therefore, the atmosphere is rather efficiently blown away but a fraction of the water may survive depending on the ice thickness at the time of impact and on impact conditions itself. Therefore, the atmosphere should be enriched in water compared to other gases. Thus, except for case 2, water is likely enriched over other gases, but the mechanism and efficiency of this enrichment are different.

The surface conditions of protoplanets depend on the time interval between giant impacts and on the distance from the central star. According to the terrestrial planet formation theory, several tens of Mars sized protoplanets are formed through the successive accretion of planetesimals in the terrestrial planet region on a timescale of 10^5 years (Kokubo and Ida 1998). After that, these protoplanets collide with each other to form terrestrial planets such as the Earth and Venus. The timescale for the Earth-sized planet to be fully grown is roughly 100 Myrs (Kokubo and Genda 2010; O'Brien et al. 2014). This timescale is consistent with the timing of the last giant impact that forms the Moon, which has been derived from Hf-W system of Earth's and lunar rocks (e.g. Touboul et al. 2007). Since the Earth and Venus experience giant impacts about 5 to 10 times during this stage (Kokubo and Genda 2010), the interval of giant impacts is roughly 10 Myrs in average. This time interval is longer than the typical cooling time of MO for type I planets (several Myr depending on their water content, see Fig. 10) reported by Lebrun et al. (2013) and Hamano et al. (2013). Another thing we have to take into account is the orbital change of protoplanets. We can expect giant impacts among protoplanets formed at various distances from the Sun. It means collisions among protoplanets with various volatile content. A protoplanet has either a solar-composition or a mixed atmosphere. Since giant impacts are intrinsically related to the orbital change of protoplanets, the time scale of orbital change is basically the same as the interval between the giant impacts. It is much longer than the time scale of the atmosphere responses such as atmosphere cooling or ocean vaporization due to insolation change (Abe et al. 2000). Hence, the place of collision is more important than the original position of protoplanets (Abe 2011).

The average time interval between giant impacts is longer than the typical cooling time of type I planets, but it is much shorter than the typical cooling time of type II planets. Therefore, we can estimate the surface conditions of protoplanets at the time of giant impacts as a function of distance from the central star as follows:

(1) If the distance is smaller than the critical distance dividing type I and type II planets, the surface is covered by an MO. Thus, gases with high solubility in silicate melt, such as water vapor, are preferentially retained in the planets. It is hard to remove water molecules dissolved in silicate melt by giant impacts, while the atmosphere would be more or less removed. Therefore, water content will become relatively enriched through the sequence of giant impacts. If the depth of the MO just before a giant impact is deep (or shallow), large (or small) fraction of water molecules should be dissolved in MO. Therefore, the enrichment factor depends on the depth of the MO, which in turn depends on the time interval between giant impacts. A longer interval results in a smaller enrichment. However, after the giant impact stage, most water should be lost into space via hydrodynamic escape induced by the intensive irradiation of XUV from the Sun before the complete MO freezing, because such a planet is a type II terrestrial planet (Hamano et al. 2013).

(2) If the distance is larger than the critical distance for global freezing, the surface of planets is likely covered by water ice at the time of giant impacts. Thus, enrichment of water is expected. If the ice layer is thick, the atmospheric loss would be slightly suppressed. Therefore, the enrichment factor would be small for longer impact interval, because a longer interval results in a thicker ice layer.

(3) If the distance lies between the critical distances dividing type I and type II planets and the global freezing, the surface of planets is likely covered by water oceans at the time of giant impacts. Thus, the loss efficiency of gases other than water vapor is enhanced and water is relatively enriched.

It is very difficult to estimate the precise value of the loss efficiency of each gas component, because it depends on the interval of impacts, which varies stochastically. Generally speaking, however, we can expect the highest loss efficiency of gases other than H_2O molecules for the last case among the 3 cased discussed above, because H_2O exists in the liquid form as a water ocean and it survives a giant impact. This is expected between the Venusian and Martian orbits (Abe 1993). Thus, planets formed near Earth orbit are likely more enriched in water than those near Venus orbit or those which experienced no giant impact.

4 Escape from Protoatmospheres

4.1 Introduction

On all terrestrial planets, the dissipation timescales of protoatmospheres are short (Montmerle et al. 2006; Lammer et al. 2014). Observations of exoplanetary atmospheres under strong stellar radiation provide excellent opportunities to verify our theoretical understandings on this key process influencing planet evolution. Shortly after the observation of an extended hydrogen cloud around transiting hot Jupiter HD 209458b (Vidal-Madjar et al. 2003) and the discovery that these planets experience most likely hydrodynamic atmospheric escape (Lammer et al. 2003), one dimensional hydrodynamic upper atmosphere models were independently developed by several research groups (e.g., Yelle 2004; García Muñoz 2007; Penz et al. 2008; Murray-Clay et al. 2009; Tian et al. 2005b) to understand this phenomenon. Despite the differences in model details (numerical solution methods, major atmosphere species, and complexity of chemistry networks, etc.), the orders of magnitude of calculated hydrogen escape rates from these models agree with each other. But unfortunately these studies did not thoroughly explore the long term evolution histories of close-in exoplanets, partially due to the perceived large uncertainties in factors influencing the model results (XUV energy flux, heating efficiency which converts absorbed photon energy into heat, geometric factors—where the stellar photons should be absorbed by the atmosphere gases, and how to treat heavier elements and molecules, etc.). More advanced models (Koskinen et al. 2010, 2013a,b; Guo 2011, 2013; Shaikhislamov et al. 2014) along this line have been developed.

On the other hand, other studies investigated the evolution of close-in exoplanets based on simple "energy-limited" escape formulae, first for hot Jupiters (Lammer et al. 2003; Baraffe et al. 2004; Lecavelier des Etangs et al. 2004), later for exoplanets with different masses, including the Neptune and super-Earth mass domain (e.g., Lecavelier des Etangs 2007; Davis and Wheatley 2009; Lammer et al. 2009; Sanz-Forcada et al. 2010; Lopez et al. 2012; Lanza 2013; Lopez and Fortney 2013; Kawahara et al. 2013; Kurosaki et al. 2014; Jin et al. 2014), the later works being successful at explaining some trends in the mass-size distribution of Kepler exoplanets. Triggered by the same Kepler observations, newer and more advanced hydrodynamic escape models (Owen et al. 2010; Owen and Wu 2013; Erkaev et al. 2013; Lammer et al. 2013, 2014) are also able to explain the Kepler data but provide somewhat different insights on the nature of physical processes controlling atmospheric escape from close-in exoplanets.

Readers familiar with works on atmospheric escape for solar system bodies such as Mars, in which the main focus is put on nonthermal escape processes, may be puzzled that very limited work regarding nonthermal escape processes is discussed in the exoplanet community. This is because the observed close-in exoplanets are under intense stellar XUV radiation and their upper atmospheres are necessarily hot, which makes hydrodynamic escape the most relevant when discussing escape from protoatmospheres.

All of these works contribute to improve our understanding of atmospheric escape. However, there are important differences among these recent works, mainly focusing on the following aspects: (1) what is the main driver of hydrodynamic escape from planetary atmospheres, the X-ray or EUV photons? (2) What is the proper heating efficiency in hydrodynamic escape from planetary atmospheres? (3) Is hydrodynamic escape from planetary atmospheres energy limited? It can be expected that future observations of exoplanets will present stronger constraints on and encourage improvements of atmospheric escape models, the applications of which will also lead to a better understanding of the paleoclimate and the evolution of solar system terrestrial planets.

In this section, we present a review on theories and observations relevant to escape from protoatmospheres. Although protoatmospheres are likely dominated by hydrogen, in this review we try to cover more general physics and thus sometimes discuss processes not limited to hydrogen atmospheres. We will first introduce basic concepts related to planetary upper atmospheres, followed by illustrations of different physical processes important to determine energy deposition and atmospheric escape processes, followed by some recent observations and theoretical results related to atmospheric escape from close-in exoplanets. After that we provide a brief discussion on the differences in recent atmospheric escape models, ending with a brief summary.

4.2 Basics Concepts

4.2.1 Planetary Upper Atmospheres

The term upper atmosphere can mean different things even to planetary atmospheric scientists. A planetary upper atmosphere can be defined as the region where the different atmospheric species begin to separate from each other due to the lack of mixing. In this definition the upper atmosphere is identical to the so called "heterosphere", in contrast to "homosphere" where atmospheric species are well mixed. An indicator of the mixing strength is the eddy diffusivity which is typically greater in the dense part of the atmosphere with a negative temperature gradient (troposphere and mesosphere in the Earth's atmosphere) and smaller in the atmosphere region with a positive temperature gradient (stratosphere of the Earth). In solar system planets, the number density of the lower boundary of the heterosphere is typically between 10^{12} and 10^{13} cm^{-3}.

Based on thermal structure one can define a planetary upper atmosphere as the region where a positive thermal gradient exists (we will soon find out that this definition is problematic). In this definition the upper atmosphere is identical to the so-called thermosphere. The thermosphere is heated by the absorption of solar or stellar soft X-ray (up to a few 100 Å) and EUV (up to 1000 Å) photons (hereafter called XUV). Since the energy of most XUV photons exceeds the ionization threshold of essentially all molecules and atoms, the absorption cross sections of gases for XUV photons are large, which in turn favors the absorption of XUV photons at low density and high altitude regions. The combination of sufficient energy deposition and a low local density makes the thermosphere hotter than the underlying atmosphere. Because of this internal link between the thermosphere density and temperature, the lower boundary of the heterosphere is normally located not far from that of the thermosphere. Thus in the following we do not distinguish between the two.

Note that atmospheric gases have smaller absorption cross sections for photons with wavelengths shorter than a few Å. These photons could penetrate deeper into the atmosphere below the upper atmosphere. Although these photons are energetic, the fluxes at these short wavelengths are orders of magnitude smaller than stellar UV radiation which forms the ozone layer in the Earth's atmosphere. In addition, the planetary atmosphere region where these energetic photons are absorbed is typically dominated by molecules, which have stronger radiative capabilities. Thus these photons do not necessarily cause significant heating of the middle atmosphere. Neither do they contribute to atmospheric escape.

Because XUV photons can ionize atmospheric species, the fraction of charged particles is higher in the thermosphere, which leads to the overlap of the ionosphere and the thermosphere. A quasi charge neutrality is reached in the thermosphere in which the total densities of ions and the density of electrons roughly balance each other at all altitudes in the thermosphere because the recombination reactions between (1) the ion species and the electrons provide a self-limiting control on the electron and ion densities; and (2) separation of charges will produce electric field which attracts opposite charges to each other.

The Earth's thermosphere temperature is ~ 900 K in solar minimum and ~ 1500 K in solar maximum with solar XUV energy flux changes by a factor of ~ 3 in a solar cycle. The most important radiative cooling agents in the Earth's thermosphere are CO_2 and NO in the lower thermosphere, and atomic oxygen in the upper part. In comparison the thermospheres of Venus and Mars are much colder, $200 \sim 300$ K respectively, and much more stable in response to variations of solar XUV radiation because CO_2 is the dominant gas in the atmospheres of these planets. On the other hand, the temperatures of hydrogen and helium dominated thermospheres of Jupiter and Saturn are on the order of 1000 K despite their large distances from the Sun because of the relative weak radiative cooling capability of hydrogen and helium. Thus a general rule for the thermosphere temperature is that CO_2 atmospheres are cooler.

There are other sources of energy and ionization sources in addition to stellar photons. In the magnetic polar regions of the Earth, the magnetic field lines are either open to the solar wind or the Earth's magnetosphere where particles can be accelerated via various processes. Thus the thermosphere, in these polar regions, is subject to collisions of these energetic charged particles and can be heated by this so called energetic particle precipitation process. In addition, the ions and electrons in the ionosphere are accelerated by the magnetic field of the planet and thus have a differential velocity in comparison to the neutral species in the thermosphere. Collisions between ions/electrons and neutrals produce so called Joule heating which is more important than solar XUV heating in the polar regions of the Earth. Globally averaged, solar XUV heating dominates these other heating mechanisms (i.e. energetic particle precipitation, Joule heating, etc.) in the thermospheres of the Earth, Mars, Venus, Titan, and giant planets.

4.2.2 Physical Processes Related to Heating

Energy deposition in planetary upper atmospheres involves some complex and interesting fundamental processes. After an XUV photon is absorbed by atmospheric species, several results can occur. For photons absorbed by molecules, if the photon energy is lower than the dissociation energy threshold of the absorbing molecules, the energy would most likely be used to excite the molecule. If the photon energy is greater than the dissociation energy threshold, photo-dissociation could occur and the daughter particles produced could each carry some kinetic energy (heat) and some potential energy (forming excited atoms or molecules). Collisional de-excitation processes would convert the potential energy of the excited species into kinetic energy and heat the atmosphere. Alternatively the potential energy could be passed partially to other species through chemical reactions and heat is generated in the process.

For photons with even higher energy, ionization could occur in which a neutral is converted into a pair of ion and an electron. An ion can transfer its potential energy and charge to a neutral species through chemical reactions. The potential energy of the reactants needs to be greater than that of the products in order for the reaction to be exothermic (heat generated) and rapid. Charge transfer reactions, a good example of which are the $H^+ + O \longleftrightarrow H + O^+$, are rapid because the two reactants have similar potential energy.

Recombination reactions are the processes through which ion species return to neutrals. Because of the fundamental laws of energy and momentum conservation, radiative recombination reactions occur for atomic ions such as H^+. However, these processes are slower than dissociative recombination (DR) reactions such as $O_2^+ + e \rightarrow O + O$. Note that these DR reactions could produce neutrals in excited states and thus the kinetic energy of the products depends on which branch the neutrals are formed. For small planets such as Mars, the kinetic energy formed in DR reactions could be sufficient for the neutrals to escape—photochemical escape. In the Earth's thermosphere the molecular ion finally produced is NO^+ because it has a low potential energy out of all ions in the Earth's ionosphere. The recombination reaction of NO^+ forms an N and an O atoms which returns the upper atmosphere to its neutrality. Carefully accounting the energy stored in ions and neutrals and those released in chemical reactions is essential to calculate the energy deposition in planetary upper atmospheres.

The electron produced in a photo-ionization reaction is called a photo-electron, which could carry much more energy than the potential energy carried by the ion if the initial photon has much greater energy than the ionization threshold of the species. In this case the photo-electron will have adequate energy to cause additional excitation, dissociation, and ionization. Electron impact cross sections are used to describe the occurrence probabilities of these events, which are measured in the laboratory and used to compute where and when the energy of photo-electrons deposit their energy. One energetic photo-electron can lead to the formation of another electron with less but still high energy, which can cause subsequent ionization. Thus one needs to trace these secondary electrons in order to correctly track the flow and deposition of energy. In the presence of a planetary magnetic field, the motion of photoelectrons and secondary electrons should be constrained, which further complicates the calculations of cascade electron transport and degradation. This is an area that multi-dimension numerical models for planetary upper atmospheres have not yet treated.

If the photon energy is exceedingly high, dissociative ionization, in which one neutral molecule is divided into one neutral and one ion, could occur. Similarly species except hydrogen can lose more than one electron when absorbing an energetic photon or when being hit by an energetic electron. A further complication is the possible multiple photon events which are unimportant in the upper atmospheres of Solar System terrestrial planets but could become non-negligible for planets under extreme strong stellar XUV radiation.

For pure hydrogen protoatmospheres, many of the above mentioned processes may not apply. However, even small amounts of other species such as CO_2, CO, H_2O, NH_3, CH_4, etc. may change the chemistry and energy network of planetary upper atmospheres. Thus it is useful to pay attention to what species and processes are involved in the deposition of photon energy in planetary upper atmospheres.

4.2.3 Escape Processes

The exobase is an altitude in the atmosphere beyond which few collisions between particles occur. Mathematically it can be defined as where the mean free path equals the atmospheric scale height. From a kinetic point of view, any particle at the exobase with outgoing velocity exceeding the escape velocity of a planet can escape. High particle velocity can be achieved in many processes. Those processes in which the high velocity is associated with the exobase temperature are called thermal escape.

Thermal Escape Although an exobase level can be defined, the collisional and collisionless regions near the exobase level cannot be divided by an arbitrary dimensionless boundary because the density distribution is a continuous function and thus a gradual transition exists between the two regions. When considering three dimensional geometry, the exobase may exist at different altitudes and pressures, because the probabilities of collisions only depend on number densities and temperature, at different longitudes and latitudes, which further complicates the problem.

In thermal escape, the percentage of escaping particles is governed by the local temperature. If the atmosphere near the exobase is in hydrostatic equilibrium, the local velocity distribution of particles follows the Maxwellian distribution and the escape is called Jeans escape (Chamberlain 1963; Opik and Singer 1961):

$$\Phi = \frac{n(z)V_0}{2\sqrt{\Pi}}\left(\frac{V_{esc}^2}{V_0^2}+1\right)\exp\left(-\frac{V_{esc}^2}{V_0^2}\right) \tag{21}$$

here Φ is the escape flux (number of escaping particles per unit area per unit time) at the exobase, $n(z)$ is the exobase number density, $V_0 = \sqrt{(2kT/m)}$ is the most probable velocity with m the mass of the escaping particle, k the Boltzmannn constant, T the local temperature, $V_{esc} = \sqrt{(2GM/r)}$ with G the gravitational constant, M the mass of the planet, and r the distance from the center of planet. The ratio between the gravitational energy and the kinetic energy of particles near the exobase is often called the Jeans escape parameter,

$$\lambda = \frac{GMm}{kT} \tag{22}$$

When this parameter is close to unity, the gravity of the planet hardly binds its atmosphere and the upper planetary atmosphere deviates from hydrostatic equilibrium, in which case the local velocity distribution of particles can be described by a shifted Maxwellian distribution (Yelle 2004; Volkov et al. 2011a,b), with the outward velocity enhancing the escape probability of particles near the exobase. Traditionally this scenario is called the blowoff or hydrodynamic escape or planetary wind, an analog to the solar wind (Parker 1964; Watson et al. 1981).

Since it is the temperature of planetary upper atmosphere which drives atmospheric escape in both cases, both Jeans escape and hydrodynamic escape are special cases of thermal escape: in Jeans escape, the escape does not change the velocity distribution of particles (in another word the temperature) near the exobase or in the upper thermosphere; in hydrodynamic escape, the temperature and velocity distributions of particles in the upper thermosphere are influenced strongly by atmospheric escape. More discussions on hydrodynamic escape are in Sect. 2.4 because it is more relevant to the protoatmospheres.

Nonthermal Escape When the velocities of escaping particles are not determined by the exobase temperature, nonthermal escape processes occur (Lammer et al. 2013, and references therein). Most nonthermal escape processes are associated with the presence of ions and their behavior in the electric and magnetic field. Photochemical escape is a nonthermal escape process important for Mars (e.g., McElroy 1972; Ip 1988; Nagy and Cravens 1988; Lammer and Bauer 1991; Fox and Hać 2009; Krestyanikova and Shematovich 2006; Chassefière et al. 2007; Chaufray et al. 2007; Gröller et al. 2014) but is unlikely to be an efficient escape mechanism on Venus (Lichtenegger et al. 2009; Gröller et al. 2010), Earth or super Earths because the maximum kinetic energy an oxygen atom can obtain from dissociative recombination reaction of O_2^+ is fixed and lower than the escape energies of planets more massive than Mars. Neutral atoms and molecules in the atmosphere of the planet may be ionized through photoionization, impact ionization, or charge exchange with solar wind ions. Once formed around a nonmagnetized planet (or a weakly magnetized one), the ions can be accelerated by solar magnetic field and some could escape the planet (e.g., Lammer et al. 2013, and references therein). This ion pickup process is less efficient on a planet with a strong intrinsic magnetic field, such as Earth, because the solar or interplanetary electric and magnetic fields are kept at distances far from the planet. In this case, the ions can be trapped by the planet's magnetosphere and do not necessarily escape the planet. However various models for hydrodynamic escape from protoatmospheres shown that the upper atmosphere under strong stellar XUV radiation could be highly expanded, with the exobase beyond a few planetary radii. In combination with the fact that the magnetopause of a protoatmosphere could be suppressed to closer distances from the planet by stronger stellar wind from a young star, the protection from planetary magnetic fields could be inefficient for protoatmospheres. When energetic ions are trapped by the magnetic field of the planet, they can escape by transferring their charges to slowly moving atoms—charge exchange. They can also impact slowly moving atmospheric species and cause the latter to escape-sputtering (e.g., Luhmann et al. 1992; Jakosky et al. 1994).

Since electrons are much lighter than ions, in the regions with open magnetic field lines (magnetic poles), there is a deviation from exact neutrality. The separation of the outgoing electrons and the ions left behind attract each other and cause a net acceleration of the ions outward which can lead to atmospheric escape—the so called polar wind. The relative significance of thermal and nonthermal escape mechanisms on present day solar system terrestrial planets is summarized in a recent review paper (Tian 2013) and is not repeated here.

4.2.4 Escape from Protoatmospheres—The Hydrodynamic Regime

The section 'Thermal Escape' briefly discussed the differences between Jeans and hydrodynamic escape (or blow-off). The Direct Simulation Monte Carlo (DSMC) model of Volkov et al. (2011a,b) shows that the transition between hydrodynamic escape and Jeans escape occurs in a rather narrow range in terms of the escape parameter. This result was recently confirmed by Erkaev et al. (2015) who found also that when the Jeans escape parameter is fixed at values less or equal than 2.5, there is no transonic solution. This is the case for a fast explosive atmospheric expansion that results in extreme thermal atmospheric escape rates. It is important to note that the negative feedback between rapid escape and energy budget throughout the thermosphere is at work in this regime (Tian 2013) and the escape parameter could be regulated by this feedback: a low escape parameter induces a faster escape which in turn lowers the temperature and leads to a greater escape parameter and hence a slower escape; so that a smooth transition from the subsonic to supersonic hydrodynamic flow could occur (Erkaev et al. 2015). It will be interesting to model and observe an alternative scenario to the steady transition, which is an oscillating planetary upper atmosphere between a more stationary state with a larger escape parameter and an explosion state with an escape parameter less or equal than 2.5. It is likely that 3D models will be needed for this.

Tian et al. (2008) showed that if the energy associated with the outflow of the upper thermosphere, which is a result of enhanced Jeans escape at the exobase, is neglected, the temperature and exobase level can grow rapidly with increasing XUV leading to blow-off. However, when the cooling associated with the outflow is included in the energy budget, blow-off is prevented until much higher XUV radiation is imposed. Thus it is important to characterize planetary thermosphere based on the importance of atmospheric escape on the energy budget: a hydrodynamic regime in which the energy budget in the thermosphere provides an ultimate limit to atmospheric escape; a hydrostatic regime in which atmospheric escape does not affect the thermosphere temperature. Traditional hydrodynamic escape (blow-off) can be considered as an extreme case of the hydrodynamic regime. Tian et al. (2008) found that even when the escape parameter does not reach the low values required for traditional hydrodynamic escape to occur, the energy budget in the upper thermosphere could nevertheless already be strongly influenced by the escape. This finding is confirmed by Erkaev et al. (2015).

The other importance of the hydrodynamic regime classification is that atmospheric escape from a planet in the hydrodynamic regime is inherently limited by energy available to drive the escape through the whole thermosphere. Thus if the amount of heating in the thermosphere remains unchanged, no matter how efficiently atmospheric escape can be at the exobase level, total escape remains unchanged. Thus the calculations of atmospheric escape and the evolution history of planetary atmospheres under strong thermosphere heating can be simplified (Tian 2013). On the other hand, nonthermal escape processes would help the transition of planetary upper atmospheres from a hydrostatic regime to a hydrodynamic regime.

It is convenient to associate incoming stellar XUV energy and the escape rate from a planet with an escape efficiency. However to calculate such an escape efficiency is not simple. As discussed in previous sections, a careful accounting is necessary in order to not only trace all energy flowing through the chemistry network, but also to consider the energy carried from one place to another by photo—and secondary electrons. Thus, the calculations of heating efficiency, the ratio between energy released to the ambient upper atmosphere as heat and the absorbed photon energy, is complicated. Most up-to-date works estimate less than 20 % heating efficiency, the conversion factor from the energy in the absorbed XUV photons to heat in the upper atmosphere, for hydrogen-dominated upper atmospheres (Shematovich et al. 2014, more discussions in the following).

In addition, when a planetary upper atmosphere is heated up, the exobase moves to greater distances which enhances the probability for atmospheric escape, which in turn increases the fraction of energy associated with escape and decreases the fraction of energy being radiated into space. This further complicates the calculations of escape efficiency. However as we shall see in the following sections, it seems nature has prepared some interesting and easy-to-apply rules to govern escape from protoatmospheres.

4.2.5 Suprathermals and Heating Efficiency

The photolysis of hydrogen-dominated protoatmospheres of the close-in exoplanets, driven by the XUV radiation of the host star, leads to the formation of suprathermal particles (particles with an excess of kinetic energy, i.e., with kinetic energies above 5–10kT, where T is the temperature of an ambient atmospheric gas and k the Boltzmann constant). These particles can be primary photoelectrons from H_2/H/He ionization or hydrogen atoms from H_2 dissociation and dissociative ionization processes. Dissociative recombination, dissociation and ionization by ultraviolet photons and electrons, and exothermic chemical reactions are accompanied by the release of energy on the order of several eV; part of this energy can be stored as the internal excitation of the products (Wayne 1991; Marov et al. 1996; Johnson et al. 2008). Charge exchange and atmospheric sputtering induced by energetic plasma ions can result in much larger energy transfers, producing hot particles with energies up to several hundred eVs (Johnson et al. 2008). If the production rate of these particles, which are typically suprathermal, is faster than thermalization, then a stable fraction of them is formed. These particles with excess kinetic energies are an important source of thermal energy in the upper atmosphere of hydrogen-rich planets. Generally the kinetics of the suprathermals is not considered in the contemporary aeronomical models because fast local thermalization of the suprathermal hydrogen atoms by elastic collisions with the ambient gas is usually assumed in order to avoid the complicated numerics caused by the solution of the non-linear kinetic Boltzmann equation.

Heating Efficiency Only some fraction of the absorbed stellar XUV energy contributes to the heating of the upper atmosphere. Therefore, it is important to calculate this fraction, and to evaluate how deep in an atmosphere stellar XUV photons, continuously release part of their energy as heat.

To estimate the effect of the XUV emission from solar-type stars of different ages an accurate description of radiative transfer and photoelectron energy deposition is required. In the studies by Cecchi-Pestellini et al. (2009) it was shown that soft X-rays provide an important contribution to the heating of hydrogen-dominated planetary atmospheres of close-in exoplanets. The flux of stellar XUV emission photons incident upon a planetary atmosphere of hydrogen-dominated composition photoionize the gas so that a flux of high-energy (suprathermal) photoelectrons can be produced, which again deposit their energy into the gas (described in Sect. 4.2.2). In a partially neutral gas, electrons ionize, excite, and dissociate atomic and molecular species, as well as heat the gas through Coulomb collisions. In determining these energy deposition, we must account for all the possible degradation histories of the energetic electrons. When the stopping medium is only partially neutral, electron-electron interactions contribute to the electron energy degradation, and a significant portion of the energetic electron energy is deposited into the stopping medium as heat. As the ionized fraction rises, more and more of the electron energy heats the gas, and the excitation and ionization yields decrease. The heating efficiency in the hydrogen-dominated upper atmosphere of the well studied "hot Jupiter" HD 209458b was calculated with the kinetic Monte Carlo model (Shematovich et al. 2014). The obtained results can also be used for any thermal escape study of hydrogen-rich protoatmospheres.

The XUV radiation of the star is absorbed by atmospheric gas and leads to the excitation, dissociation, and ionization of different components of the atmosphere. Photoionization could produce photoelectrons with energies sufficient for the subsequent ionization and excitation of atomic and molecular hydrogen (Ionov et al. 2014). The fresh electrons lose their excess kinetic energy in collisions with the ambient atmospheric particles. Their kinetics and transport is described by the kinetic Boltzmann equation (Shematovich et al. 2008) and the DSMC (Direct Simulation Monte Carlo) method is an efficient tool to solve atmospheric kinetic systems in the stochastic approximation (Marov et al. 1996; Shematovich 2010). In the numerical simulations, the evolution of the system of modeling particles due to collisional processes and particle transport is calculated from the initial to the steady state. The Monte Carlo model developed by Shematovich et al. (2008) and adapted for hydrogen atmospheres was used to calculate the collisional kinetics and the transport of photoelectrons in the atmosphere of HD 209458b (Shematovich 2010).

The heating efficiency by the stellar XUV radiation within a hydrogen-dominated atmosphere, such as that of the extrasolar gas giant HD 209458b, has been modeled (Shematovich et al. 2014). It is shown that one cannot neglect the effects caused by electron-impact processes together with the participation of suprathermal photoelectrons. By including these processes the results of the DSMC model indicate that the XUV heating efficiency $\lambda_{h\nu}$ in hydrogen-dominated planetary protoatmospheres do not exceed 20 % over most of the thermosphere. This result agrees well with the heating efficiency values assumed in the early studies by Kasting and Pollack (1983) and Chassefière (1996a). Therefore, one can conclude that atmospheric mass loss studies, that assumed $\lambda_{h\nu}$ values greater than 20 %, overestimate the hydrogen escape rates, while hydrogen escape studies that assumed $\lambda_{h\nu}$ values $\approx$ 10–15 % should likely yield accurate results.

Suprathermals in the Extended Atmospheres of Hot Jupiters Ballester et al. (2007) reported the results of observations of hot Jupiter HD 209458b with STIS (Space Telescope Imaging Spectrograph) onboard the Hubble Space Telescope (HST) in a broad UV-to-optical spectral range. They detected a new detail in the planetary atmosphere absorption spectrum at 356–390 nm that they interpreted as Balmer-continuum absorption by hot (suprathermal) neutral hydrogen atoms in the upper atmosphere of the planet. Such a fraction of suprathermal hydrogen was used (Ben-Jaffel and Hosseini 2010) to approach the observations (Vidal-Madjar et al. 2004).

As mentioned above the important sources of suprathermal hydrogen atoms in the upper atmosphere of HD 209458b are dissociation and ionization of molecular hydrogen by XUV radiation of the star that produces hydrogen atoms with an excess of kinetic energy. No detailed studies of suprathermal hydrogen atom kinetics and transfer have been performed at the molecular level (Yelle et al. 2008), because this requires the solution of the non-linear kinetic Boltzmann equation (Marov et al. 1996; Johnson et al. 2008). Accordingly, it is necessary to consider the effect of molecular hydrogen dissociation by UV radiation and the subsequent flux of photoelectrons on the suprathermal atomic hydrogen production in the $H_2 \rightarrow H$ transition region and the formation of the corresponding outflow in the upper atmosphere of HD 209458b. Recently the stochastic planetary corona model (Shematovich 2004) was used to study the kinetics and transfer of the suprathermal hydrogen atoms in the extended hydrogen-dominated protoatmosphere and estimate of the atmospheric mass-loss rate were obtained (Shematovich 2010).

The thermal regime and escape rate in the atmospheres of hot Jupiters significantly depend on the chemical composition of the atmosphere. In contrast to the planets of the Solar System that have stable atmospheric compositions over geological timescales, the composition of giant exoplanets may change rapidly as a consequence of the influence of the intense radiation field of the host star. Moreover, the thermal regime and the composition of the atmosphere are tightly connected through heating and cooling (Yelle 2004; García Muñoz 2007). Aeronomical models such as that of Yelle (2004) generally assume that reaching temperatures of several thousand Kelvin at particle densities on the order of 10^{10} cm^{-3} leads to the thermal dissociation of molecular hydrogen, $H_2 + M \rightarrow H + H + M$, and the formation of the $H_2 \rightarrow H$ transition region in the inner thermosphere of the exoplanet. In the upper thermospheric layers, the photoionization of atomic hydrogen starts to play the dominant role. Correspondingly, the makeup of the upper atmosphere changes with altitude from H_2 to H to H^+, which is an additional factor forming the extended atmosphere, since such a change in the composition is accompanied by the growth of the characteristic scale height. The molecular-to-atomic transition occurs at altitudes of about $(1\text{–}1.2) \times R_p$, where R_p is the planet radius, and the temperature of the gas attains values of about a few thousands of K. The temperature and density values in the transition region are consistent with the observations (Ballester et al. 2007). The model of the upper neutral atmosphere consisting primarily of H_2, H, and He (Yelle 2004) was used to trace the kinetic properties of suprathermal hydrogen atoms.

The dissociation processes such as photodissociation, collisional dissociation, dissociative ionization, etc., are the primary sources of thermal and suprathermal fragments of molecules in electronically excited states in the upper planetary atmospheres (Wayne 1991). Although a hydrogen molecule is relatively simple, the UV and/or electron collisional dissociation of H_2 may occur in several ways. If, as a result of the absorption of radiation, electrons are excited to unbound or antibound orbits and their excitation energy exceeds the binding energy of the molecule, the molecule may dissociate. This mechanism allows for the photodissociation of molecules excited either to a continuum bound state or immediately to an unbound (repulsive) state. The photo-dissociation cross sections for these processes are usually smooth functions dependent on the wavelength; therefore, low (0.05–0.1 nm) spectral resolution data on the incident flux and cross sections are sufficient to estimate the dissociation rate (Fox et al. 2008). Another important mechanism is a pre-dissociation, when an absorbed photon excites the molecule into a bound state of electron excitation, from which the subsequent non-radiative transition into an unbound state is possible. The dissociation rate depends on the absorption rates at the wavelengths of the selected line transitions and the pre-dissociation probabilities (Fox et al. 2008).

The excess kinetic energy ΔE_{dis} of the atoms formed through the photo-dissociation of the hydrogen molecule was calculated as the difference between the energy of the absorbed UV photon, the energy of electronically excited state, and the dissociation energy. In the case of the dissociation of H_2 by photoelectrons the distributions were calculated for several electron beam energies and include the populations of relatively slow thermal energy (less than 1 eV) and fast high-energy (1–10 eV, with a main peak at about 4 eV) atoms. Since hydrogen atoms appear in the process of dissociation with an excess of kinetic energy, their distribution in the $H_2 \rightarrow H$ transition region of the upper atmosphere of HD 209458b is determined by the Boltzmann equation with a photochemical source function due to the dissociation.

The calculated formation rates and spectra were used as source functions in the kinetic Boltzmann equation for the suprathermal fraction of atomic hydrogen (Shematovich 2010). Simulations show that the distribution functions of hot H significantly deviate from the local equilibrium distributions. At heights of the maximal production of hydrogen atoms due to the dissociation of H_2, a significant fraction of hydrogen atoms with energies high enough to escape the planetary gravitational field (about 6.9 eV) is formed, in contrast to the local equilibrium distribution.

Using the numerical stochastic model created by Shematovich (2004) for a hot planetary corona the kinetics and transfer of suprathermal hydrogen atoms in the upper atmosphere and the emergent flux of atoms evaporating from the atmosphere were investigated. The latter is estimated as 5.8×10^{12} cm^{-2} s^{-1} for a moderate stellar activity level of XUV radiation, which leads to a protoatmosphere evaporation rate of 5.8×10^9 g/s due to the process of the dissociation of H_2. This estimate is approaching the observational value of $\sim 10^{10}$ g/s for the rate of atmospheric loss of HD 209458b.

These studies show that input of the suprathermals should be taken into account in the aeronomic models to estimate the escape of the hydrogen-dominated protoatmospheres.

4.3 Observations and Models of Hot Jupiters

HD 209458b is the first observed transiting exoplanet and presents the earliest evidence (transit depth of Lyman-α) of the existence of an extended hydrogen cloud surrounding hot Jupiters (Vidal-Madjar et al. 2004). Soon it was discovered that the extended cloud surrounding the planet includes not only hydrogen atoms but also other species: O and C^+ (Vidal-Madjar et al. 2004), Si^+ and Si^{2+} (Linsky et al. 2010), and Mg (Vidal-Madjar et al. 2013). These observations provide a unique opportunity to test upper atmosphere models.

Although many workers have theoretically studied the thermosphere of HD 209458b (Yelle 2004; Tian et al. 2005b; Erkaev et al. 2007; García Muñoz 2007; Schneiter et al. 2007; Penz et al. 2008; Holmstrom et al. 2008; Murray-Clay et al. 2009; Ekenback et al. 2010; Stone and Proga 2009; Guo 2011; Trammell et al. 2011; Shaikhislamov et al. 2014; Kislyakova et al. 2014b), Koskinen et al. (2013b) developed probably the most comprehensive model to date for the thermosphere of hot Jupiter HD 209458b. The model considered photochemistry of H, He, C, N, O, Si, and their ions, and computes the stellar heating rates based on previous estimates of photoelectron heating efficiencies (the fraction of photoelectron energy that heats the thermosphere, not the overall heating efficiency). The globally averaged mass loss rate for HD 209458b in Koskinen et al. (2013b) is close to those in Yelle (2004), Yelle et al. (2008), Tian et al. (2005b), and García Muñoz (2007), but is more than 6 times greater than the value in Murray-Clay et al. (2009), which could be due to the gray assumption and a factor of 4 lower total XUV energy used in the later. Koskinen et al. (2013b) showed that the velocity and temperature differences between different species in the thermosphere of HD 209458b are small—separation of species near the homopause level is not efficient in a hydrodynamic upper atmosphere and the single fluid approximation of the momentum and energy equations is valid. Koskinen et al. (2013b) also showed that (1) H and O remain neutral up to about $3R_p$ (planetary radii), explaining the—observed transit depths in the H Lyman a and O lines (Vidal-Madjar et al. 2004, 2003); (2) C is ionized to C^+ at low altitude, explaining the observation of C^+ (Vidal-Madjar et al. 2004); (3) Si^+ is the dominant silicon species at heights less than $5R_p$ and Si^{2+} can be formed through charge exchange with H^+, explaining the observation of Si ions (Linsky et al. 2010).

Bourrier et al. have carried out a series of interesting observations and modeling of the highly extended upper atmospheres of HD 189733b and HD 209458b (Bourrier et al. 2013, 2014; Bourrier and Lecavelier des Etangs 2013). Most recently they concluded (Bourrier et al. 2015) that the Mg line at 2853 Å could provide observation opportunities to observe planets under severe mass change. Kislyakova et al. (2014a) adjusted their stellar wind upper atmosphere models in order to explain the Lyman-α transit observations of HD 209458b and found that the observations can be reproduced due to a complex interplay of the host star's radiation pressure, the hydrodynamically outward flowing upper atmosphere and the production of energetic hydrogen atoms via charge exchange around a magnetic obstacle that corresponds to a magnetic moment of the exoplanet that is about 10 % compared to that of Jupiter in the solar system.

Kislyakova et al. (2014b) applied a DSMC model on top of the upper atmosphere structures calculated in Erkaev et al. (2013) and Lammer et al. (2013) to calculate the hydrogen escape through stellar wind interaction from Kepler-11b-f. They found that the stellar wind erosion of the expanded hydrogen cloud around these super Earths enhances the total escape rates by only a few percent. However, the DSMC model suggests that high radiation pressure and frequent charge exchange between planetary wind and stellar wind result in highly asymmetric spatial distribution of the hydrogen cloud around these Kepler planets. Bisikalo et al. (2013a,b) studied the gas-dynamic interaction between the stellar wind and neutral planetary corona.

Usually the orbital velocities of hot Jupiters are supersonic relative to the stellar wind, resulting in the formation of a bow shock. Gas-dynamical modeling shows that the gaseous envelopes around hot Jupiters can belong to two classes, depending on the position of the collision point. If the collision point is inside the Roche lobe of the planet, the envelopes have almost the spherical shapes of classical atmospheres, slightly distorted by the influence of the star and interactions with the stellar-wind gas. Firstly, if the collision point is located outside the Roche lobe, outflows from the vicinity of the Lagrangian points L_1 and L_2 arise, and the envelope becomes substantially asymmetrical. The latter class of objects can also be divided into two types. If the dynamical pressure of the stellar-wind gas is high enough to stop the most powerful outflow from the vicinity of the inner Lagrangian point L_1, a closed quasi-spherical envelope with a complex shape forms in the system. Secondly, if the wind is unable to stop the outflow from L_1, an open non-spherical envelope forms. Using the typical hot Jupiter HD 209458b as an example, it was shown (Bisikalo et al. 2013a,b) that all three types of atmospheres could exist within the range of estimated parameters of this planet. Since different types of envelopes have different transit light curves, determining the type of envelope in HD 209458b could impose additional constraints on the parameters of this exoplanet.

Khodachenko et al. (2012) pointed out that hydrodynamic planetary atmospheres driven by stellar XUV radiation must necessarily lead to the formation of a plasmasphere much different from those around solar system planets. In this case, the expanding planetary plasma (planetary wind) interacts with the stellar winds and the intrinsic planetary magnetic fields, leading to the development of extended magnetodisks and the magnetic field produced by magnetodisk ring currents could dominate that of an intrinsic dipole magnetic field. As a result, the magnetospheres of planets with a hydrodynamic upper atmosphere could be more than 40 larger than those only considering a planetary dipole magnetic field. This point has important implications for estimating non-thermal escape due to erosion by the stellar wind.

Cohen and Glocer (2012) estimated that the polar wind effect, ions flowing out along open magnetic field lines, could increase the atmospheric escape rate from hot Jupiters by more than 10 times.

Tanaka et al. (2014) investigated mass loss driven by magnetohydrodynamic (MHD) waves in a 1D flux tube model and found mass loss rates similar to those inferred from observations. Valsecchi et al. (2014) studied how Roche lobe overflow could have reduced the mass of a hot Jupiter near the Roche limit of its orbit. Their model calculation results can produce small planets with masses ranging from that of the Earth to Neptune. Fortney et al. (2013) suggested that the atmospheres of sub-Neptune and Neptune-class exoplanets may experience dramatic increase in metallicity with time because of hydrogen escape.

Guo (2013) presented a two dimensional multi-fluid (H and H^+) hydrodynamic model with two-dimensional radiative transfer. They found that the planetary winds demonstrate significant horizontal movements and lower density and radial velocity are present near the polar regions. As a result, atmospheric escape is dominated by outflow from low-latitude regions. Guo (2013) also discussed the influence of two-dimensional radiative transfer on mass loss rates. They found that the 2D hydrodynamics together with the 2D radiative transfer model produced mass loss rates 7 times smaller than that of their 1D model due to a self-shielding effect (the night side upper atmosphere being shielded by the dayside and thus does not receive stellar XUV radiation). The Guo (2013) model also suggests the potential importance of multi-dimensionality in future planetary thermosphere modeling.

Volkov and Johnson (2013) studied the 1D single component hydrodynamic escape problem with a DSMC model. Since fluid models should not be used in collisionless regions, a combination of a fluid model, which requires collisions to be frequent enough, below the exobase level and a kinetic model, which does not assume frequent collisions, above this level is more physically consistent. Villarreal D'Angelo et al. (2014) investigated the dynamic response of close-in exoplanets to different stellar wind conditions and found that the mass loss rate from HD 209458b is not sensitive to the stellar wind conditions in comparison with the mass loss rates in the literature.

4.4 Evolution Models

Using a planet thermal evolution model, which studies how a giant planetary mass and radius evolve in time due to the loss of its internal heat, in combination with the energy limited formula discussed in Sect. 2 and a constant efficiency factor for atmospheric escape, Lopez et al. (2012) found that observed exoplanets could be characterized by a threshold with two controlling factors: (1) the product of planetary mass and mean density, and (2) the incident stellar bolometric flux at the top of the planetary atmosphere. They argue that planets above this threshold would not be able to maintain their hydrogen envelope because of efficient atmospheric escape driven by stellar XUV. The efficiency factor, as a free parameter in Lopez et al. (2012), is ~ 10 for all low mass low density exoplanets explored.

Lopez and Fortney (2013) further investigated the influence of the efficiency factor, total planet mass, and planet's core mass on atmospheric escape with similar methods as in Lopez et al. (2012) and suggested that the planet's core mass, not its total mass, and the efficiency factor are most important in controlling atmospheric escape of observed transiting exoplanets with masses less than 100 Earth mass. The Lopez and Fortney (2013) model can explain well the orders of density and size in multi-planet systems (Ciardi et al. 2013; Wu and Lithwick 2013). Furthermore Lopez and Fortney (2013) predicted that 1.8–4.0 Earth radii planets should be uncommon on orbits with period less than 10 days and proposed that there could be a narrow valley in the incident flux vs. planet size (between 1 and 4 Earth

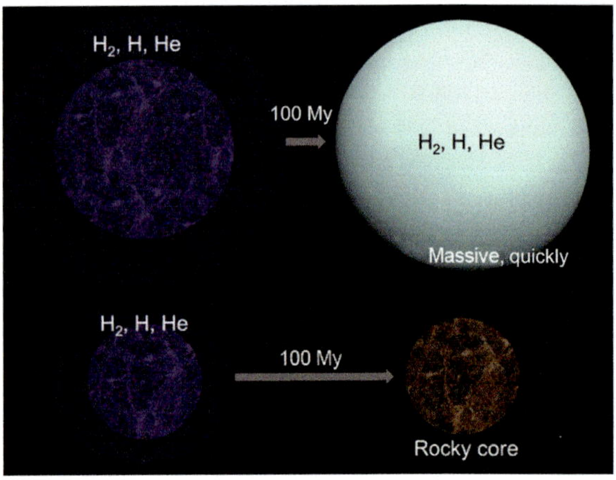

Fig. 11 *Upper panel*: a super-Earth mass core that accreted in the nebula and captured a dense H_2/He envelope may not completely lose its envelope after the nebula dissipated. As a result the planet evolves to a hydrogen-dominated mini-Neptune. *Lower panel*: a rocky core with lower mass may not accumulate enough gas envelope during the formation. Thus it may lose all of its captured H_2/He envelope and evolve to a "terrestrial" planet (Lammer et al. 2014; Luger et al. 2015)

radii) diagram where planet occurrence rate should be at a minimum. This is consistent with the physically intuitive concept that larger planets can better keep their hydrogen envelopes and smaller planets have stronger tendency to lose their hydrogen envelopes under certain stellar bolometric flux. Lopez and Fortney (2013) also proposed several uncertainty factors which could potentially smear the "occurrence valley".

Kurokawa and Nakamoto (2014) modeled the evolution of hot Jupiters around solar mass stars by considering atmospheric escape and thermal contraction. Under lower XUV conditions, they used the energy-limited escape formula of Lopez et al. (2012); under higher XUV conditions, a semianalytical model is applied following the analytical approach of Murray-Clay et al. (2009). They found that atmospheric escape leads to a dichotomy of populations: hot Jupiters keeping their envelopes and super Earths whose envelopes are completely lost, consistent with the findings of Lopez et al. (2012) and Lopez and Fortney (2013).

Kurosaki et al. (2014) studied the loss of water of hot super- and sub-Earths around solar mass stars by applying the energy limited escape formula and found that the threshold mass below which the planet has its water envelope stripped off in 1–10 Gyr is 10 times smaller than that for H and He envelope models by Lopez and Fortney (2013). In particular planets with masses less than 3 Earth masses orbiting at 0.03 AU could lose all of their oceans in 1–10 Gyrs; while water-rich planets with radii ~ 1.5 Earth radii should be common because they can keep their water envelopes against atmospheric escape.

Lammer et al. (2014) investigated how atmospheric escape influences the hydrogen envelopes of rocky planets from 0.1 to 5 Earth masses at 1 AU around solar mass stars. The model includes gravitational attraction and accumulation of nebula gas around a planet's core before the dissipation of the disk and the escape of these accumulated hydrogen-dominated protoatmospheres. A hydrodynamic upper atmosphere model is used to calculate the loss rates induced by stellar XUV radiation. As illustrated in Fig. 11, sub-Earths and Earth-mass planets can lose their captured hydrogen envelopes during the first 100 Myrs after the disc dissipated while more massive super-Earths are able to keep most of the captured hydrogen envelopes (Lammer et al. 2014). They concluded that habitable planets around solar mass stars should have core masses around 1 ± 0.5 Earth masses (corresponding to $0.8 \sim 1.15$ Earth radii). Recently, Luger et al. (2015) studied the atmospheric mass loss related transformation of hydrogen-dominated mini-Neptunes into super-Earth cores inside the habitable zone of M dwarfs and obtained similar results as Lammer et al. (2014) for similar bodies in the habitable zone of G stars. However, it is important to note that the probability that rocky cores which experienced such transition to be able to maintain an Earth-like ocean is very small as shown by Tian and Ida (2015).

A most recent development on protoatmospheric escape concerns the effect of early luminosity evolution of M dwarfs. Since the stellar luminosity of M dwarfs decreases by more than one order of magnitude during their pre-main-sequence (PMS) phase, planets in the habitable zones of main sequence M dwarfs were much hotter during the PMS phase and thus could have become dehydrated (Ramirez and Kaltenegger 2014; Tian and Ida 2015; Luger et al. 2015). The result of this escape could lead to the formation of oxygen-rich atmospheres provided that the reaction between atmospheric oxygen and surface materials is slow (Luger et al. 2015). Tian (2015) showed that the same process could potentially lead to the loss of the dense O_2 atmospheres on low mass exoplanets. In addition the reactions between magma ocean and atmospheric oxygen could become efficient enough to remove dense O_2 atmospheres.

In comparison to the simple parameterized treatment of atmospheric escape (Lopez et al. 2012; Lopez and Fortney 2013; Kurokawa and Nakamoto 2014; Jin et al. 2014; Ramirez and Kaltenegger 2014; Tian and Ida 2015; Luger et al. 2015), Owen and Wu (2013) took a more sophisticated approach. They generated a table of atmospheric escape rates as functions of ionizing flux, planetary mass, and planetary radius based on calculations using the previously developed model of Owen and Jackson (2012), and considered atmospheric escape using the general purpose MESA (Modules FOR Experiments In Stellar Astrophysics) stellar evolution code to simulate planetary structure and evolution. Thus no direct assumption on escape or heating efficiency is made in Owen and Wu (2013).

However, the Owen and Jackson (2012) model uses a temperature-ionization parameter relationship taken from Owen et al. (2010), which was originally used in calculating the photo-evaporation of discs around solar-type stars, to compute the gas temperature from the local number density of planetary atmospheres and incident X-ray for the region where heating by X-rays is important. In the region where EUV heating is important, Owen and Jackson (2012) argues that the gas should thermostat to 10^4 K in radiative equilibrium and an isothermal equation of state is employed. The ionization parameter in Owen and Jackson (2012) model includes a parameterized metallicity scaling law also originally developed for protoplanetary disc. Owen and Jackson (2012) concluded that the main driving force of atmospheric escape in hydrogen dominated planetary upper atmosphere should be stellar X-ray, and suggested that stellar EUV radiation's contribution to atmospheric escape is less than 10 %. Note that Owen and Jackson (2012) derived an escape efficiency between 5 % and 20 %, which is the source of the 10 efficiency value used in the studies using simple parameterization.

The main conclusions of Owen and Wu (2013) are that atmospheres largely determine the observed size distribution of close-in Kepler planets. Similar to Owen and Jackson (2012) and Owen and Wu (2013), Erkaev et al. (2013) and Lammer et al. (2013) developed a hydrodynamic model to investigate atmospheric escape from hydrogen dominant planetary atmospheres. Their models use constant heating efficiencies, the conversion of absorbed stellar XUV photon energy to heat, of 15 % and 60 %, and employed a two dimensional radiative transfer scheme similar to that in Tian et al. (2005a,b). Lammer et al. (2013) applied this model to transiting super-Earths with inflated radii and known masses (Kepler-11b-f, GJ 1214b and 55 Cnc e) and found that all of them except Kepler-11c can reach the blow-off state because their thermospheres reach the Roche lobes. Note that the mass loss rates predicted in Lammer et al. (2013) are one to two orders of magnitude lower than that of hot Jupiters HD 209458b and XUV heating is the main driving force for atmospheric escape in Erkaev et al. (2013) and Lammer et al. (2013).

Regarding whether an energy limited formula can be used to estimate atmospheric escape rates from hydrodynamic planetary atmospheres, Erkaev et al. (2013) and Lammer et al. (2013) showed that, by comparing the escape rates obtained by the hydrodynamic upper atmosphere model with the energy-limited equation, the overestimation of escape rates can be as small as a factor of 2 for some Kepler super Earths, which is well within parametrical uncertainties. However, application of the energy-limited equation for cooler exoplanetary atmospheres can overestimate the escape rate by one or two orders of magnitude in comparison with the results of the hydrodynamical upper atmosphere model of Chadney et al. (2015). Moreover, the energy-limited formula may also overestimate the atmospheric loss rate if one assumes a hydrogen-dominated atmosphere but in reality heavier atoms or molecules are also present. Thus again caution is required when using the energy limited formula.

Despite the debate on what the correct escape efficiency should be for protoatmospheres, the suggestion that there is a threshold of stellar XUV irradiance beyond which hydrogen envelopes of low mass exoplanets are unstable should be robust for the current level of details of physics in the models.

4.5 Summary

Future modeling of protoatmospheres of rocky planets should include fully coupled photochemistry and radiative transfer. Two or three dimensional hydrodynamical models will be important not only to better calculate atmospheric escape rates, but also to be compared against observations. Since fluid models should not be used in collisionless regions, a combination of fluid models below the exobase level and kinetic models above this level is more physically self-consistent. In addition the effect of planetary magnetic field, the influences of stellar gravity and magnetic field are all important processes for close-in exoplanets.

With more exoplanet detection missions (TESS, CHEOPS, PLATO 2.0, etc.) and facilities (NGTS, etc.) planned and about to be put in action, hundreds if not thousands of low mass exoplanets will be available for future facilities to characterize. This will start a new age of study for protoatmospheric escape by providing unprecedented opportunities to compare numerical models with direct observations. The research field of atmospheric escape has already been invigorated by exoplanet discoveries. It can be expected that with the new wave of exoplanet discoveries, atmospheric escape research will soon enter an even more exciting period.

5 Conclusion and Perspectives

The formation and evolution of protoatmospheres depend on different factors that have been reviewed in this chapter. The first section concerns the initial conditions given by disk gas accretion, such as, the timing between dissipation and accretion of disk gas and thermodynamic conditions. The structure and mass of the protoatmosphere depend on the mass of the solid protoplanet, the energy supply rate and opacity, the temperature and density of the ambient disk gas, and the composition (e.g., mean molecular weight) of the atmosphere. Depending on the position within the disk and the timing of formation, a planet of the size of Earth is thought to be able to capture an atmosphere on the order of 0.3 to 1000 water oceans. Temperature at a given location in the disk gas depends on accretion energy and thus, should also be linked to the temperature conditions at the surface at an early stage in the planet's history. Since there is also an interplay between the surface and atmospheric conditions at the end of accretion, and geodynamic regimes influencing heat flux from the planet's interior, the knowledge of surface conditions and volatile composition, is thus an important constraint to determine the future planetary evolution and this was the objective of the second section. Recent coupled atmospheric/interior models showed that two regimes exist depending on the distance from the star. For so called type II planets closer to the star than a critical distance, the surface might be covered by a magma ocean. For type I planets, farther from the star, the surface would either be covered by a water ocean or an ice layer. These first conclusions depend also on the modeling of the process of gas escape from protoatmospheres and multiple impacts during accretion will also change initial stage of the atmosphere and solid surface of a growing planets (see Sect. 4). Below, we summarize some current research directions and still open questions:

- What causes the diversity in the hydrogen content of planets?
- How much volatiles can be trapped in the planet interior?
- Required is to develop fully coupled models of photochemistry, radiative transfer and escape processes.

Aside from progressing in the knowledge of: thermal evolution of atmospheres, escape processes, convective models of planetary interiors, disk gas evolution and impact scenario, there is a need for more simple coupled models which take advantage of advances from all these fields. Applications of such models will be able to be tested in the forthcoming, promising period of further exoplanets discoveries.

Acknowledgements The authors greatly acknowledge both of the reviewers for their careful reading of our manuscript and valuable suggestions. H. Massol, P. Sarda, A. Davaille, E. Chassefière, E. Marcq and F. Leblanc are supported by the 2016 PNP program of INSU-CNRS. F. Tian is supported by the National Natural Science Foundation of China (41175039), the Startup Fund of the Ministry of Education of China, and the Tsinghua University Initiative Science Research Program (523001028). V. Shematovich acknowledges the support by the Russian Science Foundation Project No. 14-12-01048. H. Lammer acknowledges support by the FWF NFN project S11601-N16 'Pathways to Habitability: From Disks to Active Stars, Planets and Life', and the related FWF NFN subproject, S11607-N16 'Particle/Radiative Interactions with Upper Atmospheres of Planetary Bodies Under Extreme Stellar Conditions', as well as the FWF project P27256-N27 'Characterizing Stellar and Exoplanetary Environments via Modeling of Lyman-α Transit Observations of Hot Jupiters'. K. Hamano and Y. Abe are supported by a Grant-in-Aid for Scientific Research on Innovative Areas from the Ministry of Education, Culture, Sports, Science and Technology (MEXT) (No. 23103003) and K. Hamano by a Grant-in-Aid for Young Scientists (B) from Japan Society for the Promotion of Science (JSPS) (No. 26800242). Y. Hori is supported by Grant-in-Aid for Scientific Research on Innovative Areas (No. 26103711) from MEXT. M. Ikoma is supported by Grants-in-Aid for Scientific Research on Innovative Areas (No. 23103005) and Scientific Research (C) (No. 25400224) from the Ministry of Education, Culture, Sports, Science and Technology (MEXT) of Japan. Finally the authors thank the International Space Science Institute (ISSI) and ISSI-Beijing in Bern and Beijing.

References

Y. Abe, Physical state of the very early Earth. Lithos **30**(3), 223–235 (1993), the evolving Earth

Y. Abe, Thermal and chemical evolution of the terrestrial magma ocean. Phys. Earth Planet. Inter. **100**(1-4), 27–39 (1997)

Y. Abe, Protoatmospheres and surface environment of protoplanets. Earth Moon Planets **108**(1), 9–14 (2011)

Y. Abe, T. Matsui, The formation of an impact-generated H_2O atmosphere and its implications for the early thermal history of the Earth. J. Geophys. Res., Solid Earth **90**(S02), C545–C559 (1985)

Y. Abe, T. Matsui, Early evolution of the Earth: accretion, atmosphere formation, and thermal history. J. Geophys. Res., Solid Earth **91**(B13), E291–E302 (1986)

Y. Abe, T. Matsui, Evolution of an impact-generated H_2O-CO_2 atmosphere and formation of a hot protoocean on Earth. J. Atmos. Sci. **45**(21), 3081–3101 (1988)

Y. Abe, E. Ohtani, T. Okuchi, K. Righter, M. Drake, *Water in the Early Earth* (University of Arizona Press, Tucson, 2000), pp. 413–433

Y. Abe, A. Abe-Ouchi, N.H. Sleep, K.J. Zahnle, Habitable zone limits for dry planets. Astrobiology **11**, 443–460 (2011)

C.B. Agnor, R.M. Canup, H.F. Levison, On the character and consequences of large impacts in the late stage of terrestrial planet formation. Icarus **142**, 219–237 (1999)

Y. Alibert, F. Carron, A. Fortier, S. Pfyffer, W. Benz, C. Mordasini, D. Swoboda, Theoretical models of planetary system formation: mass vs. semi-major axis. Astron. Astrophys. **558**, A109 (2013)

C. Allègre, T. Staudacher, P. Sarda, Rare gas systematics: formation of the atmosphere, evolution and structure of the Earth's mantle. Earth Planet. Sci. Lett. **81**(2–3), 127–150 (1987)

C.J. Allègre, G. Manhès, C. Göpel, The major differentiation of the Earth at 4.45 Ga. Earth Planet. Sci. Lett. **267**, 386–398 (2008)

D. Andrault, N. Bolfan-Casanova, G.L. Nigro, M.A. Bouhifd, G. Garbarino, M. Mezouar, Solidus and liquidus profiles of chondritic mantle: implication for melting of the Earth across its history. Earth Planet. Sci. Lett. **304**(1–2), 251–259 (2011)

G.E. Ballester, D.K. Sing, F. Herbert, The signature of hot hydrogen in the atmosphere of the extrasolar planet HD 209458b. Nature **445**(7127), 511–514 (2007)

I. Baraffe, F. Selsis, G. Chabrier, T.S. Barman, F. Allard, P.H. Hauschildt, H. Lammer, The effect of evaporation on the evolution of close-in giant planets. Astron. Astrophys. **419**(2), L13–L16 (2004)

N.M. Batalha, Exploring exoplanet populations with NASA's Kepler mission. Proc. Natl. Acad. Sci. USA **111**, 12647–12654 (2014)

L. Ben-Jaffel, S. Hosseini, On the existence of energetic atoms in the upper atmosphere of exoplanet HD 209458b. Astrophys. J. **709**(2), 1284 (2010)

K. Berlo, J. Gardner, J. Blundy, Timescales of magma degassing, in *Timescales of Magmatic Processes: From Core to Atmosphere* (Wiley-Blackwell, Hoboken, 2011), Ch. 11

D.V. Bisikalo, P.V. Kaigorodov, D.E. Ionov, V.I. Shematovich, Types of gaseous envelopes of "hot Jupiter" exoplanets. Astron. Rep. **57**(10), 715–725 (2013a)

D. Bisikalo, P. Kaygorodov, D. Ionov, V. Shematovich, H. Lammer, L. Fossati, Three-dimensional gas dynamic simulation of the interaction between the exoplanet wasp-12b and its host star. Astrophys. J. **764**(1), 19 (2013b)

P. Bodenheimer, J.J. Lissauer, Accretion and evolution of $\sim$2.5 $M_\oplus$ planets with voluminous H/He envelopes. Astrophys. J. **791**, 103 (2014)

P. Bodenheimer, J.B. Pollack, Calculations of the accretion and evolution of giant planets. The effects of solid cores. Icarus **67**, 391–408 (1986)

V. Bourrier, A. Lecavelier des Etangs, 3d model of hydrogen atmospheric escape from HD 209458b and HD 189733b: radiative blow-out and stellar wind interactions. Astron. Astrophys. **557**, A124 (2013)

V. Bourrier, A. Lecavelier des Etangs, H. Dupuy, D. Ehrenreich, A. Vidal-Madjar, G. Hébrard, G.E. Ballester, J.-M. Désert, R. Ferlet, D.K. Sing, P.J. Wheatley, Atmospheric escape from HD 189733b observed in hi Lyman α: detailed analysis of HST/STIS September 2011 observations. Astron. Astrophys. **551**, A63 (2013)

V. Bourrier, A. Lecavelier des Etangs, A. Vidal-Madjar, Modeling magnesium escape from HD 209458b atmosphere. Astron. Astrophys. **565**, A105 (2014)

V. Bourrier, A. Lecavelier des Etangs, A. Vidal-Madjar, The Mg i line: a new probe of the atmospheres of evaporating exoplanets. Astron. Astrophys. **573**, A11 (2015)

R. Brasser, The formation of Mars: building blocks and accretion time scale. Space Sci. Rev. **174**, 11–25 (2013)

E.L. Brown, C.E. Lesher, North Atlantic magmatism controlled by temperature, mantle composition and buoyancy. Nat. Geosci. **7**(11), 820–824 (2014)

H.P. Brown, A.J. Panshin, C.C. Forsaith, *Textbook of Wood Technology* (1949)

A. Burgisser, M. Alletti, B. Scaillet, Simulating the behavior of volatiles belonging to the C–O–H–S system in silicate melts under magmatic conditions with the software d-compress. Comput. Geosci. **79**, 1–14 (2015)

R.M. Canup, Dynamics of lunar formation. Annu. Rev. Astron. Astrophys. **42**, 441–475 (2004a)

R.M. Canup, Simulations of a late lunar-forming impact. Icarus **168**(2), 433–456 (2004b)

M.R. Carroll, J.R. Holloway, *Volatiles in Magmas*. Reviews in Mineralogy, vol. 30 (Mineralogical Society of America, Washington, 1994)

C. Cecchi-Pestellini, A. Ciaravella, G. Micela, T. Penz, The relative role of EUV radiation and x-rays in the heating of hydrogen-rich exoplanet atmospheres. Astron. Astrophys. **496**(3), 863–868 (2009)

J. Chadney, M. Galand, Y. Unruh, T.T. Koskinen, J. Sanz-Forcada, XUV-driven mass loss from extrasolar giant planets orbiting active stars. Icarus **250**, 357–367 (2015)

J.W. Chamberlain, Planetary coronae and atmospheric evaporation. Planet. Space Sci. **11**(8), 901–960 (1963)

E. Chassefière, Hydrodynamic escape of hydrogen from a hot water-rich atmosphere: the case of Venus. J. Geophys. Res., Planets **101**(E11), 26039–26056 (1996a)
E. Chassefière, Hydrodynamic escape of oxygen from primitive atmospheres: applications to the cases of Venus and Mars. Icarus **124**, 537–552 (1996b)
E. Chassefière, F. Leblanc, B. Langlais, The combined effects of escape and magnetic field histories at Mars. Planet. Space Sci. **55**(3), 343–357 (2007)
J.Y. Chaufray, R. Modolo, F. Leblanc, G. Chanteur, R.E. Johnson, J.G. Luhmann, Mars solar wind interaction: formation of the martian corona and atmospheric loss to space. J. Geophys. Res., Planets **112**(E9), E09009, e09009 (2007)
D.R. Ciardi, D.C. Fabrycky, E.B. Ford, T.N. Gautier III., S.B. Howell, J.J. Lissauer, D. Ragozzine, J.F. Rowe, On the relative sizes of planets within Kepler multiple-candidate systems. Astrophys. J. **763**(1), 41 (2013)
W. Clarke, M. Beg, H. Craig, Excess ^{3}He in the sea: evidence for terrestrial primodal helium. Earth Planet. Sci. Lett. **6**(3), 213–220 (1969)
O. Cohen, A. Glocer, Ambipolar electric field, photoelectrons, and their role in atmospheric escape from hot Jupiters. Astrophys. J. Lett. **753**(1), L4 (2012)
N.R. Council, *The Limits of Organic Life in Planetary Systems* (The National Academies Press, Washington, 2007)
H. Craig, J. Lupton, Helium 3 and mantle volatiles in the ocean and the oceanic crust, in *The Oceanic Lithosphere*, vol. 7 (Wiley, New York, 1981)
E. Crawford, Arrhenius' 1896 model of the greenhouse effect in context, in *Ambio*, Stockholm, vol. 26 (1997), pp. 6–11
N. Dauphas, A. Pourmand, Hf-W-Th evidence for rapid growth of Mars and its status as a planetary embryo. Nature **473**, 489–492 (2011)
A. Davaille, A. Limare, Laboratory studies of mantle convection, in *Treatise on Geophysics (Second Edition). Vol. "Mantle Dynamics"*, ed. by G. Schubert, D. Bercovici (Elsevier, Amsterdam, 2015), pp. 73–144
T.A. Davis, P.J. Wheatley, Evidence for a lost population of close-in exoplanets. Mon. Not. R. Astron. Soc. **396**(2), 1012–1017 (2009)
A. Lecavelier des Etangs, A diagram to determine the evaporation status of extrasolar planets. Astron. Astrophys. **461**(3), 1185–1193 (2007)
A. Lecavelier des Etangs, A. Vidal-Madjar, J.C. McConnell, G. Hébrard, Atmospheric escape from hot Jupiters. Astron. Astrophys. **418**(1), L1–L4 (2004)
K.-M. Dittkrist, C. Mordasini, H. Klahr, Y. Alibert, T. Henning, Impacts of planet migration models on planetary populations. Effects of saturation, cooling and stellar irradiation. Astron. Astrophys. **567**, A121 (2014)
A. Ekenback, M. Holmstrom, M. Wurz et al., Energetic neutral atoms around HD 209458b: estimations of magnetospheric properties. Astrophys. J. **709**, 670–679 (2010)
L.T. Elkins-Tanton, Linked magma ocean solidification and atmospheric growth for Earth and Mars. Earth Planet. Sci. Lett. **271**(1), 181–191 (2008)
L.T. Elkins-Tanton, Magma oceans in the inner solar system, in *Annual Review of Earth and Planetary Sciences*, ed. by R. Jeanloz. Annual Review of Earth and Planetary Sciences, vol. 40 (2012), pp. 113–139
L. Elkins-Tanton, E. Parmentier, P. Hess, Magma ocean fractional crystallization and cumulate overturn in terrestrial planets: implications for Mars. Meteorit. Planet. Sci. **38**(12), 1753–1771 (2003)
L. Elkins-Tanton, S. Zaranek, E. Parmentier, P. Hess, Early magnetic field and magmatic activity on Mars from magma ocean cumulate overturn. Earth Planet. Sci. Lett. **236**(1-2), 1–12 (2005)
N.V. Erkaev, Y.N. Kulikov, H. Lammer, F. Selsis, D. Langmayr, G.F. Jaritz, H.K. Biernat, Roche lobe effects on the atmospheric loss from "hot Jupiters". Astron. Astrophys. **472**(1), 329–334 (2007)
N.V. Erkaev, H. Lammer, P. Odert, Y.N. Kulikov, K.G. Kislyakova et al., XUV-exposed, non-hydrostatic hydrogen-rich upper atmospheres of terrestrial planets. Part i: Atmospheric expansion and thermal escape. Astrobiology **13**(11), 1011–1029 (2013)
N.V. Erkaev, H. Lammer, L.T. Elkins-Tanton, A. Stökl, P. Odert, E. Marcq, E.A. Dorfi, K.G. Kislyakova, Y.N. Kulikov, M. Leitzinger, M. Güdel, Escape of the martian protoatmosphere and initial water inventory. Planet. Space Sci. **98**, 106–119 (2014)
N.V. Erkaev, H. Lammer, P. Odert, Y.N. Kulikov, K.G. Kislyakova, Extreme hydrodynamic atmospheric loss near the critical thermal escape regime. Mon. Not. R. Astron. Soc. **448**(2), 1916–1921 (2015)
F.P. Fanale, A case for catastrophic early degassing of the Earth. Chem. Geol. **8**(2), 79–105 (1971)
G. Fiquet, A.L. Auzende, J. Siebert, A. Corgne, H. Bureau, H. Ozawa, G. Garbarino, Melting of peridotite to 140 gigapascals. Science **329**(5998), 1516–1518 (2010)
D. Fisher, Trapped helium and argon and the formation of the atmosphere by degassing. Nature **256**(5513), 113–114 (1975)

J.J. Fortney, N. Nettelmann, The interior structure, composition, and evolution of giant planets. Space Sci. Rev. **152**, 423–447 (2010)

J. Fortney, C. Mordasini, N. Nettelmann, E.M.-R. Kempton, T.P. Greene, K. Zahnle, A framework for characterizing the atmospheres of low-mass low-density transiting planets. Astrophys. J. **775**(1), 80 (2013)

J.L. Fox, A.B. Hać, Photochemical escape of oxygen from Mars: a comparison of the exobase approximation to a Monte Carlo method. Icarus **204**(2), 527–544 (2009)

J. Fox, M. Galand, R. Johnson, Energy deposition in planetary atmospheres by charged particles and solar photons. Space Sci. Rev. **139**(1–4), 3–62 (2008)

N. Fujii, S. Uyeda, Conditions for a once molten Earth to cool. J. Phys. Earth **14**(1), 15–26 (1966)

A. García Muñoz, Physical and chemical aeronomy of HD 209458b. Planet. Space Sci. **55**(10), 1426–1455 (2007)

H. Genda, Y. Abe, Survival of a proto-atmosphere through the stage of giant impacts: the mechanical aspects. Icarus **164**, 149–162 (2003)

H. Genda, Y. Abe, Enhanced atmospheric loss on protoplanets at the giant impact phase in the presence of oceans. Nature **433**, 842–844 (2005)

G.J. Golabek, T. Keller, T.V. Gerya, G. Zhu, P.J. Tackley, J.A.D. Connolly, Origin of the martian dichotomy and Tharsis from a giant impact causing massive magmatism. Icarus **215**(1), 346–357 (2011)

C. Goldblatt, T.D. Robinson, K.J. Zahnle, D. Crisp, Low simulated radiation limit for runaway greenhouse climates. Nat. Geosci. **6**, 661–667 (2013)

D. Grinspoon, Implications of the high D/H ratio for the sources of water in Venus' atmosphere. Nature **363**, 428–431 (1993)

H. Gröller, V.I. Shematovich, H.I.M. Lichtenegger, H. Lammer, M. Pfleger, Y.N. Kulikov, W. Macher, U.V. Amerstorfer, H.K. Biernat, Venus' atomic hot oxygen environment. J. Geophys. Res., Planets **115**(E12), e12017 (2010)

H. Gröller, H. Lichtenegger, H. Lammer, V.I. Shematovich, Hot oxygen and carbon escape from the martian atmosphere. Planet. Space Sci. **98**, 93–105 (2014)

S. Grossmann, D. Lohse, Scaling in thermal convection: a unifying theory. J. Fluid Mech. **407**, 27–56 (2000)

M. Güdel, R. Dvorak, N. Erkaev, J. Kasting, M. Khodachenko, H. Lammer, E. Pilat-Lohinger, H. Rauer, I. Ribas, B.E. Wood, Astrophysical conditions for planetary habitability, in *Protostars and Planets VI* (2014), pp. 883–906

B. Guillot, P. Sarda, The effect of compression on noble gas solubility in silicate melts and consequences for degassing at mid-ocean ridges. Geochim. Cosmochim. Acta **70**(5), 1215–1230 (2006)

J.H. Guo, Escaping particle fluxes in the atmospheres of close-in exoplanets. I. Model of hydrogen. Astrophys. J. **733**(2), 98 (2011)

J.H. Guo, Escaping particle fluxes in the atmospheres of close-in exoplanets. II. Reduced mass-loss rates and anisotropic winds. Astrophys. J. **766**(2), 102 (2013)

K.E. Haisch Jr., E.A. Lada, C.J. Lada, Disk frequencies and lifetimes in young clusters. Astrophys. J. **553**, L153–L156 (2001)

A.N. Halliday, Mixing, volatile loss and compositional change during impact-driven accretion of the Earth. Nature **427**, 505–509 (2004)

Y. Hamano, M. Ozima, Earth atmosphere evolution model based on Ar isotopic data, in *Terrestrial Rare Gases* (Japan Sci. Soc., Tokyo, 1978), pp. 155–171

K. Hamano, Y. Abe, H. Genda, Emergence of two types of terrestrial planet on solidification of magma ocean. Nature **497**(7451), 607–610 (2013)

K. Hamano, H. Kawahara, Y. Abe, M. Onishi, G. Hashimoto, Lifetime and spectral evolution of a magma ocean with a steam atmosphere: its detectability by future direct imaging. Astrophys. J. **806**(2), 216 (17 pp.) (2015)

C. Hayashi, Structure of the solar nebula, growth and decay of magnetic fields and effects of magnetic and turbulent viscosities on the nebula. Prog. Theor. Phys. Suppl. **70**, 35–53 (1981)

C. Hayashi, K. Nakazawa, H. Mizuno, Earth's melting due to the blanketing effect of the primordial dense atmosphere. Earth Planet. Sci. Lett. **43**, 22–28 (1979)

C. Hayashi, K. Nakazawa, Y. Nakagawa, Formation of the solar system, in *Protostars and Planets II*, ed. by D.C. Black, M.S. Matthews (1985), pp. 1100–1153

P. Hess, E. Parmentier, A model for the thermal and chemical evolution of the Moon's interior: implications for the onset of mare volcanism. Earth Planet. Sci. Lett. **134**(3-4), 501–514 (1995)

L.A. Hillenbrand, Observational constraints on dust disk lifetimes: implications for planet formation (2005). ArXiv Astrophysics e-prints

T. Höink, J. Schmalzl, U. Hansen, Formation of compositional structures by sedimentation in vigorous convection. Phys. Earth Planet. Inter. **153**, 1–3 (2005)

M. Holmstrom, A. Ekenback, F. Selsis, T. Penz, H. Lammer, P. Wurz, Energetic neutral atoms as the explanation for the high-velocity hydrogen around HD 209458b. Nature **451**(7181), 970–972 (2008)

Y. Hori, M. Ikoma, Critical core masses for gas giant formation with grain-free envelopes. Astrophys. J. **714**, 1343–1346 (2010)
Y. Hori, S. Ida, D.N.C. Lin, Characterization of sub/super-Earths orbiting cool stars: water content and hydrogen-rich atmospheres. Astrophysical Journal (2016, submitted)
O. Hubickyj, P. Bodenheimer, J.J. Lissauer, Accretion of the gaseous envelope of Jupiter around a 5–10 Earth-mass core. Icarus **179**, 415–431 (2005)
S. Ida, J. Makino, Scattering of planetesimals by a protoplanet—slowing down of runaway growth. Icarus **106**, 210 (1993)
M. Ikoma, H. Genda, Constraints on the mass of a habitable planet with water of nebular origin. Astrophys. J. **648**, 696–706 (2006)
M. Ikoma, H. Emori, K. Nakazawa, Formation of giant planets in dense nebulae: critical core mass revisited. Astrophys. J. **553**, 999–1005 (2001)
M. Ikoma, Y. Hori, In situ accretion of hydrogen-rich atmospheres on short-period super-Earths: implications for the Kepler-11 planets. Astrophys. J. **753**, 66 (2012)
M. Ikoma, K. Nakazawa, H. Emori, Formation of giant planets: dependences on core accretion rate and grain opacity. Astrophys. J. **537**, 1013–1025 (2000)
A.P. Ingersoll, The runaway greenhouse: a history of water on Venus. J. Atmos. Sci. **26**, 1191–1198 (1969)
D.E. Ionov, D.V. Bisikalo, V.I. Shematovich, B. Huber, Ionization fraction in the thermosphere of the exoplanet HD 209458b. Sol. Syst. Res. **48**(2), 105–112 (2014)
W.H. Ip, On a hot oxygen corona of Mars. Icarus **76**(1), 135–145 (1988)
B.M. Jakosky, R.O. Pepin, R.E. Johnson, J.L. Fox, Mars atmospheric loss and isotopic fractionation by solar-wind-induced sputtering and photochemical escape. Icarus **111**(2), 271–288 (1994)
N. Jendrzejewski, T. Trull, F. Pineau, M. Javoy, Carbon solubility in mid-ocean ridge basaltic melt at low pressures (250–1950 bar). Chem. Geol. **138**(1-2), 81–92 (1997)
W.J. Jenkins, J.M. Edmond, J.B. Corliss, Excess ^{3}He and ^{4}He in Galapagos submarine hydrothermal waters. Nature **272**(5649), 156–158 (1978)
S. Jin, C. Mordasini, V. Parmentier, R. van Boekel, T. Henning, J. Ji, Planetary population synthesis coupled with atmospheric escape: a statistical view of evaporation. Astrophys. J. **795**(1), 65 (2014)
R.E. Johnson, M.R. Combi, J.L. Fox, W.H. Ip, F. Leblanc, M.A. McGrath, V.I. Shematovich, D.F. Strobel, J.H. Waite, Exospheres and atmospheric escape. Space Sci. Rev. **139**(1–4), 355–397 (2008)
L. Kamp, F. Taylor, S. Calcutt, Structure of Venus's atmosphere from modelling of night-side infrared spectra. Nature **336**, 360–362 (1988)
J. Kasting, Runaway and moist greenhouse atmospheres and the evolution of Earth and Venus. Icarus **74**, 472–494 (1988)
J.F. Kasting, J.B. Pollack, Loss of water from Venus. I. Hydrodynamic escape of hydrogen. Icarus **53**(3), 479–508 (1983)
J.F. Kasting, J.B. Pollack, T.P. Ackerman, Response of Earth's atmosphere to increases in solar flux and implications for loss of water from Venus. Icarus **57**, 335–355 (1984). doi:10.1016/0019-1035(84)90122-2
J.F. Kasting, D.P. Whitmire, R.T. Reynolds, Habitable zones around main sequence stars. Icarus **101**, 108–128 (1993)
H. Kawahara, T. Hirano, K. Kurosaki, Y. Ito, M. Ikoma, Starspots-transit depth relation of the evaporating planet candidate KIC 12557548b. Astrophys. J. Lett. **776**(1), L6 (2013)
M.L. Khodachenko, I. Alexeev, E. Belenkaya, H. Lammer, J.-M. Grießmeier, M. Leitzinger, P. Odert, T. Zaqarashvili, H.O. Rucker, Magnetospheres of "hot Jupiters": the importance of magnetodisks in shaping a magnetospheric obstacle. Astrophys. J. **744**(1), 70 (2012)
R. Kippenhahn, A. Weigert, *Stellar Structure and Evolution* (Springer, Berlin, 1994)
K.G. Kislyakova, M. Holmström, H. Lammer, P. Odert, M.L. Khodachenko, Magnetic moment and plasma environment of HD 209458b as determined from Lyα observations. Science **346**(6212), 981–984 (2014a)
K.G. Kislyakova, C.P. Johnstone, P. Odert, N.V. Erkaev, H. Lammer, T. Lüftinger, M. Holmström, M.L. Khodachenko, M. Güdel, Stellar wind interaction and pick-up ion escape of the Kepler-11 "super-Earths". Astron. Astrophys. **562**, A116 (2014b)
T. Kleine, M. Touboul, B. Bourdon, F. Nimmo, K. Mezger, H. Palme, S.B. Jacobsen, Q.-Z. Yin, A.N. Halliday, Hf-W chronology of the accretion and early evolution of asteroids and terrestrial planets. Geochim. Cosmochim. Acta **73**, 5150–5188 (2009)
E. Kokubo, H. Genda, Formation of terrestrial planets from protoplanets under a realistic accretion condition. Astrophys. J. Lett. **714**, L21–L25 (2010)
E. Kokubo, S. Ida, On runaway growth of planetesimals. Icarus **123**, 180–191 (1996)
E. Kokubo, S. Ida, Oligarchic growth of protoplanets. Icarus **131**, 171–178 (1998)

R.K. Kopparapu, R. Ramirez, J.F. Kasting, V. Eymet, T.D. Robinson, S. Mahadevan, R.C. Terrien, S. Domagal-Goldman, V. Meadows, R. Deshpande, Habitable zones around main-sequence stars: new estimates. Astrophys. J. Lett. **765**, 131 (2013)

T.T. Koskinen, R.V. Yelle, P. Lavvas, N.K. Lewis, Characterizing the thermosphere of HD 209458b with UV transit observations. Astrophys. J. **723**(1), 116 (2010)

T. Koskinen, R. Yelle, M. Harris, P. Lavvas, The escape of heavy atoms from the ionosphere of HD 209458b. II. Interpretation of the observations. Icarus **226**(2), 1695–1708 (2013a)

T.T. Koskinen, M.J. Harris, R.V. Yelle, P. Lavvas, The escape of heavy atoms from the ionosphere of HD 209458b. I. A photochemical–dynamical model of the thermosphere. Icarus **226**(2), 1678–1694 (2013b)

M.A. Krestyanikova, V.I. Shematovich, Stochastic models of hot planetary and satellite coronas: a hot oxygen corona of Mars. Sol. Syst. Res. **40**(5), 384–392 (2006)

H. Kurokawa, T. Nakamoto, Mass-loss evolution of close-in exoplanets: evaporation of hot Jupiters and the effect on population. Astrophys. J. **783**(1), 54 (2014)

K. Kurosaki, M. Ikoma, Y. Hori, Impact of photo-evaporative mass loss on masses and radii of water-rich sub/super-Earths. Astron. Astrophys. **562**, A80 (2014)

M. Kurz, W. Jenkins, J. Schilling, S. Hart, Helium isotopic variations in the mantle beneath the central North Atlantic Ocean. Earth Planet. Sci. Lett. **58**(1), 1–14 (1982)

T. La Tourette, G.J. Wasserburg, Mg diffusion in anorthite: implications for the formation of early solar planetisimals. Earth Planet. Sci. Lett. **158**, 91–108 (1998)

S. Labrosse, C. Jaupart, Thermal evolution of the Earth: secular changes and fluctuations of plate characteristics. Earth Planet. Sci. Lett. **260**(3–4), 465–481 (2007)

S. Labrosse, J.W. Hernlund, N. Coltice, A crystallizing dense magma ocean at the base of the Earth's mantle. Nature **450**(7171), 866–869 (2007)

H. Lammer, S.J. Bauer, Nonthermal atmospheric escape from Mars and Titan. J. Geophys. Res. Space Phys. **96**(A2), 1819–1825 (1991)

H. Lammer, F. Selsis, I. Ribas, E.F. Guinan, S.J. Bauer, W.W. Weiss, Atmospheric loss of exoplanets resulting from stellar X-ray and extreme-ultraviolet heating. Astrophys. J. Lett. **598**(2), L121 (2003)

H. Lammer, P. Odert, M. Leitzinger, M.L. Khodachenko, M. Panchenko, Y.N. Kulikov, T.L. Zhang, H.I.M. Lichtenegger, N.V. Erkaev, G. Wuchterl, G. Micela, T. Penz, H.K. Biernat, J. Weingrill, M. Steller, H. Ottacher, J. Hasiba, A. Hanslmeier, Determining the mass loss limit for close-in exoplanets: what can we learn from transit observations? Astron. Astrophys. **506**(1), 399–410 (2009)

H. Lammer, N.V. Erkaev, P. Odert, K.G. Kislyakova, M. Leitzinger, M.L. Khodachenko, Probing the blow-off criteria of hydrogen-rich 'super-Earths'. Mon. Not. R. Astron. Soc. **430**(2), 1247–1256 (2013)

H. Lammer, A. Stökl, N.V. Erkaev, E.A. Dorfi, P. Odert, M. Güdel, Y.N. Kulikov, K.G. Kislyakova, M. Leitzinger, Origin and loss of nebula-captured hydrogen envelopes from 'sub'- to 'super-Earths' in the habitable zone of sun-like stars. Mon. Not. R. Astron. Soc. **439**(4), 3225–3238 (2014)

M.A. Lange, T.J. Ahrens, The evolution of an impact-generated atmosphere. Icarus **51**(1), 96–120 (1982)

A.F. Lanza, Star-planet magnetic interaction and evaporation of planetary atmospheres. Astron. Astrophys. **557**, A31 (2013)

M. Le Bars, A. Davaille, Whole layer convection in a heterogeneous planetary mantle. J. Geophys. Res., Solid Earth **109**, B03403 (2004)

T. Lebrun, H. Massol, E. Chassefiere, A. Davaille, E. Marcq, P. Sarda, F. Leblanc, G. Brandeis, Thermal evolution of an early magma ocean in interaction with the atmosphere. J. Geophys. Res., Planets **118**(6), 1155–1176 (2013)

J. Leconte, F. Forget, B. Charnay, R. Wordsworth, A. Pottier, Increased insolation threshold for runaway greenhouse processes on Earth-like planets. Nature **504**, 268–271 (2013)

E.J. Lee, E. Chiang, C.W. Ormel, Make super-Earths, not Jupiters: accreting nebular gas onto solid cores at 0.1 AU and beyond. Astrophys. J. **797**, 95 (2014)

A.M. Lejeune, P. Richet, Rheology of crystal-bearing silicate melts: an experimental study at high viscosities. J. Geophys. Res., Solid Earth **100**, 4215–4229 (1995)

J.S. Lewis, R.G. Prinn, *Planets and Their Atmospheres: Origin and Evolution*, vol. 33 (Elsevier, Amsterdam, 1983)

J. Li, C. Agee, Geochemistry of mantle-core differentiation at high pressure. Nature **381**(6584), 686–689 (1996)

H.I.M. Lichtenegger, H. Gröller, H. Lammer, Y.N. Kulikov, V.I. Shematovich, On the elusive hot oxygen corona of Venus. Geophys. Res. Lett. **36**(10), L10204 (2009)

J.L. Linsky, H. Yang, K. France, C.S. Froning, J.C. Green, J.T. Stocke, S.N. Osterman, Observations of mass loss from the transiting exoplanet HD 209458b. Astrophys. J. **717**(2), 1291 (2010)

J.J. Lissauer, Timescales for planetary accretion and the structure of the protoplanetary disk. Icarus **69**, 249–265 (1987)

Y. Liu, Y. Zhang, H. Behrens, Solubility of H_2O in rhyolitic melts at low pressures and a new empirical model for mixed H_2O-CO_2 solubility in rhyolitic melts. J. Volcanol. Geotherm. Res. **143**(1–3), 219–235 (2005)
E.D. Lopez, J. Fortney, The role of core mass in controlling evaporation: the Kepler radius distribution and the Kepler-36 density dichotomy. Astrophys. J. **776**(1), 2 (2013)
E.D. Lopez, J.J. Fortney, Understanding the mass-radius relation for sub-Neptunes: radius as a proxy for composition. Astrophys. J. **792**, 1 (2014)
E.D. Lopez, J. Fortney, N. Miller, How thermal evolution and mass-loss sculpt populations of super-Earths and sub-Neptunes: application to the Kepler-11 system and beyond. Astrophys. J. **761**(1), 59 (2012)
R. Luger, R. Barnes, E. Lopez, J. Fortney, B. Jackson, V. Meadows, Habitable evaporated cores: transforming mini-Neptunes into super-Earths in the habitable zones of M dwarfs. Astrobiology **15**(1), 57–88 (2015)
J.G. Luhmann, R.E. Johnson, M.H.G. Zhang, Evolutionary impact of sputtering of the martian atmosphere by O^+ pickup ions. Geophys. Res. Lett. **19**(21), 2151–2154 (1992)
R.E. Lupu, K. Zahnle, M.S. Marley, L. Schaefer, B. Fegley, C. Morley, K. Cahoy, R. Freedman, J.J. Fortney, The atmospheres of earthlike planets after giant impact events. Astrophys. J. **784**, 27 (2014)
G.J.F. MacDonald, Calculations on the thermal history of the Earth. J. Geophys. Res. **64**(11), 1967–2000 (1959)
W.V.R. Malkus, The heat transport and spectrum of thermal turbulence. Proc. R. Soc. A, Math. Phys. Eng. Sci. (1954). doi:10.1098/rspa.1954.0197
E. Marcq, A simple 1-D radiative-convective atmospheric model designed for integration into coupled models of magma ocean planets. J. Geophys. Res., Planets **117**, 1001 (2012)
G.W. Marcy, L. Weiss, E.A. Petigura, H. Isaacson, A.W. Howard, L.A. Buchhave, Occurrence and core-envelope structure of 1-4x Earth-size planets around sun-like stars. Proc. Natl. Acad. Sci. USA **111**(35), 12655–12660 (2014)
M. Marov, V. Shematovich, D. Bisikalo, Non equilibrium aeronomic processes. Space Sci. Rev. **76**(1–2), 1–204 (1996)
D. Martin, R. Nokes, Crystal settling in a vigorously convecting magma chamber. Nature **332**, 534–536 (1988)
T. Matsui, Y. Abe, Impact-induced atmospheres and oceans on Earth and Venus. Nature **322**(6079), 526–528 (1986)
M. Maurice, N. Tosi, A.C. Plesa, D. Breuer, Evolution and consequences of magma ocean solidification, in *Goldschmidt Abstracts. No. 2060* (2015)
M.B. McElroy, Mars: an evolving atmosphere. Science **175**(4020), 443–445 (1972)
E. Miller-Ricci, M.R. Meyer, S. Seager, L. Elkins-Tanton, On the emergent spectra of hot protoplanet collision afterglows. Astrophys. J. **704**, 770–780 (2009)
H. Mizuno, K. Nakazawa, C. Hayashi, Instability of a gaseous envelope surrounding a planetary core and formation of giant planets. Prog. Theor. Phys. **60**, 699–710 (1978)
H. Mizuno, K. Nakazawa, C. Hayashi, Dissolution of the primordial rare gases into the molten Earth's material. Earth Planet. Sci. Lett. **50**, 202–210 (1980)
J. Monteux, N. Coltice, F. Dubuffet, Y. Ricard, Thermo-mechanical adjustment after impacts during planetary growth. Geophys. Res. Lett. **34**, L24201 (2007)
T. Montmerle, J.-C. Augereau, M. Chaussidon, M. Gounelle, B. Marty, A. Morbidelli, Solar system formation and early evolution: the first 100 million years. Earth Moon Planets **98**(1–4), 39–95 (2006)
C. Mordasini, Grain opacity and the bulk composition of extrasolar planets. II. An analytical model for grain opacity in protoplanetary atmospheres. Astron. Astrophys. **572**, A118 (2014)
C. Mordasini, Y. Alibert, C. Georgy, K.-M. Dittkrist, H. Klahr, T. Henning, Characterization of exoplanets from their formation. II. The planetary mass-radius relationship. Astron. Astrophys. **547**, A112 (2012a)
C. Mordasini, Y. Alibert, H. Klahr, T. Henning, Characterization of exoplanets from their formation. I. Models of combined planet formation and evolution. Astron. Astrophys. **547**, A111 (2012b)
M. Moreira, Noble gas constraints on the origin and evolution of Earth's volatiles. Geochem. Perspect. **2**, 229–403 (2013)
R. Morishima, J. Stadel, B. Moore, From planetesimals to terrestrial planets: N-body simulations including the effects of nebular gas and giant planets. Icarus **207**, 517–535 (2010)
N. Movshovitz, M. Podolak, The opacity of grains in protoplanetary atmospheres. Icarus **194**, 368–378 (2008)
N. Movshovitz, P. Bodenheimer, M. Podolak, J.J. Lissauer, Formation of Jupiter using opacities based on detailed grain physics. Icarus **209**, 616–624 (2010)
R.A. Murray-Clay, E.I. Chiang, N. Murray, Atmospheric escape from hot Jupiters. Astrophys. J. **693**(1), 23 (2009)
A. Nagy, T. Cravens, Hot oxygen atoms in the upper atmosphere of Venus and Mars. Geophys. Res. Lett. **15**(5), 433–435 (1988)

S. Nakajima, Y.-Y. Hayashi, Y. Abe, A study on the 'runaway greenhouse effect' with a one-dimensional radiative-convective equilibrium model. J. Atmos. Sci. **49**, 2256–2266 (1992)
F. Nimmo, T. Kleine, How rapidly did Mars accrete? Uncertainties in the Hf W timing of core formation. Icarus **191**, 497–504 (2007)
R. Nomura, H. Ozawa, S. Tateno, K. Hirose, J. Hernlund, S. Muto, H. Ishii, N. Hiraoka, Spin crossover and iron-rich silicate melt in the Earth's deep mantle. Nature **473**(7346), 199–202 (2011)
D.P. O'Brien, A. Morbidelli, H.F. Levison, Terrestrial planet formation with strong dynamical friction. Icarus **184**, 39–58 (2006)
D.P. O'Brien, K.J. Walsh, A. Morbidelli, S.N. Raymond, A.M. Mandell, Water delivery and giant impacts in the 'Grand Tack' scenario. Icarus **239**, 74–84 (2014)
S. Okuzumi, H. Tanaka, H. Kobayashi, K. Wada, Rapid coagulation of porous dust aggregates outside the snow line: a pathway to successful icy planetesimal formation. Astrophys. J. **752**, 106 (2012)
H.C. O'Neill, H. Palme, Composition of the silicate Earth: implications for accretion of and core formation, in *The Earth's Mantle; Composition, Structure, and Evolution* (Cambridge University Press, Cambridge, 1988), pp. 3–126
E.J. Opik, S.F. Singer, Distribution of density in a planetary exosphere. II. Phys. Fluids **4**, 221–233 (1961)
C.W. Ormel, An atmospheric structure equation for grain growth. Astrophys. J. Lett. **789**, L18 (2014)
C.W. Ormel, R. Kuiper, J.-M. Shi, Hydrodynamics of embedded planets' first atmospheres—I. A centrifugal growth barrier for 2D flows. Mon. Not. R. Astron. Soc. **446**, 1026–1040 (2015a)
C.W. Ormel, J.-M. Shi, R. Kuiper, Hydrodynamics of embedded planets' first atmospheres—II. A rapid recycling of atmospheric gas. Mon. Not. R. Astron. Soc. **447**, 3512–3525 (2015b)
J.E. Owen, A. Jackson, Planetary evaporation by UV and X-ray radiation: basic hydrodynamics. Mon. Not. R. Astron. Soc. **425**(4), 2931–2947 (2012)
J.E. Owen, Y. Wu, Kepler planets: a tale of evaporation. Astrophys. J. **775**(2), 105 (2013)
J.E. Owen, B. Ercolano, C.J. Clarke, R.D. Alexander, Radiation-hydrodynamic models of X-ray and EUV photoevaporating protoplanetary discs. Mon. Not. R. Astron. Soc. **401**(3), 1415–1428 (2010)
M. Ozima, Ar isotopes and Earth-atmosphere evolution models. Geochim. Cosmochim. Acta **39**(8), 1127–1134 (1975)
M. Ozima, K. Kudo, Excess argon in submarine basalts and an Earth-atmosphere evolution model. Nature **239**(89), 23–24 (1972)
V. Pan, J.R. Holloway, R. Hervig, The pressure and temperature-dependence of carbon-dioxide solubility in tholeiitic basalt melts. Geochim. Cosmochim. Acta **55**(6), 1587–1595 (1991)
P. Papale, Modeling of the solubility of a two-component H_2O+CO_2 fluid in silicate liquids. Am. Mineral. **84**(4), 447–492 (1999)
E.N. Parker, *Interplanetary Dynamical Processes* (Interscience, New York, 1963)
E.N. Parker, Dynamical properties of stellar coronas and stellar winds. I. Integration of the momentum equation. Astrophys. J. **139**, 72–122 (1964)
T. Penz, N.V. Erkaev, Y.N. Kulikov, D. Langmayr, H. Lammer, G. Micela, C. Cecchi-Pestellini, H.K. Biernat, F. Selsis, P. Barge, M. Deleuil, A. Léger, Mass loss from "hot Jupiters"—implications for CoRoT discoveries, part ii: Long time thermal atmospheric evaporation modeling. Planet. Space Sci. **56**(9), 1260–1272 (2008)
F. Perri, A.G.W. Cameron, Hydrodynamic instability of the solar nebula in the presence of a planetary core. Icarus **22**, 416–425 (1974)
S. Pfyffer, Y. Alibert, W. Benz, D. Swoboda, Theoretical models of planetary system formation. II. Post-formation evolution (2015). ArXiv e-prints
A.C. Plesa, N. Tosi, D. Breuer, Can a fractionally crystallized magma ocean explain the thermo-chemical evolution of Mars? Earth Planet. Sci. Lett. **403**, 225–235 (2014)
M. Podolak, The contribution of small grains to the opacity of protoplanetary atmospheres. Icarus **165**, 428–437 (2003)
J.B. Pollack, Formation of giant planets and their satellite-ring systems, in *Protostars and Planets II* (University of Arizona Press, Tucson, 1985)
J.B. Pollack, O. Hubickyj, P. Bodenheimer, J.J. Lissauer, M. Podolak, Y. Greenzweig, Formation of the giant planets by concurrent accretion of solids and gas. Icarus **124**, 62–85 (1996)
C. Priestley, *Turbulent Transfer in the Lower Atmosphere* (University of Chicago Press, Chicago, 1959)
R. Rafikov, Microwave emission from spinning dust in circumstellar disks. Astrophys. J. **646**, 288–296 (2006)
R.M. Ramirez, L. Kaltenegger, The habitable zones of pre-main-sequence stars. Astrophys. J. Lett. **797**(2), L25–L33 (2014)
S.N. Raymond, T. Quinn, J.I. Lunine, High-resolution simulations of the final assembly of Earth-like planets I. Terrestrial accretion and dynamics. Icarus **183**, 265–282 (2006)

S.N. Raymond, E. Kokubo, A. Morbidelli, R. Morishima, K.J. Walsh, Terrestrial planet formation at home and abroad, in *Protostars and Planets VI* (2014), pp. 595–618
R.T. Reynolds, P.E. Fricker, A.L. Summers, Effects of melting upon thermal models of the Earth. J. Geophys. Res. **71**(2), 573–582 (1966)
K. Righter, M. Drake, Metal/silicate equilibrium in the early Earth—new constraints from the volatile moderately siderophile elements Ga, Cu, P, and Sn. Geochim. Cosmochim. Acta **64**(20), 3581–3597 (2000)
L. Rogers, Most 1.6 Earth-radius planets are not rocky. Astrophys. J. **801**(1), 41 (2015)
W.W. Rubey, Geologic history of sea water an attempt to state the problem. Geol. Soc. Am. Bull. **62**(9), 1111–1148 (1951)
V. Safronov, The heating of the Earth during its formation. Icarus **33**, 3–12 (1978)
J. Sanz-Forcada, I. Ribas, G. Micela, A.M.T. Pollock, D. García-Álvarez, E. Solano, C. Eiroa, A scenario of planet erosion by coronal radiation. Astron. Astrophys. **511**, L8 (2010)
P. Sarda, B. Guillot, Breaking of Henry's law for noble gas and CO_2 solubility in silicate melt under pressure. Nature **436**(7047), 95–98 (2005)
P. Sarda, T. Staudacher, C.J. Allègre, $^{40}Ar/^{36}Ar$ in MORB glasses: constraints on atmosphere and mantle evolution. Earth Planet. Sci. Lett. **72**(4), 357–375 (1985)
P. Sarda, T. Staudacher, C.J. Allègre, Neon isotopes in submarine basalts. Earth Planet. Sci. Lett. **91**(1–2), 73–88 (1988)
S. Sasaki, K. Nakazawa, Metal-silicate fractionation in the growing Earth: energy source for the terrestrial magma ocean. J. Geophys. Res., Solid Earth **91**(B9), 9231–9238 (1986)
E.M. Schneiter, P. Velázquez, A. Esquivel, A.C. Raga, X. Blanco-Cano, Three-dimensional hydrodynamical simulation of the exoplanet HD 209458b. Astrophys. J. Lett. **671**(1), L57 (2007)
D. Semenov, T. Henning, C. Helling, M. Ilgner, E. Sedlmayr, Rosseland and Planck mean opacities for protoplanetary discs. Astron. Astrophys. **410**, 611–621 (2003)
I.F. Shaikhislamov, M.L. Khodachenko, Y.L. Sasunov, H. Lammer, K.G. Kislyakova, N.V. Erkaev, Atmosphere expansion and mass loss of close-orbit giant exoplanets heated by stellar XUV. I. Modeling of hydrodynamic escape of upper atmospheric material. Astrophys. J. **795**(2), 132 (2014)
H.R. Shaw, Viscosities of magmatic silicate liquids: an empirical method of prediction. Am. J. Sci. **272**, 870–893 (1972)
V.I. Shematovich, Stochastic models of hot planetary and satellite coronas. Sol. Syst. Res. **38**(1), 28–38 (2004)
V.I. Shematovich, Suprathermal hydrogen produced by the dissociation of molecular hydrogen in the extended atmosphere of exoplanet HD 209458b. Sol. Syst. Res. **44**(2), 96–103 (2010)
V.I. Shematovich, D.V. Bisikalo, J.C. Gérard, C. Cox, S.W. Bougher, F. Leblanc, Monte Carlo model of electron transport for the calculation of Mars dayglow emissions. J. Geophys. Res., Planets **113**(E2), E02011 (2008)
V.I. Shematovich, D.E. Ionov, H. Lammer, Heating efficiency in hydrogen-dominated upper atmospheres. Astron. Astrophys. **571**, A94 (2014)
E.D. Siggia, High Rayleigh number convection. Annu. Rev. Fluid Mech. **26**(1), 137–168 (1994)
J. Smith, A.T. Anderson, R. Newton, E. Olsen et al., Petrologic history of the Moon inferred from petrography, mineralogy, and petrogenesis of Apollo 11 rocks, in *Proc. Apollo 11 Lunar Sci. Conf., vol. 1*, Houston, TX, Jan. 5–8 (Pergamon, New York, 1970), pp. 897–925
V.S. Solomatov, Fluid dynamics of a terrestrial magma ocean, in *Origin of the Earth and Moon* (2000), pp. 323–338
V. Solomatov, Magma oceans and primordial mantle differentiation, in *Treatise on Geophysics*, ed. by G. Schubert (Elsevier, Amsterdam, 2007), pp. 91–119
V. Solomatov, Magma oceans and primordial mantle differentiation, in *Treatise on Geophysics*, ed. by G. Schubert (Elsevier, Amsterdam, 2009), pp. 91–119
V. Solomatov, Magma oceans and primordial mantle differentiation—magma oceans and primordial mantle differentiation, in *Treatise on Geophysics*, ed. by G. Schubert (Elsevier, Amsterdam, 2015), 81–100
V.S. Solomatov, D.J. Stevenson, Suspension in convective layers and style of differentiation of a terrestrial magma ocean. J. Geophys. Res., Planets **98**(E3), 5375–5390 (1993a)
V.S. Solomatov, D.J. Stevenson, Kinetics of crystal growth in a terrestrial magma ocean. J. Geophys. Res., Planets **98**(E3), 5407–5418 (1993b)
T. Staudacher, C.J. Allègre, Terrestrial xenology. Earth Planet. Sci. Lett. **60**(3), 389–406 (1982)
D.J. Stevenson, Formation of the giant planets. Planet. Space Sci. **30**, 755–764 (1982)
A. Stökl, E. Dorfi, H. Lammer, Hydrodynamic simulations of captured protoatmospheres around Earth-like planets. Astron. Astrophys. **576**, A87 (2015)
J.M. Stone, D. Proga, Anisotropic winds from close-in extrasolar planets. Astrophys. J. **694**(1), 205 (2009)
J. Suckale, J.A. Sethian, J-d. Yu, L.T. Elkins-Tanton, Crystals stirred up: 1. Direct numerical simulations of crystal settling in nondilute magmatic suspensions. J. Geophys. Res., Planets **117**, E08004 (2012)

H.E. Suess, Die häufigkeit der edelgase auf der erde und im kosmos. J. Geol. **57**, 600–607 (1949)

H. Tanaka, S. Ida, Growth of a migrating protoplanet. Icarus **139**, 350–366 (1999)

Y.A. Tanaka, T.K. Suzuki, S. Inutsuka, Atmospheric escape by magnetically driven wind from gaseous planets. Astrophys. J. **792**(1), 18 (2014)

G. Taylor, M. Norman, *Evidence for Magma Oceans on Asteroids, the Moon, and Earth* (Lunar Planet. Inst., Houston, 1992), pp. 58–65

E. Thommes, M. Nagasawa, D.N.C. Lin, Dynamical shake-up of planetary systems. II. N-Body simulations of solar system terrestrial planet formation induced by secular resonance sweeping. Astrophys. J. **676**, 728–739 (2008)

F. Tian, Conservation of total escape from hydrodynamic planetary atmospheres. Earth Planet. Sci. Lett. **379**, 104–107 (2013)

F. Tian, History of water loss and atmospheric O_2 buildup on rocky exoplanets near M dwarfs. Earth Planet. Sci. Lett. **432**, 126–132 (2015)

F. Tian, S. Ida, Water contents of Earth-mass planets around M dwarfs. Nat. Geosci. **8**(3), 177–180 (2015)

F. Tian, O.B. Toon, A.A. Pavlov, H. De Sterck, A hydrogen-rich early Earth atmosphere. Science **308**(5724), 1014–1017 (2005a)

F. Tian, O.B. Toon, A.A. Pavlov, H. De Sterck, Transonic hydrodynamic escape of hydrogen from extrasolar planetary atmospheres. Astrophys. J. **621**(2), 1049 (2005b)

F. Tian, J.F. Kasting, H.-L. Liu, R.G. Roble, Hydrodynamic planetary thermosphere model: 1. Response of the Earth's thermosphere to extreme solar EUV conditions and the significance of adiabatic cooling. J. Geophys. Res., Planets **113**(E5), E05008 (2008)

G. Tinetti et al., EChO. Exoplanet characterisation observatory. Exp. Astron. **34**(2), 311–353 (2012)

W. Tonks, H. Melosh, The physics of crystal settling and suspension in a turbulent magma ocean, in *Origin of the Earth* (Oxford University Press, London, 1990), pp. 151–174

W.B. Tonks, H.J. Melosh, Magma ocean formation due to giant impacts. J. Geophys. Res., Planets **98**(E3), 5319–5333 (1993)

N. Tosi, A.C. Plesa, D. Breuer, Overturn and evolution of a crystallized magma ocean: a numerical parameter study for Mars. J. Geophys. Res., Planets **118**(7), 1512–1528 (2013)

M. Touboul, T. Kleine, B. Bourdon, H. Palme, R. Wieler, Late formation and prolonged differentiation of the Moon inferred from W isotopes in lunar metals. Nature **450**, 1206–1209 (2007)

G.B. Trammell, P. Arras, Z.-Y. Li, Hot Jupiter magnetospheres. Astrophys. J. **728**(2), 152 (2011)

K.K. Turekian, S.P. Clark Jr., The non-homogeneous accumulation model for terrestrial planet formation and the consequences for the atmosphere of Venus. J. Atmos. Sci. **32**(6), 1257–1261 (1975)

H. Urey, The cosmic abundances of potassium, uranium and thorium and the heat balances of the Earth, the Moon and Mars. Proc. Natl. Acad. Sci. USA **41**, 127–144 (1955)

F. Valsecchi, F.A. Rasio, J.H. Steffen, From hot Jupiters to super-Earths via Roche lobe overflow. Astrophys. J. Lett. **793**(1), L3 (2014)

A. Vidal-Madjar, A. Lecavelier des Etangs, J.M. Desert, G.E. Ballester, R. Ferlet, G. Hebrard, M. Mayor, An extended upper atmosphere around the extrasolar planet HD 209458b. Nature **422**(6928), 143–146 (2003)

A. Vidal-Madjar, J.M. Désert, A. Lecavelier des Etangs, G. Hébrard, G.E. Ballester, D. Ehrenreich, R. Ferlet, J.C. McConnell, M. Mayor, C.D. Parkinson, Detection of oxygen and carbon in the hydrodynamically escaping atmosphere of the extrasolar planet HD 209458b. Astrophys. J. Lett. **604**(1), L69 (2004)

A. Vidal-Madjar, C.M. Huitson, V. Bourrier, J.-M. Désert, G. Ballester, A. Lecavelier des Etangs, D.K. Sing, D. Ehrenreich, R. Ferlet, G. Hébrard, J.C. McConnell, Magnesium in the atmosphere of the planet HD 209458b: observations of the thermosphere-exosphere transition region. Astron. Astrophys. **560**, A54 (2013)

C. Villarreal D'Angelo, M. Schneiter, A. Costa, P. Velázquez, A. Raga, A. Esquivel, On the sensitivity of extrasolar mass-loss rate ranges: HD 209458b a case study. Mon. Not. R. Astron. Soc. **438**(2), 1654–1662 (2014)

A. Volkov, R.E. Johnson, Thermal escape in the hydrodynamic regime: reconsideration of Parker's isentropic theory based on results of kinetic simulations. Astrophys. J. **765**(2), 90 (2013)

A. Volkov, R.E. Johnson, O.J. Tucker, J.T. Erwin, Thermally driven atmospheric escape: transition from hydrodynamic to jeans escape. Astrophys. J. Lett. **729**(2), L24 (2011a)

A.N. Volkov, O.J. Tucker, J.T. Erwin, R.E. Johnson, Kinetic simulations of thermal escape from a single component atmosphere. Phys. Fluids **23**(6), 066601 (2011b)

A.J. Watson, T.M. Donahue, J.C. Walker, The dynamics of a rapidly escaping atmosphere: applications to the evolution of Earth and Venus. Icarus **48**(2), 150–166 (1981)

R. Wayne, *Chemistry of Atmospheres* (Oxford University Press, London, 1991) 312 pp.

G.W. Wetherill, Comparison of analytical and physical modeling of planetesimal accumulation. Icarus **88**, 336–354 (1990)

G.W. Wetherill, G.R. Stewart, Accumulation of a swarm of small planetesimals. Icarus **77**, 330–357 (1989)
F. Witham, J. Blundy, S.C. Kohn, P. Lesne, J. Dixon, S.V. Churakov, R. Botcharnikov, SolEx: a model for mixed COHSCl-volatile solubilities and exsolved gas compositions in basalt. Comput. Geosci. **45**, 87–97 (2012)
J. Wood, J.S. Dickey Jr., U.B. Marvin, B.N. Powell, Lunar anorthosites and a geophysical model of the Moon, in *Proceedings of the Apollo 11 Lunar Science Conference* (Pergamon, New York, 1970), pp. 965–988
B. Wood, M. Walter, J. Wade, Accretion of the Earth and segregation of its core. Nature **441**(7095), 825–833 (2006)
Y. Wu, Y. Lithwick, Density and eccentricity of Kepler planets. Astrophys. J. **772**(1), 74 (2013)
G. Wuchterl, The critical mass for protoplanets revisited—massive envelopes through convection. Icarus **106**, 323 (1993)
G. Wuchterl, Giant planet formation. Earth Moon Planets **67**(1–3), 51–65 (1994)
G. Wuchterl, Planet masses and radii from physical principles, in *Proceedings of the International Astronomical Union 6 (Symposium S276)* (2010), pp. 76–81
J. Yang, G. Boué, D.C. Fabrycky, D.S. Abbot, Strong dependence of the inner edge of the habitable zone on planetary rotation rate. Astrophys. J. Lett. **787**, L2 (2014)
R.V. Yelle, Aeronomy of extra-solar giant planets at small orbital distances. Icarus **170**(1), 167–179 (2004)
R. Yelle, H. Lammer, W.H. Ip, Aeronomy of extra-solar giant planets. Space Sci. Rev. **139**, 437–451 (2008)
T. Yoshino, M. Walter, T. Katsura, Core formation in planetesimals triggered by permeable flow. Nature **422**(6928), 154–157 (2003)
K. Zahnle, J. Kasting, J. Pollack, Evolution of a steam atmospheres during Earth's accretion. Icarus **74**(1), 62–97 (1988)

Part II

The Products of the Coevolution of Disks and Planets

Space Sci Rev (2016) 205:255–272
DOI 10.1007/s11214-016-0249-0

Insights into Planet Formation from Debris Disks: I. The Solar System as an Archetype for Planetesimal Evolution

Brenda C. Matthews[1,2] · **JJ Kavelaars**[1,2]

Received: 23 September 2015 / Accepted: 14 March 2016 / Published online: 17 May 2016

Abstract Circumstellar disks have long been regarded as windows into planetary systems. The advent of high sensitivity, high resolution imaging in the submillimeter where both the solid and gas components of disks can be detected opens up new possibilities for understanding the dynamical histories of these systems and therefore, a better ability to place our own solar system, which hosts a highly evolved debris disk, in context. Comparisons of dust masses from protoplanetary and debris disks have revealed a stark downturn in mass in millimeter-sized grains around a stellar age of 10 Myr, ostensibly in the "transition disk" phase, suggesting a period of rapid accretion of such grains onto planetesimals. This rapid formation phase is in keeping with radionucleide studies of Kuiper Belt Objects in the solar system. Importantly, this suggests that any thermal gradients in the gas of disks of this era will be "frozen in" to the planetesimals as they rapidly accrete from the solids and ices in their vicinity. Measurements of radial gradients in thermal tracers such as DHO, DCN and other tracers can therefore provide insight into the nascent solar system's abundances. In studies of dynamical evolution of the solar system, it is tacitly assumed that such abundances can reveal the location of formation for bodies now found in the asteroid belt and Kuiper belt. Similarly, evidence of gas detected from collisional evolution in young debris disks could potentially reveal how rapidly objects have dynamically evolved in those systems, most of which will be significantly younger than the solar system.

Keywords Circumstellar disks · Planet formation

1 Introduction

Circumstellar disks are found around stars of all ages. Around forming stars, protostellar disks play a critical role in the accretion of material onto forming stars, and over time, the disks can evolve to protoplanetary disks, wherein stable locations in the midplane allows

✉ B.C. Matthews
brenda.matthews@nrc-cnrc.gc.ca

1 National Research Council of Canada, 5071 West Saanich Road, Victoria, BC V9E 2E7, Canada

2 University of Victoria, Finnerty Road, Victoria, BC V8W 3P6, Canada

agglomeration processes to assemble the solids from sub-micron sized dust to grains up to centimeters in size, crucial steps in the planet formation process (Blum and Wurm 2008; Testi et al. 2014).

As protoplanetary disks evolve, they become more tenuous, eventually beginning to erode from the inside out, creating disks with inner holes, coinciding with a decline in infrared excess emission such that half the disks are gone by 3 Myr and no disks are detectable by 6 Myr, based on a study of young clusters by Haisch et al. (2001). During the transition disk epoch, gas can continue to accrete onto planetary embryos and dust grains can continue to grow. Ultimately however, in the late stages of a transition disk, the remaining dust and gas assembled from the natal star-forming cloud or core is dispersed (Espaillat et al. 2014). The decline of near-IR emission, the loss of gas components of the disks and the changes in the dust grain size distribution are all identifiers of transitional objects, but it is not immediately clear whether these changes occur simultaneously or in a sequence (e.g., Wyatt et al. 2015). Therefore, at the end of the transition disk stage, the formation of gas giants and super Earths must be complete, although terrestrial planet formation may continue for up to 100 Myr (see the Chassefière chapter in this work). After the assembly of planetesimals and planets is complete, some parts of the disk may already be dominated by destructive collisions, rather than agglomerative processes, although it is likely destructive collisions occur even from the earliest agglomeration stages in gas-rich disks (Dullemond and Dominik 2005). When a disk is predominately populated by dust and gas produced in collisions, it is known as a debris disk. Historically, such disks were also referred to as "secondary" or "second-generation" disks owing to their content being entirely derived from collisional evolution of larger bodies, rather than any remnants of the initial protoplanetary disk.

Debris disks represent the longest lived phase of circumstellar disks, ranging from the youngest systems, many of which now are seen to still host remnant protoplanetary gas disks, to very evolved systems, including sub-giant stars (Bonsor et al. 2013, 2014) and white dwarfs, where the measured incidence rates of infrared excess are typically a few percent (e.g., Barber et al. 2012), but have been measured to be as high as 14% (Kilic and Redfield 2007). White dwarfs also show evidence of "pollution" of their atmospheres due to deposition of material, most likely from a circumstellar disk or remnant planetesimals (e.g., Zuckerman et al. 2003; Dufour et al. 2012). All these evolved systems have in common the fact that their protoplanetary disk phase and succeeded in forming planetesimals of at least 100 km in size. Such objects may be essential for planetary assembly, but not all find their way into planets. Instead, the collisions of these planetesimals provide the source of debris disk dust emission for the lifetime of the system.

The sun is a middle-aged star, and we see its debris disk components primarily through direct detection of the planetesimals (asteroids and comets) that make up the largest end of its size distribution. Currently, collisions within the Kuiper Belt occur on very long timescales, meaning its mass is relatively constant, but its dust emission is relatively low. Breakup of large objects can still occur, however, as evidenced by the asteroid P/2010 A2, which is observed to be fragmenting, either due to impact or rotationally induced breakup (Agarwal et al. 2013). This object is just one of many examples of ongoing destruction of larger bodies in our own solar system (see review of Jewitt et al. 2015).

The solar debris disk could not yet be detected in surveys of disks around other stars (e.g., with *Spitzer*, *Herschel* or *WISE*). Based on models of the solar system's dynamical evolution, Booth et al. (2009) predict its brightness variation over time. When surveys of debris disks reach comparable limits to the Kuiper belt's dust fractional luminosity ($\sim 10^{-7}$), we will be better able to judge whether many such faint systems exist.

Observations of individual planetesimals are not possible around other stars; instead, we see the dust emission produced by the collisional cascade that represents the steady state evolution of many debris disks (Wyatt et al. 2007; Wyatt 2008). The observational evidence for debris disks shows that their planetesimal belts are long-lived, existing around stars of any age, and that they can be anomalously bright at late times, suggesting that stochastic as well as steady-state processes act within them. Depending on the wavelength observed and even the resolution of the imaged region, the dust distribution can appear very different. For instance, millimeter/centimeter grains are preferentially observed near their radius of formation, since they do not migrate far before being ground down to smaller sizes. Therefore, these larger grains are the best tracers of the planetesimal belts in external debris disks because they indicate the location of active collision processes (Wyatt 2006). In contrast, scattered light imaging highlights the location of much smaller grains, which are easily pushed around in the system, moving inward due to PR drag or stellar wind drag or outward due to radiation pressure or stellar (particle) winds. Emission from warmer dust is also seen at infrared wavelengths, typically highlighting the location of asteroidal or terrestrial zones of warm dust as illustrated by Fig. 1 of Matthews et al. (2014). Generally, the timescales for the evolution of warm dust (as seen by *Spitzer* and *WISE*) are much shorter than the cold dust detected in the Kuiper belt analogues (by *IRAS*, *Spitzer* and *Herschel*), which explains why warm dust is less frequently detected, with incidence rates of a few percent, rather than the 15–25% for cold dust (Matthews et al. 2014).

Debris disks are considered signposts of planetary systems because evidence of a bright debris disk indicates that the planetesimals in the disk have been stirred, inducing collisions. For a recent review of stirring mechanisms, see Sect. 6 of Matthews et al. (2014) and references therein. Wyatt et al. (2012) and Marshall et al. (2014) find evidence for a correlation between the presence of a bright debris disk and the presence of low-mass planets, many of which were not known when the systems were targeted by *Herschel* surveys for debris disks. In parallel, the *Herschel* Survey 4 Kuiper Belts Around RV-detected Planet hosts (SKARPS) targeted 97 planet host stars and also find a higher correlation of brighter disks with low-mass planet systems (i.e., those in which the largest planet mass is $<$ a Saturn mass) than with systems in which a higher mass planet is present, though the number of low-mass planet systems are still small (Bryden et al., private communication). The limited number of low-mass planet systems prevents statistical analysis of the joint DUNES/DEBRIS dataset from firmly establishing the correlation (Moro-Martín et al. 2015), although we note that the statistics can improve in the DUNES and DEBRIS samples as more planets are identified in nearby systems.

Wyatt et al. (2015) recently authored a detailed summary of the steps that occur as a protoplanetary disk becomes a debris disk. We do not attempt to replicate that work here. In the first part of this section, we briefly discuss recent results that provide insight into the critical epoch of planetesimal formation based on observations. We then discuss the implications of the detections of gas disks around debris dust systems and the implications for interpreting measured abundances in terms of the dynamical history of those systems. We then step back from extrasolar debris disks to highlight what can be learned about the origins of the solar system from its planetesimal population. The observations of the solar system's planetesimal population are interpreted based the reasonable assumption that the chemical features observed in asteroids and comets today were inherited from the local disk at the time and place of the planetesimals' formation (i.e., the transition disk phase), and the system subsequently evolved dynamically. Finally, we highlight the deuterium fraction as a key discriminator of the solar system's planetesimal history and its potential in external debris disks.

2 Tracing the Critical Epoch of Planetesimal Formation

Planetesimals, i.e., solid bodies ranging in size from 10s–1000s of km, may form very early in protoplanetary disks. The agglomeration process must overcome the meter-size barrier to produce oligarchs capable of assembling into the cores of giant planets or terrestrial planets in their own right (e.g., see Weidenschilling 1980; Carrera et al. 2015). Observations of many planetary systems reveal that not all the planetesimals are assembled into larger bodies, however. Many asteroids and comets remain to undergo collisional evolution. Based on recent *Herschel* and *Spitzer* surveys, $\sim$25% of A stars (Thureau et al. 2014; Su et al. 2009) show evidence of a measured far-infrared excess (to the survey depths) with a detectably higher rate of detection for younger stars in the *Spitzer* sample. *Herschel* surveys of solar type (FGK) stars find a detection rate of $\sim$20% (Eiroa et al. 2013), with evidence of a significant decrease in the rate of measured incidence from $\sim$25% for F-type stars to $\sim$15% for G/K-type stars (Sibthorpe et al. 2016, in preparation). The survey data do not yet probe disks as faint as our own Kuiper Belt. The range of system ages where dust from planetesimal collisions can be observed is a strong indication of the longevity of debris disks.

In contrast to the debris disk phase, the period during which the planetesimals can be assembled must be relatively short ($<$10 Myr) since the gaseous disk components are significantly depleted for most systems after that time. This timescale is also reinforced by the planetesimal population of the solar system itself, as we discuss below. To study the assembly of planetesimals, protoplanetary and transition disks must be observed.

With the advent of the Atacama Large Millimeter/submillimeter Array (ALMA), thermal imaging of circumstellar disks with comparable resolution to scattered light imaging and high sensitivity has at last become feasible. Even in its early science phase, ALMA has observed many transition era disks, typified by the loss or strong depletion of the inner disk material (see the review by Espaillat et al. 2014); recent ALMA results establish that some gas does remain the inner regions cleared of dust (van der Marel et al. 2015b). Observations have revealed very similar morphologies in several transition disks, including dust disks that are typically much more radially confined than the gas disk components, and asymmetries are also commonly observed (e.g., HD 100546, Pineda et al. 2014; Walsh et al. 2014). Most significantly, many disks exhibit highly asymmetric dust structures, suggesting that planetesimal formation processes may be highly localized in the disk. For example, van der Marel et al. (2013) observed a significant azimuthal asymmetry in the disk around Oph IRS 48, revealing an enhanced mm-grain "dust trap" containing up to 9 Earth masses of material (assuming grains up to 4 mm in size) while the 18.7 μm emission from VISIR reveal that the smaller grains were much more uniformly distributed in the disk. The authors suggest this dust trap feature is consistent with a vortex, which confines material both radially and azimuthally. Pérez et al. (2014) note strong azimuthally asymmetric distributions of dust consistent with vorticity in the disks of SAO 206462 and SR21, and HD 142527 (Casassus et al. 2013; Fukagawa et al. 2013) exhibits a disk in which dust is confined to a horseshoe pattern peaking at the northern edge of the disk.

The composition of planets and planetesimals varies according to the constituent materials available where they form in the protoplanetary disk. Therefore, the variations in conditions across the disk have a significant impact on the larger bodies produced. For example, condensation fronts of various volatile species form boundaries between different types of compositions (Öberg et al. 2011). Icy grains are more porous and grow more quickly than pure silicates. In addition, many studies show that surface mass densities are enhanced near snow lines, creating favorable conditions to form or grow planetesimals. Zhang et al. (2015)

illustrate that the spectral index map derived from ALMA data toward the protoplanetary disk HL Tau (ALMA Partnership et al. 2015) reveals that the gaps observed in the disk correspond well to the expected location of condensation fronts of water ice, ammonia and hydrates formed from amorphous water ice and other species. Zhang et al. (2015) show that the flux ratio between 1.3 mm and 0.87 mm is more extreme in the areas of lower emission, and they suggest that the variation observed is best explained by a model with two dust populations, one of which is the product of significant grain growth within the "dips" visible in the disk.

The HL Tau data, and the theoretical models with which they are consistent, suggest that this era of disk evolution may be the most critical in the assembly of planetary masses. The assemblage of so much material concentrated in small regions of the disk is the key to pebble accretion scenarios theories (see Johansen et al. 2007). Pebble accretion allows solids, through mass accretion over a short period of time, to grow from relatively small (<1 m) bodies to planetary scale oligarchs or embryos, vaulting over the meter-size barrier, which has been an impediment to core accretion theories of planet formation.

Another window into the evolution of grains is offered by measurements of dust emission in the submillimeter and millimeter regime. Because the emission is optically thin and the opacity is relatively simply modeled (as a power law in frequency), the flux density can be viewed as a proxy for mass (in millimeter-sized grains), provided one has some information about the dust temperature. The dust masses of many disks have been measured and can be compiled to gauge how dust mass varies with age. From observations at millimeter wavelengths, grains larger than centimeter sizes will not be observable, so disks may appear to diminish in mass as agglomeration occurs. Figure 1 shows that there is a marked decline of over two orders of magnitude in the dust mass of millimeter-sized grains in disks around the 10 Myr epoch, but the decline in fact extends over a range from 3–20 Myr. This timescale for

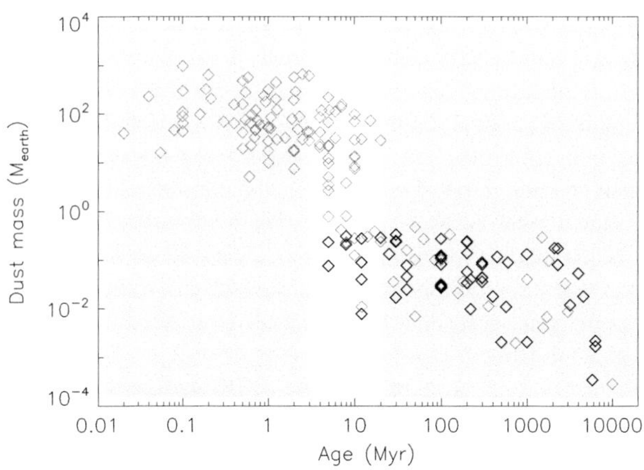

Fig. 1 Dust masses of circumstellar disks inferred from millimeter and submillimeter measurements. Data were compiled by Panić et al. (2013) to which new data from the SCUBA-2 Observations of Nearby Stars (SONS) Survey have been added (*darker symbols*). Beyond 10 Myr, disks are significantly less massive (100–1000 ×) compared to younger, protoplanetary disk systems, representing a rapid decline in the mass located in mm-sized dust grains during the transition to the debris disk phase. There is an uncertainty to any age of at least a factor of 2, though the relative ages shown should be more accurate. The broadest range of masses are observed between 3–20 Myr, during the transition disk era. We note that while typical distances of star-forming regions place detections of typical debris disk masses outside practical sensitivity limits, the dearth of high mass disks at late times is a real effect. We would be sensitive to very massive disks around nearby stars of 10+ Myr, and these are not seen. (Image courtesy of Olja Panić and the SONS Survey team.)

the decline of millimeter continuum emission is longer than the 3 Myr timescale observed by Haisch et al. (2001) during which time one half of the disks had disappeared in the infrared. This makes sense because the infrared emission traces emission closest to the star, whereas the millimeter emission preferentially traces cooler grains further from the star, which take longer to evolve or dissipate.

The fate of the millimeter grains lies in one of two paths: either they are rapidly ground down to smaller dust grains, rendering them susceptible to forces capable of rapidly removing the grains from the system via radiation pressure or, for low mass stars, stellar wind (e.g., Augereau and Beust 2006; Strubbe and Chiang 2006). Either way, their mass would be permanently removed from the disk, possibly following a short period of anomalously bright emission at infrared wavelengths and in scattered light due to the enhanced amount of small grains in the system. The alternative possibility is that the dust grains undergo a period of rapid grain growth to sizes that do not radiate efficiently at these observing wavelengths (>1 cm), rendering them invisible.

3 The Critical Role of Gas

Due to the absence of gas detections in debris disks, they were typically described as gas-poor, and the absence of gas emission was considered characteristic of debris systems. While this continues to be a generally accurate paradigm, there are now almost a dozen systems which have gas detections. The earliest detections were of CI and CO in UV absorption spectra toward β Pictoris (Roberge et al. 2000) and AU Microscopii via fluorescent H2 emission lines (France et al. 2007). More recently, detections are typically measurements of CO in emission, e.g., β Pictoris (Dent et al. 2014), 49 Ceti (Dent et al. 2005; Hughes et al. 2008; Zuckerman and Song 2012; Roberge et al. 2013, 2014), HD 21997 (Kóspál et al. 2013), HD 32297 (J.S. Greaves et al. in preparation), and several debris disk targets in the Sco–Cen association (Lieman-Sifry et al., submitted). In the case of the ALMA observations of β Pictoris, the distribution of gas mirrors that of the dust emission, with an extreme asymmetry to one side of the edge-on disk, likely at the location of a recent collision (Dent et al. 2014). In several other disks, the CO distribution is not ostensibly asymmetric and exhibits a Keplerian rotation curve. For example, the circumstellar disk of the 30 Myr old star HD 21997 has been resolved by ALMA. Its dust distribution is consistent with a debris disk (Moór et al. 2013), while the presence of the CO gas disk (Kóspál et al. 2013) in such an evolved source presents an ambiguity: is it remnant gas from a protoplanetary disk in its last throes or is the gas, as well as the dust, second generation, a product of the same collisions that produce the dust?

Kóspál et al. (2013) and Zuckerman and Song (2012) before them have pointed out that the CO mass ($4-8\times10^{-2}\ M_\oplus$) estimated in the HD 21997 disk indicates an unusually gas-rich disk *for a debris dust system*, suggesting HD 21997 (and 49 Ceti, Zuckerman and Song 2012) may just represent atypically massive protoplanetary disks. If the detected gas remains from the protoplanetary and transition disk phase, then the CO is just a proxy for H_2 emission and the mass could be as high as 26–60 $M_\oplus$. If this system is established as a true hybrid system, in which the dust is generated by collisional processing of larger bodies but the gas is not, then this object may represent an example of the final phase of a transition disk, before the remnant protoplanetary disk (small grains and gas) is dissipated and the disk becomes a true debris disk (Wyatt et al. 2015).

The possibility that gas removal is the final step in the evolution of a protoplanetary disk to a debris disk is significant because of the critical role played by gas in the final assembly of planetesimals. Gas acts to stabilize orbits of planetesimals and planets, and in its presence, small dust grains are well coupled to and co-orbiting with the gas disk. Slowing the relative velocities of dust grains enhances agglomeration properties in the disk, as described in the case of vortices in the previous section. The gas is also the primary reservoir of the magnetic field in the disk, which can act to create dead zones where mass can be focused to enhance agglomeration processes (Dzyurkevich et al. 2013).

In the solar system, the timing of gas dispersal can be constrained by the size distribution of material being delivered from the Oort Cloud. Small objects (<10 km, Brasser et al. 2007) would be more strongly affected than larger bodies by an increase in gas drag in the outer disk. If the delivery of material into the Oort cloud occurs in the presence of significant primordial gas, a reduced Oort Cloud results. In our own solar system this would be observable as a suppression in the number of 1–2 km size Oort cloud comets. Therefore, in the presence of gas, the material that would normally have arrived in a system's Oort Cloud region remains in the protoplanetary disk region, providing a substantial reservoir of short-period comets (Brasser et al. 2007). Significant gas presence during the late stage re-ordering might then provide the conditions needed for the substantial cometary number densities needed to explain the significant collisional processing seen in some systems, i.e., CO levels which could require a cometary collision every 6 seconds (e.g., 49 Ceti, Zuckerman and Song 2012).

As we will discuss below in more detail, the fact that gas chemistry is highly temperature dependent for many processes means that there will be variation in compositions in the gas throughout the disk, a radial function of temperature, and even potentially scale height. Over time, this, coupled with gas-solid chemistry, can lead to a diversity of surface properties in the planetesimals which should follow a systematic gradient as imparted from the disk. Once the gas is removed from the system, those diverse properties serve as a record of the location of those planetesimals in the disk, which should allow us to determine the final location of those planetesimals before significant dynamical evolution occurs.

Most models of the solar system are in fact predicated on this assumption, i.e., that the characteristics of the planetesimals we see today reveal that the dynamical history of the solar system must include several upheavals to account for the current observed properties of the population. The observed structure of the Kuiper belt clearly reveals that Neptune at least migrated (Malhotra 1993) and mostly likely scattered from a formation location near the present day locations of Saturn and Uranus (Levison and Morbidelli 2003). Meanwhile, the presence of Main Belt Comets and the variation among the surface properties of the various asteroid classes may imply that Jupiter and Saturn also had a complex migration history (Walsh et al. 2012). Such large scale restructuring events may be observable as chemical gradient anomalies in exo-debris disks when the gas is observable.

4 Insights from the Solar System

The debris disks of the solar system provide directly measurable examples of the suspected sources of dust seen around other stars. Observations of the current day structure of the Kuiper belt suggest that the early Kuiper belt experienced a large scale instability. Booth et al. (2009) provide a model of expected emission during such a destabilizing event in our solar system and compare that emission to detected debris disk systems. Lawler (2014) provides a model of the expected exo-debris disk emission that would be produced by the *current* Kuiper belt (see Fig. 2). The current day fractional luminosity ($\sim 10^{-7}$) is many orders of magnitude lower than the early disk predictions (a few $\times 10^{-3}$) from Booth et al.

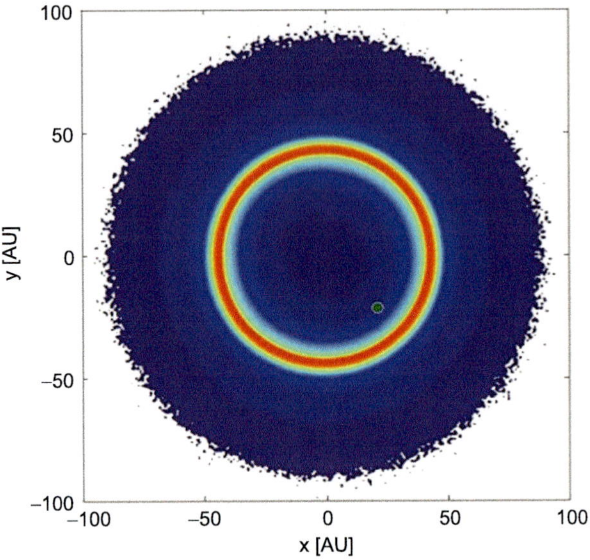

Fig. 2 Number density contour plots of the debiased Kuiper belt looking down on the Solar System from above. The position of Neptune is shown by a *green circle*. Classical objects dominate the number density, with resonant objects nearly invisible because of their much lower number density. Derived from Lawler (2014)

(2009). The orbital structure of the Kuiper belt provides an excellent motivation for the interpretation of currently observed exo-debris disks and their decay.

The solar zodiacal dust, which may be compared to warm dust around other stars, is, in the majority, produced through disruptive splitting of Jupiter family comets (JFCs) rather than through sublimation or asteroid belt collisions (Nesvorný et al. 2010). The size distribution of the zodiacal dust can be measured directly and is found to be consistent with a collisional cascade that occurs near the site of dust production (Jupiter's orbital distance) and then transported inward through PR-drag. Earlier in the evolution of our planetary system we expect that the production of JFCs would greatly exceed the current rate, due to a larger supply source in the region of the Kuiper belt, resulting in an inner warm dust belt that would be easily detected. Thus, the detection of warm dust, so called "exo-zodi" debris disk components, may be an indicator of a large cometary population that is undergoing disruption. The extrasolar disk systems in which warm dust has been detected exhibit significantly stronger emission than is seen around the Sun; see Kennedy and Wyatt (2013) for a discussion of such systems and their possible origins. (As for cold Kuiper belt analogues, we cannot yet detected warm excesses comparable to that of our own solar system.) The composition of the warm dust, having being produced predominantly by objects that formed in the more distant interplanetary region, is then not indicative of the local planetesimal formation conditions, but rather those at large distances where the temperatures, densities and chemical abundances would have been very different.

The time scales for the process of accretion are well constrained by measurements of isotopic ages of the meteorite population. Ages are ascertained by measurements of the decay products of ^{26}Al (^{26}Mg) in comparison to lead-lead dating (i.e., comparisons of ^{207}Pb or ^{206}Pb to ^{204}Pb). Comparing ages of adjacent components of a meteorite reveal the timescale for its creation, because once ^{26}Al is created, it must be bound up quickly before it undergoes decay. Applying these dating techniques to calcium–aluminum-rich inclusions (CAIs) and other components (called "grains") of H4 (among the most primordial and least heat-processed) chondrites indicate that these "grains" formed and underwent metamorphic processes within a few million years of the formation of the CAIs (Zinner and Göpel 2002).

This time scale for formation of ^{26}Mg-rich grains sets a short upper bound on the planetesimal formation timescale as the grains are taken up into the chondrites prior to the decay of ^{26}Al. These rapid (few million year) timescales are consistent with evidence that stars older than about 10 Myr are depleted in gas and do not possess protoplanetary dust disks (see Sect. 3). Models of the processes of disk evolution and planetesimal accretion must keep these formation timescales in mind.

The observable material in the asteroid and Kuiper belts have chemical properties that hold the signature of the thermal histories of those objects. Within the asteroid belt there is a clear gradient in reflected surface properties from refractory rich to volatile rich, while the Kuiper belt exhibits a diversity of processed surface ices. The gradients seen in the asteroid belt are nominally reflective of the formation locations of those objects. There are, however, anomalous objects whose properties appear more consistent with formation elsewhere within the disk, and these may indicate post-formation transport of already formed objects (Walsh et al. 2012). If the Kuiper belt ices are indicative of variations in formation location, then the diversity of surfaces reflects a large scale re-ordering of the system which would have left a detectable but very short-lived excess IR emission (Booth et al. 2009). Although collisions within the belt are currently rare, they were likely frequent during this re-ordering processes. As the transport of material was from the warmer (interior to Neptune) zone to the colder zone, the gas liberated from the ice during the collisional processing would be deficient in the volatile species that would have survived had the colliding objects formed in-situ. Detection of abundance anomalies in second generation gas in extrasolar debris disks may be a signature that such large scale re-ordering has occurred in those systems.

The observed orbital diversity in the Kuiper belt requires the giant planets to have experienced a large scale migration phase after their formation. Pluto's orbital resonance with Neptune has long been recognized as the strongest evidence of a past migration phase for Neptune (Malhotra 1993). Subsequent surveys of the Kuiper belt, see Bannister (2016, in press) for a complete listing, have revealed a rich orbital resonance structure that is not well matched by models that do not include Neptune migration (Gladman et al. 2012) while clearly indicating that smooth migration cannot have been the complete story. The Outer Solar System Origins Survey (OSSOS, Bannister et al. 2015, submitted) is specifically designed to measure the relative abundances of the various resonance populations and determine the migration history of the outer solar system. Within the non-resonant populations, the cold component of the main classical Kuiper belt (see Kavelaars et al. 2009, for definition) is composed of two distinct subcomponents: the Kernel and the Stirred (Petit et al. 2011). The Kernel component may be explained as the result of a 'skip' in Neptune's migration during a secular rearrangement of the outer solar system orbital architecture (Nesvorny 2015a). The structural rearrangements would have resulted in increased collisional cross-sections within sub-components of the Kuiper belt, likely resulting in the dust-ring type structures seen in extrasolar debris disks today.

It is possible that the Kernel subcomponent of the Kuiper belt formed in situ, rather than through migration processes from elsewhere in the disk. Then, the inferred mass of the protoplanetary disk, particularly in the region of the Kernel component, presents a challenge for 'standard' accretion scenarios. The reason in situ formation must be considered is that nearly 100% of the larger members of the cold classical Kuiper belt objects (KBOs) are binary systems (Noll et al. 2014), with a large fraction having the separation between the two components being a substantial fraction of the systems Hill radius. These wide-binary cold classical KBOs are very weakly bound and thus easily disrupted. The weakness of the binding argues in favor of their formation 'in-situ' at their current physical location (Parker and Kavelaars 2012). Indeed, the systems are so weakly bound that they would not survive in a bath of smaller objects produced during a collisional cascade (Parker and Kavelaars 2012). *These two constraints appear to require that the Kuiper belt region itself experienced in-situ planetesimal accretion that must have proceeded in a low-density environment where collisional disruption was minimal.* Formation within a dust-trap inside the protoplanetary disk may provide these conditions. Such dust-traps may have already been observed in transition disk systems (i.e., van der Marel et al. 2015a; Pinilla et al. 2015). In such a scenario, where there is no nearby massive planet, the planetesimal disk would be left relatively undisturbed, post-formation in the trap, resulting in little debris dust. This suggests that such cold accretion systems may build up, undetected. The "cold disks" reported by Krivov et al. (2013) are also suggestive of such a process.

Unlike the cold classical components, the dynamically hot, high inclination distribution component of the Kuiper belt appears to have been implanted by a large scale scattering process. The hot component appears to stretch from just beyond the orbit of Neptune out to 60+ AU (Kavelaars et al. 2009). Additionally, the inclination distribution of the hot component appears to require that the post scattering migration of Neptune was slow (Nesvorny 2015b). This component possesses very little in the way of large separation binaries (Grundy et al. 2010) suggesting a different dynamical history for this population compared to the Kernel component. The hot component has surface reflectance (Tegler and Romanishin 1998; Fraser and Brown 2012; Peixinho et al. 2015) and size distributions (Bernstein et al. 2004; Fuentes et al. 2010; Fraser et al. 2010; Petit et al. 2011) that appear to be distinct from the cold classical KBOs. The scattering event that produced the hot component would likely have resulted in significant collisional processing and dust production, especially if the event occurs in the inner solar system. Detection of sudden or very high levels of dust production in other systems (e.g., such as the detection of large amounts of warm dust by Meng et al. 2014; Melis et al. 2013) could be evidence that such large scale scattering events are common, as suggested by the solar system evolution models that contain them.

The chemical compositions of today's cometary population may provide some insight into the past restructuring processes. Observations of fractionation ratios in deuterated ice species such as DHO and DCN provide probes of the temperature at which the molecules condensed onto ice-forming grains (Mousis et al. 2000). The ice fractions and types of ice available provide additional constraints on the formation conditions (Allamandola et al. 1988). Disentangling which signatures result from interstellar origins (Mumma et al. 1996) and which might be a signature of re-arrangement of the planetesimal disk (Kavelaars et al. 2011) presents a significant challenge.

4.1 Deuterium Fractionation in the Early Solar Disk

One of the most distinctive imprints of the early disk is expected to be the fractionation of deuterium, or the "D/H ratio." In the low temperatures of the protostellar core of a forming solar system, fractionation of deuterium is expected to be enhanced in the gas phase and locked into microscopic ice particles even before the disk is formed (e.g. Kavelaars et al. 2011; Jørgensen and van Dishoeck 2010). The level of fractionation of water, defined as $f_{H_2O} = \frac{[DHO]/[H_2O]}{[DH]/[H_2]}$, is enhanced to about 35 times solar, while the level of fractionation of HCN, $f_{HCN} = \frac{[DCN]/[HCN]}{[DH]/[H_2]}$, is over 150 times solar. When the deuterium-enriched material is carried into the disk, the grains are heated and the molecules are released into the gas phase. What fraction of the deuterated species relative to the hydrogen species enters the gas phase depends on the individual molecule and the period of heating. The closer they are to the forming star, the warmer the grains become and more deuterium is released, and the fractionation can then vary based on the local gas chemistry. In the most distant parts of the disk, the grains may never get warm enough to release a given species into the gas phase (e.g., 30 K for water or HCN), in which case, the high ratio from the protostellar core should be retained. We therefore expect the D/H ratio to be large in the outer disk but drop closer to the star where temperatures are higher and deuterium enrichment is not favored.

Earth's value of f_{H_2O} is consistent with the lower values measured for primitive chondrites in the asteroid belt, an enrichment factor about 4 times solar (see, for example, Fig. 3 of Altwegg et al. 2015). Oort cloud comets, Saturn's moon Enceladus and the Rosetta measurement of comet 67P/CG show much higher values. The giant planets and the Sun have minimal enrichment factors consistent with a homogeneous protosolar nebula. Chondrites in fact show a large variation in f_{H_2O}, ranging from 4 to as high as 30, suggesting some condensed from the gas of the solar nebula, while others are consistent with formation in situ near the Earth. Several comets also show values consistent with Earth, but most show values between 10–20 with large error bars (i.e., at least 2x Earth). Interestingly, among the Jupiter family comets (JFCs), the recent result of (Altwegg et al. 2015) from Rosetta shows that comet 67P/CG has a very elevated value of f_{H_2O}, which is at odds with the other members of its class. The authors argue based on this result that the JFCs may be highly heterogeneous, reflecting diverse origins. Furthermore, the new measurement supports models that argue in favor of an asteroidal origin for Earth's oceans and its atmosphere, rather than a cometary one.

Interpretation of the variation in cometary D/H measurements in terms of the D/H ratio in the solar disk is greatly complicated by the fact that the objects observed have undergone dynamical evolution. Their D/H ratios were "locked in" at the values present in the gas phase when the ices condensed onto the asteroids and comets we observe today. We can't turn back the clock on the solar system, but we can observe young systems at this critical phase to measure how their D/H ratio varies. Since the D/H ratio in the gas phase should exhibit temperature (and hence radial) dependence, we should be able to observe a radial dependence if we can resolve such a disk.

Models, like Kavelaars et al. (2011), attempt to map the formation location of cometesimals using the observed variation of f_{H_2O} in cometary bodies. If it can be established that the D/H ratio of a protoplanetary system had a temperature dependence, then the formation distance of a given comet or asteroid can be recovered based on its relative D/H ratio. Such a tool would allow the original configuration of our solar system to be retraced, and importantly, untangle its dynamical evolution.

5 Tracing the Chemical History of Planetesimals

Where did Earth's water and atmosphere originate? This has been one of the key issues motivating the study of small bodies, comets and asteroids, in the solar system.

One of the key assumptions in the current model of the Solar system is that periods of dynamical instability occurred in the early stages of its evolution, particularly among the giant planets. This dynamical evolution was critical to Earth's habitability since these periods of upheaval were likely responsible for the delivery of the water and atmosphere we currently enjoy. Recent evidence, however, has called into question the long-accepted cometary origins for terrestrial oceans. Instead, it seems increasingly apparent that Earth's water has its likely origin in asteroids and bodies which formed much closer to the Earth (Altwegg et al. 2015; O'Brien et al. 2014; Albertsson et al. 2014). This interpretation is based on the idea that chemical evolution must also have occurred in the early solar system. As planetesimals form inside a protoplanetary disk, the chemical signatures and abundances in the disk should be recorded in those solid bodies, creating a record of the chemical diversity of the nascent solar disk.

What can this tell us about planet formation?

The position of objects in the solar system has evolved dynamically over time. Interpretation of the solar system's evolution relies on an assumption about the gas chemistry in the solar disk at the time the planets and planetesimals formed. These assumptions can be tested by observing protoplanetary disks at late stages and debris disks at early stages to directly measure the radial dependence of certain chemical tracers that are believed to have varied significantly at the time the planetesimal formation was completed. The probability is high, and it may seem obvious, that a variation in the D/H ratio was present in the solar system's disk when the planetesimals were formed, but this ratio has never been resolved in a disk of this era. Öberg et al. (2012) and Qi et al. (2008) measured the DCN abundances in the 8–12 Myr old TW Hydra disk and report a global f_{HCN} of 1.7×10^{-2}, many times the expected value in the early solar nebula, $(2.5 \pm 0.5) \times 10^{-5}$ (Mousis et al. 2000), suggesting this ratio should be detectable in other disks.

The most ideal targets would be those disks now being identified which appear to have highly evolved dust properties, but retain gas distributions that are potentially the remnants of a protoplanetary disk, e.g., HD 21997 (Kóspál et al. 2013; Moór et al. 2013). Changes to the dust grains can occur independently from the gas, since large pebbles are so poorly coupled to the gas that their disappearance would not alter the gas disk architecture. This scenario describes the situation in HD 21997 very well. Therefore it is likely that young debris systems exist in which the deuteration fraction has been imprinted onto the planetesimal population, according to the thermal chemistry at their formation location, and the gas phase is still present to measure that D/H ratio as well.

6 Summary

The study of debris disks provides direct evidence of planetesimal formation in other systems. Since debris disks must be stirred to see the bright collisionally produced dust detected thus far around 20–25% of AFGK stars, it can be presumed that in many, if not all, cases, the debris disks are also signposts of planetary systems as well. The timescales by which systems have evolved to exhibit debris disk qualities puts strong constraints on the limits of time available for giant planet formation; in actuality, the transition period from protoplanetary levels of dust and gas to debris levels of dust (and rarely detectable gas) is very abrupt. For this reason, young associations of known age are particular targets of study. Some, such as the TW Hydra moving group (~8 Myr) contain disks that are protoplanetary around some stars and still capable of forming planets (i.e., such as TW Hydra itself, Andrews et al. 2012; Bergin et al. 2013) while other stars are already debris disk hosts (e.g., HR 4796A and TWA 7, Koerner et al. 1998; Matthews et al. 2007).

The detection of gas associated with young debris disks (and of course late transition disks) has the potential to provide substantial information about the dynamical evolution of systems outside the solar system and provide important constraints on the evolution of the solar system itself. Systems in the transition epoch between protoplanetary and debris levels of mm-grain emission are ripe for detection of gas disk components. Therefore, observations of disks around the 10 Myr epoch transition period may capture the last phase of coalescence of solid material onto planetesimals, and therefore any gas detections should reflect the abundances and compositions deposited onto the planetesimals. Disks emerging from the transition as debris disks may exhibit detectable secondary gas disks, in which observed gas, like the dust, is a collisional product of the planetesimals.

The increased sensitivity and resolution of ALMA in particular is enabling the detection of secondary gas disks and even gradients in those disks, as evidenced by the recent detection of CO gas associated with several debris disks in Sco–Cen (Lieman-Sifry et al., submitted) and the detection of the secondary gas disk asymmetry of β Pictoris (Dent et al. 2014). To current sensitivity limits, $\sim$20–25% of nearby stars (less for G/K stars than A/F stars) show evidence of infrared excess emission associated with dust emission of debris disks. Despite the increased number of detected debris disks, the stronger dust and gas emission toward the shorter phase transition disks suggest that they can reveal more readily the location of planetesimal belt formation. This is particularly true at ALMA's observing wavelengths and resolution, as evidence by the strong body of work on sites of dust agglomeration already in the literature.

Wyatt et al. (2015) have identified the rapid growth of planetesimals prior to the loss of gas as one of the critical phases of the birth of a debris disk. This rapid formation phase is also supported by observations of radio-nucleotide signatures in the solar system. The radial thermal gradient in a transition disk should result in a radial variation of the refractory and volatile material being taken up into planetesimals and in predictable variations of thermally sensitive molecules such as DHO, DCN and other deuterated species. Resolved radial variations in these species will soon provide a detailed mapping of the thermal processes in the remotely observed systems.

Detailed studies of dynamical and physical properties of Kuiper Belt Objects are untangling the past evolution of giant planets, a necessary step in the understanding of giant planet formation in our solar system. Evidence of large scale re-ordering in the solar system suggests that chemical signatures in secondary gas and dust grain reflected light do not reflect the local conditions but provide evidence of transport of material in the system. Similarly, outside the solar system, detailed studies of resolved dust and gas distributions combined with known or future detections of planetary perturbers could dissect the history of the planetary system's dynamical evolution, since species trapped in planetesimal ices and released by collisions provide a record of the location of formation. This in effect would permit us to use the disk's ongoing collision evolution to trace its dynamical history.

Acknowledgements The authors acknowledge the efforts of O. Panić and W. Holland, who provided the revised SONS data of Fig. 1 prior to publication and S. Lawler for a revision of Fig. 2 for this publication. The authors also acknowledge the efforts of two referees, whose queries and suggestions improved this work.

References

J. Agarwal, D. Jewitt, H. Weaver, Dynamics of large fragments in the tail of active asteroid P/2010 A2. Astrophys. J. **769**, 46 (2013). doi:10.1088/0004-637X/769/1/46

T. Albertsson, D. Semenov, T. Henning, Chemodynamical deuterium fractionation in the early solar nebula: the origin of water on Earth and in asteroids and comets. Astrophys. J. **784**, 39 (2014). doi:10.1088/0004-637X/784/1/39

L.J. Allamandola, S.A. Sandford, G.J. Valero, Photochemical and thermal evolution of interstellar/precometary ice analogs. Icarus **76**, 225–252 (1988). doi:10.1016/0019-1035(88)90070-X

ALMA Partnership, C.L. Brogan, L.M. Pérez, T.R. Hunter, W.R.F. Dent, A.S. Hales, R.E. Hills, S. Corder, E.B. Fomalont, C. Vlahakis, Y. Asaki, D. Barkats, A. Hirota, J.A. Hodge, C.M.V. Impellizzeri, R. Kneissl, E. Liuzzo, R. Lucas, N. Marcelino, S. Matsushita, K. Nakanishi, N. Phillips, A.M.S. Richards, I. Toledo, R. Aladro, D. Broguiere, J.R. Cortes, P.C. Cortes, D. Espada, F. Galarza, D. Garcia-Appadoo, L. Guzman-Ramirez, E.M. Humphreys, T. Jung, S. Kameno, R.A. Laing, S. Leon, G. Marconi, A. Mignano, B. Nikolic, L.-A. Nyman, M. Radiszcz, A. Remijan, J.A. Rodón, T. Sawada, S. Takahashi, R.P.J. Tilanus, B. Vila Vilaro, L.C. Watson, T. Wiklind, E. Akiyama, E. Chapillon, I. de Gregorio-Monsalvo, J. Di Francesco, F. Gueth, A. Kawamura, C.-F. Lee, Q. Nguyen Luong, J. Mangum, V. Pietu, P. Sanhueza, K. Saigo, S. Takakuwa, C. Ubach, T. van Kempen, A. Wootten, A. Castro-Carrizo, H. Francke, J. Gallardo, J. Garcia, S. Gonzalez, T. Hill, T. Kaminski, Y. Kurono, H.-Y. Liu, C. Lopez, F. Morales, K. Plarre, G. Schieven, L. Testi, L. Videla, E. Villard, P. Andreani, J.E. Hibbard, K. Tatematsu, The 2014 ALMA long baseline campaign: first results from high angular resolution observations toward the HL Tau region. Astrophys. J. Lett. **808**, 3 (2015). doi:10.1088/2041-8205/808/1/L3

K. Altwegg, H. Balsiger, A. Bar-Nun, J.J. Berthelier, A. Bieler, P. Bochsler, C. Briois, U. Calmonte, M. Combi, J. De Keyser, P. Eberhardt, B. Fiethe, S. Fuselier, S. Gasc, T.I. Gombosi, K.C. Hansen, M. Hässig, A. Jäckel, E. Kopp, A. Korth, L. LeRoy, U. Mall, B. Marty, O. Mousis, E. Neefs, T. Owen, H. Rème, M. Rubin, T. Sémon, C.-Y. Tzou, H. Waite, P. Wurz, 67P/Churyumov–Gerasimenko, a Jupiter family comet with a high D/H ratio. Science **347**(27), 1261952 (2015). doi:10.1126/science.1261952

S.M. Andrews, D.J. Wilner, A.M. Hughes, C. Qi, K.A. Rosenfeld, K.I. Öberg, T. Birnstiel, C. Espaillat, L.A. Cieza, J.P. Williams, S.-Y. Lin, P.T.P. Ho, The TW Hya disk at 870 μm: comparison of CO and dust radial structures. Astrophys. J. **744**, 162 (2012). doi:10.1088/0004-637X/744/2/162

J.-C. Augereau, H. Beust, On the AU microscopii debris disk. Density profiles, grain properties, and dust dynamics. Astron. Astrophys. **455**, 987–999 (2006). doi:10.1051/0004-6361:20054250

M.T. Bannister, J.J. Kavelaars, J.-M. Petit, The outer solar system origins survey. I: design and first-quarter discoveries. Astron. J. (2015, submitted)

M.T. Bannister, Mapping the deep: discovery surveys in the outer Solar System. Planet. Space Sci. (2016, in press)

S.D. Barber, A.J. Patterson, M. Kilic, S.K. Leggett, P. Dufour, J.S. Bloom, D.L. Starr, The frequency of debris disks at white dwarfs. Astrophys. J. **760**, 26 (2012). doi:10.1088/0004-637X/760/1/26

E.A. Bergin, L.I. Cleeves, U. Gorti, K. Zhang, G.A. Blake, J.D. Green, S.M. Andrews, N.J. Evans II, T. Henning, K. Öberg, K. Pontoppidan, C. Qi, C. Salyk, E.F. van Dishoeck, An old disk still capable of forming a planetary system. Nature **493**, 644–646 (2013). doi:10.1038/nature11805

G.M. Bernstein, D.E. Trilling, R.L. Allen, M.E. Brown, M. Holman, R. Malhotra, The size distribution of trans-Neptunian bodies. Astron. J. **128**, 1364–1390 (2004). doi:10.1086/422919

J. Blum, G. Wurm, The growth mechanisms of macroscopic bodies in protoplanetary disks. Annu. Rev. Astron. Astrophys. **46**, 21–56 (2008). doi:10.1146/annurev.astro.46.060407.145152

A. Bonsor, G.M. Kennedy, J.R. Crepp, J.A. Johnson, M.C. Wyatt, B. Sibthorpe, K.Y.L. Su, Spatially resolved images of dust belt(s) around the planet-hosting subgiant κ CrB. Mon. Not. R. Astron. Soc. **431**, 3025–3035 (2013). doi:10.1093/mnras/stt367

A. Bonsor, G.M. Kennedy, M.C. Wyatt, J.A. Johnson, B. Sibthorpe, Herschel observations of debris discs orbiting planet-hosting subgiants. Mon. Not. R. Astron. Soc. **437**, 3288–3297 (2014). doi:10.1093/mnras/stt2128

M. Booth, M.C. Wyatt, A. Morbidelli, A. Moro-Martín, H.F. Levison, The history of the Solar system's debris disc: observable properties of the Kuiper belt. Mon. Not. R. Astron. Soc. **399**, 385–398 (2009). doi:10.1111/j.1365-2966.2009.15286.x

R. Brasser, M.J. Duncan, H.F. Levison, Embedded star clusters and the formation of the Oort cloud. II. The effect of the primordial solar nebula. Icarus **191**, 413–433 (2007). doi:10.1016/j.icarus.2007.05.003

D. Carrera, A. Johansen, M.B. Davies, How to form planetesimals from mm-sized chondrules and chondrule aggregates. Astron. Astrophys. **579**, 43 (2015). doi:10.1051/0004-6361/201425120

S. Casassus, G. van der Plas, S.P. M, W.R.F. Dent, E. Fomalont, J. Hagelberg, A. Hales, A. Jordán, D. Mawet, F. Ménard, A. Wootten, D. Wilner, A.M. Hughes, M.R. Schreiber, J.H. Girard, B. Ercolano, H. Canovas, P.E. Román, V. Salinas, Flows of gas through a protoplanetary gap. Nature **493**, 191–194 (2013). doi:10.1038/nature11769

W.R.F. Dent, J.S. Greaves, I.M. Coulson, CO emission from discs around isolated HAeBe and Vega-excess stars. Mon. Not. R. Astron. Soc. **359**, 663–676 (2005). doi:10.1111/j.1365-2966.2005.08938.x

W.R.F. Dent, M.C. Wyatt, A. Roberge, J.-C. Augereau, S. Casassus, S. Corder, J.S. Greaves, I. de Gregorio-Monsalvo, A. Hales, A.P. Jackson, A.M. Hughes, A.-M. Lagrange, B. Matthews, D. Wilner, Molecular gas clumps from the destruction of icy bodies in the β Pictoris debris disk. Science **343**, 1490–1492 (2014). doi:10.1126/science.1248726

P. Dufour, M. Kilic, G. Fontaine, P. Bergeron, C. Melis, J. Bochanski, Detailed compositional analysis of the heavily polluted DBZ white dwarf SDSS J073842.56+183509.06: a window on planet formation? Astrophys. J. **749**, 6 (2012). doi:10.1088/0004-637X/749/1/6

C.P. Dullemond, C. Dominik, Dust coagulation in protoplanetary disks: a rapid depletion of small grains. Astron. Astrophys. **434**, 971–986 (2005). doi:10.1051/0004-6361:20042080

N. Dzyurkevich, N.J. Turner, T. Henning, W. Kley, Magnetized accretion and dead zones in protostellar disks. Astrophys. J. **765**, 114 (2013). doi:10.1088/0004-637X/765/2/114

C. Eiroa, J.P. Marshall, A. Mora, B. Montesinos, O. Absil, J.C. Augereau, A. Bayo, G. Bryden, W. Danchi, C. del Burgo, S. Ertel, M. Fridlund, A.M. Heras, A.V. Krivov, R. Launhardt, R. Liseau, T. Löhne, J. Maldonado, G.L. Pilbratt, A. Roberge, J. Rodmann, J. Sanz-Forcada, E. Solano, K. Stapelfeldt, P. Thébault, S. Wolf, D. Ardila, M. Arévalo, C. Beichmann, V. Faramaz, B.M. González-García, R. Gutiérrez, J. Lebreton, R. Martínez-Arnáiz, G. Meeus, D. Montes, G. Olofsson, K.Y.L. Su, G.J. White, D. Barrado, M. Fukagawa, E. Grün, I. Kamp, R. Lorente, A. Morbidelli, S. Müller, H. Mutschke, T. Nakagawa, I. Ribas, H. Walker, DUst around NEarby stars. The survey observational results. Astron. Astrophys. **555**, 11 (2013). doi:10.1051/0004-6361/201321050

C. Espaillat, J. Muzerolle, J. Najita, S. Andrews, Z. Zhu, N. Calvet, S. Kraus, J. Hashimoto, A. Kraus, P. D'Alessio, An observational perspective of transitional disks, in *Protostars and Planets VI* (2014), pp. 497–520 doi:10.2458/azu_uapress_9780816531240-ch022

K. France, A. Roberge, R.E. Lupu, S. Redfield, P.D. Feldman, A low-mass H_2 component to the AU microscopii circumstellar disk. Astrophys. J. **668**, 1174–1181 (2007). doi:10.1086/521348

W.C. Fraser, M.E. Brown, The Hubble wide field camera 3 test of surfaces in the Outer Solar system: the compositional classes of the Kuiper belt. Astrophys. J. **749**, 33 (2012). doi:10.1088/0004-637X/749/1/33

W.C. Fraser, M.E. Brown, M.E. Schwamb, The luminosity function of the hot and cold Kuiper belt populations. Icarus **210**, 944–955 (2010). doi:10.1016/j.icarus.2010.08.001

C.I. Fuentes, M.J. Holman, D.E. Trilling, P. Protopapas, Trans-Neptunian objects with Hubble Space Telescope ACS/WFC. Astrophys. J. **722**, 1290–1302 (2010). doi:10.1088/0004-637X/722/2/1290

M. Fukagawa, T. Tsukagoshi, M. Momose, K. Saigo, N. Ohashi, Y. Kitamura, S.-I. Inutsuka, T. Muto, H. Nomura, T. Takeuchi, H. Kobayashi, T. Hanawa, E. Akiyama, M. Honda, H. Fujiwara, A. Kataoka, S.Z. Takahashi, H. Shibai, Local enhancement of the surface density in the protoplanetary ring surrounding HD 142527. Publ. Astron. Soc. Jpn. **65**, 14 (2013). doi:10.1093/pasj/65.6.L14

B. Gladman, S.M. Lawler, J.-M. Petit, J. Kavelaars, R.L. Jones, J.W. Parker, C. Van Laerhoven, P. Nicholson, P. Rousselot, A. Bieryla, M.L.N. Ashby, The resonant trans-Neptunian populations. Astron. J. **144**, 23 (2012). doi:10.1088/0004-6256/144/1/23

W.M. Grundy, K.S. Noll, M.W. Buie, S.D. Benecchi, H.G. Roe, S.B. Porter, F. Nimmo, J.A. Stansberry, D.C. Stephens, H.F. Levison, Transneptunian binaries: statistics of orbital properties, in *AAS/Division for Planetary Sciences Meeting Abstracts #42*. Bulletin of the American Astronomical Society, vol. 42 (2010), p. 991

K.E. Haisch Jr., E.A. Lada, C.J. Lada, Disk frequencies and lifetimes in young clusters. Astrophys. J. Lett. **553**, 153–156 (2001). doi:10.1086/320685

A.M. Hughes, D.J. Wilner, I. Kamp, M.R. Hogerheijde, A resolved molecular gas disk around the nearby a star 49 Ceti. Astrophys. J. **681**, 626–635 (2008). doi:10.1086/588520

D. Jewitt, H. Hsieh, J. Agarwal, The active asteroids. ArXiv e-prints (2015)

A. Johansen, J.S. Oishi, M.-M. Mac Low, H. Klahr, T. Henning, A. Youdin, Rapid planetesimal formation in turbulent circumstellar disks. Nature **448**, 1022–1025 (2007). doi:10.1038/nature06086

J.K. Jørgensen, E.F. van Dishoeck, The HDO/H_2O ratio in gas in the inner regions of a low-mass protostar. Astrophys. J. Lett. **725**, 172–175 (2010). doi:10.1088/2041-8205/725/2/L172

J.J. Kavelaars, R.L. Jones, B.J. Gladman, J.-M. Petit, J.W. Parker, C. Van Laerhoven, P. Nicholson, P. Rousselot, H. Scholl, O. Mousis, B. Marsden, P. Benavidez, A. Bieryla, A. Campo Bagatin, A. Doressoundiram, J.L. Margot, I. Murray, C. Veillet, The Canada–France ecliptic plane survey–L3 data release: the orbital structure of the Kuiper belt. Astron. J. **137**, 4917–4935 (2009). doi:10.1088/0004-6256/137/6/4917

J.J. Kavelaars, O. Mousis, J.-M. Petit, H.A. Weaver, On the formation location of Uranus and Neptune as constrained by dynamical and chemical models of comets. Astrophys. J. Lett. **734**, 30 (2011). doi:10.1088/2041-8205/734/2/L30

G.M. Kennedy, M.C. Wyatt, The bright end of the exo-Zodi luminosity function: disc evolution and implications for exo-Earth detectability. Mon. Not. R. Astron. Soc. **433**, 2334–2356 (2013). doi:10.1093/mnras/stt900

M. Kilic, S. Redfield, A dusty disk around WD 1150-153: explaining the metals in white dwarfs by accretion from the interstellar medium versus debris disks. Astrophys. J. **660**, 641–650 (2007). doi:10.1086/513008

D.W. Koerner, M.E. Ressler, M.W. Werner, D.E. Backman, Mid-infrared imaging of a circumstellar disk around HR 4796: mapping the debris of planetary formation. Astrophys. J. Lett. **503**, 83–87 (1998). doi:10.1086/311525

Á. Kóspál, A. Moór, A. Juhász, P. Ábrahám, D. Apai, T. Csengeri, C.A. Grady, T. Henning, A.M. Hughes, C. Kiss, I. Pascucci, M. Schmalzl, ALMA observations of the molecular gas in the debris disk of the 30 Myr old star HD 21997. Astrophys. J. **776**, 77 (2013). doi:10.1088/0004-637X/776/2/77

A.V. Krivov, C. Eiroa, T. Löhne, J.P. Marshall, B. Montesinos, C. del Burgo, O. Absil, D. Ardila, J.-C. Augereau, A. Bayo, G. Bryden, W. Danchi, S. Ertel, J. Lebreton, R. Liseau, A. Mora, A.J. Mustill, H. Mutschke, R. Neuhäuser, G.L. Pilbratt, A. Roberge, T.O.B. Schmidt, K.R. Stapelfeldt, P. Thébault, C. Vitense, G.J. White, S. Wolf, Herschel's "cold debris disks": background galaxies or quiescent rims of planetary systems? Astrophys. J. **772**, 32 (2013). doi:10.1088/0004-637X/772/1/32

S.M. Lawler, The debiased Kuiper belt: our Solar System as a debris disk, in *IAU Symposium*, ed. by M. Booth, B.C. Matthews, J.R. Graham IAU Symposium, vol. 299, 2014, pp. 232–236. doi:10.1017/S1743921313008466

H.F. Levison, A. Morbidelli, The formation of the Kuiper belt by the outward transport of bodies during Neptune's migration. Nature **426**, 419–421 (2003)

R. Malhotra, The origin of Pluto's peculiar orbit. Nature **365**, 819–821 (1993). doi:10.1038/365819a0

J.P. Marshall, A. Moro-Martín, C. Eiroa, G. Kennedy, A. Mora, B. Sibthorpe, J.-F. Lestrade, J. Maldonado, J. Sanz-Forcada, M.C. Wyatt, B. Matthews, J. Horner, B. Montesinos, G. Bryden, C. del Burgo, J.S. Greaves, R.J. Ivison, G. Meeus, G. Olofsson, G.L. Pilbratt, G.J. White, Correlations between the stellar, planetary, and debris components of exoplanet systems observed by Herschel. Astron. Astrophys. **565**, 15 (2014). doi:10.1051/0004-6361/201323058

B.C. Matthews, P.G. Kalas, M.C. Wyatt, Mass and temperature of the TWA 7 debris disk. Astrophys. J. **663**, 1103–1109 (2007). doi:10.1086/518643

B.C. Matthews, A.V. Krivov, M.C. Wyatt, G. Bryden, C. Eiroa, Observations, modeling, and theory of debris disks, in *Protostars and Planets VI*, (2014), pp. 521–544. doi:10.2458/azu_uapress_9780816531240-ch023

C. Melis, B. Zuckerman, J.H. Rhee, I. Song, S.J. Murphy, M.S. Bessell, Copious amounts of hot and cold dust orbiting the main sequence A-type stars HD 131488 and HD 121191. Astrophys. J. **778**, 12 (2013). doi:10.1088/0004-637X/778/1/12

H.Y.A. Meng, K.Y.L. Su, G.H. Rieke, D.J. Stevenson, P. Plavchan, W. Rujopakarn, C.M. Lisse, S. Poshyachinda, D.E. Reichart, Large impacts around a solar-analog star in the era of terrestrial planet formation. Science **345**, 1032–1035 (2014). doi:10.1126/science.1255153

A. Moór, A. Juhász, Á. Kóspál, P. Ábrahám, D. Apai, T. Csengeri, C. Grady, T. Henning, A.M. Hughes, C. Kiss, I. Pascucci, M. Schmalzl, K. Gabányi, ALMA continuum observations of a 30 Myr old gaseous debris disk around HD 21997. Astrophys. J. Lett. **777**, 25 (2013). doi:10.1088/2041-8205/777/2/L25

A. Moro-Martín, J.P. Marshall, G. Kennedy, B. Sibthorpe, B.C. Matthews, C. Eiroa, M.C. Wyatt, J.-F. Lestrade, J. Maldonado, D. Rodriguez, J.S. Greaves, B. Montesinos, A. Mora, M. Booth, G. Duchêne, D. Wilner, J. Horner, Does the presence of planets affect the frequency and properties of extrasolar Kuiper belts? Results from the Herschel debris and Dunes surveys. Astrophys. J. **801**, 143 (2015). doi:10.1088/0004-637X/801/2/143

O. Mousis, D. Gautier, D. Bockelée-Morvan, F. Robert, B. Dubrulle, A. Drouart, Constraints on the formation of comets from D/H ratios measured in H_2O and HCN. Icarus **148**, 513–525 (2000). doi:10.1006/icar.2000.6499

M.J. Mumma, M.A. Disanti, N. dello Russo, M. Fomenkova, K. Magee-Sauer, C.D. Kaminski, D.X. Xie, Detection of abundant ethane and methane, along with carbon monoxide and water, in comet C/1996 B2 Hyakutake: evidence for interstellar origin. Science **272**, 1310–1314 (1996). doi:10.1126/science.272.5266.1310

D. Nesvorny, Jumping Neptune can explain the Kuiper belt kernel. ArXiv e-prints (2015a)

D. Nesvorny, The evidence for slow migration of Neptune from the inclination distribution of Kuiper belt objects. ArXiv e-prints (2015b)

D. Nesvornỳ, P. Jenniskens, H.F. Levison, W.F. Bottke, D. Vokrouhlickỳ, M. Gounelle, Cometary origin of the zodiacal cloud and carbonaceous micrometeorites. Implications for hot debris disks. Astrophys. J. **713**(2), 816 (2010)

K.S. Noll, A.H. Parker, W.M. Grundy, All bright cold classical KBOs are binary, in *AAS/Division for Planetary Sciences Meeting Abstracts*, vol. 46, (2014), 507.05

K.I. Öberg, R. Murray-Clay, E.A. Bergin, The effects of snowlines on C/O in planetary atmospheres. Astrophys. J. Lett. **743**, 16 (2011). doi:10.1088/2041-8205/743/1/L16

K.I. Öberg, C. Qi, D.J. Wilner, M.R. Hogerheijde, Evidence for multiple pathways to deuterium enhancements in protoplanetary disks. Astrophys. J. **749**, 162 (2012). doi:10.1088/0004-637X/749/2/162

D.P. O'Brien, K.J. Walsh, A. Morbidelli, S.N. Raymond, A.M. Mandell, Water delivery and giant impacts in the Grand Tack scenario. Icarus **239**, 74–84 (2014). doi:10.1016/j.icarus.2014.05.009

O. Panić, W.S. Holland, M.C. Wyatt, G.M. Kennedy, B.C. Matthews, J.F. Lestrade, B. Sibthorpe, J.S. Greaves, J.P. Marshall, N.M. Phillips, J. Tottle, First results of the SONS survey: submillimeter detections of debris discs. Mon. Not. R. Astron. Soc. **435**, 1037–1046 (2013). doi:10.1093/mnras/stt1293

A.H. Parker, J.J. Kavelaars, Collisional evolution of ultra-wide trans-Neptunian binaries. Astrophys. J. **744**, 139 (2012). doi:10.1088/0004-637X/744/2/139

N. Peixinho, A. Delsanti, A. Doressoundiram, Reanalyzing the visible colors of Centaurs and KBOs: what is there and what we might be missing. Astron. Astrophys. **577**, 35 (2015). doi:10.1051/0004-6361/201425436

L.M. Pérez, A. Isella, J.M. Carpenter, C.J. Chandler, Large-scale asymmetries in the transitional disks of SAO 206462 and SR 21. Astrophys. J. Lett. **783**, 13 (2014). doi:10.1088/2041-8205/783/1/L13

J.-M. Petit, J.J. Kavelaars, B.J. Gladman, R.L. Jones, J.W. Parker, C. Van Laerhoven, P. Nicholson, G. Mars, P. Rousselot, O. Mousis, B. Marsden, A. Bieryla, M. Taylor, M.L.N. Ashby, P. Benavidez, A. Campo Bagatin, G. Bernabeu, The Canada–France ecliptic plane survey–full data release: the orbital structure of the Kuiper belt. Astron. J. **142**, 131 (2011). doi:10.1088/0004-6256/142/4/131

J.E. Pineda, S.P. Quanz, F. Meru, G.D. Mulders, M.R. Meyer, O. Panić, H. Avenhaus, Resolved images of the protoplanetary disk around HD 100546 with ALMA. Astrophys. J. Lett. **788**, 34 (2014). doi:10.1088/2041-8205/788/2/L34

P. Pinilla, N. van der Marel, L.M. Pérez, E.F. van Dishoeck, S. Andrews, T. Birnstiel, G. Herczeg, K.M. Pontoppidan, T. van Kempen, Testing particle trapping in transition disks with ALMA. Astron. Astrophys. **584**, 16 (2015). doi:10.1051/0004-6361/201526655

C. Qi, D.J. Wilner, Y. Aikawa, G.A. Blake, M.R. Hogerheijde, Resolving the chemistry in the disk of TW hydrae. I. Deuterated species. Astrophys. J. **681**, 1396–1407 (2008). doi:10.1086/588516

A. Roberge, P.D. Feldman, A.M. Lagrange, A. Vidal-Madjar, R. Ferlet, A. Jolly, J.L. Lemaire, F. Rostas, High-resolution Hubble Space Telescope STIS spectra of C I and CO in theβ Pictoris circumstellar disk. Astrophys. J. **538**, 904–910 (2000). doi:10.1086/309157

A. Roberge, I. Kamp, B. Montesinos, W.R.F. Dent, G. Meeus, J.K. Donaldson, J. Olofsson, A. Moór, J.-C. Augereau, C. Howard, C. Eiroa, W.-F. Thi, D.R. Ardila, G. Sandell, P. Woitke, Herschel observations of gas and dust in the unusual 49 Ceti debris disk. Astrophys. J. **771**, 69 (2013). doi:10.1088/0004-637X/771/1/69

A. Roberge, B.Y. Welsh, I. Kamp, A.J. Weinberger, C.A. Grady, Volatile-rich circumstellar gas in the unusual 49 Ceti debris disk. Astrophys. J. Lett. **796**, 11 (2014). doi:10.1088/2041-8205/796/1/L11

L.E. Strubbe, E.I. Chiang, Dust dynamics, surface brightness profiles, and thermal spectra of debris disks: the case of AU microscopii. Astrophys. J. **648**, 652–665 (2006). doi:10.1086/505736

K.Y.L. Su, G.H. Rieke, K.R. Stapelfeldt, R. Malhotra, G. Bryden, P.S. Smith, K.A. Misselt, A. Moro-Martin, J.P. Williams, The debris disk around HR 8799. Astrophys. J. **705**, 314–327 (2009). doi:10.1088/0004-637X/705/1/314

S.C. Tegler, W. Romanishin, Two distinct populations of Kuiper-belt objects. Nature **392**, 49 (1998). doi:10.1038/32108

L. Testi, T. Birnstiel, L. Ricci, S. Andrews, J. Blum, J. Carpenter, C. Dominik, A. Isella, A. Natta, J.P. Williams, D.J. Wilner, Dust evolution in protoplanetary disks, in *Protostars and Planets VI* (2014), pp. 339–361. doi:10.2458/azu_uapress_9780816531240-ch015

N.D. Thureau, J.S. Greaves, B.C. Matthews, G. Kennedy, N. Phillips, M. Booth, G. Duchêne, J. Horner, D.R. Rodriguez, B. Sibthorpe, M.C. Wyatt, An unbiased study of debris discs around A-type stars with Herschel. Mon. Not. R. Astron. Soc. **445**, 2558–2573 (2014). doi:10.1093/mnras/stu1864

N. van der Marel, E.F. van Dishoeck, S. Bruderer, T. Birnstiel, P. Pinilla, C.P. Dullemond, T.A. van Kempen, M. Schmalzl, J.M. Brown, G.J. Herczeg, G.S. Mathews, V. Geers, A major asymmetric dust trap in a transition disk. Science **340**, 1199–1202 (2013). doi:10.1126/science.1236770

N. van der Marel, P. Pinilla, J. Tobin, T. van Kempen, S. Andrews, L. Ricci, T. Birnstiel, A concentration of centimeter-sized grains in the Ophiuchus IRS 48 dust trap. Astrophys. J. Lett. **810**, 7 (2015a). doi:10.1088/2041-8205/810/1/L7

N. van der Marel, E.F. van Dishoeck, S. Bruderer, L. Pérez, A. Isella, Gas density drops inside dust cavities of transitional disks around young stars observed with ALMA. Astron. Astrophys. **579**, 106 (2015b). doi:10.1051/0004-6361/201525658

C. Walsh, A. Juhász, P. Pinilla, D. Harsono, G.S. Mathews, W.R.F. Dent, M.R. Hogerheijde, T. Birnstiel, G. Meeus, H. Nomura, Y. Aikawa, T.J. Millar, G. Sandell, ALMA hints at the presence of two companions in the disk around HD 100546. Astrophys. J. Lett. **791**, 6 (2014). doi:10.1088/2041-8205/791/1/L6

K.J. Walsh, A. Morbidelli, S.N. Raymond, D.P. O'Brien, A.M. Mandell, Populating the asteroid belt from two parent source regions due to the migration of giant planets: the grand tack. Meteorit. Planet. Sci. **47**, 1941–1947 (2012). doi:10.1111/j.1945-5100.2012.01418.x

S.J. Weidenschilling, Dust to planetesimals—settling and coagulation in the solar nebula. Icarus **44**, 172–189 (1980). doi:10.1016/0019-1035(80)90064-0

M.C. Wyatt, Dust in resonant extrasolar Kuiper belts: grain size and wavelength dependence of disk structure. Astrophys. J. **639**, 1153–1165 (2006). doi:10.1086/499487

M.C. Wyatt, Evolution of debris disks. Annu. Rev. Astron. Astrophys. **46**, 339–383 (2008). doi:10.1146/annurev.astro.45.051806.110525

M.C. Wyatt, R. Smith, K.Y.L. Su, G.H. Rieke, J.S. Greaves, C.A. Beichman, G. Bryden, Steady state evolution of debris disks around a stars. Astrophys. J. **663**, 365–382 (2007). doi:10.1086/518404

M.C. Wyatt, G. Kennedy, B. Sibthorpe, A. Moro-Martín, J.-F. Lestrade, R.J. Ivison, B. Matthews, S. Udry, J.S. Greaves, P. Kalas, S. Lawler, K.Y.L. Su, G.H. Rieke, M. Booth, G. Bryden, J. Horner, J.J. Kavelaars, D. Wilner, Herschel imaging of 61 Vir: implications for the prevalence of debris in low-mass planetary systems. Mon. Not. R. Astron. Soc. **424**, 1206–1223 (2012). doi:10.1111/j.1365-2966.2012.21298.x

M.C. Wyatt, O. Panić, G.M. Kennedy, L. Matrà, Five steps in the evolution from protoplanetary to debris disk. Astrophys. Space Sci. **357**, 103 (2015). doi:10.1007/s10509-015-2315-6

K. Zhang, G.A. Blake, E.A. Bergin, Evidence of fast pebble growth near condensation fronts in the HL tau protoplanetary disk. Astrophys. J. Lett. **806**, 7 (2015). doi:10.1088/2041-8205/806/1/L7

E. Zinner, C. Göpel, Aluminum-26 in H4 chondrites: implications for its production and its usefulness as a fine-scale chronometer for early solar system events. Meteorit. Planet. Sci. **37**, 1001–1013 (2002)

B. Zuckerman, I. Song, A 40 Myr old gaseous circumstellar disk at 49 Ceti: massive CO-rich comet clouds at young A-type stars. Astrophys. J. **758**, 77 (2012). doi:10.1088/0004-637X/758/2/77

B. Zuckerman, D. Koester, I.N. Reid, M. Hünsch, Metal lines in DA white dwarfs. Astrophys. J. **596**, 477–495 (2003). doi:10.1086/377492

Space Sci Rev (2016) 205:273–307
DOI 10.1007/s11214-016-0248-1

Insights into Planet Formation from Debris Disks

II. Giant Impacts in Extrasolar Planetary Systems

Mark C. Wyatt[1] · Alan P. Jackson[2]

Received: 18 September 2015 / Accepted: 14 March 2016 / Published online: 22 March 2016

Abstract *Giant impacts* refer to collisions between two objects each of which is massive enough to be considered at least a planetary embryo. The putative collision suffered by the proto-Earth that created the Moon is a prime example, though most Solar System bodies bear signatures of such collisions. Current planet formation models predict that an epoch of giant impacts may be inevitable, and observations of debris around other stars are providing mounting evidence that giant impacts feature in the evolution of many planetary systems. This chapter reviews giant impacts, focussing on what we can learn about planet formation by studying debris around other stars. Giant impact debris evolves through mutual collisions and dynamical interactions with planets. General aspects of this evolution are outlined, noting the importance of the collision-point geometry. The detectability of the debris is discussed using the example of the Moon-forming impact. Such debris could be detectable around another star up to 10 Myr post-impact, but model uncertainties could reduce detectability to a few 100 yr window. Nevertheless the 3 % of young stars with debris at levels expected during terrestrial planet formation provide valuable constraints on formation models; implications for super-Earth formation are also discussed. Variability recently observed in some bright disks promises to illuminate the evolution during the earliest phases when vapour condensates may be optically thick and acutely affected by the collision-point geometry. The outer reaches of planetary systems may also exhibit signatures of giant impacts, such as the clumpy debris structures seen around some stars.

Keywords Circumstellar disks · Planet formation

1 Introduction

The majority of the processes that contribute to the formation of a planetary system are thought to take place in the massive circumstellar disks that surround young stars called

✉ M.C. Wyatt
wyatt@ast.cam.ac.uk

[1] Institute of Astronomy, University of Cambridge, Madingley Road, Cambridge CB3 0HA, UK

[2] School of Earth and Space Exploration, Arizona State University, Tempe, AZ 85287, USA

protoplanetary disks. Such disks last for up to around 10 Myr before dispersing through mechanisms that are still debated, leaving a planetary system and a debris disk (Wyatt et al. 2015). The debris disk is made up of the components of the planetary system that are not massive enough to be classed as planets, i.e., planetesimals, dust and gas (see Matthews et al. 2014). For example, the Asteroid Belt and Kuiper Belt account for the majority of Solar System's debris disk, along with the Zodiacal Cloud and the Oort Cloud (see e.g., Matthews and Kavelaars 2016).

The motivation for saying that planet formation processes are largely complete by the time the protoplanetary disk is dispersed is based on the absence of large quantities of gas and dust at later stages from which to form planets (e.g., Haisch et al. 2001). However, while the disappearance of the protoplanetary gas disk does make the formation of gas giant planets like Jupiter rather problematic, this does not necessarily prevent the continued growth of solid rocky or icy planets. From an observational perspective we can only say that the mass present in less than cm-sized dust cannot be significant after the protoplanetary disk has dispersed (Panić et al. 2013). The majority of the solid mass in the system could instead be locked up in planets (or planetary embryos) that continue to grow. Indeed this is the basis of some models for planetary system formation and evolution that aim to explain some debris disk observations (Kenyon and Bromley 2008).

Isotopic compositions indicate that the parent bodies of many of the meteorites formed within the first few Myr of the Solar System's history (e.g. Kleine et al. 2009), i.e., while the protoplanetary disk would have still been present. However, a lengthier timescale for the formation of solid rocky planets is inferred from isotopic measurements of the terrestrial planets, which shows that they formed over a range of timescales, with Earth taking up to ~ 100 Myr to reach its final mass (e.g. Halliday 2008). The long formation timescale for the Earth is readily understood within the context of favoured models for planet formation. In such models, the protoplanetary disk is suffused with planetesimals that grow via mutual collisions to eventually become planets, with a system of terrestrial planets that bears many similarities to our own inner Solar System a likely outcome for the right initial conditions (e.g., Hansen 2009; Walsh et al. 2011). Of relevance to the current discussion is that planetesimal growth results in many planetary embryos (Kokubo and Ida 1998), and in the final stage the number of embryos is whittled down to a few as they undergo repeated chaotic close encounters and eventually giant collisions, until the few that remain are sufficiently separated to be dynamically stable (e.g., Goldreich et al. 2004; Kenyon and Bromley 2006). One consequence of this scenario is that the formation of the Earth's moon in a giant impact ~ 100 Myr after the Solar System formed is not an atypical occurrence.

The detailed structures of other inner Solar System bodies also bear witness to a past epoch of giant (i.e., planetary scale) impacts in the inner Solar System (e.g., Benz et al. 1988; Marinova et al. 2011). There are more uncertainties about the exact formation mechanism of the outer Solar System (Goldreich et al. 2004; Helled and Bodenheimer 2014), but there are several lines of evidence to suggest that giant impacts were also a feature of the more distant reaches of the Solar System (e.g., Safronov 1972; Canup 2005). There is also growing evidence for giant impacts having occurred recently in some extrasolar planetary systems, from observations of what could be impact-generated debris (Song et al. 2005; Rhee et al. 2008). Stars with anomalously high levels of dust in their regions that would be analogous to the location of terrestrial planets in the inner Solar System are often interpreted within the context of terrestrial planet formation (e.g., Lisse et al. 2008; Melis et al. 2010), particularly if they are at an age at which planet formation models predict such giant impacts to be taking place (i.e., $\ll 100$ Myr). Clearly, studying this debris and how frequently it is detected around nearby stars has the potential to inform about terrestrial planet formation processes more generally, as well as the fraction of stars that host such planets (Jackson and Wyatt 2012). Some stars also show evidence that could point to recent giant impacts in the more distant reaches of their planetary systems (Telesco et al. 2005; Dent et al. 2014; Stark et al. 2014), but conclusions remain tentative for now.

The aim of this chapter is to review the topic of giant impacts, focussing on the potential of observations of giant impact debris to inform on planet formation processes. It starts with evidence for giant impacts in the Solar System (Sect. 2) before discussing the various possible observable consequences of such impacts in extrasolar systems (Sect. 3). Generic aspects of the evolution of impact-generated debris are then discussed in Sect. 4. The observability of debris from the Moon-forming giant impact is considered in Sect. 5 and then compared with observations of extrasolar debris disks (Sect. 6). Giant impacts at larger distances from the star are discussed in Sect. 7, followed by a discussion of the implications from debris observations for the formation of the ubiquitous super-Earth class of planetary system (Sect. 8). Conclusions are given in Sect. 9.

2 Giant Impacts in the Solar System

In discussing the role of giant impacts in the formation of planets it is beneficial to turn first to our own Solar System, as here we have detailed information about the final end products—the planets and satellites themselves. Aside from Jupiter and possibly Saturn, true twins for the Solar System planets (in both size and orbital location) remain tantalisingly outside detection limits of searches for extrasolar planets at present. While future missions will certainly extend these detection limits and allow us to begin detecting true Solar System twins, we will presumably always have more information about our own backyard than other star systems. As is common in discussions of planet formation we divide our overview of the evidence for giant impacts in the Solar System into the inner Solar System (interior to Jupiter) and the outer Solar System (exterior to and including Jupiter).

2.1 Inner Solar System

The inner Solar System is replete with evidence for giant impacts. Each of the terrestrial planets has features for which a giant impact explanation has been suggested, with the exception of Venus, and it is probably no coincidence that Venus is also the terrestrial planet whose surface and geological history are least well understood.

Mercury In comparison with the other terrestrial planets Mercury is unusually rich in iron, with a very large core. This extreme iron enrichment can be reproduced by a collision between a roughly 2.25 M_{Mercury} chondritic composition proto-Mercury and a roughly 0.4 M_{Mercury} impactor at speeds of around 30 $\mathrm{km\,s^{-1}}$ (e.g. Benz et al. 1988; Cameron et al. 1988; Anic 2006; Benz et al. 2007). This impact speed is around 6 times the escape velocity of proto-Mercury and the resulting impact is extremely violent, almost destroying the planet. While commonly thought of as a 'mantle-stripping' event, in physical terms it is gravitational re-accumulation that is responsible for bringing the core of the former planet back together, rather than the outer layers being peeled away to reveal the exposed core. It is also notable that proto-Mercury in this scenario is very similar in mass to Mars, being around 16 per cent more massive.

Debris is a key component of the formation of Mercury via a giant impact, accounting for a larger fraction of the initial mass than the final planet. As noted by Gladman and Coffey (2009) the fate of this debris, and in particular the fraction that is re-accreted, is essential to the final success or failure of a giant impact model to reproduce Mercury.

Earth-Moon System Perhaps the most striking thing about Earth (from a dynamical perspective) is the large mass of the Moon. At 1.2 per cent of Earth's mass the Moon is, by a substantial margin, the largest satellite relative to the size of its parent planet in the Solar System.

To find a comparable system one must look to the dwarf planet binary Pluto-Charon with a mass ratio of 0.12, and which is also believed to be the result of a giant impact (Sect. 2.2). The angular momentum of the Earth-Moon system is also unusually high—equivalent to a 4 hour rotation period in an isolated Earth, much faster than the other terrestrial planets. Meanwhile tracing back the tidal evolution of the Moon suggests that it would have been much closer to Earth in the early Solar System, and compositionally the Moon is substantially depleted in both iron and volatile elements relative to Earth.

All of these factors led to the suggestion that the Moon might be the product of a giant impact between proto-Earth and another body (e.g. Cameron and Ward 1976; Canup 2004a). Continued work on the giant impact hypothesis eventually resulted in what has become known as the Canonical model (e.g. Canup and Asphaug 2001; Canup 2004a,b), in which a roughly Mars mass body (known as Theia) collides with a nearly fully formed proto-Earth at an oblique angle of $\sim 45^\circ$ and a speed near the mutual escape velocity. Such impacts would have been quite common in the late stages of terrestrial planet formation (e.g. Agnor et al. 1999), and the Canonical model is very successful at reproducing many aspects of the Earth-Moon system.

In recent years the Canonical model has faced something of a crisis in the near indistinguishability of Earth and Moon in isotope ratios, which has spawned a re-think of the giant impact scenario (Asphaug 2014). Some have invoked the evection resonance to permit the Moon to be created with collision parameters that are quite different to the Canonical model, satisfying both the dynamical and isotopic constraints (e.g. Canup 2012; Ćuk and Stewart 2012; Reufer et al. 2012), while others have argued that the Canonical model parameters still satisfy the isotopic constraints, e.g., if some equilibration of the accreted material occurs in the protolunar disk (Pahlevan and Stevenson 2007; Salmon and Canup 2012) or if the impactor formed sufficiently close to the Earth (Mastrobuono-Battisti et al. 2015), but the basic idea of the origin of the Moon in a giant impact does not seem to be in doubt.

Martian Hemispheric Dichotomy The largest topographic feature on Mars is the difference in elevation by $\sim$5 km between the lower northern and higher southern hemispheres (e.g. Smith et al. 1999) commonly referred to as the hemispheric dichotomy. The possibility of the northern lowlands representing a massive impact basin was first suggested by Wilhelms and Squyres (1984). With more recent work (e.g. Andrews-Hanna et al. 2008; Nimmo et al. 2008; Marinova et al. 2008, 2011) this suggestion has become more concrete, and a giant impact appears to be the best explanation for the hemispheric dichotomy.

A giant impact is also suggested as a possible origin for the Martian moons Phobos and Deimos (e.g. Rosenblatt 2011; Rosenblatt and Charnoz 2012; Citron et al. 2015). The Borealis basin impact might thus be responsible for both the hemispheric dichotomy and the formation of the Martian moons.

Asteroid Families and (4) Vesta The Hirayama asteroid families are the result of large collisions within the asteroid belt (e.g. Durda et al. 2007). Though material strength begins to play an important role in asteroid sized bodies large impacts between asteroids share many of the features of giant impacts between planetary bodies (e.g. Jutzi 2015). In the case of (4) Vesta and the associated Vesta family asteroids there is particularly strong link. The second largest asteroid (mean diameter 525 km), Vesta is believed to be differentiated and possess a largely intact crust, and has been suggested as an intact proto-planet, or even 'the smallest terrestrial planet' (e.g. Keil 2002; Russell et al. 2012). In addition the numerous Vesta family asteroids can be traced to the formation of the massive Rheasilvia basin, which together with the underlying Veneneia basin dominate the southern hemisphere of Vesta. Models of the formation of Rheasilvia indicate a ~66 km diameter impactor (Jutzi et al. 2013). The formation of the Rheasilvia basin can thus be thought of in some senses as a recent ($\lesssim$1 Gyr ago, Binzel et al. e.g. 1997) 'mini' giant impact.

2.2 Outer Solar System

Pluto-Charon As mentioned above the Pluto-Charon system is the closest analogue to the Earth-Moon system, despite lying on average over 40 times further from the Sun and having a substantially different composition, being rich in icy material. Pluto-Charon is well described as a binary, having a mass ratio of around 8.5:1, both bodies being tidally locked to one another, and with the system barycentre lying outside Pluto. The angular momentum in the Pluto-Charon system is larger than could be contained in a single body, suggesting a collisional origin (e.g. McKinnon 1989) and Canup (2005) showed that a giant impact is indeed a good explanation for the Pluto-Charon binary. The smaller satellites in the Pluto-Charon system may also have formed in the giant impact that produced Charon (e.g. Canup 2011), but explaining the complex resonant structure of the small satellites presents some difficulties (e.g. Showalter and Hamilton 2015).

Haumea Collisional Family Analogous to the Hirayama families in the asteroid belt there is presently one collisional family known in the Kuiper belt, that associated with the dwarf planet Haumea (Brown et al. 2007). This collisional family has a significantly lower velocity dispersion than the escape velocity of Haumea, and it has thus been suggested that the collisional family may have been created in an impact with a precursor of the current two satellites of Haumea (Schlichting and Sari 2009; Ćuk et al. 2013) rather than Haumea itself.

Middle-Sized Moons of Saturn It has been suggested that the formation of satellites around the gas giant planets is in many ways like a miniature version of the formation of the Solar System (Canup and Ward 2002, 2006). This would include giant impacts between proto-satellites, and such impacts may explain the origin of the otherwise mysterious middle-sized moons (those ~500–1500 km in diameter) of Saturn (Sekine and Genda 2012; Asphaug and Reufer 2013). In this scenario Saturn would have started out with a system of larger satellites like the Galilean moons of Jupiter, but whereas in the Jovian system these moons remained locked in a stable Laplace resonance, in the Saturnian system they became unstable and collided, with the present middle-sized moons representing the remnants of this former satellite population.

Obliquity of Uranus One of the important features of both the Earth-Moon and Pluto-Charon systems that is well explained by a giant impact scenario is their large angular momentum. Given the large amount of angular momentum that can be imparted by a giant impact, if the impact orientation is substantially different to the spin axis of the planet it is possible for the impact to lead to a large change in the planet's obliquity. It has long been recognised that an impact, involving an impactor around the size of Earth, represents a possible explanation for the large obliquity of Uranus (e.g. Safronov 1972; Parisi and Brunini 1997). An impact is not the only means by which Uranus can be tilted; tidal evolution of a massive retrograde satellite has also been proposed (Greenberg 1974; Kubo-Oka and Nakazawa 1995), though the moon required would have had to be greater than 1.5 times the mass of Mars and thus within an order of magnitude of the mass of proposed impactors.

3 Giant Impact Extrasolar Observables

One conclusion from Sect. 2 is that there is abundant evidence for giant impacts occurring in the Solar System. Indeed, the majority of the planets show some evidence for giant impacts. It is also the case that evidence for the impacts is manifested in many different ways, in terms of the physical or dynamical properties of the planets and their environments. The various consequences of giant impacts are summarised in Table 1, along with the evidence in the Solar System for these outcomes and the potential for observing them in extrasolar systems which is discussed further below.

The formation of moons is one of the most obvious lines of evidence for giant impacts, since this has emerged as the favoured explanation for our own Moon, with other Solar System moons also explained in this way. Given the small mass of any extrasolar moon, it seems like an impossible task to identify them. However, planets as small as 0.066 $M_\oplus$ have been detected in orbit around nearby stars (Jontof-Hutter et al. 2015). The mechanism to find such low mass planets involves looking for small perturbations to the timing of the transit of another large planet in front of the star due to the dynamical perturbation of the smaller object (so-called transit timing variations, or *TTVs*). This may be the most promising way of detecting extrasolar moons, although none are known yet (Kipping et al. 2014). The formation of a moon through giant impact requires an intermediate stage of a circumplanetary disk. There may be evidence for the passage of a circumplanetary ring system in front of its host star 1SWASP J140747.93-394542.6 (Mamajek et al. 2012; Kenworthy and Mamajek 2015). While the nature of that object remains unsure, this may provide evidence for satellite formation processes similar to those discussed in Sect. 2.2 for Saturn's moons. The thermal emission from circumplanetary disks may also enhance the brightness of an extrasolar planet, as has been suggested as the explanation for the spectrum of the Fomalhaut-b planet (Kalas et al. 2013), and it is suggested that rings may also be detectable in transit observations (Zuluaga et al. 2015).

A moon is not always an outcome of a giant impact, since in some cases very little material is placed into orbit around the planet, one example being the loss of Mercury's mantle in a violent collision to leave a high density planet (Sect. 2.1). The number of extrasolar planets for which densities have been measured has increased dramatically recently due to

Table 1 Summary of the consequences of a planetary system being subjected to giant impacts: the evidence for this from the Solar System and the possible manifestation in the observable properties of extrasolar systems

What do giant impacts do?	Evidence in Solar System	Potential extrasolar observable
Formation of moons	Earth-Moon, Pluto-Charon, Saturn?	Transit timing variations from exomoons
Modify internal composition	Mercury's Fe-rich composition	Exoplanet densities inferred from radial velocities and transits
Modify planet surface and atmosphere	Mars' hemispheric dichotomy, past magma ocean	Surface effects subtle, but hot planet spectrum
Modify planet spin	Uranus tilt	Exoplanet spins are measurable
Create debris	Hirayama asteroid families; dust bands; craters on small bodies	Infrared signature

the combination of high sensitivity transit observations with Kepler (that measure the size of the planet) and radial velocity observations (that determine the mass of the planet). In some cases the transit measurements also determine the mass of the planet, providing there are nearby planets from which to measure the TTVs as mentioned above. Such densities are used to infer the composition of the planets, with densities possibly as high as 14 g/cm^3 being recorded (Marcy et al. 2014). This is a promising avenue, but one which has yet to be used to infer a collisional history for the planets, rather this information has been used to assess the presence or absence of an atmosphere.

The effect on the planetary surface is more subtle. For example, the Martian hemispheric dichotomy is only apparent to us because its surface height has been measured at high precision; the level of difference is 0.07 % the diameter of Mars. Variations in surface height may be measured from the shape of the transit of an exoplanet, but the precision required is beyond current instrumentation. However, if the collision is energetic enough, the surface would be melted in the collision resulting in a magma ocean at 1000–3000 K for rocky planets. This molten surface would only be temporary since the heat would be radiated away (e.g., Earth's magma ocean may have survived 2 Myr; Zahnle et al. 2007), but could have a dramatic effect on the emission spectrum of the planet. Whether the hot surface itself is observable depends on the planet's atmospheric properties, which would in turn be modified by the impact; e.g., the early nebula atmosphere could be stripped by impacts (Schlichting et al. 2015; Inamdar and Schlichting 2015), and the atmosphere could also be replenished by magma ocean outgassing. However, it is clear that the excess energy must be radiated somehow, and this can increase the detectability of the planets following impacts (Miller-Ricci et al. 2009; Lupu et al. 2014). Such impact afterglows would be easier to detect at larger distance from the star due to favourable contrast with the stellar emission. This lead to the suggestion that the companion found at 55 au from 2M1207 was in fact the result of an impact between a 7 $M_\oplus$ and 74 $M_\oplus$ planets (Mamajek and Meyer 2007). While more detailed analysis shows that the properties of this system are better explained by the atmospheric properties of a low mass companion without any collision required (Barman et al. 2011), this at least demonstrates that impact afterglows are in the realm of current observational capabilities.

Conservation of angular momentum requires a planet's spin to be modified as a result of a giant impact. In some cases this can be significant, perhaps explaining the tilt of Uranus' spin relative to its orbital plane. Exoplanet spins are already measurable in observations of the velocity profile from lines in their atmospheres (Snellen et al. 2014). For now such observations have been applied to one of the most massive known extrasolar planets ($\sim 8M_{\rm jup}$ β Pic-b), showing it to spin much faster than its Solar System counterparts, likely due to its higher mass as opposed to its collisional history. There are also prospects to measure the orientation of an exoplanet's spin axis with respect to its orbital plane using observations of its radial velocity during secondary eclipse in a manner analogous to the Rossiter-McLaughlin effect (Nikolov and Sainsbury-Martinez 2015).

It is a general outcome of a collision that some fraction of the mass of the two bodies does not remain bound to either one at the end of this process. Indeed, the loss of Mercury's mantle in a giant collision not only modified the composition of that planet, but also released large quantities of debris. That debris, made up of solid material ranging in size from μm-sized dust grains to km-sized planetesimals, as well as vaporised rock if the collision was energetic enough, ended up orbiting the Sun. There is abundant evidence for the creation of debris in collisions in the asteroid belt, ranging from witnessing dust released in an impact which occurred in the last few years (Jewitt et al. 2011), to clustering in asteroid orbital elements showing these to have been created in an event a few Myr in the past (Nesvorný et al. 2002), evidence for which is also present in the dust bands in the inner Solar System (Grogan et al. 2001), to the Hirayama asteroid families which may have been much more violent events Gyr-ago (Hirayama 1918). Evidence for debris released in the giant impacts proposed to have shaped the Solar System's planets is harder to find since these collisions likely occurred long ago, shortly after the Solar System formed. Nevertheless such evidence may exist in the cratering record and chemical composition of asteroids and meteorites (Bottke et al. 2015).

From an extrasolar perspective, giant impact debris is expected to be readily detectable, because even relatively small quantities of dust (e.g., that corresponding to an asteroid of a few 10s of km in diameter ground into μm-sized dust) can be detected around nearby stars due to its thermal emission. The dust is heated by the star and then re-radiates that emission in the infrared at levels that can exceed the star's own photospheric flux (Wyatt 2008). Some 20 % of nearby stars show evidence for circumstellar dust originating in the break-up of planetesimals. While a large fraction of this dust is likely produced in the steady state grinding down of extrasolar analogues to the Kuiper belt, since such an interpretation explains the statistics of disk detection as a function of age (Wyatt et al. 2007b), some fraction of these may instead be the signature of a recent giant impact, since giant impact debris is expected to emit at potentially detectable levels (e.g., Jackson and Wyatt 2012).

To summarise, there are many ways in which evidence for giant impacts may be found in extrasolar systems. While there has been much progress in identifying these signatures, for now the most promising evidence for giant impacts in extrasolar systems lies in the creation of debris, and it is to this aspect that the remainder of this chapter will be devoted.

4 Evolution of Debris from Giant Impacts

As noted in Sect. 3, one of the key observable features produced by a giant impact around another star is debris. Though the character of the observable features will vary somewhat depending on the nature of the impact, for example the mass of the colliding bodies, or the orbital distance at which the impact occurs, the underlying processes which shape these features and how they evolve possess many similarities. Here we briefly outline a general framework within which the evolution of giant impact debris may be understood (for a more detailed discussion see Jackson et al. 2014).

4.1 Dynamics of Giant Impact Debris

The dynamical evolution of debris released in a giant impact can be considered as beginning from a single point with velocities relative to the star given by the progenitor's Keplerian orbital velocity v_k plus some velocity dispersion. While the velocity dispersion is likely to be asymmetric, it is convenient to model this as isotropic given that the orientation of any asymmetry would be specific to a given impact, and is likely to be random. The velocity dispersion of debris in giant impact simulations is approximately Gaussian with a width of σ_v (e.g. Jackson and Wyatt 2012), where σ_v is comparable to the escape velocity of the progenitor v_{esc}. The subsequent dynamical evolution and morphology of the debris depends on the ratio σ_v/v_k which is shown in Fig. 1 for different progenitors as a function of orbital distance, using the assumption that $\sigma_v \approx 0.46 v_{esc}$ as found for the Moon-forming impact (Jackson and Wyatt 2012). Note that the velocities of the debris being comparable to the escape velocity of the progenitor (regardless of the progenitor mass) holds largely independently of the velocity distribution of the debris particles. This is because to go into

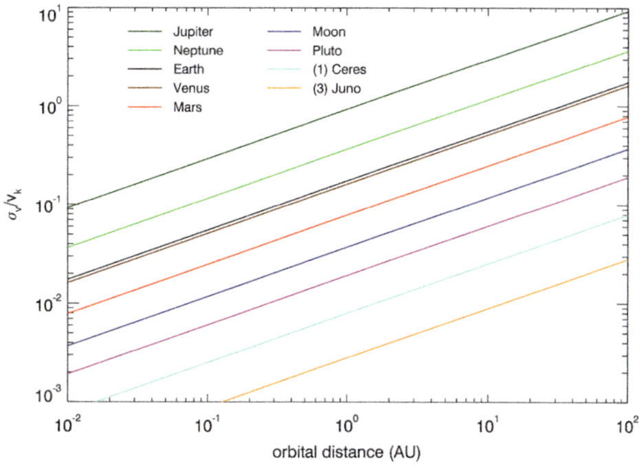

Fig. 1 The scaled velocity dispersion imparted to giant impact debris, $\sigma_v/v_{\rm k}$, as a function of orbital distance of the progenitor for a selection of Solar System bodies ranging in mass from Neptune to the asteroid (3) Juno, under the assumption that $\sigma_v = 0.46v_{\rm esc}$. This ratio is important for the morphology and subsequent evolution of the debris; e.g., much of the debris may be placed onto orbits that are unbound from the star if $\sigma_v/v_{\rm k} \gg 1$

heliocentric orbit the debris must have sufficient energy to escape the progenitor body, and once the energy required to escape is exceeded the resulting relative speed post-escape then rapidly becomes comparable to the escape velocity. For example a particle that exceeds the escape velocity by only 5 % at launch will then have a relative speed post-escape of $0.32v_{\rm esc}$.

To understand the evolution and morphology of the debris it is important to remember that immediately after the impact the debris occupies what is essentially, on the scale of the wider planetary system, a single point at the location of the impact, which we will refer to as the *collision-point* (Jackson and Wyatt 2012; Jackson et al. 2014). This geometry means that the orbits of all debris fragments must pass through this point, which thus has a high density. A small density enhancement is also found on the opposite side of the star from the collision-point, since all debris orbits must pass through the plane of the progenitors' orbit somewhere along what we will refer to as the *anti-collision line* (Jackson and Wyatt 2012; Jackson et al. 2014). Since debris passes through the collision-point once per orbit, the resulting repeat encounters can also have implications for the rate at which debris encounters the progenitor and indeed other debris.

Common to the dynamical evolution of all (bound) giant impact debris are four key morphological phases: (i) After the impact, both the debris and the progenitor move away from the collision-point. The debris initially forms an expanding clump that moves with the progenitor body around its orbit (see upper left panel of Fig. 2). (ii) Since fragments that have been placed onto orbits with smaller semi-major axes than the progenitor will have shorter orbital periods than those with larger semi-major axes, the clump will be sheared as it continues around the orbit. Once the shearing has progressed enough that fragments on smaller orbits have reached the collision-point before those on larger orbits have left it, the clump will become a spiral structure, which then continues to coil as Keplerian shearing continues (see upper right panel of Fig. 2). (iii) Eventually the spiral structure of the debris is no longer visible and the debris forms a smooth but highly asymmetric disk, with a strong pinch at the collision-point (see bottom left panel of Fig. 2). The density enhancement at the collision-point, as well as how spread out the disk is away from the collision-point, both increase with velocity dispersion (i.e., with $\sigma_v/v_{\rm k}$, for more detail see Jackson et al. 2014). (iv) The orbits of debris fragments precess due to interactions with massive bodies, including the progenitor, at a rate that depends on the semi-major axis of the fragments. The resulting differential precession eventually smears out the asymmetry producing an axisymmetric disk (see bottom right panel of Fig. 2).

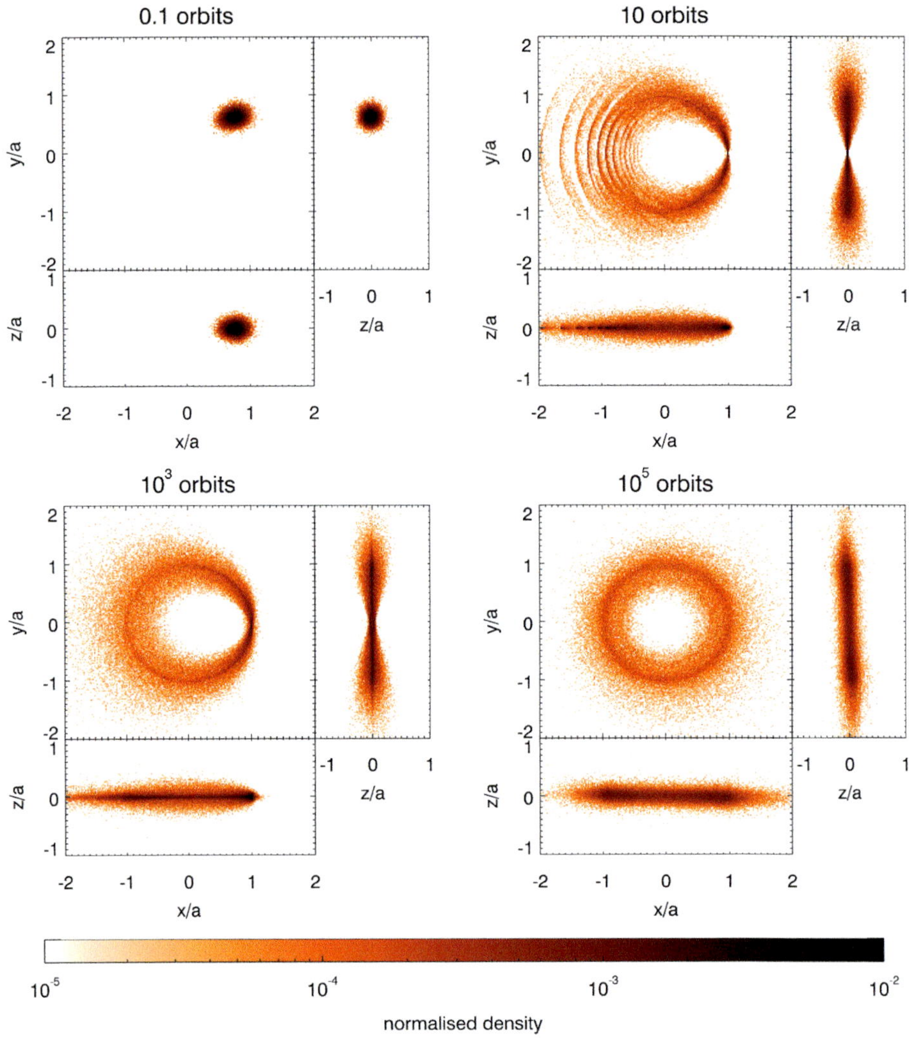

Fig. 2 Snapshots of the distribution of debris ejected by a giant impact with $\sigma_v/v_{\rm K} = 0.18$, showing the four key stages of its dynamical evolution: (*top left*) clump, (*top right*) spiral, (*bottom left*) asymmetric disk and (*bottom right*) axisymmetric disk. For each stage the *top left panel* shows a face-on view of the disk with the star at the origin and the collision-point at $(1, 0)$, while the *panels to the right and below* this show edge-on views. The axis scale is in units of the progenitor semi-major axis, a, while the snapshot times are in units of the progenitor's orbital period. The *colour* corresponds to the disk density normalised to an integrated value of 1 at time zero

The timescale for the shearing process, and so for the evolution from clump to asymmetric disk, depends on $\sigma_v/v_{\rm K}$. The shearing is quite rapid in Fig. 2 since $\sigma_v/v_{\rm K} = 0.18$; the early spiral stage is visible after 1 progenitor orbit, and after 100 orbits the spiral is no longer distinguishable. In the case of an impact with a smaller $\sigma_v/v_{\rm K}$ the rate of Keplerian shearing is slower, since the velocity difference across the debris distribution is smaller. While this would result in slightly longer clump and spiral phases, in general the clump phase is expected to end within ten orbits, and the spiral phase within a few hundred orbits after the impact. Since the morphology of the disk, and the rate at which the disk progresses through the early phases of the evolution, depend on σ_v/v_k, this means that impacts involving progenitors of the same mass can have substantially different morphologies and different early stage evolution if they occur at different orbital distances (see Fig. 1).

The smooth asymmetric phase lasts much longer, since it is determined by the precession timescale, rather than by the dynamical timescales that set the clump and spiral phases. Precession rates depend on the system architecture, i.e., the masses and orbits of other planets in the system, although the progenitor itself may also be massive enough to set the precession period. In general the collision-point may be expected to have been smeared out after a few thousand orbits and the disk to have achieved total axisymmetry after a few tens of thousands of orbits, timescales that apply to precession due to a Jupiter mass planet 5 times more distant than the impact (see Eq. 23 of Jackson et al. 2014), or due to an Earth-mass progenitor (Jackson and Wyatt 2012).

The morphological evolution shown in Fig. 2 will occur whatever the wider architecture of the planetary system in which the progenitor and debris reside. However, the progenitor and other planets play a potentially much more important role beyond setting the precession timescale of the debris, since close debris-planet approaches can lead to the debris being accreted onto the planet or ejected from the system. Which of these outcomes is more likely depends on the ratio of the escape velocity of the planet to its Keplerian velocity, v_{esc}/v_k: ejection is more likely for $v_{esc}/v_k \gg 1$ and accretion more likely for $v_{esc}/v_k \ll 1$ (see Fig. 1). The rate at which debris is accreted onto a planet of radius R_p can be estimated as

$$R_{col} = n\sigma_{col} v_{rel}, \tag{1}$$

where n is the density of debris in the vicinity of the planet, v_{rel} is the relative velocity of encounters between the planet and the debris, and $\sigma_{col} = \pi R_p^2[1 + (v_{esc}/v_{rel})^2]$ is the collision cross-section.[1] However, note that this rate may be significantly underestimated at early times for re-accretion onto the progenitor if the geometrical effect of the collision-point is not taken into account; during the early phases of dynamical evolution, most re-accretion onto the progenitor will occur when it passes through the collision-point, and the accretion rate will be substantially raised during this early period (Jackson and Wyatt 2012, see also discussion in Sect. 4.2).

4.2 Collisional Evolution

In addition to evolving through dynamical interactions with massive bodies in the system, the debris will also evolve as a result of mutual collisions within the debris population. Such collisions lead to the break-up of larger fragments into smaller ones, redistributing mass down the debris size distribution. Eventually the debris is ground down into dust grains that are small enough to be removed by radiation pressure on orbital timescales. The size at which this occurs is called the blow-out size and for dust of density ρ is given approximately by (e.g. Wyatt 2008)

$$D_{bl} = 0.8\left(\frac{L_*}{L_\odot}\right)\left(\frac{M_\odot}{M_*}\right)\left(\frac{2700\ \mathrm{kg\,m^{-3}}}{\rho}\right)\ \mu\mathrm{m}. \tag{2}$$

[1] Note that the gravitational focussing factor for re-accretion onto the progenitor is independent of the progenitor's properties, since v_{rel} scales with σ_v and so v_{esc}. This factor is found to be close to 10 for the Moon-forming giant impact debris.

The rate at which debris fragments collide with each other is determined by an equation that is analogous to Eq. (1). This calculation is complicated by the fact that any given debris particle will interact with particles with a range of sizes, with the outcomes of such collisions ranging from cratering to complete pulverisation. For this reason the collisional evolution of debris populations is often studied numerically (Thébault et al. 2003; Krivov et al. 2005; Bottke et al. 2005). However, this is not necessary to understand what happens, because for most common assumptions about collisional outcomes, once the size distribution has reached steady state it tends to a shape that can be readily calculated (O'Brien and Greenberg 2003; Wyatt et al. 2011). If it can be assumed that the strength of a debris fragment is independent of its size, the size distribution tends to a power law

$$n(D) \propto D^{-\alpha}, \tag{3}$$

where $\alpha = 7/2$. Such a size distribution has most of the mass in the largest objects, while most of its cross-sectional area is in the smallest objects (i.e., those close to the blow-out limit in Eq. (2)).

It is important to understand how both mass and cross-sectional area evolve within the debris distribution. It is the cross-sectional area to which the observations discussed in Sect. 3 are sensitive (Wyatt 2008). However that cross-section resides in small grains that are relatively short-lived, and so must be replenished from the larger debris fragments which make up the majority of the mass and have longer collisional lifetimes. The discussion is simplified if the size distribution has a fixed shape, because this implies that cross-sectional area scales with mass by a constant ratio, though this is only true at all times if the debris size distribution starts with the size distribution it will tend to in steady state. Typically the total mass of debris created in a giant impact is well known. For example, for large bodies (those with $\sigma_v/v_k > 0.1$), impacts generally fall in the hit-and-run or partial accretion/erosion regimes with debris releases of around 3–5 % of the colliding mass (Leinhardt and Stewart 2012; Stewart and Leinhardt 2012). This is because for large σ_v/v_k, the impact speed must be a larger fraction of the orbital speed to push the impact into a violent outcome state. Smaller bodies (those with $\sigma_v/v_k \ll 0.1$) are more likely to undergo violent impacts where debris masses can be a much higher fraction of the colliding mass, but if the impact parameters are fairly well constrained the debris mass can also be determined with reasonable confidence. On the other hand the size distribution is not well constrained, as discussed further in Sect. 5.2. This is unfortunate, as for a fixed mass of giant impact debris, both its cross-sectional area and (as we will see below) the timescale of its evolution, are strongly dependent on the size of object where most of the mass is created.

Nevertheless, the above discussion motivates a simple prescription with which to predict the evolution of cross-sectional area of the giant impact debris. It is only necessary to consider the evolution of the mass, which will deplete due to the dynamical processes discussed in Sect. 4.1, and due to the collisional erosion of the largest fragments. A quick estimate for the rate at which the largest fragments deplete by collisions can be made using Eq. (1) and assuming that all the debris is in fragments of the same size. While more accurate calculations are readily available (see Wyatt et al. 2007a), this illustrates how the collisional depletion rate is proportional to the mass of largest fragments. This leads to the mass remaining constant until the largest objects have come to collisional equilibrium, following which the mass decays inversely with time, with dynamical processes further accelerating the depletion process.

There are several caveats to the above picture. For example, it is known that the strength of a debris fragment is dependent on its size, and the debris size distribution may not be

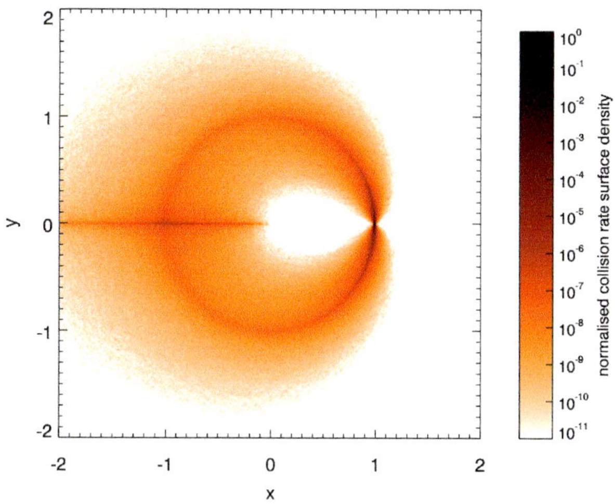

Fig. 3 Image showing how the rate of mutual debris collisions varies with position during the asymmetric phase of evolution of the debris produced by a giant impact with $\sigma_v/v_k = 0.3$. The rates are normalised such that the maximum collision rate is 1. Orbital motion is in the anti-clockwise sense. The geometry of the debris orbits results in an increased density along $y = 0$ (see Fig. 2), and so a higher collision rate at the collision point and along the anti-collision line

expected to start with the shape it tends to in steady state. Both are valid criticisms, but their consequences can be readily accounted for (Löhne et al. 2008; Wyatt et al. 2011). Furthermore, it is found that the simple prescription given above is usually accurate to within a factor of a few and so is acceptable for many applications.

A more complicated caveat to address is the geometrical effect of the collision-point. In the early phases of the dynamical evolution of the disk (prior to axisymmetrisation) the number density varies substantially around the orbits of the debris fragments; in particular the collision-point is a region of exceptionally high density. Figure 3 shows how this results in the collision-point having a collision rate that is many orders of magnitude higher than elsewhere. This effect is such that, to a first approximation, all collisions that a debris particle experiences occur close to the collision-point. It is also such that the collision rate averaged over the particle's orbit is around two orders of magnitude higher than would be the case for an axisymmetric disk; i.e., the rate of collisional evolution will be raised during the early phases of dynamical evolution. The density enhancement along the anti-collision line also leads to a slightly increased collision rate there, though this is much less dramatic. These geometrical effects can only be accounted for using numerical techniques (e.g., Kral et al. 2015), and have further implications which will be discussed in Sect. 7.

5 Observability of Debris from the Moon-Forming Giant Impact

Though recent work has put the Canonical model in some doubt (see Sect. 2.1), the Moon-forming giant impact remains the most well studied example of a giant impact, and thus provides a good starting point for discussions of the behaviour and observability of debris released by giant impacts (for more detail see Jackson and Wyatt 2012).

5.1 Dynamical Evolution

From smoothed-particle hydrodynamics simulations conducted by Marcus et al. (2009) the velocity dispersion of the debris released by the Moon-forming giant impact is well fit by a Gaussian with a width of $\sigma_v = 5.2\ \mathrm{km\,s^{-1}}$. At an orbital distance of 1 au this corresponds to $\sigma_v/v_k = 0.18$ (the value used in Fig. 2). Once it has been launched from Earth the debris goes

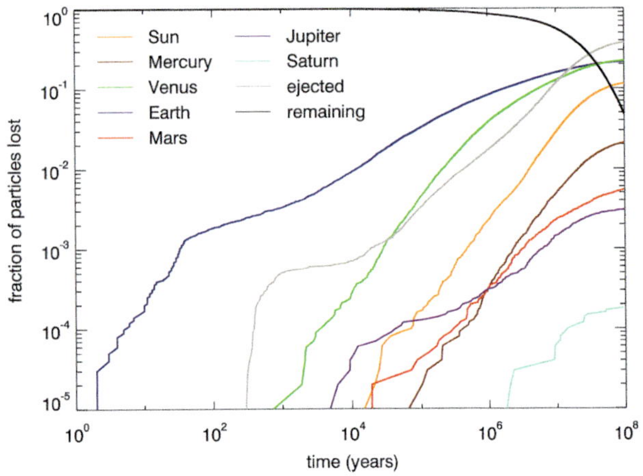

Fig. 4 The fates of particles in a 100 Myr 10^5 particle N-body simulation of debris from the Moon-forming giant impact. Debris particles are massless test particles, and no account is made for collisional erosion. The present day Solar System architecture is used. *Lines for different planets* indicate the fraction of particles accreted onto that planet

into heliocentric orbit and the debris distribution then evolves dynamically as described in Sect. 4.1, with the proviso that the general discussion therein only gave brief consideration to planets in the system other than the progenitor, which were not included in Fig. 2.

The wider architecture of the planetary system becomes important as the giant impact debris leaves the asymmetric stage. The two most important planets aside from Earth in determining the dynamical evolution of the debris are Venus and Jupiter. Venus draws the debris inwards such that, rather than being narrowly peaked around the orbit of Earth, the debris occupies a broad band encompassing the orbits of both Earth and Venus. Jupiter meanwhile truncates the outer edge of the disk by ejecting fragments that encounter it from the Solar System. The differing effects of Venus and Jupiter are in line with expectations from Fig. 1, since Venus has $v_{\mathrm{esc}}/v_{\mathrm{K}} \ll 1$ while Jupiter has $v_{\mathrm{esc}}/v_{\mathrm{K}} \gg 1$. Figure 4 shows the ultimate fate of debris created in the Moon-forming impact, which shows that the three fates of accretion by Earth, accretion by Venus, and ejection (mostly by Jupiter) account for the overwhelming majority of the dynamical losses of the debris. The next most likely fate, collision with the Sun, is in fact also a secondary effect of Jupiter; in systems lacking Jupiter the quantity of debris which impacts the Sun is much lower. In total half of the dynamical particles have been lost after 13 Myr, while just 5 % of the dynamical particles remain after 100 Myr.

While of course other planetary systems may have very different orbital architectures to the Solar System, we can expect that some general features of the evolution of the debris from Moon-formation will apply to all systems. In particular we can expect that the most important other bodies in the system will be nearby terrestrial planets of comparable (or larger) mass than the progenitor, and the nearest giant planet. More distant giant planets have limited influence, since the nearest one will almost always eject particles before they have a chance to interact with more distant giant planets (i.e., as long as $v_{\mathrm{esc}}/v_{\mathrm{K}} \gg 1$).

5.2 Predicted Brightness Evolution

As discussed in Sect. 4.2, the brightness of giant impact debris depends on the quantity of small dust present, which in turn depends on how the debris size distribution evolves due to collisions. Several issues with calculating the collisional evolution were discussed in Sect. 4.2, where it was noted that there may be inherent uncertainties arising from how well the initial size distribution is known, and most collision rate calculations also suffer from

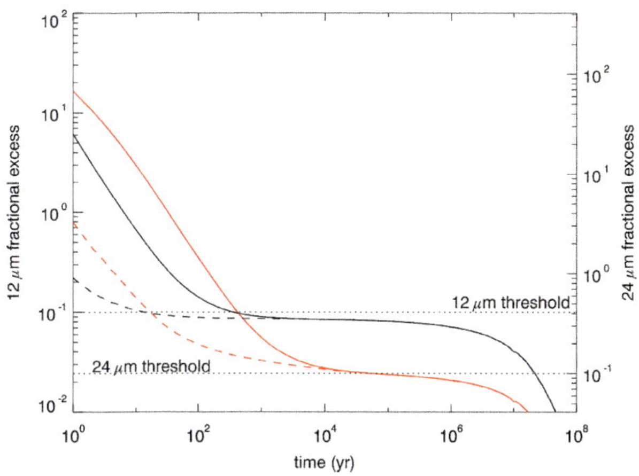

Fig. 5 Evolution of the fractional excess of debris released by a Moon-forming impact around a Sun-like star for different assumptions about its initial size distribution and about its collisional evolution. The fractional excess is defined as the flux from the debris relative to the flux from the stellar photosphere, shown at 12 μm *on the left axis* and 24 μm *on the right axis*; the *dotted lines* show typical detection limits at fractional excesses of 10 % at both wavelengths. The *black lines* show the evolution of a two component size distribution with 30 % vapour condensates up to 1 cm in size and 70 % boulders up to 500 km in size. The *red lines* show the evolution of a three component size distribution with 30 % vapour condensates up to 1 cm in size, 50 % intermediate material up to 7.5 km in size and 20 % large boulders up to 500 km in size. *Solid lines* calculate collision rates assuming the debris to be axisymmetric at all times, while *dashed lines* show the results incorporating some aspects of the collision-point geometry in this calculation (see text for details)

an inaccurate treatment of the geometry of the collision-point. For this reason Fig. 5, which shows the predicted brightness evolution of giant impact debris at two wavelengths 12 and 24 μm at which stars are commonly observed, shows that evolution for two different initial size distributions and two different treatments of the collisional evolution.

There are additional uncertainties involved in translating dust cross-sectional area into the emergent emission spectrum. In all cases on Fig. 5 the emission from the dust is calculated assuming the grains to act like simple blackbodies at the orbital distance of the progenitor. This has the benefit of resulting in a constant ratio between the 12 and 24 μm emission so that both can be shown on the same plot. However, the small grains that dominate the emission are expected to be hotter than black body, while at the same time emitting inefficiently, and potentially exhibiting spectral features due to their composition (e.g., Sect. 6.2). These effects can be properly accounted for, but require additional assumptions (e.g., about the grain composition). For simplicity these are not included in Fig. 5, which should thus be considered to have additional factors of a few or more uncertainty.

As mentioned in Sect. 4.2 the size distribution of debris released by a giant impact is poorly constrained. SPH simulations of giant impacts are unable to resolve anything but the largest debris fragments ($\gg$100 km for Earth-mass impactors; Genda et al. 2015). The main constraint from such simulations is the total mass of the debris, which is around 0.016 $M_\oplus$ ($\sim 10^{23}$ kg) in the Canonical model. However, SPH simulations of giant impacts do indicate that, due to the enormous energy involved in the collision of two planetary-sized bodies, a significant fraction of the debris material is released in the form of vapour. As the cloud of vapour expands it cools and condenses, forming small droplets with typical sizes in the mm-cm range (e.g., Melosh and Vickery 1991; Johnson et al. 2012; Johnson and Melosh 2014). For impacts involving Earth-sized bodies, typically a few tens of percent of the debris is released in the form of vapour; Canup (2008) estimated ~10–30 % for the Canonical

Moon-forming impact, though Kraus et al. (2012) suggest that the equations of state in use in current hydrodynamical codes may underestimate vapour production. The production of vapour alongside material that is not vaporised suggests that the size distribution should consist of (at least) two components, the vapour condensates and unvaporised *boulder* material that extends to larger sizes.

The simplest assumption for the initial boulder size distribution is that it starts with the shape it would tend to in steady state (e.g., Eq. (3)) and extends up to some maximum size D_{max}. This was the assumption made in Jackson and Wyatt (2012), which is reproduced in Fig. 5 for $D_{max} = 500$ km with the solid black line. This illustrates the properties common to all models for the brightness evolution. In the early phases the vapour condensates shine brightly at readily detectable levels. However, these are rapidly depleted by collisional evolution to non-detectable levels within a few hundred years. Dust arising from the break-up of the boulders starts at a lower level, but is maintained for much longer, eventually starting to deplete after a few Myr due to a combination of collisional erosion and dynamical erosion (see Fig. 4). Taken at face value, the boulder distribution remains undetectable throughout the evolution, and the total emission from both vapour condensates and boulders drops below the typical 12 μm detection threshold after a few hundred years. However, the predicted excess level is so close to the typical detection threshold that, given the various factor of a few uncertainties mentioned elsewhere, the giant impact debris could be detectable at 12 μm up to 10 Myr, which is its duration of detectability at 24 μm.

An alternative choice for the initial size distribution of the boulder material is that of Bottke et al. (2015), basing it instead on the size distribution of the Vesta family asteroids, with additional constraints from Lunar cratering. The size distribution of the Vesta family is considerably steeper than a self-similar cascade, with more mass in smaller bodies.[2] Bottke et al. (2015) suggest that the boulder material may itself have two components, one a steep power law ($\alpha \sim 5$ in Eq. (3)) extending to a largest size of ~80–100 km and an additional less massive population of larger bodies. The red lines in Fig. 5 show the predicted brightness evolution for a three-component size distribution intended to mimic that of Bottke et al. (2015). The resulting evolution is very similar to that of the two-component distribution considered in Jackson and Wyatt (2012). In particular, it is noticeable that the brightness evolution exhibits just two rather than three distinct phases (i.e., a drop over ~ 100 years followed by another after ~ 10 Myr), which is because the emission from the grinding of intermediate-sized (~ 10 km) material overlaps with the peak from the vapour condensates (since this population is both bright and has a very short collisional lifetime). The relatively low mass in the largest boulders with this assumption makes it less likely that the debris remains detectable up to 10 Myr after the impact event, but any such conclusion would bear the caveat of the factors of few uncertainty in brightness.

The solid lines on Fig. 5 made the same assumption as Jackson and Wyatt (2012) in calculating the collision rate in assuming the debris to be distributed in an axisymmetric torus. The assumption of axisymmetry is common in calculations of collision rates (Wetherill 1967; Wyatt et al. 2010), since precession causes orbits to precess on long timescales. However, in the early phases of evolution the orbital planes and pericentres have yet to be

[2]It is unclear how appropriate it is to extrapolate the Vesta family size distribution to planetary scale giant impacts. For example, the Rheasilvia impact speed of $\sim 15 v_{esc}$ is much faster than typical giant impact speeds, while at ~0.1 % of the mass of Vesta the 66 km impactor is unusually small (Jutzi et al. 2013). The largest of the Vesta family asteroids (aside from (4) Vesta itself) is only ~8 km in size, and gravitational re-accumulation may become more efficient at larger sizes. Additionally, the size distribution of Vesta family asteroids may have undergone collisional evolution in the ~1 Gyr since the Rheasilvia impact. Nevertheless, it may be the only observable example of a giant impact-like debris distribution.

randomised due to precession, and moreover in the clump phase it is incorrect to assume that particles are evenly distributed along their orbits. While such effects are automatically accounted for in the N-body simulations for accretion of debris onto planets, these simulations do not follow debris-debris collisions due to the large number of particles that would be required. It is possible to make a more accurate calculation of the collision rates using the output of the N-body simulations. The dashed lines on Fig. 5 show the evolution using collision rates calculated from the output of the N-body simulations in which the orbital planes and pericentre orientations are no longer assumed to be random (see Fig. 3 and Jackson et al. 2014). As expected from the discussion in Sect. 4.2, the decay in brightness occurs much faster due to the geometry of the collision-point. However, there is a caveat in that this calculation still assumed that particles are evenly spread around their orbits. This means that the collision rate is overestimated during the clump phase, and the true evolution is likely to lie somewhere between the solid and dashed lines on Fig. 5.

A further caveat that applies during the earliest clump and spiral phases is the prospect for the debris to have high optical depth. All of the brightness calculations discussed so far assumed the debris to be optically thin, whereas in fact the dust arising from the break-up of vapour condensates is predicted to be so abundant that its distribution may be optically thick. This could have several consequences, by changing the detectable cross-sectional area, the dust temperature, and whether it is removed from the system by radiation pressure. These effects would also vary on orbital timescales. Clearly further work is needed to improve our understanding of the evolution of giant impact debris during the first tens of orbits. Progress has been made in incorporating collisional evolution into N-body simulations of the evolution of giant impact debris (Kral et al. 2015), however the small volume of the collision-point means that it remains a challenge to accurately model its effect due to the spatial resolution required to do so.

6 Comparison with Observed Systems

6.1 Photometric Fluxes

From Sect. 5.2 it is clear that if we were observing another star shortly after its growing terrestrial planets had undergone an impact similar to that which created the Earth-Moon system, then this event would have had a significant effect on the level of infrared emission observed. Moreover, with multiple such events occurring, it seems that the Solar System would have shone brightly in the mid-IR throughout its infancy. Given the other evidence for collisions occurring throughout the history of the Solar System (see Sect. 2), Rieke et al. (2005) suggested that the scatter in levels of 24 μm emission seen in their Spitzer survey of nearby A stars (i.e., stars slightly more massive and luminous than the Sun), and their higher levels in the first 150 Myr of the stars' lives, could be a result of recent collisions. In that interpretation, high levels of mid-IR emission are attained stochastically following a giant impact, with levels fading back to quiescent levels thereafter. A similar interpretation of 24 μm emission levels was applied to a Spitzer survey of Sun-like stars (Meyer et al. 2008), wherein it was assumed that this emission is indicative of ongoing terrestrial planet formation processes.

However, the situation is complicated, because 24 μm emission can arise for a number of reasons unconnected with collisions or planet formation. It was later shown that both the scatter seen in the Rieke et al. (2005) observations and the decay in emission levels could be explained by the steady state erosion of extrasolar Kuiper belts (Wyatt et al. 2007b); the level observed simply reflects the initial mass in the belt and its distance from the star, which was far enough that the emission was generally cold and so also explained the observations at 70 μm toward the same stars (Su et al. 2006). The same issue affected the interpretation of Sun-like stars, since it was found that the emission spectrum for the majority of these stars was rising toward longer wavelengths implying that the 24 μm emission arises in a belt far enough from the star (> 10 au) to be explained by steady state processes (Carpenter et al. 2009). That is, while recent giant impacts are not ruled out as the origin of the 24 μm emission, a steady state interpretation is more likely for most systems, and regardless the relatively cool temperature of the emission implies that this arises from a location outside that typically associated with the formation of terrestrial planets.

The problem is simply that while 24 μm emission should accompany a giant impact (see Fig. 5), it is a flawed proxy for such an impact. Nevertheless, the large samples of young stars observed by Spitzer mean that the evolution of a star's (or more accurately a population of stars') 24 μm emission is well characterised (e.g., Siegler et al. 2007), and as such has often been used to assess the frequency of giant impacts (Jackson and Wyatt 2012). More recently it has become clear that a better proxy for such collisions is a star's excess 12 μm emission. Such hot emission is seen less frequently and cannot arise from a more distant Kuiper belt, rather the requisite temperature to emit at such short wavelengths places the dust at a temperature compatible with the location of the terrestrial planets in the Solar System. This does not preclude the presence of a more distant cold dust belt in the system, and indeed there are many examples of debris disks with both cold outer belts and hot inner dust (Su et al. 2013; Kennedy and Wyatt 2014); the hot dust in some of the younger such systems has been attributed to ongoing planet formation processes (Smith et al. 2009a).

Early searches for 12 μm emission from dust around nearby stars with IRAS were not particularly successful (Aumann and Probst 1991), though several good candidates were re-discovered in the last decade and followed up with higher resolution instruments to confirm the emission was associated with the star and to characterise its spectrum (e.g., Rhee et al. 2008; Melis et al. 2010). These hot excesses were picked out on a case-by-case basis, and interpreted again within the context of terrestrial planet formation, based on the fact that the stars with bright emission were predominantly found around < 100 Myr stars.

A significant advance in the statistical analysis of the frequency of stars exhibiting excess 12 μm emission came with WISE, which surveyed the whole sky at that wavelength with a spatial resolution and sensitivity far superior to IRAS. This enabled Kennedy and Wyatt (2013) to quantify the rarity of bright excesses by correlating Sun-like stars in the Hipparcos database with WISE. They found that detectable excesses (those for which 12 μm dust emission is at levels above $\sim 10\ \%$ that of the stellar photosphere) occur around 1:1000 stars. While it is not feasible to derive the ages of all Hipparcos stars, an age dependence for the excess emission was evident because the majority of the 22 stars with excess were known to be members of young associations. To make a rough estimate of the age dependence, Kennedy and Wyatt (2013) made the assumption that stars in the Hipparcos catalogue have ages that are randomly distributed up to the main sequence age for their spectral type. This led to the conclusion that for stars in the 10–120 Myr age range, the frequency of detectable emission is closer to 3 %; i.e., the excess rate declines with age, just as it does at other wavelengths.

Eventually the WISE data should be mined to determine the excess distributions in all young clusters, similar to the analysis carried out for Spitzer. For now, this has been applied to a few regions such as 10–20 Myr Upper Sco (Luhman and Mamajek 2012) and 3–5 Myr Taurus (Esplin et al. 2014). The interpretation is complicated by the presence of long-lived protoplanetary disk systems among the excess candidates. However, based on this work

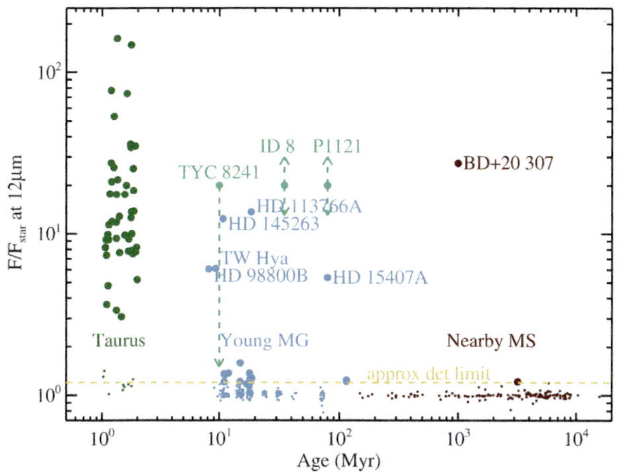

Fig. 6 Evolution of 12 µm emission around Sun-like stars. *Large circles* are those for which an excess above photospheric levels are detected, while *small circles* are those for which no excess is discernable; the approximate detection threshold is given by the *horizontal yellow line*, though this varies from star to star. The *different colours* indicate different surveys: *green* is Taurus (Esplin et al. 2014), *light blue* are young moving groups such as Sco Cen (Rizzuto et al. 2012) and our SED fitting of bona fide members from Malo et al. (2013), *brown* are our SED fits of nearby members from the DEBRIS sample (Phillips et al. 2010). Individually identified 12 µm excess sources are also included from Kennedy and Wyatt (2013). The *vertical dashed lines* show the level of variability for sources discussed in Sect. 6.3 (Melis et al. 2012; Meng et al. 2014, 2015)

we estimate the fraction of Sun-like stars in Upper Sco with detectable 12 µm excesses is ∼ 10 %, in broad agreement with the findings of Kennedy and Wyatt (2013), particularly if one considers that the fraction is likely to decrease over the 10–120 Myr age range (and noting that some of the Upper Sco excesses are from protoplanetary disks).

The 12 µm excesses are plotted on Fig. 6 as a function of stellar age for a number of Sun-like stars. Some of these were identified in statistical samples from which the rate of excesses was determined above, but others were simply excess detections reported in the literature such that care should be taken with the interpretation of this plot. Nevertheless, this illustrates the level at which excess is detected, and how that level is typically higher around younger stars. In fact the level of emission around main sequence stars can reach the level seen around T Tauri stars, meaning that the full emission spectrum needs to be considered when interpreting a 12 µm excess, as it is not always clear whether this should be interpreted as evidence for a recent collision or for a protoplanetary disk (Schneider et al. 2013; Kennedy et al. 2014).

6.2 Compositional Constraints

The information provided by a detection of excess emission extends beyond the photometric measurement of the level of excess. For those systems with detections, the mid-IR spectrum provides information about the composition of the dust (e.g., Chen et al. 2006). A prime example of this is the 12 Myr A5V star HD172555 at 29 pc for which the mid-IR spectrum is dominated by emission from amorphous silica dust (Lisse et al. 2009). Dust of different compositions leaves signatures in the spectrum at specific wavelengths; e.g., the silica feature at ∼ 9.3 µm allows its presence to be picked out by eye for HD172555. However, extracting more detailed information about the dust composition is not trivial, since the spectral features are relatively broad so that those from different compositions overlap, and the exact shape of the spectrum depends on the dust size distribution and the temperatures attained by different sizes.

Debris disk modellers commonly employ Mie theory for interpretation of debris disk observations (e.g., Lebreton et al. 2013), however the approximations in this approach mean that such detailed spectra are often interpreted using optical properties obtained from laboratory measurements of astrophysical materials (Lisse et al. 2009). In the case of HD172555 it was the latter approach which was used, resulting in the conclusion that around half of the dust is composed of silica, with a total dust mass equivalent to a 150–200 km asteroid. Since silica is produced in high energy impacts, this points to an origin for the dust in a recent collision at > 10 km/s between two massive protoplanets each at least the mass of Mercury. This interpretation was supported by a feature in the spectrum at 8 μm. While this was originally believed to indicate the presence of SiO gas created in the collision (Lisse et al. 2009), the short lifetime of such gas due to photodissociation suggested an alternative origin for the feature in solid SiO created by condensing vaporised silicate (Johnson et al. 2012). In summary, the composition is similar to that which would have been anticipated following the Earth-Moon forming collision. This analogy is reinforced by the dust temperature which places it at a distance of ~ 6 au from the star, in keeping with spatial constraints from imaging and interferometry (Smith et al. 2012), a location that is comparable in terms of incident energy to the Solar System's terrestrial planets given the higher luminosity of this star.

One puzzling aspect of the dust orbiting HD172555 is its inferred size distribution, since the prominence of the silica feature requires there to be abundant sub-μm dust, whereas such dust is small enough that it would be expected to be blown away from the star by radiation pressure on orbital timescales of a few years (see Eq. (2)). This is in contrast to the persistence of the mid-IR emission at the same level to within 4 % over 27 years (from IRAS to WISE), suggesting that the dust is created more rapidly than it is removed, or that radiation pressure is somehow ineffective. Johnson et al. (2012) preferred the latter explanation, and found that for certain compositions it is possible for this star to retain a long-lived population of $\ll 0.1$ μm grains even if those 0.1–1 μm are significantly depleted; this is because the effect of radiation pressure is strongest for 0.1–1 μm grains, and for certain compositions such grains can have $\beta > 1$ (and so must be unbound) while smaller (and larger) grains have $\beta < 0.5$ (and so can be bound). However, the origin of the size distribution remains unclear, and may be related to the issues discussed in Sect. 6.3, and may also affect the assumed detectability of giant impact debris discussed in Sect. 5.2. Another relevant aspect of this system that we are just beginning to learn about is its gas content; for now it is known that OI is present (Riviere-Marichalar et al. 2012), but its location relative to the hot silica dust is not yet known, as are its origin and impact on the hot dust.

While this section has focussed on just one system, there are many other systems for which compositional constraints exist, not all of which exhibit the same silica composition, even if they are interpreted within the framework of giant collisions (Lisse et al. 2008; Currie et al. 2011; Ballering et al. 2014).

6.3 Time Variability

Whereas the hot dust emission toward HD172555 was concluded to not be variable (see Sect. 6.2), the same is not true for all stars with hot dust mid-IR emission. For example, that of BD+20307 brightened by 7 % over the course of the 6 months between two WISE observations (Kennedy and Wyatt 2013). The case of TYC8241 is particularly interesting, because the emission persisted at a high level of excess ($\sim$ 10 times brighter than the star at 12 μm) for around 25 years before dropping by an order of magnitude to a negligible level over the course of 2 years (Melis et al. 2012). A good explanation for the disappearance of the hot dust is lacking. One of the ideas proposed is that the dust was removed in a collisional avalanche, a runaway process wherein the input of large quantities of small grains that get blown out by radiation pressure break-up larger debris on their way out (Grigorieva et al. 2007). If so, the previously high levels of excess will return once the population of small dust is replenished in collisions between the bigger debris that was unaffected by the avalanche.

However, given the discussion in Sect. 6.2, perhaps it is not the disappearance of the dust which is surprising, rather it is the earlier persistence of the dust in the face of the star's radiation pressure which is more surprising. Indeed, a two year timescale for the emission to decline is compatible with the orbital removal timescale expected for radiation pressure. The question then would be why radiation pressure was ineffective for so long, and what then caused it to turn on so suddenly?

Another case that is particularly illuminating is that of 35 Myr-old Sun-like (G6V) star ID8 for which the 24 μm emission was noted to be variable in Meng et al. (2012). Later observations showed that the near-IR (3–5 μm) emission to this star underwent a substantial brightening followed by decay on a year timescale, with quasi-periodic modulations (with period $\sim$ 30 days) in the disk flux (Meng et al. 2014). This was interpreted as the result of a giant collision in which silicate vapour is created out of which small particles condense and subsequently collisionally deplete. The geometry of the collision-point during the early phases of the debris evolution explains the periodic modulation (see Sect. 5.1), implying that the collision was located at $\sim$ 0.33 au. Subsequent monitoring of 5 other systems with large levels of mid-IR emission showed that 5/6 of the full sample show significant variations on timescales less than a year (Meng et al. 2015).

Regardless of the origin of this variability, there is an empirical implication for how bright mid-IR emission is interpreted, since its level does not evolve in the manner that the colder dust in debris disks is thought to evolve. That is, the evolution is more complicated than that outlined in Sects. 4.2 and 5.2, in which the dust emission was assumed to decay slowly as the largest objects are ground into dust that is removed by radiation pressure (Eq. (2)). Indeed, the shape of what Kennedy and Wyatt (2013) called the exozodi luminosity function, that is the fraction of stars with 12 μm excess levels above a given level, may already bear witness to that; the luminosity function is relatively flat compared with the expectation that the brightest known excesses will decay due to collisions such that their brightness scales inversely with time.

6.4 How Common Are Late Giant Impacts?

In Jackson and Wyatt (2012) the predicted evolution of 24 μm emission from the debris following the Moon-forming impact was compared with the statistics for nearby stars to estimate the fraction of stars that form terrestrial planets. This section carries out a similar calculation, but for the reasons described in Sect. 6.1 is based on the evolution of the 12 μm emission. The question is also framed slightly differently, so that we seek to determine the fraction of stars that undergo late giant impacts

$$f_{\mathrm{gi}} = f_{12} * f_{12\mathrm{gi}}/f_{12\mathrm{gidet}}. \tag{4}$$

That is simply the fraction of 10–120 Myr stars with bright 12 μm excess, f_{12}, multiplied by the fraction of those bright excesses that originate in giant impacts rather than bright asteroid belts, f_{12gi}, divided by the fraction of the 10–120 Myr timeframe that it would be expected the giant impact debris would be at a detectable level, $f_{12gidet}$.

One of these parameters is well constrained, since Sect. 6.1 showed that $f_{12} \approx 3$ %, though this may be around a factor of 3 higher at the youngest end of the age range. While f_{12gi} is not well known, it can be said to be constrained to be less than 100 %. One might be mistaken for thinking that this is close to 100 % from the literature, since mid-IR emission is usually interpreted as the result of giant impacts, but there are not many cases where it is possible to rule out that the dust is simply the bottom end of the collisional cascade of a massive asteroid belt, at least for systems that are in this age range for which collisional processes have yet to have a significant effect on depleting any remnant asteroid belt (Wyatt et al. 2007a).

The hardest parameter to estimate is $f_{12gidet}$. The most pessimistic assumption would be that there is just one giant impact following the dispersal of the protoplanetary disk (i.e., the Moon-forming impact), and that the dust created from the grinding of boulders is below the detection threshold. This would result in $f_{12gidet} \approx 10^{-5}$ (i.e., the impact debris is detectable for around 1000 years out of the $\sim$ 100 Myr timeframe). In this case, the only way to explain the high detection rate of 12 μm excesses would be for the majority of these to originate in asteroid belts rather than giant impacts (i.e., $f_{12gi} < 3 \times 10^{-4}$). However, the conclusion is quite different if the dust from boulders starts out bright enough to be detectable, since the debris from a single impact would remain detectable for $\sim$ 10 Myr, and so $f_{12gidet} \approx 0.1$. This would mean that $f_{gi} < 0.3$. This calculation starts to become more constraining if one considers that terrestrial planet formation simulations predict that multiple giant impacts occur in the 10–100 Myr timeframe. Indeed, it is usually assumed that $f_{12gidet} = 0.5$–1 (Kenyon and Bromley 2004; Jackson and Wyatt 2012; Genda et al. 2015). If this is the case, then the fraction of systems that undergo a late giant impact must be less than 6 %. This would provide a useful constraint for planet formation models, by indicating that either the formation of Solar System-like terrestrial planets is relatively rare, or that planet growth is usually complete by the time the protoplanetary disk disperses; i.e., terrestrial planet formation is not normally accompanied by a late Moon-forming impact, which means planet formation models need to be revised. However, given the caveats on the brightness evolution in Sect. 5.2, strong conclusions cannot yet be drawn.

7 Collisions at Larger Separations

While giant impacts are perhaps most associated with terrestrial planet formation, there is substantial evidence from the Solar System that such events can take place in the outer reaches of planetary systems too (Sect. 2.2). Debris from giant impacts at larger separations undergoes the same evolution as described in Sect. 4. The one point of note is that, for a given progenitor, σ_v/v_k will be higher (and so the debris ends up in a broader annulus) for impacts at larger separations (see Fig. 1). However, the considerations for the detectability of debris from large separation impacts are slightly different than for impacts closer to the star (see Jackson et al. 2014).

One obvious difference is that dust is colder at larger separations, and so it is much more difficult to detect its emission at 12 μm. This is compounded by the more general point that a larger mass of dust (and so a larger progenitor mass) is required for a detectable level of emission at larger separations. The emission from giant impact debris at 10s of au would be most readily detected in the far-IR. Cold dust is quite commonly detected around nearby stars, with ~ 20 % of nearby Sun-like stars having far-IR emission from circumstellar debris (e.g., Eiroa et al. 2013). However, this is not interpreted as giant impact debris. Rather, this dust is believed to be the product of steady state collisional grinding of planetesimal belts that are analogous to the Kuiper Belt in the Solar System (Wyatt et al. 2007b; Löhne et al. 2008; Gáspár et al. 2013). At the large orbital separations inferred for these belts, collisional lifetimes are long enough that their dust can persist several Gyr above the detection threshold. Since observations of how debris levels vary with stellar age are consistent with the collisional erosion of Kuiper belt analogues, this interpretation is favoured over a giant impact origin for the dust (Wyatt et al. 2007b). Thus it is not possible to use the presence of cold dust from photometry to infer a recent giant impact as it was for hot dust found closer to the star (Kenyon and Bromley 2005).

However, as is evident in Fig. 2, the asymmetric morphology of a debris disk may be used to point to a recent giant impact. The detectability of such asymmetries is favoured at large separations, because the orbital timescales there are much longer than those in the terrestrial planet region, which means that the early asymmetric phases of the disk evolution last much longer. For example, while in the terrestrial planet region the typical lifetime of the asymmetric disk is a few thousand years, this lifetime may rise to of order a million years at 10s of au. In addition it is easier to resolve the morphology of debris in the outer reaches of a system due to its larger angular size (notwithstanding the longer wavelength of the emission). The most intuitive evidence of a giant impact would be the detection of an expanding clump (e.g., Wyatt and Dent 2002). However, the clump phase is so short-lived that it is much less likely to be witnessed than the asymmetric disk phase which is much longer lived.

In addition to the asymmetry in the distribution of debris seen in Fig. 2, the dominance of the collision-point in setting the collision rate of the debris means it plays a further important role in enhancing the asymmetry observed during the asymmetric disk phase at wavelengths that trace the distribution of certain collision products. Such products include the very smallest dust grains that are placed on orbits that are significantly modified by radiation pressure (i.e., those close to or below the blow-out limit given by Eq. (2)). This is because such grains are put on orbits with pericentres that are close to the site of their production, which means that when such grains are closest to the star they are predominantly found close to the collision-point (see Fig. 3). This leads to an excess of hot emission at this point, which may stand out further because such grains are also heated to temperatures above black body equilibrium (and so are much hotter than the rest of the debris); their apocentre distribution also means that their spatial distribution extends to much larger distance on the side of the star opposite the collision-point.

Another collision product with an enhanced asymmetry due to the collision-point is carbon monoxide (CO) gas. Cold bodies at large distances from their host star may contain CO ice (e.g., Solar System comets are typically ~1–10 % CO by mass), and when they are broken up in collisions this might be released as gas, promptly through vaporisation of the ice in the collision, or perhaps delayed through desorption from freshly exposed surfaces. CO gas has a comparatively short lifetime of around 150 years to photo-dissociation by interstellar ultraviolet radiation (e.g. Visser et al. 2009). This is comparable to the orbital timescale at tens of au from the host star, and thus the CO will undergo substantial decay over the course of the orbit after its production, which is predominantly at the collision-point (see Fig. 3). Thus the asymmetry in the CO distribution will be enhanced relative to the parent body population, since not only does the disk geometry reduce the CO density away from the collision-point, but so does the photo-dissociation.

One system in which this effect might be visible is the well known edge-on disk of β Pic. Telesco et al. (2005) revealed a large brightness asymmetry in the SW portion of the disk at

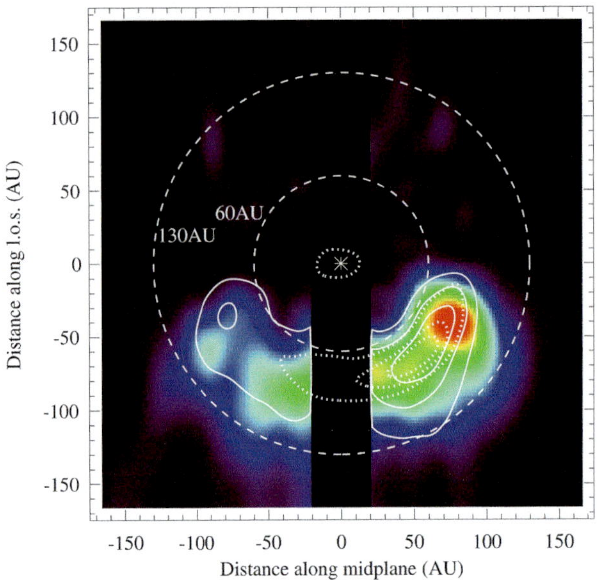

Fig. 7 Image of the face-on distribution of CO gas in the β Pic disk, derived from spatial and velocity information in ALMA observations of this edge-on disk (Dent et al. 2014). Note this is just one possible de-projection, since the observations are insensitive to reflections in the x-axis (i.e., the gas has the same line-of-sight velocity whether it is *in front* or *behind the star*). The *dashed circles* denote the inner and outer radii of the gas disk at 60–130 au, and the *asterisk* represents the location of the star. The distance along the line of sight (l.o.s.) is relative to the stellar position and is positive for gas on the far side of the star; orbital motion is clockwise. Overlaid on the *colour map* of the observed CO are *contours* of a model for CO produced at the collision-point of a giant impact involving Mars-size embryos at 85 au (Jackson et al. 2014). The *solid contours* show the model distribution after being run through the same de-projection process as the ALMA data, while the *dotted contours* show the true distribution

a projected separation of 52 au in the mid-infrared, and Dent et al. (2014) showed that this asymmetry is also present at a lower level in sub-millimetre continuum images. Dent et al. (2014) also found that an asymmetry is present in CO line observations at the same location, but with a significantly greater contrast. The CO line observations also provide information on the line-of-sight velocity of the gas, allowing the edge-on images of the gas distribution to be de-projected to get a face-on view of the disk, under the assumption that the gas is in Keplerian rotation. Figure 7 shows one possible de-projection in which most of the CO is in a clump at around 85 au from the star, with a tail extending 1/3 of an orbit in front of the clump (Dent et al. 2014). In optical scattered light the asymmetry is reversed, with the dust extending around 25 % farther in the NE compared with the SW (Larwood and Kalas 2001).

While it is possible that the clump in β Pic was created in the break-up of a planetesimal within the last 50 years (Telesco et al. 2005), we are many orders of magnitude more likely to witness the same impact in the asymmetric disk phase than in the clump phase. Moreover the morphology, and in particular its wavelength dependence outlined above, are exactly that expected if the asymmetry in the clump is the result of a giant impact witnessed in the asymmetric disk phase; e.g., Fig. 7 shows contours corresponding to a model for CO gas produced in a giant impact onto a Mars-sized progenitor occurring at 85 au from the star (based on the model presented in Jackson et al. 2014). However, an alternative explanation exists for the asymmetry in the β Pic disk that invokes a population of debris trapped in resonance with an unseen planet (Dent et al. 2014). Such resonant populations have a clumpy distribution (e.g. Wyatt 2003), and the collision rates would be enhanced in the clumps (Wyatt 2006), leading to similar asymmetries in the distributions of both CO and small grains to those of the asymmetric disk phase of a giant impact.

One way to distinguish between the two interpretations is to look for motion of the clump relative to the star, since the collision-point in the giant impact model is stationary on orbital timescales (only moving on much longer precession timescales), whereas resonant clumps would move with the planet. Li et al. (2012) re-imaged the mid-infrared clump and found tentative evidence for motion, but further observations are needed to definitively confirm or reject its motion. If the clump were found to be stationary, and so the giant impact interpretation favoured, this would have significant implications for our understanding of planet formation processes in the outer reaches of planetary systems. This would imply that sufficient Mars-sized embryos exist at a time of ~ 20 Myr after the star formed for us to be likely to witness a collision between them in a ~ 10 Myr timeframe. This would set constraints on the total mass of such embryos and how they are distributed in the disk and could be compared with models for the formation of such embryos (e.g., Kenyon and Bromley 2008). It would also boost the prospects for observing such collisions in other disks, as has been already been proposed for HD181327 (Stark et al. 2014).

8 Giant Collisions in Super-Earth Systems

As pointed out in Sect. 7, while giant impacts are an integral part of the process of forming bodies like our own terrestrial planets, they are not exclusive to the formation of Solar System-like terrestrial planets. One of the most interesting outcomes of exoplanet searches in recent years is the emergence of a new class of *super-Earth* planets. Such planets lie between roughly Earth and Neptune in size and/or mass and orbit in the inner regions of their planetary systems. There are no analogues to these planets in our Solar System, and yet they make up a substantial portion of the population of exoplanets discovered to date, thought to exist around 30–50 % of stars (Howard et al. 2010; Mayor et al. 2011). As such the formation mechanism of these planets is the subject of intense study and debate.

The key question regarding the origin of super-Earths is whether they formed in-situ at $\ll 1$ au (Hansen and Murray 2012, 2013; Chiang and Laughlin 2013) or at larger distance then migrated in through interaction with the protoplanetary disk (Alibert et al. 2006; Lopez et al. 2012; Kenyon and Bromley 2014). Knowing the internal composition of the planets would help, e.g. with a more distant formation location favoured for volatile-rich planets. A significant question is thus whether these planets are 'super-Earths' that are largely rocky, or 'sub-Neptunes' that are rich in volatile elements. It is becoming clear that planets with radii larger than around 1.5 $R_\oplus$ cannot be purely rocky (e.g. Weiss and Marcy 2014; Rogers 2015), however the nature and quantity of the lighter component of planets inferred to have low density is unclear. Planets with radii of around 2–3 $R_\oplus$ can be well fit by a rocky core with a hydrogen-helium envelope that comprises $\sim$0.1–5 per cent of the planet mass (e.g. Wolfgang and Lopez 2015), however they could equally well have less hydrogen and helium, but be predominantly composed of water (e.g. Seager et al. 2007; Howe et al. 2014). Thus far atmospheric studies of such planets have been inconclusive (e.g. Knutson et al. 2014; Kreidberg et al. 2014).

This section explores the possibility that constraints can be set on the origin and evolution of super-Earth planets by considering the debris released in giant impacts involving such planets. If rocky super-Earths form in situ in the same way as smaller terrestrial planets, they would be expected to undergo giant impacts with concomitant debris releases during their

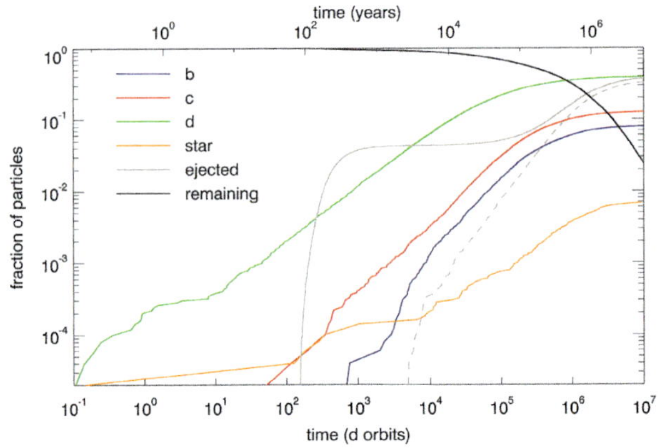

Fig. 8 Fate of particles in a 50000 particle N-body simulation of debris released in a giant impact onto the planet HD 69830d. Time is given in terms of d orbital periods (197 days). By 10^7 d orbits after the impact >95 per cent of the debris has been lost. *Both dashed and solid grey lines* correspond to particles that are ejected, however for the *dashed line* the prompt ejections (i.e., those particles placed on unbound orbits in the impact itself) are removed to make the longer term evolution more visible. The *black line* shows the number of particles remaining in the simulation

formation. A super-Earth planetary system formed by inward migration may also undergo a late phase of giant impacts after the protoplanetary disk dispersed, for example if this promoted instability in the system. Regardless of its origin a super-Earth planetary system may also undergo late giant impacts with any remnant planetary embryos in the system. While the rate of giant impacts experienced by a super-Earth system has yet to be discussed in the literature, the discussion in Sect. 6.4 shows that it should at least be possible to set constraints on that rate from observations.

To focus the discussion of giant impact debris in super-Earth systems, here we consider the HD 69830 system. This system contains three super-Earth planets with minimum masses of 10.2, 11.8 and 18.1 $M_\oplus$ at orbital distances of 0.079, 0.19 and 0.63 au respectively (Lovis et al. 2006) as well as a bright hot debris disk (Beichman et al. 2005). Modelling of the dust emission spectrum suggests that the debris is centred at around 1 au (Lisse et al. 2007), though this indirect constraint on its radial location has a factor of a few uncertainty; direct measurements of the radial location are consistent with a 1 au location, but again have a factor of a few uncertainty (Smith et al. 2009b). This system is particularly interesting as collisional modelling indicates that the dust cannot be produced by the steady-state grinding of a planetesimal belt that has existed at 1 au since the formation of the system ~2 Gyr ago, but rather must have been created more recently (Wyatt et al. 2007b; Heng and Tremaine 2010). The fact that the predicted location of the debris disk is very close to, and could be coincident with, the outermost planet HD 69830d also points to the possibility that the two are related.

While there are other explanations for the dust that do not involve HD 69830d, such as a recent collision within a stable asteroid belt at 1 au (Wyatt et al. 2010) or a super-comet (Beichman et al. 2005), consideration of giant impact debris from HD 69830d will highlight the differences and similarities to terrestrial planet giant impact debris that will be useful for a more general discussion of the detectability of super-Earth giant impact debris. Figure 8 shows results from a simulation of the dynamical evolution of debris released by a hypothetical impact involving HD 69830d. For the purposes of determining the velocity dispersion of the debris it was assumed that the planet has the same density as Neptune

(1.64 g/cm^3), which leads to $\sigma_v/v_k = 0.32$. This shows that accretion onto all of the planets (b, c and d) is rapid, as is ejection, with around 98 % of the particles having been lost by 10^7 d orbits (5 Myr) after the impact (with roughly twice as many lost to accretion as to ejection). This contrasts with the Solar System case of debris from Moon-formation where only 44 % of the particles had been lost to dynamical effects at 10 Myr (see Fig. 4).

A substantial part of this difference is due to the larger radii of the HD 69830 planets, since a Neptune-like HD 69830d has a radius of almost 4 $R_\oplus$ which would result in an accretion rate 16 times larger than Earth (Eq. (1), noting that re-accretion onto progenitors has a similar gravitational focussing factor). A more Earth-like composition, with a higher bulk density, would result in a smaller radius and a lower accretion rate, and so a longer dynamical evolution than shown in Fig. 8. Another difference with debris from the Moon-forming impact is that the accretion curves of all three planets around HD 69830 are roughly parallel. This is because the larger mass (and escape velocity) of HD 69830d compared with Earth leads to the debris being much more broadly distributed across the system. As a result, and due to the relatively closely-packed nature of the system, all three planets interact strongly with the debris from the beginning of the simulation. A higher density for HD 69830d would enhance this effect due to the corresponding increase in σ_v.

In addition to the more rapid dynamical evolution for super-Earth impact debris, the collisional evolution will also be faster, since the launch (and so collision) velocities will be higher than for debris from an Earth-like planet. A larger fraction of the debris may also be released as vapour, since the larger escape velocities of super-Earths imply that material ejected in their giant impacts will have been subject to more violent shocks. Both of these effects would add to the tendency for debris from a super-Earth giant impact to last for less time than one involving an Earth-size planet. While the initial mass of debris released from a more massive planet will be larger, this cannot offset the more rapid collisional evolution of debris from a super-Earth giant impact and so the duration of detectability would be unaffected, though this would mean that the disk is brighter while it is present.

This section only scratches the surface of the topic of giant impacts in systems of super-Earths. However, it is clear that there are potential implications for our understanding of the formation and composition of these worlds. For example, Eq. (4) can be rewritten as

$$t_{\mathrm{gise12}} = 100\ \mathrm{Myr} * f_{12} * f_{12\mathrm{gise}}/f_{\mathrm{se}}, \tag{5}$$

where $f_{\mathrm{se}} \approx 0.3$ is the fraction of stars that form super-Earths, $f_{12} \approx 0.03$ is the fraction of 10–120 Myr stars that have 12 μm excess, $f_{12\mathrm{gise}} < 1$ is the fraction of those 12 μm excesses that arise from giant impacts onto super-Earths, and t_{gise12} is the duration of detectability of giant impact debris in the 10–100 Myr window from super-Earths at 12 μm. This calculation benefits from several parameters being well constrained, leading to $t_{\mathrm{gise12}} < 10$ Myr. This excludes a long period of late giant impacts which make long-lived detectable debris, which could suggest most of the formation processes takes place in the protoplanetary disk, and/or that only a small fraction of the impact-generated debris is in the form of boulders. However, the 0.06 au radial location of the dust around ~ 60 Myr-old V488 Per was noted to be similar to that of transiting planets discovered by the Kepler satellite (Zuckerman et al. 2012), suggesting a possible origin in super-Earth giant impact debris. If confirmed this could indicate that such debris persists for a non-negligible duration (e.g., $t_{\mathrm{gise12}} > 0.1$ Myr).

9 Conclusions

There is abundant evidence for giant impacts in the history of the Solar System. There are many ways in which such impacts could be manifested in observations of extrasolar systems, including in the properties of any extrasolar planets (e.g., their spin, surface, density, atmosphere, and moons), but the most promising method at the moment for detecting giant impacts is through debris released in the collisions and the resulting infrared emission from dust that is heated by the star. This dust is likely to be readily detectable long after the event, up to 10–40 Myr in the case of the Earth-Moon forming impact. However, there remain significant uncertainties in the duration of detectability because of the unknown size distribution of impact-generated debris.

There are 10s of candidate young Sun-like stars with bright mid-IR emission from dust at levels expected following a giant collision like that which created the Moon. One of the best candidates for giant impact debris is around the star HD172555, since its spectrum shows dust with a silica composition that suggests an origin in a hypervelocity impact. Other systems also show complex variability in their emission, similar to that expected due to the dynamical and physical evolution of impact generated debris, which could provide a valuable window into these events. Despite the growing number of examples of potential giant impact debris, the occurrence rate of such emission is relatively low, at a few %, which either implies that late stage giant impacts are rare (i.e., planet formation is largely complete by the time the protoplanetary disk has dispersed), or that the debris from impacts rapidly grinds itself down to undetectable levels.

Giant impacts in terrestrial planet forming regions are those most readily identified, since these inner regions are usually clear of dust following the dispersal of the protoplanetary disk. However, giant impacts may also occur at larger distances from the stars (at 10s of au). At such distances, larger quantities of debris are required to be detected photometrically, not least because emission from cold debris is relatively common at this distance (~ 20 % for Sun-like stars), so evidence for giant impacts must rest on the morphology of the disk. A clump observed in the β Pictoris debris disk may be evidence of a collision between icy Mars-sized embryos. If so this would require the outer regions of this system to be host to multiple such embryos to give a reasonable chance of witnessing a collision. The ubiquitous class of super-Earth planetary system may also be subject to giant impacts. This chapter provides an initial exploration of this topic by showing how the dynamical evolution differs from that of Moon-forming debris, and it is shown that studying such impacts could provide important clues to the formation of super-Earth systems.

Acknowledgements M.C. Wyatt is grateful for support from the European Union through ERC grant number 279973.

References

C.B. Agnor, R.M. Canup, H.F. Levison, On the character and consequences of large impacts in the late stage of terrestrial planet formation. Icarus **142**, 219–237 (1999)

Y. Alibert, I. Baraffe, W. Benz, G. Chabrier, C. Mordasini, C. Lovis, M. Mayor, F. Pepe, F. Bouchy, D. Queloz, S. Udry, Formation and structure of the three Neptune-mass planets system around HD 69830. Astron. Astrophys. **455**, 25–28 (2006)

J.C. Andrews-Hanna, M.T. Zuber, W.B. Banerdt, The Borealis basin and the origin of the martian crustal dichotomy. Nature **453**, 1212–1215 (2008)

A. Anic, Giant collisions in the early solar system. Ph.D. thesis, Univ. Bern, 2006, http://space.unibe.ch/fileadmin/media/pdf/taps/diss-augustina.pdf

E. Asphaug, Impact origin of the Moon? Annu. Rev. Earth Planet. Sci. **42**, 551–578 (2014)
E. Asphaug, A. Reufer, Late origin of the Saturn system. Icarus **223**, 544–565 (2013)
H.H. Aumann, R.G. Probst, Search for Vega-like nearby stars with 12 micron excess. Astrophys. J. **368**, 264–271 (1991)
N.P. Ballering, G.H. Rieke, A. Gáspár, Probing the terrestrial regions of planetary systems: Warm debris disks with emission features. Astrophys. J. **793**, 57 (2014)
T.S. Barman, B. Macintosh, Q.M. Konopacky, C. Marois, The young planet-mass object 2M1207b: A cool, cloudy, and methane-poor atmosphere. Astrophys. J. Lett. **735**, 39 (2011)
C.A. Beichman, G. Bryden, T.N. Gautier, K.R. Stapelfeldt, M.W. Werner, K. Misselt, G. Rieke, J. Stansberry, D. Trilling, An excess due to small grains around the nearby K0 V Star HD 69830: Asteroid or cometary debris? Astrophys. J. **626**, 1061–1069 (2005)
W. Benz, W.L. Slattery, A.G.W. Cameron, Collisional stripping of Mercury's mantle. Icarus **74**, 516–528 (1988)
W. Benz, A. Anic, J. Horner, J.A. Whitby, The origin of Mercury. Space Sci. Rev. **132**, 189–202 (2007)
R.P. Binzel, M.J. Gaffey, P.C. Thomas, B.H. Zellner, A.D. Storrs, E.N. Wells, Geologic mapping of Vesta from 1994 Hubble space telescope images. Icarus **128**, 95–103 (1997)
W.F. Bottke, D.D. Durda, D. Nesvorný, R. Jedicke, A. Morbidelli, D. Vokrouhlický, H. Levison, The fossilized size distribution of the main asteroid belt. Icarus **175**, 111–140 (2005)
W.F. Bottke, D. Vokrouhlický, S. Marchi, T. Swindle, E.R.D. Scott, J.R. Weirich, H. Levison, Dating the Moon-forming impact event with asteroidal meteorites. Science **348**, 321–323 (2015)
M.E. Brown, K.M. Barkume, D. Ragozzine, E.L. Schaller, A collisional family of icy objects in the Kuiper belt. Nature **446**, 294–296 (2007)
A.G.W. Cameron, W.R. Ward, The origin of the Moon, in *Abstracts of the Lunar and Planetary Science Conference*, vol. 7, 1976, p. 120
A.G.W. Cameron, W. Benz, B. Fegley Jr., W.L. Slattery, The strange density of Mercury—Theoretical considerations, in *Mercury*, ed. by F. Vilas, C.R. Chapman, M.S. Matthews (1988), pp. 692–708
R.M. Canup, Dynamics of lunar formation. Annu. Rev. Astron. Astrophys. **42**, 441–475 (2004a)
R.M. Canup, Simulations of a late lunar-forming impact. Icarus **168**, 433–456 (2004b)
R.M. Canup, A giant impact origin of Pluto-Charon. Science **307**, 546–550 (2005)
R.M. Canup, Lunar-forming collisions with pre-impact rotation. Icarus **196**, 518–538 (2008)
R.M. Canup, On a giant impact origin of Charon, Nix, and Hydra. Astron. J. **141**, 35 (2011)
R.M. Canup, Forming a Moon with an Earth-like composition via a giant impact. Science **338**, 1052 (2012)
R.M. Canup, E. Asphaug, Origin of the Moon in a giant impact near the end of the Earth's formation. Nature **412**, 708–712 (2001)
R.M. Canup, W.R. Ward, Formation of the Galilean satellites: Conditions of accretion. Astron. J. **124**, 3404–3423 (2002)
R.M. Canup, W.R. Ward, A common mass scaling for satellite systems of gaseous planets. Nature **441**, 834–839 (2006)
J.M. Carpenter, J. Bouwman, E.E. Mamajek, M.R. Meyer, L.A. Hillenbrand, D.E. Backman, T. Henning, D.C. Hines, D. Hollenbach, J.S. Kim, A. Moro-Martin, I. Pascucci, M.D. Silverstone, J.R. Stauffer, S. Wolf, Formation and evolution of planetary systems: Properties of debris dust around solar-type stars. Astrophys. J. Suppl. Ser. **181**, 197–226 (2009)
C.H. Chen, B.A. Sargent, C. Bohac, K.H. Kim, E. Leibensperger, M. Jura, J. Najita, W.J. Forrest, D.M. Watson, G.C. Sloan, L.D. Keller, Spitzer IRS spectroscopy of IRAS-discovered debris disks. Astrophys. J. Suppl. Ser. **166**, 351–377 (2006)
E. Chiang, G. Laughlin, The minimum-mass extrasolar nebula: In situ formation of close-in super-Earths. Mon. Not. R. Astron. Soc. **431**, 3444–3455 (2013)
R.I. Citron, H. Genda, S. Ida, Formation of Phobos and Deimos via a giant impact. Icarus **252**, 334–338 (2015)
M. Ćuk, S.T. Stewart, Making the Moon from a fast-spinning Earth: A giant impact followed by resonant despinning. Science **338**, 1047 (2012)
M. Ćuk, D. Ragozzine, D. Nesvorný, On the dynamics and origin of Haumea's Moons. Astron. J. **146**, 89 (2013)
T. Currie, C.M. Lisse, A. Sicilia-Aguilar, G.H. Rieke, K.Y.L. Su, Spitzer infrared spectrograph spectroscopy of the 10 Myr old EF Cha debris disk: Evidence for phyllosilicate-rich dust in the terrestrial zone. Astrophys. J. **734**, 115 (2011)
W.R.F. Dent, M.C. Wyatt, A. Roberge, J.-C. Augereau, S. Casassus, S. Corder, J.S. Greaves, I. de Gregorio-Monsalvo, A. Hales, A.P. Jackson, A.M. Hughes, A.-M. Lagrange, B. Matthews, D. Wilner, Molecular gas clumps from the destruction of icy bodies in the β Pictoris debris disk. Science **343**, 1490–1492 (2014)

D.D. Durda, W.F. Bottke, D. Nesvorný, B.L. Enke, W.J. Merline, E. Asphaug, D.C. Richardson, Size-frequency distributions of fragments from SPH/ N-body simulations of asteroid impacts: Comparison with observed asteroid families. Icarus **186**, 498–516 (2007)
C. Eiroa, J.P. Marshall, A. Mora, B. Montesinos, O. Absil, J.C. Augereau, A. Bayo, G. Bryden, W. Danchi, C. del Burgo, S. Ertel, M. Fridlund, A.M. Heras, A.V. Krivov, R. Launhardt, R. Liseau, T. Löhne, J. Maldonado, G.L. Pilbratt, A. Roberge, J. Rodmann, J. Sanz-Forcada, E. Solano, K. Stapelfeldt, P. Thébault, S. Wolf, D. Ardila, M. Arévalo, C. Beichmann, V. Faramaz, B.M. González-García, R. Gutiérrez, J. Lebreton, R. Martínez-Arnáiz, G. Meeus, D. Montes, G. Olofsson, K.Y.L. Su, G.J. White, D. Barrado, M. Fukagawa, E. Grün, I. Kamp, R. Lorente, A. Morbidelli, S. Müller, H. Mutschke, T. Nakagawa, I. Ribas, H. Walker, DUst around NEarby Stars. The survey observational results. Astron. Astrophys. **555**, 11 (2013)
T.L. Esplin, K.L. Luhman, E.E. Mamajek, A WISE survey of circumstellar disks in Taurus. Astrophys. J. **784**, 126 (2014)
A. Gáspár, G.H. Rieke, Z. Balog, The collisional evolution of debris disks. Astrophys. J. **768**, 25 (2013)
H. Genda, H. Kobayashi, E. Kokubo, Warm debris disks produced by giant impacts during terrestrial planet formation. Astrophys. J. **810**, 136 (2015)
B. Gladman, J. Coffey, Mercurian impact ejecta: Meteorites and mantle. Meteorit. Planet. Sci. **44**, 285–291 (2009)
P. Goldreich, Y. Lithwick, R. Sari, Final stages of planet formation. Astrophys. J. **614**, 497–507 (2004)
R. Greenberg, Outcomes of tidal evolution for orbits with arbitrary inclination. Icarus **23**, 51–58 (1974)
A. Grigorieva, P. Artymowicz, P. Thébault, Collisional dust avalanches in debris discs. Astron. Astrophys. **461**, 537–549 (2007)
K. Grogan, S.F. Dermott, D.D. Durda, The size-frequency distribution of the zodiacal cloud: Evidence from the solar system dust bands. Icarus **152**, 251–267 (2001)
K.E. Haisch Jr., E.A. Lada, C.J. Lada, Disk frequencies and lifetimes in young clusters. Astrophys. J. Lett. **553**, 153–156 (2001)
A.N. Halliday, A young Moon-forming giant impact at 70–110 million years accompanied by late-stage mixing, core formation and degassing of the Earth. Philos. Trans. R. Soc. A, Math. Phys. Eng. Sci. **366**, 4163–4181 (2008)
B.M.S. Hansen, Formation of the terrestrial planets from a narrow annulus. Astrophys. J. **703**, 1131–1140 (2009)
B.M.S. Hansen, N. Murray, Migration then assembly: Formation of Neptune-mass planets inside 1 AU. Astrophys. J. **751**, 158 (2012)
B.M.S. Hansen, N. Murray, Testing in situ assembly with the Kepler planet candidate sample. Astrophys. J. **775**, 53 (2013)
R. Helled, P. Bodenheimer, The formation of Uranus and Neptune: Challenges and implications for intermediate-mass exoplanets. Astrophys. J. **789**, 69 (2014)
K. Heng, S. Tremaine, Long-lived planetesimal discs. Mon. Not. R. Astron. Soc. **401**, 867–889 (2010)
K. Hirayama, Groups of asteroids probably of common origin. Astron. J. **31**, 185–188 (1918)
A.W. Howard, G.W. Marcy, J.A. Johnson, D.A. Fischer, J.T. Wright, H. Isaacson, J.A. Valenti, J. Anderson, D.N.C. Lin, S. Ida, The occurrence and mass distribution of close-in super-Earths, Neptunes, and Jupiters. Science **330**, 653 (2010)
A.R. Howe, A. Burrows, W. Verne, Mass-radius relations and core-envelope decompositions of super-Earths and sub-Neptunes. Astrophys. J. **787**, 173 (2014)
N.K. Inamdar, H.E. Schlichting, The formation of super-Earths and mini-Neptunes with giant impacts. Mon. Not. R. Astron. Soc. **448**, 1751–1760 (2015)
A.P. Jackson, M.C. Wyatt, Debris from terrestrial planet formation: The Moon-forming collision. Mon. Not. R. Astron. Soc. **425**, 657–679 (2012)
A.P. Jackson, M.C. Wyatt, A. Bonsor, D. Veras, Debris from giant impacts between planetary embryos at large orbital radii. Mon. Not. R. Astron. Soc. **440**, 3757–3777 (2014)
D. Jewitt, H. Weaver, M. Mutchler, S. Larson, J. Agarwal, Hubble space telescope observations of Main-belt Comet (596) Scheila. Astrophys. J. Lett. **733**, 4 (2011)
B.C. Johnson, H.J. Melosh, Formation of melt droplets, melt fragments, and accretionary impact lapilli during a hypervelocity impact. Icarus **228**, 347–363 (2014)
B.C. Johnson, C.M. Lisse, C.H. Chen, H.J. Melosh, M.C. Wyatt, P. Thebault, W.G. Henning, E. Gaidos, L.T. Elkins-Tanton, J.C. Bridges, A. Morlok, A self-consistent model of the circumstellar debris created by a giant hypervelocity impact in the HD 172555 system. Astrophys. J. **761**, 45 (2012)
D. Jontof-Hutter, J.F. Rowe, J.J. Lissauer, D.C. Fabrycky, E.B. Ford, The mass of the Mars-sized exoplanet Kepler-138 b from transit timing. Nature **522**, 321–323 (2015)
M. Jutzi, SPH calculations of asteroid disruptions: The role of pressure dependent failure models. Planet. Space Sci. **107**, 3–9 (2015)

M. Jutzi, E. Asphaug, P. Gillet, J.-A. Barrat, W. Benz, The structure of the asteroid 4 Vesta as revealed by models of planet-scale collisions. Nature **494**, 207–210 (2013)
P. Kalas, J.R. Graham, M.P. Fitzgerald, M. Clampin, STIS coronagraphic imaging of Fomalhaut: Main belt structure and the orbit of Fomalhaut b. Astrophys. J. **775**, 56 (2013)
K. Keil, Geological history of Asteroid 4 Vesta: The "Smallest Terrestrial Planet", in *Asteroids III* (2002), pp. 573–584
G.M. Kennedy, M.C. Wyatt, The bright end of the exo-Zodi luminosity function: Disc evolution and implications for exo-Earth detectability. Mon. Not. R. Astron. Soc. **433**, 2334–2356 (2013)
G.M. Kennedy, M.C. Wyatt, Do two-temperature debris discs have multiple belts? Mon. Not. R. Astron. Soc. **444**, 3164–3182 (2014)
G.M. Kennedy, S.J. Murphy, C.M. Lisse, F. Ménard, M.L. Sitko, M.C. Wyatt, D.D.R. Bayliss, F.E. DeMeo, K.B. Crawford, D.L. Kim, R.J. Rudy, R.W. Russell, B. Sibthorpe, M.A. Skinner, G. Zhou, Evolution from protoplanetary to debris discs: The transition disc around HD 166191. Mon. Not. R. Astron. Soc. **438**, 3299–3309 (2014)
M.A. Kenworthy, E.E. Mamajek, Modeling giant extrasolar ring systems in eclipse and the case of J1407b: Sculpting by exomoons? Astrophys. J. **800**, 126 (2015)
S.J. Kenyon, B.C. Bromley, Detecting the dusty debris of terrestrial planet formation. Astrophys. J. Lett. **602**, 133–136 (2004)
S.J. Kenyon, B.C. Bromley, Prospects for detection of catastrophic collisions in debris disks. Astron. J. **130**, 269–279 (2005)
S.J. Kenyon, B.C. Bromley, Terrestrial planet formation. I. The transition from oligarchic growth to chaotic growth. Astron. J. **131**, 1837–1850 (2006)
S.J. Kenyon, B.C. Bromley, Variations on debris disks: Icy planet formation at 30–150 AU for 1–3 M_{solar} main-sequence stars. Astrophys. J. Suppl. Ser. **179**, 451–483 (2008)
S.J. Kenyon, B.C. Bromley, Coagulation calculations of icy planet formation around 0.1–0.5 $M_{\odot}$ stars: Super-Earths from large planetestimals. Astrophys. J. **780**, 4 (2014)
D.M. Kipping, D. Nesvorný, L.A. Buchhave, J. Hartman, G.Á. Bakos, A.R. Schmitt, The hunt for exomoons with Kepler (HEK). IV. A search for Moons around eight M dwarfs. Astrophys. J. **784**, 28 (2014)
T. Kleine, M. Touboul, B. Bourdon, F. Nimmo, K. Mezger, H. Palme, S.B. Jacobsen, Q.-Z. Yin, A.N. Halliday, Hf-W chronology of the accretion and early evolution of asteroids and terrestrial planets. Geochim. Cosmochim. Acta **73**, 5150–5188 (2009)
H.A. Knutson, B. Benneke, D. Deming, D. Homeier, A featureless transmission spectrum for the Neptune-mass exoplanet GJ436b. Nature **505**, 66–68 (2014)
E. Kokubo, S. Ida, Oligarchic growth of protoplanets. Icarus **131**, 171–178 (1998)
Q. Kral, P. Thébault, J.-C. Augereau, A. Boccaletti, S. Charnoz, Signatures of massive collisions in debris discs. A self-consistent numerical model. Astron. Astrophys. **573**, 39 (2015)
R.G. Kraus, S.T. Stewart, D.C. Swift, C.A. Bolme, R.F. Smith, S. Hamel, B.D. Hammel, D.K. Spaulding, D.G. Hicks, J.H. Eggert, G.W. Collins, Shock vaporization of silica and the thermodynamics of planetary impact events. J. Geophys. Res., Planets **117**, 9009 (2012)
L. Kreidberg, J.L. Bean, J.-M. Désert, B. Benneke, D. Deming, K.B. Stevenson, S. Seager, Z. Berta-Thompson, A. Seifahrt, D. Homeier, Clouds in the atmosphere of the super-Earth exoplanet GJ1214b. Nature **505**, 69–72 (2014)
A.V. Krivov, M. Sremčević, F. Spahn, Evolution of a Keplerian disk of colliding and fragmenting particles: A kinetic model with application to the Edgeworth Kuiper belt. Icarus **174**, 105–134 (2005)
T. Kubo-Oka, K. Nakazawa, Gradual increase in the obliquity of Uranus due to tidal interaction with a hypothetical retrograde satellite. Icarus **114**, 21–32 (1995)
J.D. Larwood, P.G. Kalas, Close stellar encounters with planetesimal discs: The dynamics of asymmetry in the β Pictoris system. Mon. Not. R. Astron. Soc. **323**, 402–416 (2001)
J. Lebreton, R. van Lieshout, J.-C. Augereau, O. Absil, B. Mennesson, M. Kama, C. Dominik, A. Bonsor, J. Vandeportal, H. Beust, D. Defrère, S. Ertel, V. Faramaz, P. Hinz, Q. Kral, A.-M. Lagrange, W. Liu, P. Thébault, An interferometric study of the Fomalhaut inner debris disk. III. Detailed models of the exozodiacal disk and its origin. Astron. Astrophys. **555**, 146 (2013)
Z.M. Leinhardt, S.T. Stewart, Collisions between gravity-dominated bodies. I. Outcome regimes and scaling laws. Astrophys. J. **745**, 79 (2012)
D. Li, C.M. Telesco, C.M. Wright, The mineralogy and structure of the inner debris disk of β Pictoris. Astrophys. J. **759**, 81 (2012)
C.M. Lisse, C.A. Beichman, G. Bryden, M.C. Wyatt, On the nature of the dust in the debris disk around HD 69830. Astrophys. J. **658**, 584–592 (2007)
C.M. Lisse, C.H. Chen, M.C. Wyatt, A. Morlok, Circumstellar dust created by terrestrial planet formation in HD 113766. Astrophys. J. **673**, 1106–1122 (2008)

C.M. Lisse, C.H. Chen, M.C. Wyatt, A. Morlok, I. Song, G. Bryden, P. Sheehan, Abundant circumstellar silica dust and SiO gas created by a giant hypervelocity collision in the $\sim$ 12 Myr HD172555 system. Astrophys. J. **701**, 2019–2032 (2009)
T. Löhne, A.V. Krivov, J. Rodmann, Long-term collisional evolution of debris disks. Astrophys. J. **673**, 1123–1137 (2008)
E.D. Lopez, J.J. Fortney, N. Miller, How thermal evolution and mass-loss sculpt populations of super-Earths and sub-Neptunes: Application to the Kepler-11 system and beyond. Astrophys. J. **761**, 59 (2012)
C. Lovis, M. Mayor, F. Pepe, Y. Alibert, W. Benz, F. Bouchy, A.C.M. Correia, J. Laskar, C. Mordasini, D. Queloz, N.C. Santos, S. Udry, J.-L. Bertaux, J.-P. Sivan, An extrasolar planetary system with three Neptune-mass planets. Nature **441**, 305–309 (2006)
K.L. Luhman, E.E. Mamajek, The disk population of the upper Scorpius association. Astrophys. J. **758**, 31 (2012)
R.E. Lupu, K. Zahnle, M.S. Marley, L. Schaefer, B. Fegley, C. Morley, K. Cahoy, R. Freedman, J.J. Fortney, The atmospheres of earthlike planets after giant impact events. Astrophys. J. **784**, 27 (2014)
L. Malo, R. Doyon, D. Lafrenière, É. Artigau, J. Gagné, F. Baron, A. Riedel, Bayesian analysis to identify new star candidates in nearby young stellar kinematic groups. Astrophys. J. **762**, 88 (2013)
E.E. Mamajek, M.R. Meyer, An improbable solution to the underluminosity of 2M1207B: A hot protoplanet collision afterglow. Astrophys. J. Lett. **668**, 175–178 (2007)
E.E. Mamajek, A.C. Quillen, M.J. Pecaut, F. Moolekamp, E.L. Scott, M.A. Kenworthy, A. Collier Cameron, N.R. Parley, Planetary construction zones in occultation: Discovery of an extrasolar ring system transiting a young Sun-like star and future prospects for detecting eclipses by circumsecondary and circumplanetary disks. Astron. J. **143**, 72 (2012)
R.A. Marcus, S.T. Stewart, D. Sasselov, L. Hernquist, Collisional stripping and disruption of super-Earths. Astrophys. J. Lett. **700**, 118–122 (2009)
G.W. Marcy, H. Isaacson, A.W. Howard, J.F. Rowe, J.M. Jenkins, S.T. Bryson, D.W. Latham, S.B. Howell, T.N. Gautier III, N.M. Batalha, L. Rogers, D. Ciardi, D.A. Fischer, R.L. Gilliland, H. Kjeldsen, J. Christensen-Dalsgaard, D. Huber, W.J. Chaplin, S. Basu, L.A. Buchhave, S.N. Quinn, W.J. Borucki, D.G. Koch, R. Hunter, D.A. Caldwell, J. Van Cleve, R. Kolbl, L.M. Weiss, E. Petigura, S. Seager, T. Morton, J.A. Johnson, S. Ballard, C. Burke, W.D. Cochran, M. Endl, P. MacQueen, M.E. Everett, J.J. Lissauer, E.B. Ford, G. Torres, F. Fressin, T.M. Brown, J.H. Steffen, D. Charbonneau, G.S. Basri, D.D. Sasselov, J. Winn, R. Sanchis-Ojeda, J. Christiansen, E. Adams, C. Henze, A. Dupree, D.C. Fabrycky, J.J. Fortney, J. Tarter, M.J. Holman, P. Tenenbaum, A. Shporer, P.W. Lucas, W.F. Welsh, J.A. Orosz, T.R. Bedding, T.L. Campante, G.R. Davies, Y. Elsworth, R. Handberg, S. Hekker, C. Karoff, S.D. Kawaler, M.N. Lund, M. Lundkvist, T.S. Metcalfe, A. Miglio, V. Silva Aguirre, D. Stello, T.R. White, A. Boss, E. Devore, A. Gould, A. Prsa, E. Agol, T. Barclay, J. Coughlin, E. Brugamyer, F. Mullally, E.V. Quintana, M. Still, S.E. Thompson, D. Morrison, J.D. Twicken, J.-M. Désert, J. Carter, J.R. Crepp, G. Hébrard, A. Santerne, C. Moutou, C. Sobeck, D. Hudgins, M.R. Haas, P. Robertson, J. Lillo-Box, D. Barrado, Masses, radii, and orbits of small Kepler planets: The transition from gaseous to rocky planets. Astrophys. J. Suppl. Ser. **210**, 20 (2014)
M.M. Marinova, O. Aharonson, E. Asphaug, Mega-impact formation of the Mars hemispheric dichotomy. Nature **453**, 1216–1219 (2008)
M.M. Marinova, O. Aharonson, E. Asphaug, Geophysical consequences of planetary-scale impacts into a Mars-like planet. Icarus **211**, 960–985 (2011)
A. Mastrobuono-Battisti, H.B. Perets, S.N. Raymond, A primordial origin for the compositional similarity between the Earth and the Moon. Nature **520**, 212–215 (2015)
B.C. Matthews, J.J. Kavelaars, Insights into planet formation from debris disks. I. The solar system as an archetype for planetesimal evolution. Space Sci. Rev. (2016, this issue)
B.C. Matthews, A.V. Krivov, M.C. Wyatt, G. Bryden, C. Eiroa, Observations, modeling, and theory of debris disks, in *Protostars and Planets VI* (2014), pp. 521–544
M. Mayor, M. Marmier, C. Lovis, S. Udry, D. Ségransan, F. Pepe, W. Benz, J.-L. Bertaux, F. Bouchy, X. Dumusque, G. Lo Curto, C. Mordasini, D. Queloz, N.C. Santos, The HARPS search for southern extra-solar planets XXXIV. Occurrence, mass distribution and orbital properties of super-Earths and Neptune-mass planets, ArXiv e-prints (2011). arXiv:1109.2497
W.B. McKinnon, On the origin of the Pluto-Charon binary. Astrophys. J. Lett. **344**, 41–44 (1989)
C. Melis, B. Zuckerman, J.H. Rhee, I. Song, The age of the HD 15407 system and the epoch of final catastrophic mass accretion onto terrestrial planets around Sun-like stars. Astrophys. J. Lett. **717**, 57–61 (2010)
C. Melis, B. Zuckerman, J.H. Rhee, I. Song, S.J. Murphy, M.S. Bessell, Rapid disappearance of a warm, dusty circumstellar disk. Nature **487**, 74–76 (2012)
H.J. Melosh, A.M. Vickery, Melt droplet formation in energetic impact events. Nature **350**, 494–497 (1991)

H.Y.A. Meng, G.H. Rieke, K.Y.L. Su, V.D. Ivanov, L. Vanzi, W. Rujopakarn, Variability of the infrared excess of extreme debris disks. Astrophys. J. Lett. **751**, 17 (2012)
H.Y.A. Meng, K.Y.L. Su, G.H. Rieke, D.J. Stevenson, P. Plavchan, W. Rujopakarn, C.M. Lisse, S. Poshyachinda, D.E. Reichart, Large impacts around a solar-analog star in the era of terrestrial planet formation. Science **345**, 1032–1035 (2014)
H.Y.A. Meng, K.Y.L. Su, G.H. Rieke, W. Rujopakarn, G. Myers, M. Cook, E. Erdelyi, C. Maloney, J. McMath, G. Persha, S. Poshyachinda, D.E. Reichart, Planetary collisions outside the solar system: Time domain characterization of extreme debris disks. Astrophys. J. **805**, 77 (2015)
M.R. Meyer, J.M. Carpenter, E.E. Mamajek, L.A. Hillenbrand, D. Hollenbach, A. Moro-Martin, J.S. Kim, M.D. Silverstone, J. Najita, D.C. Hines, I. Pascucci, J.R. Stauffer, J. Bouwman, D.E. Backman, Evolution of mid-infrared excess around Sun-like stars: Constraints on models of terrestrial planet formation. Astrophys. J. Lett. **673**, 181–184 (2008)
E. Miller-Ricci, M.R. Meyer, S. Seager, L. Elkins-Tanton, On the emergent spectra of hot protoplanet collision afterglows. Astrophys. J. **704**, 770–780 (2009)
D. Nesvorný, W.F. Bottke Jr., L. Dones, H.F. Levison, The recent breakup of an asteroid in the main-belt region. Nature **417**, 720–771 (2002)
N. Nikolov, F. Sainsbury-Martinez, Radial velocity eclipse mapping of exoplanets. ArXiv e-prints (2015)
F. Nimmo, S.D. Hart, D.G. Korycansky, C.B. Agnor, Implications of an impact origin for the martian hemispheric dichotomy. Nature **453**, 1220–1223 (2008)
D.P. O'Brien, R. Greenberg, Steady-state size distributions for collisional populations:. analytical solution with size-dependent strength. Icarus **164**, 334–345 (2003)
K. Pahlevan, D.J. Stevenson, Equilibration in the aftermath of the lunar-forming giant impact. Earth Planet. Sci. Lett. **262**, 438–449 (2007)
O. Panić, W.S. Holland, M.C. Wyatt, G.M. Kennedy, B.C. Matthews, J.F. Lestrade, B. Sibthorpe, J.S. Greaves, J.P. Marshall, N.M. Phillips, J. Tottle, First results of the SONS survey: Submillimetre detections of debris discs. Mon. Not. R. Astron. Soc. **435**, 1037–1046 (2013)
M.G. Parisi, A. Brunini, Constraints to Uranus' great collision—II. Planet. Space Sci. **45**, 181–187 (1997)
N.M. Phillips, J.S. Greaves, W.R.F. Dent, B.C. Matthews, W.S. Holland, M.C. Wyatt, B. Sibthorpe, Target selection for the SUNS and DEBRIS surveys for debris discs in the solar neighbourhood. Mon. Not. R. Astron. Soc. **403**, 1089–1101 (2010)
A. Reufer, M.M.M. Meier, W. Benz, R. Wieler, A hit-and-run giant impact scenario. Icarus **221**, 296–299 (2012)
J.H. Rhee, I. Song, B. Zuckerman, Warm dust in the terrestrial planet zone of a Sun-like Pleiades star: Collisions between planetary embryos? Astrophys. J. **675**, 777–783 (2008)
G.H. Rieke, K.Y.L. Su, J.A. Stansberry, D. Trilling, G. Bryden, J. Muzerolle, B. White, N. Gorlova, E.T. Young, C.A. Beichman, K.R. Stapelfeldt, D.C. Hines, Decay of planetary debris disks. Astrophys. J. **620**, 1010–1026 (2005)
P. Riviere-Marichalar, D. Barrado, J.-C. Augereau, W.F. Thi, A. Roberge, C. Eiroa, B. Montesinos, G. Meeus, C. Howard, G. Sandell, G. Duchêne, W.R.F. Dent, J. Lebreton, I. Mendigutía, N. Huélamo, F. Ménard, C. Pinte, HD 172555: Detection of 63 μm [OI] emission in a debris disc. Astron. Astrophys. **546**, 8 (2012)
A.C. Rizzuto, M.J. Ireland, D.B. Zucker, WISE circumstellar discs in the young Sco-Cen association. Mon. Not. R. Astron. Soc. **421**, 97–101 (2012)
L.A. Rogers, Most 1.6 Earth-radius planets are not rocky. Astrophys. J. **801**, 41 (2015)
P. Rosenblatt, The origin of the Martian moons revisited. Astron. Astrophys. Rev. **19**, 44 (2011)
P. Rosenblatt, S. Charnoz, On the formation of the martian moons from a circum-martian accretion disk. Icarus **221**, 806–815 (2012)
C.T. Russell, C.A. Raymond, A. Coradini, H.Y. McSween, M.T. Zuber, A. Nathues, M.C. De Sanctis, R. Jaumann, A.S. Konopliv, F. Preusker, S.W. Asmar, R.S. Park, R. Gaskell, H.U. Keller, S. Mottola, T. Roatsch, J.E.C. Scully, D.E. Smith, P. Tricarico, M.J. Toplis, U.R. Christensen, W.C. Feldman, D.J. Lawrence, T.J. McCoy, T.H. Prettyman, R.C. Reedy, M.E. Sykes, T.N. Titus, Dawn at Vesta: Testing the protoplanetary paradigm. Science **336**, 684 (2012)
V.S. Safronov, *Evolution of the Protoplanetary Cloud and Formation of the Earth and Planets*. 1972
J. Salmon, R.M. Canup, Lunar accretion from a Roche-interior fluid disk. Astrophys. J. **760**, 83 (2012)
H.E. Schlichting, R. Sari, The creation of Haumea's collisional family. Astrophys. J. **700**, 1242–1246 (2009)
H.E. Schlichting, R. Sari, A. Yalinewich, Atmospheric mass loss during planet formation: The importance of planetesimal impacts. Icarus **247**, 81–94 (2015)
A. Schneider, I. Song, C. Melis, B. Zuckerman, M. Bessell, T. Hufford, S. Hinkley, The nearby, young, isolated, dusty star HD 166191. Astrophys. J. **777**, 78 (2013)
S. Seager, M. Kuchner, C.A. Hier-Majumder, B. Militzer, Mass-radius relationships for solid exoplanets. Astrophys. J. **669**, 1279–1297 (2007)

Y. Sekine, H. Genda, Giant impacts in the Saturnian system: A possible origin of diversity in the inner mid-sized satellites. Planet. Space Sci. **63**, 133–138 (2012)

M.R. Showalter, D.P. Hamilton, Resonant interactions and chaotic rotation of Pluto's small moons. Nature **522**, 45–49 (2015)

N. Siegler, J. Muzerolle, E.T. Young, G.H. Rieke, E.E. Mamajek, D.E. Trilling, N. Gorlova, K.Y.L. Su, Spitzer 24 μm observations of open cluster IC 2391 and debris disk evolution of FGK stars. Astrophys. J. **654**, 580–594 (2007)

D.E. Smith, M.T. Zuber, S.C. Solomon, R.J. Phillips, J.W. Head, J.B. Garvin, W.B. Banerdt, D.O. Muhleman, G.H. Pettengill, G.A. Neumann, F.G. Lemoine, J.B. Abshire, O. Aharonson, C.D. Brown, S.A. Hauck, A.B. Ivanov, P.J. McGovern, H.J. Zwally, T.C. Duxbury, The global topography of Mars and implications for surface evolution. Science **284**, 1495 (1999)

R. Smith, L.J. Churcher, M.C. Wyatt, M.M. Moerchen, C.M. Telesco, Resolved debris disc emission around η Telescopii: A young solar system or ongoing planet formation? Astron. Astrophys. **493**, 299–308 (2009a)

R. Smith, M.C. Wyatt, C.A. Haniff, Resolving the hot dust around HD69830 and η Corvi with MIDI and VISIR. Astron. Astrophys. **503**, 265–279 (2009b)

R. Smith, M.C. Wyatt, C.A. Haniff, Resolving the terrestrial planet forming regions of HD 113766 and HD 172555 with MIDI. Mon. Not. R. Astron. Soc. **422**, 2560–2580 (2012)

I.A.G. Snellen, B.R. Brandl, R.J. de Kok, M. Brogi, J. Birkby, H. Schwarz, Fast spin of the young extrasolar planet β Pictoris b. Nature **509**, 63–65 (2014)

I. Song, B. Zuckerman, A.J. Weinberger, E.E. Becklin, Extreme collisions between planetesimals as the origin of warm dust around a Sun-like star. Nature **436**, 363–365 (2005)

C.C. Stark, G. Schneider, A.J. Weinberger, J.H. Debes, C.A. Grady, H. Jang-Condell, M.J. Kuchner, Revealing asymmetries in the HD 181327 debris disk: A recent massive collision or interstellar medium warping. Astrophys. J. **789**, 58 (2014)

S.T. Stewart, Z.M. Leinhardt, Collisions between gravity-dominated bodies. II. The diversity of impact outcomes during the end stage of planet formation. Astrophys. J. **751**, 32 (2012)

K.Y.L. Su, G.H. Rieke, J.A. Stansberry, G. Bryden, K.R. Stapelfeldt, D.E. Trilling, J. Muzerolle, C.A. Beichman, A. Moro-Martin, D.C. Hines, M.W. Werner, Debris disk evolution around a stars. Astrophys. J. **653**, 675–689 (2006)

K.Y.L. Su, G.H. Rieke, R. Malhotra, K.R. Stapelfeldt, A.M. Hughes, A. Bonsor, D.J. Wilner, Z. Balog, D.M. Watson, M.W. Werner, K.A. Misselt, Asteroid belts in debris disk twins: Vega and Fomalhaut. Astrophys. J. **763**, 118 (2013)

C.M. Telesco, R.S. Fisher, M.C. Wyatt, S.F. Dermott, T.J.J. Kehoe, S. Novotny, N. Mariñas, J.T. Radomski, C. Packham, J. De Buizer, T.L. Hayward, Mid-infrared images of β Pictoris and the possible role of planetesimal collisions in the central disk. Nature **433**, 133–136 (2005)

P. Thébault, J.C. Augereau, H. Beust, Dust production from collisions in extrasolar planetary systems. The inner beta Pictoris disc. Astron. Astrophys. **408**, 775–788 (2003)

R. Visser, E.F. van Dishoeck, J.H. Black, The photodissociation and chemistry of CO isotopologues: Applications to interstellar clouds and circumstellar disks. Astron. Astrophys. **503**, 323–343 (2009)

K.J. Walsh, A. Morbidelli, S.N. Raymond, D.P. O'Brien, A.M. Mandell, A low mass for Mars from Jupiter's early gas-driven migration. Nature **475**, 206–209 (2011)

L.M. Weiss, G.W. Marcy, The mass-radius relation for 65 exoplanets smaller than 4 Earth radii. Astrophys. J. Lett. **783**, 6 (2014)

G.W. Wetherill, Collisions in the asteroid belt. J. Geophys. Res. **72**, 2429 (1967)

D.E. Wilhelms, S.W. Squyres, The martian hemispheric dichotomy may be due to a giant impact. Nature **309**, 138–140 (1984)

A. Wolfgang, E. Lopez, How rocky are they? The composition distribution of Kepler's sub-Neptune planet candidates within 0.15 AU. Astrophys. J. **806**, 183 (2015)

M.C. Wyatt, Resonant trapping of planetesimals by planet migration: Debris disk clumps and Vega's similarity to the solar system. Astrophys. J. **598**, 1321–1340 (2003)

M.C. Wyatt, Dust in resonant extrasolar Kuiper belts: Grain size and wavelength dependence of disk structure. Astrophys. J. **639**, 1153–1165 (2006)

M.C. Wyatt, Evolution of debris disks. Annu. Rev. Astron. Astrophys. **46**, 339–383 (2008)

M.C. Wyatt, W.R.F. Dent, Collisional processes in extrasolar planetesimal discs - dust clumps in Fomalhaut's debris disc. Mon. Not. R. Astron. Soc. **334**, 589–607 (2002)

M.C. Wyatt, R. Smith, J.S. Greaves, C.A. Beichman, G. Bryden, C.M. Lisse, Transience of hot dust around Sun-like stars. Astrophys. J. **658**, 569–583 (2007a)

M.C. Wyatt, R. Smith, K.Y.L. Su, G.H. Rieke, J.S. Greaves, C.A. Beichman, G. Bryden, Steady state evolution of debris disks around a stars. Astrophys. J. **663**, 365–382 (2007b)

M.C. Wyatt, M. Booth, M.J. Payne, L.J. Churcher, Collisional evolution of eccentric planetesimal swarms. Mon. Not. R. Astron. Soc. **402**, 657–672 (2010)
M.C. Wyatt, C.J. Clarke, M. Booth, Debris disk size distributions: Steady state collisional evolution with Poynting-Robertson drag and other loss processes. Celest. Mech. Dyn. Astron. **111**, 1–28 (2011)
M.C. Wyatt, O. Panić, G.M. Kennedy, L. Matrà, Five steps in the evolution from protoplanetary to debris disk. Astrophys. Space Sci. **357**, 103 (2015)
K. Zahnle, N. Arndt, C. Cockell, A. Halliday, E. Nisbet, F. Selsis, N.H. Sleep, Emergence of a Habitable planet. Space Sci. Rev. **129**, 35–78 (2007)
B. Zuckerman, C. Melis, J.H. Rhee, A. Schneider, I. Song, Stellar membership and dusty debris disks in the α persei cluster. Astrophys. J. **752**, 58 (2012)
J.I. Zuluaga, D.M. Kipping, M. Sucerquia, J.A. Alvarado, A novel method for identifying exoplanetary rings. Astrophys. J. Lett. **803**, 14 (2015)

Space Sci Rev (2016) 205:309–325
DOI 10.1007/s11214-016-0246-3

The Architecture of Exoplanets

Artie P. Hatzes[1]

Received: 31 August 2015 / Accepted: 14 March 2016 / Published online: 17 May 2016

Abstract Prior to the discovery of exoplanets our expectations of their architecture were largely driven by the properties of our solar system. We expected giant planets to lie in the outer regions and rocky planets in the inner regions. Planets should probably only occupy orbital distances 0.3–30 AU from the star. Planetary orbits should be circular, prograde and in the same plane. The reality of exoplanets have shattered these expectations. Jupiter-mass, Neptune-mass, Superearths, and even Earth-mass planets can orbit within 0.05 AU of the stars, sometimes with orbital periods of less than one day. Exoplanetary orbits can be eccentric, misaligned, and even in retrograde orbits. Radial velocity surveys gave the first hints that the occurrence rate increases with decreasing mass. This was put on a firm statistical basis with the *Kepler* mission that clearly demonstrated that there were more Neptune- and Superearth-sized planets than Jupiter-sized planets. These are often in multiple, densely packed systems where the planets all orbit within 0.3 AU of the star, a result also suggested by radial velocity surveys. Exoplanets also exhibit diversity along the main sequence. Massive stars tend to have a higher frequency of planets (≈ 20–$25~\%$) that tend to be more massive ($M \approx 5$–$10~M_{\mathrm{Jup}}$). Giant planets around low mass stars are rare, but these stars show an abundance of small (Neptune and Superearth) planets in multiple systems. Planet formation is also not restricted to single stars as the Kepler mission has discovered several circumbinary planets. Although we have learned much about the architecture of planets over the past 20 years, we know little about the census of small planets at relatively large ($a > 1$ AU) orbital distances. We have yet to find a planetary system that is analogous to our own solar system. The question of how unique are the properties of our own solar system remains unanswered. Advancements in the detection methods of small planets over a wide range of orbital distances is needed before we gain a complete understanding of the architecture of exoplanetary systems.

Keywords Exoplanetary systems

✉ A.P. Hatzes
artie@tls-tautenburg.de

[1] Thüringer Landessternwarte, Sternwarte 5, 07778 Tautenburg, Germany

1 Introduction

The field of exoplanet research has been one "full of surprises" starting with the detection of 51 Peg b and continuing through the stunning discoveries of the *Kepler* mission. These surprises, however, largely stemmed from the biases presented by our own solar system. As our only known example of a planetary system, it dictated not only our expectations, but early search strategies (e.g. Walker et al. 1992). From our solar system we expected that:

- Rocky planets were in the inner regions (semi-major axis, $a < 2$ AU)
- Giant gaseous planets were in the outer regions ($a > 5$ AU)
- Planets moved around their host stars in circular prograde orbits which were generally in the same plane
- No planets, rocky or otherwise, were found within distances of 0.3 AU
- No giant planets were found beyond 30 AU.

This architecture could easily be understood in terms of our rudimentary understanding of star formation. Both stars and planets were born from the collapsing molecular cloud that eventually formed the proto-planetary disk. Because the angular momentum vectors of the disk and proto-star were in the same direction, planet orbits should all be prograde. The planets are formed in the same disk, so all orbits should be roughly in the same plane. Finally, viscosity in the disk would quickly damp out any orbital eccentricities leaving planets in nearly circular orbits.

The presence of giant planets in the outer regions of our solar system could be understood in the context of the so-called 'ice line'. A giant planet first needs to form a massive core ($\sim 10\ M_\oplus$) in order to accrete a gaseous envelope. The formation of this core requires the presence of significant amounts of icy material which can only be found in the outer regions of the disk at distances greater than about 3 AU. The inner region of the protoplanetary disk is hot with less solid particles, thus only small rocky planets can form.

The discovery of exoplanets have shown that the architecture of planetary systems can be considerably more diverse than our solar system led us to believe. Our understanding of exoplanet architecture can be divided into two "eras": The pre-*Kepler* mission era that was largely dominated by the discoveries made by the Doppler method and ground-based transit searches for giant planets and the post-*Kepler* era that was dominated by discoveries made with photometric transits measured from space.

2 The Architecture of Exoplanets Before *Kepler*

Figure 1 represents many of the exoplanet discoveries in the mass–semi-major axis (m–a) plane. The different symbols indicate the different detection methods that have been employed. These include 1) The radial velocity (RV) method; 2) The transit method from both ground- and space-based surveys; 3) Timing variations exemplified for example by the pulsar planets (Wolszczan and Frail 1992); 4) Direct imaging discoveries; 5) Gravitational micro-lensing which uses the gravitational bending of light to infer the presence of companions orbiting the lensing star; and 6) Astrometric detections.

Note that each method has a different sensitivity to orbital distance and planet mass and all are required to fill the parameter space. The RV method covers the largest area in the m–a plane. The transit method is sensitive to close-in planets, although space-based missions like CoRoT and *Kepler* are extending these discoveries to planets at larger orbital distances. Micro-lensing and timing variations can be quite sensitive to the lowest mass

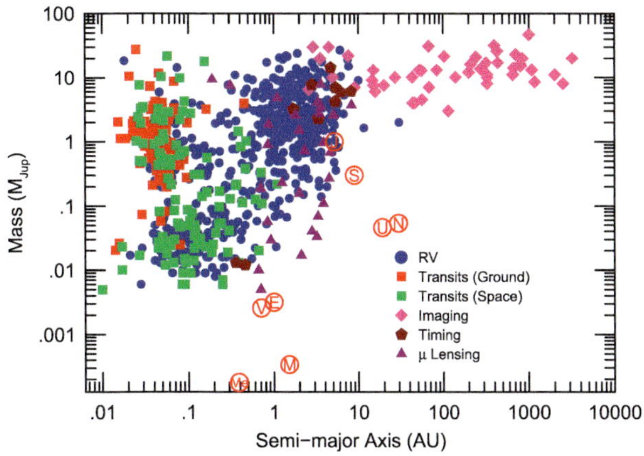

Fig. 1 The mass–semi-major axis discovery space for exoplanets with representative discoveries shown. *Symbols* represent the different detection methods: radial velocity (*dots*), ground-based transit detections (*squares*), space-based transit detections (*green squares*), direct imaging (*diamonds*), timing variations, but excluding transit timing variations (*pentagons*), and micro-lensing (*triangles*). Only transits for which the planet mass has been determined are shown. The planet mass shown is a mixture of true mass (from transits) and minimum mass (from RV surveys)

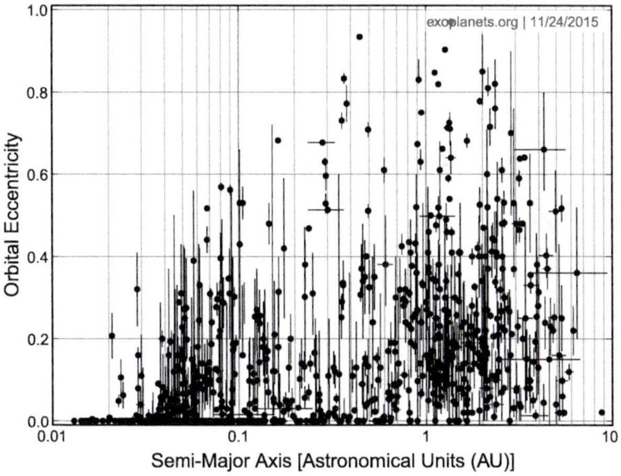

Fig. 2 The eccentricity versus semi-major axis for planets discovered by the radial velocity method. (From exoplanets.org)

planets, but only in the region ≈ 1 AU from the star. The number of discoveries from microlensing, however, lags the other methods like RV and transits. Direct imaging, to date, is only sensitive to massive planets at a large orbital distances. Astrometry has yet to discover exoplanets, although it has detected the astrometric signature of known exoplanets (Benedict et al. 2002, 2006; Martioli et al. 2010).

From Fig. 1 it is immediately apparent that the architecture of exoplanets is more diverse than the solar system led us to believe. Close-in planets with orbital distances less than a < 0.05 can have masses (radii) ranging from several times that of Jupiter to that of Neptune and even down to Earth mass (size). Giant planets can also be found well beyond the boundary defined by our outermost planet, Neptune. Note that the region of the parameter space occupied by the planets of our own solar system is empty. We have only found Jupiter analogs, but no Uranus or Neptune analogs. The detection of a planetary *system* with prop-

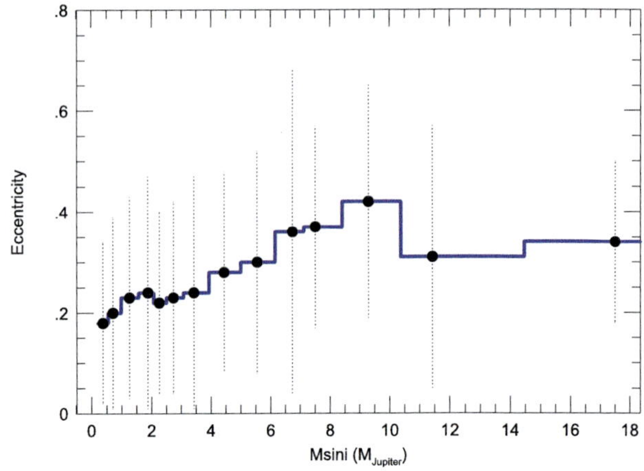

Fig. 3 The mean eccentricity as a function of planet mass. The "*error*" *bars* represent the standard deviation of the eccentricities for each bin average

erties like the solar system still eludes us. The lack of these systems does not mean they are rare, but only reflects the difficulty in detecting such exoplanets. As the precision of current detection methods increases this region of the m–a parameter space will eventually fill up. Until then our understanding of the architecture of exoplanets will be largely incomplete.

2.1 Orbital Eccentricities

Unlike the planets in our solar system, exoplanet orbits show a wide range of eccentricities in the eccentricity–semi-major axis (e–a) plane (Fig. 2). If we only consider planets with $a > 0.3$, in order to make a more direct comparison to our solar system, we find that planets have a mean orbital eccentricity of $\approx 0.26 \pm 0.20$. One could argue that circular orbits are the exception rather than the rule. The orbital eccentricities occupy the entire e–a plane except at the shortest orbital distances. The lack of planets in highly eccentric orbits close to the star is expected since tidal effects would quickly circularize any initial eccentric orbit.

One way to reconcile the observed orbital eccentricities with our view of planet formation in a viscous proto-planetary disk is if the eccentric orbits were not primordial. Several mechanisms have been proposed to pump-up the eccentricity of planets that initially had circular orbits. If the host star is a member of a binary system, then both the eccentricity and the inclination of the planet would oscillate between low and high values via the Kozai mechanism, if the orbital of the planet is initially inclined with respect to the binary orbit. This mechanism was invoked to explain the high eccentricity of 16 Cyg Bb (Holman et al. 1997), a planet in one member of a visual binary system.

Planet-planet interactions could also alter the orbital eccentricities of exoplanets. In particular, if more than one giant planet is formed in the system, then planet-planet interaction may cause one giant planet to be scattered out of the system leaving behind a giant planet in an eccentric orbit in the system (Lin and Ida 1997).

There is some evidence that more massive planets tend to be in more eccentric orbits, consistent with this scenario. Figure 3 shows the mean eccentricity for planets discovered by the radial velocity method as a function of planet mass. Averages were taken over bin sizes of $\approx 0.5\ M_{\mathrm{Jup}}$, except at larger planet masses where bin sizes of 1–3 M_{Jup} were used. The "error" bar represents the standard deviation of eccentricities of planets for each bin. Although there is a large scatter in the eccentricities of each bin, on average higher mass planets tend to be in more eccentric orbits.

Alternatively, the proto-planetary disk itself may be the source of eccentric planets. This can be either via a disk–planet interaction (e.g. Moorhead and Adams 2008) or eccentric disks increasing the planet eccentricity (e.g. D'Angelo et al. 2006). However, some theoretical work concludes that planet-disc interaction is not a viable mechanism for producing the high eccentricity of exoplanet orbits (Bitsch and Kley 2010). It may well be that several mechanisms are responsible for the observed eccentricity distribution of exoplanet orbits.

2.2 Orbital Alignments

Information about the orbital alignment of exoplanets comes predominantly from measurements of the Rossiter-McClaughlin (R-M) effect in transiting exoplanets. The R-M effect produces an apparent distortion to the orbital motion of a transiting planet and was first observed in eclipsing binaries (Rossiter 1924; McLaughlin 1924). As a transiting planet crosses in front a rotating star it first obscures the approaching, or blue-shifted part of the star resulting in an observed net redshift. After mid-transit the receding part of the star is obscured resulting in net blue-shift. The R-M effect yields information about the projected angle between the spin axis of the star and the orbital axis of the exoplanet. It can also tell us if a transiting planet orbits in a retrograde orbit.

The first R-M measurements for transiting exoplanets showed low obliquities for the orbit (Queloz et al. 2000). This was even the case for exoplanets in highly eccentric orbits where one might expect oblique orbits (Winn et al. 2007). This was a bit of a surprise because if the Kozai mechanism was responsible for the high eccentricity, then it would be highly likely that the orbit would also be mis-aligned, which was not the case.

Subsequently, it was found that exoplanets could indeed be in retrograde orbits (Narita et al. 2009; Winn et al. 2009; Triaud et al. 2010), highly oblique orbits (e.g. Pont et al. 2010), or even polar orbits (Addison et al. 2013). The distribution of orbital obliquities of exoplanet orbits may offer hints as to the formation mechanism of hot Jupiters. Planet-planet interactions or the Kozai mechanism could bring in giant planets formed beyond the ice line close to the star (Nagasawa et al. 2008; Fabrycky and Tremaine 2007). These should change the relative orientation between the spin and orbital axes. Planet migration via interaction with the disk (Lin et al. 1996) should result in low obliquity angles.

Winn et al. (2010) found that short-period exoplanets around stars with effective temperature, $T_{\rm eff} = 6250$ K tended to have higher obliquity angles than exoplanets around stars cooler than this temperature. Albrecht et al. (2012) confirmed this trend in an analysis of 53 transiting systems (Fig. 4). An effective temperature of 6250 K marks the boundary where stars develop an extensive convective envelope. Stars hotter than this tend to rotate more rapidly (Fig. 4). Albrecht et al. argued that hot Jupiters formed with a wide range of orbital obliquities, either through planet-planet interactions, or the Kozai mechanism and that the axes became aligned through tidal interactions.

Astrometric measurements can give us information about the co-planarity of multi-planet systems, but so far this has only been done for one case. Using the Fine Guidance Sensor of the Hubble Space Telescope McArthur et al. (2010) determined the astrometric orbit for υ And c ($a = 0.86$ AU) and d ($a = 0.86$ AU). They found a mutual inclination of 29.9 degrees between the two planets. The astrometric mission Gaia (Casertano et al. 2008; Perryman et al. 2014) has the potential of measuring the mutual inclination of a large number of planetary systems.

2.3 Low Mass Planets and Compact Systems

Radial velocity surveys gave the first indications that the number of planets increases with decreasing planet mass. Howard et al. (2010) estimated that based on a survey of 166 stars

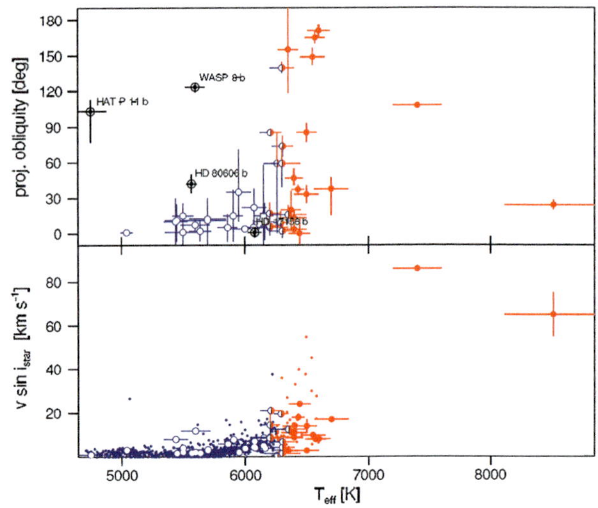

Fig. 4 (*Upper panel*) Measurements of the projected obliquity of exoplanet orbits as function of effective temperature of the host star. *Filled symbols* are stars with temperatures higher than 6250 K and *open circles* less than 6250 K. (*Lower panel*) Projected rotational velocities of the host stars. (Reproduced from Albrecht et al. 2012)

the occurrence rate for Neptune-like planets in the mass range 3–10 $M_\oplus$ was $11.8^{+4.3}_{-3.5}$ %. This estimate was consistent with Mayor et al. (2011) who found an occurrence rate of 16.6 ± 4.4 % for this same mass range which was significantly higher than the ≈ 1 % for planets with masses 30–100 $M_\oplus$. This result was based on a larger survey of 367 stars using the HARPS spectrograph.

However, the results from radial velocity surveys were only suggestive as one had to apply a large correction factor to correct for the detection efficiency of the Doppler method. Low mass planets produce small radial velocity amplitudes which can be easily missed due to measurement error, too few data points, or poor sampling. Both surveys had to apply rather large corrections to account for the "missed" planets. The lowest mass bin in the Howard et al. (2010) had only 5 detected planets, 2 candidate planets, and ≈ 10 "missed" planets. This is a correction of a factor of two for the detection efficiency. Without this correction the occurrence rate is only about 3 %.

The HARPS survey had a larger sample, and better overall precision, but in this case the corrections for the detection efficiency were even larger. The detected planets showed a peak at ≈ 10 M_{Earth} and these decreased to lower mass planets. Only after accounting for the detection efficiency using simulations did one see a rise in the occurrence rate for low-mass planets. For example, the lowest two mass bins had approximately 80 and 60 planets after correcting for the detection efficiency, but only 5 and 2 detected planets, respectively, or correction factors of approximately 10–30. These results would be on a firmer statistical ground if the number of undetected planets represented only a small fraction of the mass bin.

The RV surveys also gave us the early hints of the compact multiple systems that the Kepler mission later found. The first, HD 40307 has three superearths with periods of 4.3, 9.6, and 20.5 d (Mayor et al. 2009a). HD 10180 that had seven planets with periods up to 2149 d (Lovis et al. 2011). Five of these had relatively short periods ($P < 125$ d). Most of the multiple systems detected by RV surveys have 2–3 planets (Mayor et al. 2011), but as more data are obtained and found analysis tools refined undoubtedly more will be detected. For example, Jenkins et al. (2014) a six planet in GL 587 to complement the other known planets (Delfosse et al. 1998; Marcy et al. 2001; Rivera et al. 2005, 2010; Correia et al. 2010).

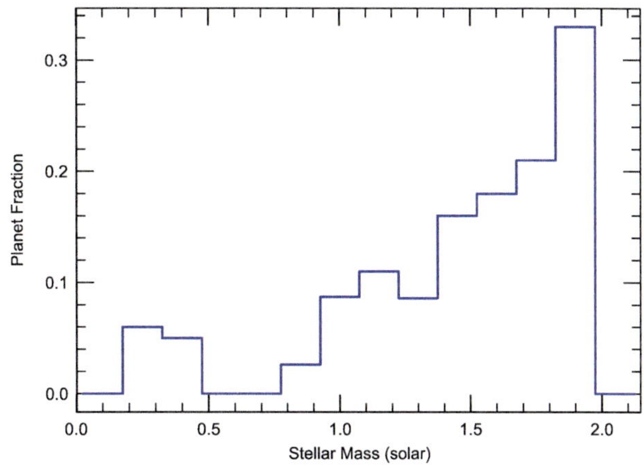

Fig. 5 The fraction of stars having giant planets as a function of stellar mass. (Based on Johnson et al. 2010)

2.4 Exoplanet Architecture Around Intermediate Mass Stars

Our knowledge of planets around intermediate mass (IM) stars comes predominantly from Doppler measurements of evolved stars. IM stars on the main sequence are early-type stars with high effective temperatures and thus a paucity of stellar lines. Early-type stars also show high rotation rates of several hundreds of $\mathrm{km\,s^{-1}}$. Because of the few and rotationally broadened spectral lines one can rarely achieve an RV precision of less than several hundreds of $\mathrm{m\,s^{-1}}$, insufficient for the detection of exoplanets.

IM stars that have evolved up the giant branch, on the other hand, have cool effective temperatures that result in a dense forest of spectral lines. Furthermore, the rotation rate of the star slows as it ascends the giant branch, thus the spectral lines are quite narrow. As a result an excellent Doppler precision is possible. The problem for evolved stars, however, is in determining the mass of the host star. Evolving stars covering a wide range of masses all converge to the same region of the color-magnitude diagram occupied by the giant branch. Thus one has to rely on evolutionary tracks to obtain the stellar mass. These are not only model dependent, but tracks for a wide range of masses converge to the same region of the color magnitude diagram. Still, Doppler measurements of evolved stars are still the best way to study the occurrence of planets around stars more massive than the sun.

The first hints of planets around evolved stars date to the early 1990s. Hatzes and Cochran (1993) using precise RV measurements found evidence for exoplanets around the giant stars β Gem and α Tau. Both were eventually confirmed as planets (Hatzes et al. 2006, 2015). The presence of planets around giant stars was firmly established with the discovery of the giant planet around ι Dra (Frink et al. 2002). Since then a large number of programs have been surveying giant stars for planets (e.g. Setiawan et al. 2003; Döllinger et al. 2007; Sato et al. 2008; Niedzielski et al. 2009; Johnson et al. 2010; Wittenmeyr et al. 2011; Jones et al. 2011). To date approximately 100 evolved IM stars have been found to host giant planets. Although much fewer than the over 1500 known exoplanets discovered by RV surveys of main sequence stars, this is a sufficiently large sample to show that differences in the architecture may exist between solar-type and more massive stars.

There appears to be a tendency for stars more massive than the Sun to have a higher frequency of planets (Fig. 5). This occurrence rate increases to about 25–30% for stars with masses $M > 1.5\ M_{\odot}$.

IM stars tend to show a lack of planets with semi-major axes between 0.1 and 0.7 AU (Fig. 6). IM giant stars have radii of 0.05 AU or more so the lack of planets at $a < 0.1$ AU is

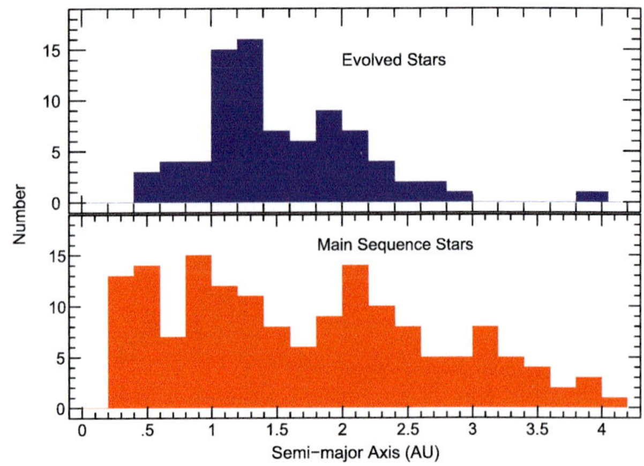

Fig. 6 (*Top*) The semi-major axis distribution for giant planets ($M > 0.5\ M_{\rm Jup}$) around evolved stars with $M_{\rm star} > 1.1\ M_\odot$. (*Bottom*) The semi-major axis distribution for main sequence stars with $M_{\rm star} < 1.1\ M_\odot$. Semi-major axis bins start with $a > 0.2$

probably due to the engulfment of the planet by the star as it moves up the giant branch. But this cannot explain the giant planets found at larger orbital distances. The lack of planets out to 0.7 AU does not seem to be an observational bias (Johnson et al. 2010; Bowler et al. 2010) One might hypothesize that the evolution of the central star somehow disrupts those planets orbiting with 0.7 AU. Sato et al. (2008) showed that clump stars, helium core burning stars that have already had their first ascent up the giant branch, may indeed disrupt planetary orbits. However, there is not a large influence on the planet's orbit for host stars that have recently left the main sequence. It is still open whether this lack of close-in planets results from the formation process, or the evolution of the host star.

The vast majority of planets around IM stars are single, so we have little knowledge about planetary systems around stars more massive than the sun. A hint to the architecture of planetary systems around IM stars comes from the planetary system to HR 8799. This star is a pulsating γ Dor star with a mass of 1.5 $M_\odot$. Using differential imaging observations Marois et al. (2008, 2010) found four giant planets with masses of 7–10 $M_{\rm Jup}$ located at orbital distances of 14.5–68 AU (Fig. 7). This planetary system has an architecture analogous to the outer planets of our solar system. Scaling the HR 8799 planetary system to our solar system shows that its four giant planets would have the same relative separation as the outer giants of the solar system. However, all the giant planets of HR 8799 have masses several times that of Jupiter. The planetary system of HR 8799 cannot be easily explained by current models of in-situ planet formation and thus represents a unique system for the study of giant planet formation at large orbital distances from the star.

2.5 Exoplanet Architecture Around M Dwarfs

If higher mass stars tend to have larger mass planets, then one might naively expect low mass stars to have lighter planets. This appears to be the case as Doppler surveys of M dwarf stars has yielded relatively few giant planets (Endl et al. 2006; Bonfils et al. 2013; Montet et al. 2014). In a survey of 102 M dwarfs Bonfils et al. (2013) found that close-in giant planets ($P = 1$–10 d) have a frequency, f, less than 1 %, For giant planets in wider orbits ($P = 10$–100 d) this only increases to $f = 0.02^{+0.03}_{-0.1}$. One of the few exceptions is the M5 dwarf, GL 876, which hosts two giant planets with masses of 1.9 $M_{\rm Jup}$ and 0.6 $M_{\rm Jup}$ with periods of $\sim$ 60 and 30 days, respectively (Marcy et al. 1998; Delfosse et al. 1998; Marcy et al. 2001). The outer planet, GL 876b was the first planet whose true mass was determined

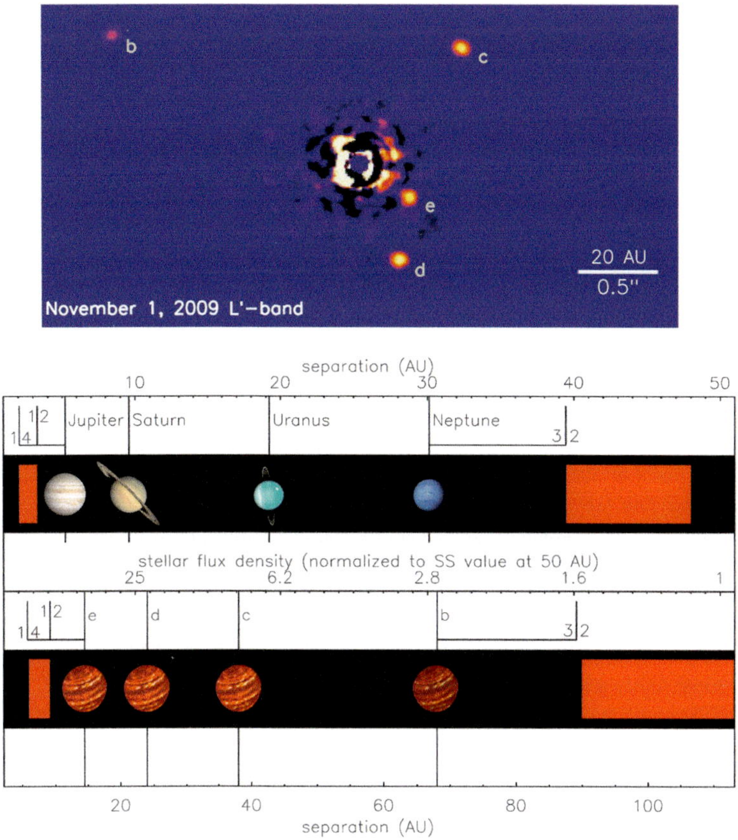

Fig. 7 (*Top*) Image of the 4-planet system to HR 8799. (*Bottom*) The HR 8799 system compared to the outer planets of the solar system. (Reproduced from Marois et al. 2010)

astrometrically at $M_b = 1.89 \pm 0.34\ M_{\rm Jup}$ using the Fine Guidance Sensor of the Hubble Space Telescope (Benedict et al. 2002). However, Correia et al. (2010) determined a slightly higher mass of $2.64 \pm 0.04\ M_{\rm Jup}$ by modeling the effect of planet-planet interactions on the radial velocities.

Although lacking in gas giant planets, M dwarfs seem to be abundant with small planets. Bonfils et al. (2013) estimated that the frequency of superearths in mass range 1–10 $M_\oplus$ in short period orbits ($P = 1$–10 d) was $f = 0.36^{+0.25}_{-0.10}$. In longer period orbits ($P = 10$–100 d) the frequency is higher with $f = 0.52^{+0.50}_{-0.16}$. Although these estimates have large errors, these are largely in line with the occurrence rate found by the *Kepler* mission (see below).

2.5.1 Planets in the Habitable Zone of M Dwarfs

Recently, M dwarf stars have become attractive objects for exoplanet searches because they offer us a "fast track" to finding terrestrial planets in the habitable zone of stars. M dwarf stars have low effective temperatures which means that the habitable zone is much closer to the star ($a \sim 0.1$ AU). Furthermore, the low mass of the host star ($M \sim 0.1\ M_\odot$) results in a higher reflex motion of the star. A terrestrial planet in the habitable zone of an M dwarf will produce a barycentric motion of the host star of $\sim 1\ \mathrm{m\,s^{-1}}$, a value measurable by current radial velocity techniques.

Most of the flux from an M-star is in the infrared (IR) and consequently several high resolution spectrographs are being constructed with the scientific goal of finding terrestrial planets in the habitable zone of M dwarfs. These include the CARMENES spectrograph for Calar Alto (Quirrenbach et al. 2014), the upgraded CRIRES+ project for the VLT (Follert et al. 2014), the Pathfinder Instrument (Ramsey 2010), and SPIRou (Artigau et al. 2014).

A major challenge to finding terrestrial planets in the habitable zone of M dwarfs is the activity of the star. Surface structure in the form of spots, plage, convection cells, etc. produce their own radial velocity signal that varies with the rotational period of the star or its harmonics. This activity "jitter" can either hinder the detection of low mass planets, or even create a false planet signal. The rotational period of an M dwarf can be $\sim$ tens of days, or comparable to the orbital period of a planet in the habitable zone. Thus activity can present serious challenges to the detection of these habitable zone planets.

A case in point is the M dwarf star GL 581. Mayor et al. (2009b) first reported five planets orbiting this star with periods less than 130 days. Vogt et al. (2010) claimed six planet candidates with periods up to 433 d. GL 581d and GL 581g had orbital distances in the habitable zone of the star. Subsequent RV measurements (Forveille et al. 2011) could not confirm the presence of GL 581g. The second habitable planet, GL 581d also may be an artifact of stellar activity. Robertson et al. (2014) showed that the RV variations of GL 581d correlated with the strength of the Balmer Hα line, an indication that the RV signal of GL 581d was most likely a harmonic of the 130 d rotation period of the star. Spectrographs like CARMENES which will make simultaneous RV measurements in the optical and infrared should be able to distinguish between RV signals due to activity and help in our search for bona fide terrestrial planets in the habitable zone of M dwarfs.

This activity RV jitter also is a concern for Doppler searches for planets in the habitable zone of G-type stars. The barycentric velocity of a solar-mass star due to a planet orbiting at 1 AU is $\approx 10\ \mathrm{cm\,s^{-1}}$. The RV jitter due to stellar activity thus represents the single largest obstacle to our discovery of terrestrial planets in the habitable zone of G-type stars using the Doppler method. Such detections will require ways of correcting the RV for the activity signal—not easily done considering the stars are unresolved, or a large number of RV measurements in order to extract the planet signal from the activity noise.

2.6 Planets in Binary Systems

There have been a number of investigations of the role of stellar binarity on the presence of exoplanets (Desidera and Barbieri 2007; Eggenberger et al. 2007; Mugrauer and Neuhäuser 2009). Planets in binary systems can often offer us important clues to the process of planet formation. For example, as mentioned before, the binary nature of 16 Cyg B (binary separation $\approx$ 800 AU) may be responsible for the high eccentricity of this planet. This system is also interesting because both stellar components are nearly identical with the same mass, effective temperature, age, chemical abundance, and presumably the same birth environment. The only distinguishing characteristic of 16 Cyg B is depleted in lithium by a factor of six compared to 16 Cyg A (King et al. 1997). The component 16 Cyg A was also monitored extensively with radial velocity measurements and it shows no indication of hosting a giant planet. It is not clear what environmental factors favored the formation of a giant planet around 16 Cyg B, but not 16 Cyg A.

Planets have also been found in binary systems that are "dynamically challenged". The components of the binary system ν Oct have masses of 1.4 $M_{\odot}$ and 0.5 $M_{\odot}$ and orbit the barycenter with a period of 1050 d. Precise stellar RV measurements revealed an additional perturbation of ν Oct A with a period of 417 d that was consistent with the presence of a planetary mass companion of 2.4 $M_\oplus$ (Ramm et al. 2009). The only problem with ν Oct Ab is that it is dynamically unstable, thus this planet should not exist! However, the system can be stable under very special configurations. Eberle and Cuntz (2010) demonstrated that if the planet's orbit is retrograde with respect to the binary motion then it can have a long term stability.

Alternatively the RV perturbation at 417 d may just be due to the rotational modulation of stellar surface structure (e.g. cool or hot spots). If true, then many of the giant planets around evolved IM stars may be fake planets due to rotational modulation of surface structure. Many of these planets have orbital distances of $a = 1$–1.5 AU (Fig. 6) which corresponds to orbital periods of ≈ 400–600 d. These are comparable to the expected rotation periods of the star.

Ramm (2015) measured line depth ratios of temperature sensitive lines in ν Oct from the same spectral data used for the RV measurements. He found that the standard deviation of any temperature variations in ν Oct A was only 4.2 K. Furthermore, by combining Hipparcos photometry he was able to show that this star has been photometrically stable for 15 years. It seems unlikely that starspots are the cause of the 417 d RV perturbation of ν Oct A. Currently, the most likely explanation is that this is a giant planet in a retrograde S-type orbit where the planet orbits one component of the binary.

The shortest period binary certain to host an exoplanet is γ Cep. The primary is orbited by a 0.40 $M_\odot$ M dwarf with a period of 67.5 years and it also hosts a 1.46 M_{Jup} planet in a 903 d orbit (Hatzes et al. 2003). This system is of historical interest because it is one of the first exoplanets whose presence was inferred by Doppler measurements (Campbell et al. 1988). It is also a key object for theoretical studies because the relative shortness of the binary period makes it very challenging to form this planet with the standard theory (Kley and Nelson 2008).

3 Architecture of Exoplanets During the *Kepler* Mission

The *Kepler* Space Mission established that the frequency of exoplanets increases as the size decreases and that there are more close-in Neptune and Superearths that Jupiter-sized planets (Borucki et al. 2011). In fact, *Kepler* has discovered only about 200 of such objects, roughly in line with the 1 % frequency of short-period Jupiters detected by RV surveys of nearby field stars.

The frequency of small sized planets apparently decreases from a peak at Neptune-sized objects (Fig. 8). As is the case with the RV surveys this decreases is due to the lower efficiency of detecting these objects. For example, the transit depth of an Earth-sized planet is $\sim 10^{-5}$ and is often difficult to detect in the presence of intrinsic stellar noise of activity, even when using space-based photometry.

Howard et al. (2012) assessed the detectability of planets in the Kepler light curves as a function of radius, taking into account the detection efficiency for each star. They showed that the planet occurrence rate indeed increases for small planets consistent with a power law distribution, $N \approx 0.23\ R^{-1.66}$ (Fig. 8). Note that *Kepler* has only a 50 % detection efficiency for the smallest planets ($R < 1\ R_\oplus$).

The frequency of small planets is strongly dependent on the temperature, and thus mass of the host star. Howard et al. (2012) concluded that planets in the radius range 2–4 $R_\oplus$ were seven times more abundant around stars with $T_{\mathrm{eff}} = 3600$–4100 K than for stars with $T_{\mathrm{eff}} = 6600$–7100 K. In other words, lower mass stars have a higher frequency of low mass planets. This is consistent with the Doppler results from M dwarf stars.

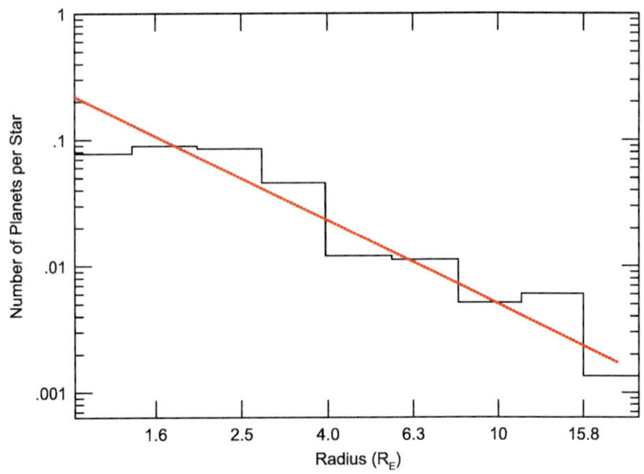

Fig. 8 Planet occurrence as a function of planet radius for *Kepler* planets with orbital periods less than 50 days. The *red line* represents a power law fit, $N \approx 0.23\ R^{-1.66}$ showing that the planet occurrence rate increases substantially for small planets. (Based on Howard et al. 2012)

3.1 Densely-Packed Multiple Systems

RV surveys first found the presence of multiple systems with up to seven planets in the systems with periods out to several hundreds of days (see Mayor et al. 2011). These surveys also detected the first ultra-compact system, HD 40307, shown to host three super-earths all with periods less than 21 d (Mayor et al. 2009a, 2009b). Space transit missions showed that these ultra-compact systems are common.

One of these was CoRoT-7 system. The CoRoT space mission (Baglin et al. 2009) detected the first transiting rocky planet, CoRoT-7b, with an orbital period of 0.85 d and a mass of 6.9 $M_\oplus$ (Léger et al. 2009). Follow-up radial velocity measurements established the presence of two (Queloz et al. 2009) and most likely a third planet in the system (Hatzes et al. 2010). The three planets of CoRoT-7 have periods 0.85–9 days and masses of 7–20 $M_\oplus$. This system appears to be densely packed system in that the Hill exclusion zones for these three planets do not overlap. Adding an additional planet in the orbital space of the planets would cause it to be unstable.

CoRoT-7b was also significant in that it demonstrated that small planets can occur on ultra-short period orbits ($P < 1$ d). A virtual twin to CoRoT-7b was found in Kepler-10b, a rocky planet also in a 0.85 d orbit (Batalha et al. 2011). The current short-period record holder of a fully characterized transiting planet is Kepler-78, an Earth-mass planet with an 8 hr orbital period (Sanchis-Ojeda et al. 2013; Howard et al. 2013; Pepe et al. 2013).

The *Kepler* mission found that approximately 30 % of late-type stars have densely packed multiple planet systems. The first of these was Kepler-11 system which has six planets with masses ranging from 2–13 $M_\oplus$ and all at orbital periods of 10–118 d (Lissauer et al. 2011).

The type of planetary system a host star has depends on whether a giant planet is in the system. One of the important early results was that for the single systems, i.e. stars for which only one transiting planet was detected, the planets came in all sizes ranging from 1 $R_\oplus$ to 1 R_{Jup} (Fig. 9). However, for multiple systems the planet sizes tended to be Neptune-size or smaller (Latham et al. 2011). It seems that the presence of a Jupiter-mass planet either prevents their formation or disrupts their orbits after they formed.

One of the few Kepler multiple systems that has planets ranging in size from Jupiter to that of the Earth is Kepler 90 (= KIC 11442793). This system has seven planets ranging in size from Jupiter-sized down to Earth-sized (Cabrera et al. 2014). Comparing Kepler-90 to our solar system (Fig. 10) shows that it shares a similar property: the giant planets are in

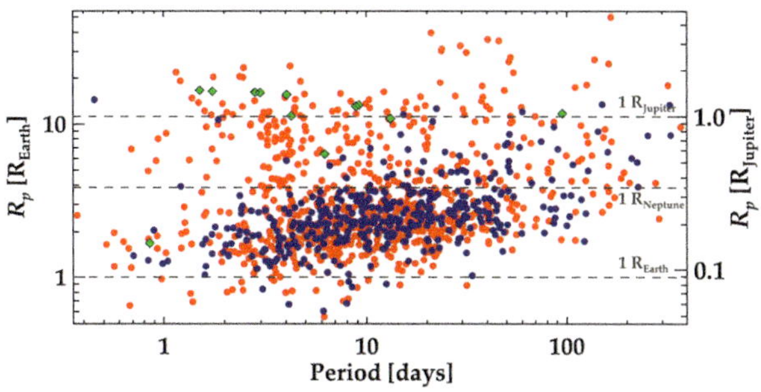

Fig. 9 The planet radius vs. orbital period. *Red dots* represent planet candidates that are singles, whereas the *blue dots* represent planets found in multiple systems. *Green dots* represent planets discovered by CoRoT. (Reproduced from Latham et al. 2011)

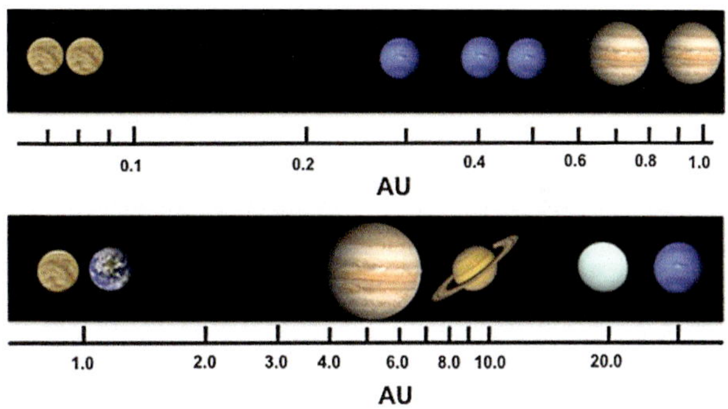

Fig. 10 (*Top*) The planetary system of Kepler-90 consisting of two giant planets, three Neptune-size planets, and two terrestrial planets. (*Bottom*) Comparison to inner terrestrial and outer giant planets of the solar system. (Planet images courtesy of NASA)

the outer regions, while the small planets are in the inner regions. In this respect Kepler-90 can be considered a compact Solar analog, counter to the "extended" analog of HR 8799. However, the Neptune-sized planets lie in-between the rocky planets and gas giants, rather than beyond the gas giants like in our solar system.

3.2 *Kepler's* Circumbinary Planets

Until *Kepler*, all the exoplanets found in binary systems were in S-type systems. In this case planet formation could proceed more or less as for an isolated stars. What about circumbinary planets?

Theoretical work established that planets can form around binary stars (e.g. Pierrens and Nelson 2008). The possible birthplace for such planets have already been observed in the form of circumbinary disks (see Krist et al. 2002).

Ground-based efforts to look for circumbinary planets around eclipsing binary systems using eclipse timing variations pre-date the *Kepler* mission and have produced mixed results. Deeg et al. (2000) found possible periodic variations in the eclipse timings of the binary CM Dra that might be due to a 1.5–3 $M_{\rm Jup}$ companion orbiting ~ 1 AU from the binary barycenter. This has not been confirmed and a concern is the fact that CM Dra is a spotted BY Dra-type star and this activity may influence the eclipse timings. Eclipse timing variations revealed two giant planets around the cataclysmic variable NN Ser (Beuermann et al. 2010) and a giant planet in a 28-year orbit around DP Leo (Beuermann et al. 2011). These planets would have had to survive the common envelope phase of the binary evolution. However, dynamical studies of many of the claimed circumbinary planets around post main sequence stars have shown these to be unstable (Hinse et al. 2014; Zorotovic and Schreiber 2013;

Table 1 Kepler's first five circumbinary planets

Parameter	Kepler 16	Kepler 34	Kepler 35	Kepler 38	Kepler 47
M_A [$M_\odot$]	0.69	1.05	0.89	0.95	1.04
M_B [$M_\odot$]	0.20	1.02	1.02	0.26	0.36
P_{AB} [d]	41.1	27.8	20.7	18.8	7.45
P_b [d]	228.7	288.8	131.4	105.6	49.5
r_b [$R_{\rm Jup}$]	0.75	0.76	0.73	0.39	0.17
m_b [$M_{\rm Jup}$]	0.33	0.22	0.13	–	–
P_c [d]	–	–	–	–	202.2
r_c [$R_{\rm Jup}$]	–	–	–	–	0.42
m_c [$M_{\rm Jup}$]	–	–	–	–	–

Horner et al. 2012, 2013; Wittenmeyr et al. 2012). The eclipse timing variations observed from these systems may not arise from planetary companions.

With the discovery of Kepler-16 (AB)b the Kepler mission established without doubt that circumbinary planets could exist (Doyle et al. 2011). The stellar components (K5V + M5V) have an orbital period of 41 d. These are orbited by a giant planet with a period 229 d. To date Kepler has discovered ten transiting circumbinary planets (Welsh et al. 2014). The first five discoveries are summarized in Table 1 to give an overall impression of the typical orbital properties. The smallest circumbinary planet is the Neptune-sized planet Kepler-38(AB)b (Orosz et al. 2012).

All the binary stars have relatively short periods ranging from 7.4 to 41 d, and the planets are in relatively long period orbits of greater than 100 d. The exception is Kepler 47 (AB)b with a period of 49.5 d. Of particular interest is Kepler 47, the first multiple binary system. The binary hosts three exoplanets, one with period of 303 d that places it near the habitable zone of the binary system. Based on the Kepler data Armstrong et al. (2014) estimate that $10^{+18}_{-6.5}$ % occurrence rate for circumbinary planets with $R_p > 6\ R_\oplus$ and periods < 300 d.

The first circumbinary planets had configurations such that the planet transits the host stars every orbit, but this will not last. Due to secular changes in the planet's orbit Kepler-16(AB)b will no longer transit the secondary star in 2014, and cease transiting the primary star in 2018. The last transit of Kepler-35 (AB)b stopped in 2012 and will not start again for a decade (Welsh et al. 2012). It is fortunate that *Kepler* just happened to be observing these systems at the right time.

Many of the circumbinary planets have orbital planes that are well-aligned with that of the binary stars. Kepler-413 (AB)b, however, is a slightly misaligned Neptune-sized circumbinary planet (Kostov et al. 2014). The binary components have masses of 0.82 $M_\odot$ and 0.54 $M_\odot$ and an orbital period of 10 d. The planet, with a radius of 4.3 $R_\oplus$, orbits the binary host stars every 66 d on an slightly eccentric orbit ($e = 0.118 \pm 0.002$). The orbital plane of the binary is inclined by 2°.5 to the binary plane. As a result of this configuration hundreds of days pass when no transits occur.

The circumbinary planets found by Kepler as well as those to be discovered with upcoming space missions will provide key data to theories on the formation and evolution of planets in multiple star systems.

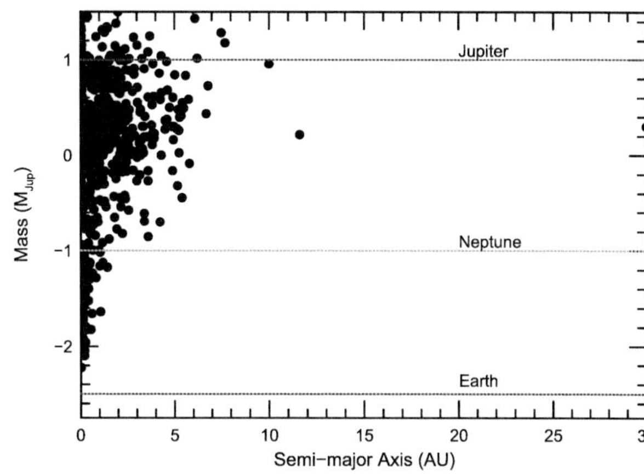

Fig. 11 The discovery space for exoplanets but plotted on a linear scale out to 30 AU, the orbital distance of Neptune

4 Towards a Complete Understanding of Architecture of Exoplanets

Although tremendous progress has been made in the past 20 years in understanding the architecture of exoplanets, this knowledge is still incomplete. By plotting orbital distance on a logarithm scale Fig. 1 gives us a bit of a false impression. The same figure on a linear scale (Fig. 11) shows just how incomplete our knowledge truly is. Kepler has given us a relatively good understanding of exoplanet architecture down to small planets, but only for orbital distances $a < 0.5$ AU. For Earth-sized and Neptune-sized planets beyond 1–2 AU, and even giant gas planets beyond 5 AU our knowledge is woefully inadequate. We have probed less than 5 % of the available parameter space for exoplanets. Unfortunately, upcoming space missions such as TESS (Ricker et al. 2015) and PLATO (Rauer et al. 2014) will not help matters as these will still only find the close in planets out to ≈ 1 AU. Probing the region of the parameter space at larger orbital distances requires increased sensitivity of the radial velocity, astrometric, and direct imaging techniques.

The radial velocity method might be able to detect Earth-mass planets orbiting at 1 AU to a G-type star, but it will be seriously challenged in detecting Neptune-mass planets out to ~ 10 AU. Doppler programs would require a several decades commitment of telescope resources to monitor stars. Another issue is naturally the enthusiasm and drive of the astronomers since the scientific rewards will be reaped by future generation of astronomers. Such legacy programs, however, are important.

Astrometry not only provides us with the true mass of the companion, but it is complementary to the radial velocity method in that it is most sensitive to planets at large orbital distances. The astrometric space mission GAIA should make a significant contribution to the architecture of exoplanets, but only for giant planets. Finding the small planets with large semi-major axes requires space-based astrometric measurements.

Arguably, direct imaging is the most efficient way to probe the planet census at large orbital radii since relatively few measurements are required. The recent commissioning of high contrast imaging instruments such as the Gemini Planet Imager (Macintosh et al. 2014) and Sphere (Beuzit et al. 2008) should explore the parameter space of giant planets with $M \sim 1\ M_{\mathrm{Jup}}$ to within a few AU from the star. However, to find the Neptunes at 10–20 AU, or terrestrial planets at ≈ 1 AU would require space-based telescopes using coronographs or nulling interferometry. The James Webb Telescope is a good first start to such measurements, but this is a multi-purpose facility. Dedicated imaging space telescopes will be required to gain the full census of planets over the full parameter space.

Our current understanding of planet architecture has largely been driven by a "Golden Age" of Doppler surveys, followed by one for transit surveys. Ushering in a "Golden Age" of astrometry and direct imaging will be needed for a full understanding of the architecture of exoplanetary systems.

References

B.C. Addison, C.J. Tinney, D.J. Wright et al., Astrophys. J. **774**, 9 (2013)
S. Albrecht, J.N. Winn, J.A. Johnson et al., Astrophys. J. **757**, 18 (2012)
D.J. Armstrong, H.P. Obsborn, D.J.A. Brown et al., Mon. Not. R. Astron. Soc. **444**, 1873 (2014)
É. Artigau, D. Kouach, J-F. Donati, Proc. SPIE **9147**, 15 (2014)
A. Baglin et al., Proc. Int. Astron. Union **253**, 71–81 (2009)
N. Batalha, W.J. Borucki, S.T. Bryson et al., Astrophys. J. **729**, 27 (2011)
G.F. Benedict, B.E. McArthur, T. Forveille et al., Astrophys. J. **581**, L115 (2002)
G.F. Benedict, B.E. McArthur, G. Gatewood et al., Astron. J. **132**, 2206 (2006)
K. Beuermann, F.V. Hessmann, S. Dreizler et al., Astron. Astrophys. **521**, 60 (2010)
K. Beuermann, J. Buhlmann, J. Diese et al., Astron. Astrophys. **526**, 53 (2011)
J-L. Beuzit et al., Proc. SPIE **7014**, 18 (2008)
B. Bitsch, W. Kley, Astron. Astrophys. **523**, A30 (2010)
X. Bonfils, X. Delfosse, S. Udry et al., Astron. Astrophys. **549**, 109 (2013)
W.J. Borucki et al., Astrophys. J. **728**, 117 (2011)
B.P. Bowler et al., Astrophys. J. **709**, 396 (2010)
J. Cabrera, Sz. Csizmadia, H. Lehmann et al., Astron. J. **781**, 18 (2014)
B. Campbell, G.A.H. Walker, S. Yang, Astrophys. J. **331**, 902 (1988)
S. Casertano et al., Astron. Astrophys. **482**, 699 (2008)
A.C.M. Correia, J. Couetdic, J. Laskar et al., Astron. Astrophys. **511**, 21 (2010)
G. D'Angelo, S.H. Lubow, M.R. Bate, Astrophys. J. **652**, 1698 (2006)
H.J. Deeg et al., Astron. Astrophys. **358**, 5 (2000)
X. Delfosse et al., Astron. Astrophys. **338**, L67 (1998)
S. Desidera, M. Barbieri, Astron. Astrophys. **462**, 345 (2007)
M. Döllinger et al., Astron. Astrophys. **472**, 649 (2007)
L.R. Doyle, J.A. Carter, D.C. Fabrycky et al., Science **333**, 6049 (2011)
J. Eberle, M. Cuntz, Astron. Astrophys. **420**, 357 (2010)
A. Eggenberger, S. Udry, G. Chavin et al., Astron. Astrophys. **474**, 273 (2007)
M. Endl, W.D. Cochran, M. Kürster, Astrophys. J. **649**, 436 (2006)
D. Fabrycky, S. Tremaine, Astrophys. J. **669**, 1298 (2007)
R. Follert et al., Proc. SPIE **9147**, 19F (2014)
T. Forveille et al., eprint arXiv:1109.2505 (2011)
S. Frink, D.S. Mitchell, A. Quirrenbach, D.A. Fischer, G.W. Marcy, R.P. Butler, Astrophys. J. **576**, 478 (2002)
A.P. Hatzes, W.D. Cochran, Astrophys. J. **413**, 339 (1993)
A.P. Hatzes et al., Astrophys. J. **599**, 1383 (2003)
A.P. Hatzes et al., Astron. Astrophys. **457**, 670 (2006)
A.P. Hatzes et al., Astron. Astrophys. **520**, 93 (2010)
A.P. Hatzes, W.D. Cochran, M. Endl et al., Astron. Astrophys. **580**, 31 (2015)
T.C. Hinse, J. Horner, J.W. Lee et al., Mon. Not. R. Astron. Soc. **438**, 307 (2014)
M. Holman, J. Touma, S. Tremaine, Nature **386**, 254 (1997)
J. Horner, R.A. Wittenmeyer, T.C. Hinse, C.G. Tinney, Mon. Not. R. Astron. Soc. **425**, 749 (2012)
J. Horner, R.A. Wittenmeyer, T.C. Hinse et al., Mon. Not. R. Astron. Soc. **435**, 2033 (2013)
A.W. Howard, J.A. Johnson, G.W. Marcy et al., Science **330**, 653 (2010)
A.W. Howard, G.W. Marcy, S.T. Bryson et al., Astrophys. J. Suppl. Ser. **201**, 15 (2012)
A.W. Howard, R. Sanchis-Ojedo, G.W. Marcy et al., Nature **503**, 381 (2013)
J.S. Jenkins, N. Yoma, R. Becerra et al., Mon. Not. R. Astron. Soc. **441**, 2253 (2014)
J.A. Johnson, K.M. Aller, A.W. Howard, J.R. Crepp, Publ. Astron. Soc. Pac. **122**, 905 (2010)
M.I. Jones, J.S. Jenkins, P. Rojo, C.H.F. Melo, Astron. Astrophys. **536**, 71 (2011)
J.R. King et al., Astron. J. **113**, 1871 (1997)
W. Kley, R.P. Nelson, Astron. Astrophys. **486**, 617 (2008)
V.B. Kostov et al., Astrophys. J. **784**, 14 (2014)

J.E. Krist, K.R. Stapelfeldt, A.M. Watson, Astrophys. J. **570**, 570 (2002)
D.W. Latham et al., Astrophys. J. **732**, 24 (2011)
A. Léger, D. Rouan, J. Schneider et al., Astron. Astrophys. **506**, 287 (2009)
D.N.C. Lin, S. Ida, Astrophys. J. **477**, 781 (1997)
D.N.C. Lin, P. Bodenheimer, D.C. Richardson, Nature **380**, 606 (1996)
J.J. Lissauer, D.C. Fabrycky, E.B. Ford et al., Nature **470**, 53 (2011)
C. Lovis, D. Ségransam, M. Mayor et al., Astron. Astrophys. **528**, 112 (2011)
B.A. Macintosh, A. Anthony, J. Atwood et al., Proc. SPIE **9148**, 91480J (2014)
G.W. Marcy, R.P. Butler, S.S. Vogt et al., Astrophys. J. **505**, L147 (1998)
G.W. Marcy, R.P. Butler, D. Fischer et al., Astrophys. J. **556**, 296 (2001)
C. Marois, B. Zuckerman, Q.M. Konopacky et al., Science **322**, 1348 (2008)
C. Marois, B. Macintosh, T. Barman et al., Nature **468**, 1348 (2010)
E. Martioli, B.E. McArthur, G.F. Benedict et al., Astrophys. J. **708**, 625 (2010)
M. Mayor, X. Bonfils, T. Forveille, Astron. Astrophys. **493**, 639 (2009a)
M. Mayor, X. Bonfils, T. Forveille, Astron. Astrophys. **507**, 487 (2009b)
M. Mayor, M. Marmier, C. Lovis et al., eprint arXiv:1109.2497 (2011)
B. McArthur et al., Astrophys. J. **560**, 907 (2010)
D.B. McLaughlin, Astrophys. J. **60**, 22 (1924)
B. Montet, J.R. Crepp, J.A. Johnson et al., Astron. J. **781**, 28 (2014)
A.V. Moorhead, F.C. Adams, Icarus **193**, 475 (2008)
M. Mugrauer, R. Neuhäuser, Astron. Astrophys. **494**, 373 (2009)
M. Nagasawa, S. Ida, T. Bessho, Astrophys. J. **249**, 279 (2008)
N. Narita, B. Sato, T. Hurano, M. Tamura, Publ. Astron. Soc. Jpn. **61**, 35 (2009)
A. Niedzielski, G. Novak, M. Adamów, A. Wolszczan, Astrophys. J. **707**, 768 (2009)
J.A. Orosz, W.F. Welsh, J.A. Carter et al., Astrophys. J. **758**, 87 (2012)
F. Pepe, A.C. Cameron, D.W. Latham et al., Nature **503**, 337 (2013)
M. Perryman, K. Hartmann, G. Bakos, L. Lindegren, Astrophys. J. **797**, 14 (2014)
A. Pierrens, R.P. Nelson, Astron. Astrophys. **483**, 633 (2008)
F. Pont, M. Endl, W.D. Cochran et al., Mon. Not. R. Astron. Soc. **402**, L1 (2010)
D. Queloz et al., Astron. Astrophys. **359**, 13 (2000)
D. Queloz et al., Astron. Astrophys. **506**, 303 (2009)
A. Quirrenbach et al., Proc. SPIE **9147**, 91471F (2014)
D.J. Ramm, Mon. Not. R. Astron. Soc. **449**, 4428 (2015)
D.J. Ramm et al., Mon. Not. R. Astron. Soc. **394**, 1695 (2009)
L.W. Ramsey, Proc. SPIE **7735**, 71R (2010)
H. Rauer et al., Exp. Astron. **38**, 249 (2014)
G.R. Ricker, J.N. Winn, R. Vanderspeck et al., J. Astron. Telesc. Instrum. Syst. **1**, 014003 (2015)
E.J. Rivera, J.J. Lissauer, R.P. Butler et al., Astrophys. J. **634**, 625 (2005)
E.J. Rivera, G. Laughlin, R.P. Butler et al., Astrophys. J. **719**, 890 (2010)
P. Robertson, S. Mahadevan, M. Endl, R. Arpita, Science **345**, 440 (2014)
R.A. Rossiter, Astron. J. **60**, 15 (1924)
R. Sanchis-Ojeda et al., Astrophys. J. **775**, 54 (2013)
B. Sato et al., Publ. Astron. Soc. Jpn. **60**, 539 (2008)
J. Setiawan et al., Astron. Astrophys. **397**, 1151 (2003)
A.H.M. Triaud et al., Astron. Astrophys. **524**, 25 (2010)
S.S. Vogt, R.P. Butler, E.J. Rivera et al., Astrophys. J. **723**, 954 (2010)
G.A.H. Walker et al., Icarus **116**, 359 (1992)
W.F. Welsh et al., eprint arXiv:1409.1605 (2014)
W.F. Welsh, J.A. Orosz, J.A. Carter et al., Nature **481**, 475 (2012)
J.N. Winn, J.A. Johnson, K.M.G. Peek et al., Astrophys. J. **665**, 167 (2007)
J.N. Winn et al., Astrophys. J. **703**, 99 (2009)
J.N. Winn, D. Fabrycky, S. Albrecht, J.A. Johnson, Astrophys. J. **718**, 145 (2010)
R.A. Wittenmeyr, J. Horner, J.P. Marshall, O.W. Butters, C.G. Tinney, Mon. Not. R. Astron. Soc. **419**, 3258 (2012)
R.A. Wittenmeyr, M. Endl, L. Wang et al., Astrophys. J. **743**, 184 (2011)
A. Wolszczan, D.A. Frail, Nature **355**, 145 (1992)
M. Zorotovic, M.R. Schreiber, Astron. Astrophys. **549**, 95 (2013)

Space Sci Rev (2016) 205:327–390
DOI 10.1007/s11214-016-0254-3

Exoplanetary Atmospheres—Chemistry, Formation Conditions, and Habitability

Nikku Madhusudhan[1] · Marcelino Agúndez[2] · Julianne I. Moses[3] · Yongyun Hu[4]

Received: 29 July 2015 / Accepted: 11 April 2016 / Published online: 12 May 2016

Abstract Characterizing the atmospheres of extrasolar planets is the new frontier in exoplanetary science. The last two decades of exoplanet discoveries have revealed that exoplanets are very common and extremely diverse in their orbital and bulk properties. We now enter a new era as we begin to investigate the chemical diversity of exoplanets, their atmospheric and interior processes, and their formation conditions. Recent developments in the field have led to unprecedented advancements in our understanding of atmospheric chemistry of exoplanets and the implications for their formation conditions. We review these developments in the present work. We review in detail the theory of atmospheric chemistry in all classes of exoplanets discovered to date, from highly irradiated gas giants, ice giants, and super-Earths, to directly imaged giant planets at large orbital separations. We then review the observational detections of chemical species in exoplanetary atmospheres of these various types using different methods, including transit spectroscopy, Doppler spectroscopy, and direct imaging. In addition to chemical detections, we discuss the advances in determining chemical abundances in these atmospheres and how such abundances are being used to constrain exoplanetary formation conditions and migration mechanisms. Finally, we review

✉ N. Madhusudhan
nmadhu@ast.cam.ac.uk

M. Agúndez
marcelino.agundez@icmm.csic.es

J.I. Moses
jmoses@SpaceScience.org

Y. Hu
yyhu@pku.edu.cn

1 Institute of Astronomy, University of Cambridge, Madingley Road, Cambridge CB3 0HA, UK

2 Instituto de Ciencia de Materiales de Madrid, CSIC, C/Sor Juana Inés de la Cruz 3, 28049 Cantoblanco, Spain

3 Space Science Institute, 4750 Walnut Street, Suite 205, Boulder, CO 80301, USA

4 Laboratory for Climate and Ocean-Atmosphere Sciences, Department of Atmospheric and Oceanic Sciences, School of Physics, Peking University, Beijing 100871, China

recent theoretical work on the atmospheres of habitable exoplanets, followed by a discussion of future outlook of the field.

Keywords Exoplanets · Exoplanetary atmospheres · Atmospheric chemistry · Planet formation · Habitability

1 Introduction

The study of extrasolar planets (or 'exoplanets') is one of the most dynamic frontiers of modern astronomy. A few thousand exoplanets are now known. The numerous exoplanet discoveries via different methods have shown that exoplanetary systems are very common and are extremely diverse in their macroscopic properties. Recent statistics from observational surveys are suggesting that almost every star in the solar neighborhood hosts at least one planet around it, and that terrestrial-size exoplanets are amongst the most numerous (Howard et al. 2012; Fressin et al. 2013). The exoplanets detected to date span a diverse range in masses, radii, temperatures, and orbital parameters (orbital periods, separations, eccentricities, inclinations, etc.). The extreme diversity of these macroscopic properties of exoplanets has for the first time placed the solar system in cosmic context, thereby opening a plethora of new questions with far reaching implications.

We are now entering a new era in exoplanetary science. Going beyond planet detections, the field is now moving towards detailed characterization of exoplanetary atmospheres through spectroscopic observations. Encoded within a spectrum of an exoplanet is information about the chemical composition and manifold physical processes in its atmosphere. State-of-the-art observations are now beginning to provide both the high sensitivity and long spectral baseline required to place detailed constraints on the various physicochemical properties of exoplanetary atmospheres. Such observations in recent years are already providing good constraints on the chemical compositions of exoplanetary atmospheres. The most observed exoplanets to date are hot giant planets whose large scale-heights and high temperatures ($\sim$800–3000 K) make them particularly conducive to atmospheric observations. Several prominent molecules of carbon and oxygen and several atomic species (e.g. Na, K) are expected to be abundant and observable in these atmospheres, making giant exoplanets rich laboratories for understanding atmospheric chemistry. The atmospheric chemical compositions are in turn beginning to provide the first insights into the possible formation conditions and migration mechanisms of exoplanets.

Atmospheric observations have been reported for a variety of exoplanets detected via transits, direct imaging, as well as the radial velocity method. Spitzer observations have been obtained for about 20 transiting exoplanets in at least four photometric bands (3.6, 4.5, 5.8, 8 μm) and about 50 in two (3.6 and 4.5 μm), primarily for close-in hot Jupiters (Madhusudhan et al. 2014c). Photometric observations have also been obtained in the near infrared using large ground-based telescopes. Recently, high S/N near-infrared transit spectroscopy has become possible thanks to the HST Wide Field Camera 3 (WFC3) leading to the first high-confidence detections of H_2O in several transiting exoplanets (e.g. Deming et al. 2013). Multi-wavelength datasets are also providing the long spectral baseline and high precision required to derive joint constraints on the chemical compositions and temperature profiles of exoplanetary atmospheres. Over 400 HST orbits have been recently allocated for high S/N spectra of over a dozen transiting exoplanets using the HST instruments in the visible and near-infrared. High resolution spectroscopic observations have also been reported for several directly imaged planets in the near-infrared, and dedicated surveys are expected to pursue the same for dozens more. Finally, it has also now become possible to detect molecules in the atmospheres of close-in planets using very high resolution ($R = 10^5$) Doppler spectroscopy (Snellen et al. 2010). These observational advancements have led to the detection of chemical species in a sizable ensemble of planets and have motivated rapid progress in our theoretical understanding of atmospheric chemistry in exoplanets.

In the present work, we review the latest developments in our understanding of atmospheric chemistry of exoplanets. We begin with a brief introduction to observational methods for exoplanet detection and atmospheric characterization in Sect. 2. We then discuss the theoretical developments in our understanding of atmospheric chemistry in exoplanets (in Sect. 3), with particular emphasis on large gas giant planets (in Sect. 4) as well as ice giants and super-Earths (in Sect. 5). We then review, in Sect. 6, observational inferences of chemical species in exoplanetary atmospheres of various class using different observational methods. In Sect. 7, we review developments towards a theoretical framework for using atmospheric chemical abundances to constrain exoplanetary formation conditions and migration pathways. In Sect. 8, we discuss theoretical developments in our expectations for the habitability of terrestrial exoplanets. We conclude with a discussion of the future outlook for the field.

2 Observational Methods

The thousands of exoplanets known to date have been detected via a variety of observational methods (see Fischer et al. 2014, for a review). By far the majority of the planets have been detected using the transit method and the radial velocity method, followed by some detections using various other methods such as pulsar timing, gravitational microlensing, and direct imaging. On the other hand, observations of exoplanetary atmospheres present much more stringent requirements on the instrumental capabilities than planet detection. A combination of significantly higher sensitivity and higher spectral resolution are required for reliable atmospheric detections. Amongst the various planet finding methods, those amenable for atmospheric characterization are the transit method, direct imaging, and radial velocity method, with the transit method being the most successful to date. A detailed review of all these observational methods for atmospheric characterization can be found in Madhusudhan et al. (2014c). Here we briefly review the key attributes of each method.

One of the most successful methods for observing exoplanetary atmospheres has been the transit method. When the planet transits in front of the host star, part of the star light traverses through the day-night terminator region of the planetary atmosphere before reaching the observer. The resulting 'transmission spectrum', obtained by subtracting the in-transit spectrum from the out-of-transit spectrum, contains absorption features imprinted on the starlight by chemical species in the planetary atmosphere. Therefore, transmission spectra probe the chemical composition and temperature structure at the day-night terminator of the planet. On the other hand, when the planet is at full phase before being occulted by the star (i.e. 'secondary eclipse') the thermal emission and reflection spectrum from the planet is observed along with the stellar spectrum which can be subtracted out later; during secondary eclipse only the stellar spectrum is observed. Thus, secondary eclipse spectra probe composition and thermal structure of the dayside atmosphere of the planet. Spectra for exoplanets have been obtained in both transmission and at secondary eclipse, and in a wide range of wavelengths. While transmission spectra have been obtained from UV to mid-infrared wavelengths, thermal emission spectra have been obtained predominantly in the near to mid infrared. Besides the transit method, atmospheric observations have also been obtained from direct imaging of planets as well as by high-resolution Doppler spectroscopy, which are both discussed briefly in their respective sections below.

3 Factors Influencing Atmospheric Chemistry of Exoplanets

The advent of observational techniques able to characterize atmospheres of extrasolar planets has been a major breakthrough in astronomy. The discovery of a great diversity of exoplanets, many of them with no analogue in the solar system, suggests that there is an exoplanet zoo out there with a continuum of possible sizes (from Jupiter-sized planets to sub-Earth bodies), atmospheric temperatures (from thousands of degrees Kelvin down to a few tens), and elemental compositions (from H/He-dominated atmospheres with a solar composition to atmospheres dominated by heavy molecules such as H_2O, CO_2, or N_2). This diversity poses a great challenge to theorists who seek an understanding of the variety of existing planetary climates.

A theoretical approach to the chemistry of exoplanet atmospheres must start by posing a question of the following type. How can we infer the chemical composition of a planetary atmosphere from a reduced set of parameters related to the planet? Of course it is first necessary to identify such parameters and to understand how each of them affect the atmospheric chemistry. Some of these parameters can be obtained in a straightforward way from observations, although there are some others for which we can just have guesses, albeit in some cases with formidable uncertainties. Let's have a look to each of these parameters.

Gravity—This parameter follows directly from the mass and radius of the planet, which in turn can be derived from radial velocity and transit techniques, respectively. Whether or not a planet can retain a substantial amount of atmosphere is largely related to its gravity and the X-ray and EUV flux of the host star (see e.g. chapter on protoatmospheres in Massol et al. 2016, this issue). Moreover, depending on the mass and radius of the planet, light elements such as hydrogen and helium may or not escape from the planet, with strong implications for the elemental composition of the atmosphere. The gravity is also important in that, together with the mean mass of particles and temperature, it sets the scale height of the atmosphere, i.e., how compact or extended the atmosphere is.

Elemental composition—The relative abundances of the different elements is one of the most important aspects that determine the atmospheric composition, although unfortunately these are rather difficult to infer from observations. Based on the study of solar system planets and on existing theories of planet formation (Lissauer 1993; Hubbard et al. 2002; Goldreich et al. 2004; Alibert et al. 2005; Papaloizou and Terquem 2006; Blum and Wurm 2008; Chiang and Youdin 2010; Morbidelli et al. 2012), it is expected that giant planets will from by core accretion, becoming massive enough at an early stage of planet formation to efficiently capture the nebular gas, and thus will retain a thick H/He-dominated atmosphere (see e.g. chapter on protoatmospheres in Massol et al. 2016, this issue). Terrestrial planets, on the other hand, will not become massive enough to efficiently accrete or retain H_2/He from the nebular gas, and their atmospheres, whether thick or thin, will depend on their mass and evolutionary history (e.g., orbital evolution, stellar/disk evolution, impact history, interior outgassing, atmospheric escape, climate evolution, surface-atmosphere interactions, magnetospheric interactions). Terrestrial-planet atmospheres are expected to be typically dominated by secondary products such as H_2O, CO_2, CO, N_2, Ne, Ar, Kr, SO_2, SiO_2 that were outgassed from the solid planetesimals that formed the planets or that were supplied by later impacts by solid bodies (see e.g. chapter on planet formation in Bai et al. 2016, this issue). In between giant and terrestrial planets there is a regime of super-Earths/sub-Neptunes whose elemental composition is largely unknown (e.g., Haghighipour 2013; Fortney et al. 2013; Lopez and Fortney 2014; Marcy et al. 2014; Lee et al. 2014; Rogers 2015). By comparing the outcomes of chemical models and observations, it is in principle possible to put constraints on the elemental composition of the planetary atmosphere (e.g., Madhusudhan et al. 2014c), which in turn can provide clues to the formation and evolution of the planet itself.

Insolation—The amount of energy received at the top of the atmosphere per unit time and unit area depends essentially on the luminosity of the host star and the orbital distance, two properties that can be relatively well constrained by observations. Of course, the way the stellar luminosity is divided across the electromagnetic spectrum, i.e., the spectral type of the star, has a great importance in affecting atmospheric characteristics. Incoming visible and infrared photons take care of the heating of the atmosphere, especially at pressures greater than $\sim$1 microbar (e.g., Seager and Sasselov 1998; Sudarsky et al. 2003; Barman et al. 2005; Iro et al. 2005; Fortney et al. 2005; Burrows et al. 2006), while ultraviolet photons heat the high-altitude thermosphere and lead to ionization and dissociation of atmospheric constituents and induce photochemistry at a variety of altitudes (e.g., Liang et al. 2003; Yelle 2004; García Muñoz 2007; Zahnle et al. 2009a; Line et al. 2010; Moses et al. 2011; Venot et al. 2012; Agúndez et al. 2012; Hu et al. 2012; Miller-Ricci Kempton et al. 2012; Koskinen et al. 2013; Lavvas et al. 2014; Shaikhislamov et al. 2014; Rimmer and Helling 2015).

Internal heating—The interiors of planets can provide an important source of heat for the atmosphere, depending on the planet's age and tidal interactions with the host star or other planets in the system. The internal heating can be estimated from theoretical models of the evolution of the planet interior and its reaction to tidal forces (Guillot 2005; Baraffe et al. 2010; Ogilvie 2014). However, uncertainties in some key magnitudes, such as the dissipation properties of the bulk material of the planet interior and the age of the star (and thus of the planetary system), keep the internal heating poorly constrained in most cases.

A self-consistent model of a planetary atmosphere should in principle be able to describe the physical and chemical state of the atmosphere from the parameters described above. The situation resembles that of stellar atmospheres (e.g., Kurucz 1979; Hauschildt et al. 1999), where models need just three parameters (gravity, metallicity, and effective temperature)—external insolation being not important. Of course, in certain types of planets, parameters or processes different from those listed above, such as, e.g., exchange of matter with the surface in the case of terrestrial planets (e.g., Leconte et al. 2014) or grain formation and resulting opacity effects (e.g., Tsuji et al. 1996; Marley et al. 1996; Allard et al. 1997), may be also of great relevance in establishing the physical and chemical atmospheric properties.

In practice, models of planetary atmospheres lack a full self-consistency and tend to focus on some particular aspects, adopting approximations for some others. On the one hand, there are one-dimensional models that concentrate on solving the radiative transfer and thermal balance in the vertical direction. These models result in a pressure-temperature (P-T) profile for which the temperature gradient (dT/dP) is subadiabatic in the radiative zone of the atmosphere and adiabatic in the deeper convective regions (e.g., McKay et al. 1989; Burrows et al. 1997, 2008; Seager et al. 2005). These radiative-convective equilibrium models usually make some assumptions about the chemical composition of the atmosphere, either adopting a particular one suitable for the modeled planet or assuming that chemical equilibrium holds in the atmosphere, and neglecting any fluid dynamics. For an assumed composition, in addition to the P-T profiles these non-gray numerical models also calculate the line-by-line thermal emission and reflectance spectra of the planetary atmospheres. Recent studies have also reported analytic and semi-analytic P-T profiles derived for irradiated planetary atmospheres in radiative equilibrium under the assumptions of gray/non-gray opacities (Hansen 2008; Guillot 2010; Robinson and Catling 2012; Heng et al. 2012; Parmentier and Guillot 2014; Parmentier et al. 2015).

Other type of models, the so-called general circulation models (GCMs), were originally developed to study the climate of the Earth (e.g., Phillips 1956; Mellor and Yamada 1982) and are now routinely used to study atmospheres of other solar system planets (e.g., Forget et al. 1999) as well as exoplanets such as hot Jupiters (e.g., Showman and Guillot 2002; Showman et al. 2009), hot Neptunes (e.g., Lewis et al. 2010), super-Earths (e.g., Castan and Menou 2011) and terrestrial planets (e.g., Showman et al. 2013). These are three-dimensional models which solve the Navier-Stokes equations or a reduced set of "primitive equations" (see e.g. Showman et al. 2010; Heng and Showman 2015) and end up with a three-dimensional view of the circulation and thermal structure of the atmosphere. Some of the limitations of GCMs have to do with the approximations used to deal with the radiative transfer (e.g., Showman et al. 2009) and the chemical processes (Cooper and Showman 2006). Another type of models are those focused on the atmospheric chemistry, in which we concentrate hereafter.

Chemical models of planetary atmospheres aim at describing how the atmospheric constituents are distributed, usually in the vertical direction. To build such a model it is necessary to have information on some basic parameters, among them three of the four aforementioned: (1) *gravity*, (2) *elemental composition*, and (3) *ultraviolet/X-ray irradiation*. It is also necessary to have information on a couple of additional parameters.

(4) *Thermal structure*—The spatial distribution of the atmospheric temperature, at least in the vertical direction, can be retrieved from radiative-convective models and GCMs (see above). In the case of these latter models it is possible to get a three-dimensional map of the temperature. The atmospheric temperature has an enormous influence on the chemical composition because depending on whether the atmosphere is hot or cold the major constituents and the type of condensates formed (if any) are completely different.

(5) *Strength of transport processes*—An atmosphere is in essence a fluid and therefore a variety of processes such as advection, diffusion, and turbulent motions can occur at different scales, having as consequence the transport and mixing of material between different regions. These processes are an important source of disequilibrium because they can transport molecules that were originally formed in a given location to other atmospheric regions. In particular, the vertical chemical structure can be strongly affected by transport processes, the strength of which is usually parameterized through an eddy diffusion coefficient (K_{zz} if referred to the vertical direction). It is, however, difficult to obtain realistic estimates of the coefficient K_{zz}. In the case of solar system planets such as Jupiter and Saturn, if observations are able to determine the abundance of certain species at a given altitude, it is possible to put constraints on the eddy diffusion coefficient (e.g., Atreya 1986). In the case of extrasolar planets, information on the strength of mixing processes must be obtained from theoretical models such as GCMs (e.g., Parmentier et al. 2013).

If one wants to build a chemical model of a planetary atmosphere, these five ingredients (*gravity*, *elemental composition*, *ultraviolet/X-ray irradiation*, *thermal structure*, and *strength of transport processes*) should be in principle enough to provide a good description of the chemical structure of the atmosphere.

The characterization of exoplanet atmospheres through observations has been to date restricted to hot Jupiters, some hot Neptunes, and a few super-Earths, most of them characterized by transit techniques, and to a few young and self-luminous gas giant planets with an orbital distance sufficiently large to be accessible by direct imaging. It is therefore not surprising that theoretical studies of exoplanet atmospheres carried out to date have been to a large extent biased toward these types of extrasolar planets. A general theoretical characterization of habitable-zone exoplanet atmospheres has also been a popular topic, due to the high intrinsic interest in the potential for life (and its detectability) outside of Earth.

4 Theory of Atmospheric Chemistry of Gas Giant Exoplanets

Gas giant exoplanets are expected to have H/He-dominated atmospheres based on their formation mechanism and on the bulk densities inferred from observations (Lissauer and

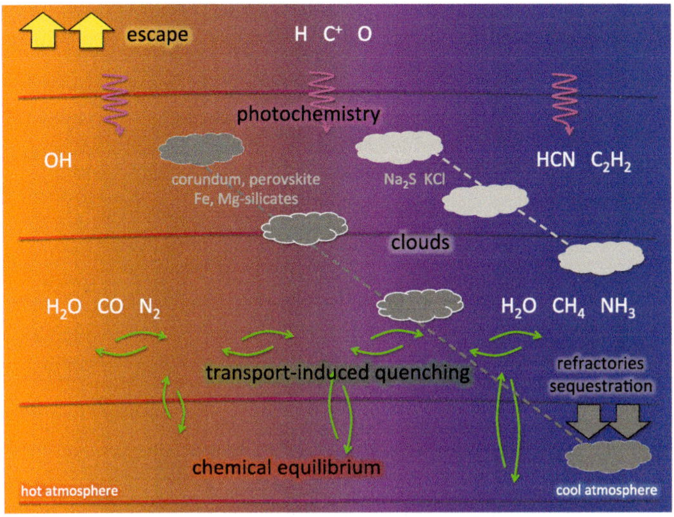

Fig. 1 Sketch of the structure and processes at work in H/He-dominated atmospheres. The chemical composition in deep layers is controlled by chemical equilibrium, by transport-induced quenching in upper layers, and by photochemistry in still upper layers. Formation of clouds is an issue of increasing importance as the atmosphere becomes cooler. Escape can also be an important process in highly irradiated atmospheres (i.e., those with hot thermospheres), while sequestration of refractory elements becomes important in the cooler atmospheres

Stevenson 2007; Baraffe et al. 2010). Their chemical composition depends on various parameters, perhaps the most important being the temperature. A sketch of the structure of such an atmosphere is depicted in Fig. 1, where various important processes at work are indicated. The chemical composition is controlled by chemical equilibrium in deep atmospheric layers, by transport-induced quenching in upper layers, and by photochemistry in still upper layers. Cloud formation becomes increasingly important as the atmosphere is cooler. Atmospheric escape can be an important issue for highly irradiated giant planets (e.g., Koskinen et al. 2014), while sequestration of refractory elements plays an important role in cool atmospheres (e.g., Lodders 2010).

4.1 Chemical Equilibrium

Under the chemical equilibrium assumption, the chemical composition does only depend on temperature, pressure, and elemental abundances, and can be calculated in a relatively straightforward way by minimizing the Gibbs free energy of the system. Chemical equilibrium can accurately describe the composition of an atmosphere provided it is sufficiently hot and dense to ensure that chemical reactions occur faster than any other process at work (dynamics, interaction with energetic radiation, etc.). This can be the case for atmospheres of cool stars and brown dwarfs, where chemical equilibrium has been routinely applied (Tsuji 1973; Fegley and Lodders 1996; Allard et al. 1996; Burrows and Sharp 1999; Lodders and Fegley 2002), and of hot exoplanets with atmospheric temperatures in excess of $\sim$2000 K (Moses et al. 2011; Venot et al. 2012). In the case of cooler planets, it can still be a good starting point to have a general idea of which could be the main atmospheric constituents.

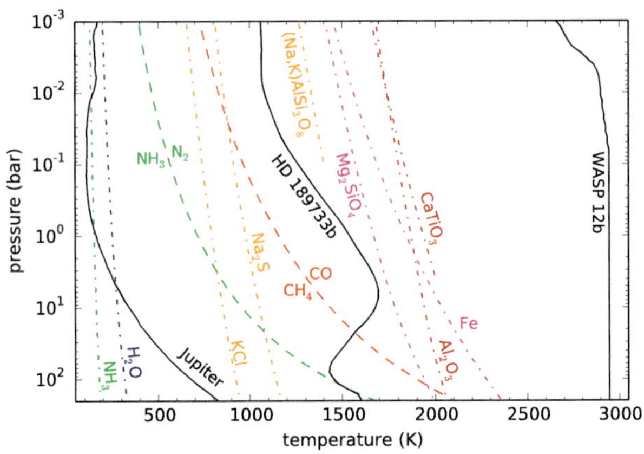

Fig. 2 Pressure-temperature profiles for three gas giant planets covering a broad range of atmospheric temperatures: Jupiter (Lindal 1992), the widely studied hot Jupiter HD 189733b (Agúndez et al. 2014a), and the highly irradiated gas giant WASP-12b (Stevenson et al. 2014a). *Dashed lines* delimitate the regions where either CH_4 or CO (*red curve*) or NH_3 or N_2 (*green curve*) are the major carbon or nitrogen reservoirs, respectively. *Dot-dashed lines* indicate the condensation curves for water, ammonia, and various refractory species containing Ca, Ti, Al, Fe, Mg, Na, and K

The atmospheric temperatures of gas giant planets may span over a very broad range, from a few hundreds of kelvins up to some thousands (see Fig. 2). Gas giant planets are expected to have a nearly solar elemental composition, although some elements heavier than helium could be enriched or depleted by a factor of a few relative to the solar composition, as occurs in Jupiter (Mahaffy et al. 2000; Wong et al. 2004). In any case, gas giant atmospheres are expected to be dominated by hydrogen and helium, with heavier elements like oxygen, carbon, and nitrogen being present at a lower level. Figure 2 shows various curves, calculated under chemical equilibrium for a solar elemental composition, which delimitate different regions of interest from a chemical view point. The red dashed curve indicates where CO and CH_4 have the same abundance. To the right of this curve (at high temperatures/low pressures) we have the region of stability of CO while to the left (at low temperatures/high pressures) methane is the main carbon reservoir. Similarly, in the case of nitrogen chemical equilibrium indicates that to the right of the green dashed curve N_2 is the main nitrogen reservoir while to the left it is ammonia that dominates. Thus, a first lesson to be learnt from chemical equilibrium is that at high temperatures and/or low pressures CO and N_2 are the main reservoirs of carbon and nitrogen, respectively, while at low temperatures and/or high pressures the hydrides CH_4 and NH_3 dominate. In the case of oxygen, water vapor remains a major reservoir over most of the temperature regime of interest. This molecule locks either the excess of oxygen not locked into CO if the atmosphere resides in the stability region of CO (oxygen has a solar abundance about twice than carbon), or directly locks most of the available oxygen if the atmosphere resides in the stability region of CH_4. Only at very low temperatures (see blue dot-dashed line) can water vapor be depleted from the gas phase because of condensation, something that can occur also for ammonia, albeit at even lower temperatures (see green dot-dashed line). That is, according to chemical equilibrium, depending on the thermal profile, the atmosphere of a giant planet may be dominated by (apart from H_2 and He) a mixture of $H_2O/CO/N_2$, $H_2O/CH_4/N_2$, $H_2O/CH_4/NH_3$, or a modification of the latter in which first H_2O, then NH_3, and finally CH_4 are progressively removed from the gas phase as the atmosphere gets cooler (e.g., Burrows et al. 1997; Sudarsky et al. 2000; Lodders and Fegley 2002).

4.2 Disequilibrium Chemistry Driven by Transport Processes

The atmospheric composition of gas giant exoplanets is governed by chemical equilibrium only in the hottest regions, usually in the deepest, densest regions (or the bulk of the atmosphere for very hot, strongly irradiated planets). In cooler upper regions, when chemical reactions become slower than dynamic processes (advection, diffusion, or turbulence), the bulk composition can be significantly driven out of equilibrium. The transport of material between different regions has as a consequence a redistribution (homogenization) of heat and chemical composition, with a net flow from hot to cooler regions.

The vertical distribution of the atmospheric constituents can be strongly affected by the so-called chemical quenching, in which as material moves upwards chemical reactions become slower, and at a certain level the abundances of the different species are progressively quenched. This phenomenon was first described by Prinn and Barshay (1977) to explain the presence of CO in the troposphere of Jupiter as a result of upward mixing from deeper and hotter levels, where CO is thermochemically stable. In gas giant exoplanets, chemical quenching regulates to a large extent the composition of the layers probed by observations, especially if their pressure-temperature profiles cross the boundaries of stability of the pairs CH_4/CO or NH_3/N_2 (see Fig. 2). A quantitative evaluation of chemical quenching (i.e., the quench level and resulting quenched abundance) requires a good knowledge of both the chemical kinetics of the interconversion scheme of interest and the vertical eddy mixing coefficient.

The main issue when dealing with quenching and chemical conversion schemes is the identification of the elementary reactions involved and the "limiting step", whose rate controls the kinetics of the overall conversion. The interconversion CO $\rightleftharpoons$ CH_4 is of great importance in cool to moderately warm gas giant planets because exoplanet temperature-pressure profiles often cross the stability boundary between CO and CH_4 (red dashed curve in Fig. 2). Understanding how the conversion between CO $\rightleftharpoons$ CH_4 proceeds kinetically and which reaction is the rate-limiting step is thus of primordial importance (see Visscher and Moses 2011 for more details). In the same vein, the interconversion $N_2 \rightleftharpoons NH_3$ is also of great relevance for the atmospheres of gas giant planets, N_2 and NH_3 being the main reservoirs of nitrogen at high and low temperatures (or low and high pressures), respectively. Although there are still important uncertainties in the reaction schemes and the rate-limiting step, the conversion $N_2 \rightarrow NH_3$ is likely intrinsically slower than the conversion CO $\rightarrow$ CH_4, resulting in a deeper quench level for the nitrogen species (see Moses 2014 for more details). The identification of interconversion reaction schemes is interesting in that it permits implementation of simple prescriptions to deal with chemical quenching in high-demanding computational models of atmospheres such as GCMs (Cooper and Showman 2006).

For 1D exoplanetary atmosphere models, the eddy diffusion coefficient must be quantified from 3D theoretical models like GCMs, or constrained from the observations themselves. Although mixing in planetary atmospheres can occur through large-scale advection, atmospheric waves, eddies of a variety of scales, and other transport processes that are not diffusive in a rigorous sense, vertical mixing can typically be well represented by a diffusion equation (e.g. Lindzen 1981; Strobel 1981; Brasseur et al. 1999). A common practice has been to estimate the vertical diffusion coefficient K_{zz} as the root mean square of the vertical velocity (as extracted from a GCM) times the vertical scale height (Line et al. 2010; Moses et al. 2011). A better approach when 3D circulation models are available, as outlined by Parmentier et al. (2013), is (1) to follow the behavior of passive tracers in a GCM simulation and fit the resulting planet-averaged vertical tracer profiles via a 1D diffusion equation with an effective diffusion coefficient and/or (2) to determine the diffusive flux and resulting K_{zz}

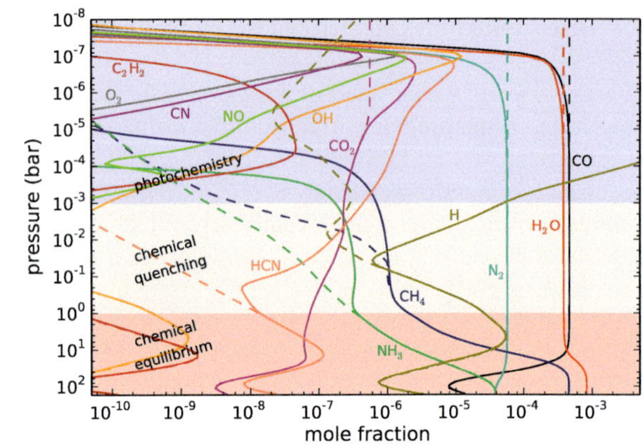

Fig. 3 Vertical distribution of abundances at the substellar point of HD 189733b. *Dashed lines* correspond to chemical equilibrium and *solid lines* to a one-dimensional vertical model including thermochemical kinetics and photochemistry (Agúndez et al. 2014a). The atmosphere can be schematically divided into three regions where the composition is controlled by either chemical equilibrium, chemical quenching, or photochemistry

profile that best matches the horizontally averaged vertical flux in the GCMs. These latter methods have been shown to result in K_{zz} values significantly lower than those obtained by representing K_{zz} as the root-mean-square vertical velocity times the scale height (Parmentier et al. 2013; Agúndez et al. 2014a). The exact value of the eddy diffusion coefficient, especially at pressures greater than $\sim$0.1 bar, has a direct impact on the location of the quench level of each species (the higher the K_{zz}, the deeper the quench level) and the abundance at which each species gets quenched in the upper atmosphere. If the quench point falls within the radiative region of the atmosphere, K_{zz} is expected to vary with the inverse square root of atmospheric pressure near this quench point (Lindzen 1981; Parmentier et al. 2013). If the quench point falls within the deeper convective region of the atmosphere, mixing-length theory and free-convection theory can be used to estimate the magnitude of K_{zz} near the quench point (Visscher et al. 2010b), or expressions based on laboratory studies of turbulent rotating convection can be developed (Wang et al. 2015). However, the appropriate mixing length to use for the K_{zz} estimates is not straightforward (e.g., Smith 1998), and observations of quenched species may themselves provide our best means of estimating the magnitude of convective mixing at depth.

Chemical quenching in the vertical direction of hot Jupiter atmospheres has been studied through timescale arguments (Visscher et al. 2006; Line et al. 2010; Bilger et al. 2013) and using more robust chemical networks (in which endothermic reactions are included and reverse reaction rates are computed from thermochemical grounds) suitable to model high temperature environments (Zahnle et al. 2009a,b; Moses et al. 2011, 2013b; Venot et al. 2012; Kopparapu et al. 2012; Agúndez et al. 2014a; Miguel and Kaltenegger 2014; see also Fig. 3). Most of these studies focus on the atmospheres of the widely observed hot Jupiters HD 189733b and HD 209458b, where it is predicted that, assuming nearly solar elemental abundances, CO and N_2 are the major reservoirs of carbon and nitrogen, respectively, H_2O is very abundant (it takes up most of the excess of oxygen not locked into CO), and methane and ammonia are present at a lower level, with CO/CH_4 and N_2/NH_3 abundance ratios in the ranges 50–500 and 5–200, respectively, in HD 189733b, and in the ranges 1000–10000 and 100–1000, respectively, in the hotter HD 209458b (Moses et al. 2011; Venot et al. 2012; Agúndez et al. 2014a; Miguel and Kaltenegger 2014). The abundance of carbon dioxide is found to be somewhat low for atmospheres with an assumed solar elemental composition (see Fig. 3). It is interesting to note that when CO is the dominant carbon component, CO_2 is much less affected by chemical quenching than many other molecules because the various reactions involved in the interconversion between H_2O-CO-CO_2 are rapid enough (Moses

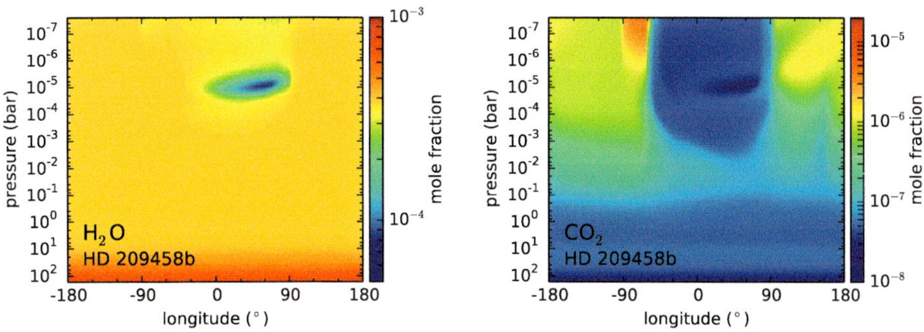

Fig. 4 Calculated distributions of H_2O and CO_2 as a function of longitude and pressure in the atmosphere of HD 209458b, from Agúndez et al. (2014a). Note that H_2O maintains a rather homogeneous distribution while CO_2 experiences important abundance variations with longitude and height

et al. 2011; Venot et al. 2012). The chemistry of elements other than C, N, and O, although more difficult to constrain through observations and more uncertain due to a lack of robust reaction rate measurements under appropriate conditions, has been also studied theoretically. It has been pointed out, for example, that chemical quenching leads to PH_3 becoming a major reservoir of phosphorus in hot Jupiter atmospheres (Visscher et al. 2006).

Chemical quenching can also occur horizontally driven by winds that move material from hot to cooler atmospheric regions. In hot Jupiters, which are usually tidally locked (with permanent day and night sides), the uneven heating of the planet may result in important temperature contrasts, and thus in possible variations in the composition, between the day and night sides. On the other hand, circulation dominated by a strong superrotating equatorial jet tends to homogenize both the temperature and chemical composition between the different planetary sides (Showman and Guillot 2002; Cooper and Showman 2006; Knutson et al. 2007). The main effect of horizontal chemical quenching is that molecular abundances are quenched horizontally to values typical of the hottest dayside regions, making the cooler nightside to be highly contaminated by the warmer dayside regions (Cooper and Showman 2006; Agúndez et al. 2012, 2014a). The effect is however more marked for molecules such as H_2O, CO, and N_2, which show rather uniform abundances, than for others such as CH_4, NH_3, CO_2, and HCN, which may still show important abundance gradients among the different planetary sides (see Fig. 4). The horizontal homogenization of abundances becomes more important as the atmosphere gets cooler. The distribution of the atmospheric constituents has implications for transit observations, which probe the terminator region, and for phase curves probing the different planetary sides.

4.3 Photochemistry

Hot Jupiters are strongly irradiated by their host stars and thus receive a high ultraviolet flux, which is absorbed in the upper atmosphere and induces a more or less rich photochemistry. That is, the parent molecules transported from deep atmospheric regions are photodissociated in upper layers producing radicals that react to form new species. The study of the photochemistry of exoplanet atmospheres has benefited from decades of study of solar system atmospheres such as those of Jupiter (Gladstone et al. 1996), Saturn (Moses et al. 2000), and Titan (Yung et al. 1984; Lara et al. 1996). However, some of the chemical kinetics and photoabsorption cross section data still need to be adapted to the high temperatures of hot Jupiters (Venot et al. 2013). The type of molecules formed by photochemistry depends on the composition of the precursor material transported from deeper layers, which in turn depends on the elemental composition (see Sect. 4.4) and thermal atmospheric structure (see Sects. 4.1 and 4.2). The cooler the planet the larger the extent of photochemistry because, on the one hand, high temperature chemistry counterbalances the action of photochemistry, and on the other, low temperatures favor the presence of CH_4 and NH_3, which are more active photochemical precursors than their respective high-temperature counterparts CO and N_2 (Moses 2014).

There are various species produced by photochemistry in the atmospheres of gas giant planets (see Fig. 3). These atmospheres being hydrogen-rich, a major species produced by photochemistry is atomic hydrogen, which is formed by photolysis and thermal decomposition of H_2 as well as by catalytic photolysis of H_2O (Liang et al. 2003). The higher the irradiation of the planet the more abundant becomes H. Another important photochemical species is HCN, which is efficiently formed in the photochemical layer through schemes starting with the photodissociation of CH_4 and NH_3 (provided the atmosphere is not too hot to exclude these two hydrides as important constituents), and is able to diffuse to deeper regions (Zahnle et al. 2009a; Moses et al. 2011). The photodissociation of CH_4 also triggers the formation of larger hydrocarbons such as C_2H_2, which may then polymerize to form soots (Zahnle et al. 2009a; Bilger et al. 2013). Another important photochemical product is the radical OH, which results from the photodissociation of water and act as a key intermediate in the synthesis of other O-bearing molecules such as O_2 and NO (Moses et al. 2011). As concerns sulfur compounds, the major photochemical products of the dominant equilibrium sulfur species, H_2S, are expected to be S, HS, S_2, SO, and SO_2 (Zahnle et al. 2009b). Lavvas et al. (2014) show that for planets hot enough that refractory elements are not tied up in condensates, atomic neutrals and ions such as Mg, Mg^+, Fe, Fe^+, Ca, Ca^+, Na, Na^+, K, K^+, Si^+, Al^+ could be important photochemical and equilibrium species, although some dominant equilibrium molecular forms (e.g., SiO, SiS, AlOH, FeH) can also survive in the photosphere on hot planets (see also Visscher et al. 2010a), and molecular species like NaCl and KCl may form from both photochemistry and thermochemistry.

On highly irradiated hot Jupiters, most photochemical products are confined to relatively high altitudes (e.g., above $\sim$0.1 mbar, see Fig. 3). Therefore, photochemical products are likely to have a limited impact on the infrared emission spectrum of the planet, whose photosphere is located at deeper layers, although they may leave their imprints on the transmission spectrum, which can probe higher atmospheric layers. The cooler the planet, however, the deeper the photochemical "layers" extend, and the higher the likelihood that the photochemical products affect the infrared emission spectra. On our own solar-system giant planets, for example, methane photochemical products such as C_2H_2 and C_2H_6 survive throughout the stratosphere, dominating the mid-IR spectra and enabling more effective cooling of the stratospheres of these planets. In the case of HCN, its extended vertical distribution and relatively high abundance in cool to moderately warm gas giant planets makes it a very likely candidate for detection (e.g., Moses et al. 2011).

One-dimensional models indicate that in still upper layers, molecules are completely photodestroyed and atoms become the major constituents. In the absence of a hydrodynamic escape flow, molecular diffusion at a high-altitude homopause layer (typically in the 10^{-6} to 10^{-9} bar range for hot Jupiters) will cause the atmospheric constituents to become layered according to their weight, following their own density scale heights. Because of both this molecular diffusion and the strong X-ray/ultraviolet irradiation from the host star (which strongly heats the high-altitude thermosphere), atomic hydrogen becomes the only remaining dominant neutral constituent at high altitudes on close-in giant planets (e.g., Moses et al. 2011). If the upper atmosphere were to remain cool, the molecular region could extend to very high altitudes (e.g., Agúndez et al. 2014a). However, thermospheric photochemical models that include X-ray and extreme ultraviolet photolysis and photodissociative ionization of H_2, ion chemistry, and realistic atmospheric escape processes (García Muñoz 2007; Koskinen et al. 2013) demonstrate that hydrodynamic winds fueled by the escaping gas can drag heavy atoms and ions to regions as high as the exosphere in strongly irradiated planets, but that molecules (including H_2) cannot survive the large thermospheric temperatures on such planets. In fact, various types of atoms, from the lightest hydrogen to heavy metals, have been detected in the vertically extended, escaping atmospheres of hot Jupiters (Vidal-Madjar et al. 2003, 2004, 2013; Lecavelier Des Etangs et al. 2010; Fossati et al. 2010; Linsky et al. 2010).

Ionization of atmospheric constituents by galactic cosmic rays provides an additional source of disequilibrium chemistry on exoplanets. Rimmer and Helling (2013) and Rimmer et al. (2014) have explored the effects of cosmic rays on the atmospheric compositions of extrasolar giants planets. Charging of mineral grains, and potential subsequent electrical discharges, are another possible source of disequilibrium chemistry on extrasolar giant planets (see Helling et al. 2013; Bailey et al. 2014; Stark et al. 2014).

4.4 Sensitivity to Bulk Elemental Composition

Gas giant planets are expected to have nearly solar elemental compositions. However some may show a different metallicity, i.e., the abundances of elements heavier than helium scaled up or down by a similar factor, and/or elemental abundance ratios different from those found in the Sun. These deviations from the solar elemental composition, whatever their origin (see Sect. 6), can have a significant impact on the atmospheric composition.

In general, an enhancement of metallicity in H/He-dominated atmospheres of gas giant planets favors an increase in the abundances of molecules that contain multiple heavy atoms, i.e., CO and N_2 are favored over CH_4 and NH_3. Molecules with more than two heavy atoms are even more favored, as occurs in the case of CO_2, whose abundance increases as the square of metallicity (Zahnle et al. 2009a,b). See more details on this subject in Sects. 5.1 and 5.2.

In warm H/He-dominated atmospheres, one of the most critical elemental ratios is C/O. The issue is well known in the study of evolved stars, some of which can bring out to the surface enough carbon to revert the C/O abundance ratio, which in the Sun is 0.55 (Asplund et al. 2009). The high stability of CO under these conditions makes it lock almost all the limiting reactant, either C or O, allowing for the reactant in excess to form O-bearing molecules such as H_2O when $C/O < 1$ and C-bearing molecules such as HCN and C_2H_2 when $C/O > 1$ (Tsuji 1973). The observational claim of a carbon-rich atmosphere in the hot Jupiter WASP-12b (Madhusudhan et al. 2011b), although still subjected to debate (Cowan et al. 2012a; Crossfield et al. 2012; Madhusudhan 2012; Swain et al. 2013; Mandell et al. 2013; Stevenson et al. 2014a,b; Kreidberg et al. 2015), has opened a window on the possible existence of carbon-rich giant extrasolar planets. Whatever their origin (see Sect. 6), such planets would show a chemical composition dramatically different from the more traditional gas giants with a nearly solar C/O ratio (i.e., oxygen-rich). In moderately warm and hot carbon-rich atmospheres, water is no longer an abundant constituent, the C-bearing molecules C_2H_2, CH_4, and HCN become major constituents, and CO_2 vanishes to a negligible level (Madhusudhan 2012; Kopparapu et al. 2012; Moses et al. 2013a,b; Venot et al. 2015; Heng and Lyons 2016). If the atmosphere is cool enough (<1000 K) then CH_4 and H_2O become the main reservoirs of carbon and oxygen, respectively, at the expense of other carbon- and oxygen-bearing molecules (Madhusudhan 2012; Moses et al. 2013a,b; Venot et al. 2015).

4.5 Clouds and Hazes

The terms *cloud* and *haze* are often used interchangeably, but here we follow the spirit of Marley et al. (2013) and use these terms to refer to two fundamentally different types of condensates that may appear in the atmospheres of gas giant planets. By *cloud* we refer to the typical "cooling clouds" (e.g., Rossow 1978) that form when a volume of gas is cooled dynamically or radiatively, such that the partial pressure of a constituent or constituents exceeds its saturation vapor pressure and condensation ensues. The typical equilibrium condensates expected along the cloud-condensation sequence in a hydrogen-dominated atmosphere (e.g., Lodders 2010; Marley et al. 2013; Morley et al. 2013) are thus considered clouds, regardless of their optical depth, horizontal extent, etc. We restrict the term "hazes" to refer to aerosols that form in situ by the action of photochemistry or any other disequilibrium chemical process (e.g., the so-called "soots" discussed by Zahnle et al. 2009a) or from photochemically produced gases flowing dynamically into cooler regions, where they can condense.

In the rainout scenario of cloud formation, a condensate forms and gravitationally settles in the atmospheric layer where the temperature equals its condensation temperature, and the atmosphere above the cloud becomes depleted in the elements that take part in that condensate (Lewis 1969; Fegley and Lodders 1994, 1996; Burrows and Sharp 1999; Lodders 1999; Lodders and Fegley 2002; Sudarsky et al. 2003). Thus, as one moves from the deep and hot atmosphere to upper and cooler regions, elements are progressively removed from the gas phase according to their refractory character. Chemical equilibrium calculations including gas and condensed species are very useful to identify the most plausible cloud-forming species (see Fig. 2). In the rainout approach, chemical equilibrium is solved in order of decreasing temperature and, when a given species is found to condense the elements which take part in that condensate are removed (in the corresponding stoichiometric proportions) before continuing to solve chemical equilibrium at lower temperatures. Thus, it is not only necessary to know the possible condensates but also the condensation sequence and the main condensates that deplete each element.

Unlike the case of Jupiter, whose atmosphere is so cold that water and ammonia condense to form tropospheric clouds, most extrasolar giant planets characterized to date are sufficiently hot to ensure the survival of water vapor and other volatiles in their atmospheres. In hot Jupiters only the most refractory species can condense to form clouds. Although the exact temperature-pressure profile of the planet in question controls which refractory species will condense first (i.e., at the deepest pressures; see Lodders 2010), aluminum, titanium, and calcium are expected to be the first elements to be removed from the gas phase, condensing as corundum (Al_2O_3), Ca-aluminates such as hibonite ($CaAl_{12}O_{19}$), and Ca-titanates such as perovskite ($CaTiO_3$). Therefore, atmospheres hotter than about 2000 K can maintain titanium in the gas phase to form oxides such as TiO, while cooler atmospheres would deplete most of this element in the form of perovskite or other Ca-titanate (Burrows and Sharp 1999; Lodders 2002). In fact, it has been proposed that in the atmospheres of very hot giant exoplanets, the survival in the gas phase of species such as TiO and VO can provide a sufficiently high opacity at optical wavelengths as to induce a thermal inversion (Hubeny et al. 2003; Fortney et al. 2008a; Spiegel et al. 2009). The detections of TiO and VO in brown dwarf atmospheres at high temperatures (Kirkpatrick 2005) provide support to this proposition. However, TiO and VO have not yet been unambiguously identified in the atmospheres of strongly irradiated hot Jupiters (Désert et al. 2008; Huitson et al. 2013; Mancini et al. 2013; Swain et al. 2013; Sing et al. 2013; Gibson et al. 2013; Stevenson et al. 2014b; Schlawin et al. 2014; Hoeijmakers et al. 2015), though recently a potential detection of TiO has been reported for one of the most irradiated hot Jupiters WASP-33b (Haynes et al. 2015).

At temperatures below ~1800 K, iron is expected to condense homogeneously, and magnesium, together with a good fraction of silicon, are expected to deplete in the form of silicates such as Mg_2SiO_4 and $MgSiO_3$. The presence of such type of clouds has been inferred from transit observations of the hot Jupiter HD 189733b (Pont et al. 2008, 2013; Lecavelier Des Etangs et al. 2008a; Sing et al. 2009; Gibson et al. 2012). The alkali metals Na and K can be depleted in the form of the aluminosilicates $NaAlSi_3O_8$ and $KAlSi_3O_8$ at temperatures around 1500 K (Burrows and Sharp 1999), although in the rainout scenario it is likely that at these temperatures most of the aluminum is no longer available in the gas phase, in which case Na and K would condense at ~1000 K as Na_2S and KCl, respectively (Lodders 1999). In sufficiently hot atmospheres neutral Na and K atoms can survive in the gas phase and can be readily detected through the Na I doublet at 589.0 nm and 589.6 nm and the K I doublet at 766.5 nm and 769.9 nm (Charbonneau et al. 2002; Redfield et al. 2008; Wood et al. 2011; Sing et al. 2011a, 2012; Nikolov et al. 2014; Murgas et al. 2014).

The rainout scenario provides a useful methodology to predict the composition of the clouds that may be present in a given atmosphere and the base level at which each cloud forms. However, the computation of the size and concentration of particles above this level and the horizontal distribution of the clouds, i.e., how particles nucleate and grow and are affected by atmospheric dynamics, still remains a formidable challenge. Nevertheless, efforts to this end have been undertaken by various groups (Ackerman and Marley 2001; Helling et al. 2008; Freytag et al. 2010; Parmentier et al. 2013; Morley et al. 2013; Lee et al. 2015).

The photochemical formation of organic hazes in the atmospheres of hot Jupiters has been addressed in a couple of theoretical studies (Liang et al. 2004; Zahnle et al. 2009a). The general view provided by these studies is that in hot atmospheres, most of the carbon remains locked into CO, while in cooler atmospheres the larger amount of carbon stored into CH_4 can be photochemically driven to larger hydrocarbons, eventually producing hydrocarbon aerosols or soots. Cooler planets would therefore be expected to have more prevalent photochemical hazes. This scenario agrees with the presence of hazes in the cold atmospheres of Jupiter, Saturn, and Titan (e.g., Tomasko et al. 2005; Zhang et al. 2013), although it is in contradiction with the inference of aerosols in the very hot atmosphere of WASP-12b (Sing et al. 2013), unless these aerosols result from the most refractory species along the cloud-condensation sequence (e.g., Al_2O_3) at the cooler limb of the planet (Sing et al. 2013), or unless the planet is indeed carbon-rich (and thus contains the more photochemically active molecules HCN and C_2H_2, e.g., Kopparapu et al. 2012 and Moses et al. 2013b). It is clear that more work is needed in both the theoretical and observational sides to better understand the formation and role of hazes in extrasolar giant planets.

4.6 Hot Jupiters Versus Directly Imaged Planets

One of the most recent and formidable successes achieved in the field of exoplanets has been the detection by direct imaging of young and self-luminous gas giant planets, opening the way to characterize their atmospheres by direct spectroscopy (Janson et al. 2010, 2013; Hinz et al. 2010; Bowler et al. 2010; Currie et al. 2011; Barman et al. 2011; Skemer et al. 2012; Oppenheimer et al. 2013; Konopacky et al. 2013; Chilcote et al. 2015). The few planets that have been characterized by this method are more massive than Jupiter and have effective temperatures in the range 600–1700 K. Thus, they have some similarities with brown dwarfs and free-floating planets (Zapatero Osorio et al. 2000; Burrows et al. 2001, 2003). Directly imaged planets share also some characteristics with hot Jupiters, as both are gas giants and hot. However, unlike hot Jupiters, directly imaged planets orbit far from their host star and are young, so that they are heated predominantly from the interior rather than irradiated by the star. This difference leads to qualitative differences in the dynamical and thermal structure of the atmosphere. In hot Jupiters, the high irradiation causes the atmospheres to be radiative in almost the entire observable atmosphere, and the temperature profile is isothermal in the lower atmosphere, at pressures above $\sim$1 bar (see e.g. Burrows et al. 2008), before convection dominates in the deep atmosphere ($\gtrsim$100 bar). On the other hand, the atmospheres of self-luminous planets and brown dwarfs are largely driven by convection, leading to adiabatic temperature gradients even in the observable atmospheres.

The atmospheres of hot Jupiters and directly imaged planets have temperatures of the same order and are expected to have a nearly solar elemental composition, so that one would expect a similar atmospheric chemistry in both types of planets. There are, however, a couple of major differences. First, the atmospheres of directly imaged planets are likely less affected by photochemistry than in the case of hot Jupiters because of their much larger orbital distances; however, directly imaged planets tend to orbit young stars, and young stars tend to have high ultraviolet output, so photochemistry will not be negligible on these planets. And second, the source of heat being located in the interior of the planet rather than outside imprints differences in the atmospheric thermal profile. According to the recent theoretical study by Zahnle and Marley (2014), the higher temperatures in the deep atmosphere of self-luminous planets favor CO over CH_4 in deep layers, but also in the upper observable atmosphere as a consequence of upward mixing. Methane is therefore predicted to be a minor atmospheric constituent in most self-luminous planets characterized to date, except for a couple of planets with cool atmospheres whose near-infrared spectra show evidence for CH_4 absorption, such as GJ 504b (Janson et al. 2013), 51 Eri b (Macintosh et al. 2015), and HR 8799b (Barman et al. 2015). Years of study of brown dwarf atmospheres (see e.g., Marley and Robinson 2015), which share many similarities with the atmospheres of self-luminous planets, provide an invaluable basis to aid in the understanding of atmospheric chemistry of self-luminous planets, although further observations are still needed.

5 Theory of Atmospheric Chemistry of Exo-Neptunes and Super-Earths

The same chemical processes discussed above for giant planets—thermochemical equilibrium, disequilibrium quenching due to transport, and disequilibrium photochemistry—also affect smaller planets, but the basic ingredients available to the atmospheres of smaller planets differ from those of gas giants. These differences emerge early on in the planet's evolution, with such factors as the dust/planetesimal surface density distribution within the protoplanetary disk (see chapters on protoplanetary disk evolution in Oberg et al. 2016, this issue; Fang et al. 2016, this issue), the planet's initial formation and feeding-zone location within the disk, and the planet's migration history being important parameters that control how massive the planet becomes, what atmospheric volatiles are collected, and how much atmospheric hydrogen and helium are retained (e.g., Pollack et al. 1996; Lissauer and Stevenson 2007; D'Angelo et al. 2010; Dodson-Robinson and Bodenheimer 2010; Marboeuf et al. 2014; Ali-Dib et al. 2014; Madhusudhan et al. 2014a; Helling et al. 2014). In the gravitational-instability theory for giant-planet formation (e.g., Boss 1997), giant planets are expected to end up with a metallicity similar to the host star. In the core-accretion theory of giant-planet formation (e.g. Pollack et al. 1996), protoplanetary cores form from the accretion of solid planetesimals, and gas accretion rates are initially slow. As the solid protoplanetary core grows, it can accrete more and more of the surrounding hydrogen-rich nebular gas (Lammer et al. 2014; Stökl et al. 2015, also see chapter by Massol et al. 2016, this issue). When the core becomes massive enough, with a critical mass of order ~10 Earth masses (but cf. Venturini et al. 2015), the protoplanet can experience a runaway gas-accretion phase in which the nebular gas is rapidly accreted onto the protoplanetary core. Planets that reach this runaway gas-accretion stage become hydrogen- and helium-rich gas-giant planets. Planets that don't reach this stage contain less H and He.

Whether a protoplanet can reach this runaway gas-accretion stage or not depends largely on the accretion rate of solids in comparison to the time scale for dissipation of gas from the disk. Just beyond condensation fronts in the disk, such as the water-ice "snow line", the surface density of solid material becomes large, leading to more rapid accretion of solids and a reduced time scale for formation of massive protoplanetary cores (e.g., Stevenson and Lunine 1988; Ciesla and Cuzzi 2006; Dodson-Robinson et al. 2009). The formation of giant planets is thus expected to be particularly efficient near the disk snow line. In lower-surface-density regions of the disk farther out from the snow line, the solid accretion rate is slower, and a protoplanet may never reach this runaway gas-accretion phase, leading to less accumulation of the nebular gas. In this classical picture, the formation of Uranus and Neptune occurred too slowly to allow these planets to fully reach the runaway gas-accretion phase, leading to "ice giant" planets whose total mass is dominated by heavier volatiles such H_2O rather than hydrogen and helium (e.g., Lissauer and Stevenson 2007; Fortney and Nettelmann 2010). The overall initial mass fraction of H/He in a planet's atmosphere is a sensitive function of the disk characteristics and lifetime, with small changes in disk properties or evolutionary history having a significant impact on the resulting mass and composition of planets that are intermediate in size between gas giants and solid terrestrial planets (e.g., Helled and Bodenheimer 2014; Lammer et al. 2014; Luger et al. 2015). While generally atmospheres of ice-giants are expected to be H_2-rich, recent theoretical studies have also suggested the possibility of He-dominated atmospheres (Hu et al. 2015b).

In this section, we discuss the atmospheric chemistry of these intermediate-sized planets, the so-called super-Earths, mini-Neptunes, and Neptune-sized planets, with radii spanning ~1–6 $R_\oplus$, that are observed to constitute a large percentage of the known planets in our galaxy (e.g., Borucki et al. 2011; Mayor et al. 2011; Howard et al. 2012; Batalha et al. 2013; Fressin et al. 2013; Burke et al. 2014; Rowe et al. 2015). These planets will typically contain less hydrogen and helium (and thus a higher atmospheric metallicity) than giant planets because of the less efficient accretion of nebular gas during their formation and evolution, as well as the higher likelihood of the escape of light gases over time. Outgassing from the planetary interior is expected to contribute additional volatiles—a component of the atmosphere that will become increasingly important for smaller planets—and other evolutionary processes such as atmospheric escape, impact delivery or erosion, atmosphere-surface exchange, weathering, and sequestration of volatiles into the interior can have and major influence on atmospheric composition and chemistry. The stochastic nature of the different possible evolutionary pathways is expected to lead to highly diverse atmospheric properties for intermediate and small planets (e.g. Pepin 2006; Dodson-Robinson and Bodenheimer 2010; Rogers et al. 2011; Lopez and Fortney 2013; Fortney et al. 2013; Haghighipour 2013; Leconte et al. 2014). Here, we consider the atmospheric chemistry of exo-Neptunes and super-Earths with widely diverse volatile contents, as well as the chemistry of outgassed atmospheres of smaller, hot, rocky planets. The atmospheric chemistry of terrestrial planets near the habitable zone is briefly discussed in Sect. 8.

5.1 Chemical Equilibrium in Exo-Neptune and Super-Earth Atmospheres

Thermochemical equilibrium can be maintained in high-temperature, high-pressure regions of exoplanet atmospheres (see Sect. 4.1), so equilibrium conditions are appropriate to consider to first order for any super-Earths or exo-Neptunes with thick, hot atmospheres. This statement is true whether the planet is a low-density fluid "ice giant", like Neptune itself, or has a thick atmosphere overlying a solid surface, like Venus. Even secondary atmospheres produced from the outgassing of interior volatiles during and after the accretion phase can lead to thick, high-pressure atmospheres that remain hot at depth due to accretional energy, radioactive decay in the interior, tidal heating, strong stellar irradiation, and/or a greenhouse effect.

Our own terrestrial planets have demonstrated the importance of secondary outgassing of interior volatiles in shaping the atmospheric properties of solid-surface planets (Pepin 2006). The theoretical equilibrium composition of super-Earth atmospheres dominated by such an outgassing source has been explored by several investigators. For example, Elkins-Tanton and Seager (2008) have examined the wide range of atmospheric masses and compositions that can result from the degassing of different meteoritic compositions during the planet's accretion phase. Schaefer and Fegley (2010) have performed similar more detailed calculations, albeit with a focus on the early Earth, to investigate the chemistry of potential steam atmospheres, such as those predicted to be associated with magma oceans (e.g., Abe and Matsui 1985) thought to form during the accretion of the Earth and other terrestrial (exo-)planets. Both Schaefer and Fegley (2010) and Elkins-Tanton and Seager (2008) find that the resulting atmospheric composition is a sensitive function of the assumed composition of the meteoritic material being accreted—water-dominated steam atmospheres occur only for CI and CM chondritic starting material. For other assumed meteoritic starting compositions, the atmospheres can be dominated by CO_2, N_2, H_2, CH_4, or CO, depending on the starting material composition and atmospheric temperatures (Schaefer and Fegley 2010). Schaefer and Fegley (2011) have also pursued the potential atmosphere-surface buffering of hot atmospheres in equilibrium with planetary surfaces.

The possible formation of exotic (by solar-system standards) silicate atmospheres on hot super-Earths, in which volatile elements such as H, C, N, S, and Cl have already escaped from the planet, is explored by Schaefer and Fegley (2009) and Miguel et al. (2011). In these calculations, the atmosphere is assumed to be in gas-melt equilibrium with a volatile-free magma ocean or partially molten lithosphere. Their results indicate that such atmospheres can be composed largely of atomic Na, O_2, O, with SiO, Fe, and/or Mg, depending on the planet's orbital distance and resulting atmospheric temperature. Ito et al. (2015) have performed a similar set of equilibrium calculations for super-Earth atmospheric compositions over a volatile-free magma ocean of various assumed compositions; they discuss the resulting thermal structure, spectroscopy, and detectability of such atmospheres. They find that SiO, in particular, affects the atmospheric opacity, causing thermal inversions and notable infrared emission signatures.

Schaefer et al. (2012) have extended their earlier 2010 investigation to consider a wider range of parameter space that might be relevant to outgassed atmospheres of hot terrestrial exoplanets and super-Earths, including considerations of variations in the relative abundances of elements such as H, C, and O. Again, these authors emphasize that the equilibrium atmospheric composition of such secondary outgassed atmospheres depends strongly on the source material and temperatures. Outgassing from accreted material with the composition of the bulk silicate Earth or the terrestrial crust would result in a water-dominated atmosphere under conditions relative to GJ 1214b, with CO_2 as an important secondary component (see Fig. 5), whereas O_2 and Na could dominate in the hotter, lower-density atmosphere

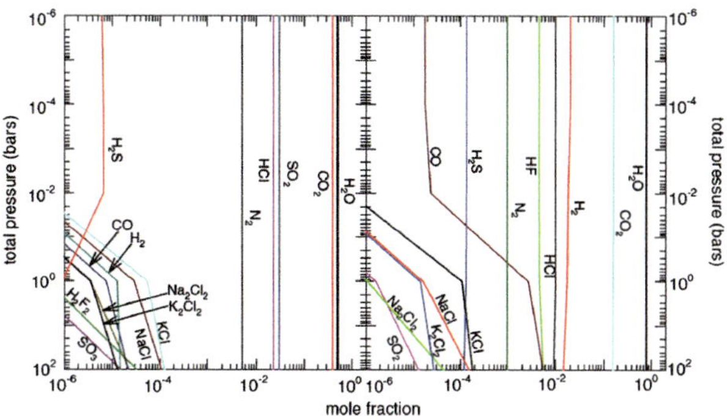

Fig. 5 Equilibrium atmospheric composition for a GJ 1214b-like exoplanet, assuming the atmosphere results from outgassing from a high-temperature felsic silicate terrestrial crustal composition (*left*) or a more mafic bulk-Earth silicate composition (*right*). Figure from Schaefer et al. (2012)

of CoRoT-7b (Schaefer et al. 2012; see also Léger et al. 2011). The chemistry of possible hot equilibrium atmospheres of Earth-like planets after giant impact events is explored by Lupu et al. (2014), who find that the main consituents are water and CO_2, along with smaller amounts of HCl, H_2, HF, CO, N_2, alkali halides, SO_2, and H_2S.

The equilibrium atmospheric composition of warm super-Earths like GJ 581c is investigated by Miller-Ricci et al. (2009), for an assumed cometary-like complement of volatiles with variable hydrogen content. For hydrogen-rich atmospheres, the dominant constituents are H_2, H_2O, CH_4, and NH_3. For the hydrogen-poor situation, the resulting Venus-like atmosphere has dominant constituents CO_2 and N_2, with much less H_2O. For the "intermediate" hydrogen case, the atmosphere has an interesting mixture of dominant constituents like H_2O, H_2, CO_2, and CH_4. The overall hydrogen content therefore has a strong influence on the resulting composition.

Observations of exoplanets for which both mass and radius have been determined show a transition in planetary bulk densities near radii of $\sim$1.5–2 $R_\oplus$; the smallest planets tend to be dense and presumably rocky, with at best a tiny mass fraction of light volatiles like hydrogen and helium, and larger planets tend to require increasingly larger mass fractions of H/He envelopes to explain the bulk density (cf. Weiss et al. 2013; Weiss and Marcy 2014; Marcy et al. 2014; Wu and Lithwick 2013; Hadden and Lithwick 2014; Rogers 2015). Population-synthesis models and other theoretical arguments suggest that the trend of decreasing hydrogen and helium content with decreasing planetary size is a natural consequence of planetary formation and evolution (Miller and Fortney 2011; Rogers et al. 2011; Lopez et al. 2012; Fortney et al. 2013; Owen and Wu 2013; Lopez and Fortney 2013, 2014; Benz et al. 2014; Wolfgang and Lopez 2015). Regardless of whether close-in exo-Neptunes and super-Earths with comparatively large H/He contents formed farther out in the disk and migrated/scattered inward (e.g., Ida and Lin 2005; Alibert et al. 2006; Terquem and Papaloizou 2007; Kennedy and Kenyon 2008; Mordasini et al. 2012; Inamdar and Schlichting 2015) or whether they formed in situ (Hansen and Murray 2012, 2013; Chiang and Laughlin 2013; Chatterjee and Tan 2014), the amount of hydrogen in the atmospheric envelope has a strong influence on the resulting atmospheric composition. The metallicity (or bulk mole fraction of H, in general) of the atmosphere is therefore an important parameter controlling the chemistry of low-density super-Earths and exo-Neptunes.

The effect of metallicity on the equilibrium composition, temperature structure, and/or spectra of exoplanet atmospheres has been studied by Lodders and Fegley (2002), Visscher et al. (2006, 2010a), and Fortney et al. (2008b) for metallicities up to 3–5$\times$ solar; Zahnle et al. (2009a,b), Spiegel et al. (2010), Fortney et al. (2010), Lewis et al. (2010), Line et al. (2011) for metallicities up to 30–50$\times$ solar; Agúndez et al. (2014b) and Venot et al. (2014) for metallicities up to 100$\times$ solar; and Moses et al. (2013a), Hu and Seager (2014), and

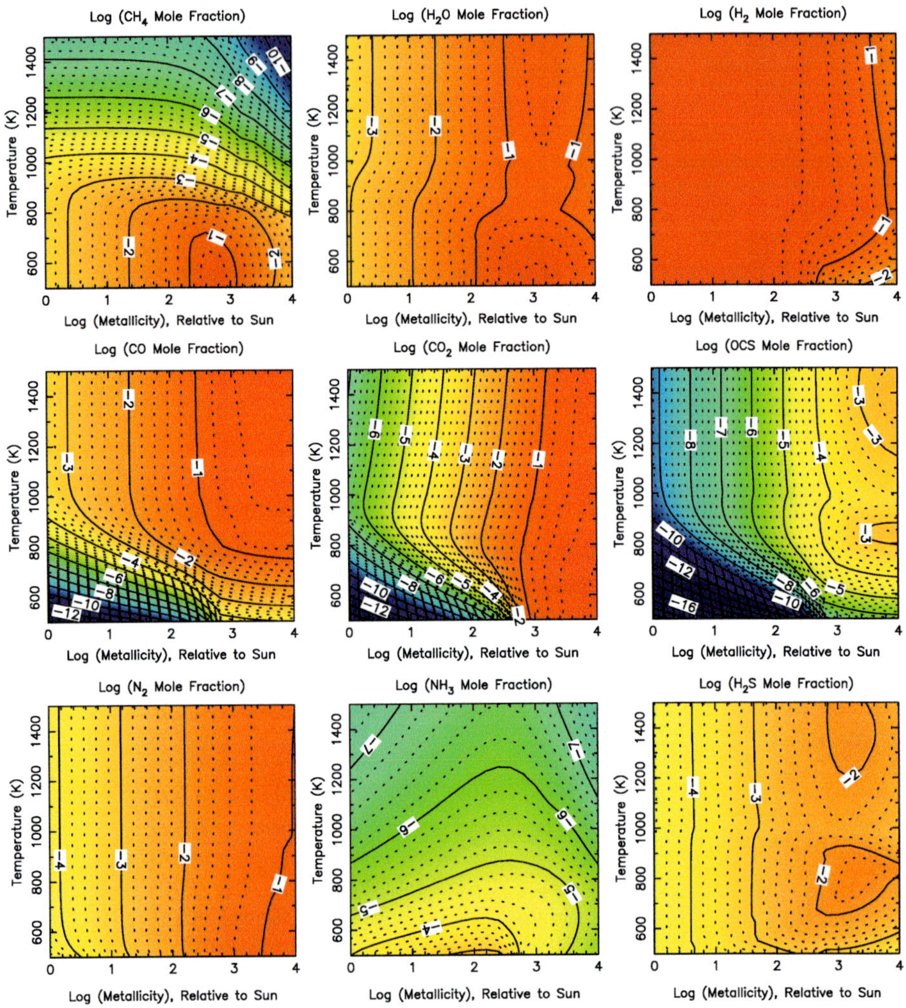

Fig. 6 Equilibrium mixing ratios for various atmospheric constituents as a function of temperature and metallicity for an assumed typical photospheric pressure of 0.1 bar. Figure adapted from Moses et al. (2013a)

Miguel et al. (2015) for very high metallicities (e.g., up to and beyond 1000× solar). These investigations reach concensus on several general trends. First, high-metallicity planets will have higher temperatures at lower pressures than otherwise similar low-metallicity planets, due to greater atmospheric opacity from the heavy (i.e., non-H_2) molecular constituents; that is, the photosphere will shift to higher altitudes. Second, the CO/CH_4 and N_2/NH_3 equal-abundance curves shift to lower temperatures with higher metallicities, leading to a higher likelihood that CO and N_2 will be important carbon and nitrogen components, respectively, of a high-metallicity planetary atmosphere. Third, molecules with multiple heavy elements, such as CO, CO_2, N_2, CS, S_2 become favored at the expense of molecules that just contain one heavy element, like CH_4, NH_3, H_2S, as the metallicity increases—an effect that is particularly notable for molecules that contain more than two heavy elements, such as CO_2 and OCS (see Fig. 6). At sufficiently high metallicities, CO_2 will even replace H_2 as the dominant atmospheric constituent for an otherwise solar-composition atmosphere. Molecular hydrogen remains a major constituent of the atmosphere under the conditions studied in Fig. 6, but it ceases to dominate at the highest metallicities considered (e.g., several thousand times solar). Water increases roughly linearly with metallicity until metallicities of ~1000 times solar, at which point the lower H mole fraction begins to adversely impact all H-bearing molecules. The expected depletion of hydrogen in super-Earths and exo-Neptunes due to inefficient accretion of the nebular gas and/or efficient escape of hydrogen at small orbital distances is therefore expected to lead to increased atmospheric mean molecular weights and a wide variety of interesting atmospheric compositions.

The sensitivity-to-metallicity calculations in Fig. 6 and in most of the investigations described in the previous paragraph have been performed assuming that the relative abundances of all elements other than H, He (and sometimes Ne) remain in solar proportion. Of course, that is not likely going to be true for exoplanetary atmospheres, given the different formation scenarios and evolutionary process at work. The sensitivity of the chemical equilibrium composition to changes in the C/O ratio at near-solar metallicities is discussed in Sect. 4.4. Moses et al. (2013a) examine the more general case of the sensitivity of the equilibrium atmospheric composition to both the C/O ratio and metallicity (i.e., to the relative abundances of H, C, and O) as a function of temperature, and Hu and Seager (2014) perform a similar general analysis considering disequilibrium processes like photochemistry (see also Sect. 5.2). Here is where the diversity of potential heavy-element-rich super-Earths and exo-Neptunes really stands out. The dominant equilibrium atmospheric constituent on intermediate-sized planets will typically be H_2 at low-enough metallicities (e.g., less than several hundred times solar) but can become H_2O at moderately high metallicities and subsolar C/O ratios, can become CO_2 at solar-like and subsolar C/O ratios and high metallicities, can potentially become CO at high metallicities and C/O ratios near unity (depending on graphite stability), and can even become O_2 at very low C/O ratios and high metallicities. At high C/O ratios and low metallicities, CH_4 is an important atmospheric component at low temperatures, while HCN and C_2H_2 become more important carbon phases at high temperatures. Hu and Seager (2014) suggest that hydrocarbons like C_2H_2 and C_2H_4 can even become dominant atmospheric constituents at bulk C/O ratios greater than ~2, but they have ignored graphite formation in their calculations. At high C/O ratios, the graphite stability field for equilibrium conditions is greatly expanded (see Moses et al. 2013a), and for a large range of temperature and metallicity conditions, graphite will condense out and sequester a significant fraction of the carbon when bulk C/O ratios are greater than ~1 (or even C/O <1 for high metallicities), leaving the remaining gas much less carbon-rich. Therefore, high-metallicity atmospheres can achieve an unusual state where the dominant gas is CO_2, even for bulk C/O ratios greatly exceeding unity (Moses et al. 2013a). It is also possible that atmospheric O may be lost preferentially by photodissociation of H_2O and subsequent hydrodynamic escape of O, thereby changing the C/O ratio (Chassefière 1996; Luger and Barnes 2015).

Between the potential for hot, silicate- and metal-rich outgassed atmospheres and the variety of volatile compositions available from inefficient accretion of nebular gas, the super-Earths and exo-Neptune population can be expected to have a rich diversity of atmospheric compositions from thermochemical equilibrium considerations, and disequilibrium processes (below) simply augment this possible diversity.

5.2 Disequilibrium Chemistry in Exo-Neptune and Super-Earth Atmospheres

Both photochemistry and transport-induced quenching can affect the atmospheric composition of intermediate-size planets, just as on giant planets (see Sects. 4.2 and 4.3). The first photochemical models specifically applied to intermediate-sized planets were those of Line et al. (2011) for the exo-Neptune GJ 436b and Miller-Ricci Kempton et al. (2012) for the super-Earth GJ 1214b, although the generic, higher-metallicity, hot-Jupiter models studied by Zahnle et al. (2009a,b) should also have relevance to some exoplanets in the intermediate-size range. Thermochemistry, photochemistry, and transport-induced quenching are considered in the Miller-Ricci Kempton et al. (2012) and Line et al. (2011) models, and the planets are assumed to have H_2-rich atmospheres with metallicities up to 30–50× solar. Given that both GJ 1214b and GJ 436b are expected to be relatively cool transiting planets, the results from both models are qualitatively similar. Methane is expected to be the dominant equilibrium carbon species for both planets up to 30–50× solar metallicity for the other conditions considered. At the quench point where interconversion between CH_4 and CO shuts down, methane is the main carbon component. However, transport-induced quenching causes CO to be more abundant than it otherwise would have been in equilibrium. This situation represents the opposite of the case for hotter planets, where quenching in the CO-dominated regime causes CO to be the major carbon component, with methane then being a less-abundant, but still important, quenched component (see Sect. 4.2). Photolysis of methane at high altitudes leads to the production of C_2H_x hydrocarbons and, because of interactions with water photolysis products, the photochemical production of CO and CO_2. Coupled methane-ammonia photochemistry causes the production of HCN. However, methane is not removed from the photospheric region of either planet due to photochemistry (as was suggested as a possibility by Madhusudhan and Seager 2011), which is problematic, given that cloudless H_2-rich models with equilibrium methane abundances do not reproduce transit and eclipse observations of these planets (e.g., Stevenson et al. 2010; Désert et al. 2011; Bean et al. 2011; Berta et al. 2012; Kreidberg et al. 2014a).

This model-data mismatch led Moses et al. (2013a) to suggest that intermediate-sized planets could have much higher metallicities than previously considered, as long as their overall H content remains consistent with constraints supplied by the planet's bulk density. As discussed above, higher metallicities lead to hotter photospheres at lower pressures and shift the atmosphere toward the CO and CO_2 stability fields and away from the CH_4 stability field. Figure 7 shows how the atmospheric composition and photochemistry change for GJ 436b as the metallicity is assumed to increase from 1× solar to 10000× solar. The atmosphere transitions from being hydrogen-dominated, with abundant hydrogen-saturated components like H_2O, CH_4, and NH_3 and photochemically produced hydrocarbons and nitriles at low metallicities, to becoming CO_2-, CO-, H_2O-, and N_2-rich, with more oxidized photochemical products like O_2 and NO at high metallicities. Condensed graphite is a likely cloud component in the very-high-metallicity scenarios. Moses et al. (2013a) find that GJ 436b models with metallicities in the 230–2000× solar range provide the best overall consistency with both the planet's inferred interior structure (see Nettelmann et al. 2010, and references therein) and its apparent CO-rich, CH_4-poor dayside atmosphere and relatively flat transmission spectrum (e.g., Stevenson et al. 2010; Lanotte et al. 2014; Knutson et al. 2014a).

Venot et al. (2013) also consider the thermo/photochemistry of a presumed 100× solar metallicity atmosphere for GJ 436b, focusing in particular on how their new measurements of the CO_2 ultraviolet absorption cross sections at high temperatures (up to 800 K) affect the model abundances, and how stellar type affects the results. Venot et al. (2013) find that their results for the predicted abundances of NH_3, CO_2, and CO are especially affected by the temperature-sensitive CO_2 cross sections, with models that are irradiated by hotter stars (with their corresponding higher near-ultraviolet fluxes) exhibiting the biggest change in abundance due to use of high-temperature CO_2 cross sections.

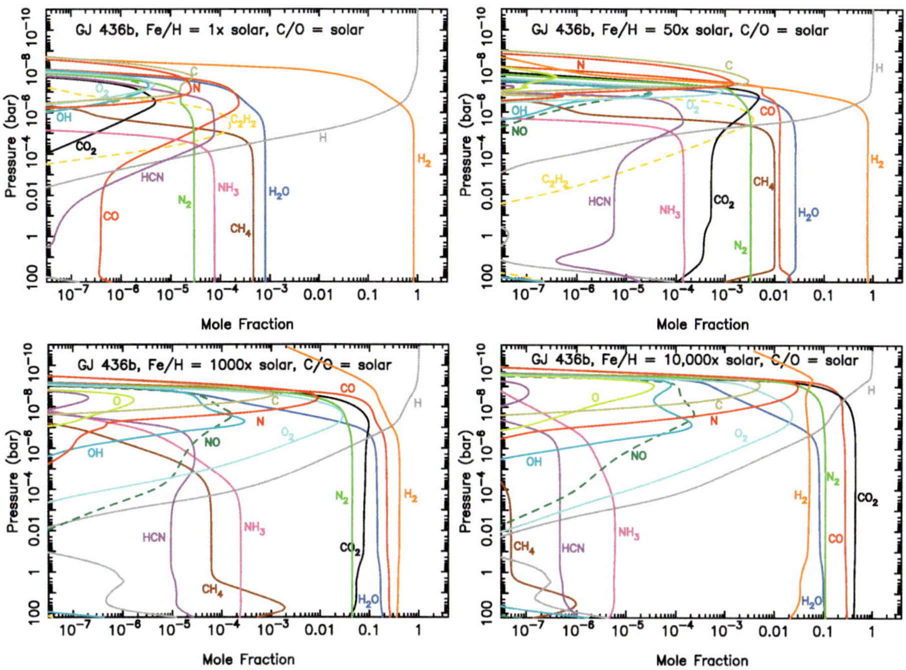

Fig. 7 Vertical mixing-ratio profiles for several major atmospheric constituents in disequilibrium thermo/photochemical models of GJ 436b, for assumed metallicities of 1× solar (*top left*), 50× solar (*top right*), 1000× solar (*bottom left*), and 10000× solar (*bottom right*). Note the major change in the abundance of CO_2, CO, and CH_4 with the increase in metallicity (figure from Moses et al. 2013a)

Thermo/photochemical models for GJ 436b are also presented by Agúndez et al. (2014b) and Miguel et al. (2015). Agúndez et al. (2014b) explore the influence of tidal heating and metallicity on the thermal structure and resulting disequilibrium chemistry of planet, calculating the thermal structure self-consistently. Both greater tidal heating at depth and higher metallicities help shift the thermal structure into the CO dominated regime, again helping favor higher CO/CH_4 ratios and a greater CO_2 abundance. Miguel et al. (2015) examine the sensitivity of the atmospheric composition of GJ 436b to the flux in the stellar Lyman alpha line for both low-metallicity and high-metallicity models. They find that constituent abundances in the upper stratosphere at pressures less than ~0.1 mbar are affected by Lyman alpha, with stronger fluxes favoring the destruction of molecules and the resulting dominance of atomic species. Note, however, that when one considers the formation and presence of an extended hot thermosphere (García Muñoz 2007; Koskinen et al. 2013), which is the likely consequence of strong X-ray and EUV radiation received by close-in extrasolar planets—including GJ 436b (see Sanz-Forcada et al. 2011; Koskinen et al. 2014)—much or all of the Lyman alpha flux itself could be absorbed within the thermosphere (Lavvas et al. 2011), never reaching the stratosphere. On the other hand, other strong X-ray and EUV lines could have a similar effect to that described in Miguel et al. (2015), ultimately moving the base of the atomic-dominated hot thermosphere to deeper pressures for stronger EUV and X-ray fluxes.

Hu and Seager (2014) have examined the sensitivity of the disequilibrium atmospheric composition of several intermediate-sized exoplanets (e.g., GJ 1214b, HD 97658b, and 55 Cnc e) to the general variation of H, O, and C elemental abundances, using a

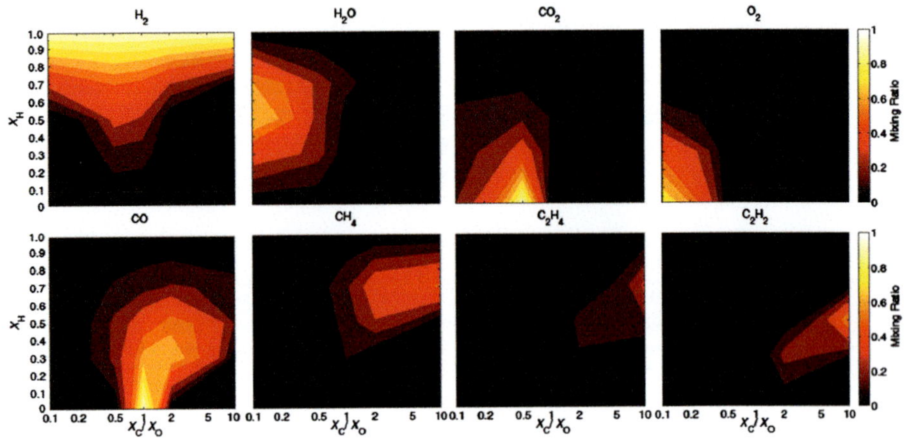

Fig. 8 Mixing ratios for important atmospheric constituents vertically averaged over the pressure range 1–100 mbar from a disequilibrium thermo/photochemistry model of the super-Earth GJ 1214b, as a function of the bulk atmospheric H mole fraction (X_H, where smaller values correspond to higher metallicities) and the C/O ratio (X_C/X_O). Figure from Hu and Seager (2014)

thermo/photochemistry model that self-consistently calculates temperatures. Figure 8 shows their results for GJ 1214b. These results are similar to the equilibrium results shown in Fig. 6 in that the atmospheric composition can be highly diverse, depending on the relative abundances of the key volatile elements H, C, and O. At low metallicities, H_2 dominates, but GJ 1214b's atmosphere could also be dominated by H_2O, CO_2, CO, or O_2, depending on the overall H mole fraction and C/O ratio. Hu and Seager (2014) suggest that the atmosphere could be dominated by hydrocarbons such as C_2H_2 and C_2H_4 for high C/O ratios and moderately high metallicities; however, graphite condensation has been neglected in their calculations. Unless graphite formation is somehow kinetically inhibited (see Moses et al. 2013a), condensation of graphite will tie up much of the excess carbon, causing the remaining vapor to have lower C/O ratios, disfavoring hydrocarbons as the dominant constituents. Hu and Seager (2014) also show how the results for intermediate-sized planets can change with stellar flux (and thus temperature), and they suggest classification schemes for super-Earths with thick atmospheres based on the dominant constituents that appear for different H mole fractions and C/O ratios. In an earlier study, Hu et al. (2013) examine the fate of sulfur species from photochemistry in atmospheres of different dominant compositions relevant to super-Earths.

Venot et al. (2014) have investigated the disequilibrium chemistry of the warm exo-Neptune GJ 3470b, using a grid of disequilibrium models with various metallicities, thermal profiles, incident ultraviolet fluxes, and eddy diffusion coefficients (specifying the strength of atmospheric mixing) to determine the possible sensitivity of the atmospheric composition to these parameters. They find that, like GJ 1214b and GJ 436b, methane is favored over CO as the dominant atmospheric constituent, except when both metallicities (up to $100\times$ solar) and temperatures are high.

Conspicuously absent from the current literature is a study of how ion chemistry affects the gas-phase composition and possible formation of hazes on super-Earths and exo-Neptunes. Given that ion chemistry initiates the formation of high-molecular-weight organics in the N_2- and CH_4-rich upper atmosphere of Titan (Waite et al. 2007), it seems likely that ion chemistry would be interesting on intermediate-sized exoplanets (especially the cooler ones).

5.3 Clouds and Hazes in Exo-Neptune and Super-Earth Atmospheres

Basic concepts of cloud and haze formation in extrasolar-planet atmospheres are discussed in Sect. 4.5. The same physics and chemistry that was described for giant planets is relevant to intermediate-sized planets, but the different starting ingredients can lead to different possible aerosol compositions. Morley et al. (2013) provide a good discussion of the possible equilibrium condensates along the standard cloud condensation sequence for H_2-rich super-Earths and exo-Neptunes. In order from the hottest to the coldest dominant condensates, H_2-rich atmospheres could typically contain equilibrium clouds of Al-Ca-Ti oxides and silicates, Fe metal, Mg silicates, Cr metal, MnS, Na_2S, ZnS, KCl, $NH_4H_2PO_4$, H_2O, NH_4SH, NH_3, and CH_4. For a solar-metallicity atmosphere, the total available mass for some of these clouds (e.g., ZnS, KCl) is pretty sparse, suggesting that they would not be very optically thick in the vertical; however, Morley et al. (2013) demonstrate that this conclusion changes as the metallicity is increased. They find that ZnS and KCl clouds could obscure the transit spectra of a metal-rich GJ 1214b if the particles are lofted to sufficiently high altitudes and have sedimentation times that are sufficiently long (e.g., due to small particle sizes).

Other possibilities for aerosols on moderately H_2-rich planets include high-molecular-weight organics or "soots" that form through photochemical processes (e.g., Zahnle et al. 2009a; Moses et al. 2011; Miller-Ricci Kempton et al. 2012; Morley et al. 2013), or graphite that is stable in equilibrium for metal-rich atmospheres or ones with high C/O ratios (Moses et al. 2013a). Graphite can also be stable for a range of other compositions relevant to outgassed atmospheres of intermediate-sized planets (Schaefer and Fegley 2009, 2010), as can various other equilibrium condensates, such as alkali salts (Schaefer et al. 2012). Formation of sulfuric acid (H_2SO_4) clouds is likely through photochemical processes under a wide variety of conditions for high-metallicity or CO_2-rich atmospheres that have sufficient SO_2 (e.g., Schaefer and Fegley 2011; Schaefer et al. 2012; Hu et al. 2013), and elemental sulfur aerosols (e.g. S8) can be photochemically produced (e.g., Hu et al. 2013). However, the detailed formation pathways of photochemical hazes within the diverse atmospheres of intermediate-size planets has received relatively little attention and is still poorly understood.

6 Observational Inferences of Chemical Compositions

The last decade has witnessed substantial progress in observational inferences of chemical species in the atmospheres of giant exoplanets through a variety of methods. The planets for which such detections have been made are mostly hot giant planets, either in close-in orbits (i.e. 'hot Jupiters') or directly-imaged young objects at large orbital separations. The methods employed include differential photometry and spectroscopy of transiting hot Jupiters, high-resolution Doppler spectroscopy of transiting and non-transiting hot Jupiters, and high-resolution spectroscopy of directly-imaged planets. Whereas atomic species have been detected across the ultraviolet and visible, molecular species have been detected primarily in the near infrared. These detections were made thanks to pioneering observations using *Spitzer*, *HST*, and ground-based facilities. In what follows, we review the developments in each of these various areas.

6.1 Chemical Detections via Transit Spectroscopy

Atmospheric observations have been reported for over 50 transiting exoplanets to date, mostly as broadband/narrowband photometry or low resolution spectra. But, given the limited spectral resolution of the observations of most planets, molecular compositions have been inferred for only a few exoplanets to date. The inferred molecules typically include the most abundant and spectroscopically dominant molecules expected in hot atmospheres e.g. H_2O, CO, CH_4, and CO_2. On the other hand, inferences have also been made of several atomic species and the presence of clouds/hazes in some atmospheres.

6.1.1 Detections of Atomic Species

In high-temperature gas giant atmospheres ($T \gtrsim 1000$ K) in chemical equilibrium, alkali metals sodium (Na) and potassium (K) are expected to exist predominantly in atomic form thereby causing significant absorption in the visible via strong resonance lines at 589 nm and 770 nm, respectively (Seager and Sasselov 2000; Brown 2001). The strong line cores and wide pressure broadened wings are observable in optical transmission spectra of hot Jupiters. Thus the first Na detection was made using a visible band transmission spectrum of the hot Jupiter HD 209458b obtained using the HST STIS spectrograph (Charbonneau et al. 2002). However, while unambiguous, the spectrum revealed lower Na absorption than predicted by equilibrium models, indicating either lower atomic abundances than assumed or non-equilibrium processes such as rainout of condensed species and photoionization of Na/K (Fortney et al. 2003; Barman 2007).

The Na resonance doublet was also detected from ground in the optical transmission spectrum of the hot Jupiter HD 189733b (Redfield et al. 2008; Wyttenbach et al. 2015). Additionally, HST STIS observations of the transmission spectrum of HD 189733 also detected the Na absorption line core (Huitson et al. 2012). However, the ensemble of observations spanning UV to visible showed that the Na line is significantly weaker than that observed for HD 209458b, with a clear lack of broad line wings. Besides the weak Na I line, the spectrum was found to be largely featureless with a blue-ward slope which was consistent with the presence of strong scattering due to a thick haze of condensate grains (Vidal-Madjar et al. 2011; Pont et al. 2013). In recent years, Na I has been inferred in visible transmission spectra of a few other hot Jupiters, e.g. XO-2b (Sing et al. 2012), WASP-17b (Wood et al. 2011; Zhou and Bayliss 2012), HAT-P-1b (Nikolov et al. 2014), and WASP-12b (Burton et al. 2015).

Visible transmission spectra of hot Jupiters have also revealed other atomic species. Similar to Na, several studies have also detected the K resonance double at 770 nm using transmission spectra of hot Jupiters from both space and ground-based instruments, e.g., XO-2b (Sing et al. 2011a), and HD 80606 (Colón et al. 2012), HAT-P-1b (Wilson et al. 2015), WASP-31b (Sing et al. 2015). Other atomic species inferred in the visible include H α at 656 nm (Jensen et al. 2012), Ca and Sc (Astudillo-Defru and Rojo 2013).

Several atomic species have also been detected in the exospheres of hot Jupiters using UV transmission spectroscopy. The first such observations in Ly α revealed an extended envelope of escaping H in the hot Jupiter HD 209458b (Vidal-Madjar et al. 2003). Subsequent observations have revealed a rich population of atomic species in several hot Jupiter exospheres, e.g. H (Lecavelier Des Etangs et al. 2010; Bourrier et al. 2013), O and C (Vidal-Madjar et al. 2003), Si (Linsky et al. 2010; Schlawin et al. 2010), Mg (Fossati et al. 2010; Vidal-Madjar et al. 2013). Most recently, Ehrenreich et al. (2015) reported a detection of escaping atomic hydrogen in the exosphere of the hot Neptune GJ 436b.

6.1.2 Early Molecular Inferences

Early inferences of molecules in exoplanetary atmospheres were based on few channels of photometry or low-resolution spectra obtained using then available instruments on *Spitzer* and *HST*. For example, some early studies used 2–3 near-infrared photometric observations in transmission to infer the presence of H_2O at the day-night terminator regions of HD 209458b (Barman 2007) and HD 189733b Tinetti et al. (2007), but cf. Beaulieu et al. (2008) and Désert et al. (2009). Early attempts were also made to detect molecules in a handful of hot Jupiters using infrared spectroscopy with *HST* and *Spitzer*. For example, several studies used near-infrared spectra of hot Jupiters in transmission and/or emission obtained with the *HST* NICMOS instrument (1.8–2.3 μm) to report detections of H_2O, CH_4, CO, and/or CO_2 in the hot Jupiters HD 189733b (Swain et al. 2008, 2009b; Madhusudhan and Seager 2009), HD 209458b (Swain et al. 2009a), and XO-1b (Tinetti et al. 2010). However, the uncertainties on the NICMOS observations have since been extensively debated in the literature leading to different molecular detection significances claimed by different teams, ranging from confident detections to no detections at the nominal 3-σ significance (Gibson et al. 2011, 2012; Waldmann et al. 2013; Swain et al. 2014).

Similar molecular inferences were also made using observations over a longer spectral baseline using *Spitzer* photometry and spectroscopy. Grillmair et al. (2008) used *Spitzer IRS* spectroscopy to infer the presence of H_2O in the dayside atmosphere of HD 189733b using chemical equilibrium models (also see Todorov et al. 2014), whereas (Madhusudhan and Seager 2009) were able to place only an upper-limit on the possible abundances of H_2O using the same dataset. Using six-channel photometric and/or spectroscopic observations from *Spitzer* and/or *HST*, Madhusudhan and Seager (2009) also inferred the presence of H_2O, CH_4, CO, and/or CO_2 in the hot Jupiters HD 189733b and HD 209458b. However, several of the *Spitzer* photometric observations have also since been revised drastically (Knutson et al. 2012; Diamond-Lowe et al. 2014), thanks to our improved understanding of the instrumental systematics and, hence, the observational uncertainties (but cf. Hansen et al. 2014). Therefore, early inferences based on such observations are not currently substantiated.

The above early inferences, were followed by more molecular inferences from multi-band photometry for smaller and/or more distant planets. Stevenson et al. (2010) and Madhusudhan and Seager (2011) used six-channel *Spitzer* photometry in thermal emission to infer the presence of CO and/or CO_2 and the absence of CH_4 in the dayside atmosphere of the hot Neptune GJ 436b. Given the relatively lower equilibrium temperature of the planet, the presence of CO and absence of CH_4 was suggested to be indicative of strong chemical disequilibrium and high metallicity in the atmosphere. While these data, and hence the molecular inferences, were originally contested by Beaulieu et al. (2011), recent comprehensive and independent analyses by Lanotte et al. (2014) support the original inferences of a methane poor and CO/CO_2 rich atmosphere in GJ 436b.

Another major observational advancement that followed was the possibility of detecting thermal emission from hot Jupiters in the near-infrared from ground (e.g. Sing and López-Morales 2009; Croll et al. 2010). The advantage of these measurements was that their spectral range, between ~0.9 μm and 2.3 μm, complemented that typically available from *Spitzer* photometry (3.6–8 μm) to provide a long spectral baseline to retrieve chemical species from the combined data. In the first such instance, observations of thermal emission in seven photometric bandpasses (between ~1.2–8 μm) were used to infer the presence of significant CO and CH_4 and lack of H_2O, and hence a C/O ratio ≥ 1, in the hot Jupiter WASP-12b (Madhusudhan et al. 2011a). Again, the *Spitzer* photometric observations were a subject of intense debate initially (Crossfield et al. 2012; Cowan et al. 2012a). However, subsequent multi-epoch *Spitzer* observations in the same bandpasses have reinstated the original data and the conclusions (Stevenson et al. 2014a) for the dayside atmosphere of WASP-12b.

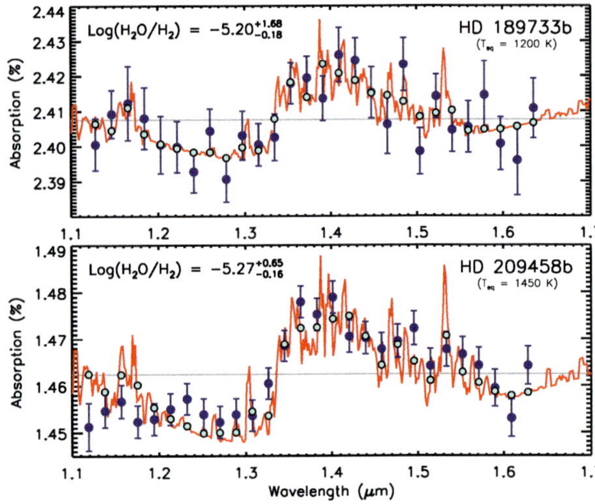

Fig. 9 Transmission spectra of two hot Jupiters observed with HST WFC3 (adapted from Madhusudhan et al. 2014b). The *vertical axis* shows absorption (transit depth). The *blue circles* show the data: HD 209458b from Deming et al. (2013) and HD 189733b from McCullough et al. (2014). The *red curves* shows the best-fit model spectra, and the *cyan circles* show the models binned to the same resolution as the data. The *peaks* around 1.4 μm show H_2O absorption

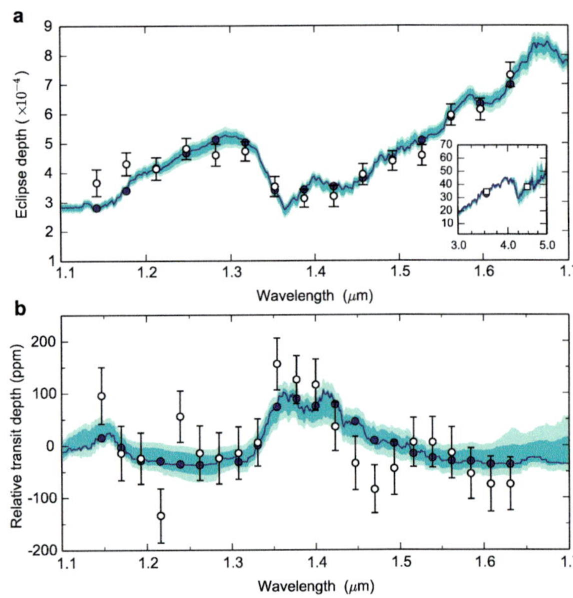

Fig. 10 Spectra of the hot Jupiter WASP-43b observed with HST WFC3 (from Kreidberg et al. 2014b). *Top*: eclipse spectrum. *Bottom*: transmission spectrum. In both cases, the *black circles with error bars* show the data and *solid curves* show best-fit model spectra. The *filled dark blue circles* show the best-fit model binned to the same resolution as the data

6.1.3 Molecular Detections with HST WFC3

In recent years, near-infrared *HST* spectroscopy has led to detections of H_2O in the atmospheres of several transiting hot Jupiters. This has been made possible by the *HST* Wide Field Camera 3 (WFC3) spectrograph (McCullough and MacKenty 2012) which operates in the near-infrared range 1.1–1.8 μm, where water vapor has a strong absorption band, as shown in Figs. 9 and 10. Since H_2O is one of the most abundant molecules expected in giant exoplanetary atmospheres, the HST WFC3 instrument provides a unique opportunity to constrain H_2O abundances in such atmospheres. As shown in Fig. 9, Deming et al. (2013) reported the first WFC3 detection of H_2O in the transiting hot Jupiter HD 209458b using transmission spectroscopy, which probed the atmosphere at the day-night terminator region of the planet. The observations led to a clear 10-σ detection of H_2O in a single transit thanks to the high brightness of the host star ($V = 7.6$).

Several subsequent studies reported single-event transmission spectra of transiting hot Jupiters orbiting less brighter stars with varied levels of success in H_2O detections. Some of the notable H_2O detections, albeit less significant than that for HD 209458b, were reported for hot Jupiters WASP-17b (Mandell et al. 2013), WASP-19b (Huitson et al. 2013), HAT-P-1b (Wakeford et al. 2013), WASP-12b (Stevenson et al. 2014a,b), and HD 189733b (McCullough et al. 2014), and a hot Neptune HAT-P-11b (Fraine et al. 2014), whereas similar observations of several other planets resulted in featureless spectra within the observed uncertainties (e.g. Swain et al. 2013; Mandell et al. 2013; Ranjan et al. 2014). These efforts demonstrate that single transits are inadequate to make high-confidence H_2O detections for most hot Jupiters currently known with the exception of those transiting the brightest stars such as HD 209458b. Consequently, more recent observations using multiple visits to reduce the observational uncertainties have led to good quality transmission spectra with clear H_2O detections for hot Jupiters orbiting even moderate brightness stars (Kreidberg et al. 2014a, 2015).

Several studies have also used *HST* WFC3 to observe thermal emission spectra of hot Jupiters at secondary eclipse. In addition to the presence of molecular abundances, thermal emission spectra also provide constraints on the disk-averaged temperature profile of the dayside atmosphere. However, with the exception of a few cases WFC3 observations to date have generally revealed thermal spectra with subdued molecular features, if any. The first such spectrum was observed for the very hot Jupiter WASP-12b ($T_{\rm eq} \sim 2500$ K) which revealed a featureless thermal spectrum consistent with a blackbody spectrum (Swain et al. 2013; Madhusudhan 2012; Stevenson et al. 2014a). Considered on its own, the spectrum is consistent with an isothermal temperature structure in the atmosphere. On the other hand, considering together with existing photometric observations in the range 0.9–8 μm the observations require an atmosphere with a temperature profile decreasing outward and significantly depleted in H_2O as expected for a carbon-rich atmosphere, $C/O \geq 1$ (Madhusudhan 2012; Stevenson et al. 2014a). Subsequent studies have observed WFC3 thermal emission spectra for several other hot Jupiters (Wilkins et al. 2014; Ranjan et al. 2014; Stevenson et al. 2014a; Crouzet et al. 2014; Haynes et al. 2015).

Robust detections of molecular features in thermal emission spectra using HST WFC3 have been reported for the dayside atmospheres of only two hot Jupiters to date. Stevenson et al. (2014c) and Kreidberg et al. (2014b) reported unambiguous detection of H_2O in the dayside atmosphere of WASP-43b, as shown in Fig. 10. Most recently, Haynes et al. (2015) reported a detection of H_2O and the first detection of TiO in the dayside atmosphere of the extremely irradiated hot Jupiter WASP-33b.

6.1.4 Abundance Estimations Using Retrieval Methods

Going beyond detecting the presence of molecules, the exoplanetary spectra discussed above have also been used to derive the molecular abundances which in turn can be used to constrain atmospheric processes, elemental abundances, bulk compositions, and formation conditions. Molecular abundances are derived from exoplanetary spectra using detailed atmospheric retrieval methods which lead to joint statistical constraints on the chemical composition and temperature profile of an exoplanetary atmosphere given an observed spectrum; see e.g. Madhusudhan et al. (2014c) for a detailed review of atmospheric retrieval techniques for exoplanets. Atmospheric retrieval methods for exoplanets typically comprise of a 1-D atmospheric model coupled with an optimization algorithm to estimate the free parameters of the model given the data. The molecular abundances and the pressure-temperature (P-T) profile are free parameters in the model; typically there are over ten free parameters depending on the number of molecules included and the adopted parametrization of the temperature profile. For the optimization algorithm, a number of methods have been tried over time ranging from grid-search in the early days (Madhusudhan and Seager 2009) to Bayesian approaches such as the Markov Chain Monte Carlo (MCMC) method in subsequent years (Madhusudhan et al. 2011a; Line et al. 2012; Benneke and Seager 2013; Waldmann et al. 2015) and gradient descent methods (Lee et al. 2012).

Chemical abundances have been retrieved for several giant exoplanets to date. The majority of these planets are transiting hot Jupiters for which high-precision transmission and/or emission spectra have been obtained using the HST WFC3 spectrograph and, in some cases, Spitzer and/or ground-based photometry. Initial statistical constraints on atmospheric abundances (e.g. Madhusudhan and Seager 2009; Lee et al. 2012; Line et al. 2012) in the pre-WFC3 era were based on photometric data and/or low-resolution spectra, the data and uncertainties of which have since been revised as discussed above. On the other hand, while robust HST WFC3 spectra have been observed for over a dozen planets now, most of them have uncertainties large enough to be consistent with featureless spectra (Mandell et al. 2013; Ranjan et al. 2014). Therefore, in what follows we review only a selection of observed transiting exoplanets for which non-flat spectra have been observed at $\geq$3-σ significance and for which abundance constraints have been reported. This includes five transiting hot Jupiters (HD 209458b, HD 189733b, WASP-12b, WASP-33b, WASP-43b), and a hot Neptune (HAT-P-11b).

Currently, the molecule with the best abundance estimates available is H_2O, thanks to high-precision spectra with HST WFC3. As discussed above, the HST WFC3 G141 grism with a spectral range of 1.1–1.8 μm covers a strong H_2O absorption band near 1.4 μm thereby allowing H_2O measurements from exoplanetary spectra. H_2O has been detected in all the six giant exoplanets listed above, albeit with different abundances. One important feature in almost all the transmission spectra with robust H_2O detections to date is that the amplitude of the H_2O absorption feature is typically smaller than model predictions for a clear (i.e. free of clouds/hazes) solar-abundance atmosphere. This was first noted for the hot Jupiter with the most-precise transmission spectrum, HD 209458b (Deming et al. 2013). For such spectra, assuming a clear atmosphere results in an H_2O abundance at the day-night terminator that is markedly sub-solar: 20–100$\times$ sub-solar for HD 209458b and 3–200$\times$ sub-solar for HD 189733b (Madhusudhan et al. 2014b). However, the derived low H_2O abundances are degenerate with the possibility of clouds or hazes at the terminator in these atmospheres (Sing et al. 2013; Benneke 2015), i.e. the same spectra can be explained with solar or super-solar abundances if high altitude clouds or hazes are allowed in the atmosphere. Transmission spectra of several other hot Jupiters were also found to be consistent with solar abundances when clouds/hazes are invoked or when the uncertainties are large enough to preclude high-precision abundance estimates (Huitson et al. 2013; Line et al. 2013; Barstow et al. 2014; Benneke 2015). Therefore, the possibility of clouds present a fundamental challenge in deriving chemical abundances from transmission spectra, which may be mitigated if optical transmission spectra are available (discussed in the following section) and/or where abundance estimates are available for the dayside atmospheres using thermal spectra.

Constraints on molecular abundances have also been reported in the dayside atmospheres of hot Jupiters using thermal emission spectra. An early example in this regard is the hot Jupiter WASP-12b, for which broadband photometric observations originally suggested a 3-σ upper-limit on the H_2O abundance that was $\sim$100$\times$ sub-solar, and implied a C/O ratio of $\geq$1 (Madhusudhan et al. 2011a). While the veracity of the Spitzer photometric observations of WASP-12b were a subject of substantial debate in the intervening years (Cowan et al. 2012a; Crossfield et al. 2012), latest multi-epoch Spitzer observations together with high-precision HST WFC3 spectra are consistent with the lack of significant H_2O in the dayside atmosphere of WASP-12b (Madhusudhan 2012; Stevenson et al. 2014b). Most recently, H_2O was detected at the day-night terminator region of the planetary atmosphere using a HST WFC3 transmission spectrum, however the abundance of H_2O detected was subject to model assumptions as discussed above (Kreidberg et al. 2015). Similar to the other datasets discussed above, a cloud-free atmosphere required a sub-solar H_2O abundance whereas the inclusion of clouds/hazes and/or the imposition of chemical equilibrium allowed the data to be consistent with solar abundance H_2O.

In another example, joint constraints on the H_2O abundance at both the terminator as well as the dayside atmosphere were obtained using transmission and emission spectra for the hot Jupiter WASP-43b resulting in an H_2O abundance of 0.3–3$\times$ solar, i.e. both sub-solar and marginally super-solar abundances are allowed by the data (Kreidberg et al. 2014b). Finally, abundance estimates for the dayside atmosphere of WASP-33b, the most irradiated hot Jupiter observed with HST WFC3 to date, also revealed a marginally sub-solar ($\sim$0.5$\times$) H_2O abundance (Haynes et al. 2015). HST WFC3 thermal spectra for a few other hot Jupiters have also revealed relatively low-amplitude or non-existent H_2O features, hinting at the possibility of low H_2O abundances (e.g. Wilkins et al. 2014; Crouzet et al. 2014). While in principle clouds or hazes could also cause featureless (blackbody-like) spectra in thermal emission, the observed brightness temperatures of the spectra are quite high, indicative of the lower atmospheres, implying that they were unlikely to have been impeded by clouds/hazes in the atmosphere.

The observed molecular abundances can be used to constrain the O/H, C/H, and C/O ratio of the atmosphere (e.g. Madhusudhan et al. 2011a), which in turn are important to constrain exoplanetary formation conditions, as discussed in Sect. 6. However, estimating C/O ratios is non-trivial. Reliably estimating the C/O ratio for an exoplanetary atmosphere requires that the observations are able to constrain the molecular abundances of all the dominant O and C bearing molecules in a given atmosphere. As discussed in Sect. 3, for hot giant exoplanets, the dominant molecules are H_2O, CO, CH_4, and to a lesser extent CO_2, C_2H_2, and HCN, depending on the C/O ratio. Simultaneously constraining the abundances of all these molecules requires a long spectral baseline with observable bandpasses where these molecules have strong spectral features. This is, in principle, possible with the combination of HST WFC3 and the Spitzer IRAC Channels 3.6–8 μm (Madhusudhan 2012). Indeed, such constraints were placed for the few planets where such data are available in thermal emission allowing constraints on the C/O ratios of their dayside atmospheres. While a carbon-rich dayside atmosphere (C/O $\geq$ 1) has been reported for the hot Jupiter WASP-12b (Madhusudhan 2012; Stevenson et al. 2014b), oxygen-rich dayside atmospheres consistent with a solar C/O of 0.5 were reported for the hot Jupiters WASP-43b (Kreidberg et al. 2014b) and WASP-33b (Haynes et al. 2015).

On the other hand, determining the C/O ratio at the day-night terminator from transmission spectra is challenging and prone to misinterpretation. One of the main reason is that typically only WFC3 spectra are available in transmission, implying that stringent constraints are possible only on the H_2O abundance. In principle, the abundances of CH_4 and HCN may also be constrained in the WFC3 spectral range, however the most dominant O and C bearing molecule in hot Jupiters is CO, which is unconstrained by WFC3 observations. Consequently, with WFC3 data alone the C/H and hence the C/O ratio cannot be reliably constrained. For example, if a sub-solar H_2O abundance is observed in a hot Jupiter (assuming a cloud-free atmosphere), the low H_2O can be caused either by a low overall

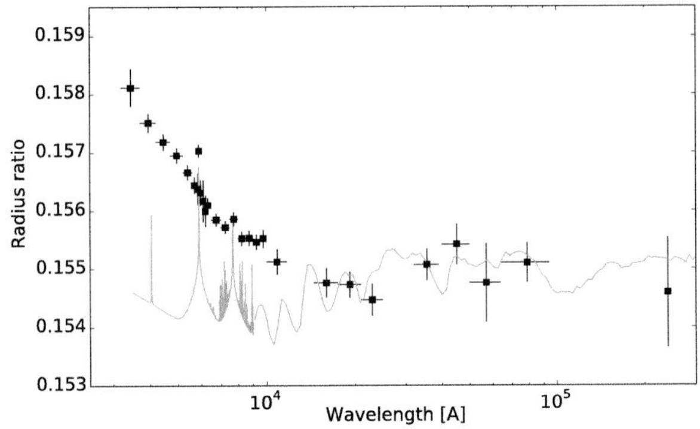

Fig. 11 Transmission spectrum of the hot Jupiter HD 189733b from UV to near-infrared (from Pont et al. 2013). The *black circles with error bars* show the data available at that time, observed using multiple instruments on *HST* and *Spitzer*. The *grey curve* shows a model spectrum for a dust-free atmosphere. The *steep blue-ward rise* in the spectrum was indicative of strong scattering due to haze/dust in the atmosphere which was not consistent with the expected model spectrum for a haze-free atmosphere

metallicity (i.e. low O/H) with a solar-like C/O ratio or by a high metallicity but a high C/O ratio ($\sim$1), depending on the temperature. Therefore, studies which attempt to derive C/O ratios from WFC3 transmission spectra alone do so by enforcing constraints of equilibrium chemistry or temperature profiles in radiative equilibrium or both (Benneke 2015; Kreidberg et al. 2015), neither of which is strictly justified at the day-night terminator regions of strongly irradiated atmospheres with strong atmospheric circulation. On the other hand, the presence of clouds only makes the inference more degenerate. Therefore, multi-wavelength observations besides WFC3 spectra are essential for reliable constraints on C/O ratios from transmission spectra.

6.1.5 Degeneracy with Clouds and Hazes

The presence of high-temperature condensates as hazes and clouds in hot exoplanetary atmospheres may contribute significantly to the optical depth in transmission spectra (Seager and Sasselov 2000; Fortney 2005). Therefore, determinations of chemical compositions from transmission spectra can be strongly degenerate with the presence of clouds or hazes in the atmospheres. In transmission spectra, the amplitude of a spectral feature is directly proportional to the height of the atmospheric annulus through which the star light traverses. The presence of an opaque cloud deck at a given height in the atmosphere means that only the layers above the cloud deck contribute to the transmission spectrum. Thus for a cloud deck located high enough in the atmosphere the amplitude of a spectral feature can be significantly diminished thereby confounding estimates of the corresponding molecular abundances.

High-precision observations of several transiting exoplanets have revealed such spectra with diminished features. At the extreme end are recent HST WFC3 observations of remarkably flat spectra, despite extremely high precisions, for super-Earths GJ 1214b (Kreidberg et al. 2014a) and HD 97658b (Knutson et al. 2014b), exo-Neptune GJ 436b (Knutson et al. 2014a), and exo-Uranus GJ 3470b (Ehrenreich et al. 2014). The featureless transmission spectra in these low-mass, low-temperature (T below $\sim$800 K), planets have been attributed to the presence of thick high-altitude clouds with cloud-top pressures below a mbar. On the other hand, even for hot Jupiters ($T \sim$ 1200–3000 K) which show unambiguous H_2O features the amplitude of the spectral features are significantly diminished, corresponding to only $\sim$1–2 scale heights (Deming et al. 2013; Kreidberg et al. 2015), compared to expectations of 5–8 scale heights (Madhusudhan and Redfield 2015). These diminished spectral features may be attributed to the presence of clouds/hazes in these hot Jupiters, but as discussed above the same features may also be attributed to significantly lower H_2O abundances than previously expected (Madhusudhan et al. 2014c; Benneke 2015).

Scattering due to haze particles have also been inferred from visible transmission spectra of hot Jupiters which show an inverse power-law dependence on wavelength with parameters different from that due to pure gaseous Rayleigh scattering. The archetype for such inferences is the hot Jupiter HD 189733b (Lecavelier Des Etangs et al. 2008b; Pont et al. 2008; Sing et al. 2011b) where observations with HST revealed a power law slope in the UV and optical indicating the presence of hazes, as shown in Fig. 11. Similar inferences have been made for several other hot Jupiters in recent years (Sing et al. 2013, 2015; Nikolov et al. 2015). A potential degeneracy in such inferences arises from the fact that the presence of star spots may also cause similar optical spectra with inverse power law slopes (e.g. McCullough et al. 2014). However, in addition to the power-law slopes observations of partially muted alkali (Na and/or K) absorption lines in the optical have also indicated the presence of cloud decks in some atmospheres (Sing et al. 2013, 2015).

6.1.6 Albedos and Optical Phase Curves

The presence of clouds/hazes have also been inferred using reflection spectra of hot Jupiters obtained at occultation. Early observations presented little evidence for clouds/hazes and suggested that hot Jupiters are dark, potentially due to strong absorption in the optical due to alkali metals Na and K. Such studies reported rather low geometric albedos for several hot Jupiters in the optical, e.g. 0.038 ± 0.045 for HD 209458b (Rowe et al. 2008) and 0.0136 ± 0.0027 for TrES-2 (Kipping and Spiegel 2011). However, recent new observations are providing evidence for high geometric albedos and evidence for clouds/hazes in dayside spectra of hot Jupiters. Demory et al. (2011, 2013) reported a high visible geometric albedo of 0.35 ± 0.02 for the hot Jupiter Kepler-7b observed in the Kepler photometric bandpass ($\sim$0.4–1.0 μm). Furthermore, based on the asymmetry in the visible phase curve they also suggested the presence of an inhomogeneous cloud cover in the atmosphere of Kepler-7b. In another study, Evans et al. (2013) reported a low-resolution albedo spectrum of the hot Jupiter HD 189733b in the 0.29–0.57 μm range using HST which nominally showed the albedo increasing blue-ward and inferred the presence of reflective clouds in the atmosphere; also see Barstow et al. (2014). Most recently, Martins et al. (2015) reported a high-resolution spectroscopic detection of reflected light for the non-transiting planet 51 Peg b which suggests a high geometric albedo ($\sim$0.5) of the planet. Several other recent studies have used photometric observations to understand the trends and/or report constraints on albedos and/or optical phase curves for a sizeable sample of exoplanets (e.g. Heng and Demory 2013; Esteves et al. 2015; Hu et al. 2015a; Schwartz and Cowan 2015).

While it is apparent from the above observations that clouds/hazes likely exist in at least a subset of irradiated atmospheres, the compositions of cloud forming species in currently unconstrained. However, as discussed in Sects. 3–4, theoretical studies over the past decade show that a wide range of high-temperature refractory condensates can form in such atmospheres in the 500–2000 K range. Such condensates range from NaCl, KCl, ZnS for T below 1000 K to silicates, Fe, and Al_2O_3 at $\sim$1000–2000 K temperatures (e.g. Sudarsky et al. 2003; Marley et al. 2013; Morley et al. 2013).

6.2 Molecular Detections via High-Resolution Doppler Spectroscopy

Robust detection of molecules in exoplanetary atmospheres have also been made using high-resolution Doppler spectroscopy in the near infrared. This technique involves the detection of molecular lines in the planetary spectrum that are shifted in wavelength due to the radial velocity of the planet (Brogi et al. 2012; Birkby et al. 2013). For close-in hot Jupiters, the orbital velocities are $\sim$ km s^{-1} whereas the stellar orbital velocities are significantly lower (e.g. below $\sim$100 m s^{-1}). Thus the spectrally shifted molecular lines of the planetary atmosphere are easily identifiable compared to those in the stellar spectrum as well as those in the telluric spectrum which is static. A template planetary spectrum including the sought after molecule is cross-correlated with the observed spectrum to detect the Doppler shift in the molecular lines with phase thereby revealing the presence of the molecule. Critical to this method, however, is the high resolution of the observed spectrum so that individual molecular lines can be resolved. Consequently, most of the success in this area has been achieved using the CRIRES instrument on the Very Large Telescope (VLT) with a spectral resolving power of 100000 in the near-infrared (Snellen et al. 2010). Moreover, as the planetary signal diminishes with the increased resolution the method has been successfully applied only to planets orbiting the brightest stars.

This method has been used to robustly detect CO and H_2O in several hot Jupiter atmospheres. The detections of these particular molecules are favored by the fact that they are expected to be the most dominant O and C bearing molecules in hot Jupiter atmospheres, especially for $T > 1300$ K, and also have detectable absorption lines in the range $\sim$2–3 μm where these atmospheres are most conducive to observe from ground. This wavelength range offers the optimal conditions because in the near-infrared the planet-star flux ratio increases with wavelength, however the background noise also increases with wavelength; therefore the K-band (around 2.1 μm) typically offers an optimal choice for ground based transit/eclipse spectroscopy. Snellen et al. (2010) reported the first detection of CO using this technique in the hot Jupiter HD 209458b in transit which in turn also led to a constraint on the day-night wind velocity in the planetary atmosphere. Subsequent observations have led to the detection of CO and H_2O in the dayside atmospheres of several transiting and non-transiting planets: τ Boo (Brogi et al. 2012; Rodler et al. 2012; Lockwood et al. 2014), 51 Peg b (Brogi et al. 2013), HD 189733b (de Kok et al. 2013; Birkby et al. 2013; Rodler et al. 2013), and HD 179949b (Brogi et al. 2014).

6.3 Directly Imaged Gas Giants

Direct imaging offers another avenue to characterize atmospheric compositions of exoplanets and works preferentially for a complementary region in planetary parameter space. Exoplanets detected to date via direct imaging are all young gas giants at large orbital separations. Their young ages (below $\sim$100 Myr) implies high effective temperatures ($\sim$1000 K) and, hence, high planet-star flux contrasts in the near-infrared. On the other hand, their large orbital separations (beyond $\sim$10 AU), which are set by the inner working angles of the instruments used, help in minimizing the stellar glare. Together these factors make young gas giants particularly favorable to near-infrared spectroscopy via direct imaging.

The advantage of direct imaging is that the detection of a planet simultaneously results in observation of its atmospheric thermal emission spectrum, i.e. atmospheric characterization can be pursued simultaneously with detection. This is in contrast to transiting exoplanets where the detections are generally made using photometric transit surveys and RV confirmations, while atmospheric spectra are obtained for optimal targets using follow-up observations with specialized instruments. On the other hand, the challenge with directly imaged planets is that generally only the atmospheric spectrum is available with little information about any other planet property; the mass, radius, and hence gravity, temperature, age, are all unknowns in modeling the planets. Therefore, typically, planetary evolution models are required along with atmospheric models to robustly constrain the atmospheric and physical parameters of the planets. Moreover, spectra are available at only one point of the orbital phase which means thermal phase curves are not observable precluding constraints on atmospheric properties with longitude. Nevertheless, the possibility of high-resolution absolute near-infrared spectroscopy for directly-imaged planets means that chemical signatures can be robustly detected in emission spectra using ground-based instruments.

Chemical detections have been reported for a few directly-imaged exoplanets in recent years. Given the expected temperature range of young giant planets, the dominant molecular species expected are H_2O, CO, and CH_4. All these molecules have been detected in one or more directly-imaged planets. Most of the detections have been reported for planets in the well characterized and nearby HR 8799 system. H_2O, CO, and CH_4 have been unambiguously detected in the planet HR 8799b using high-resolution spectroscopy in the H and K band (Barman et al. 2011, 2015), though the CH_4 absorption is weaker than expected from chemical equilibrium (Bowler et al. 2010; Barman et al. 2015). H_2O and CO have also been detected in high-resolution spectra of HR 8799c (Konopacky et al. 2013). The presence of CH_4 and other candidate molecules (e.g. NH_3, C_2H_2, CO_2, and/or HCN) have also been suggested using lower resolution spectra for the four planets in the HR 8799 system (Oppenheimer et al. 2013). Janson et al. (2013) reported an H-band detection of CH_4 in GJ 504b. More recently, CO was also detected unambiguously in β Pic b (Snellen et al. 2014) which also provided the first measurement of the spin period of an exoplanet.

In addition to molecular detections, spectra of directly imaged planets have also been use to place constraints on their elemental abundance ratios just as have been pursued for transiting hot Jupiters. However, determining elemental abundance ratios for directly imaged planets is challenging since most of the planetary properties (e.g. mass, radius, gravity, temperature) are not known a priori and hence need to be fit for, often requiring evolutionary models, in addition to the usual complexities introduced by the presence of clouds. Nevertheless, several studies have attempted to fit static models to spectra of directly-imaged planets to report nominal constraints on the elemental abundances. Using a model grid to fit spectroscopic observations of HR 8799 c, Konopacky et al. (2013) suggested a C/O ratio of $0.65^{+0.10}_{-0.05}$ which is slightly higher than a solar value of 0.54. Lee et al. (2013) performed an atmospheric retrieval analysis for HR 8799b to suggest a high metallicity, with a mean molecular weight of 3.8 compared to 2.4 for solar abundance, and a C/O ratio $\sim$1 in its atmosphere. More recently, Todorov et al. (2015) performed an atmospheric retrieval analysis to constrain the molecular abundances in the sub-stellar companion κ and b and reported an H_2O abundance of $\sim 10^{-4}$, nearly consistent with a solar abundance composition and those of several transiting hot Jupiters.

7 Implications for Formation Conditions of Exoplanets

Atmospheric elemental abundances of solar-system giant planets have led to important constraints on the origins of the solar system. In Jupiter's atmosphere, the abundances of C, N, S, Ar, Kr, and Xe, relative to H have been measured to be 2–3$\times$ solar values (Owen et al. 1999; Atreya and Wong 2005) which suggest substantial accretion of solids during its formation, and have been used as evidence for its metal-rich interior and its formation by core accretion. However, the abundance of O is not known for Jupiter. Given the low temperatures ($\lesssim$125 K) in Jupiter's observable troposphere, H_2O is condensed down to deeper layers of the atmosphere at pressures greater than ~10 bar. As the Galileo probe descended into Jupiter's atmosphere, its last measurement showed an H_2O abundance, and hence an O abundance, of 0.3× solar, which is 10× lower than the other elements (Atreya and Wong 2005). However, this measurement of Jupiter's H_2O abundance is considered to be a lower-limit since the region the probe entered is thought to be an anomalously dry spot in Jupiter's atmosphere. On the other hand, indirect constraints based on kinetics models explaining the observed CO abundance suggest an O abundance of 0.3–7.3× solar (Visscher et al. 2010b), which allows for a wide range of C/O ratios.

Taking the currently available lower-limit on O at its true abundance, implying more carbon than oxygen, would require unusual formation conditions in the early solar system. Lodders (2004) studied the hypothetical possibility and suggested that tarry planetesimals may have dominated Jupiter's accretion history, instead of planetesimals dominant in water-ice as expected from compositions of minor bodies in the solar system. Instead, using core accretion models Mousis et al. (2012) suggested that considering a solar-composition nebula predicts an O abundance of 3–7× solar. Therefore, a significantly lower observed O abundance would require substantial depletion of H_2O ice in Jupiter's formation region, unless the planetesimals are unusually carbon-rich. Thus, accurately measuring the O abundance is critical to constrain the formation conditions of Jupiter, and of the outer solar system in general. The H_2O abundance is similarly unknown for any other giant planet in the solar system. The upcoming Juno mission to Jupiter (Matousek 2007), therefore, aims to constrain its O abundance by measuring the H_2O abundance in Jupiter's atmosphere. In particular, the H_2O abundances play a central role in constraining exoplanetary formation conditions. Since O is cosmically the most abundant heavy element it is expected that H_2O is one of the most dominant volatile in interstellar and planet-forming environments (van Dishoeck et al. 2014). Thus, the H_2O abundances in giant planetary atmospheres are of fundamental importance to constrain the inventory of O that was available in planet's formation environments.

The O/H and C/O ratios are easier to measure for hot giant exoplanets than they are for solar-system giant planets (Madhusudhan 2012). The vast majority of extrasolar gas giants known have $T \sim 600$–3000 K, thus hosting gaseous H_2O in their observable atmospheres accessible to spectroscopic observations. Other detectable gases include CH_4, CO, CO_2, and NH_3, depending on the temperature and incident irradiation. Measurements of such molecular abundances allow estimations of elemental abundances ratios involving H, C, O, and N. Such elemental abundances can in turn provide crucial clues regarding exoplanetary atmospheric processes, interior compositions, and formation mechanisms, just as pursued for solar system planets. This fortuitous opportunity makes hot giant exoplanets the perfect laboratories to investigate the origins of giant planets and the diversity of their atmospheres and interiors. Indeed, as discussed above, recent observations are already leading to unprecedented detections of chemical species in giant exoplanetary atmospheres. Such spectra have already led to clear H_2O detections in atmospheres of several transiting hot Jupiters, as discussed above (e.g. Deming et al. 2013; Kreidberg et al. 2014b; Madhusudhan et al. 2014b). H_2O, CO, and/or CH_4 have also been detected using high-resolution ground-based spectroscopy of hot Jupiters (e.g. Brogi et al. 2014), as well as directly imaged planets (e.g. Konopacky et al. 2013). Upcoming large facilities such as the *JWST* and *E-ELT* will further revolutionize the field.

New studies are beginning to investigate the influence of formation and migration histories of giant exoplanets on their observable chemical compositions. As discussed in previous sections, observations are suggesting the possibility of carbon-rich ($C/O \geq 1$) as well as oxygen-rich (C/O $<$ 1) atmospheres in giant exoplanets (Madhusudhan et al. 2011a; Madhusudhan 2012) which are in turn motivating new ideas on the formation conditions of planetary systems. One of the key questions in this regard is about how C-rich gas giants can form around O-rich sun-like stars; the solar C/O is 0.5 (Asplund et al. 2009). As discussed above, an early investigation into this question was pursued in the context of Jupiter in the solar-system for which, as discussed above, only a lower limit on the O/H is known, which allows for a C/O $>$ 1 in Jupiter. More recently, following the inference of C/O $\geq$ 1 in WASP-12b (Madhusudhan et al. 2011a), Öberg et al. (2011) suggested that C/O ratios in giant planetary envelopes depend on the formation location of the planets in the disk relative to the snow lines of major C and O bearing volatile species, such as H_2O, CO, and CO_2, since the C/O ratio of the gas approaches 1 outside the CO and CO_2 snow lines. By predominantly accreting such C-rich gas, more so than O-rich planetesimals, gas giants could, in principle, host C-rich atmospheres even when orbiting O-rich stars. On the other hand, it may also be possible that inherent inhomogeneities in the C/O ratios of the disk itself may contribute to higher C/O ratios of the planets relative to the host stars (Kuchner and Seager 2005; Madhusudhan et al. 2011b; Mousis et al. 2012; Moses et al. 2013b; Ali-Dib et al. 2014). The compositions of gas and solids accreted also depend on the physicochemical properties of the disk at the given location which are time-dependent (Helling et al. 2014; Marboeuf et al. 2014). Furthermore, some of the volatile elements can also be depleted due to dust formation and settling in the atmosphere (Moses et al. 2013b; Helling et al. 2014). These various scenarios predict different limits on the metallicities and C/O ratios of the giant planets, which high-precision observations of their atmospheres will be able to test in the near future.

Besides the various factors discussed above, the main formation mechanisms also contribute significantly to the planetary composition. Giant planets are thought to form via one of two primary mechanisms, core accretion (CI) versus gravitational instability (GI). In the CI model (Pollack et al. 1996), the planetary embryos start out as 10 Earth-mass cores in the protoplanetary disk that subsequently undergo runaway accretion of a large volume of gas and planetesimals to form a massive gaseous envelope. On the other hand, a GI in a young disk can cause rapid collapse of a large volume of ambient gas and solids to form a giant planet (Boss 2000). Both scenarios occur in planet-forming disks, but at different orbital separations. While CA is favored closer to the snow-line (within $\sim$2–10 AU) because cores take too long to form at larger distances and only reach large masses after the disk has dispersed, GI is favored at larger distances ($\gtrsim$10 AU) where the disk can cool sufficiently on orbital timescales to fragment. In this regard, GI may be the favored mechanism for the formation of distant gas giants detected via direct imaging. However, because the different formation mechanisms are favorable at different orbital separations in the disk they may be expected to probe different chemical compositions in the disk. In principle, both CA and GI can cause significant metallicity enhancements or depletions depending on the specific accretion history (Helled and Bodenheimer 2010), but the relative elemental abundances (e.g. C, O, N, etc.) may be different depending on the formation location corresponding to the respective snow lines.

The existence of hot Jupiters in very close-in orbits (e.g. $\lesssim$0.1 AU) presents a key challenge to theories of giant planet formation. Neither GI nor CA discussed above is thought to operate in such a way that allows hot Jupiters to form in situ at their current locations close to the host stars. The disk cannot fragment at those distances (Gammie 2001; Rafikov 2005), and cores with sufficient mass to attract significant envelopes cannot form. Therefore, the existence of hot Jupiters requires some form of "migration" from their original formation locations to their present orbits. Two competing hypotheses suggest that the planets migrated either through interaction with the protoplanetary disk during their formation (Papaloizou et al. 2007), or by 'disk-free' mechanisms such as gravitational interactions with a third body (e.g. Rasio and Ford 1996; Fabrycky and Tremaine 2007). Measurements of spit-orbit misalignment (or stellar obliquity) for a large sample of hot Jupiters over the past decade (e.g. Triaud et al. 2010; Albrecht et al. 2012) have been advocated as possible metrics to distinguish between the two hypotheses for migration. In a simplistic view, migration through the disk was thought to likely align the orbital angular momentum vector of the planet with the stellar spin axis whereas disk-free migration via dynamical encounters could lead to spin-orbit misalignment. A significant number of large spin-orbit misalignments (or stellar obliquity) observed in hot Jupiter systems initially supported disk-free migration mechanisms (Winn et al. 2010). However, recent studies show that spin-orbit misalignments may also be caused by planet migration through disks that are themselves misaligned and due to star-disk interactions (Crida and Batygin 2014; Lai 2014). Consequently, observed dynamical properties of hot Jupiters have been unable to conclusively constrain their migration pathways.

Instead, recent studies suggest that chemical abundances of hot Jupiters could provide stronger constraints on their formation and migration pathways. As discussed above, the O/H, C/H, and C/O ratios can change substantially with their formation locations in the protoplanetary disk. Madhusudhan et al. (2014a) suggested that atmospheric metallicities of hot Jupiter atmospheres could potentially constrain their migration mechanisms as migration through the disk is more likely to cause metal enrichment due to planetesimal accretion compared to disk-free migration mechanisms. The results lead to three key predictions as evident from Fig. 12. Firstly, planets migrating through the disk always accrete solids efficiently enough to result in solar or super-solar abundances (top-right quadrant in Fig. 12). Secondly, the C and O abundances are enhanced or depleted together, i.e. no cases in the top-left or bottom-right quadrants. Thirdly, planets with sub-stellar O and C abundances could not result from disk-migration. These findings imply that elemental abundances of hot Jupiters could potentially provide important constraints not only on their local formation environments but also on the migration pathways. Current observations are already providing the first constraints on the O and C abundances which in turn are being used to constrain formation and migration mechanisms (Madhusudhan et al. 2014a). These efforts will be further bolstered in the future as the spectral range of JWST instruments will contain strong molecular features of several molecules which will allow precise abundance estimates for a wider range of elements (e.g. O, C, N, P, S, Si, etc.).

8 Theory: Climates of Habitable Exoplanets

8.1 Habitability of Exoplanets

Existence of life requires many kinds of conditions. One of the critically necessary conditions for a planet to harbor life is permanent existence of liquid water on the planet's surface. To maintain liquid water, the planetary surface temperature has to be between the freezing point of water (0 °C) and the boiling point (100 °C), i.e., at 1 bar,

$$0\,^{\circ}\mathrm{C} < T_s < 100\,^{\circ}\mathrm{C}. \tag{1}$$

In reality, the upper limit of surface temperature has to be lower than the boiling point of water. Otherwise, the planet would fall into the runaway greenhouse state. We will return to

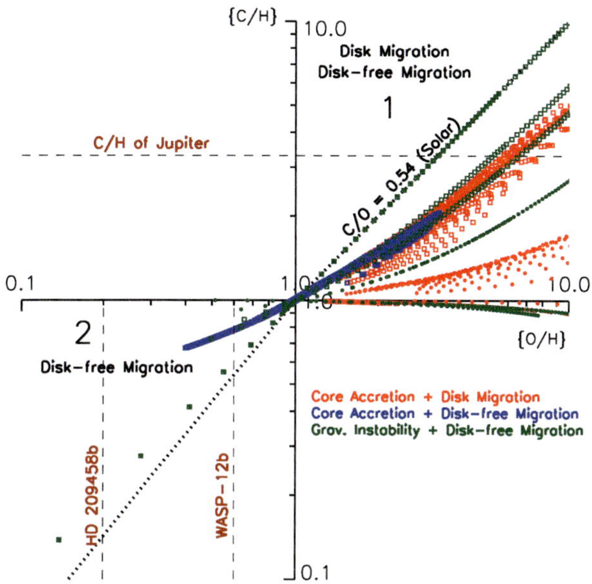

Fig. 12 Predicted elemental O and C abundances in hot Jupiters from semi-analytic chemical models of hot Jupiter formation and migration (from Madhusudhan et al. 2014a). The *axes* show O/H and C/H abundances relative to solar values, and the *colored symbols* represent models with different formation-migration histories. *Each dot* represents a different model realization and shows the final O/H and C/H of the planet resulting from that model. The final elemental abundances in a planet depend critically on the amount of gas and solids accreted by the planet as well the region from which that material is accreted relative to the H_2O, CO_2, and CO *snow lines*, giving rise to the distinct loci of models in the O-C plane as shown. The results suggest that different migration mechanisms can cause different patterns of elemental enhancements in hot Jupiters

this issue in the next section. To retain a fairly dense atmosphere and large amount of water against escape processes, the planetary mass has to be large enough so that the planet has sufficiently strong gravity. However, the planetary mass cannot be too large. Otherwise, it could have accreted a massive H_2-He envelope whose greenhouse effect along with the high pressure of the envelope would warm the surface and prevent water from being liquid, or it could be a Neptune-like ice-giant planet (Selsis et al. 2007; Venturini et al. 2015). Specifically, a habitable planet has to be a solid planet, and its mass should be in an approximate range given by:

$$0.5M_\oplus < M < 10M_\oplus, \tag{2}$$

where, $M_\oplus$ denotes Earth mass (Selsis et al. 2007).

Based on the above two conditions, an Earth Similarity Index (ESI) has been proposed (Schulze-Makuch et al. 2011) which attempts to quantify the similarity of a given property of an exoplanet to that of the Earth and ranges between 0.0 and 1.0. There are about 30 exoplanets that have ESI greater than 0.5. In other words, there are about 30 potentially habitable exoplanets among nearly 2000 discovered exoplanets. Gl 581c is the first exoplanet that had ever been thought to be habitable (Udry et al. 2007). However, radiative-convective model simulations indicated that Gl 581c is too hot to maintain liquid water (Selsis et al. 2007; Hu and Ding 2011). Especially, Hu and Ding (2011) showed that the climate of Gl 581c can readily fall into runaway greenhouse state even if the CO_2 level is extremely low and cloud albedo is extremely high. Gl 581d had also been considered a habitable super-

Earth (Udry et al. 2007; Selsis et al. 2007; Von Bloh et al. 2007), and it is probably the first exoplanet whose habitability has drawn intensive studies. Wordsworth et al. (2010), Hu and Ding (2011), and Von Paris et al. (2011) all showed that at least 7 bars of CO_2 are required to maintain the surface temperature of Gl 581d above the freezing point of water. Kepler 186f, which has a size similar to Earth's, was also suggested to be a habitable exoplanet. However, calculations indicate that the surface temperature of Kepler 186f would be much colder than that of Gl 581d for same CO2 concentration because Kepler 186f receives less stellar radiation than Gl 581d does. So far, none of these potentially habitable exoplanets has been confirmed to be habitable.

8.2 Habitable Zones

Condition (1) defines a circumstellar zone in which a terrestrial planet can hold permanent liquid water on its surface. Such a zone is the so-called Habitable Zone (HZ). The HZ was first proposed by Shapley (1953) as the region around a star in which liquid water could exist on a planet's surface. The term 'habitable zone' was first introduced by Huang (1960), in the context of planetary habitability and extraterrestrial life. The HZ was more precisely defined in later works (Kasting et al. 1993; Kopparapu 2013), based on results from one-dimensional radiative-convective climate models. In these works, various factors that influence the width of the HZ, such as surface albedo, atmospheric compositions, and stellar radiation spectra, were considered. Especially, Kasting et al. (1993) extensively addressed how the inner and outer edges of HZ are constrained by the positive water-vapor feedback and by the saturation limit of the maximum CO_2 concentration, respectively. One might expect that the inner edge of HZ is at the distance where planetary surface temperature is equal the boiling point of water (100 °C). However, as shown by Ingersoll (1969), all liquid water on the planetary surface would completely evaporate into the atmosphere due to the positive water-vapor feedback once the surface temperature reaches about 70 °C, and the planet falls into the runaway greenhouse climate state. Then, water is lost throughout photolysis and hydrogen escape. The outer edge is constrained by the limit of CO_2 saturation, with which CO_2 condensation begins to take place. Thus, further increasing CO_2 does not enhance the greenhouse effect. Instead, latent heat release by CO_2 condensation even causes weakened greenhouse effect due to the decrease of lapse rate. Using the one-dimensional climate model, Kasting et al. (1993) found that conservative estimates of the inner and outer edges of HZ in our own Solar System is about 0.95 and 1.37 AU, respectively. Thus, the width of the solar HZ is about 0.42 AU. It is wider than the mean spacing between the four terrestrial planets in the Solar System, which is about 0.35 AU. It suggests that, statistically speaking, at least one of the terrestrial planets ought to be in the solar HZ. Kasting (2010) pointed out that the possibility that at least one rocky planet will be in the HZ should be high if such a statistical result can be applied to other exoplanetary systems. For darker stars like M dwarfs that have effective temperatures of about 3500 K, the HZ is much closer to the stars, at about 0.1 AU.

One uncertainty of the HZ width calculated from radiative-convective models is cloud radiative effects that are not included in these one-dimensional models. It was thought that the inner edge could be pushed closer to stars as the negative feedback of cloud reflection of stellar radiation is considered (Kasting et al. 1993; Selsis et al. 2007). Three-dimensional general circulation models (GCMs) include cloud radiative effects because GCMs have self-adjusted dynamical and physical processes of cloud formation. Several recent works have studied the HZ width with GCMs. However, Leconte et al. (2013) demonstrated that the inner edge of the solar HZ calculated from their GCM is nearly the same as that predicted by one-dimensional models. The reason why the inner edge does not move closer to Sun in their GCM is because of two important factors. One is that water-vapor absorption of solar radiation is higher in their model than previously assumed. The other one is that for high surface temperatures and water-rich atmosphere conditions high cirrus clouds increase in fraction faster than low stratus clouds. The former consists of ice particles and have greenhouse effects, and the later mainly reflects solar radiation and cools the surface. On the other hand, Yang et al. (2013) used a GCM to show that deep convective clouds develop around the substellar point for slow-rotating or tidally-locked exoplanets. These clouds largely reflect stellar radiation and cool the surface. As a result, the inner edge of the HZ is pushed much closer to the star, and the HZ width is nearly doubled.

It is worthwhile to point out that the cloud radiative effect on the inner edge of the HZ has large uncertainties. Parameterized clouds in GCMs are the primary source of uncertainty even in simulating modern climates where we have a wealth of direct observations. Thus, it is not clear to what extent cloud parameterizations in Earth GCMs can be applied to studying exoplanetary atmospheres. Using Earth GCMs to study the outer edge of the HZ also has great challenges. There is also lack of parameterizations of CO_2 clouds. Radiative effects of CO_2 clouds are not well understood (Forget and Pierrehumbert 1997).

8.3 Climates and Habitability of Tidal-Locking Exoplanets in the HZ of M Dwarfs

M dwarfs are the most common stars in the Universe (Rodono 1986). Thus, the probability of finding habitable exoplanets around M dwarfs is much higher than around other types of stars. Moreover, it is much easier to find habitable exoplanets closer to stars than that with greater distances with current observation techniques. Therefore, it is very likely to first discover habitable exoplanets around M dwarfs. Since M dwarfs have much weaker luminosities than Sun-like stars, the HZ around an M dwarf is typically about 0.1 AU, much closer than that for a solar-type star. However, the short distance of the HZ could also cause exoplanets in the HZ around M dwarfs to be uninhabitable, as discussed below.

First, exoplanets within such short distance to M dwarfs can be exposed to high levels of X-ray and extreme ultraviolet radiation (X-EUV) and strong particle fluxes from stellar winds or coronal mass ejections (Lammer et al. 2011). Such strong emissions are due to M dwarfs magnetic activity. It can generate significant atmospheric escapes to space and can even potentially cause atmosphere erosion or strip the whole atmosphere. For solar-type stars, such strong X-EUV emissions may be limited to the first few hundred millions years. However, the extreme irradiation could last several Gyr for M-type stars. Thus, exoplanets in the HZ of M-type stars may receive X-EUV fluxes that are 10–100 times higher than those in the HZ of solar-type stars of the same age (Selsis et al. 2007). X-EUV flux could cause a CO_2 dominant atmosphere with modest amount of free oxygen (O_2). Rapid escape of carbon from a CO_2 dominant atmosphere could lead to the formation of O_2 even without the presence of biological O_2 production (Tian 2009). Photochemical models also suggest that O_3 and CH_4 can also be produced abiotically for different stellar characteristics and levels of volcanic outgassing (Domagal-Goldman et al. 2014). On the other hand, several other compounds, e.g. of sulfur, could act as potential biosignatures (Domagal-Goldman et al. 2011). At the current stage, more observational and quantitative modeling works are needed for understanding the effects of the extreme irradiation on atmospheres of exoplanets in the HZ around M dwarfs (see e.g. Segura et al. 2010).

Second, habitable exoplanets orbiting M-type stars are very likely to be tidally locked to their primaries due to strong forcing of gravitational gradients. Tidal-locking exoplanets receive very uneven stellar heating because their one side permanently faces their stars and the other side remains dark. While the dayside can be warm enough to sustain liquid water, the nightside could be so cold that any gases condense out there. If there exists atmosphere collapse, exoplanets in the HZ of M dwarfs would be uninhabitable. There were works that studied whether atmospheric heat transports could prevent atmosphere from collapse on the nightside using simplified or sophisticate models (Haberle et al. 1996; Joshi et al. 1997; Joshi 2003; Pierrehumbert 2011). They all showed that for a sufficiently dense atmosphere (surface air pressure greater than 0.1 bar) atmospheric heat transports would be able to warm the nightside and maintain nightside surface temperature above the condensation point of CO_2 (-78.5 °C at 1 bar of CO_2 partial pressure, and much lower at lower partial pressures). Therefore, atmosphere collapse can hardly happen as long as exoplanets are able to attract a fairly dense atmosphere envelope. However, more recent works suggest that the threshold of air pressure to prevent atmosphere collapse requires further investigation (Heng and Kopparla 2012; Wordsworth 2015).

The above studies have not taken ocean into account. If an exoplanet has an extensive ocean, its habitability also involves ocean heat transports and the positive sea-ice albedo feedback (Hu and Yang 2014). On the one hand, it is well known that ocean heat transports are equally important in Earth's climate (Peixoto and Oort 1992). In the presence of sea ice, ocean heat transports are likely to be especially important, since it is known from studies of the Snowball Earth phenomenon in Earth-like conditions that ocean heat transports are very effective in holding back the advance of the sea-ice margin (Pierrehumbert 2011; Yang et al. 2012a,b,c). On the other hand, in the presence of sea ice there exists the possibility that the exoplanet could be locked in a globally glaciated Snowball state due to the positive ice-albedo feedback. Note that the distribution of ice on tidally locked exoplanets is a particularly interesting issue only for M-type stars, since exoplanets orbiting hotter stars in orbits close enough to yield tidal-locking are likely to be too hot to permit ice, and may even be too hot to retain liquid water. Using a fully coupled atmospheric-oceanic GCM (AOGCM), Hu and Yang (2014) carried out simulations and demonstrated that ocean heat transports substantially extend the area of open water along the equator, and that the open-ocean area shows a lobster-like spatial pattern (Fig. 13), instead of an eyeball as shown in Pierrehumbert (2011). Figure 13 also shows that as CO_2 concentration increases, the dayside temperature does not increase very much, but the nightside temperature increases a lot, indicating that most heat is transported from the dayside to the nightside by ocean currents. They also showed that ocean heat transports can even lead to complete deglaciation of the nightside as greenhouse gas concentration is sufficiently high or stellar radiation is sufficiently strong. By contrast, the open-ocean area is not expanded very much by atmospheric heat transports alone for the same high level of CO_2. They also showed that it is more ready for aqua-planets to fall into runaway greenhouse and runaway freezing at the inner and outer edges, respectively, compared with simulation results without a dynamic ocean. It implies that the HZ would be narrower for aqua-planets or exoplanets with extensive oceans. These results indicate that ocean heat transports play critically important roles in determining the climate state and habitability of exoplanets.

The third problem is whether water could be completely trapped on the nightside of tidal-locking exoplanets in the HZ of M dwarfs (Menou 2013). Although Fig. 13 shows that the lowest surface temperature on the nightside is about -40 °C, well above the condensation point of CO_2, it is still far below the freezing point of water. Therefore, there is the possibility of water trap on the nightside. Hu and Yang (2014) showed that as long as the ocean is not completely frozen, wind stresses transport sea ice toward the dayside and the ocean carries heat toward the nightside. As a result, sea-ice thickness on the nightside remains thin, less than 10 meters. Yang et al. (2014) studied that a case with a super-continent on the nightside

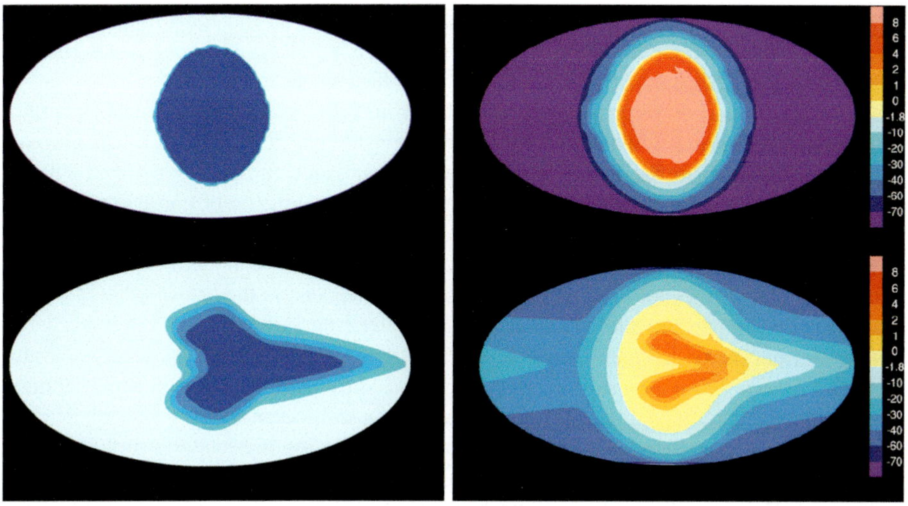

Fig. 13 Spatial patterns of open-ocean areas and surface temperatures. *Left panels*: sea-ice fraction. *Blue colors* indicate open-ocean areas, and *light-gray* indicates ice coverage. *Right panels*: surface temperatures, *color bar* unit is °C. *Top panels*: with only atmospheric heat transport, and *bottom panels*: heat transports by both atmosphere and ocean. The substellar point is *at the center of each panel*. After Hu and Yang Hu and Yang (2014)

Fig. 14 Ice sheets over a supercontinent spanning the entire nightside. *Color shading* indicates ice-sheet thickness, and *arrows* indicate ice-sheet flow velocity. Geothermal flux is 0.1 W m^{-2} in the simulation. Adapted from Yang et al. (2014)

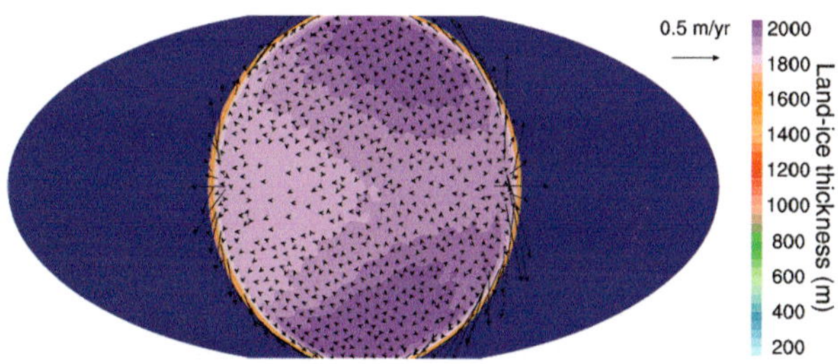

and an ocean on the dayside. In this case, ocean heat transports from the dayside to the nightside are ceased. It is found that ice sheets over the nightide supercontinent can grow 2000 m thick if geothermal heat flux is close to Earth's or smaller (Fig. 14). It suggests that if the dayside ocean is deeper than 2000 m, water will not be completely trapped on the nightside continent. Only exoplanets with a geothermal heat flux lower than Earths, and much of their surface covered by continents would be susceptible to complete water trapping.

8.4 Observable Signatures of Tidally-Locked Habitable Exoplanets

The above climate patterns have important implications for future observations. Showman et al. (2009), Showman and Polvani (2011), Heng and Showman (2015) showed that an equatorial superrotating jet stream develops in the atmosphere of tidal-locking exoplanets. The hot spot of tidal-locking exoplanets is advected to the downstream of the substellar point by the equatorial superrotating jet stream. Their simulation results are confirmed by observations for hot Jupiters (Knutson et al. 2009). It was shown that the hot spot is shifted

20°–30° east of the substellar point of hot Jupiter HD 189733b. The equatorial superrotating flow is a common feature for tidal-locking exoplanets (Heng and Vogt 2011; Hu and Ding 2013). Thus, the hot-spot shift shall also be observed from the thermal phase curve of terrestrial exoplanets. In fact, the climate pattern in Fig. 13 also suggests eastward shift of the hot spot. Especially, as a dynamic ocean is considered, the equatorial ocean current, together with the equatorial atmospheric jet stream, can cause further downstream shift of the hot spot. Cowan et al. (2012b) studied how surface thermal inertia of Earth-like exoplanets leads to distinguishable thermal phase curve for observations. They studied two types of climate states. One is the temperate climate which is just like the modern Earth with low- and middle-latitude oceans and polar ice-caps. The other one is the Snowball state. They found that the former has a relatively flat thermal phase curve because its relatively large surface thermal inertia tends to damp out the amplitude of the thermal phase curve. Wang et al. (2014) argued that terrestrial exoplanets around M dwarfs could have different climate states. They use an atmospheric GCM coupled with a slab ocean to show that exoplanets with nonzero eccentricities could have different spin-orbit resonance states different from the synchronous rotation state, and these exoplanets should have a striped-ball climate pattern, with a global belt of open ocean at low and middle latitudes and ice caps over both polar regions. This is in contrast to synchronous rotating habitable exoplanets around M dwarfs that have an eyeball climate pattern a limited region of open water on the dayside and ice on the rest of the planet (Pierrehumbert 2011; Hu and Yang 2014). They pointed out that the striped-ball climate state should be common for habitable exoplanets in large eccentric orbits around M dwarfs. They further suggested that these different climate patterns can be observed by future exoplanet detection missions. Overall, the features of climate patterns provide future observations of habitable exoplanets with additional constraints and information.

9 Future Prospects

Major observational advancements are happening in two key directions: (a) detections of exoplanets around bright stars, and (b) high precision observations of exoplanetary spectra. Firstly, bright exoplanet host stars are important to obtain precise measurements of exoplanetary masses, radii, and atmospheric spectra. Upcoming space missions, CHEOPS (Broeg et al. 2014) and TESS (Ricker et al. 2015), as well as various ground-based surveys (e.g. NGTS, SPECULOOS, MEARTH, etc.) are expected to find thousands of transiting exoplanets, including hundreds of super-Earths, orbiting nearby bright stars within the next five years (also see chapter on future landscape by Fridlund et al. 2016, this issue). During the same time, major ground-based direct imaging platforms (e.g. SPHERE, GPI, etc.) are also expected to find at least dozens of exoplanets at wide orbital separations orbiting nearby young stars. On a longer run, early 2020s, the recently selected ESA mission, PLATO (Rauer et al. 2014), will discover numerous transiting exoplanets in habitable zones of nearby stars. Secondly, major parallel efforts are also being pursued for spectroscopic observations of exoplanetary atmospheres. In this regard, enormous amount of time is currently being dedicated on the Hubble and Spitzer space telescopes, as well as major ground-based telescopes (e.g. VLT, Keck, Gemini, Magellan, CFHT, etc.). In addition, the James Webb Space Telescope (*JWST*) scheduled for launch in 2018 will revolutionize exoplanetary spectroscopy. By early 2020s, next generation ground-based telescopes, such as the European-Extremely Large Telescope (E-ELT), will further revolutionize the field. Future spectra with *JWST* would be of unprecedented precision and resolution which will enable us to derive precise chemical abundances for transiting exoplanets. In addition, JWST would also be able to observe atmospheres of much cooler low-mass planets that are possible to observe today.

Acknowledgements This review was conceived during the 2014 ISSI workshop in Beijing on "The Disk in Relation to the Formation of Planets and their Protoatmospheres". N. Madhusudhan and M. Agúndez thank the International Space Science Institute (ISSI) and ISSI-Beijing for the hospitality during the workshop and the participants of the workshop for the lively discussions on exoplanetary atmospheres. M. Agúndez thanks K. Lodders for interesting discussions on atmospheric condensates. M. Agúndez acknowledges funding support from Spanish MINECO through grants CSD2009-00038, AYA2009-07304, and AYA2012-32032 and from the European Research Council (ERC Grant 610256: NANOCOSMOS). J. Moses thanks the NASA Exoplanet Research program NNX15AN82G for support. Y. Hu is supported by the National Natural Science Foundation of China 435 (NSFC) under grants 41375072 and 41530423.

References

Y. Abe, T. Matsui, The formation of an impact-generated H_2O atmosphere and its implications for the early thermal history of the Earth. J. Geophys. Res. **90**, 545 (1985). doi:10.1029/JB090iS02p0C545

A.S. Ackerman, M.S. Marley, Precipitating condensation clouds in substellar atmospheres. Astrophys. J. **556**, 872–884 (2001). doi:10.1086/321540

M. Agúndez, O. Venot, N. Iro, F. Selsis, F. Hersant, E. Hébrard, M. Dobrijevic, The impact of atmospheric circulation on the chemistry of the hot Jupiter HD 209458b. Astron. Astrophys. **548**, 73 (2012). doi:10.1051/0004-6361/201220365

M. Agúndez, V. Parmentier, O. Venot, F. Hersant, F. Selsis, Pseudo 2D chemical model of hot-Jupiter atmospheres: Application to HD 209458b and HD 189733b. Astron. Astrophys. **564**, 73 (2014a). doi:10.1051/0004-6361/201322895

M. Agúndez, O. Venot, F. Selsis, N. Iro, The puzzling chemical composition of GJ 436b's atmosphere: Influence of tidal heating on the chemistry. Astrophys. J. **781**, 68 (2014b). doi:10.1088/0004-637X/781/2/68

S. Albrecht, J.N. Winn, J.A. Johnson, A.W. Howard, G.W. Marcy, R.P. Butler, P. Arriagada, J.D. Crane, S.A. Shectman, I.B. Thompson, T. Hirano, G. Bakos, J.D. Hartman, Obliquities of hot Jupiter host stars: Evidence for tidal interactions and primordial misalignments. Astrophys. J. **757**, 18 (2012). doi:10.1088/0004-637X/757/1/18

M. Ali-Dib, O. Mousis, J.-M. Petit, J.I. Lunine, Carbon-rich planet formation in a solar composition disk. Astrophys. J. **785**, 125 (2014). doi:10.1088/0004-637X/785/2/125

Y. Alibert, C. Mordasini, W. Benz, C. Winisdoerffer, Models of giant planet formation with migration and disc evolution. Astron. Astrophys. **434**, 343–353 (2005). doi:10.1051/0004-6361:20042032

Y. Alibert, I. Baraffe, W. Benz, G. Chabrier, C. Mordasini, C. Lovis, M. Mayor, F. Pepe, F. Bouchy, D. Queloz, S. Udry, Formation and structure of the three Neptune-mass planets system around HD 69830. Astron. Astrophys. **455**, 25–28 (2006). doi:10.1051/0004-6361:20065697

F. Allard, P.H. Hauschildt, I. Baraffe, G. Chabrier, Synthetic spectra and mass determination of the brown dwarf Gl 229B. Astrophys. J. Lett. **465**, 123 (1996). doi:10.1086/310143

F. Allard, P.H. Hauschildt, D.R. Alexander, S. Starrfield, Model atmospheres of very low mass stars and brown dwarfs. Annu. Rev. Astron. Astrophys. **35**, 137–177 (1997). doi:10.1146/annurev.astro.35.1.137

M. Asplund, N. Grevesse, A.J. Sauval, P. Scott, The chemical composition of the Sun. Annu. Rev. Astron. Astrophys. **47**, 481–522 (2009). doi:10.1146/annurev.astro.46.060407.145222

N. Astudillo-Defru, P. Rojo, Ground-based detection of calcium and possibly scandium and hydrogen in the atmosphere of HD 209458b. Astron. Astrophys. **557**, 56 (2013). doi:10.1051/0004-6361/201219018

S.K. Atreya, *Atmospheres and Ionospheres of the Outer Planets and Their Satellites* (1986)

S.K. Atreya, A.-S. Wong, Coupled clouds and chemistry of the giant planets—A case for multiprobes. Space Sci. Rev. **116**, 121–136 (2005). doi:10.1007/s11214-005-1951-5

Bai et al., Space Sci. Rev. (2016, this issue)

R.L. Bailey, C. Helling, G. Hodosán, C. Bilger, C.R. Stark, Ionization in atmospheres of brown dwarfs and extrasolar planets VI: Properties of large-scale discharge events. Astrophys. J. **784**, 43 (2014). doi:10.1088/0004-637X/784/1/43

I. Baraffe, G. Chabrier, T. Barman, The physical properties of extra-solar planets. Rep. Prog. Phys. **73**(1), 016901 (2010). doi:10.1088/0034-4885/73/1/016901

T. Barman, Identification of absorption features in an extrasolar planet atmosphere. Astrophys. J. Lett. **661**, 191–194 (2007). doi:10.1086/518736

T.S. Barman, P.H. Hauschildt, F. Allard, Phase-dependent properties of extrasolar planet atmospheres. Astrophys. J. **632**, 1132–1139 (2005). doi:10.1086/444349

T.S. Barman, B. Macintosh, Q.M. Konopacky, C. Marois, Clouds and chemistry in the atmosphere of extrasolar planet HR8799b. Astrophys. J. **733**, 65 (2011). doi:10.1088/0004-637X/733/1/65

T.S. Barman, Q.M. Konopacky, B. Macintosh, C. Marois, Simultaneous detection of water, methane, and carbon monoxide in the atmosphere of exoplanet HR8799b. Astrophys. J. **804**, 61 (2015). doi:10.1088/0004-637X/804/1/61

J.K. Barstow, S. Aigrain, P.G.J. Irwin, T. Hackler, L.N. Fletcher, J.M. Lee, N.P. Gibson, Clouds on the hot Jupiter HD189733b: Constraints from the reflection spectrum. Astrophys. J. **786**, 154 (2014). doi:10.1088/0004-637X/786/2/154

N.M. Batalha, J.F. Rowe, S.T. Bryson, T. Barclay, C.J. Burke, D.A. Caldwell, J.L. Christiansen, F. Mullally, S.E. Thompson, T.M. Brown, A.K. Dupree, D.C. Fabrycky, E.B. Ford, J.J. Fortney, R.L. Gilliland, H. Isaacson, D.W. Latham, G.W. Marcy, S.N. Quinn, D. Ragozzine, A. Shporer, W.J. Borucki, D.R. Ciardi, T.N. Gautier III, M.R. Haas, J.M. Jenkins, D.G. Koch, J.J. Lissauer, W. Rapin, G.S. Basri, A.P. Boss, L.A. Buchhave, J.A. Carter, D. Charbonneau, J. Christensen-Dalsgaard, B.D. Clarke, W.D. Cochran, B.-O. Demory, J.-M. Desert, E. Devore, L.R. Doyle, G.A. Esquerdo, M. Everett, F. Fressin, J.C. Geary, F.R. Girouard, A. Gould, J.R. Hall, M.J. Holman, A.W. Howard, S.B. Howell, K.A. Ibrahim, K. Kinemuchi, H. Kjeldsen, T.C. Klaus, J. Li, P.W. Lucas, S. Meibom, R.L. Morris, A. Prša, E. Quintana, D.T. Sanderfer, D. Sasselov, S.E. Seader, J.C. Smith, J.H. Steffen, M. Still, M.C. Stumpe, J.C. Tarter, P. Tenenbaum, G. Torres, J.D. Twicken, K. Uddin, J. Van Cleve, L. Walkowicz, W.F. Welsh, Planetary candidates observed by Kepler. III. Analysis of the first 16 months of data. Astrophys. J. Suppl. Ser. **204**, 24 (2013). doi:10.1088/0067-0049/204/2/24

J.L. Bean, J.-M. Désert, P. Kabath, B. Stalder, S. Seager, E. Miller-Ricci Kempton, Z.K. Berta, D. Homeier, S. Walsh, A. Seifahrt, The optical and near-infrared transmission spectrum of the super-Earth GJ 1214b: Further evidence for a metal-rich atmosphere. Astrophys. J. **743**, 92 (2011). doi:10.1088/0004-637X/743/1/92

J.P. Beaulieu, S. Carey, I. Ribas, G. Tinetti, Primary transit of the planet HD 189733b at 3.6 and 5.8 μm. Astrophys. J. **677**, 1343–1347 (2008). doi:10.1086/527045

J.-P. Beaulieu, G. Tinetti, D.M. Kipping, I. Ribas, R.J. Barber, J.Y.-K. Cho, I. Polichtchouk, J. Tennyson, S.N. Yurchenko, C.A. Griffith, V. Batista, I. Waldmann, S. Miller, S. Carey, O. Mousis, S.J. Fossey, A. Aylward, Methane in the atmosphere of the transiting hot Neptune GJ436B? Astrophys. J. **731**, 16 (2011). doi:10.1088/0004-637X/731/1/16

B. Benneke, Strict upper limits on the carbon-to-oxygen ratios of eight hot Jupiters from self-consistent atmospheric retrieval. ArXiv e-prints (2015)

B. Benneke, S. Seager, How to distinguish between cloudy mini-Neptunes and water/volatile-dominated super-Earths. Astrophys. J. **778**, 153 (2013). doi:10.1088/0004-637X/778/2/153

W. Benz, S. Ida, Y. Alibert, D. Lin, C. Mordasini, Planet population synthesis, in *Protostars and Planets VI*, ed. by H. Beuther, R.S. Klessen, C.P. Dullemond, T. Henning (2014), pp. 691–713. doi:10.2458/azu_uapress_9780816531240-ch030

Z.K. Berta, D. Charbonneau, J.-M. Désert, E. Miller-Ricci Kempton, P.R. McCullough, C.J. Burke, J.J. Fortney, J. Irwin, P. Nutzman, D. Homeier, The flat transmission spectrum of the super-Earth GJ1214b from wide field camera 3 on the Hubble Space Telescope. Astrophys. J. **747**, 35 (2012). doi:10.1088/0004-637X/747/1/35

C. Bilger, P. Rimmer, C. Helling, Small hydrocarbon molecules in cloud-forming brown dwarf and giant gas planet atmospheres. Mon. Not. R. Astron. Soc. **435**, 1888–1903 (2013). doi:10.1093/mnras/stt1378

J.L. Birkby, R.J. de Kok, M. Brogi, E.J.W. de Mooij, H. Schwarz, S. Albrecht, I.A.G. Snellen, Detection of water absorption in the day side atmosphere of HD 189733 b using ground-based high-resolution spectroscopy at 3.2 μm. Mon. Not. R. Astron. Soc. **436**, 35–39 (2013). doi:10.1093/mnrasl/slt107

J. Blum, G. Wurm, The growth mechanisms of macroscopic bodies in protoplanetary disks. Annu. Rev. Astron. Astrophys. **46**, 21–56 (2008). doi:10.1146/annurev.astro.46.060407.145152

W.J. Borucki, D.G. Koch, G. Basri, N. Batalha, T.M. Brown, S.T. Bryson, D. Caldwell, J. Christensen-Dalsgaard, W.D. Cochran, E. DeVore, E.W. Dunham, T.N. Gautier III, J.C. Geary, R. Gilliland, A. Gould, S.B. Howell, J.M. Jenkins, D.W. Latham, J.J. Lissauer, G.W. Marcy, J. Rowe, D. Sasselov, A. Boss, D. Charbonneau, D. Ciardi, L. Doyle, A.K. Dupree, E.B. Ford, J. Fortney, M.J. Holman, S. Seager, J.H. Steffen, J. Tarter, W.F. Welsh, C. Allen, L.A. Buchhave, J.L. Christiansen, B.D. Clarke, S. Das, J.-M. Désert, M. Endl, D. Fabrycky, F. Fressin, M. Haas, E. Horch, A. Howard, H. Isaacson, H. Kjeldsen, J. Kolodziejczak, C. Kulesa, J. Li, P.W. Lucas, P. Machalek, D. McCarthy, P. MacQueen, S. Meibom,

T. Miquel, A. Prsa, S.N. Quinn, E.V. Quintana, D. Ragozzine, W. Sherry, A. Shporer, P. Tenenbaum, G. Torres, J.D. Twicken, J. Van Cleve, L. Walkowicz, F.C. Witteborn, M. Still, Characteristics of planetary candidates observed by Kepler. II. Analysis of the first four months of data. Astrophys. J. **736**, 19 (2011). doi:10.1088/0004-637X/736/1/19

A.P. Boss, Giant planet formation by gravitational instability. Science **276**, 1836–1839 (1997). doi:10.1126/science.276.5320.1836

A.P. Boss, Possible rapid gas giant planet formation in the solar nebula and other protoplanetary disks. Astrophys. J. Lett. **536**, 101–104 (2000). doi:10.1086/312737

V. Bourrier, A. Lecavelier des Etangs, H. Dupuy, D. Ehrenreich, A. Vidal-Madjar, G. Hébrard, G.E. Ballester, J.-M. Désert, R. Ferlet, D.K. Sing, P.J. Wheatley, Atmospheric escape from HD 189733b observed in H I Lyman-α: Detailed analysis of HST/STIS September 2011 observations. Astron. Astrophys. **551**, 63 (2013). doi:10.1051/0004-6361/201220533

B.P. Bowler, M.C. Liu, T.J. Dupuy, M.C. Cushing, Near-infrared spectroscopy of the extrasolar planet HR 8799 b. Astrophys. J. **723**, 850–868 (2010). doi:10.1088/0004-637X/723/1/850

G.P. Brasseur, J.J. Orlando, G.S. Tyndall, *Atmospheric Chemistry and Global Change* (1999)

C. Broeg, W. Benz, N. Thomas (Cheops Team), The CHEOPS mission. Contrib. Astron. Obs. Skaln. Pleso **43**, 498 (2014)

M. Brogi, I.A.G. Snellen, R.J. de Kok, S. Albrecht, J. Birkby, E.J.W. de Mooij, The signature of orbital motion from the dayside of the planet τ Boötis b. Nature **486**, 502–504 (2012). doi:10.1038/nature11161

M. Brogi, I.A.G. Snellen, R.J. de Kok, S. Albrecht, J.L. Birkby, E.J.W. de Mooij, Detection of molecular absorption in the dayside of exoplanet 51 Pegasi b? Astrophys. J. **767**, 27 (2013). doi:10.1088/0004-637X/767/1/27

M. Brogi, R.J. de Kok, J.L. Birkby, H. Schwarz, I.A.G. Snellen, Carbon monoxide and water vapor in the atmosphere of the non-transiting exoplanet HD 179949 b. Astron. Astrophys. **565**, 124 (2014). doi:10.1051/0004-6361/201423537

T.M. Brown, Transmission spectra as diagnostics of extrasolar giant planet atmospheres. Astrophys. J. **553**, 1006–1026 (2001). doi:10.1086/320950

C.J. Burke, S.T. Bryson, F. Mullally, J.F. Rowe, J.L. Christiansen, S.E. Thompson, J.L. Coughlin, M.R. Haas, N.M. Batalha, D.A. Caldwell, J.M. Jenkins, M. Still, T. Barclay, W.J. Borucki, W.J. Chaplin, D.R. Ciardi, B.D. Clarke, W.D. Cochran, B.-O. Demory, G.A. Esquerdo, T.N. Gautier III, R.L. Gilliland, F.R. Girouard, M. Havel, C.E. Henze, S.B. Howell, D. Huber, D.W. Latham, J. Li, R.C. Morehead, T.D. Morton, J. Pepper, E. Quintana, D. Ragozzine, S.E. Seader, Y. Shah, A. Shporer, P. Tenenbaum, J.D. Twicken, A. Wolfgang, Planetary candidates observed by Kepler IV: Planet sample from q1–q8 (22 months). Astrophys. J. Suppl. Ser. **210**, 19 (2014). doi:10.1088/0067-0049/210/2/19

A. Burrows, C.M. Sharp, Chemical equilibrium abundances in brown dwarf and extrasolar giant planet atmospheres. Astrophys. J. **512**, 843–863 (1999). doi:10.1086/306811

A. Burrows, M. Marley, W.B. Hubbard, J.I. Lunine, T. Guillot, D. Saumon, R. Freedman, D. Sudarsky, C. Sharp, A nongray theory of extrasolar giant planets and brown dwarfs. Astrophys. J. **491**, 856–875 (1997)

A. Burrows, W.B. Hubbard, J.I. Lunine, J. Liebert, The theory of brown dwarfs and extrasolar giant planets. Rev. Mod. Phys. **73**, 719–765 (2001). doi:10.1103/RevModPhys.73.719

A. Burrows, D. Sudarsky, J.I. Lunine, Beyond the T dwarfs: Theoretical spectra, colors, and detectability of the coolest brown dwarfs. Astrophys. J. **596**, 587–596 (2003). doi:10.1086/377709

A. Burrows, D. Sudarsky, I. Hubeny, Theory for the secondary eclipse fluxes, spectra, atmospheres, and light curves of transiting extrasolar giant planets. Astrophys. J. **650**, 1140–1149 (2006). doi:10.1086/507269

A. Burrows, J. Budaj, I. Hubeny, Theoretical spectra and light curves of close-in extrasolar giant planets and comparison with data. Astrophys. J. **678**, 1436–1457 (2008). doi:10.1086/533518

J.R. Burton, C.A. Watson, P. Rodríguez-Gil, I. Skillen, S.P. Littlefair, S. Dhillon, D. Pollacco, Defocused transmission spectroscopy: A potential detection of sodium in the atmosphere of WASP-12b. Mon. Not. R. Astron. Soc. **446**, 1071–1082 (2015). doi:10.1093/mnras/stu2149

T. Castan, K. Menou, Atmospheres of hot super-Earths. Astrophys. J. Lett. **743**, 36 (2011). doi:10.1088/2041-8205/743/2/L36

D. Charbonneau, T.M. Brown, R.W. Noyes, R.L. Gilliland, Detection of an extrasolar planet atmosphere. Astrophys. J. **568**, 377–384 (2002). doi:10.1086/338770

E. Chassefière, Hydrodynamic escape of oxygen from primitive atmospheres: Applications to the cases of Venus and Mars. Icarus **124**, 537–552 (1996). doi:10.1006/icar.1996.0229

S. Chatterjee, J.C. Tan, Inside-out planet formation. Astrophys. J. **780**, 53 (2014). doi:10.1088/0004-637X/780/1/53

E. Chiang, G. Laughlin, The minimum-mass extrasolar nebula: In situ formation of close-in super-Earths. Mon. Not. R. Astron. Soc. **431**, 3444–3455 (2013). doi:10.1093/mnras/stt424

E. Chiang, A.N. Youdin, Forming planetesimals in solar and extrasolar nebulae. Annu. Rev. Earth Planet. Sci. **38**, 493–522 (2010). doi:10.1146/annurev-earth-040809-152513

J. Chilcote, T. Barman, M.P. Fitzgerald, J.R. Graham, J.E. Larkin, B. Macintosh, B. Bauman, A.S. Burrows, A. Cardwell, R.J. De Rosa, D. Dillon, R. Doyon, J. Dunn, D. Erikson, D. Gavel, S.J. Goodsell, M. Hartung, P. Hibon, P. Ingraham, P. Kalas, Q. Konopacky, J. Maire, F. Marchis, M.S. Marley, C. Marois, M. Millar-Blanchaer, K. Morzinski, A. Norton, R. Oppenheimer, D. Palmer, J. Patience, M. Perrin, L. Poyneer, L. Pueyo, F.T. Rantakyrö, N. Sadakuni, L. Saddlemyer, D. Savransky, A. Serio, A. Sivaramakrishnan, I. Song, R. Soummer, S. Thomas, J.K. Wallace, S. Wiktorowicz, S. Wolff, The first H-band spectrum of the giant planet β Pictoris b. Astrophys. J. Lett. **798**, 3 (2015). doi:10.1088/2041-8205/798/1/L3

F.J. Ciesla, J.N. Cuzzi, The evolution of the water distribution in a viscous protoplanetary disk. Icarus **181**, 178–204 (2006). doi:10.1016/j.icarus.2005.11.009

K.D. Colón, E.B. Ford, S. Redfield, J.J. Fortney, M. Shabram, H.J. Deeg, S. Mahadevan, Probing potassium in the atmosphere of HD 80606b with tunable filter transit spectrophotometry from the Gran Telescopio Canarias. Mon. Not. R. Astron. Soc. **419**, 2233–2250 (2012). doi:10.1111/j.1365-2966.2011.19878.x

C.S. Cooper, A.P. Showman, Dynamics and disequilibrium carbon chemistry in hot Jupiter atmospheres, with application to HD 209458b. Astrophys. J. **649**, 1048–1063 (2006). doi:10.1086/506312

N.B. Cowan, P. Machalek, B. Croll, L.M. Shekhtman, A. Burrows, D. Deming, T. Greene, J.L. Hora, Thermal phase variations of WASP-12b: Defying predictions. Astrophys. J. **747**, 82 (2012a). doi:10.1088/0004-637X/747/1/82

N.B. Cowan, A. Voigt, D.S. Abbot, Thermal phases of Earth-like planets: Estimating thermal inertia from eccentricity, obliquity, and diurnal forcing. Astrophys. J. **757**(1), 80 (2012b)

A. Crida, K. Batygin, Spin-orbit angle distribution and the origin of (mis)aligned hot Jupiters. Astron. Astrophys. **567**, 42 (2014). doi:10.1051/0004-6361/201323292

B. Croll, L. Albert, D. Lafreniere, R. Jayawardhana, J.J. Fortney, Near-infrared thermal emission from the hot Jupiter TrES-2b: Ground-based detection of the secondary eclipse. Astrophys. J. **717**, 1084–1091 (2010). doi:10.1088/0004-637X/717/2/1084

I.J.M. Crossfield, T. Barman, B.M.S. Hansen, I. Tanaka, T. Kodama, Re-evaluating WASP-12b: Strong emission at 2.315 μm, deeper occultations, and an isothermal atmosphere. Astrophys. J. **760**, 140 (2012). doi:10.1088/0004-637X/760/2/140

N. Crouzet, P.R. McCullough, D. Deming, N. Madhusudhan, Water vapor in the spectrum of the extrasolar planet HD 189733b. II. The eclipse. Astrophys. J. **795**, 166 (2014). doi:10.1088/0004-637X/795/2/166

T. Currie, A. Burrows, Y. Itoh, S. Matsumura, M. Fukagawa, D. Apai, N. Madhusudhan, P.M. Hinz, T.J. Rodigas, M. Kasper, T.-S. Pyo, S. Ogino, A combined Subaru/VLT/MMT 1-5 μm study of planets orbiting HR 8799: Implications for atmospheric properties, masses, and formation. Astrophys. J. **729**, 128 (2011). doi:10.1088/0004-637X/729/2/128

G. D'Angelo, R.H. Durisen, J.J. Lissauer, Giant planet formation, in *Exoplanets*, ed. by S. Seager (2010), pp. 319–346

R.J. de Kok, M. Brogi, I.A.G. Snellen, J. Birkby, S. Albrecht, E.J.W. de Mooij, Detection of carbon monoxide in the high-resolution day-side spectrum of the exoplanet HD 189733b. Astron. Astrophys. **554**, 82 (2013). doi:10.1051/0004-6361/201321381

D. Deming, A. Wilkins, P. McCullough, A. Burrows, J.J. Fortney, E. Agol, I. Dobbs-Dixon, N. Madhusudhan, N. Crouzet, J.-M. Desert, R.L. Gilliland, K. Haynes, H.A. Knutson, M. Line, Z. Magic, A.M. Mandell, S. Ranjan, D. Charbonneau, M. Clampin, S. Seager, A.P. Showman, Infrared transmission spectroscopy of the exoplanets HD 209458b and XO-1b using the wide field camera-3 on the Hubble Space Telescope. Astrophys. J. **774**, 95 (2013). doi:10.1088/0004-637X/774/2/95

B.-O. Demory, S. Seager, N. Madhusudhan, H. Kjeldsen, J. Christensen-Dalsgaard, M. Gillon, J.F. Rowe, W.F. Welsh, E.R. Adams, A. Dupree, D. McCarthy, C. Kulesa, W.J. Borucki, D.G. Koch, The high albedo of the hot Jupiter Kepler-7 b. Astrophys. J. Lett. **735**, 12 (2011). doi:10.1088/2041-8205/735/1/L12

B.-O. Demory, J. de Wit, N. Lewis, J. Fortney, A. Zsom, S. Seager, H. Knutson, K. Heng, N. Madhusudhan, M. Gillon, T. Barclay, J.-M. Desert, V. Parmentier, N.B. Cowan, Inference of inhomogeneous clouds in an exoplanet atmosphere. Astrophys. J. Lett. **776**, 25 (2013). doi:10.1088/2041-8205/776/2/L25

J.-M. Désert, A. Vidal-Madjar, A. Lecavelier Des Etangs, D. Sing, D. Ehrenreich, G. Hébrard, R. Ferlet, TiO and VO broad band absorption features in the optical spectrum of the atmosphere of the hot-Jupiter <ASTROBJ>HD 209458b</ASTROBJ>. Astron. Astrophys. **492**, 585–592 (2008). doi:10.1051/0004-6361:200810355

J.-M. Désert, A. Lecavelier des Etangs, G. Hébrard, D.K. Sing, D. Ehrenreich, R. Ferlet, A. Vidal-Madjar, Search for carbon monoxide in the atmosphere of the transiting exoplanet HD 189733b. Astrophys. J. **699**, 478–485 (2009). doi:10.1088/0004-637X/699/1/478

J.-M. Désert, J. Bean, E. Miller-Ricci Kempton, Z.K. Berta, D. Charbonneau, J. Irwin, J. Fortney, C.J. Burke, P. Nutzman, Observational evidence for a metal-rich atmosphere on the super-Earth GJ1214b. Astrophys. J. Lett. **731**, 40 (2011). doi:10.1088/2041-8205/731/2/L40
H. Diamond-Lowe, K.B. Stevenson, J.L. Bean, M.R. Line, J.J. Fortney, New analysis indicates no thermal inversion in the atmosphere of HD 209458b. Astrophys. J. **796**, 66 (2014). doi:10.1088/0004-637X/796/1/66
S.E. Dodson-Robinson, P. Bodenheimer, The formation of Uranus and Neptune in solid-rich feeding zones: Connecting chemistry and dynamics. Icarus **207**, 491–498 (2010). doi:10.1016/j.icarus.2009.11.021
S.E. Dodson-Robinson, K. Willacy, P. Bodenheimer, N.J. Turner, C.A. Beichman, Ice lines, planetesimal composition and solid surface density in the solar nebula. Icarus **200**, 672–693 (2009). doi:10.1016/j.icarus.2008.11.023
S.D. Domagal-Goldman, V.S. Meadows, M.W. Claire, J.F. Kasting, Using biogenic sulfur gases as remotely detectable biosignatures on anoxic planets. Astrobiology **11**, 419–441 (2011). doi:10.1089/ast.2010.0509
S.D. Domagal-Goldman, A. Segura, M.W. Claire, T.D. Robinson, V.S. Meadows, Abiotic ozone and oxygen in atmospheres similar to prebiotic Earth. Astrophys. J. **792**, 90 (2014). doi:10.1088/0004-637X/792/2/90
D. Ehrenreich, X. Bonfils, C. Lovis, X. Delfosse, T. Forveille, M. Mayor, V. Neves, N.C. Santos, S. Udry, D. Ségransan, Near-infrared transmission spectrum of the warm-Uranus GJ 3470b with the wide field camera-3 on the Hubble Space Telescope. Astron. Astrophys. **570**, 89 (2014). doi:10.1051/0004-6361/201423809
D. Ehrenreich, V. Bourrier, P.J. Wheatley, A. Lecavelier des Etangs, G. Hébrard, S. Udry, X. Bonfils, X. Delfosse, J.-M. Désert, D.K. Sing, A. Vidal-Madjar, A giant comet-like cloud of hydrogen escaping the warm Neptune-mass exoplanet GJ 436b. Nature **522**, 459–461 (2015). doi:10.1038/nature14501
L.T. Elkins-Tanton, S. Seager, Ranges of atmospheric mass and composition of super-Earth exoplanets. Astrophys. J. **685**, 1237–1246 (2008). doi:10.1086/591433
L.J. Esteves, E.J.W. De Mooij, R. Jayawardhana, Changing phases of alien worlds: Probing atmospheres of Kepler planets with high-precision photometry. Astrophys. J. **804**, 150 (2015). doi:10.1088/0004-637X/804/2/150
T.M. Evans, F. Pont, D.K. Sing, S. Aigrain, J.K. Barstow, J.-M. Désert, N. Gibson, K. Heng, H.A. Knutson, A. Lecavelier des Etangs, The deep blue color of HD 189733b: Albedo measurements with Hubble Space Telescope/space telescope imaging spectrograph at visible wavelengths. Astrophys. J. Lett. **772**, 16 (2013). doi:10.1088/2041-8205/772/2/L16
D. Fabrycky, S. Tremaine, Shrinking binary and planetary orbits by Kozai cycles with tidal friction. Astrophys. J. **669**, 1298–1315 (2007). doi:10.1086/521702
Fang et al., Space Sci. Rev. (2016, this issue)
B. Fegley Jr., K. Lodders, Chemical models of the deep atmospheres of Jupiter and Saturn. Icarus **110**, 117–154 (1994). doi:10.1006/icar.1994.1111
B. Fegley Jr., K. Lodders, Atmospheric chemistry of the brown dwarf Gliese 229B: Thermochemical equilibrium predictions. Astrophys. J. Lett. **472**, 37 (1996). doi:10.1086/310356
D.A. Fischer, A.W. Howard, G.P. Laughlin, B. Macintosh, S. Mahadevan, J. Sahlmann, J.C. Yee, Exoplanet detection techniques, in *Protostars and Planets VI* (2014), pp. 715–737. doi:10.2458/azu_uapress_9780816531240-ch031
F. Forget, R.T. Pierrehumbert, Warming early Mars with carbon dioxide clouds that scatter infrared radiation. Science **278**(5341), 1273–1276 (1997)
F. Forget, F. Hourdin, R. Fournier, C. Hourdin, O. Talagrand, M. Collins, S.R. Lewis, P.L. Read, J.-P. Huot, Improved general circulation models of the Martian atmosphere from the surface to above 80 km. J. Geophys. Res. **104**, 24155–24176 (1999). doi:10.1029/1999JE001025
J.J. Fortney, The effect of condensates on the characterization of transiting planet atmospheres with transmission spectroscopy. Mon. Not. R. Astron. Soc. **364**, 649–653 (2005). doi:10.1111/j.1365-2966.2005.09587.x
J.J. Fortney, N. Nettelmann, The interior structure, composition, and evolution of giant planets. Space Sci. Rev. **152**, 423–447 (2010). doi:10.1007/s11214-009-9582-x
J.J. Fortney, D. Sudarsky, I. Hubeny, C.S. Cooper, W.B. Hubbard, A. Burrows, J.I. Lunine, On the indirect detection of sodium in the atmosphere of the planetary companion to HD 209458. Astrophys. J. **589**, 615–622 (2003). doi:10.1086/374387
J.J. Fortney, M.S. Marley, K. Lodders, D. Saumon, R. Freedman, Comparative planetary atmospheres: Models of TrES-1 and HD 209458b. Astrophys. J. Lett. **627**, 69–72 (2005). doi:10.1086/431952
J.J. Fortney, K. Lodders, M.S. Marley, R.S. Freedman, A unified theory for the atmospheres of the hot and very hot Jupiters: Two classes of irradiated atmospheres. Astrophys. J. **678**, 1419–1435 (2008a). doi:10.1086/528370

J.J. Fortney, M.S. Marley, D. Saumon, K. Lodders, Synthetic spectra and colors of young giant planet atmospheres: Effects of initial conditions and atmospheric metallicity. Astrophys. J. **683**, 1104–1116 (2008b). doi:10.1086/589942

J.J. Fortney, M. Shabram, A.P. Showman, Y. Lian, R.S. Freedman, M.S. Marley, N.K. Lewis, Transmission spectra of three-dimensional hot Jupiter model atmospheres. Astrophys. J. **709**, 1396–1406 (2010). doi:10.1088/0004-637X/709/2/1396

J.J. Fortney, C. Mordasini, N. Nettelmann, E.M.-R. Kempton, T.P. Greene, K. Zahnle, A framework for characterizing the atmospheres of low-mass low-density transiting planets. Astrophys. J. **775**, 80 (2013). doi:10.1088/0004-637X/775/1/80

L. Fossati, C.A. Haswell, C.S. Froning, L. Hebb, S. Holmes, U. Kolb, C. Helling, A. Carter, P. Wheatley, A. Collier Cameron, B. Loeillet, D. Pollacco, R. Street, H.C. Stempels, E. Simpson, S. Udry, Y.C. Joshi, R.G. West, I. Skillen, D. Wilson, Metals in the exosphere of the highly irradiated planet WASP-12b. Astrophys. J. Lett. **714**, 222–227 (2010). doi:10.1088/2041-8205/714/2/L222

J. Fraine, D. Deming, B. Benneke, H. Knutson, A. Jordán, N. Espinoza, N. Madhusudhan, A. Wilkins, K. Todorov, Water vapour absorption in the clear atmosphere of a Neptune-sized exoplanet. Nature **513**, 526–529 (2014). doi:10.1038/nature13785

F. Fressin, G. Torres, D. Charbonneau, S.T. Bryson, J. Christiansen, C.D. Dressing, J.M. Jenkins, L.M. Walkowicz, N.M. Batalha, The false positive rate of Kepler and the occurrence of planets. Astrophys. J. **766**, 81 (2013). doi:10.1088/0004-637X/766/2/81

B. Freytag, F. Allard, H.-G. Ludwig, D. Homeier, M. Steffen, The role of convection, overshoot, and gravity waves for the transport of dust in M dwarf and brown dwarf atmospheres. Astron. Astrophys. **513**, 19 (2010). doi:10.1051/0004-6361/200913354

Fridlund et al., Space Sci. Rev. (2016, this issue)

C.F. Gammie, Nonlinear outcome of gravitational instability in cooling, gaseous disks. Astrophys. J. **553**, 174–183 (2001). doi:10.1086/320631

A. García Muñoz, Physical and chemical aeronomy of HD 209458b. Planet. Space Sci. **55**, 1426–1455 (2007). doi:10.1016/j.pss.2007.03.007

N.P. Gibson, F. Pont, S. Aigrain, A new look at NICMOS transmission spectroscopy of HD 189733, GJ-436 and XO-1: No conclusive evidence for molecular features. Mon. Not. R. Astron. Soc. **411**, 2199–2213 (2011). doi:10.1111/j.1365-2966.2010.17837.x

N.P. Gibson, S. Aigrain, F. Pont, D.K. Sing, J.-M. Désert, T.M. Evans, G. Henry, N. Husnoo, H. Knutson, Probing the haze in the atmosphere of HD 189733b with Hubble Space Telescope/WFC3 transmission spectroscopy. Mon. Not. R. Astron. Soc. **422**, 753–760 (2012). doi:10.1111/j.1365-2966.2012.20655.x

N.P. Gibson, S. Aigrain, J.K. Barstow, T.M. Evans, L.N. Fletcher, P.G.J. Irwin, The optical transmission spectrum of the hot Jupiter HAT-P-32b: Clouds explain the absence of broad spectral features? Mon. Not. R. Astron. Soc. **436**, 2974–2988 (2013). doi:10.1093/mnras/stt1783

G.R. Gladstone, M. Allen, Y.L. Yung, Hydrocarbon photochemistry in the upper atmosphere of Jupiter. Icarus **119**, 1–52 (1996). doi:10.1006/icar.1996.0001

P. Goldreich, Y. Lithwick, R. Sari, Planet formation by coagulation: A focus on Uranus and Neptune. Annu. Rev. Astron. Astrophys. **42**, 549–601 (2004). doi:10.1146/annurev.astro.42.053102.134004

C.J. Grillmair, A. Burrows, D. Charbonneau, L. Armus, J. Stauffer, V. Meadows, J. van Cleve, K. von Braun, D. Levine, Strong water absorption in the dayside emission spectrum of the planet HD189733b. Nature **456**, 767–769 (2008). doi:10.1038/nature07574

T. Guillot, The interiors of giant planets: Models and outstanding questions. Annu. Rev. Earth Planet. Sci. **33**, 493–530 (2005). doi:10.1146/annurev.earth.32.101802.120325

T. Guillot, On the radiative equilibrium of irradiated planetary atmospheres. Astron. Astrophys. **520**, 27 (2010). doi:10.1051/0004-6361/200913396

R.M. Haberle, C.P. McKay, D. Tyler, R.T. Reynolds, Can synchronously rotating planets support an atmosphere?, in *Circumstellar Habitable Zones*, vol. 1, 1996, p. 29

S. Hadden, Y. Lithwick, Densities and eccentricities of 139 Kepler planets from transit time variations. Astrophys. J. **787**, 80 (2014). doi:10.1088/0004-637X/787/1/80

N. Haghighipour, The formation and dynamics of super-Earth planets. Annu. Rev. Earth Planet. Sci. **41**, 469–495 (2013). doi:10.1146/annurev-earth-042711-105340

B.M.S. Hansen, On the absorption and redistribution of energy in irradiated planets. Astrophys. J. Suppl. Ser. **179**, 484–508 (2008). doi:10.1086/591964

B.M.S. Hansen, N. Murray, Migration then assembly: Formation of Neptune-mass planets inside 1 AU. Astrophys. J. **751**, 158 (2012). doi:10.1088/0004-637X/751/2/158

B.M.S. Hansen, N. Murray, Testing in situ assembly with the Kepler planet candidate sample. Astrophys. J. **775**, 53 (2013). doi:10.1088/0004-637X/775/1/53

C.J. Hansen, J.C. Schwartz, N.B. Cowan, Features in the broad-band eclipse spectra of exoplanets: Signal or noise? Mon. Not. R. Astron. Soc. **444**, 3632–3640 (2014). doi:10.1093/mnras/stu1699

P.H. Hauschildt, F. Allard, E. Baron, The NextGen model atmosphere grid for $3000 \leq T_{eff} \leq 10000$ K. Astrophys. J. **512**, 377–385 (1999). doi:10.1086/306745
K. Haynes, A.M. Mandell, N. Madhusudhan, D. Deming, H. Knutson, Spectroscopic evidence for a temperature inversion in the dayside atmosphere of hot Jupiter WASP-33b. Astrophys. J. **806**, 146 (2015). doi:10.1088/0004-637X/806/2/146
R. Helled, P. Bodenheimer, Metallicity of the massive protoplanets around HR 8799 if formed by gravitational instability. Icarus **207**, 503–508 (2010). doi:10.1016/j.icarus.2009.11.023
R. Helled, P. Bodenheimer, The formation of Uranus and Neptune: Challenges and implications for intermediate-mass exoplanets. Astrophys. J. **789**, 69 (2014). doi:10.1088/0004-637X/789/1/69
C. Helling, A. Ackerman, F. Allard, M. Dehn, P. Hauschildt, D. Homeier, K. Lodders, M. Marley, F. Rietmeijer, T. Tsuji, P. Woitke, A comparison of chemistry and dust cloud formation in ultracool dwarf model atmospheres. Mon. Not. R. Astron. Soc. **391**, 1854–1873 (2008). doi:10.1111/j.1365-2966.2008.13991.x
C. Helling, M. Jardine, C. Stark, D. Diver, Ionization in atmospheres of brown dwarfs and extrasolar planets. III. Breakdown conditions for mineral clouds. Astrophys. J. **767**, 136 (2013). doi:10.1088/0004-637X/767/2/136
C. Helling, P. Woitke, P.B. Rimmer, I. Kamp, W.-F. Thi, R. Meijerink, Disk evolution, element abundances and cloud properties of young gas giant planets. Life **4**, 142–173 (2014). doi:10.3390/life4020142
K. Heng, B.-O. Demory, Understanding trends associated with clouds in irradiated exoplanets. Astrophys. J. **777**, 100 (2013). doi:10.1088/0004-637X/777/2/100
K. Heng, P. Kopparla, On the stability of super-Earth atmospheres. Astrophys. J. **754**(1), 60 (2012). doi:10.1088/0004-637X/754/1/60
K. Heng, J.R. Lyons, Carbon dioxide in exoplanetary atmospheres: Rarely dominant compared to carbon monoxide and water in hot, hydrogen-dominated atmospheres. Astrophys. J. **817**, 149 (2016)
K. Heng, A.P. Showman, Atmospheric dynamics of hot exoplanets. Annu. Rev. Earth Planet. Sci. **43**, 509–540 (2015). doi:10.1146/annurev-earth-060614-105146
K. Heng, S.S. Vogt, Gliese 581g as a scaled-up version of Earth: Atmospheric circulation simulations. Mon. Not. R. Astron. Soc. **415**(3), 2145–2157 (2011)
K. Heng, W. Hayek, F. Pont, D.K. Sing, On the effects of clouds and hazes in the atmospheres of hot Jupiters: Semi-analytical temperature-pressure profiles. Mon. Not. R. Astron. Soc. **420**, 20–36 (2012). doi:10.1111/j.1365-2966.2011.19943.x
P.M. Hinz, T.J. Rodigas, M.A. Kenworthy, S. Sivanandam, A.N. Heinze, E.E. Mamajek, M.R. Meyer, Thermal infrared MMTAO observations of the HR 8799 planetary system. Astrophys. J. **716**, 417–426 (2010). doi:10.1088/0004-637X/716/1/417
H.J. Hoeijmakers, R.J. de Kok, I.A.G. Snellen, M. Brogi, J.L. Birkby, H. Schwarz, A search for TiO in the optical high-resolution transmission spectrum of HD 209458b: Hindrance due to inaccuracies in the line database. Astron. Astrophys. **575**, 20 (2015). doi:10.1051/0004-6361/201424794
A.W. Howard, G.W. Marcy, S.T. Bryson, J.M. Jenkins, J.F. Rowe, N.M. Batalha, W.J. Borucki, D.G. Koch, E.W. Dunham, T.N. Gautier III, J. Van Cleve, W.D. Cochran, D.W. Latham, J.J. Lissauer, G. Torres, T.M. Brown, R.L. Gilliland, L.A. Buchhave, D.A. Caldwell, J. Christensen-Dalsgaard, D. Ciardi, F. Fressin, M.R. Haas, S.B. Howell, H. Kjeldsen, S. Seager, L. Rogers, D.D. Sasselov, J.H. Steffen, G.S. Basri, D. Charbonneau, J. Christiansen, B. Clarke, A. Dupree, D.C. Fabrycky, D.A. Fischer, E.B. Ford, J.J. Fortney, J. Tarter, F.R. Girouard, M.J. Holman, J.A. Johnson, T.C. Klaus, P. Machalek, A.V. Moorhead, R.C. Morehead, D. Ragozzine, P. Tenenbaum, J.D. Twicken, S.N. Quinn, H. Isaacson, A. Shporer, P.W. Lucas, L.M. Walkowicz, W.F. Welsh, A. Boss, E. Devore, A. Gould, J.C. Smith, R.L. Morris, A. Prsa, T.D. Morton, M. Still, S.E. Thompson, F. Mullally, M. Endl, P.J. MacQueen, Planet occurrence within 0.25 AU of solar-type stars from Kepler. Astrophys. J. Suppl. Ser. **201**, 15 (2012). doi:10.1088/0067-0049/201/2/15
Y. Hu, F. Ding, Radiative constraints on the habitability of exoplanets Gliese 581c and Gliese 581d. Astron. Astrophys. **526**, 135 (2011)
Y. Hu, F. Ding, Atmospheric circulations and climate of tidal-locking exoplanets. Sci. Sinica Phys. Mech. Astron. **43**(10), 1356 (2013). doi:10.1360/132013-252
R. Hu, S. Seager, Photochemistry in terrestrial exoplanet atmospheres. III. Photochemistry and thermochemistry in thick atmospheres on super Earths and mini Neptunes. Astrophys. J. **784**, 63 (2014). doi:10.1088/0004 637X/784/1/63
Y. Hu, J. Yang, Role of ocean heat transport in climates of tidally locked exoplanets around m dwarf stars. Proc. Natl. Acad. Sci. **111**(2), 629–634 (2014)
R. Hu, S. Seager, W. Bains, Photochemistry in terrestrial exoplanet atmospheres. I. Photochemistry model and benchmark cases. Astrophys. J. **761**, 166 (2012). doi:10.1088/0004-637X/761/2/166
R. Hu, S. Seager, W. Bains, Photochemistry in terrestrial exoplanet atmospheres. II. H_2S and SO_2 photochemistry in anoxic atmospheres. Astrophys. J. **769**, 6 (2013). doi:10.1088/0004-637X/769/1/6

R. Hu, B.-O. Demory, S. Seager, N. Lewis, A.P. Showman, A semi-analytical model of visible-wavelength phase curves of exoplanets and applications to Kepler- 7 b and Kepler- 10 b. Astrophys. J. **802**, 51 (2015a). doi:10.1088/0004-637X/802/1/51

R. Hu, S. Seager, Y.L. Yung, Helium atmospheres on warm Neptune- and sub-Neptune-sized exoplanets and applications to GJ 436b. Astrophys. J. **807**, 8 (2015b). doi:10.1088/0004-637X/807/1/8

S.-S. Huang, Life-supporting regions in the vicinity of binary systems, in *Publications of the Astronomical Society of the Pacific* (1960), pp. 106–114

W.B. Hubbard, A. Burrows, J.I. Lunine, Theory of giant planets. Annu. Rev. Astron. Astrophys. **40**, 103–136 (2002). doi:10.1146/annurev.astro.40.060401.093917

I. Hubeny, A. Burrows, D. Sudarsky, A possible bifurcation in atmospheres of strongly irradiated stars and planets. Astrophys. J. **594**, 1011–1018 (2003). doi:10.1086/377080

C.M. Huitson, D.K. Sing, A. Vidal-Madjar, G.E. Ballester, A. Lecavelier des Etangs, J.-M. Désert, F. Pont, Temperature-pressure profile of the hot Jupiter HD 189733b from HST sodium observations: Detection of upper atmospheric heating. Mon. Not. R. Astron. Soc. **422**, 2477–2488 (2012). doi:10.1111/j.1365-2966.2012.20805.x

C.M. Huitson, D.K. Sing, F. Pont, J.J. Fortney, A.S. Burrows, P.A. Wilson, G.E. Ballester, N. Nikolov, N.P. Gibson, D. Deming, S. Aigrain, T.M. Evans, G.W. Henry, A. Lecavelier des Etangs, A.P. Showman, A. Vidal-Madjar, K. Zahnle, An HST optical-to-near-IR transmission spectrum of the hot Jupiter WASP-19b: Detection of atmospheric water and likely absence of TiO. Mon. Not. R. Astron. Soc. **434**, 3252–3274 (2013). doi:10.1093/mnras/stt1243

S. Ida, D.N.C. Lin, Toward a deterministic model of planetary formation. III. Mass distribution of short-period planets around stars of various masses. Astrophys. J. **626**, 1045–1060 (2005). doi:10.1086/429953

N.K. Inamdar, H.E. Schlichting, The formation of super-Earths and mini-Neptunes with giant impacts. Mon. Not. R. Astron. Soc. **448**, 1751–1760 (2015). doi:10.1093/mnras/stv030

A.P. Ingersoll, The runaway greenhouse: A history of water on Venus. J. Atmos. Sci. **26**(6), 1191–1198 (1969)

N. Iro, B. Bézard, T. Guillot, A time-dependent radiative model of HD 209458b. Astron. Astrophys. **436**, 719–727 (2005). doi:10.1051/0004-6361:20048344

Y. Ito, M. Ikoma, H. Kawahara, H. Nagahara, Y. Kawashima, T. Nakamoto, Theoretical emission spectra of atmospheres of hot rocky super-Earths. Astrophys. J. **801**, 144 (2015). doi:10.1088/0004-637X/801/2/144

M. Janson, C. Bergfors, M. Goto, W. Brandner, D. Lafrenière, Spatially resolved spectroscopy of the exoplanet HR 8799 c. Astrophys. J. Lett. **710**, 35–38 (2010). doi:10.1088/2041-8205/710/1/L35

M. Janson, T.D. Brandt, M. Kuzuhara, D.S. Spiegel, C. Thalmann, T. Currie, M. Bonnefoy, N. Zimmerman, S. Sorahana, T. Kotani, J. Schlieder, J. Hashimoto, T. Kudo, N. Kusakabe, L. Abe, W. Brandner, J.C. Carson, S. Egner, M. Feldt, M. Goto, C.A. Grady, O. Guyon, Y. Hayano, M. Hayashi, S. Hayashi, T. Henning, K.W. Hodapp, M. Ishii, M. Iye, R. Kandori, G.R. Knapp, J. Kwon, T. Matsuo, M.W. McElwain, K. Mede, S. Miyama, J.-I. Morino, A. Moro-Martín, T. Nakagawa, T. Nishimura, T.-S. Pyo, E. Serabyn, T. Suenaga, H. Suto, R. Suzuki, Y. Takahashi, M. Takami, N. Takato, H. Terada, D. Tomono, E.L. Turner, M. Watanabe, J. Wisniewski, T. Yamada, H. Takami, T. Usuda, M. Tamura, Direct imaging detection of methane in the atmosphere of GJ 504 b. Astrophys. J. Lett. **778**, 4 (2013). doi:10.1088/2041-8205/778/1/L4

A.G. Jensen, S. Redfield, M. Endl, W.D. Cochran, L. Koesterke, T. Barman, A detection of Hα in an exoplanetary exosphere. Astrophys. J. **751**, 86 (2012). doi:10.1088/0004-637X/751/2/86

M. Joshi, Climate model studies of synchronously rotating planets. Astrobiology **3**(2), 415–427 (2003)

M. Joshi, R. Haberle, R. Reynolds, Simulations of the atmospheres of synchronously rotating terrestrial planets orbiting m dwarfs: Conditions for atmospheric collapse and the implications for habitability. Icarus **129**(2), 450–465 (1997)

J. Kasting, *How to Find a Habitable Planet.* Science Essentials (Princeton University Press, Princeton, 2010). 9780691138053

J.F. Kasting, D.P. Whitmire, R.T. Reynolds, Habitable zones around main sequence stars. Icarus **101**(1), 108–128 (1993)

G.M. Kennedy, S.J. Kenyon, Planet formation around stars of various masses: Hot super-Earths. Astrophys. J. **682**, 1264–1276 (2008). doi:10.1086/589436

D.M. Kipping, D.S. Spiegel, Detection of visible light from the darkest world. Mon. Not. R. Astron. Soc. **417**, 88–92 (2011). doi:10.1111/j.1745-3933.2011.01127.x

J.D. Kirkpatrick, New spectral types L and T. Annu. Rev. Astron. Astrophys. **43**, 195–245 (2005). doi:10.1146/annurev.astro.42.053102.134017

H.A. Knutson, D. Charbonneau, L.E. Allen, J.J. Fortney, E. Agol, N.B. Cowan, A.P. Showman, C.S. Cooper, S.T. Megeath, A map of the day-night contrast of the extrasolar planet HD 189733b. Nature **447**, 183–186 (2007). doi:10.1038/nature05782

H.A. Knutson, D. Charbonneau, N.B. Cowan, J.J. Fortney, A.P. Showman, E. Agol, G.W. Henry, The 8 μm phase variation of the hot Saturn hd 149026b. Astrophys. J. **703**(1), 769 (2009)

H.A. Knutson, N. Lewis, J.J. Fortney, A. Burrows, A.P. Showman, N.B. Cowan, E. Agol, S. Aigrain, D. Charbonneau, D. Deming, J.-M. Désert, G.W. Henry, J. Langton, G. Laughlin, 3.6 and 4.5 μm phase curves and evidence for non-equilibrium chemistry in the atmosphere of extrasolar planet HD 189733b. Astrophys. J. **754**, 22 (2012). doi:10.1088/0004-637X/754/1/22

H.A. Knutson, B. Benneke, D. Deming, D. Homeier, A featureless transmission spectrum for the Neptune-mass exoplanet GJ436b. Nature **505**, 66–68 (2014a). doi:10.1038/nature12887

H.A. Knutson, D. Dragomir, L. Kreidberg, E.M.-R. Kempton, P.R. McCullough, J.J. Fortney, J.L. Bean, M. Gillon, D. Homeier, A.W. Howard, Hubble Space Telescope near-IR transmission spectroscopy of the super-Earth HD 97658b. Astrophys. J. **794**, 155 (2014b). doi:10.1088/0004-637X/794/2/155

Q.M. Konopacky, T.S. Barman, B.A. Macintosh, C. Marois, Detection of carbon monoxide and water absorption lines in an exoplanet atmosphere. Science **339**, 1398–1401 (2013). doi:10.1126/science.1232003

R.K. Kopparapu, A revised estimate of the occurrence rate of terrestrial planets in the habitable zones around Kepler m-dwarfs. Astrophys. J. Lett. **767**(1), 8 (2013)

R.k. Kopparapu, J.F. Kasting, K.J. Zahnle, A photochemical model for the carbon-rich planet WASP-12b. Astrophys. J. **745**, 77 (2012). doi:10.1088/0004-637X/745/1/77

T.T. Koskinen, M.J. Harris, R.V. Yelle, P. Lavvas, The escape of heavy atoms from the ionosphere of HD209458b. I. A photochemical-dynamical model of the thermosphere. Icarus **226**, 1678–1694 (2013). doi:10.1016/j.icarus.2012.09.027

T.T. Koskinen, P. Lavvas, M.J. Harris, R.V. Yelle, Thermal escape from extrasolar giant planets. Philos. Trans. R. Soc. A, Math. Phys. Eng. Sci. **372**, 30089 (2014). doi:10.1098/rsta.2013.0089

L. Kreidberg, J.L. Bean, J.-M. Désert, B. Benneke, D. Deming, K.B. Stevenson, S. Seager, Z. Berta-Thompson, A. Seifahrt, D. Homeier, Clouds in the atmosphere of the super-Earth exoplanet GJ1214b. Nature **505**, 69–72 (2014a). doi:10.1038/nature12888

L. Kreidberg, J.L. Bean, J.-M. Désert, M.R. Line, J.J. Fortney, N. Madhusudhan, K.B. Stevenson, A.P. Showman, D. Charbonneau, P.R. McCullough, S. Seager, A. Burrows, G.W. Henry, M. Williamson, T. Kataria, D. Homeier, A precise water abundance measurement for the hot Jupiter WASP-43b. Astrophys. J. Lett. **793**, 27 (2014b). doi:10.1088/2041-8205/793/2/L27

L. Kreidberg, M.R. Line, J.L. Bean, K.B. Stevenson, J.-M. Désert, N. Madhusudhan, J.J. Fortney, J.K. Barstow, G.W. Henry, M.H. Williamson, A.P. Showman, A detection of water in the transmission spectrum of the hot Jupiter WASP-12b and implications for its atmospheric composition. Astrophys. J. **814**, 66 (2015). doi:10.1088/0004-637X/814/1/66

M.J. Kuchner, S. Seager, Extrasolar carbon planets. ArXiv Astrophysics e-prints (2005)

R.L. Kurucz, Model atmospheres for G, F, A, B, and O stars. Astrophys. J. Suppl. Ser. **40**, 1–340 (1979). doi:10.1086/190589

D. Lai, Star-disc-binary interactions in protoplanetary disc systems and primordial spin-orbit misalignments. Mon. Not. R. Astron. Soc. **440**, 3532–3544 (2014). doi:10.1093/mnras/stu485

H. Lammer, V. Eybl, K. Kislyakova, J. Weingrill, M. Holmström, M. Khodachenko, Y.N. Kulikov, A. Reiners, M. Leitzinger, P. Odert, et al., Uv transit observations of euv-heated expanded thermospheres of Earth-like exoplanets around m-stars: Testing atmosphere evolution scenarios. Astrophys. Space Sci. **335**(1), 39–50 (2011)

H. Lammer, A. Stökl, N.V. Erkaev, E.A. Dorfi, P. Odert, M. Güdel, Y.N. Kulikov, K.G. Kislyakova, M. Leitzinger, Origin and loss of nebula-captured hydrogen envelopes from 'sub'- to 'super-Earths' in the habitable zone of Sun-like stars. Mon. Not. R. Astron. Soc. **439**, 3225–3238 (2014). doi:10.1093/mnras/stu085

A.A. Lanotte, M. Gillon, B.-O. Demory, J.J. Fortney, N. Astudillo, X. Bonfils, P. Magain, X. Delfosse, T. Forveille, C. Lovis, M. Mayor, V. Neves, F. Pepe, D. Queloz, N. Santos, S. Udry, A global analysis of Spitzer and new HARPS data confirms the loneliness and metal-richness of GJ 436 b. Astron. Astrophys. **572**, 73 (2014). doi:10.1051/0004-6361/201424373

L.M. Lara, E. Lellouch, J.J. López-Moreno, R. Rodrigo, Vertical distribution of Titan's atmospheric neutral constituents. J. Geophys. Res. **101**, 23261–23283 (1996). doi:10.1029/96JE02036

P. Lavvas, T. Koskinen, R.V. Yelle, Chemical composition of extrasolar giant planets, in *EPSC-DPS Joint Meeting*, 2011, 2011, p. 645

P. Lavvas, T. Koskinen, R.V. Yelle, Electron densities and alkali atoms in exoplanet atmospheres. Astrophys. J. **796**, 15 (2014). doi:10.1088/0004-637X/796/1/15

A. Lecavelier Des Etangs, F. Pont, A. Vidal-Madjar, D. Sing, Rayleigh scattering in the transit spectrum of HD 189733b. Astron. Astrophys. **481**, 83–86 (2008a). doi:10.1051/0004-6361:200809388

A. Lecavelier Des Etangs, A. Vidal-Madjar, J.-M. Désert, D. Sing, Rayleigh scattering by H_2 in the extrasolar planet HD 209458b. Astron. Astrophys. **485**, 865–869 (2008b). doi:10.1051/0004-6361:200809704

A. Lecavelier Des Etangs, D. Ehrenreich, A. Vidal-Madjar, G.E. Ballester, J.-M. Désert, R. Ferlet, G. Hébrard, D.K. Sing, K.-O. Tchakoumegni, S. Udry, Evaporation of the planet HD 189733b observed in H I Lyman-α. Astron. Astrophys. **514**, 72 (2010). doi:10.1051/0004-6361/200913347

J. Leconte, F. Forget, B. Charnay, R. Wordsworth, A. Pottier, Increased insolation threshold for runaway greenhouse processes on Earth-like planets. Nature **504**(7479), 268–271 (2013)

J. Leconte, F. Forget, H. Lammer, On the (anticipated) diversity of terrestrial planet atmospheres. Exp. Astron. (2014). doi:10.1007/s10686-014-9403-4

J.-M. Lee, L.N. Fletcher, P.G.J. Irwin, Optimal estimation retrievals of the atmospheric structure and composition of HD 189733b from secondary eclipse spectroscopy. Mon. Not. R. Astron. Soc. **420**, 170–182 (2012). doi:10.1111/j.1365-2966.2011.20013.x

J.-M. Lee, K. Heng, P.G.J. Irwin, Atmospheric retrieval analysis of the directly imaged exoplanet HR 8799b. Astrophys. J. **778**, 97 (2013). doi:10.1088/0004-637X/778/2/97

E.J. Lee, E. Chiang, C.W. Ormel, Make super-Earths, not Jupiters: Accreting nebular gas onto solid cores at 0.1 AU and beyond. Astrophys. J. **797**, 95 (2014). doi:10.1088/0004-637X/797/2/95

G. Lee, C. Helling, I. Dobbs-Dixon, D. Juncher, Modelling the local and global cloud formation on HD 189733b. Astron. Astrophys. **580**, 12 (2015). doi:10.1051/0004-6361/201525982

A. Léger, O. Grasset, B. Fegley, F. Codron, A.F. Albarede, P. Barge, R. Barnes, P. Cance, S. Carpy, F. Catalano, C. Cavarroc, O. Demangeon, S. Ferraz-Mello, P. Gabor, J.-M. Grießmeier, J. Leibacher, G. Libourel, A.-S. Maurin, S.N. Raymond, D. Rouan, B. Samuel, L. Schaefer, J. Schneider, P.A. Schuller, F. Selsis, C. Sotin, The extreme physical properties of the CoRoT-7b super-Earth. Icarus **213**, 1–11 (2011). doi:10.1016/j.icarus.2011.02.004

J.S. Lewis, Observability of spectroscopically active compounds in the atmosphere of Jupiter. Icarus **10**, 393–409 (1969). doi:10.1016/0019-1035(69)90094-3

N.K. Lewis, A.P. Showman, J.J. Fortney, M.S. Marley, R.S. Freedman, K. Lodders, Atmospheric circulation of eccentric hot Neptune GJ436b. Astrophys. J. **720**, 344–356 (2010). doi:10.1088/0004-637X/720/1/344

M.-C. Liang, C.D. Parkinson, A.Y.-T. Lee, Y.L. Yung, S. Seager, Source of atomic hydrogen in the atmosphere of HD 209458b. Astrophys. J. Lett. **596**, 247–250 (2003). doi:10.1086/379314

M.-C. Liang, S. Seager, C.D. Parkinson, A.Y.-T. Lee, Y.L. Yung, On the insignificance of photochemical hydrocarbon aerosols in the atmospheres of close-in extrasolar giant planets. Astrophys. J. Lett. **605**, 61–64 (2004). doi:10.1086/392509

G.F. Lindal, The atmosphere of Neptune—An analysis of radio occultation data acquired with Voyager 2. Astron. J. **103**, 967–982 (1992). doi:10.1086/116119

R.S. Lindzen, Turbulence and stress owing to gravity wave and tidal breakdown. J. Geophys. Res. **86**, 9707–9714 (1981). doi:10.1029/JC086iC10p09707

M.R. Line, M.C. Liang, Y.L. Yung, High-temperature photochemistry in the atmosphere of HD 189733b. Astrophys. J. **717**, 496–502 (2010). doi:10.1088/0004-637X/717/1/496

M.R. Line, G. Vasisht, P. Chen, D. Angerhausen, Y.L. Yung, Thermochemical and photochemical kinetics in cooler hydrogen-dominated extrasolar planets: A methane-poor GJ436b? Astrophys. J. **738**, 32 (2011). doi:10.1088/0004-637X/738/1/32

M.R. Line, X. Zhang, G. Vasisht, V. Natraj, P. Chen, Y.L. Yung, Information content of exoplanetary transit spectra: An initial look. Astrophys. J. **749**, 93 (2012). doi:10.1088/0004-637X/749/1/93

M.R. Line, A.S. Wolf, X. Zhang, H. Knutson, J.A. Kammer, E. Ellison, P. Deroo, D. Crisp, Y.L. Yung, A systematic retrieval analysis of secondary eclipse spectra. I. A comparison of atmospheric retrieval techniques. Astrophys. J. **775**, 137 (2013). doi:10.1088/0004-637X/775/2/137

J.L. Linsky, H. Yang, K. France, C.S. Froning, J.C. Green, J.T. Stocke, S.N. Osterman, Observations of mass loss from the transiting exoplanet HD 209458b. Astrophys. J. **717**, 1291–1299 (2010). doi:10.1088/0004-637X/717/2/1291

J.J. Lissauer, Planet formation. Annu. Rev. Astron. Astrophys. **31**, 129–174 (1993). doi:10.1146/annurev.aa.31.090193.001021

J.J. Lissauer, D.J. Stevenson, Formation of giant planets, in *Protostars and Planets V* (2007), pp. 591–606

A.C. Lockwood, M.E. Brown, J. Stansberry, The size and shape of the oblong dwarf planet Haumea. Earth Moon Planets **111**, 127–137 (2014). doi:10.1007/s11038-014-9430-1

K. Lodders, Alkali element chemistry in cool dwarf atmospheres. Astrophys. J. **519**, 793–801 (1999). doi:10.1086/307387

K. Lodders, Titanium and vanadium chemistry in low-mass dwarf stars. Astrophys. J. **577**, 974–985 (2002). doi:10.1086/342241

K. Lodders, Jupiter formed with more tar than ice. Astrophys. J. **611**, 587–597 (2004). doi:10.1086/421970

K. Lodders, in *Exoplanet Chemistry*, ed. by R. Barnes (2010), p. 157. doi:10.1002/9783527629763.ch8

K. Lodders, B. Fegley, Atmospheric chemistry in giant planets, brown dwarfs, and low-mass dwarf stars. I. Carbon, nitrogen, and oxygen. Icarus **155**, 393–424 (2002). doi:10.1006/icar.2001.6740

E.D. Lopez, J.J. Fortney, The role of core mass in controlling evaporation: The Kepler radius distribution and the Kepler-36 density dichotomy. Astrophys. J. **776**, 2 (2013). doi:10.1088/0004-637X/776/1/2

E.D. Lopez, J.J. Fortney, Understanding the mass-radius relation for sub-Neptunes: Radius as a proxy for composition. Astrophys. J. **792**, 1 (2014). doi:10.1088/0004-637X/792/1/1

E.D. Lopez, J.J. Fortney, N. Miller, How thermal evolution and mass-loss sculpt populations of super-Earths and sub-Neptunes: Application to the Kepler-11 system and beyond. Astrophys. J. **761**, 59 (2012). doi:10.1088/0004-637X/761/1/59

R. Luger, R. Barnes, Extreme water loss and abiotic O2Buildup on planets throughout the habitable zones of M dwarfs. Astrobiology **15**, 119–143 (2015). doi:10.1089/ast.2014.1231

R. Luger, R. Barnes, E. Lopez, J. Fortney, B. Jackson, V. Meadows, Habitable evaporated cores: Transforming mini-Neptunes into super-Earths in the habitable zones of M dwarfs. Astrobiology **15**, 57–88 (2015). doi:10.1089/ast.2014.1215

R.E. Lupu, K. Zahnle, M.S. Marley, L. Schaefer, B. Fegley, C. Morley, K. Cahoy, R. Freedman, J.J. Fortney, The atmospheres of earthlike planets after giant impact events. Astrophys. J. **784**, 27 (2014). doi:10.1088/0004-637X/784/1/27

B. Macintosh, J.R. Graham, T. Barman, R.J. De Rosa, Q. Konopacky, M.S. Marley, C. Marois, E.L. Nielsen, L. Pueyo, A. Rajan, J. Rameau, D. Saumon, J.J. Wang, J. Patience, M. Ammons, P. Arriaga, E. Artigau, S. Beckwith, J. Brewster, S. Bruzzone, J. Bulger, B. Burningham, A.S. Burrows, C. Chen, E. Chiang, J.K. Chilcote, R.I. Dawson, R. Dong, R. Doyon, Z.H. Draper, G. Duchêne, T.M. Esposito, D. Fabrycky, M.P. Fitzgerald, K.B. Follette, J.J. Fortney, B. Gerard, S. Goodsell, A.Z. Greenbaum, P. Hibon, S. Hinkley, T.H. Cotten, L.-W. Hung, P. Ingraham, M. Johnson-Groh, P. Kalas, D. Lafreniere, J.E. Larkin, J. Lee, M. Line, D. Long, J. Maire, F. Marchis, B.C. Matthews, C.E. Max, S. Metchev, M.A. Millar-Blanchaer, T. Mittal, C.V. Morley, K.M. Morzinski, R. Murray-Clay, R. Oppenheimer, D.W. Palmer, R. Patel, M.D. Perrin, L.A. Poyneer, R.R. Rafikov, F.T. Rantakyrö, E.L. Rice, P. Rojo, A.R. Rudy, J.-B. Ruffio, M.T. Ruiz, N. Sadakuni, L. Saddlemyer, M. Salama, D. Savransky, A.C. Schneider, A. Sivaramakrishnan, I. Song, R. Soummer, S. Thomas, G. Vasisht, J.K. Wallace, K. Ward-Duong, S.J. Wiktorowicz, S.G. Wolff, B. Zuckerman, Discovery and spectroscopy of the young Jovian planet 51 Eri b with the Gemini Planet Imager. Science **350**, 64–67 (2015). doi:10.1126/science.aac5891

N. Madhusudhan, C/O ratio as a dimension for characterizing exoplanetary atmospheres. Astrophys. J. **758**, 36 (2012). doi:10.1088/0004-637X/758/1/36

N. Madhusudhan, S. Redfield, Optimal measures for characterizing water-rich super-Earths. Int. J. Astrobiol. **14**, 177–189 (2015). doi:10.1017/S1473550414000421

N. Madhusudhan, S. Seager, A temperature and abundance retrieval method for exoplanet atmospheres. Astrophys. J. **707**, 24–39 (2009). doi:10.1088/0004-637X/707/1/24

N. Madhusudhan, S. Seager, High metallicity and non-equilibrium chemistry in the dayside atmosphere of hot-Neptune GJ 436b. Astrophys. J. **729**, 41 (2011). doi:10.1088/0004-637X/729/1/41

N. Madhusudhan, A. Burrows, T. Currie, Model atmospheres for massive gas giants with thick clouds: Application to the HR 8799 planets and predictions for future detections. Astrophys. J. **737**, 34 (2011a). doi:10.1088/0004-637X/737/1/34

N. Madhusudhan, J. Harrington, K.B. Stevenson, S. Nymeyer, C.J. Campo, P.J. Wheatley, D. Deming, J. Blecic, R.A. Hardy, N.B. Lust, D.R. Anderson, A. Collier-Cameron, C.B.T. Britt, W.C. Bowman, L. Hebb, C. Hellier, P.F.L. Maxted, D. Pollacco, R.G. West, A high C/O ratio and weak thermal inversion in the atmosphere of exoplanet WASP-12b. Nature **469**, 64–67 (2011b). doi:10.1038/nature09602

N. Madhusudhan, M.A. Amin, G.M. Kennedy, Toward chemical constraints on hot Jupiter migration. Astrophys. J. Lett. **794**, 12 (2014a). doi:10.1088/2041-8205/794/1/L12

N. Madhusudhan, N. Crouzet, P.R. McCullough, D. Deming, C. Hedges, H_2O abundances in the atmospheres of three hot Jupiters. Astrophys. J. Lett. **791**, 9 (2014b). doi:10.1088/2041-8205/791/1/L9

N. Madhusudhan, H. Knutson, J.J. Fortney, T. Barman, Exoplanetary atmospheres, in *Protostars and Planets VI* (2014c), pp. 739–762. doi:10.2458/azu_uapress_9780816531240-ch032

P.R. Mahaffy, H.B. Niemann, A. Alpert, S.K. Atreya, J. Demick, T.M. Donahue, D.N. Harpold, T.C. Owen, Noble gas abundance and isotope ratios in the atmosphere of Jupiter from the Galileo probe mass spectrometer. J. Geophys. Res. **105**, 15061–15072 (2000). doi:10.1029/1999JE001224

L. Mancini, S. Ciceri, G. Chen, J. Tregloan-Reed, J.J. Fortney, J. Southworth, T.G. Tan, M. Burgdorf, S. Calchi Novati, M. Dominik, X.-S. Fang, F. Finet, T. Gerner, S. Hardis, T.C. Hinse, U.G. Jørgensen, C. Liebig, N. Nikolov, D. Ricci, S. Schäfer, F. Schönebeck, J. Skottfelt, O. Wertz, K.A. Alsubai, V. Bozza, P. Browne, P. Dodds, S.-H. Gu, K. Harpsøe, T. Henning, M. Hundertmark, J. Jessen-Hansen, N. Kains, E. Kerins, H. Kjeldsen, M.N. Lund, M. Lundkvist, N. Madhusudhan, M. Mathiasen, M.T. Penny, S. Prof, S. Rahvar, K. Sahu, G. Scarpetta, C. Snodgrass, J. Surdej, Physical properties, transmission and emission spectra of the WASP-19 planetary system from multi-colour photometry. Mon. Not. R. Astron. Soc. **436**, 2–18 (2013). doi:10.1093/mnras/stt1394

A.M. Mandell, K. Haynes, E. Sinukoff, N. Madhusudhan, A. Burrows, D. Deming, Exoplanet transit spectroscopy using WFC3: WASP-12 b, WASP-17 b, and WASP-19 b. Astrophys. J. **779**, 128 (2013). doi:10.1088/0004-637X/779/2/128

U. Marboeuf, A. Thiabaud, Y. Alibert, N. Cabral, W. Benz, From stellar nebula to planetesimals. Astron. Astrophys. **570**, 35 (2014). doi:10.1051/0004-6361/201322207

G.W. Marcy, H. Isaacson, A.W. Howard, J.F. Rowe, J.M. Jenkins, S.T. Bryson, D.W. Latham, S.B. Howell, T.N. Gautier III, N.M. Batalha, L. Rogers, D. Ciardi, D.A. Fischer, R.L. Gilliland, H. Kjeldsen, J. Christensen-Dalsgaard, D. Huber, W.J. Chaplin, S. Basu, L.A. Buchhave, S.N. Quinn, W.J. Borucki, D.G. Koch, R. Hunter, D.A. Caldwell, J. Van Cleve, R. Kolbl, L.M. Weiss, E. Petigura, S. Seager, T. Morton, J.A. Johnson, S. Ballard, C. Burke, W.D. Cochran, M. Endl, P. MacQueen, M.E. Everett, J.J. Lissauer, E.B. Ford, G. Torres, F. Fressin, T.M. Brown, J.H. Steffen, D. Charbonneau, G.S. Basri, D.D. Sasselov, J. Winn, R. Sanchis-Ojeda, J. Christiansen, E. Adams, C. Henze, A. Dupree, D.C. Fabrycky, J.J. Fortney, J. Tarter, M.J. Holman, P. Tenenbaum, A. Shporer, P.W. Lucas, W.F. Welsh, J.A. Orosz, T.R. Bedding, T.L. Campante, G.R. Davies, Y. Elsworth, R. Handberg, S. Hekker, C. Karoff, S.D. Kawaler, M.N. Lund, M. Lundkvist, T.S. Metcalfe, A. Miglio, V. Silva Aguirre, D. Stello, T.R. White, A. Boss, E. Devore, A. Gould, A. Prsa, E. Agol, T. Barclay, J. Coughlin, E. Brugamyer, F. Mullally, E.V. Quintana, M. Still, S.E. Thompson, D. Morrison, J.D. Twicken, J.-M. Désert, J. Carter, J.R. Crepp, G. Hébrard, A. Santerne, C. Moutou, C. Sobeck, D. Hudgins, M.R. Haas, P. Robertson, J. Lillo-Box, D. Barrado, Masses, radii, and orbits of small Kepler planets: The transition from gaseous to rocky planets. Astrophys. J. Suppl. Ser. **210**, 20 (2014). doi:10.1088/0067-0049/210/2/20

M.S. Marley, T.D. Robinson, On the cool side: Modeling the atmospheres of brown dwarfs and giant planets. Annu. Rev. Astron. Astrophys. **53**, 279–323 (2015). doi:10.1146/annurev-astro-082214-122522

M.S. Marley, D. Saumon, T. Guillot, R.S. Freedman, W.B. Hubbard, A. Burrows, J.I. Lunine, Atmospheric, evolutionary, and spectral models of the brown dwarf Gliese 229 B. Science **272**, 1919–1921 (1996). doi:10.1126/science.272.5270.1919

M.S. Marley, A.S. Ackerman, J.N. Cuzzi, D. Kitzmann, Clouds and hazes in exoplanet atmospheres, in *Comparative Climatology of Terrestrial Planets*, ed. by S.J. Mackwell, A.A. Simon-Miller, J.W. Harder, M.A. Bullock (2013), pp. 367–391. doi:10.2458/azu_uapress_9780816530595-ch15

J.H.C. Martins, N.C. Santos, P. Figueira, J.P. Faria, M. Montalto, I. Boisse, D. Ehrenreich, C. Lovis, M. Mayor, C. Melo, F. Pepe, S.G. Sousa, S. Udry, D. Cunha, Evidence for a spectroscopic direct detection of reflected light from <ASTROBJ>51 Pegasi b</ASTROBJ>. Astron. Astrophys. **576**, 134 (2015). doi:10.1051/0004-6361/201425298

Massol et al., Space Sci. Rev. (2016, this issue)

S. Matousek, The Juno new frontiers mission. Acta Astronaut. **61**, 932–939 (2007). doi:10.1016/j.actaastro.2006.12.013

M. Mayor, M. Marmier, C. Lovis, S. Udry, D. Ségransan, F. Pepe, W. Benz, J.-. Bertaux, F. Bouchy, X. Dumusque, G. Lo Curto, C. Mordasini, D. Queloz, N.C. Santos, The HARPS search for southern extra-solar planets XXXIV. Occurrence, mass distribution and orbital properties of super-Earths and Neptune-mass planets. ArXiv e-prints (2011)

P. McCullough, J. MacKenty, Considerations for using Spatial Scans with WFC3. Technical report, 2012

P.R. McCullough, N. Crouzet, D. Deming, N. Madhusudhan, Water vapor in the spectrum of the extrasolar planet HD 189733b. I. The transit. Astrophys. J. **791**, 55 (2014). doi:10.1088/0004-637X/791/1/55

C.P. McKay, J.B. Pollack, R. Courtin, The thermal structure of Titan's atmosphere. Icarus **80**, 23–53 (1989). doi:10.1016/0019-1035(89)90160-7

G.L. Mellor, T. Yamada, Development of a turbulence closure model for geophysical fluid problems. Rev. Geophys. Space Phys. **20**, 851–875 (1982). doi:10.1029/RG020i004p00851

K. Menou, Water-trapped worlds. Astrophys. J. **774**(1), 51 (2013)

Y. Miguel, L. Kaltenegger, Exploring atmospheres of hot mini-Neptunes and extrasolar giant planets orbiting different stars with application to HD 97658b, WASP-12b, CoRoT-2b, XO-1b, and HD 189733b. Astrophys. J. **780**, 166 (2014). doi:10.1088/0004-637X/780/2/166

Y. Miguel, L. Kaltenegger, B. Fegley, L. Schaefer, Compositions of hot super-Earth atmospheres: Exploring Kepler candidates. Astrophys. J. Lett. **742**, 19 (2011). doi:10.1088/2041-8205/742/2/L19

Y. Miguel, L. Kaltenegger, J.L. Linsky, S. Rugheimer, The effect of Lyman α radiation on mini-Neptune atmospheres around M stars: Application to GJ 436b. Mon. Not. R. Astron. Soc. **446**, 345–353 (2015). doi:10.1093/mnras/stu2107

N. Miller, J.J. Fortney, The heavy-element masses of extrasolar giant planets, revealed. Astrophys. J. Lett. **736**, 29 (2011). doi:10.1088/2041-8205/736/2/L29

E. Miller-Ricci Kempton, K. Zahnle, J.J. Fortney, The atmospheric chemistry of GJ 1214b: Photochemistry and clouds. Astrophys. J. **745**, 3 (2012). doi:10.1088/0004-637X/745/1/3

E. Miller-Ricci, S. Seager, D. Sasselov, The atmospheric signatures of super-Earths: How to distinguish between hydrogen-rich and hydrogen-poor atmospheres. Astrophys. J. **690**, 1056–1067 (2009). doi:10.1088/0004-637X/690/2/1056
A. Morbidelli, J.I. Lunine, D.P. O'Brien, S.N. Raymond, K.J. Walsh, Building terrestrial planets. Annu. Rev. Earth Planet. Sci. **40**, 251–275 (2012). doi:10.1146/annurev-earth-042711-105319
C. Mordasini, Y. Alibert, C. Georgy, K.-M. Dittkrist, H. Klahr, T. Henning, Characterization of exoplanets from their formation. II. The planetary mass-radius relationship. Astron. Astrophys. **547**, 112 (2012). doi:10.1051/0004-6361/201118464
C.V. Morley, J.J. Fortney, E.M.-R. Kempton, M.S. Marley, C. Visscher, K. Zahnle, Quantitatively assessing the role of clouds in the transmission spectrum of GJ 1214b. Astrophys. J. **775**, 33 (2013). doi:10.1088/0004-637X/775/1/33
J.I. Moses, Chemical kinetics on extrasolar planets. Philos. Trans. R. Soc. A, Math. Phys. Eng. Sci. **372**, 30073 (2014). doi:10.1098/rsta.2013.0073
J.I. Moses, B. Bézard, E. Lellouch, G.R. Gladstone, H. Feuchtgruber, M. Allen, Photochemistry of Saturn's atmosphere. I. Hydrocarbon chemistry and comparisons with ISO observations. Icarus **143**, 244–298 (2000). doi:10.1006/icar.1999.6270
J.I. Moses, C. Visscher, J.J. Fortney, A.P. Showman, N.K. Lewis, C.A. Griffith, S.J. Klippenstein, M. Shabram, A.J. Friedson, M.S. Marley, R.S. Freedman, Disequilibrium carbon, oxygen, and nitrogen chemistry in the atmospheres of HD 189733b and HD 209458b. Astrophys. J. **737**, 15 (2011). doi:10.1088/0004-637X/737/1/15
J.I. Moses, M.R. Line, C. Visscher, M.R. Richardson, N. Nettelmann, J.J. Fortney, T.S. Barman, K.B. Stevenson, N. Madhusudhan, Compositional diversity in the atmospheres of hot Neptunes, with application to GJ 436b. Astrophys. J. **777**, 34 (2013a). doi:10.1088/0004-637X/777/1/34
J.I. Moses, N. Madhusudhan, C. Visscher, R.S. Freedman, Chemical consequences of the C/O ratio on hot Jupiters: Examples from WASP-12b, CoRoT-2b, XO-1b, and HD 189733b. Astrophys. J. **763**, 25 (2013b). doi:10.1088/0004-637X/763/1/25
O. Mousis, J.I. Lunine, N. Madhusudhan, T.V. Johnson, Nebular water depletion as the cause of Jupiter's low oxygen abundance. Astrophys. J. Lett. **751**, 7 (2012). doi:10.1088/2041-8205/751/1/L7
F. Murgas, E. Pallé, M.R. Zapatero Osorio, L. Nortmann, S. Hoyer, A. Cabrera-Lavers, The GTC exoplanet transit spectroscopy survey. I. OSIRIS transmission spectroscopy of the short period planet WASP-43b. Astron. Astrophys. **563**, 41 (2014). doi:10.1051/0004-6361/201322374
N. Nettelmann, U. Kramm, R. Redmer, R. Neuhäuser, Interior structure models of GJ 436b. Astron. Astrophys. **523**, 26 (2010). doi:10.1051/0004-6361/200911985
N. Nikolov, D.K. Sing, F. Pont, A.S. Burrows, J.J. Fortney, G.E. Ballester, T.M. Evans, C.M. Huitson, H.R. Wakeford, P.A. Wilson, S. Aigrain, D. Deming, N.P. Gibson, G.W. Henry, H. Knutson, A. Lecavelier des Etangs, A.P. Showman, A. Vidal-Madjar, K. Zahnle, Hubble Space Telescope hot Jupiter transmission spectral survey: A detection of Na and strong optical absorption in HAT-P-1b. Mon. Not. R. Astron. Soc. **437**, 46–66 (2014). doi:10.1093/mnras/stt1859
N. Nikolov, D.K. Sing, A.S. Burrows, J.J. Fortney, G.W. Henry, F. Pont, G.E. Ballester, S. Aigrain, P.A. Wilson, C.M. Huitson, N.P. Gibson, J.-M. Désert, A.L.d. Etangs, A.P. Showman, A. Vidal-Madjar, H.R. Wakeford, K. Zahnle, HST hot-Jupiter transmission spectral survey: Haze in the atmosphere of WASP-6b. Mon. Not. R. Astron. Soc. **447**, 463–478 (2015). doi:10.1093/mnras/stu2433
K.I. Öberg, R. Murray-Clay, E.A. Bergin, The effects of snowlines on C/O in planetary atmospheres. Astrophys. J. Lett. **743**, 16 (2011). doi:10.1088/2041-8205/743/1/L16
Oberg et al., Space Sci. Rev. (2016, this issue)
G.I. Ogilvie, Tidal dissipation in stars and giant planets. Annu. Rev. Astron. Astrophys. **52**, 171–210 (2014). doi:10.1146/annurev-astro-081913-035941
B.R. Oppenheimer, C. Baranec, C. Beichman, D. Brenner, R. Burruss, E. Cady, J.R. Crepp, R. Dekany, R. Fergus, D. Hale, L. Hillenbrand, S. Hinkley, D.W. Hogg, D. King, E.R. Ligon, T. Lockhart, R. Nilsson, I.R. Parry, L. Pueyo, E. Rice, J.E. Roberts, L.C. Roberts Jr., M. Shao, A. Sivaramakrishnan, R. Soummer, T. Truong, G. Vasisht, A. Veicht, F. Vescelus, J.K. Wallace, C. Zhai, N. Zimmerman, Reconnaissance of the HR 8799 exosolar system. I. Near-infrared spectroscopy. Astrophys. J. **768**, 24 (2013). doi:10.1088/0004-637X/768/1/24
J.E. Owen, Y. Wu, Kepler planets: A tale of evaporation. Astrophys. J. **775**, 105 (2013). doi:10.1088/0004-637X/775/2/105
T. Owen, P. Mahaffy, H.B. Niemann, S. Atreya, T. Donahue, A. Bar-Nun, I. de Pater, A low-temperature origin for the planetesimals that formed Jupiter. Nature **402**, 269–270 (1999). doi:10.1038/46232
J.C.B. Papaloizou, C. Terquem, Planet formation and migration. Rep. Prog. Phys. **69**, 119–180 (2006). doi:10.1088/0034-4885/69/1/R03
J.C.B. Papaloizou, R.P. Nelson, W. Kley, F.S. Masset, P. Artymowicz, Disk-planet interactions during planet formation, in *Protostars and Planets V* (2007), pp. 655–668

V. Parmentier, T. Guillot, A non-grey analytical model for irradiated atmospheres. I. Derivation. Astron. Astrophys. **562**, 133 (2014). doi:10.1051/0004-6361/201322342

V. Parmentier, A.P. Showman, Y. Lian, 3D mixing in hot Jupiters atmospheres. I. Application to the day/night cold trap in HD 209458b. Astron. Astrophys. **558**, 91 (2013). doi:10.1051/0004-6361/201321132

V. Parmentier, T. Guillot, J.J. Fortney, M.S. Marley, A non-grey analytical model for irradiated atmospheres. II. Analytical vs. numerical solutions. Astron. Astrophys. **574**, 35 (2015). doi:10.1051/0004-6361/201323127

J. Peixoto, A.H. Oort, *Physics of Climate* (Am. Inst. of Phys., New York, 1992). 520 pp.

R.O. Pepin, Atmospheres on the terrestrial planets: Clues to origin and evolution. Earth Planet. Sci. Lett. **252**, 1–14 (2006). doi:10.1016/j.epsl.2006.09.014

N.A. Phillips, The general circulation of the atmosphere: A numerical experiment. Q. J. R. Meteorol. Soc. **82**, 123–164 (1956). doi:10.1002/qj.49708235202

R.T. Pierrehumbert, A palette of climates for Gliese 581g. Astrophys. J. Lett. **726**(1), 8 (2011)

J.B. Pollack, O. Hubickyj, P. Bodenheimer, J.J. Lissauer, M. Podolak, Y. Greenzweig, Formation of the giant planets by concurrent accretion of solids and gas. Icarus **124**, 62–85 (1996). doi:10.1006/icar.1996.0190

F. Pont, H. Knutson, R.L. Gilliland, C. Moutou, D. Charbonneau, Detection of atmospheric haze on an extrasolar planet: The 0.55-1.05 μm transmission spectrum of HD 189733b with the Hubble Space Telescope. Mon. Not. R. Astron. Soc. **385**, 109–118 (2008). doi:10.1111/j.1365-2966.2008.12852.x

F. Pont, D.K. Sing, N.P. Gibson, S. Aigrain, G. Henry, N. Husnoo, The prevalence of dust on the exoplanet HD 189733b from Hubble and Spitzer observations. Mon. Not. R. Astron. Soc. **432**, 2917–2944 (2013). doi:10.1093/mnras/stt651

R.G. Prinn, S.S. Barshay, Carbon monoxide on Jupiter and implications for atmospheric convection. Science **198**, 1031–1034 (1977). doi:10.1126/science.198.4321.1031

R.R. Rafikov, Can giant planets form by direct gravitational instability? Astrophys. J. Lett. **621**, 69–72 (2005). doi:10.1086/428899

S. Ranjan, D. Charbonneau, J.-M. Désert, N. Madhusudhan, D. Deming, A. Wilkins, A.M. Mandell, Atmospheric characterization of five hot Jupiters with the wide field camera 3 on the Hubble Space Telescope. Astrophys. J. **785**, 148 (2014). doi:10.1088/0004-637X/785/2/148

F.A. Rasio, E.B. Ford, Dynamical instabilities and the formation of extrasolar planetary systems. Science **274**, 954–956 (1996). doi:10.1126/science.274.5289.954

H. Rauer, C. Catala, C. Aerts, T. Appourchaux, W. Benz, A. Brandeker, J. Christensen-Dalsgaard, M. Deleuil, L. Gizon, M.-J. Goupil, M. Güdel, E. Janot-Pacheco, M. Mas-Hesse, I. Pagano, G. Piotto, D. Pollacco, Ċ. Santos, A. Smith, J.-C. Suárez, R. Szabó, S. Udry, V. Adibekyan, Y. Alibert, J.-M. Almenara, P. Amaro-Seoane, M.A.-v. Eiff, M. Asplund, E. Antonello, S. Barnes, F. Baudin, K. Belkacem, M. Bergemann, G. Bihain, A.C. Birch, X. Bonfils, I. Boisse, A.S. Bonomo, F. Borsa, I.M. Brandão, E. Brocato, S. Brun, M. Burleigh, R. Burston, J. Cabrera, S. Cassisi, W. Chaplin, S. Charpinet, C. Chiappini, R.P. Church, S. Csizmadia, M. Cunha, M. Damasso, M.B. Davies, H.J. Deeg, R.F. Díaz, S. Dreizler, C. Dreyer, P. Eggenberger, D. Ehrenreich, P. Eigmüller, A. Erikson, R. Farmer, S. Feltzing, F. de Oliveira Fialho, P. Figueira, T. Forveille, M. Fridlund, R.A. García, P. Giommi, G. Giuffrida, M. Godolt, J. Gomes da Silva, T. Granzer, J.L. Grenfell, A. Grotsch-Noels, E. Günther, C.A. Haswell, A.P. Hatzes, G. Hébrard, S. Hekker, R. Helled, K. Heng, J.M. Jenkins, A. Johansen, M.L. Khodachenko, K.G. Kislyakova, W. Kley, U. Kolb, N. Krivova, F. Kupka, H. Lammer, A.F. Lanza, Y. Lebreton, D. Magrin, P. Marcos-Arenal, P.M. Marrese, J.P. Marques, J. Martins, S. Mathis, S. Mathur, S. Messina, A. Miglio, J. Montalban, M. Montalto, M.J.P.F.G. Monteiro, H. Moradi, E. Moravveji, C. Mordasini, T. Morel, A. Mortier, V. Nascimbeni, R.P. Nelson, M.B. Nielsen, L. Noack, A.J. Norton, A. Ofir, M. Oshagh, R.-M. Ouazzani, P. Pápics, V.C. Parro, P. Petit, B. Plez, E. Poretti, A. Quirrenbach, R. Ragazzoni, G. Raimondo, M. Rainer, D.R. Reese, R. Redmer, S. Reffert, B. Rojas-Ayala, I.W. Roxburgh, S. Salmon, A. Santerne, J. Schneider, J. Schou, S. Schuh, H. Schunker, A. Silva-Valio, R. Silvotti, I. Skillen, I. Snellen, F. Sohl, S.G. Sousa, A. Sozzetti, D. Stello, K.G. Strassmeier, M. Švanda, G.M. Szabó, A. Tkachenko, D. Valencia, V. Van Grootel, S.D. Vauclair, P. Ventura, F.W. Wagner, N.A. Walton, J. Weingrill, S.C. Werner, P.J. Wheatley, K. Zwintz, The PLATO 2.0 mission. Exp. Astron. **38**, 249–330 (2014). doi:10.1007/s10686-014-9383-4

S. Redfield, M. Endl, W.D. Cochran, L. Koesterke, Sodium absorption from the exoplanetary atmosphere of HD 189733b detected in the optical transmission spectrum. Astrophys. J. Lett. **673**, 87–90 (2008). doi:10.1086/527475

G.R. Ricker, J.N. Winn, R. Vanderspek, D.W. Latham, G.Á. Bakos, J.L. Bean, Z.K. Berta-Thompson, T.M. Brown, L. Buchhave, N.R. Butler, R.P. Butler, W.J. Chaplin, D. Charbonneau, J. Christensen-Dalsgaard, M. Clampin, D. Deming, J. Doty, N. De Lee, C. Dressing, E.W. Dunham, M. Endl, F. Fressin, J. Ge, T. Henning, M.J. Holman, A.W. Howard, S. Ida, J. Jenkins, G. Jernigan, J.A. Johnson, L. Kaltenegger, N. Kawai, H. Kjeldsen, G. Laughlin, A.M. Levine, D. Lin, J.J. Lissauer, P. MacQueen, G. Marcy, P.R. McCullough, T.D. Morton, N. Narita, M. Paegert, E. Palle, F. Pepe, J. Pepper, A. Quirrenbach, S.A. Rinehart, D. Sasselov, B. Sato, S. Seager, A. Sozzetti, K.G. Stassun, P. Sullivan, A. Szentgyorgyi, G.

Torres, S. Udry, J. Villasenor, Transiting Exoplanet Survey Satellite (TESS). J. Astron. Telesc. Instrum. Syst. **1**(1), 014003 (2015). doi:10.1117/1.JATIS.1.1.014003
P.B. Rimmer, C. Helling, Ionization in atmospheres of brown dwarfs and extrasolar planets. IV. The effect of cosmic rays. Astrophys. J. **774**, 108 (2013). doi:10.1088/0004-637X/774/2/108
P.B. Rimmer, C. Helling, A chemical kinetics network for lightning and life in planetary atmospheres. ArXiv e-prints (2015)
P.B. Rimmer, C. Helling, C. Bilger, The influence of galactic cosmic rays on ion-neutral hydrocarbon chemistry in the upper atmospheres of free-floating exoplanets. Int. J. Astrobiol. **13**, 173–181 (2014). doi:10.1017/S1473550413000487
T.D. Robinson, D.C. Catling, An analytic radiative-convective model for planetary atmospheres. Astrophys. J. **757**, 104 (2012). doi:10.1088/0004-637X/757/1/104
F. Rodler, M. Lopez-Morales, I. Ribas, Weighing the non-transiting hot Jupiter τ Boo b. Astrophys. J. Lett. **753**, 25 (2012). doi:10.1088/2041-8205/753/1/L25
F. Rodler, M. Kürster, J.R. Barnes, Detection of CO absorption in the atmosphere of the hot Jupiter HD 189733b. Mon. Not. R. Astron. Soc. **432**, 1980–1988 (2013). doi:10.1093/mnras/stt462
M. Rodono, The atmospheres of m dwarfs: Observations. NASA Spec. Publ. **492**, 409–453 (1986)
L.A. Rogers, Most 1.6 Earth-radius planets are not rocky. Astrophys. J. **801**, 41 (2015). doi:10.1088/0004-637X/801/1/41
L.A. Rogers, P. Bodenheimer, J.J. Lissauer, S. Seager, Formation and structure of low-density exo-Neptunes. Astrophys. J. **738**, 59 (2011). doi:10.1088/0004-637X/738/1/59
W.B. Rossow, Cloud microphysics—Analysis of the clouds of Earth, Venus, Mars, and Jupiter. Icarus **36**, 1–50 (1978). doi:10.1016/0019-1035(78)90072-6
J.F. Rowe, J.M. Matthews, S. Seager, E. Miller-Ricci, D. Sasselov, R. Kuschnig, D.B. Guenther, A.F.J. Moffat, S.M. Rucinski, G.A.H. Walker, W.W. Weiss, The very low albedo of an extrasolar planet: MOST space-based photometry of HD 209458. Astrophys. J. **689**, 1345–1353 (2008). doi:10.1086/591835
J.F. Rowe, J.L. Coughlin, V. Antoci, T. Barclay, N.M. Batalha, W.J. Borucki, C.J. Burke, S.T. Bryson, D.A. Caldwell, J.R. Campbell, J.H. Catanzarite, J.L. Christiansen, W. Cochran, R.L. Gilliland, F.R. Girouard, M.R. Haas, K.G. Hełminiak, C.E. Henze, K.L. Hoffman, S.B. Howell, D. Huber, R.C. Hunter, H. Jang-Condell, J.M. Jenkins, T.C. Klaus, D.W. Latham, J. Li, J.J. Lissauer, S.D. McCauliff, R.L. Morris, F. Mullally, A. Ofir, B. Quarles, E. Quintana, A. Sabale, S. Seader, A. Shporer, J.C. Smith, J.H. Steffen, M. Still, P. Tenenbaum, S.E. Thompson, J.D. Twicken, C. Van Laerhoven, A. Wolfgang, K.A. Zamudio, Planetary candidates observed by Kepler. V. Planet sample from Q1-Q12 (36 months). Astrophys. J. Suppl. Ser. **217**, 16 (2015). doi:10.1088/0067-0049/217/1/16
J. Sanz-Forcada, G. Micela, I. Ribas, A.M.T. Pollock, C. Eiroa, A. Velasco, E. Solano, D. García-Álvarez, Estimation of the XUV radiation onto close planets and their evaporation. Astron. Astrophys. **532**, 6 (2011). doi:10.1051/0004-6361/201116594
L. Schaefer, B. Fegley Jr., Atmospheric chemistry of Venus-like exoplanets. Astrophys. J. **729**, 6 (2011). doi:10.1088/0004-637X/729/1/6
L. Schaefer, B. Fegley, Chemistry of silicate atmospheres of evaporating super-Earths. Astrophys. J. Lett. **703**, 113–117 (2009). doi:10.1088/0004-637X/703/2/L113
L. Schaefer, B. Fegley, Chemistry of atmospheres formed during accretion of the Earth and other terrestrial planets. Icarus **208**, 438–448 (2010). doi:10.1016/j.icarus.2010.01.026
L. Schaefer, K. Lodders, B. Fegley, Vaporization of the Earth: Application to exoplanet atmospheres. Astrophys. J. **755**, 41 (2012). doi:10.1088/0004-637X/755/1/41
E. Schlawin, E. Agol, L.M. Walkowicz, K. Covey, J.P. Lloyd, Exoplanetary transits of limb-brightened lines: Tentative Si IV absorption by HD 209458b. Astrophys. J. Lett. **722**, 75–79 (2010). doi:10.1088/2041-8205/722/1/L75
E. Schlawin, M. Zhao, J.K. Teske, T. Herter, A 0.8-2.4 μm transmission spectrum of the hot Jupiter CoRoT-1b. Astrophys. J. **783**, 5 (2014). doi:10.1088/0004-637X/783/1/5
D. Schulze-Makuch, A. Méndez, A.G. Fairén, P. Von Paris, C. Turse, G. Boyer, A.F. Davila, M.R.d.S. António, D. Catling, L.N. Irwin, A two-tiered approach to assessing the habitability of exoplanets. Astrobiology **11**(10), 1041–1052 (2011)
J.C. Schwartz, N.B. Cowan, Balancing the energy budget of short-period giant planets: Evidence for reflective clouds and optical absorbers. Mon. Not. R. Astron. Soc. **449**, 4192–4203 (2015). doi:10.1093/mnras/stv470
S. Seager, D.D. Sasselov, Extrasolar giant planets under strong stellar irradiation. Astrophys. J. Lett. **502**, 157–161 (1998). doi:10.1086/311498
S. Seager, D.D. Sasselov, Theoretical transmission spectra during extrasolar giant planet transits. Astrophys. J. **537**, 916–921 (2000). doi:10.1086/309088
S. Seager, L.J. Richardson, B.M.S. Hansen, K. Menou, J.Y.-K. Cho, D. Deming, On the dayside thermal emission of hot Jupiters. Astrophys. J. **632**, 1122–1131 (2005). doi:10.1086/444411

A. Segura, L.M. Walkowicz, V. Meadows, J. Kasting, S. Hawley, The effect of a strong stellar flare on the atmospheric chemistry of an Earth-like planet orbiting an M dwarf. Astrobiology **10**, 751–771 (2010). doi:10.1089/ast.2009.0376

F. Selsis, J. Kasting, B. Levrard, J. Paillet, I. Ribas, X. Delfosse, Habitable planets around the star Gliese 581? Astron. Astrophys. **476**(3), 1373–1387 (2007)

I.F. Shaikhislamov, M.L. Khodachenko, Y.L. Sasunov, H. Lammer, K.G. Kislyakova, N.V. Erkaev, Atmosphere expansion and mass loss of close-orbit giant exoplanets heated by stellar XUV. I. Modeling of hydrodynamic escape of upper atmospheric material. Astrophys. J. **795**, 132 (2014). doi:10.1088/0004-637X/795/2/132

H. Shapley, *Climatic Change: Evidence, Causes, and Effects* (1953)

A.P. Showman, T. Guillot, Atmospheric circulation and tides of "51 Pegasus b-like" planets. Astron. Astrophys. **385**, 166–180 (2002). doi:10.1051/0004-6361:20020101

A.P. Showman, L.M. Polvani, Equatorial superrotation on tidally locked exoplanets. Astrophys. J. **738**(1), 71 (2011)

A.P. Showman, J.J. Fortney, Y. Lian, M.S. Marley, R.S. Freedman, H.A. Knutson, D. Charbonneau, Atmospheric circulation of hot Jupiters: Coupled radiative-dynamical general circulation model simulations of HD 189733b and HD 209458b. Astrophys. J. **699**, 564–584 (2009). doi:10.1088/0004-637X/699/1/564

A.P. Showman, J.Y.-K. Cho, K. Menou, Atmospheric circulation of exoplanets, in *Exoplanets*, ed. by S. Seager (2010), pp. 471–516

A.P. Showman, R.D. Wordsworth, T.M. Merlis, Y. Kaspi, Atmospheric circulation of terrestrial exoplanets, in *Comparative Climatology of Terrestrial Planets*, ed. by S.J. Mackwell, A.A. Simon-Miller, J.W. Harder, M.A. Bullock (2013), pp. 277–326. doi:10.2458/azu_uapress_9780816530595-ch12

D.K. Sing, M. López-Morales, Ground-based secondary eclipse detection of the very-hot Jupiter OGLE-TR-56b. Astron. Astrophys. **493**, 31–34 (2009). doi:10.1051/0004-6361:200811268

D.K. Sing, J.-M. Désert, A. Lecavelier Des Etangs, G.E. Ballester, A. Vidal-Madjar, V. Parmentier, G. Hebrard, G.W. Henry, Transit spectrophotometry of the exoplanet HD 189733b. I. Searching for water but finding haze with HST NICMOS. Astron. Astrophys. **505**, 891–899 (2009). doi:10.1051/0004-6361/200912776

D.K. Sing, J.-M. Désert, J.J. Fortney, A. Lecavelier Des Etangs, G.E. Ballester, J. Cepa, D. Ehrenreich, M. López-Morales, F. Pont, M. Shabram, A. Vidal-Madjar, Gran Telescopio Canarias OSIRIS transiting exoplanet atmospheric survey: Detection of potassium in XO-2b from narrowband spectrophotometry. Astron. Astrophys. **527**, 73 (2011a). doi:10.1051/0004-6361/201015579

D.K. Sing, F. Pont, S. Aigrain, D. Charbonneau, J.-M. Désert, N. Gibson, R. Gilliland, W. Hayek, G. Henry, H. Knutson, A. Lecavelier Des Etangs, T. Mazeh, A. Shporer, Hubble Space Telescope transmission spectroscopy of the exoplanet HD 189733b: High-altitude atmospheric haze in the optical and near-ultraviolet with STIS. Mon. Not. R. Astron. Soc. **416**, 1443–1455 (2011b). doi:10.1111/j.1365-2966.2011.19142.x

D.K. Sing, C.M. Huitson, M. Lopez-Morales, F. Pont, J.-M. Désert, D. Ehrenreich, P.A. Wilson, G.E. Ballester, J.J. Fortney, A. Lecavelier des Etangs, A. Vidal-Madjar, GTC OSIRIS transiting exoplanet atmospheric survey: Detection of sodium in XO-2b from differential long-slit spectroscopy. Mon. Not. R. Astron. Soc. **426**, 1663–1670 (2012). doi:10.1111/j.1365-2966.2012.21938.x

D.K. Sing, A. Lecavelier des Etangs, J.J. Fortney, A.S. Burrows, F. Pont, H.R. Wakeford, G.E. Ballester, N. Nikolov, G.W. Henry, S. Aigrain, D. Deming, T.M. Evans, N.P. Gibson, C.M. Huitson, H. Knutson, A.P. Showman, A. Vidal-Madjar, P.A. Wilson, M.H. Williamson, K. Zahnle, HST hot-Jupiter transmission spectral survey: Evidence for aerosols and lack of TiO in the atmosphere of WASP-12b. Mon. Not. R. Astron. Soc. **436**, 2956–2973 (2013). doi:10.1093/mnras/stt1782

D.K. Sing, H.R. Wakeford, A.P. Showman, N. Nikolov, J.J. Fortney, A.S. Burrows, G.E. Ballester, D. Deming, S. Aigrain, J.-M. Désert, N.P. Gibson, G.W. Henry, H. Knutson, A. Lecavelier des Etangs, F. Pont, A. Vidal-Madjar, M.W. Williamson, P.A. Wilson, HST hot-Jupiter transmission spectral survey: Detection of potassium in WASP-31b along with a cloud deck and Rayleigh scattering. Mon. Not. R. Astron. Soc. **446**, 2428–2443 (2015). doi:10.1093/mnras/stu2279

A.J. Skemer, P.M. Hinz, S. Esposito, A. Burrows, J. Leisenring, M. Skrutskie, S. Desidera, D. Mesa, C. Arcidiacono, F. Mannucci, T.J. Rodigas, L. Close, D. McCarthy, C. Kulesa, G. Agapito, D. Apai, J. Argomedo, V. Bailey, K. Boutsia, R. Briguglio, G. Brusa, L. Busoni, R. Claudi, J. Eisner, L. Fini, K.B. Follette, P. Garnavich, R. Gratton, J.C. Guerra, J.M. Hill, W.F. Hoffmann, T. Jones, M. Krejny, J. Males, E. Masciadri, M.R. Meyer, D.L. Miller, K. Morzinski, M. Nelson, E. Pinna, A. Puglisi, S.P. Quanz, F. Quiros-Pacheco, A. Riccardi, P. Stefanini, V. Vaitheeswaran, J.C. Wilson, M. Xompero, First light LBT AO images of HR 8799 bcde at 1.6 and 3.3 μm: New discrepancies between young planets and old brown dwarfs. Astrophys. J. **753**, 14 (2012). doi:10.1088/0004-637X/753/1/14

M.D. Smith, Estimation of a length scale to use with the quench level approximation for obtaining chemical abundances. Icarus **132**, 176–184 (1998). doi:10.1006/icar.1997.5886
I.A.G. Snellen, R.J. de Kok, E.J.W. de Mooij, S. Albrecht, The orbital motion, absolute mass and high-altitude winds of exoplanet HD209458b. Nature **465**, 1049–1051 (2010). doi:10.1038/nature09111
I.A.G. Snellen, B.R. Brandl, R.J. de Kok, M. Brogi, J. Birkby, H. Schwarz, Fast spin of the young extrasolar planet β Pictoris b. Nature **509**, 63–65 (2014). doi:10.1038/nature13253
D.S. Spiegel, K. Silverio, A. Burrows, Can TiO explain thermal inversions in the upper atmospheres of irradiated giant planets? Astrophys. J. **699**, 1487–1500 (2009). doi:10.1088/0004-637X/699/2/1487
D.S. Spiegel, A. Burrows, L. Ibgui, I. Hubeny, J.A. Milsom, Models of Neptune-mass exoplanets: Emergent fluxes and albedos. Astrophys. J. **709**, 149–158 (2010). doi:10.1088/0004-637X/709/1/149
C.R. Stark, C. Helling, D.A. Diver, P.B. Rimmer, Electrostatic activation of prebiotic chemistry in substellar atmospheres. Int. J. Astrobiol. **13**, 165–172 (2014). doi:10.1017/S1473550413000475
D.J. Stevenson, J.I. Lunine, Rapid formation of Jupiter by diffuse redistribution of water vapor in the solar nebula. Icarus **75**, 146–155 (1988). doi:10.1016/0019-1035(88)90133-9
K.B. Stevenson, J. Harrington, S. Nymeyer, N. Madhusudhan, S. Seager, W.C. Bowman, R.A. Hardy, D. Deming, E. Rauscher, N.B. Lust, Possible thermochemical disequilibrium in the atmosphere of the exoplanet GJ 436b. Nature **464**, 1161–1164 (2010). doi:10.1038/nature09013
K.B. Stevenson, J.L. Bean, N. Madhusudhan, J. Harrington, Deciphering the atmospheric composition of WASP-12b: A comprehensive analysis of its dayside emission. Astrophys. J. **791**, 36 (2014a). doi:10.1088/0004-637X/791/1/36
K.B. Stevenson, J.L. Bean, A. Seifahrt, J.-M. Désert, N. Madhusudhan, M. Bergmann, L. Kreidberg, D. Homeier, Transmission spectroscopy of the hot Jupiter WASP-12b from 0.7 to 5 μm. Astron. J. **147**, 161 (2014b). doi:10.1088/0004-6256/147/6/161
K.B. Stevenson, J.-M. Désert, M.R. Line, J.L. Bean, J.J. Fortney, A.P. Showman, T. Kataria, L. Kreidberg, P.R. McCullough, G.W. Henry, D. Charbonneau, A. Burrows, S. Seager, N. Madhusudhan, M.H. Williamson, D. Homeier, Thermal structure of an exoplanet atmosphere from phase-resolved emission spectroscopy. Science **346**, 838–841 (2014c). doi:10.1126/science.1256758
A. Stökl, E. Dorfi, H. Lammer, Hydrodynamic simulations of captured protoatmospheres around Earth-like planets. Astron. Astrophys. **576**, 87 (2015). doi:10.1051/0004-6361/201423638
D.F. Strobel, Parameterization of linear wave chemical transport in planetary atmospheres by eddy diffusion. J. Geophys. Res. **86**, 9806–9810 (1981). doi:10.1029/JC086iC10p09806
D. Sudarsky, A. Burrows, P. Pinto, Albedo and reflection spectra of extrasolar giant planets. Astrophys. J. **538**, 885–903 (2000). doi:10.1086/309160
D. Sudarsky, A. Burrows, I. Hubeny, Theoretical spectra and atmospheres of extrasolar giant planets. Astrophys. J. **588**, 1121–1148 (2003). doi:10.1086/374331
M.R. Swain, G. Vasisht, G. Tinetti, The presence of methane in the atmosphere of an extrasolar planet. Nature **452**, 329–331 (2008). doi:10.1038/nature06823
M.R. Swain, G. Tinetti, G. Vasisht, P. Deroo, C. Griffith, J. Bouwman, P. Chen, Y. Yung, A. Burrows, L.R. Brown, J. Matthews, J.F. Rowe, R. Kuschnig, D. Angerhausen, Water, methane, and carbon dioxide present in the dayside spectrum of the exoplanet HD 209458b. Astrophys. J. **704**, 1616–1621 (2009a). doi:10.1088/0004-637X/704/2/1616
M.R. Swain, G. Vasisht, G. Tinetti, J. Bouwman, P. Chen, Y. Yung, D. Deming, P. Deroo, Molecular signatures in the near-infrared dayside spectrum of HD 189733b. Astrophys. J. Lett. **690**, 114–117 (2009b). doi:10.1088/0004-637X/690/2/L114
M. Swain, P. Deroo, G. Tinetti, M. Hollis, M. Tessenyi, M. Line, H. Kawahara, Y. Fujii, A.P. Showman, S.N. Yurchenko, Probing the extreme planetary atmosphere of WASP-12b. Icarus **225**, 432–445 (2013). doi:10.1016/j.icarus.2013.04.003
M.R. Swain, M.R. Line, P. Deroo, On the detection of molecules in the atmosphere of HD 189733b using HST NICMOS transmission spectroscopy. Astrophys. J. **784**, 133 (2014). doi:10.1088/0004-637X/784/2/133
C. Terquem, J.C.B. Papaloizou, Migration and the formation of systems of hot super-Earths and Neptunes. Astrophys. J. **654**, 1110–1120 (2007). doi:10.1086/509497
F. Tian, Thermal escape from super Earth atmospheres in the habitable zones of m stars. Astrophys. J. **703**(1), 905 (2009)
G. Tinetti, A. Vidal-Madjar, M.-C. Liang, J.-P. Beaulieu, Y. Yung, S. Carey, R.J. Barber, J. Tennyson, I. Ribas, N. Allard, G.E. Ballester, D.K. Sing, F. Selsis, Water vapour in the atmosphere of a transiting extrasolar planet. Nature **448**, 169–171 (2007). doi:10.1038/nature06002
G. Tinetti, P. Deroo, M.R. Swain, C.A. Griffith, G. Vasisht, L.R. Brown, C. Burke, P. McCullough, Probing the terminator region atmosphere of the hot-Jupiter XO-1b with transmission spectroscopy. Astrophys. J. Lett. **712**, 139–142 (2010). doi:10.1088/2041-8205/712/2/L139
K.O. Todorov, D. Deming, A. Burrows, C.J. Grillmair, Updated Spitzer emission spectroscopy of bright transiting hot Jupiter HD 189733b. Astrophys. J. **796**, 100 (2014). doi:10.1088/0004-637X/796/2/100

K.O. Todorov, M.R. Line, J.E. Pineda, M.R. Meyer, S.P. Quanz, S. Hinkley, J.J. Fortney, The water abundance of the directly imaged substellar companion {\kappa} and b retrieved from a near infrared spectrum. ArXiv e-prints (2015)

M.G. Tomasko, B. Archinal, T. Becker, B. Bézard, M. Bushroe, M. Combes, D. Cook, A. Coustenis, C. de Bergh, L.E. Dafoe, L. Doose, S. Douté, A. Eibl, S. Engel, F. Gliem, B. Grieger, K. Holso, E. Howington-Kraus, E. Karkoschka, H.U. Keller, R. Kirk, R. Kramm, M. Küppers, P. Lanagan, E. Lellouch, M. Lemmon, J. Lunine, E. McFarlane, J. Moores, G.M. Prout, B. Rizk, M. Rosiek, P. Rueffer, S.E. Schröder, B. Schmitt, C. See, P. Smith, L. Soderblom, N. Thomas, R. West, Rain, winds and haze during the Huygens probe's descent to Titan's surface. Nature **438**, 765–778 (2005). doi:10.1038/nature04126

A.H.M.J. Triaud, A. Collier Cameron, D. Queloz, D.R. Anderson, M. Gillon, L. Hebb, C. Hellier, B. Loeillet, P.F.L. Maxted, M. Mayor, F. Pepe, D. Pollacco, D. Ségransan, B. Smalley, S. Udry, R.G. West, P.J. Wheatley, Spin-orbit angle measurements for six southern transiting planets. New insights into the dynamical origins of hot Jupiters. Astron. Astrophys. **524**, 25 (2010). doi:10.1051/0004-6361/201014525

T. Tsuji, Molecular abundances in stellar atmospheres. II. Astron. Astrophys. **23**, 411–431 (1973)

T. Tsuji, K. Ohnaka, W. Aoki, T. Nakajima, Evolution of dusty photospheres through red to brown dwarfs: How dust forms in very low mass objects. Astron. Astrophys. **308**, 29–32 (1996)

S. Udry, X. Bonfils, X. Delfosse, T. Forveille, M. Mayor, C. Perrier, F. Bouchy, C. Lovis, F. Pepe, D. Queloz, et al., The harps search for southern extra-solar planets ⋆, ⋆⋆. Astron. Astrophys. **469**, 43–47 (2007)

E.F. van Dishoeck, E.A. Bergin, D.C. Lis, J.I. Lunine, Water: From clouds to planets, in *Protostars and Planets VI* (2014), pp. 835–858. doi:10.2458/azu_uapress_9780816531240-ch036

O. Venot, E. Hébrard, M. Agúndez, M. Dobrijevic, F. Selsis, F. Hersant, N. Iro, R. Bounaceur, A chemical model for the atmosphere of hot Jupiters. Astron. Astrophys. **546**, 43 (2012). doi:10.1051/0004-6361/201219310

O. Venot, N. Fray, Y. Bénilan, M.-C. Gazeau, E. Hébrard, G. Larcher, M. Schwell, M. Dobrijevic, F. Selsis, High-temperature measurements of VUV-absorption cross sections of CO_2 and their application to exoplanets. Astron. Astrophys. **551**, 131 (2013). doi:10.1051/0004-6361/201220945

O. Venot, M. Agúndez, F. Selsis, M. Tessenyi, N. Iro, The atmospheric chemistry of the warm Neptune GJ 3470b: Influence of metallicity and temperature on the CH_4/CO ratio. Astron. Astrophys. **562**, 51 (2014). doi:10.1051/0004-6361/201322485

O. Venot, E. Hébrard, M. Agúndez, L. Decin, R. Bounaceur, New chemical scheme for studying carbon-rich exoplanet atmospheres. Astron. Astrophys. **577**, A33 (2015)

J. Venturini, Y. Alibert, W. Benz, M. Ikoma, Critical core mass for enriched envelopes: The role of H_2O condensation. Astron. Astrophys. **576**, 114 (2015). doi:10.1051/0004-6361/201424008

A. Vidal-Madjar, A. Lecavelier des Etangs, J.-M. Désert, G.E. Ballester, R. Ferlet, G. Hébrard, M. Mayor, An extended upper atmosphere around the extrasolar planet HD209458b. Nature **422**, 143–146 (2003). doi:10.1038/nature01448

A. Vidal-Madjar, J.-M. Désert, A. Lecavelier des Etangs, G. Hébrard, G.E. Ballester, D. Ehrenreich, R. Ferlet, J.C. McConnell, M. Mayor, C.D. Parkinson, Detection of oxygen and carbon in the hydrodynamically escaping atmosphere of the extrasolar planet HD 209458b. Astrophys. J. Lett. **604**, 69–72 (2004). doi:10.1086/383347

A. Vidal-Madjar, D.K. Sing, A. Lecavelier Des Etangs, R. Ferlet, J.-M. Désert, G. Hébrard, I. Boisse, D. Ehrenreich, C. Moutou, The upper atmosphere of the exoplanet HD 209458 b revealed by the sodium D lines. Temperature-pressure profile, ionization layer, and thermosphere. Astron. Astrophys. **527**, 110 (2011). doi:10.1051/0004-6361/201015698

A. Vidal-Madjar, C.M. Huitson, V. Bourrier, J.-M. Désert, G. Ballester, A. Lecavelier des Etangs, D.K. Sing, D. Ehrenreich, R. Ferlet, G. Hébrard, J.C. McConnell, Magnesium in the atmosphere of the planet HD 209458 b: Observations of the thermosphere-exosphere transition region. Astron. Astrophys. **560**, 54 (2013). doi:10.1051/0004-6361/201322234

C. Visscher, J.I. Moses, Quenching of carbon monoxide and methane in the atmospheres of cool brown dwarfs and hot Jupiters. Astrophys. J. **738**, 72 (2011). doi:10.1088/0004-637X/738/1/72

C. Visscher, K. Lodders, B. Fegley Jr., Atmospheric chemistry in giant planets, brown dwarfs, and low-mass dwarf stars. II. Sulfur and phosphorus. Astrophys. J. **648**, 1181–1195 (2006). doi:10.1086/506245

C. Visscher, K. Lodders, B. Fegley Jr., Atmospheric chemistry in giant planets, brown dwarfs, and low-mass dwarf stars. III. Iron, magnesium, and silicon. Astrophys. J. **716**, 1060–1075 (2010a). doi:10.1088/0004-637X/716/2/1060

C. Visscher, J.I. Moses, S.A. Saslow, The deep water abundance on Jupiter: New constraints from thermochemical kinetics and diffusion modeling. Icarus **209**, 602–615 (2010b). doi:10.1016/j.icarus.2010.03.029

W. Von Bloh, C. Bounama, M. Cuntz, S. Franck, The habitability of super-earths in Gliese 581. Astron. Astrophys. **476**(3), 1365–1371 (2007)

P. Von Paris, S. Gebauer, M. Godolt, H. Rauer, B. Stracke, Atmospheric studies of habitability in the Gliese 581 system. Astron. Astrophys. **532**, 58 (2011)

J.H. Waite, D.T. Young, T.E. Cravens, A.J. Coates, F.J. Crary, B. Magee, J. Westlake, The process of tholin formation in Titan's upper atmosphere. Science **316**, 870–875 (2007). doi:10.1126/science.1139727

H.R. Wakeford, D.K. Sing, D. Deming, N.P. Gibson, J.J. Fortney, A.S. Burrows, G. Ballester, N. Nikolov, S. Aigrain, G. Henry, H. Knutson, A. Lecavelier des Etangs, F. Pont, A.P. Showman, A. Vidal-Madjar, K. Zahnle, HST hot Jupiter transmission spectral survey: Detection of water in HAT-P-1b from WFC3 near-IR spatial scan observations. Mon. Not. R. Astron. Soc. **435**, 3481–3493 (2013). doi:10.1093/mnras/stt1536

I.P. Waldmann, G. Tinetti, P. Deroo, M.D.J. Hollis, S.N. Yurchenko, J. Tennyson, Blind extraction of an exoplanetary spectrum through independent component analysis. Astrophys. J. **766**, 7 (2013). doi:10.1088/0004-637X/766/1/7

I.P. Waldmann, G. Tinetti, M. Rocchetto, E.J. Barton, S.N. Yurchenko, J. Tennyson, Tau-REx I: A next generation retrieval code for exoplanetary atmospheres. Astrophys. J. **802**, 107 (2015). doi:10.1088/0004-637X/802/2/107

Y. Wang, F. Tian, Y. Hu, Climate patterns of habitable exoplanets in eccentric orbits around m dwarfs. Astrophys. J. Lett. **791**(1), 12 (2014)

D. Wang, P.J. Gierasch, J.I. Lunine, O. Mousis, New insights on Jupiter's deep water abundance from disequilibrium species. Icarus **250**, 154–164 (2015). doi:10.1016/j.icarus.2014.11.026

L.M. Weiss, G.W. Marcy, The mass-radius relation for 65 exoplanets smaller than 4 Earth radii. Astrophys. J. Lett. **783**, 6 (2014). doi:10.1088/2041-8205/783/1/L6

L.M. Weiss, G.W. Marcy, J.F. Rowe, A.W. Howard, H. Isaacson, J.J. Fortney, N. Miller, B.-O. Demory, D.A. Fischer, E.R. Adams, A.K. Dupree, S.B. Howell, R. Kolbl, J.A. Johnson, E.P. Horch, M.E. Everett, D.C. Fabrycky, S. Seager, The mass of KOI-94d and a relation for planet radius, mass, and incident flux. Astrophys. J. **768**, 14 (2013). doi:10.1088/0004-637X/768/1/14

A.N. Wilkins, D. Deming, N. Madhusudhan, A. Burrows, H. Knutson, P. McCullough, S. Ranjan, The emergent 1.1-1.7 μm spectrum of the exoplanet CoRoT-2b as measured using the Hubble Space Telescope. Astrophys. J. **783**, 113 (2014). doi:10.1088/0004-637X/783/2/113

P.A. Wilson, D.K. Sing, N. Nikolov, A. Lecavelier des Etangs, F. Pont, J.J. Fortney, G.E. Ballester, M. López-Morales, J.-M. Désert, A. Vidal-Madjar, GTC OSIRIS transiting exoplanet atmospheric survey: Detection of potassium in HAT-P-1b from narrow-band spectrophotometry. Mon. Not. R. Astron. Soc. **450**, 192–200 (2015). doi:10.1093/mnras/stv642

J.N. Winn, D. Fabrycky, S. Albrecht, J.A. Johnson, Hot stars with hot Jupiters have high obliquities. Astrophys. J. Lett. **718**, 145–149 (2010). doi:10.1088/2041-8205/718/2/L145

A. Wolfgang, E. Lopez, How rocky are they? The composition distribution of Kepler's sub-Neptune planet candidates within 0.15 AU. Astrophys. J. **806**, 183 (2015)

M.H. Wong, P.R. Mahaffy, S.K. Atreya, H.B. Niemann, T.C. Owen, Updated Galileo probe mass spectrometer measurements of carbon, oxygen, nitrogen, and sulfur on Jupiter. Icarus **171**, 153–170 (2004). doi:10.1016/j.icarus.2004.04.010

P.L. Wood, P.F.L. Maxted, B. Smalley, N. Iro, Transmission spectroscopy of the sodium 'D' doublet in WASP-17b with the VLT. Mon. Not. R. Astron. Soc. **412**, 2376–2382 (2011). doi:10.1111/j.1365-2966.2010.18061.x

R. Wordsworth, Atmospheric heat redistribution and collapse on tidally locked rocky planets. Astrophys. J. **806** (2015). doi:10.1088/0004-637X/806/2/180

R. Wordsworth, F. Forget, F. Selsis, J.-B. Madeleine, E. Millour, V. Eymet, Is Gliese 581d habitable? Some constraints from radiative-convective climate modeling. Astron. Astrophys. **522**, 22 (2010)

Y. Wu, Y. Lithwick, Density and eccentricity of Kepler planets. Astrophys. J. **772**, 74 (2013). doi:10.1088/0004-637X/772/1/74

A. Wyttenbach, D. Ehrenreich, C. Lovis, S. Udry, F. Pepe, Spectrally resolved detection of sodium in the atmosphere of HD 189733b with the HARPS spectrograph. Astron. Astrophys. **577**, 62 (2015). doi:10.1051/0004-6361/201525729

J. Yang, W.R. Peltier, Y. Hu, The initiation of modern soft and hard snowball Earth climates in ccsm4. Clim. Past **8**(3), 907–918 (2012a). doi:10.5194/cp-8-907-2012

J. Yang, W.R. Peltier, Y. Hu, The initiation of modern soft snowball and hard snowball climates in ccsm3. part i: The influences of solar luminosity, co2 concentration, and the sea ice/snow albedo parameterization. J. Climate **25**(8), 2711–2736 (2012b)

J. Yang, W.R. Peltier, Y. Hu, The initiation of modern soft snowball and hard snowball climates in ccsm3. part ii: Climate dynamic feedbacks. J. Climate **25**(8), 2737–2754 (2012c)

J. Yang, N.B. Cowan, D.S. Abbot, Stabilizing cloud feedback dramatically expands the habitable zone of tidally locked planets. Astrophys. J. Lett. **771**(2), 45 (2013)

J. Yang, Y. Liu, Y. Hu, D.S. Abbot, Water trapping on tidally locked terrestrial planets requires special conditions. Astrophys. J. Lett. **796**(2), 22 (2014)

R.V. Yelle, Aeronomy of extra-solar giant planets at small orbital distances. Icarus **170**, 167–179 (2004). doi:10.1016/j.icarus.2004.02.008

Y.L. Yung, M. Allen, J.P. Pinto, Photochemistry of the atmosphere of Titan—Comparison between model and observations. Astrophys. J. Suppl. Ser. **55**, 465–506 (1984). doi:10.1086/190963

K. Zahnle, M.S. Marley, Methane, carbon monoxide, and ammonia in brown dwarfs and self-luminous giant planets. Astrophys. J. **797**, 41 (2014). doi:10.1088/0004-637X/797/1/41

K. Zahnle, M.S. Marley, J.J. Fortney, Thermometric soots on warm Jupiters? ArXiv e-prints (2009a)

K. Zahnle, M.S. Marley, R.S. Freedman, K. Lodders, J.J. Fortney, Atmospheric sulfur photochemistry on hot Jupiters. Astrophys. J. Lett. **701**, 20–24 (2009b). doi:10.1088/0004-637X/701/1/L20

M.R. Zapatero Osorio, V.J.S. Béjar, E.L. Martín, R. Rebolo, D. Barrado y Navascués, C.A.L. Bailer-Jones, R. Mundt, Discovery of young, isolated planetary mass objects in the σ orionis star cluster. Science **290**, 103–107 (2000). doi:10.1126/science.290.5489.103

X. Zhang, R.A. West, D. Banfield, Y.L. Yung, Stratospheric aerosols on Jupiter from Cassini observations. Icarus **226**, 159–171 (2013). doi:10.1016/j.icarus.2013.05.020

G. Zhou, D.D.R. Bayliss, Detection of sodium absorption in WASP-17b with Magellan. Mon. Not. R. Astron. Soc. **426**, 2483–2488 (2012). doi:10.1111/j.1365-2966.2012.21817.x

Part III

Future Projects and Their Scientific Perspectives

Space Sci Rev (2016) 205:393–416
DOI 10.1007/s11214-016-0247-2

The Way Forward

Malcolm Fridlund[1,2,3] · **Artie Hatzes**[4] · **René Liseau**[3]

Received: 11 September 2015 / Accepted: 14 March 2016 / Published online: 12 May 2016

Abstract For the last few decades the study of disks around stars young and old and of different types have progressed significantly. During the same time a completely new discipline—the study of exoplanets, planets orbiting stars other than our Sun—have emerged. Both these fields, which are interconnected, have benefited from the development of new instrumentation, and especially by telescopes and detectors deployed in space. In this chapter we are describing the state of the art of such instruments and make an inventory of what is being currently developed. We also state some of the requirements of the next steps and what type of instruments will lead the way forward.

Keywords Exoplanets · Starforming disks · Debris disks · Instrumentation: photometers · Instrumentation: space

1 Introduction

Our Solar System has a complex architecture that contains (1) Planets of several types, from giant gas giants to rocky terrestrial planets, (2) Debris disks, e.g. the zodiacal dust belt and the Kuiper-Edgeworth belt in the Solar System or disks of the β Pic or AU Mic type, and (3) Planetesimals or remnants thereof—e.g. Comets, Asteroids or Dwarf planets. There is also the Oort cloud of comets stretching tens of thousands of astronomical units from the Sun, but the understanding of the structure of the Oort cloud in the context of star formation is a separate issue and will not be discussed here. One of the ultimate goals of the research into both exoplanets and circumstellar disks is to be able to place this complex picture of our own Solar System into a proper context.

✉ M. Fridlund
fridlund@strw.leidenuniv.nl

1 Max-Planck-Institute für Astronomie, Königstuhl 17, 69117 Heidelberg, Germany

2 Leiden Observatory, P.O. Box 9513, 2300 RA, Leiden, The Netherlands

3 Department of Earth and Space Sciences, Chalmers University of Technology, Onsala Space Observatory, 439 92, Onsala, Sweden

4 Thüringer Landessternwarte Tautenburg, Sternwarte 5, 07778 Tautenburg, Germany

Answering questions like "Is our system a run-of-the-mill kind of system" or is it of a "rarer" kind—or even "unique" is of course central to this research. The answer to such questions is related to the overarching issue of whether we are alone in the Universe. The question of whether life exists elsewhere, than on the Earth, has been discussed by mankind since the beginning of recorded history, or even before, and it is still a fundamental question among scientists and laypersons alike today.[1]

The understanding of how disks and planetary bodies relate to each other is the key to how planetary systems form and evolve. What kind of planets is the more common? What is the architecture of these planetary systems? And under what conditions were they formed? It must be a consequence of the formation- and evolution-processes that lead to the structure and physics of a specific system or an individual planet. This is very likely due to the conditions leading up to the formation of an accreting star and its associated disk. We can be fairly certain that the existence of planetary systems is a consequence of the process that forms stars. This understanding has not changed significantly since the time when we only had knowledge about only one such system—our own.

Our own solar system is of course very important since it is the only place in the Universe where we know that life exists. It is also the only system where we can study individual objects in situ and in exquisite detail. This is evidenced by the fact that we have more than half a dozen working spacecrafts in orbit or on the surface of Mars. As of this writing the New Horizons mission had just made its flyby of the dwarf planet Pluto on its way out into the Solar Systems outer debris disk—the Kuiper-Edgeworth belt. The pioneering Voyager 2 spacecraft has already crossed the heliopause and has become the first human spacecraft to be traveling in interstellar space. And, for more than a year the Rosetta spacecraft has been flying in tandem with Comet 67P/Churyumov-Gerasimenko, giving us valuable, in situ knowledge of a comet/planetesimal. These are indeed exciting times for solar system research.

The understanding of star forming and debris disks of different structures and sizes is thus the key to understanding planet formation, specifically the types and masses of planets and the range of orbital distances that can form. Not only must a planet be located in the Habitable Zone (HZ, usually defined as the range of distance from the host star within liquid water can exist). Once such planets have formed, their orbits must be stable for a long enough time to provide the physical conditions for the formation of life.

Eventually, the central issue we wish to address is whether we are indeed alone in the Universe. The uniqueness or commonality of the Earth is fundamental to a proper understanding of such central issues as the properties needed by a planet in order for life to arise, the origin and evolution of life, and of course how common life is in the Universe. To correctly understand these issues we must also understand the context within which life on Earth exists and our place in the Universe. This thought, which has been with us throughout recorded history, can now be explored within the structure of a scientific inquiry. Is the Earth the only one of uncountable planets where life has arisen and evolved to the point that one of its life forms has the ability to formulate such a question? Or are the Earth, and its humans unique? Possibly the Earth formed out of random processes that were quite rare and very special circumstances were needed for higher life forms to evolve. The conditions may have been extremely rare and unlikely, but given the shear number of stars in our Galaxy perhaps it can occur at least once every several billions of years. It could also be that we may be alone in our galaxy, but certainly not alone in the vast Universe. Or....

[1] As demonstrated by e.g. The recent announcement that "the Breakthrough Prize Foundation", an organization created by investor Yuri Milner, has announced that it will commit $100 million over the next ten years to a new, large-scale SETI initiative.

The answer to such questions should be based on sound scientific inquiry and not just speculation.

1.1 The Study of Disks

Interstellar disks and the process of stellar and planetary formation and evolution have now been studied in detail for more than 60 years. Through advances in Infrared and mm/sub-mm astronomy a consistent picture has slowly been built up. The Milky Way Galaxy, like all galaxies, is made up of stars, stellar remnants, planets, and an interstellar medium. This Inter Stellar Medium (ISM) is composed of atomic and molecular gas as well as solid matter in the form of dust grains. The densest part of the ISM is mainly located in the plane of our Galaxy and it has different components, one of which is (giant) molecular clouds with a much higher (10^3–10^4 times higher) density than the "general" diffuse medium. These molecular clouds consist mainly of H_2, but with a significant enrichment of metals that supposedly originated in stars during the later stages of stellar evolution. These were then ejected into interstellar space through mass loss and supernova explosions to form the parental environment benevolent to the forming of stars and planets. While elements of the clouds collapse into protostars, some of the molecular material forms a protoplanetary disk, due to the conservation of angular momentum. This momentum preserves an enormous amount of rotational energy that the protostar has to lose before it can settle into a stable configuration. This is done through bipolar outflows (atomic and molecular), or jets.

Although most of the disk mass and its angular momentum is lost from the system through these processes, a large amount of material is nevertheless left behind in remnant disks. This remaining material disperses slowly during the stellar lifetime either through collisions that form dust and debris disks or through ejections via interactions with the planetary bodies. However some of these debris disks remain as evidenced by the dust and large number of small bodies in the Solar System. We have begun to observe more and more of this material in other stellar systems (Eiroa et al. 2013) and essentially all of the stages outlined above have been detected either from the ground or space. Most recently the ESA space observatory Herschel has observed many aspects of these processes for the first time and in greater detail than ever before. On the ground the "first light" of the Atacama Large MM/sub-mm Array (ALMA) have detected the gaps in protoplanetary disks presumably resulting from the "sweeping up" of material by planets being formed (Partnership ALMA 2015). While many of the details in these processes remain to be observed and clarified it appears that at least the broad elements of the picture are understood.

One important remaining step is to connect the understanding of the distribution of exoplanets (type of planets, type of orbits, type of host stars) with this picture of the early stages of a planetary system. As is believed today (see below) planets are extremely common, and if not ubiquitous, anyhow important somehow to the star formation process. Nevertheless, one of the surprising results of the recent large surveys of exoplanets (e.g. ground based and/or space) is their diversity concerning radii, masses, orbital characteristics and their distribution among types of stars. The extreme dynamism of exoplanetary systems is also not what one expects from our understanding of the Solar System. According to numerical simulation the Solar System has remained stable enough for life to develop on the Earth, for more than 4 Gyrs although the major planets may have moved significantly with even Uranus and Neptune exchanging their order (Gomes et al. 2005) and it may remain so for at least as long although there is a finite possibility that the orbits of either Mars or Mercury (or both) may go haywire in the distant future with potentially disastrous consequences for the Earth (Batygin and Laughlin 2008;

Laskar 2008). This stability may or may not be the norm as what concerns other planetary systems. Further, it has becoming increasingly clear that many planets are found currently far away from where they must have formed. The dynamicism, particularly during the early phases of planetary formation need to be explored more. In order to understand these facts it will be necessary to bring the picture of how stars and planets form together with what is actually observed of exoplanetary systems—never forgetting that we need to explain our own Solar System in the same picture.

1.2 The study of exoplanets

The issue of how common planets are in our Galaxy and by inference in the Universe was totally open until about 20 years ago. While the first 'modern' proposal for a 'proper' search for exoplanets date to more than 60 years ago (Struve 1952), the first discoveries of bodies with masses comparable to those within the Solar System were made during the late 1980s and beginning of the 1990s, most spectacularly with the detection of Earth-mass planets orbiting pulsars (Wolszczan and Frail 1992). These early investigations culminated with the paper by Mayor and Queloz (1995) reporting a planet orbiting the solar type star 51 Pegasi, located about 50 light years away from the Sun. This latter planet was found to be Jupiter-like, with a minimum mass of about 0.5 Jupiter masses, but with an orbital period of 4.23 days equivalent to an orbital distance of 0.05 astronomical units. The discovery of a planet with these characteristics was unexpected by the scientific community, with the possible exception of Struve (1952) who first speculated on the existence of such hot Jupiters.

Since 51 Peg b was found with the radial velocity method (see below) which is biased towards finding massive planets in short period orbits, it was first written off as an unusual object. Soon thereafter Marcy and Butler (1996), Butler and Marcy (1996) discovered two more planets (7.4 and 2.4 Jupiter masses respectively) hosted by two solar analogues (70 Virginis and 47 Ursa Majoris) and with orbital periods of 116 d and 2.98 years. None of these objects resembled planets found in our own Solar System and it seemed that giant planets with shorter orbital periods than in the Solar System were not so unusual. These first discoveries opened up the search for exoplanets that has now been going on actively for more than 20 years. Today, close to 2 000 objects are characterized as confirmed planets, while a further ∼4 000 objects (detected by NASA's Kepler space mission—see below) are classified as Kepler Objects of Interest (KOI). The European CoRoT space mission, while confirming more than 35 planets, also has around 200 objects which are of interest but where the host star is too faint for proper follow-up and confirmation with current equipment. These have to await ongoing developments in instrumentation (e.g. with the ESPRESSO ground based spectrograph with first light due ∼2016). As mentioned before, the first confirmed exoplanets were found orbiting around pulsars through radio pulse timing variations (Wolszczan and Frail 1992). The formation process of these objects is not clear, but they presumably accreted out of debris produced by the supernova explosion that created the pulsar. Most exoplanets around normal stars have however, been discovered through one or more of three other methods, viz., the radial velocity method, the transit method or through gravitational lensing. All of these techniques were first applied from ground-based observatories and only later from space (gravitational lensing observations will be implemented on NASA's WFIRST mission and possibly also on ESA's EUCLID spacecraft. Radial velocity measurement with a precision high enough for planetary detection albeit only for high mass planets, is now flying on GAIA. A detailed description of these and the other methods can be found in Perryman (2011), or Haswell (2010). Briefly, the radial velocity method measures the radial velocity of the host star's reflex motion about the center of mass of the system due to the influence of the companion. Since it only measures the velocity component along the line of sight only the minimum planet mass is obtained ($m_p \times \sin i$ where i is the orbital inclination relative to the line of sight to the host star). Assuming a value for the stellar mass, we get an estimate of the minimum mass of the planet, but not its true mass. The transit method is also very simple in that it searches for exoplanets whose orbital inclination is such that the planet passes between us, and the disk of its host star, causing the stellar flux to drop periodically.

It must be recognized, however, that the current instruments and methods are limited by observational bias. For instance, the current generation of high resolution Echelle spectrographs are limited by systematic stability (i.e. between observing epochs), by having the best and most stable spectrographs deployed on (relatively) small telescopes, and by having reached somewhat the possible inherent precision possible with current calibration methods (i.e. lamp spectrum calibration or gas cell transmission calibration). Further radial velocity observations have reached the precision where they are impacted by the inherent stellar "noise", i.e. activity, where both granulation and p-mode (astroseismic) variations produce a noise term. The previous bias of being mostly sensitive to large planets orbiting close to the host star is going away by time and several large planets in orbits of 10 years period or longer are being identified or proposed (Bedell et al. 2015).

While the radial velocity method gives us the (minimum) mass, the transit method provides the radius of the exoplanet in terms of the stellar radius (which usually is unknown by factors of about 10 %). Because for transiting planets we are viewing the orbit nearly within the orbital plane, the orbital inclination is nearly 90 deg (and can be measured more precisely by fitting the transit light curve with an appropriate model). This means that radial velocity measurements now give us the true planetary mass, and consequently its average density can be calculated. As will become clear, the accurate measurement of ages, radii and masses of large numbers of exoplanets of different types represents the next advance in exoplanetology. The problem with the current data is that the accuracy of the planetary parameters of mass and radius are fully dependent on the accuracy of the same parameters for the host star. The current errors in the host star values are of order >20 %, >20 %, and 50–100 % for the radii, masses and ages respectively. This will be further discussed below in the context of the TESS and PLATO future space missions.

The detection of exoplanets through the detection of microlensing events is very different from the transit and radial velocity methods. A microlensing event takes place when one star passes exactly (to within a fraction of a milliarcsecond) in front of another star and this then gravitationally boosts the intensity level of the background star like a magnifying glass would. If the hither, i.e. 'lensing' star has a planet orbiting around it, the planets gravitational field may add a 'blip' on the light curve. It is intuitively clear that such events must be rare and that very large numbers of stars have to be observed, or rather monitored, regularly. When such a 'blip' is observed, the resulting light curve can be modeled and the planetary nature determined. Unfortunately, to have a high confidence in the model, one requires the stellar parameters for the lensing star, and this is 'blended' with the background object for decades after the event took place. Although more than 3 dozen exoplanet candidates have been recorded through this method, only in one case have it recently been possible to confirm it through observations of the host star (using HST as well as ground based observations: Batista et al. 2015; Bennet et al. 2015), 8.21 years after the original lensing event.

The method that originally was considered to be the most difficult, namely imaging, have recently produced interesting results, mainly due to the improvement in coronography and

Adaptive Optics (AO), with instruments like e.g. NACO (http://www.eso.org/sci/facilities/paranal/instruments/naco.html) and SPHERE (http://www.eso.org/sci/facilities/paranal/instruments/sphere.html) operated by the European Southern Observatory (ESO), or the GEMINI Planetary Imager (http://planetimager.org). Such instruments operating on the worlds largest (i.e. 8–10 m class) telescopes have over the last few years managed to look for planets down to 1 AU away from the host star (Vigan et al. 2015; Zurlo et al. 2015; Rapson et al. 2015)

1.3 The Hubble Space Telescope

The Hubble Space Telescope (HST) has a very long history. First mentioned by the space pioneers Oberth, Goddard and Tsiolkovsky, its direct history goes back to a paper by Spitzer (1946), pointing out the high spatial resolution (the telescope being essentially diffraction limited), as well as the access to short and long wavelengths not possible to observe from the ground. After launch on the space shuttle Discovery, in 1990, both of these aspects have made the HST arguably the most productive telescope ever. Especially within the two topics discussed in this paper—star- and planet-forming disks, as well as the direct and indirect observation of exoplanets themselves the telescope have contributed significantly.

One of these discoveries were the detection of proto-planetary disks (the so called "proplyds") in Orion (O'Dell et al. 1993) which demonstrated the presence of dense dusty structures of Solar System dimensions that were visible since the surrounding gas had been 'blown away' by stellar wind and radiation pressure from the O- and B-type stars forming in the vicinity.

Another fundamental observation is the detailed study of jets, originating at the center of the dusty, gaseous disk the star is drawing on for its raw material while forming. The jets are now thought to play a major part in star formation, perhaps by removing angular momentum from the disk and so driving the accretion onto the star (e.g. Fridlund and Liseau 1998).

Indirect and direct observations of the exoplanets themselves e.g. through observing transits of the exoplanets across the surface of their host star and photometrically measured the light drop cause by the planet with such a high precision that significant conclusions could be drawn both about the planet and its atmosphere (Charbonneau et al. 2002), as well as properties of the host star e.g. activity, Brown et al. (2001).

Direct observation of the exoplanets themselves have followed. HST direct imaging of Fomalhaut, a bright star 7.7 parsecs away, displays a belt of cold dust with a structure consistent with the gravitational influence of an orbiting planet. The optical observations of the exoplanet candidate, Fomalhaut b found about 119 AU away from the star and 18 AU away from the dust belt, at the position where it had been predicted due to the shape of the dust disk. HST observed the system at times separated by 1.73 years which revealed orbital motion, best (dynamically) modeled by a planet with a mass at most three times that of Jupiter (Kalas et al. 2008).

HST remain a most valuable asset today, particularly in the connection between the research on the disks and the exoplanets formed/located within them.

1.4 The Spitzer Space Telescope

The Spitzer Space Telescope (SST), formerly the Space Infrared Telescope Facility (SIRTF), is an infrared space observatory launched in 2003. The mission was designed for an operational lifetime of 2.5 years, with the possibility of extending it until the liquid Helium coolant boiled off, an event that took place after about 6 years in 2009. The instruments (imagers and spectrographs) designed for the longer wavelengths are no longer operational since the telescope and focal plane have stabilized at temperatures (29 K) not compatible with the required noise figures. Nevertheless, the two shortest channels, operating at 3.6 and 4.5 μm wavelength remain operational and have given the space telescope a new lease of life, among other things in the area of exoplanetology. During the 'cold' part of the mission, the objective was studying the equally cold parts of the Universe, i.e. gas and dust at very low temperatures such as star forming disks and the central rings in the Andromeda galaxy. In the 'warm' phase, *Spitzer* could be put to creative work studying exoplanets thanks to the possibility to changing the operations modes of its hardware. The photometric stability could be increased and the "peak-up" camera could be brought to bear on the problem. Further providing 'maps' of the sensors at a sub-pixel level increased the sensitivity. Spitzer observes e.g. exoplanetary transits, and obtain (admittedly at very low resolution) transmission spectra of the (larger) planets during this eclipses.

1.5 CoRoT and Kepler

The scientific case for a space mission to detect small transiting planets orbiting other stars began in earnest around the first half of the 1980s (see Roxburgh 2006a, 2006b; Borucki 2010 and references therein). This work led eventually to the launch in December 2006 of the European-Brazilian space mission CoRoT (Convection, Rotation and exoplanetary Transits; Baglin and Fridlund 2006). CoRoT was the first space mission dedicated to the detection of exoplanets by searching for exoplanetary transits around solar-like stars. CoRoT utilized a relatively small (30 cm) telescope with a limited field of view (4 deg^2). Due to its low-Earth orbit (900 km), the longest the spacecraft could remain pointed towards the same target field was about 160 d. Observing more than 20 individual fields, CoRoT, during a 6 year mission life time (nominally $2\frac{1}{2}$ years), found more then 35 new planets (now published with another 10–15 being currently under confirmation) and acquired more than 160 000 light curves. CoRoT was the second spacecraft designed for asteroseismology that was successfully launched and increased significantly the number of, and the precision with which asteroseismology is carried out from space (The Canadian microsatellite MOST was the pioneer in this context). CoRoT's planet yield include the first transiting terrestrial planet (Legér et al. 2009) that could be confirmed with follow-up radial velocity measurements of its host star (Queloz et al. 2009). This planet, CoRoT-7b has a diameter of 1.58 Earth-radii and a mass of 7.42 Earth-masses (Hatzes et al. 2011). The average density is of the same order as the terrestrial type planets in the Solar System. CoRoT-7b however, has a very unusual orbit. It orbits its G9V primary in 20.2 h, clearly an object that has no analogue in our own system.

The next fundamental step was taken by NASA's Kepler mission launched in 2009. As a planet search mission Kepler currently holds the record with 1028 confirmed planets. Kepler consists of a one-meter aperture telescope that monitored about 160 000 stars (about the same number as CoRoT) in a fixed 110 deg^2 field of view in the sky for a duration of approximately 4 years. The many thousands of transit candidates still require ground based (mainly radial velocity) follow up observations. Kepler was able to follow a significantly larger number of brighter stars than was possible for CoRoT. This allowed researchers to carry out asteroseismology on number of host stars and generally it was easier to make follow-up radial velocity observations of stars with smaller planet candidates. In May 2013, a second reaction wheel onboard Kepler failed, making it impossible to stabilize Kepler to the precision required for its initial mission. Since all other systems on board were working NASA issued a request for White Papers suggesting ways to use the space craft in its more limited mode. As a result of this, the K2 mission was developed.

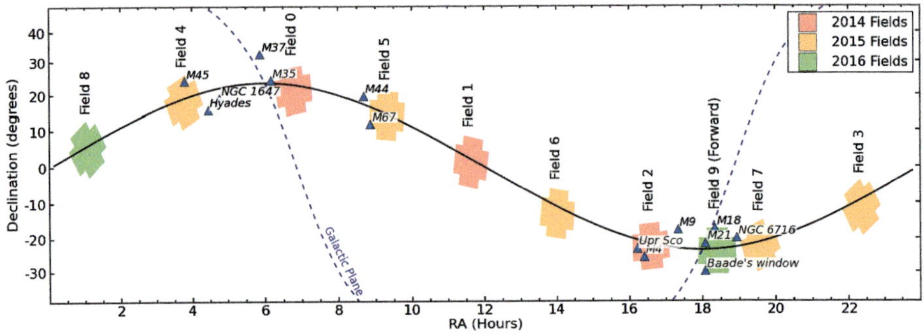

Fig. 1 The target fields for the 2-year K2 space mission

By observing in the ecliptic which minimized the force from the solar wind on the spacecraft and using its thrusters to compensate for the lost reaction wheel Kepler could continue to collect data. In this mode K2 observes targets in the plane of the Earth's orbit and in a direction away from the Sun. K2 can observe each target field for a maximum of 75 days before re-positioning for the next individual target field. This allows it to focus on bright stars and search for transiting planets that can be well defined in follow-up observations (see below). The K2 mission started in June 2014. Test observations demonstrated that a photometric precision of 50 ppm can be achieved for a 12 magnitude G-type star in 6.5 h. All K2 targets are proposed by the community through the Guest Observer Program and the data is immediately available to all interested parties. Although K2's inheritance is based on exoplanetary research, the K2 mission welcomes all proposals including, but not exclusive to, exoplanet, stellar, extragalactic and solar system science. The mission is funded for 2 years, and will observe a total of about 10 fields (each 100 deg^2) for 80 days. Each field will have about 10 000 targets. The selected fields are shown in Fig. 1 and it is immediately noticeable that several important (in terms of stellar evolution) clusters will be observed (in November 2015 a presentation at Extreme Solar Systems III meeting in Hawaii, data from the globular cluster M4 was presented showing a "Christmas tree" of blinking RR Lyrae stars).

The first results from K2 have now been published and it seems like the predictions before implementing the mission have been met or even superseded (Barros et al. 2015; Narita et al. 2015; Armstrong et al. 2015; Becker et al. 2015).

Although the original Kepler mission is over, the analysis and follow-up of the material from its 4 year targeting towards the 110 deg^2 field in Cygnus is ongoing, as evidenced by the recent announcement of a super-Earth orbiting a G2V star within its habitable zone. This exoplanet, Kepler 452b orbits its host every 385 d (Jenkins et al. 2015). The planet have a diameter of $\sim$1.6 times that of the Earth, but whether it is really Earth-like remains to be seen. At this distance from its host none of the currently operating spectrographs can obtain the radial velocity signature of its influence onto the host stars orbit, the planet being to small and the star to faint. As is explained below this is the problem with the current status of exoplanetology, that to progress downwards in radii and masses of the studied planets we will need at least a new generation of ground based and space based instrumentation.

Recently the seventh release of Kepler data netted 521 new planet candidates including 12 with a diameter twice that of the Earth that orbits within their host stars habitable zone. The tally of confirmed planets have risen to more than 1030 of which a few dozens orbit within the Habitable Zone.

2 The Important Next Step

The conclusion that one can draw from these two first exploratory space missions is that the brighter the host star, the better one can constrain the exoplanetary parameters, and more precise work both as what concerns exoplanetary physics and the physics of the exo-atmospheres, can be carried out. And as has been shown (e.g. Rauer et al. 2014), high precision measurements of the planetary physical parameters are necessary in order to take exoplanetology to the next step. One of the most interesting things in this capacity will be the study of exo-atmospheres during eclipses in the visible as well as in the infrared.

The last 20 years have been very exciting when considering the discovery of literally thousands of exoplanetary systems. One of the most fundamental discoveries central to this wealth of data has been the realization of the enormous diversity displayed by these planets and systems. Whole new classes of planets have been discovered, orbiting in both very strange orbits and being of new types—compared to our own Solar System. An example demonstrating this is the aforementioned CoRoT-7b which is clearly a super-Earth with an Earth-like density. In the same system radial velocity observations have also demonstrated the presence of another Super-Earth in a 3 d orbit, as well as a Neptune size planet orbiting in a 9 d orbit and exterior to the other two planetary bodies. The discovery of similar objects (e.g. Kepler-10b; Batalha et al. 2011) thus indicate the existence of a class of objects much different to our own Solar System. The radius of CoRoT-7b was determined using the best possible photometric data (from space). Currently, the largest error on the density comes from the mass determination based on radial velocity measurements. The star CoRoT-7 may be bright by CoRoT standards, but it is relatively faint for radial velocity measurements. However, even if we could measure the radial velocity amplitude of CoRoT-7 with a very small error we would still not know if the structure of the planet is like (density wise) Mercury, or the Earth (Hatzes et al. 2011) since we do not know the 'true' mass of the star CoRoT-7 itself. The same is true with the Kepler-78b object. This approximately Earth-sized planet orbits its host in a mere 8 hrs (Sanchis-Ojeda et al. 2013). Radial velocity measurements have established a planet mass of about one Earth mass (Pepe et al. 2013; Howard et al. 2013). Again, with zero error on the radial velocity amplitude we would still not know if the structure of this planet was like Mercury, the Earth, or the Moon (Hatzes 2014) since we would not know the accurate stellar mass. The problem is that the ultimate planet radius and mass hinges on the accuracy of the stellar mass and radius. If we are to do precise exoplanetology to the point where we know whether a planet has a structure like the Moon, Earth, or Mercury we need to know the stellar parameters to a much better accuracy than we know now. The new instruments and satellites described below are in part intended to remedy this by making more exact observations of mainly transiting planets where both radii and masses can then be determined. Nevertheless, we will get both these parameters expressed in terms of the masses and radii of the host stars. And these are not known to a precision higher than maybe at best 10 % which will introduce errors of the same order into the planetology. Our only way forward is to implement the new technique of asteroseismology, as intended with PLATO (see below) and as has been exquisitely demonstrated by Kepler (e.g. Metcalfe et al. 2014). It has been shown (e.g. Rauer et al. 2014) that asteroseismology can provide an accuracy of 1–2 % in radius, of better than 5 % in mass and around 10 % in age for solar-type main sequence stars as long as the photometric precision is good enough and the cadence is high. These errors are intrinsic, and e.g. the specifications for the PLATO spacecraft (e.g. jitter) have been given in terms of not exceeding these factors.

2.1 The Near Perspective: Ground

2.1.1 Radial Velocities

At the moment, the "gold standard" in radial velocity measurements is held by the two HARPS (High Accuracy Radial velocity Planet Searcher) spectrographs, HARPS-N (https://plone.unige.ch/HARPS-N/) deployed at the Telescope Nazionale Galileo (TNG) in La Palma, Canary Islands, Spain. It is a 3.58 m Alt-Az telescope equipped with an active optics system. HARPS-S (http://www.eso.org/sci/facilities/lasilla/instruments/harps.html) has been in regular use since 2003 at the 3.6 m telescope at the European Southern Observatory's La Silla Facility in Chile. These ultra-stable Echelle spectrographs, covering the spectral range of 380 to 690 nm with a spectral resolution of up to $R = 115\,000$, have claimed a radial velocity precision of around 0.5 $\mathrm{m\,s^{-1}}$ under the best circumstances.

La Palma hosts the 2.6 m Nordic Optical Telescope, equipped with the FIES (FIbre-fed Echelle Spectrograph—http://www.not.iac.es/instruments/fies) with a resolution of up to $R = 67\,000$. By obtaining simultaneous Thorium-Argon spectra a very high stability is achieved and a radial velocity measurement precision better than 10 $\mathrm{m\,s^{-1}}$ has been obtained.

Similarly, to the NOT/FIES combination, the ESO VLT/UVES (http://eso.org/sci/facilities/paranal/instruments/uves.html) combination can achieve relatively high spectral resolution ($\sim$5 $\mathrm{m\,s^{-1}}$), and because of the aperture (8.4 m) for significantly fainter stars. The same is true for the 10 m Keck/HIRES combination, used with an iodine absorption cell as a wavelength calibration that achieves about 2–5 $\mathrm{m\,s^{-1}}$.

The Automated Planet Finder Telescope (APF, commissioned in 2013 http://exoplanets.org/rpf.html) is a robotic 2.4-meter optical telescope at the Lick Observatory in California USA. It is designed to search for exoplanets in the 5 to 20 Earth mass range. It does so by performing high cadence observations on a sample of relatively few bright stars. A small number of bright stars are observed every night for 6 months. Although this telescope has a relatively small aperture, it is competitive in terms of the number of observations possible on a dedicated telescope. The philosophy is that it requires a large number of observations to beat down the stellar noise in order to detect really low mass planets.

All of these telescopes deployed today are either relatively small and therefore the magnitude range available for exoplanetary work is limited, or, when a large aperture is present, the spectrographs have not been designed with the ultra-stable requirements necessary in mind. Further, the spectral range has so far been limited to the visual and therefore not very suitable for the searching and study of small (super Earth size) planets orbiting red dwarf stars that are both faint and very cool (e.g. Scalo et al. 2006). This latter requirement to study specifically red dwarf stars in this context has been introduced by the realization that if a red dwarf star would be a host to a Super-Earth planet orbiting within its Habitable Zone (HZ), it would be possible to characterize its atmosphere spectroscopically with already existing (or soon to be commissioned) equipment. There are some interesting instruments delivering significant performance improvements that will be available soon.

The first such instrument is CARMENES (Calar Alto high-Resolution search for M dwarfs with Exoplanets with Near-infrared and optical Echelle Spectrographs http://carmenes.caha.es/index.html) which will be deployed at the 3.5 m telescope at Calar Alto in Spain. Built by a German-Spanish consortium it will provide a spectral resolution of $R \sim 82\,000$ in two bands: One from 0.5 μm to 1.0 μm, and a second one from 1.0 μm to 1.7 μm. It is specifically going to survey at least 300 red dwarf stars during 5 years, searching for super-Earths (about 5 Earth-masses). For this purpose, 600 *clear* nights have been reserved for the instrument in the 5 years starting from "first light", an event that is expected at the end of 2015. The instrument is specifically designed to be able to measure the radial velocity of the target star in both wavelength bands. False positives due to stellar activity will manifest themselves with different velocity amplitudes in the optical and infrared.

The second instrument is *ESPRESSO* (Echelle SPectrograph for Rocky Exoplanet and Stable Spectroscopic Observations https://www.eso.org/sci/facilities/develop/instruments/espresso.html) that will be deployed at one of the 8.2 m Very Large Telescopes (VLT) at the ESO Paranal facility in Chile. This ultra-stable spectrograph which will be sensitive between 350 and 720 nm will have a spectral resolution of 140 000. The stated goal is to reach a precision of 0.1 $m\,s^{-1}$. First light is expected in 2016.

In order to further improve on the stability of spectrographs and thus the precision (eventually to better than 10 $cm\,s^{-1}$) the technology of Laser frequency Combs is being developed for radial velocity work (Lo Curto et al. 2012), but may not reach the maturity level for a usable instrument (in the visual wavelength range) before 2020. In the infrared, however, the recently funded upgrade of NIRSPEC (http://www2.keck.hawaii.edu/inst/nirspec/) on the Keck telescopes in Hawaii intend to implement laser combing as a reference, aiming for a velocity resolution of 1 $m\,s^{-1}$ and first light in 2016/2017 (Beichman, private communication).

2.1.2 *Transits*

On the ground a large number of new networks of telescopes searching for exoplanetary transits of smaller and smaller planets are being deployed. These include:

The "Next Generation Transit Survey" (NGTS) which builds on the previously so successful SuperWasp program that utilized robotic telescopes at two sites. WASP has a proven record of detecting transiting Jupiter size planets, whereas NGTS has as its objective the detection of down to Neptune size exoplanets. NGTS employs an array of robotic small telescopes operating in the 600–900 nm band, thereby maximizing sensitivity to bright but relatively small and cool host stars (K and early-M spectral type). NGTS will survey the brighter stars that CoRoT and Kepler could not access. This will provide scores of prime targets for high quality characterization by the next generation of telescope/instrument combinations such as the VLT/ESPRESSO, the E-ELT and the space missions JWST and CHEOPS. Bright host stars provide the key element to this follow-up, particularly to the radial velocity measurements. But since the discovered exoplanets will all be transiting, the brightness of the host star will also enhance the possibility of investigating their atmospheres spectroscopically.

NGTS is located at a site with excellent photometric conditions, namely the ESO E-ELT site at Cerro Armazones, about 20 km away from the Paranal facility (location of VLT). The science requirement (0.1 % photometric precision across a wide field) is more technically demanding than in the case of previous ground based transit surveys, and the key technologies has been demonstrated with a prototype system. The NGTS telescopes are 200 mm, f/2.8 high-quality commercial telescopes and 12 such telescopes are equatorially mounted on individual piers. Each telescope has a field of view of 8 deg^2. The telescopes are working together, and the total field covered thus is 96 deg^2. Each telescope is equipped with red-sensitive ($2k \times 2k$) CCDs designed to observe with optimal sensitivity in the 600 to 900 nm range in order to match the peak emission of the primary K and early M stars targets.

MASCARA, The Multi-site All-Sky CAmeRA is a unique concept to monitor the whole sky looking for exoplanets around the brightest stars in the sky (<8 mag in V). Previous surveys like the aforementioned SuperWasp, as well as HATNET and TrES targeted stars with a magnitude fainter than $V = 8$ due to saturation limits on their detectors and exoplanets around the brighter stars have up to now only been detectable via radial velocity measurements. MASCARA is a fully funded network consisting of 5 stations across the globe, with each station composed of a battery of cameras to monitor the near-entire sky at each location. Once all stations have been installed, MASCARA will be able to provide a nearly 24-hr coverage of the complete sky, down to magnitude 8, at a sub-minute cadence. Its main purpose is to find the brightest transiting exoplanets, expected in the $V = 4$–8 magnitude range, but by surveying so many bright stars MASCARA will also allow for a wealth of secondary science. The design of each station with its cameras is especially interesting. The basic principle is to be able to use modified CCD based cameras developed for the amateur astronomy market, mated to commercial camera lens systems. The concept has been tested and the required precisions have been confirmed.

2.2 The Near Perspective: Space

Kepler and CoRoT demonstrated that space is the best place for transit studies. In this respect the situation looks bright with two upcoming missions. After CoRoT and Kepler/K2, the next two space missions are TESS and CHEOPS, both of which will be studying the transit light curve of exoplanets. Both are going to follow the recommendation based on the results of their predecessors that in order to be able to study the physics of exoplanets we need to observe the light curve of transiting exoplanets orbiting *bright stars* in order to be able to carry out also the follow-up observations with the necessary precision.

2.2.1 TESS

The Transiting Exoplanet Survey Satellite (TESS), planned for a launch in 2017, is a NASA all sky survey to be conducted over 2 years. It will monitor the brightness of a total of 1/2 million stars and study 100 000 of the brightest stars in detail. It will map the northern half of the sky in the first year and switch to the southern half in year 2. The spacecraft will be using an array of 4 wide-field 10 cm cameras scanning the sky. Each camera features a low-noise, low-power 16.8 megapixel CCD detector with a 576 deg^2 field. Due to the scan pattern, most targets will be followed for about 30 d. A small percentage will be followed for a longer period up to about 160 d. TESS will focus on relatively bright G- and K-type stars where the transits will be excellent targets for further study with the James Webb Space Telescope (JWST). There will also be a prioritization on all available red dwarf stars. These stars have a habitable zone quite close to the star and radial velocity confirmation of rocky planets in this zone is supposedly relatively easy. TESS will also continue the observations of the photometric variations due to stellar oscillations ("asteroseismology") begun by CoRoT and Kepler. The result of those two pioneering missions indicate that TESS will detect p-mode oscillations (variations where the stochastically induced acoustical waves are restored by pressure) for about 6000 of the brightest stars (brighter than $V = 7.5$). These objects include about 2000 stars on the upper main sequence as well as subgiants, and all giant stars within the magnitude range. While TESS will offer opportunities to detect stellar variability for its (in total) 500 000 targets, its limited time base of 30 days will prevent the high-precision seismic age determination for stars with the exception of those located at the ecliptic poles.

2.2.2 CHEOPS

CHaracterizing ExOPlanet Satellite is the first mission dedicated to searching for exoplanetary transits by performing ultra-high precision photometry on bright stars already known to host planets. The mission's main science goals are to measure the average density of super-Earths and Neptunes orbiting bright stars and thus to provide suitable targets for future in-depth characterization studies of exoplanetary atmospheres. It is the first of ESA's "small missions", carried out together with a European consortium led by Switzerland, and is planned for a launch in 2018. The main difference between CHEOPS and previous exoplanet missions is that one will already know if a planet is present from radial velocity observations and one will likewise have a reliable ephemeris for when to expect the transit. The data will allow the determination of the densities of the planets studied, the errors only being dominated by the assumed/modeled host star mass and radius. CHEOPS will have the capability to measure a photometric signal with a precision of 150 ppm/min for a $V = 9$ magnitude star. This allows the detection of the transit of an Earth-sized planet orbiting with a period of 60 days around a star of 0.9 R_{sun} and with a S/N of the transit greater than 10. CHEOPS uses a single frame-transfer back-side illuminated CCD mated to a 32 cm on-axis Ritchey-Chretien telescope with the image of the target star being de-focused onto the detector. Special care has been taken to achieve a design that minimizes stray light entering the telescope and using a dedicated field stop and a baffling system. Calculations and testing demonstrate that the design meets the requirement of less than 10 ppm stray light falling onto the detector even in the worst observational scenario. Thermal stabilization of the detector to within 10 mK is obtained by radiating surplus heat to cold space. The mission critical technology of CHEOPS have already been space qualified since it was flown onboard the CoRoT space craft where the baffling/field stop combination surpassed the strict requirements by an order-of-magnitude in a similar orbit. On CoRoT, the 1-sigma noise levels reached less than 1 ppm per $\text{mHz}^{1/2}$ for the brightest stars (around 6:th magnitude). For an 11.5 magnitude star the noise level was measured to be 2.00×10^{-4} per 512 s integration (Auvergne et al. 2009). These numbers are probably indicative for the levels reachable by CHEOPS.

A summary of the main science objective of the CHEOPS mission is:

- The study of exoplanets with sizes ranging between one and six Earth radii and the following goals:
  - Deriving the mass-radius relation in a planetary mass range for which only a handful of data exist and with a precision not previously achieved
  - Identifying planets with significant atmospheres over a wide range of planetary masses
  - Placing constraints on possible planet migration paths followed during the formation and evolution of planets
  - Probing the atmospheres of known hot Jupiters
  - Providing unique targets for future ground-based (e.g. E-ELT) and space-based (e.g. JWST) observatories with spectroscopic capabilities.

2.3 Astrometry and Gaia

Apart from when we observe a planet transit its host star, the orbital inclination and thus the true planet mass as discerned from radial velocity measurements is not known. And in the case of transiting planets the inclination is usually only known for small star-planet distances due to the lower probability of a transit occurring for planets with large orbital radii. This became an interesting complication when the first exoplanets were detected with the radial velocity technique and one thus had only a minimum mass of the planet itself. This led to interpretations where the planets could easily be considered to be over the limit (then usually stated as being 13 M_{Jupiter}) that would be transforming them into brown dwarfs (e.g. Latham et al. 1989).

A clear example of this problem is the object found orbiting the star HD 33636, where radial velocities gave a minimum mass $M \times \sin i = 10.2\ M_{\text{Jupiter}}$. Astrometric measurements using the Fine Guidance Sensor (FGS) of the Hubble Space Telescope (HST) determined an orbital inclination of only 4 degrees which results in a true companion mass of 142 M_{Jupiter} or 0.14 M_{Sun} (Bean et al. 2007). Astrometric measurements with FGS have over the years been able to measure the true mass of a few exoplanets. These include GL 876b (Benedict et al. 2002), ε Eri (Benedict et al. 2006), and the brown dwarf companion around HD 136118 (Martioli et al. 2010). In the case of 55 Cnc astrometric orbits gave the mass and orbital inclination for two of the planets in the system, 55 Cnc c and d. The orbits of these two planets are mis-aligned having a mutual inclination of 29 degrees (McArthur et al. 2010). However, due to the limited sensitivity of performing such astrometric measurements with the FGS on HST such important observations have only been done in a very few cases.

This situation is about to change. In December 2013, after more than a decade of development, ESA launched its successor to the HIPPARCOS satellite (which flew 1989–1993), Gaia. The name was originally an acronym for Global Astrometric Interferometer for Astrophysics, reflecting its inheritance from ESA's program for Space Interferometry. Astrometry (GAIA) together with exoplanetology (Darwin) originated in ESA's Horizon 2000 program initiated by the then Director of Science, Roger Bonnet, that identified Interferometry carried out from Space as a so-called 'Green Dream' topic, i.e. something clearly to be carried out when technology had matured enough. However, the first studies of GAIA carried out industrially demonstrated clearly that it would be much more (cost-)effective to base the new spacecraft on the principles used already for the previously successfully flown HIPPARCOS, taking advantage of advances in technology but without the added complexities of interferometry. It was demonstrated that this goal to achieve 2–3 orders of magnitude improvement in astrometric precision as well as observing more than 1000 times the number of stars than what was achieved by HIPPARCOS, was feasible. Thus the name of the mission changed to Gaia. Gaia has thus begun to take a census of roughly 1 billion stars or about 1 % of all stars in the Milky Way Galaxy. The spacecraft has a nominal lifetime of 5 years and during this period each star should be observed about 70 times. During each observation the position, brightness and color are recorded. From these data, the stars distance and proper motions can be calculated with an unprecedented precision.

Gaia has three instruments. The most important is the astrometric detector that records the passing of a star across a grid, projected onto the hitherto largest CCD assembly flown in a space instrument, as the spacecraft rotates. The second instrument is a spectrograph that records the radial velocity of each star. This third dimension together with the x- and y-position on the sky will provide a three-dimensional map of the Galaxy. A third instrument will measure the color of the stars observed so as to derive effective temperatures. Gaia will study exoplanets in several fashions. It will discover giant exoplanets by detecting the wobble in the plane of the sky that the orbital motion of the exoplanet induces in its host star. Although it makes relatively few photometric observations, by virtue of the large number of stars that will be observed Gaia will also be able detect some transits of planets as they pass between us and their host star (see Perryman et al. 2014). Finally Gaia may also be able to confirm massive giant planets and brown dwarfs through the radial velocity measurements. Gaia will also add to the exoplanetary science by providing an "input catalogue" for ESA's M3 PLATO mission (see above). Gaia's highly precise stellar distances of all of future PLATO target stars will also greatly help that later mission by improving the asteroseismic data that PLATO will obtain through the removal of some of the properties of the exoplanetary host stars as free parameters. Accurate distances, magnitudes, and effective temperatures can fix the stellar radius of the object in the retrieval of the asteroseismic parameters. This will provide an increase in the precision of the planet's physical data.

2.4 The Next Steps: Ground

2.4.1 E-ELT (2024)

At the moment several Extremely Large Telescopes are being considered or are under development worldwide as one of the highest priorities in ground-based astronomy. While such telescopes will also study and vastly advance astrophysical knowledge in many disciplines such as the first objects in the Universe, super-massive black holes, and maybe contribute to the issue about the nature and distribution of the dark matter and dark energy what we are most concerned with here is the major advances that will be possible concerning planets around other stars. Of these instruments the E-ELT (European Extremely Large Telescope), being developed by the European Southern Observatory (ESO) is the largest. This revolutionary new ground-based telescope concept will have a 39-meter main mirror and will be the largest optical/near-infrared telescope in the world. The E-ELT program was approved in 2012 and green light for construction was given at the end of 2014. First light is targeted for 2024. One of the stated goals for this instrument is "to track down Earth-like planets around other stars in the "habitable zones" where life could exist". While the detailed calculations showing what can be done in this context is lacking, it is clear that the aperture alone of this instrument will allow the detection of enough photons. Nevertheless there are significant problems to be clarified. While it is clear also that the telescope can in principle resolve the 0.1 arcsec required to separate the Earth from a G2V star at 10 pc, the issue of how to correct for speckle noise in a telescope with many hundreds of segments at the relevant timescales in order to diminish the contrast has to be clarified. Spectroscopy appear more directly feasible at this time, either as what concerns radial velocity detection or the direct detection of spectroscopic features in exoplanetary atmospheres. It of course assumes that a suitable spectrograph is made available at the appropriate time.

2.5 The Next Steps: Space

2.5.1 JWST (2018)

The James Webb Space Telescope (JWST) observatory consists of a passively cooled telescope that is optimized for diffraction-limited performance in the near-infrared (2–5 μm) region, but with extensions to either side into the visible (0.6–2 μm) and mid-infrared (5–28 μm). This is in contrast compared to the Hubble Space Telescope's (HST) 0.1–2.5 μm (ultraviolet to the near infrared) wavelength range. The JWST observatory includes three main elements, the Integrated Science Instrument Module (ISIM), the Optical Telescope Element (OTE) and the Spacecraft Element which comprises the spacecraft bus and the sun-shield. The primary mirror is 6.5 m in diameter and consist of 18 mirror elements made of gold-coated beryllium It will have a giant sun-shield (22×12 m) protecting the telescope and the instruments from the light and heat of the Sun. The total mass of the JWST will be 6500 kg and the spacecraft will be launched with a European Ariane 5 rocket into a Lagrangian L2 orbit located approximately 1.5 million km away from the Earth. This makes its operation and pointing/stability requirements both much simpler and more efficient, in comparison with Hubble. Of course refurbishment missions to JWST will be impossible in this orbit.

The science objectives of the JWST falls into several broad categories, *viz*. the observation of:

- The phase of the evolution of the Universe when the first generation of stars were formed and started to shine, thereby re-ionizing the Universe.
- The first formation of early galaxies and clusters of galaxies.
- Detailed observations of star formation and the accompanying process of planetary system formation.
- "Planetary systems and the origin of life" taken to mean the first high sensitivity spectroscopic observations of exoplanetary atmospheres, as well as complex molecules in the interstellar medium accompanying the star formation process.

2.5.2 Millimetron (2019/2022?)

The 12 m MILLIMETRON is a radio astronomical observatory that will be launched by Russia and is equipped with Bolometers operating in the 200–400 μm and 700–1400 μm ranges, having a sensitivity at 300 μm and 1 hour integration that is 1 nanoJy. It will also be equipped with 3 spectrometers for the 50–300 μm ($R = 10^6$), 30–800 μm ($R = 1000$) and 20–2000 μm ($R = 3$). The idea is also to use it as an interferometer (like its predecessor Radioastron), using other radio telescopes on the ground as the other nodes and creating a very long baselines.

It is also quite clear, that a 10 m-class telescope, operating in space and with the quoted sensitivities in the FIR wavelength range would fill a gap between previous space missions and ground based interferometric installations such as ALMA. The strength of the instrument would be high resolution, long wavelength observations of very cold disks.

2.5.3 PLATO (2024)

After CHEOPS and TESS, Europe will return to the study of exoplanets from space with the PLATO mission which will take the results of those missions to the next level. The PLATO spacecraft (Rauer et al. 2014) was selected, in February 2014, as the European Space Agency, ESA's, medium class mission for the third (M3) launch opportunity upcoming in 2024 in the Cosmic Vision program.

It addresses very fundamental questions relating to the study of exoplanets such as:

1. How do planets and planetary systems form and evolve?
2. Are there other systems with planets like ours?
3. What kind of planets do we find within the Habitable Zone around different types of stars?
4. What is the role of the host star in the context of the formation and evolution of exoplanets?

PLATO is based on the experiences of both the CoRoT and Kepler missions, and consists of a spacecraft bus, equipped with 34 individual 12 cm aperture telescopes, thus providing both a wide field-of-view, yet a large photometric magnitude range.

Two of the telescopes are equipped with red- and blue-filters and are used both for the pointing and the stability control, as well as in order to observe the very brightest of the target stars, giving the system as a whole a very large dynamical range. PLATO targets bright stars, primarily in order to detect and characterize planets down to Earth-size or smaller by photometric transits and whose masses can then be well determined by ground-based radial-velocity follow-up measurements.

PLATO will be the first mission with sufficient time base, precision, and cadence to derive masses, radii, ages, and internal rotation profiles with very high accuracy for 100 000s of stars across the Hertzsprung-Russel Diagram utilizing the new tool of asteroseismology (e.g., Aerts et al. 2010 with updated reviews in Chaplin and Miglio 2013 and Aerts 2015). The asteroseismology will be used for the determination of the stellar fundamental parameters for all stars brighter than about magnitude 11.5. The combination of bright targets and asteroseismology results in high accuracy for the planet parameters, quantifiable as better than 2 %, 4–10 % and 10 % for planet radii, masses and ages, respectively. The foreseen baseline observing strategy includes two long pointings (2–3 years each) to detect and characterize planets down to and beyond Earth-size and reaching into the habitable zone (HZ) of solar-like stars, pushing the border beyond previous missions. An additional so-called step-and-stare phase with shorter pointings (1–5 months each and in special cases just a few days to confirm earlier long term detections) is filling up the remaining lifetime of the spacecraft. It is estimated that the survey will cover at least about 50 % of the sky during the nominal 6 year mission. For a potential extended mission of up to 8 years and beyond it is possible to cover more than 70–80 % of the sky.

During the nominal mission PLATO will observe up to 1 000 000 stars and will discover and characterize hundreds of planets in the Earth to Super-Earth category, and thousands of planets in the Neptune to gas giant regime in orbits out to the HZ, as well as detecting many more for which partial characterization will be possible. PLATO will therefore provide the first large-scale catalogue of planets with radii, masses, mean densities and ages, accurate enough to carry out a detailed analyses of each planet. This catalogue will include Earth-like planets at intermediate orbital distances, where surface temperatures are moderate. Coverage of this parameter range, with a statistical large number of characterized planets, is unique to PLATO.

PLATO will provide a census for small (low-mass) planets. The results of the mission will help complete the knowledge of planet diversity for both low- and high-mass objects (at orbital distances $<$ a few AU), and correlate the planet mean density as a function of orbital distance. These parameters can be compared to predictions from planet formation theories. Other investigations that are possible to carry out include the constraining of theories for planet migration and scattering, and to specify how planet and system parameters change with host star characteristics, such as stellar mass, metallicity and age. The catalogue will also allow us to study planets and planetary systems at different evolutionary phases. This will serve to identify objects which are retaining their primordial hydrogen atmosphere and in general the typical characteristics of planets in the low-mass, low-density range. Since PLATO observes mainly bright stars, many of the discovered planets will be targets for future spectroscopic investigations exploring their atmospheres. Furthermore, the mission has the potential to detect exo-moons, planetary rings, binary planets, and Trojan planets.

The planetary science possible with PLATO is complemented by its impact on stellar and galactic science. Asteroseismology will provide us with fundamental stellar parameters including stellar ages for a large number of stars spread throughout the Galaxy. Precision light curves will be obtained for all kinds of variable stars, including those in stellar clusters of different ages. This will allow us to improve stellar models and to have a better understanding of stellar activity as a function of stellar age. A large number of well-known ages and masses of red giant stars will probe the structure and evolution of our Galaxy. Asteroseismic ages of bright stars for different phases of stellar evolution will allow us to calibrate stellar age-rotation relationships. Together with the results of ESA's Gaia mission, the results of PLATO will provide a huge legacy to planetary, stellar and galactic science. Most importantly, with its high accuracy and sensitivity, PLATO will make the first unique observations that will take the field of exoplanetology from its infancy to a mature phase where we will be able to carry out geophysics on Earth-like worlds and compare these observations with data from our own Solar System. At the same time, observations of the host stars will give us accurate ages for the exo-systems, allowing us for the first time to study the evolution of planetary systems and thus understand the context of our own world among these systems.

2.6 The Further Future

2.6.1 SPICA (Beyond 2025?)

SPICA, The Space Infrared Telescope for Cosmology and Astrophysics is a 3.5 m cryogenically cooled space telescope proposed by the Japanese community through its space agency JAXA/ISAS and intended to be carried out in collaboration with ESA and NASA. It has been considered to be equipped with a coronograph for high resolution, high dynamical range observations of disks and exoplanets, as well as being equipped with high- and low-resolution spectrometers in the 4 μm to 210 μm wavelength range. This will complement the wavelength ranges studied with JWST as well as Herschel (which are/were not cryogenically cooled) bringing high spatial resolution to the cool Universe (20–40 K).

The telescope has also been considered to be cooled to below 6 K allowing unprecedented sensitivity in the long wavelength infrared regime. SPICA would then be an essential pre-cursor to more futuristic facilities such as the proposed large space based infrared interferometers as it would demonstrate many of the technologies required for these missions. SPICA should have a guaranteed three year lifetime, but, because the cooling of the telescope and instruments is achieved using closed cycle mechanical coolers, it would be expected that the actual operational life could be significantly longer than this. SPICA's status at the moment is that it is awaiting a new proposal to the relevant space agencies.

The instrument suite for the SPICA satellite could comprise the following instrument capabilities:

Mid-infrared camera and spectrometers: 5–37 μm imaging and spectroscopy. The grating spectrometers will have a resolving power of typically a few thousand and will cover the full MIR including the 27–37 μm band not accessible to JWST. As an option there would also be two immersion grating spectrometers operating in the short wavelength MIR (5–15 μm) with resolving powers of up to 30 000. These would allow, for instance, the study of the dynamics of protostellar disks.

Far infrared spectrometer and camera: Utilized for 34–210 μm imaging and spectroscopy. This instrument will be a Mach-Zehnder configuration Fourier transform spectrometer to provide imaging spectroscopy from 34–210 μm covering from the rest frame [Si] 34 μm to [NII] 205 μm lines over a 2×2 arcmin field of view.

Guidance camera: NIR and visible focal plane camera with a scientific imaging and low resolution spectroscopy capability.

MIR coronagraph: 5–27 μm coronagraph with medium to low resolution spectroscopy. This instrument will have an inner working angle of $3.3\lambda/D$ allowing it to image and take modest resolution ($R = 10$–200) spectra of young gas giant planets orbiting as close as about 10 AU around stars at 10 pc distance. There is also an option to extend the wavelength range down to 3.5 μm.

SPICA would have the capability to directly detect and undertake detailed characterization of a number of hot young gas giant planets using its coronagraph. Additionally the MIR spectrometers will be capable of undertaking transit spectroscopy on many targets with high efficiency and high sensitivity. The coronagraphic capabilities of SPICA could, in principle, exceed those of JWST as the PSF will be very much cleaner and the dedicated instrument would have a rather better contrast (10^{-6} compared to 10^{-4} for JWST). A unique feature of the coronagraph would be the ability to take direct spectra of a few target exo-planets. The far infrared region is unexplored territory for exo-planet research and it is anticipated that both transit photometry and, possibly, spectroscopy will be possible on some targets allowing us to probe to different depths into planetary atmospheres and opening a new discovery space in characterization. Also directly related to exo-planets, the detection and characterization of exo-Zodi dust clouds is one of the prime goals for the SPICA FIR instrument. It will follow up on the results of Spitzer and Herschel, and will allow a much more complete picture of planetary formation scenarios and the role of characterization.

2.6.2 *FIRI (2030?)*

The European Space Agency, ESA, first began to look into a Infrared technology interferometer in the context of the nulling interferometer for the Darwin study in 1997. In 2006, a study began of a more general purpose instrument based on constructive interference in the Infrared wavelength region. The Far InfraRed Interferometer (FIRI) Technology Reference Study investigated the feasibility of an interferometer to achieve an angular resolution of less than 1 arcsecond at wavelengths between 25 μm and 300 μm. The selected baseline concept was based on a single spacecraft with two cryogenically cooled telescopes on two deployable booms and with a central hub, forming a Michelson interferometer. This would be quite different to the previously studied 'nulling' interferometer Darwin, which consisted of 3–4 telescopes (passively cooled), and free-flying in space allowing very large baselines. FIRI would address the same wavelength region as SPICA, but with significantly higher spatial resolution.

A later study (http://arxiv.org/pdf/0707.1822v1.pdf), utilized the Darwin concept with instead 3–4 telescopes and with the telescopes and the beam-combiner in free-flying mode.

Since the FIR is a critically important wavelength range for studying the origin and evolution of planetary systems and most solar systems, including our own, are pervaded by dust, which is very bright at FIR the utilization of the FIR makes great sense. FIRI would observe the structure and dynamics of this dust, with many times the effective sensitivity of ALMA, providing information on how such systems formed. FIRI could also infer the presence and orbital motions of very young—maybe still coalescing—planets, whose formation process disturbs the distribution of dust and gas. FIRI would also be able to study the distribution of the organic molecules that are important in the understanding of the chemistry of the of the interstellar medium (ISM). By mapping out—in detail—the contents and chemistry of the ISM in our own and nearby galaxies, it would teach us a lot about the possibilities of life in the Universe.

2.7 Exoplanet and Disk Studies in the Future

To continue the work on star- and planet-forming disks, on debris disks, and the equivalent of the zodiacal- and Kuiper-Edgeworth disks it appears that a number of possible important instruments are either already present (ALMA), in development (JWST and the E-ELT) or suggested (SPICA, or FIRI). The importance of high signal-to-noise, high spatial resolution

Fig. 2 The contrast at different wavelengths for the case of the Sun and a few of the Solar System planets

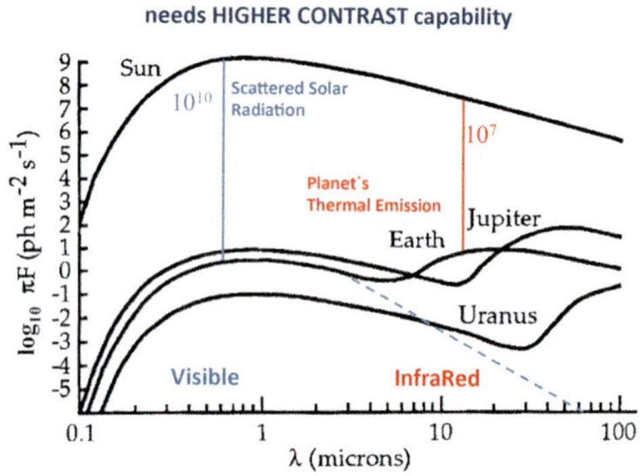

spectroscopic IR data can not be underestimated in this context, but the situation appear to be well in hand.

However, one of the more important problems to be addressed in the long term time schedule concerns direct spectroscopy of habitable planets with the capability to detect signs of biological activity at interstellar distances, and here the situation is more unclear. o that it is to early to start designing such an instrument, since we do not know, at the moment, exact what (type of) target(s) to design for. Either Super-Earths orbiting in the HZ around a very late M-dwarf, or the a 'normal' Earth analogue orbiting at 1 AU around a G2V star would require tow completely different instruments, particularly considering that the former would likely be at a distance less than 5 pc (in order to be discovered at all), while the latter would very likely be much further away (there are only a handful of G-type main sequence stars within 10 pc; Kaltenegger et al. 2010). Regardless of which type of target will be the first objective, it is a formidable problem, since, that while a reasonable spectral resolving power for measuring the spectrum of an Earth-sized planet in a habitable orbit with a V-mag of 35 is only 10–100, the main problem is one of contrast and spatial resolution. The host star is $>10^{10}$ brighter than the planet in the visual, and located with a spatial separation of <0.1 arcsec (0.1 arcsec at 10 pc for a planet orbiting a G2V star at 1 AU separation and progressively much closer for more distant objects or for fainter host stars). This can be achieved either through imaging—with or without coronography—or through (destructive, so-called 'nulling') interferometry. A third option being pursued mainly in the USA is the "Starshade" or "Occulter", which is basically a variety of coronography but where the mask is flown in space at a great distance (many tens of thousands of km) away from the actual telescope.

In obtaining a spectrum of a habitable exo-earth one needs to choose to observe the appropriate region of the electromagnetic spectrum. The contrast is several orders of magnitude lower if one selects the near- to mid-infrared spectral region and scattered light from the host star will be (somewhat) less of a problem (see Fig. 2). At the same time, in order to achieve the required spatial resolution, one requires larger optical elements in the infrared, which have historically been one of the drivers for using an interferometer in space. Finally and ultimately, obtaining spectroscopy of a "habitable" planet should indeed also demonstrate not only if it is indeed 'habitable' but also whether it is inhabited by life forms. One need thus to determine and select so-called 'biomarkers' in the spectrum that can be observed in the appropriate wavelength range. With this criterion one has entered the field of Astrobiology (Fig. 3).

Taking all these complications into account, it appears that the development of the different technologies required should be started now. The development time of flight hardware is

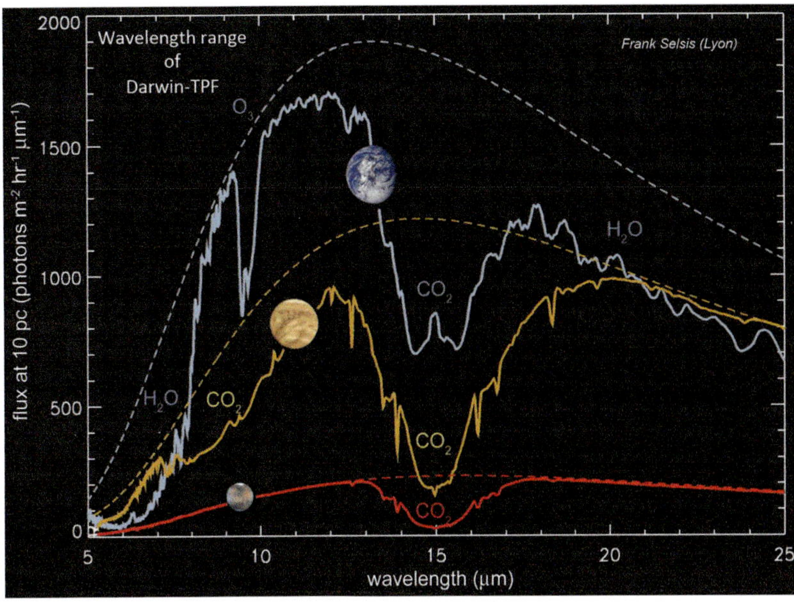

Fig. 3 An example of spectral features that could be considered to be 'biomarkers'—at least in our own Solar System

so long that it will be barely ready in the time when we are likely to have identified suitable targets. And the space agencies of the world have realized this long time ago. In 1996 ESA and NASA began parallel but coordinated studies of how such an instrument could be defined, developed and deployed in the shortest possible time. The ESA study went under the name of Darwin, while the NASA effort was designated Terrestrial Planet Finder or TPF for short.

Although suggestions for a possible technology, as well as a definition of a realistic scientific case had been put forward much earlier (e.g. Angel et al. 1986; Burke 1988; Bracewell and MacPhie 1979), this was the first time the problem was studied systematically by the worlds two largest space agencies and in conjunction with the worlds leading aerospace industries. At the time—1996—only 2–3 exoplanets had been found and were still questionable in many scientists mind. Particularly since the newly discovered planets were of types (e.g hot Jupiters and sometimes with wildly eccentric orbits) not present in our Solar System. The Darwin and TPF studies were therefore targeted towards clones of our own Solar System, i.e. early G-type stars orbited by terrestrial planets in the Habitable zone, with gas giants in the region beyond (later, the Kepler mission was targeted towards the same objective). The design specifications were thus aimed towards detecting a 1 Earth-radii planet orbiting 1 astronomical unit away from a G2V star. Between 1996 and 2007 ESA and NASA developed space components that could detect such a system out to 33 pc distance. The ESA study was based on a so-called nulling interferometer, while NASA also studied the Coronography and eventually also the occulter options. In 2000 ESA decided to continue the technology development of the nulling interferometer and this work progressed until 2007 when it was closed down. By then ESA had validated all technologies required in a free flying interferometric telescope based on destructive interference (Fig. 4). Specifically the free-flying of a constellation of spacecraft, based on the "Darwin-technology" was

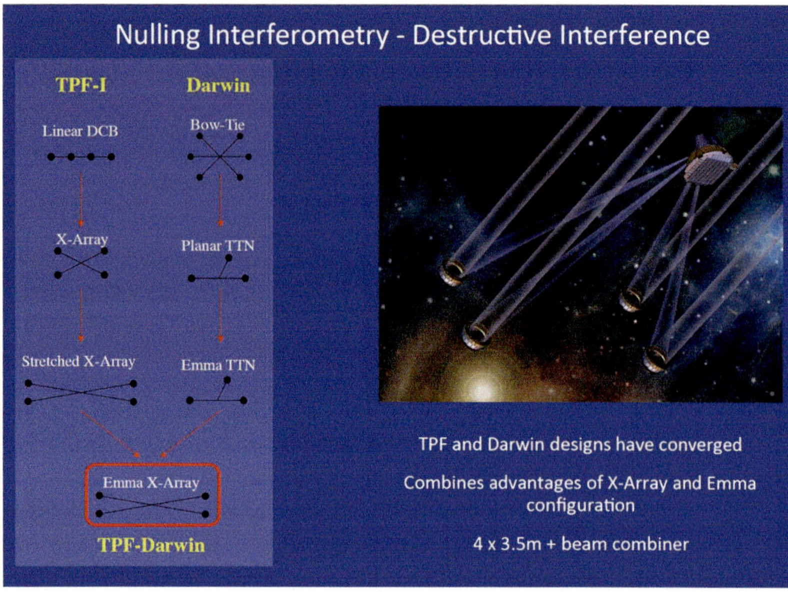

Fig. 4 Different constellation concepts evaluated in ESA's Darwin study, as well as an artists rendering of a 4 telescope array plus beam combiner. The telescopes in this array are passively cooled 3.5 m telescopes

also later space qualified by Sweden in the PRISMA experiment that was free flying two nano-satellites, utilizing this technology, in low earth orbit.

Following the definition and technical studies of these major mission concepts during the period 1996–2007, very little activity has taken place in Europe. The final conclusion of the studies were that several technologies exist with the capability of studying the direct light from small planets in habitable orbits around stars at least out to a distance of 30–50 pc. Which technology to use, depends, however, on the target in question. After 2007, some work has been carried out, mainly in the US, but the extensive technological effort aimed at a launch anytime soon has slowed down significantly. There are some studies being carried out w.r.t. coronography by NASA. This is also true for the so-called New Worlds Missions, where studies of the deployment of star shades, i.e. external occulters intended to fly many tens of thousands of km away from the telescope have been carried out utilizing literally hundreds of nano-satellites brought together in orbit to assemble the large structures required (Seager, *private communication*).

At the moment exoplanetology has reached an impasse. This may seem like a provocative statement, but what is meant is simply that we are limited at the moment by instrumentation and methods in progressing downwards in radii and mass of exoplanets. In spite of the more than 4000 exoplanet candidates we currently have in our catalogues, none is exactly like our own Solar System. A handful of super-Earths have been tentatively identified in what could be designated as 'habitable' orbits around the stars in question and even one case, Kepler 452b (Jenkins et al. 2015) with such a planet at roughly one AU from a 'true' Solar analogue. Since essentially all objects for which we have data both on the radii and the mass tend to be located many hundreds of parsecs away from the Earth (the CoRoT and Kepler planets) they would not be observable with the Darwin or TPF interferometers/coronographs (typically too small separation between host star and planet). Therefore, we at this moment do not know what kind of instrument to select for the objective of obtaining direct spectroscopy of

Earth analogues. Even if such planets turn out to be relatively 'common' the observable targets may be few since most may be located at large distances. The instruments—especially if it is necessary to deploy them in space—will likely turn out to be complicated, long-term and expensive. It is therefore absolutely required that we continue the present work that has been described in this chapter in order to eventually identify an as large as possible assemblage of targets before we commit to the appropriate instrument. Regardless of if this then turns out to be an interferometer, a coronographic large space telescope or a system with a free-flying occulter, the scientific case is strong enough to motivate a great effort.

Luckily, the situation with the new swath of instruments, new networks of telescopes, the E-ELT and its American cousins, CHEOPS, TESS and PLATO being either under development or selected give a clear promise that such Earth-analogue targets will be identified within the near future. Stressing the importance of working with the newest ground based instrumentation in cooperation with the data from current and next-generation space missions the future appears bright and important results are going to be made. The design and implementation of the appropriate instrumentation for the detailed characterization of their properties, especially the habitability, may then go ahead.

Acknowledgements Two anonymous reviewers made some very constructive comments for which the authors are grateful.

MF is also very grateful for a grant from the Max Planck Gesellshaft which allowed him to spend a year at MPIA in Heidelberg, Germany during which time this article was written.

References

C. Aerts, Astron. Nachr. **336**, 477 (2015)
C. Aerts, J. Christensen-Dalsgaard, D.W. Kurtz, *Asteroseismology, Astronomy and Astrophysics Library*. (Springer, Dordrecht, 2010). ISBN 978-1-4020-5178-4
J.R.P. Angel, A.Y.S. Cheng, N.J. Woolf, Nature **322**, 341 (1986)
D.J. Armstrong, A. Santerne, D. Veras et al., Astron. Astrophys. **582**, 33 (2015)
M. Auvergne, P. Bodin, L. Boisnard, J.-T. Buey et al., Astron. Astrophys. **506**, 411 (2009)
A. Baglin, M. Fridlund, *ESA SP-1306* (2006), p. 11
S.C.C. Barros, J.M. Almenara, O. Demangeon et al., Mon. Not. R. Astron. Soc. **454**, 4267 (2015)
N.M. Batalha, W.J. Borucki, S.T. Bryson, L.A. Buchhave et al., Astrophys. J. **729**, 27 (2011)
V. Batista, J.-P. Beaulieu, D.P. Bennet et al., Astrophys. J. **808**, 170 (2015)
K. Batygin, G. Laughlin, Astrophys. J. **683**, 1207 (2008)
J. Bean, B. McArthur, G.F. Benedict et al., Astron. J. **134**, 749 (2007)
M. Bedell, J. Melendéz, J.L. Bean et al., Astron. Astrophys. **581**, 34 (2015)
J.C. Becker, A. Vanderburg, F.C. Adams et al., Astrophys. J. **812**, L18 (2015)
G.F. Benedict, B.E. McArthur, T. Forveille et al., Astrophys. J. **581**, 115 (2002)
G.F. Benedict, B.E. McArthur, G. Gatewood et al., Astron. J. **132**, 2206 (2006)
D.P. Bennet, A. Bhattacharya, J. Anderson et al., Astrophys. J. **808**, 169 (2015)
W.J. Borucki (2010). http://kepler.nasa.gov/Mission/QuickGuide/history/
R.N. Bracewell, R.H. MacPhie, Icarus **38**, 136 (1979)
T.M. Brown, D. Charbonneau, R. Gilliland, R.W. Noyes, A. Burrows, Astrophys. J. **552**, 699 (2001)
B.F. Burke, in *Bioastronomy—the Next Steps; Proceedings of the Ninety-Ninth IAU Colloquium*, Balaton, Hungary, June 22–27, 1987 (Kluwer/Academic Publishers, Dordrecht, 1988), pp. 139–144, A88-5520124-88
P.R. Butler, G.W. Marcy, Astrophys. J. **464**, L153 (1996)
W.J. Chaplin, A. Miglio, Annu. Rev. Astron. Astrophys. **51**, 353 (2013)
D. Charbonneau, T.M. Brown, R.W. Noyes, R. Gilliland, Astrophys. J. **568**, 377 (2002)

C. Eiroa, J.P. Marshall, A. Mora, B. Montesinos et al., Astron. Astrophys. **555**, 11 (2013)
C.V.M. Fridlund, R. Liseau, Astrophys. J. **499**, L75 (1998)
R. Gomes, H.F. Levison, K. Tsiganis, A. Morbidelli, Nature **435**, 466 (2005)
C. Haswell, *Transiting Exoplanets* (Cambridge University Press, Cambridge, 2010). ISBN 978-0-521-13938-0
A.P. Hatzes, Astron. Astrophys. **568**, 84 (2014)
A.P. Hatzes, M. Fridlund, G. Nachmani, T. Mazeh et al., Astrophys. J. **743**, 75 (2011)
A. Howard, R. Sanchis-Ojeda, G.W. Marcy et al., Nature **503**, 381 (2013)
J. Jenkins, J.D. Twicken, N.M. Batalha et al., Astron. J. **150**, 56 (2015)
P. Kalas, J.R. Graham, E. Chiang, M.P. Fitzgerald, M. Clampin et al., Science **322**, 1345 (2008)
L. Kaltenegger, C. Eiroa, C.V.M. Fridlund, Astrophys. Space Sci. **326**, 233 (2010)
J. Laskar, Icarus **196**, 1 (2008)
D.W. Latham, R.P. Stefanik, T. Mazeh, M. Mayor, G. Burki, Nature **339**, 38 (1989)
A. Lèger, D. Rouan, J. Schneider et al., Astron. Astrophys. **506**, 287 (2009)
G. Lo Curto, A. Manesu, G. Avila, L. Pasquini et al., SPIE J. **8446**, 1 (2012)
B.E. McArthur, F.G. Benedict, R. Barnes et al., Astrophys. J. **715**, 1203 (2010)
T.S. Metcalfe, O.L. Creevey, G. Dogan, S. Mathur et al., Astrophys. J. Suppl. Ser. **214**, 27 (2014)
G.W. Marcy, P.R. Butler, Astrophys. J. **464**, L147 (1996)
E. Martioli, B.E. McArthur, G.F. Benedict et al., Astrophys. J. **708**, 625 (2010)
M. Mayor, D. Queloz, Nature **378**, 355 (1995)
N. Narita, H. Teruyaki, A. Fukui et al. (2015). arXiv:1510.01060
C.R. O'Dell, Z. Wen, X. Hu, Astrophys. J. **410**, 696 (1993)
Partnership ALMA, C.L. Brogan, L.M. Pérez et al., Astrophys. J. **808**, 3 (2015)
F. Pepe, A.C. Cameron, D.W. Latham et al., Nature **503**, 377 (2013)
M.A.C. Perryman, *The Exoplanet Handbook* (Cambridge University Press, Cambridge, 2011). ISBN 978-0-521-76559-6
M. Perryman, J. Hartman, G.A. Bakos, L. Linegren, Astrophys. J. **797**, 14 (2014)
D. Queloz, F. Bouchy, C. Moutou et al., Astron. Astrophys. **506**, 303 (2009)
V.A. Rapson, J.H. Kastner, M.A. Millar-Blanchaer, R. Dong (2015). arXiv:1512.01865
H. Rauer, C. Catala, C. Aerts, T. Appourchaux et al., Exp. Astron. **38**, 249 (2014)
I. Roxburgh, in *Proceedings of "The CoRoT Mission: Pre-launch Status – Stellar Seismology and Planet Finding" (ESA SP-1306)*, ed. by M. Fridlund, A. Baglin, J. Lochard, L. Conroy (2006a), p. 521. ISBN 92-9092-465-9
I.W. Roxburg ESA SP-1306, 521 (2006b)
R. Sanchis-Ojeda, S. Rappaport, J. Winn et al., Astrophys. J. **774**, 545 (2013)
J. Scalo, L. Kaltenegger, A. Segura, M. Fridlund et al., Astrobiology **7**(1), 85–166 (2006)
L. Spitzer Jr., 1946, "Report to Project Rand: Astronomical Advantages of an Extra-Terrestrial Observatory", reprinted in *NASA SP-2001-4407: Exploring the Unknown*, Chapter 3, Document III-1, p. 546
O. Struve, Observatory **72**, 199–200 (1952)
A. Vigan, C. Gry, G. Salter et al., Mon. Not. R. Astron. Soc. **454**, 129 (2015)
A. Wolszczan, D.A. Frail, Nature **355**, 145 (1992)
A. Zurlo, A. Vigan, R. Galicher et al. (2015). arXiv:1511.04082

Printed by Printforce, the Netherlands